Integrative Omics

Concept, Methodology and Application

Integrative Omics

Concept, Methodology and Application

Edited by

Manish Kumar Gupta
Department of Biotechnology, Veer Bahadur Singh Purvanchal University, Jaunpur, Uttar Pradesh, India

Pramod Katara
Centre of Bioinformatics, University of Allahabad, Prayagraj, Uttar Pradesh, India

Sukanta Mondal
National Institute of Animal Nutrition and Physiology, Adugodi, Bangalore, Karnataka, India

Ram Lakhan Singh
Department of Biochemistry, Dr. Rammanohar Lohia Avadh University, Ayodhya, Uttar Pradesh, India

Academic Press is an imprint of Elsevier
125 London Wall, London EC2Y 5AS, United Kingdom
525 B Street, Suite 1650, San Diego, CA 92101, United States
50 Hampshire Street, 5th Floor, Cambridge, MA 02139, United States

Notices

Knowledge and best practice in this field are constantly changing. As new research and experience broaden our understanding, changes in research methods, professional practices, or medical treatment may become necessary.

Practitioners and researchers must always rely on their own experience and knowledge in evaluating and using any information, methods, compounds, or experiments described herein. In using such information or methods they should be mindful of their own safety and the safety of others, including parties for whom they have a professional responsibility.

To the fullest extent of the law, neither the Publisher nor the authors, contributors, or editors, assume any liability for any injury and/or damage to persons or property as a matter of products liability, negligence or otherwise, or from any use or operation of any methods, products, instructions, or ideas contained in the material herein.

ISBN: 978-0-443-16092-9

For information on all Academic Press publications visit our website at https://www.elsevier.com/books-and-journals

Publisher: Stacy Masucci
Acquisitions Editor: Linda Buschman
Editorial Project Manager: Billie Jean Fernandez
Production Project Manager: Jayadivya Saiprasad
Cover Designer: Matt Limbert

Typeset by TNQ Technologies

Dedication

Dedicated to our teachers and families

Contents

23. Omics technology policy and society research

Manjusa Singh, Athaven Sukunathan, Swati Jain, Sunil Kumar Gupta, Ram Lakhan Singh and Manish Kumar Gupta

List of contributors

Shivaleela Biradar, Laboratory of Natural Compounds and Drug Discovery, Department of Bioinformatics, Karnataka State Akkamahadevi Women University, Vijayapura, Karnataka, India

Lydia Bogomolnaya, Marshall University, Huntington, WV, United States

Bhaskar Chittoori, Boise State University, Boise, ID, United States

James Denvir, Marshall University, Huntington, WV, United States

Ajay Singh Dhama, Division of Biomedical Informatics, Indian Council of Medical Research, New Delhi, India

Archana Gupta, Department of Biotechnology, Faculty of Science, Veer Bahadur Singh Purvanchal University, Jaunpur, Uttar Pradesh, India

Manish Kumar Gupta, Department of Biotechnology, Faculty of Science, Veer Bahadur Singh Purvanchal University, Jaunpur, Uttar Pradesh, India

Payal Gupta, Department of Biotechnology, Graphic Era University, Dehradun, Uttarakhand, India

Sunil Kumar Gupta, Department of Pharmacoinformatics, National Institute of Pharmaceutical Education and Research, Hyderabad, Telangana, India

Ravi Kumar Gutti, Department of Biochemistry, School of Life Sciences, University of Hyderabad, Hyderabad, Telangana, India

Atifa Hafeez, Department of Biotechnology, Faculty of Science, Veer Bahadur Singh Purvanchal University, Jaunpur, Uttar Pradesh, India

Tejaswini Hipparagi, Laboratory of Natural Compounds and Drug Discovery, Department of Bioinformatics, Karnataka State Akkamahadevi Women University, Vijayapura, Karnataka, India

Swati Jain, North Middlesex University Hospital, London, United Kingdom

Madhulika Jha, Department of Biotechnology, Graphic Era University, Dehradun, Uttarakhand, India

Swarna Kanchan, Marshall University, Huntington, WV, United States; Boise State University, Boise, ID, United States

Shruti Kapoor, Department of Genetics, University of Alabama, Birmingham, AL, United States

Hiren Karathia, Greenwood Genetic Centre, Greenwood, SC, United States

Pramod Katara, Computational Omics Lab, Centre of Bioinformatics, IIDS, University of Allahabad, Prayagraj, Uttar Pradesh, India

Jasmine Kaur, Division of Biomedical Informatics, Indian Council of Medical Research, New Delhi, India

Sahista Zulfikar Keshavani, Department of Biotechnology, Atmiya University, Rajkot, Gujarat, India

Minu Kesheri, Marshall University, Huntington, WV, United States; Boise State University, Boise, ID, United States

Sanjay Kumar, Bioinformatics Center, Biotech Park, Lucknow, Uttar Pradesh, India

Raj Kumar, Department of Mathematics, Faculty of Engineering & Technology, Veer Bahadur Singh Purvanchal University, Jaunpur, Uttar Pradesh, India

Navin Kumar, Department of Biotechnology, Graphic Era University, Dehradun, Uttarakhand, India

Tapobrata Lahiri, Department of Applied Sciences, Indian Institute of Information Technology, Allahabad, Prayagraj, Uttar Pradesh, India

Deepika Lakhwani, ICAR-National Bureau of Plant Genetic Resources (ICAR-NBPGR), New Delhi, India

Smriti Mall, Molecular Plant Pathology Laboratory, Department of Botany, Deen Dayal Upadhayay Gorakhpur University, Gorakhpur, Uttar Pradesh, India

Rajkrishna Mondal, Department of Biotechnology, Nagaland University, Dimapur, Nagaland, India

Sapna Pandey, Department of Computational Biology & Bioinformatics, Jacob Institute of Biotechnology & Bio-Engineering, Sam Higginbottom University of Agriculture, Technology and Science (SHUATS), Allahabad, Uttar Pradesh, India

Donald A. Primerano, Marshall University, Huntington, WV, United States

Babu R.L., Laboratory of Natural Compounds and Drug Discovery, Department of Bioinformatics, Karnataka State Akkamahadevi Women University, Vijayapura, Karnataka, India

Rajesh P. Rastogi, Ministry of Environment, Forest, and Climate Change, IPB, New Delhi, India

Ratnaprabha Ratna-Raj, Boise State University, Boise, ID, United States; Texas A&M University, College Station, TX, United States

Swasti Rawal, SRC Division of Biophysics, Medical University of Graz, Graz, Austria

Srushti S.C., Laboratory of Natural Compounds and Drug Discovery, Department of Bioinformatics, Karnataka State Akkamahadevi Women University, Vijayapura, Karnataka, India

Sarika Sahu, ICAR-Indian Agricultural Statistics Research Institute, New Delhi, India

Rajesh Sharma, Department of Biotechnology, Veer Bahadur Singh Purvanchal University, Jaunpur, Uttar Pradesh, India

Soumya Sharma, ICAR-Indian Agricultural Statistics Research Institute, New Delhi, India

Naveen Sharma, Division of Biomedical Informatics, Indian Council of Medical Research, New Delhi, India

Aalok Shiv, ICAR-Indian Institute of Sugarcane Research, Lucknow, Uttar Pradesh, India

Bharti Shree, ICAR-National Bureau of Plant Genetic Resources (ICAR-NBPGR), New Delhi, India

Manjusa Singh, Department of Biotechnology, Faculty of Science, Veer Bahadur Singh Purvanchal University, Jaunpur, Uttar Pradesh, India

Ajay Kumar Singh, Department of Bioinformatics, Central University of South Bihar, Gaya, Bihar, India

Amit Kumar Singh, ICAR-National Bureau of Plant Genetic Resources (ICAR-NBPGR), New Delhi, India

Dev Bukhsh Singh, Department of Biotechnology, Siddharth University, Kapilvastu, Uttar Pradesh, India

Ram Lakhan Singh, Department of Biochemistry, Dr. Rammanohar Lohia Avadh University, Ayodhya, India

Satendra Singh, Department of Computational Biology and Bioinformatics, Sam Higginbottom University of Agriculture, Technology and Sciences, Allahabad, Uttar Pradesh, India

Sanjay Kumar Singh, Division of Biomedical Informatics, Indian Council of Medical Research, New Delhi, India

Harpreet Singh, Division of Biomedical Informatics, Indian Council of Medical Research, New Delhi, India

Rajeshwar P. Sinha, Banaras Hindu University, Varanasi, Uttar Pradesh, India

Upasna Srivastava, University of California, San Diego, CA, United States; School of Medicine, Yale University, New Haven, CT, United States

Apoorva Srivastava, Molecular Plant Pathology Laboratory, Department of Botany, Deen Dayal Upadhayay Gorakhpur University, Gorakhpur, Uttar Pradesh, India

Athaven Sukunathan, North Middlesex University Hospital, London, United Kingdom

Puru Supriya, ICAR-National Academy of Agricultural Research Management, Hyderabad, Telangana, India

Asmita Tripathi, Department of Applied Sciences, Indian Institute of Information Technology, Allahabad, Prayagraj, Uttar Pradesh, India

Shivani Tyagi, Computational Omics Lab, Centre of Bioinformatics, IIDS, University of Allahabad, Prayagraj, Uttar Pradesh, India

Akhouri Vaishampayan, Banaras Hindu University, Varanasi, Uttar Pradesh, India

Deepak Verma, School of Medicine, Johns Hopkins University, Baltimore, MD, United States

Pulkit Verma, Division of Biomedical Informatics, Indian Council of Medical Research, New Delhi, India

Nutan Prakash Vishwakarma, Department of Biotechnology, Atmiya University, Rajkot, Gujarat, India

Arvind Kumar Yadav, Institute of Life Sciences (ILS), Bhubaneswar, Odisha, India; ICAR-National Bureau of Plant Genetic Resources (ICAR-NBPGR), New Delhi, India

Keerti Kumar Yadav, Department of Bioinformatics, Central University of South Bihar, Gaya, Bihar, India

Shiv Kumar Yadav, Department of Mathematics, Faculty of Engineering & Technology, Veer Bahadur Singh Purvanchal University, Jaunpur, Uttar Pradesh, India

Nidhi Yadav, Department of Chemistry, SS Khanna Girls Degree College, University of Allahabad, Allahabad, Uttar Pradesh, India

Ravi Kumar Yadav, Department of Botany, Kashi Naresh Government Post Graduate College, Gyanpur, Uttar Pradesh, India

Tara Chand Yadav, Department of Computer Science and Biosciences, Marwadi University, Rajkot, Gujarat, India; Department of Electronics, Electric, and Automatic Engineering, Rovira I Virgili University (URV), Tarragona, Spain

Preface

The term '*-ome'* addresses the 'whole' (universe) and commonly used in biological sciences, such as genome, transcriptome, proteome, and metabolome. The word *omics* refers to the study of whole field 'ome', in biology; nowadays multiple omics are commonly in practice, the major of them are genomics (the structural and functional analysis of genes), transcriptomics (to study the gene expression profiling), proteomics (large-scale study of proteomes), and metabolomics (analysis of metabolome identification in particular conditions). Biological processes are very dynamic and a wide range of interactions between the genome, epigenome, transcriptome, proteome, metabolome, and phenome are necessary for both their regulation and functionality. Individual omics provide picture of single biological layer but unable to provide connection between them which, thus unable to provide complete picture of any biological system. To understand any biological process completely, it is essential to analyze how these biological layers interact with one another in addition to understanding them individually.

Availability of high-throughput techniques, such as next-generation sequencing, microarray, mass spectrometry etc., produced a huge amount of data mainly about genome, transcriptome, proteome, and metabolome. Emergence of bioinformatics resources created the opportunity to handle process and interpret these data in integrative manner. As a result, a new age in biological studies utilizing many of omics approaches in integrative manner (multiomics), which require the integration of distinct biological information types. Multiomics approach assists to construct a conception among the physiology and disease as well as to discover new connections among biological units and potential biomarker. The high-throughput experimental techniques, such NGS, and microarray are utilized to study the genomes, transcriptomes, and their epitomes, while the mass spectrometry techniques are used to examine the proteome and metabolome.

While there is hope that multiomics may be able to uncover hidden biological pathways, there is also an evident need for a "multilevel" biological analysis. Despite the advancements in high-throughput technology and the increasing amount of data available, the integration of omics remains a difficult topic. In addition to the limited accessibility of multiomics datasets for identical samples, technological constraints impede the integration procedure. Although the notion of integrative omics is not new, it has been used for the past decade the majority of existing applications only combine two or more omics with specific datasets. High-level integrative omics is now required due to the exponential growth in data generation across all omics levels (all level of omics with voluminous dataset). For computational biologists, creating such an algorithm and the necessary tools to deliver the necessary assistance will be extremely difficult. Big data analytics and artificial intelligence are the two fields that are currently being used to solve a variety of these kinds of issues, including those in astronomy, stock exchange, and marketing and business. Computational biologists have recently experimented with them and discovered their great potential to aid in computational omics. This book presents omics technology and its approaches and scope of their use in an integrative manner.

Organization of the book

This book *Integrative Omics: Concepts, Methodology and Application* covers all the recent aspects of 'omics and integrative omics,' which are structured in 23 chapters.

Chapter 1: ***From omic to multi-integrative omics approach*** traces the scope and of omics and integrative biology for holistic and systems effect and from omics to multiintegrative omics approach.

Chapter 2: ***Types of omics data: genomics, metagenomics, epigenomics, transcriptomics, proteomics, metabolomics and phenomics*** covers the types, methods, and advanced high-throughput data generation techniques such as microarray, genomics, metagenomics, epigenomics, transcriptomics, proteomics, metabolomics, glycomics, lipidomics, and phenomics high-throughput data generation by advanced technologies.

Chapter 3: ***Biological omics databases and tools*** describes the multiomics data generated from various high-throughput technologies and kept in systematics and organized format along with how the multiomics datasets will be integrated with the various biological tools available today.

Chapter 4: ***Systematic benchmarking of computational tools*** represents a systematic assessment of software tool performance, methods for the study of benchmarking study set, computational tools of systematic benchmarking processing of omics data analysis and challenges and limitations of benchmarking studies, case studies about the concept of how benchmarking results can be applied in specific omics domains.

Chapter 5: ***Pharmacogenomics, nutrigenomics and microbial omics*** are three areas have the capability to completely transform the way that personalized medicine, nutrition and microbiology are practiced.

Chapter 6: ***Proteomics: present and future prospective*** covers the comprehensive study of protein positioning, organization, posttranslational modification and also discusses the label-free quantitative proteomics, techniques used in proteomics, present prospective of proteomics, future prospective of proteomics, and advance tools used for proteomics.

Chapter 7: ***Foodomics: integrated omics for the food and nutrition science*** deals with the utilization of food and nutrition science to improve human nutrition, health, and exhaustively understand the functionality of food component through integrative omics approach.

Chapter 8: ***Vaccinomics*** aims to investigate and comprehend the core idea behind vaccinations, procedure of vaccine antigen designing, immunoinformatics in specific vaccine development, and limitations for vaccine design in order to clarify the underlying causes of the variety of immune reactions to vaccines.

Chapter 9: ***Integrative omics approach for identification of genes associated with disease*** covers the omics data of GWAS (Genome Wide Association Study), TWAS (Transcriptome Wide Association Study), and EWAS (Epigenome Wide Association Study) and how the multilevel data are integrated with the help of biological tool.

Chapter 10: ***Integrative omics approaches for identification of biomarkers*** includes the detection and discovery of biomarkers and molecular signatures of complex diseases through integrative methods across different multiomics datasets and pathway enrichment analysis of metabolites.

Chapter 11: ***Omics approach for personalized and diagnostics medicine*** covers the integration of clinical data of personal and group with patient-specific omics dataset to develop personalized and diagnostic medicine in various populations utilizing the translation approach, understanding the disease prognosis, influence on the approach to personalized medicine, and the challenges, opportunities, and future prospects.

Chapter 12: ***Role of bioinformatics in genome analysis*** highlights various bioinformatic tools for the analysis of genome sequence data including whole genome sequencing (WGA), next generation sequencing (NGS) technique, and methods to process raw data from genome-wide mRNA expression studies (microarrays and RNA-seq) including data normalization, differential expression, clustering, and enrichment analysis.

Chapter 13: ***Data management in cross omics*** covers the data management approach for volume and diversity, reproducibility, storing, updating, merging, inference, semantic search, and data workflow management of heterogenous omics profile dataset.

Chapter 14: ***Omics and clinical data integration and data warehousing*** covers the data security, protection, and privacy. It also covers the medical and translational research of clinical data through patient data management system and along with the biological research data integrated and stored in one unit database.

Chapter 15: ***Integrative omics data mining: challenges and opportunities*** represents high-throughput multiomics in human health and diseases, languages and database resources for multiomics analysis, methods, as well as challenges of multiomics data integration and data mining.

Chapter 16: ***Data science and analytics, modeling, simulation and issues of omics dataset*** incorporates the basic fundamentals of data science and analysis of integrated omics data, hardware and software requirement for data analytics, biological tools used for modeling and simulation of the constructed integrated network data, and also covers the issues associated with the omics data generated by high-throughput technologies.

Chapter 17: ***Emerging trends in translational omics*** highlights the utilization of various multiomics areas in clinical practice, translational omics, omics-based precision medicine and tests and its development process, discovery phase of omics-based test development, and issues and limitations of translational omics analysis.

Chapter 18: ***Omics technology for crop improvement*** represents the subject of food security, the use of omics technologies to improve agricultural yields and nutritional content, the analysis of multiple technical challenges and complexities throughout the implementation of omics technologies, as well as advances in sequencing, omics technologies with artificial intelligence to improve crop quality along with quantity and gene expression networks in crops.

Chapter 19: ***Ecology and environmental omics*** includes environmental and genetic factors, toxicity mechanism concept in response to chemical, application of high-throughput molecular technologies in environmental sciences and molecular profiles for identifying toxic mechanisms, toxicity signatures, biomarkers, and pathways after exposure to environmental chemicals.

Chapter 20: ***Current trends and approaches in clinical metagenomics*** includes the techniques, analysis of clinical metagenomics, metagenome annotation, pipelines for metagenomic data analysis, metagenomic database of clinical importance, metagenomic database of clinical importance, role of clinical metagenomics in human health, concern and issues of clinical data handling, and applications of clinical metagenomics.

Chapter 21: ***Biomolecular networks*** covers network measurers, node and hub nodes, robustness, modularity, subgraph, motifs, motif cluster, modules, betweenness centrality, and network topology. gene regulatory network, gene coexpression network, protein–protein interaction network, RNA network, metabolic network, cell signaling network, neuronal network, etc. It also includes the scope and application of molecular network in different areas of biological sciences.

Chapter 22: ***Machine learning fundamentals to explore complex OMICS data*** includes the machine learning methods for integrating multilevel omics data and machine learning (ML)–based data mining methods i.e. (i) classification: supervised classification, (ii) clustering, and (iii) probabilistic graphical models; the chapter will also include deep learning approach, artificial intelligence (AI).

Chapter 23: ***Omics technology policy and society research*** covers consistent tools, methods, technology, catalogue of resources, tools and repositories, harmonization, interoperable systems, and the ethical, legal, and social implications of both practice and their products along with the principle of transparency, privacy, regulation, informed consent, counseling, etc.

Manish Kumar Gupta
Department of Biotechnology, Veer Bahadur Singh Purvanchal University, Jaunpur, Uttar Pradesh, India

Pramod Katara
Centre of Bioinformatics, University of Allahabad, Prayagraj, Uttar Pradesh, India

Sukanta Mondal
National Institute of Animal Nutrition and Physiology, Adugodi, Bangalore, Karnataka, India

Ram Lakhan Singh
Department of Biochemistry, Dr. Rammanohar Lohia Avadh University, Ayodhya,Uttar Pradesh, India

Chapter 1

From omic to multi-integrative omics approach

Pramod Katara and Shivani Tyagi
Computational Omics Lab, Centre of Bioinformatics, IIDS, University of Allahabad, Prayagraj, Uttar Pradesh, India

1. Introduction

Biomolecules are the substances that are produced by cells and living organisms. Biomolecules are generally small size molecules with different sizes and structures and play a wide range of bio functions. The four major types of biomolecules are carbohydrates, lipids, nucleic acids, and proteins. Though all biomolecules have their own functions which are crucial for the cell systems, nucleic acids and proteins are the key molecules. Nucleic acids in the form of DNA and RNA play a central role and have the unique function of storing an organism's genetic code, and because of that it is also considered as an informative molecule of the system. It carries the information in the form of the sequence of nucleotides that determines the amino acid sequence of proteins. Protein is another molecule that is very crucial and critical for life, almost all specific functions within the cells are performed by the proteins (mainly in the form of the structural and enzymatic proteins), and thus it's considered as functional molecule of the system. Sequence patterns of the 20 different amino acids which act as a building block of the proteins play a fundamental role in shaping protein structure and function. In combination, nucleic acid (DNA/RNA) and proteins make a central dogma of the life, which decides the flow of the genetic information required for the life. Traditionally nucleic acids and proteins were studied to understand the genetic information and functionality of any cell or living organism for various purposes. Growth of the science and technology come up with various advance concepts and techniques which allow advanced studies at a high-throughput level about these biomolecules and gave birth to "-omics", a branch of technologies (Hasin et al., 2017).

1.1 Omics

The term "-ome" was initially introduced by bioinformaticians and molecular biologists. "Ome" and omics concepts are now widely utilized by biologists especially when they deal with high throughput experimental analyses to get a holistic view of a particular molecular system. Omics can be defined as a collection of various technologies, i.e., next-generation sequencing (NGS), DNA and cDNA microarray, nuclear magnetic resonance (NMR), and mass spectrometry (MS), to understand and examine the roles, relationships, and actions of the biomolecules. Omics mainly lay with the flow of central dogma that includes genomics, transcriptomics, and proteomics, which deal with the study of the genome, transcriptome, and proteome. As technologies and molecular information are increasing day by day, the number of "-ome" and "-omics" are increasing exponentially (Vailati-Riboni et al., 2017).

1.2 Genomics

According to the *Central Dogma*, the genome makes the basal layer of the molecular information flow. Every cell has its own genome which is static in nature and all cells and tissues of the organism contain identical DNA content unless there is no mutation. Genomics is the most understood, practiced, and explored omics which utilizes various techniques including genome sequencing, genetic engineering, and bioinformatics to determine the nucleotide patterns, function, mapping, and editing of the genome (Del Giacco & Cattaneo, 2012). The emergence of NGS starts to produce huge data at an

Integrative Omics. https://doi.org/10.1016/B978-0-443-16092-9.00001-1

exponential rate, which increases the possibilities for various genomic investigations as well as created data management and mining issues (van Dijk et al., 2014). In the recent past, some object-specific genomics has also emerged, e.g., metagenomics, pharmacogenomics, phylogenomics, etc (Hugenholtz & Tyson, 2008).

1.3 Transcriptomics

The study of the transcriptome—the complete set of RNA, i.e., transcripts that are produced by the cell at any given time—is considered transcriptomics (Table 1.1). The major high-throughput approaches which are dealing with transcriptome are cDNA microarray and RNA-Seq analysis. By definition, transcriptome comprises all kinds of transcript (RNA) practically transcriptomics mainly focuses on protein-coding RNA, i.e., mRNA. Transcriptome reflects expressed genes in given conditions, and thus acts as an essential key to deducing a functional set of genes involved in any given condition including diseases (Borrageiro et al., 2018; Katara, 2020). Transcriptomics was the first technology that deals with multiomics studies; it correlates the genome and transcriptome by providing gene expression information through transcripts (Stahl et al., 2012).

1.4 Proteomics

The proteome is the entire set of proteins of the cell at a particular given time. Proteomics involves the extensive study of the proteome (e.g., structure and physiological role under a given set of conditions). MS technology, 2D-PAGE,

TABLE 1.1 Major objectives and intrinsic challenges for individual omics.

Omics	Objective	Challenges
Genomics	• Attempts to understand the structural organization and function of genomic elements. • Find genomic variations in the given DNA sequence (e.g., SNPs, CNV) of individuals and determine their significance. • Develop and apply genome-based strategies for the early detection, diagnosis, and treatment of disease.	• DNA sequence information is static by nature and is not directly informative about biological mechanisms encoded within it. • Biological problems are more often investigated on a single-cell level due to cellular heterogeneity, it's difficult to get a realistic conclusion.
Transcriptomics	• To quantify and catalog all species of the transcript, including mRNAs, noncoding RNAs, and small RNAs. • To quantify the differential gene expression under different conditions and their biological significance.	• Transcripts are regulated by different biological processes and molecules (e.g., miRNA) thus transcriptomics alone is not capable to provide a clear picture of gene expression.
Proteomics	• To understand the structure and function of the proteins in various biological processes. • To determine the effect of genetic modifications and biological stress on protein structure and functions • To discover the protein interaction, i.e., protein–protein interaction, DNA–protein interaction, and protein–drug interactions.	• Isolation and purification of most of the proteins are not easy. • Proteomics alone is not capable to deal with all these aspects.
Metabolomics	• To perform metabolic profiling and fingerprinting in different biological conditions i.e., disease, stress, and infection. • Identification of new disease indicators (markers). • On the basis of the metabolic profiling and fluxes, it attempts to establish the connection between genotype and phenotype.	• It's almost impossible to draw any conclusion by only considering metabolome data. • To develop a true and robust marker, need intense data.
Epigenomics	• To observe nongenetic (epigenetic) variations in the genome and their impact. • To establish an association between epigenetic modifications and biological circumstances.	• Limited experimental data. • Lack of standardized protocol and algorithm.

chromatographic techniques, and bioinformatics are the major technologies that facilitate various dimensions of proteomics (Zhang et al., 2014). Experimental technique such as MS, X-ray crystallography, etc. generates the primary information about the proteome (sequence, structure, interaction, etc.), and bioinformatics helps in the maintenance, processing, and analysis of the proteome information and data (Table 1.2). Proteome is the main bimolecular component but due to some technical limitations, proteomics alone is not able to provide the complete picture (Kumar et al., 2022). To understand the protein function and its involvement in various cellular processes, proteomics needs the use of multiomics and data integration (Altelaar et al., 2013).

1.5 Metabolomics

The set of low molecular weight organic compounds (mass range of 50−1500 Da) are considered as metabolome. These compounds are not directly encoded by the genome and are formed as a product of the functional spectrum of the proteome. Therefore, the metabolome constitutes a "phenotype" of the cell (Schrimpe-Rutledge et al., 2016). Metabolomics was most recently introduced among the omics to analyze low molecular weight organic compounds in different biological systems and fields of research. As metabolomics provides insights into metabolic reactions and biochemical processes of living organisms that define phenotype, it is used across multiple research areas. The major research areas are cancer, metabolic and cardiovascular disease, immunology, neurology, genomics, animal biology, and plants (Johnson et al., 2016).

TABLE 1.2 Omics techniques and major bioinformatics resources.

S. No.	Omics	Experimental techniques	Data resources	Software
1	Genomics	Sequencing technology, i.e., NGS (WGS, WES)	SRA, ENA, human genome project, HGNC database, GenAtlas, HGVbase, HapMap Consortium, 1000 genome browser	BLAST, Galaxy, GATK, Freebayes, BWA, Bowtie, MEGA, GCTA, PLINK
2	Transcriptomics	cDNA-microarray, RNA-Seq. ESTs, SAGE	GEO, SRA, dbESTs, UniGene	Bioconductor, Cufflinks, TopHat, WGCNA, FASTQC, Samtools, bowtie2
3	Proteomics	X-ray crystallography, NMR, mass spectroscopy, protein microarray, homology modeling	PDB, PIR, Swiss-Prot, STRING	Modeler, Gromacs, Autodock, Amber, Glide, SwissModel, Pymol, Cytoscape
4	Metabolomics	Mass spectrometry (GC-MS, LC-MS, CE-MS), NMR	HMDB, MetaboLights, MetaCyc, METLIN, MassBank, META-GENE, MetaboLights, Reactome	sFAME, GEMSiRV, MEMOSys, merlin, RAVEN Toolbox, SuBliMinaL
5	Epigenome	DNA-modification through NGS, DNA-methylation, DNA−protein interaction	The epigenome atlas, DeepBlue	Epigenome Browser
6	Fluxomics	NMR, mass spectrometry (GC-MS, LC-MS, CE-MS	PeptideAtlas, PRIDE, METLIN	OpenFLUX, VistaFlux, Flux-P, geoRge
7	Interactomics	Phage display, molecular network, Docking	STRING, Intact	Cytoscape, Gephi,
8	Pharmacogenomics	NGS (WGS, WES, RNA-Seq), SNPs	PharmGKB, ClinVar, Drug bank	g-Nomic, PGRN
9	Metagenomics	NGS, RNA-Seq, 16S-RNA technology	ENA, SRA	MetaWRAP, MEGAN-LR
10	Phenomics	Infrared imaging, 3D imaging, magnetic resonance imaging	Phenomicsdb	Human Phenotype Ontology (HPO)
11	Foodomics	Mass spectrometry (GC-MS, LC-MS, CE-MS), NMR	Foodomics database, FooDB, MetaboLights	MetaboAnalyst, SimMet

1.6 Epigenomics

Epigenomics is the study of genome-wide characterization of heritable changes in gene expression that do not involve changes to the underlying DNA sequence—a change in phenotype without a change in genotype. Those changes can be influenced both by genetic and environmental factors, can be long-lasting, and are sometimes heritable. Epigenetic mechanisms are mediated by either chemical modifications of the DNA itself or by modifications of proteins such as chromatin that are closely associated with DNA. Some of the best-characterized epigenetic modifications thought to initiate and sustain epigenetic changes include DNA methylation, chromatin remodeling, histone modification, and noncoding RNA (ncRNA) associated mechanism (Wang & Chang, 2018). Though epigenomics studies did not remain fancy for the scientific world, reported epigenome-wide association studies in biological processes and disease development processes are evidence of their significance and potential. In the recent past, epigenomics studies are trying to explore the association between epigenetic modifications and diseases (Gupta & Hawkins, 2015). It has great potential to enhance our functional interpretation and development of epigenetic markers associated with disease independently of genetic variation.

1.7 Other potential omics

1.7.1 Fluxomics

Fluxomics is considered as a branch of metabolomics that deals with the rates of all intracellular fluxes in the metabolism of biological systems. To understand the fluxes and their correlation, it gathers data from different -omics fields. It helps to investigate the whole picture of molecular interactions and provides metabolic phenotypes. As fluxomics reveals the functioning of multimolecular metabolic pathways, nowadays it is routinely applied in pharmacology and clinical science (Emwas et al., 2022; Winter & Krömer, 2013).

1.7.2 Interactomics

Interactomics deals with the interaction between any of the small and large macromolecules and provides comprehensive information of that particular organization at molecular levels, e.g., gene–gene interaction, gene-regulatory network, protein–protein interaction network, and protein-small molecules (protein-metabolites) interaction network. Interactomics provides a systemic picture of molecular interaction, thus allowing for getting a better understanding of the flow of signals among the molecules (James et al., 2022; Verma et al., 2020).

1.7.3 Metagenomics

Metagenomics, which is also known as community genomics or ecogenomics, is the study of genetic material recovered directly from environmental samples. It provides a better understanding of the microbial ecology in conditions of interest (e.g., disease), which is otherwise not possible (Wajid et al., 2022).

1.7.4 Pharmacogenomics

Pharmacogenomics is the study of the role of the genome in drug response. It attempts to correlate genomic variation and drug response. It analyzes how the genetic makeup of an individual affects their response to drugs and may cause variable drug response, which sometimes converts into an adverse drug response as well (Katara & Yadav, 2019; Yadav et al., 2023).

1.7.5 Phenomics

Phenomics deals with the study and measurement of the phenomes, where a phenome is a set of phenotypes (physical and biochemical traits). The considered phenome can be produced by a given organism over the course of development and in response to surrounding influences (genetic mutation and environmental influences) (Houle et al., 2010).

1.7.6 Foodomics

Foodomics is connected with food compounds (foodome). It attempts to explore the connection and impact of foodome with respect to living biological systems. Foodomics utilizes transcriptomics, proteomics, and metabolomics, together with biostatistics, chemometrics, and bioinformatics (Rodríguez-Carrasco, 2022).

2. From omics to multi-integrative omics

The central dogma of molecular information itself indicates that there is an influence of each -omic on the other. Individual omics are not capable to explain any biological process even in single-cell organisms. There is a need to consider multiple molecular components in a holistic manner to get the real picture of any biological process (Balmer et al., 2015). System's biology emerged as a field that aims to consider a system and its functions as a whole rather than an individual component and attempts to model biological systems and simulate their outcomes. Biological processes are tremendously vivacious, and their regulation as well as functionality involves multiple biomolecules. To understand and model such a process, a multitude of interactions between -ome are required (Fig. 1.1). System's biology clearly indicates that to understand any biological process comprehensively it is critical not only to understand elements of individual -omics, but to dissect how they interact with one another across omics. For example, to understand the effect of nonsynonymous SNP on protein structure and functions, one has to rely on genomics as well as proteomics (Pandey et al., 2020). An unprecedented pace in the development of "omic" technologies has greatly increased access to multiomics data across biosciences. Studies involving multiple of these techniques (multiomics) have given rise to a new era in systems biology, which needs to integrate and combine very diverse types of biological information to understand any biological mechanism completely (Subramanian et al., 2020; Yadav et al., 2020). The fundamental theory behind multi-integrative omics is that multiple types of molecular profiles (e.g., CNV [copy number variation], DNA methylation, and protein) might provide a more rational and comprehensive signature of any biological process (Katara et al., 2023; Li et al., 2020).

3. Potential of multi-integrative omics

Integration of multiomics data helps us to get understood any kind of physiological process, e.g., diseases, pathophysiological conditions, etc. There are various dimensions and probable scopes for integrative -omics, and medical science is a field where it is explored most. Almost every -omics produced clinical, disease, and pathophysiology-related data. The development of various databases makes these data available to scientists in a user-friendly manner. Bioinformatics software and algorithm make it feasible to use multiomics-based integrative concepts on these data which can provide a clear and comprehensive understanding of health and disease-related physiological process (Goh, 2018; Schneider & Orchard, 2011). On the basis of genome-wide association and other studies (protein and metabolic fingerprinting), integrative omics projected various disease-diagnosis markers for a range of diseases and few of them are already in practice (Heo et al., 2021).

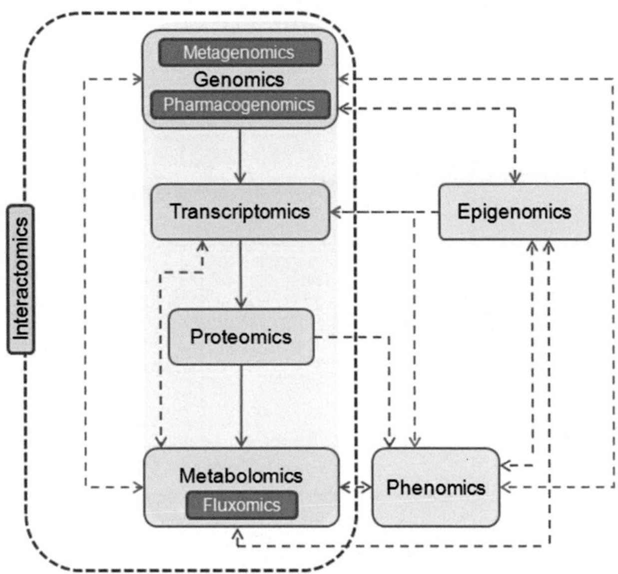

FIGURE 1.1 Interdependencies of -omics. Omics have different types which lay surrounding the central dogma of molecular information. They all are dependent on each other in a community manner.

4. Integration and interdependencies of omics

Data integration is the very first step to practicing integrative omics for any purpose. Due to the diverse nature of each "-omics" there is a need for an array of experimental techniques, different protocols, handling, and interpretation by diverse scientists or specialists, because of all these factors, data integration has been characterized as one of the key bottlenecks to all multiomics studies (Schneider & Orchard, 2011). As per the objective, different data integration approaches are utilized by scientists to interpret the multiomics data (Pinu et al., 2019). These approaches can be categorized broadly into the following two types: (1) postanalysis data integration approaches, and (2) integrated data analysis approaches.

4.1 Postanalysis data integration approaches

Integrative omics approaches are relatively new, thus most of the data which is generated in the past and available at data resources are derived through individual omics. In such conditions, postanalysis data integration approaches allow scientists to first analyze -omics data individually then on the basis of the individual analysis, their key outcome features are integrated together and finally come out with a significant conclusion (Fig. 1.2). Though, theoretically, these kinds of conclusions are multiomics based, in reality, due to several reasons (e.g., data generation with a different purpose, different scales of omics data), mostly, they compromise with coverage, sensitivity, and specificity of the complete data. At the same time, this approach relies on manual involvement almost at every stage, e.g., data selection, selection of key features for integration, data explanation, etc., thus being sensitive to human biases (Urbanski et al., 2019).

4.2 Integrated data analysis approaches

In contrast to postanalysis data integration, the integrated data analysis approach utilized software with integrated pipelines which facilitate the handling and merging of different omics datasets before the analysis and interpretation process (Fig. 1.2). As integrative pipelines provide automated data handling, integration, and analysis process, there are very less chances of human biases. Due to its integrative workflow, it is expected that these kinds of approaches are more reliable for multiomics analysis. Though integrated data analysis approaches can provide a more realistic advantage of multiomics, due to the lack of purpose-specific pipelines, it is currently not preferably used by scientists (Reel et al., 2021).

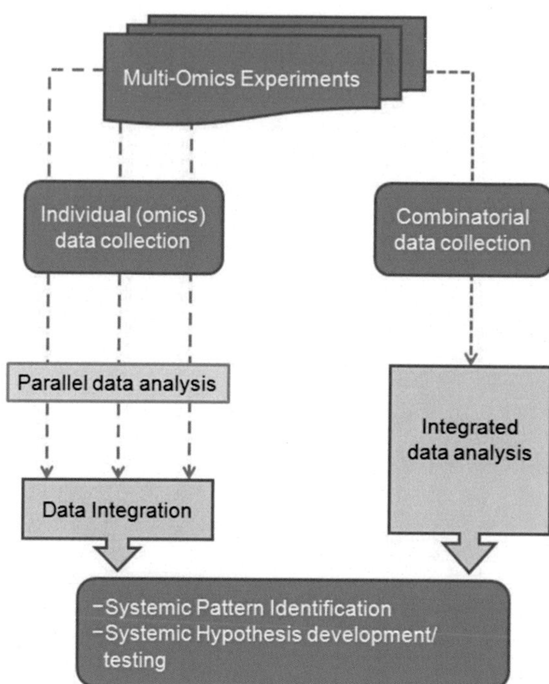

FIGURE 1.2 Data integration and analysis approach for integrative omics studies to get a systemic understanding, as shown in the figure, data integration can proceed in two ways. Though the postanalysis data integration method is relatively easy, it misses a lot of primary information thus integrated data analysis approaches are recommendable.

5. Data for multi-integrative omics

Generally, data are of specific types which are mainly on the basis of information classified as primary and secondary databases. Mostly these databases belong to a particular information type, i.e., GenBank—store data of gene sequence (genomics), and Protein Data Bank (PDB)—store data belonging to protein structure (Proteomics) (Burley et al., 2017). These individual omics databases are good with their information, but not supportive of integrative analysis. The user has to access different databases for a different type of information on the same molecules (gene/protein).

5.1 Multiomics data

Multiomics data broadly cover the data generated from the genome, proteome, transcriptome, metabolome, and epigenome. Multiomics data generated for the same set of samples can provide useful insights into the flow of biological information at multiple levels and thus can help in unraveling the mechanisms underlying the biological condition of interest (Table 1.3).

5.2 Integration of multiomics data

Composite databases contain information from a variety of primary databases (genomics/transcriptomics/proteomics etc.), which eliminates the need to search each one separately. As there is a need of an integrative approach, with the help of high-performance computational platforms, these databases have become important in providing the infrastructure needed for biological research, from data preparation to data extraction.

6. Data mining and exploitation for omics data

In the last 2 decades, an unprecedented amount of data from different omics have been produced, and most of them are stored and available in the primary or secondary form in a range of biological databases. As most of these data are produced by high-throughput techniques (e.g., microarray and NGS), before use, they need data-specific mining and a rigorous preprocessing process. Just like any other data, -omics data mining needs to follow some specific steps before they are used for a specific purpose (Wu et al., 2021). Though these mining steps are standard, sometimes on the basis of the data-mining objectives few of them can be modified.

The major steps for data mining are as follows:

1. *Data cleaning/Noise removal:* It removes unwanted, unexpected (noise), and incomplete/missing data. As the presence of noise or low data quality can affect the entire knowledge extraction process, data cleaning is always considered as a first necessary step.
2. *Data integration:* As most of the data come from different data sources they need to be aggregated.
3. *Data selection:* On the basis of the data mining objective, relevant data are selected.
4. *Data transformation:* Data transformation needs to remove the effect of variability and convert the data into a specific range. Sometimes it also integrated new attributes or features, useful for the process of data mining.

TABLE 1.3 List of multiomics data repositories.

S. No.	Data repository	Description	Types of multiomics data available
1	KEGG	Mainly focused on biological pathways	Genomics, transcriptomics, proteomics, and metabolomics
2	Omics discovery index	Consolidated datasets from 11 repositories in a uniform framework	Genomics, transcriptomics, proteomics, and metabolomics
3	PharmGKB	Pharmacogenomic data	Genomics, transcriptomics, proteomics, and metabolomics
4	TARGET	Focused on pediatric cancers	Gene expression, miRNA expression, copy number, and sequencing data
5	The cancer genome atlas (TCGA)	Cancer-related data	RNA-Seq, DNA-Seq, miRNA-Seq, SNV, CNV, DNA methylation, and RPPA

5. *Data mining:* It searches for the patterns of interest through different mining algorithms (classification, clustering, decision trees, regression, etc.).
6. *Data Visualization:* Mining approaches provide data patterns (e.g., data clusters) that need to be visualized for better interpretation.
7. *Pattern evaluation:* Once patterns are extracted they need evaluation for their reliability and fitting for further use.
8. *Knowledge presentation:* The extracted knowledge and information are presented.

7. Data mining tools/software

To handle and perform the data mining task, especially when data are huge in size, as almost in all omics cases, there is a need for specific data mining tools. Several data mining tools/software are available nowadays, but most graphical user interface (GUI)-based tools are purpose specific and have their own advantages and limitations. Among data analysts, "R" is one of the most preferable open access tools, which is specifically designed for statistical computing and graphics. "R" supports a wide range of pattern mining methods and can easily handle data for classical statistical tests, time-series analysis, classification, and graphical techniques. "R" offers bioconductor which exclusively offers effective data handling and storage facility, for -omics related data (Rajasundaram & Selbig, 2016). The availability of object-oriented packages in bioconductor is the reason that it is one of the preferable tools for bioinformaticians and computational biologists (Fig. 1.3). Along with the packages, it also offers extensive interfaces to other open resources databanks (e.g., GEO, Array-Express, Biomart, genome browsers, GO, KEGG), and diverse annotation sources (www.bioconductor.org/). Weka is another open-source tool for data mining; it offers a collection of machine-learning algorithms that can be applied directly to the data. Users can use Weka for regression, clustering, association, data preprocessing, classification, and visualization purposes (https://www.cs.waikato.ac.nz/ml/weka/).

8. Integrative data mining challenges and possibilities

The rapid advancement of high-throughput technologies provides opportunities to perform research experiments in a rapid manner at a low cost, which resulted in an enormous amount of data production. As there is a variation in the nature of each -omics, the generated data raise various challenges for integration (Naqa et al., 2018). Data scientists are trying to provide compatible analysis tools and other resources to read -omics data in a machine-readable format. At some level, scientists

FIGURE 1.3 Software, tools used for the analysis of -omics data in an integrative manner a range of computational resources is available which offer -omics data analysis. The selection of these software and tools depends on the user's requirements.

S.No.	↓Software | Omics→	G	T	P	M	F	Mg
1	3Omics			√	√		
2	BioCyc/ MetaCyc	√		√	√		
3	COBRA		√	√	√	√	
4	E-Cell		√	√	√		
5	Gaggle	√	√	√	√		
6	IMPaLA		√	√	√		
7	MarVis-Pathway		√		√		√
8	MetaboAnalyst	√	√	√	√		
9	PaintOmics		√		√		
10	Reactome		√	√	√	√	
	R Packages for multi Integrative-Omics						
11	mixOmics (R)		√	√	√		√
12	Omickriging (R)		√	√	√	√	
13	OmicsEV (R)		√	√			
14	OmicsPLS (R)		√	√	√		√
15	PathwayMultiomics (R)	√	√		√	√	

#F: Fluxomics, G: Genomics, T: Transcriptomics, P: Proteomics, M: Metabolomics, Mg: Metagenomics

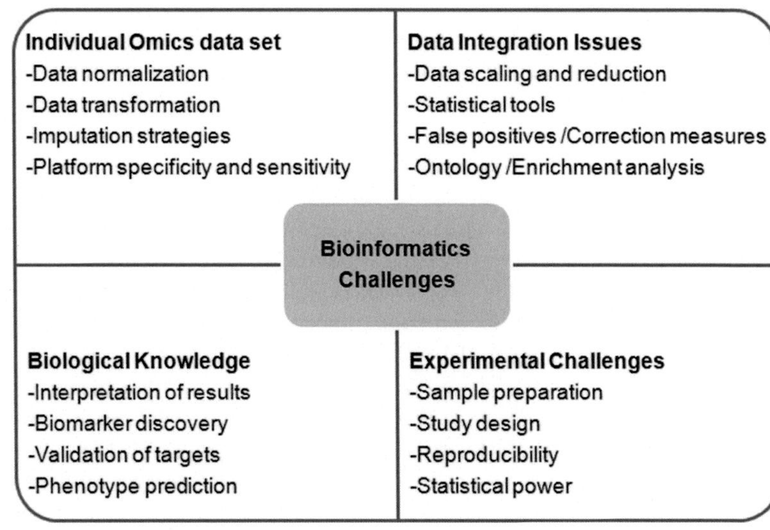

Individual Omics data set
-Data normalization
-Data transformation
-Imputation strategies
-Platform specificity and sensitivity
Data Integration Issues
-Data scaling and reduction
-Statistical tools
-False positives /Correction measures
-Ontology /Enrichment analysis
Bioinformatics Challenges
Biological Knowledge
-Interpretation of results
-Biomarker discovery
-Validation of targets
-Phenotype prediction
Experimental Challenges
-Sample preparation
-Study design
-Reproducibility
-Statistical power

FIGURE 1.4 Bioinformatics challenges for integrative data mining. Bioinformatics challenges play a very crucial role in multiomics integration. As these challenges are inherent in biological data, with the help of current data mining approaches it's a challenging task to overcome them.

succeeded still various factors are there that challenge the data integration and obstruct further progress in multiomics integration (Choi & Pavelka, 2012; Karczewski & Snyder, 2018). A few of the major challenges are as follows:

1. *Integration of qualitative and quantitative data:* Most of the "omics" data are only qualitative in nature, making it very hard to reproduce and even harder to compare (Fig. 1.4). When only qualitative data are available, multiomics integration, particularly from multiple sources, becomes difficult (Pinu et al., 2019).
2. *Lack of sufficient meta-data:* Meta-data provides the details about the experiment; it is very crucial in data pooling; lack of sufficient meta-data affects purpose-specific data integration
3. *Selections of database and tools:* With the exponential growth of bioinformatics, hundreds of software and tools are now available for each purpose. At the same time, it is very difficult for the user to differentiate between reliable and nonreliable, which cause confusion in selection for integration purpose
4. *Variable data formats:* The availability of the data in different formats, especially in nonstandard formats creates problems in data integration.
5. *Noninteroperable tools:* Lots of programs/tools work on nonstandard formats and sometimes also provide output data in such a format that is incompatible with other programs
6. *Voluminous Data:* Next-generation experimental methods (e.g., WGS, RNA-Seq, etc.) are working in high-throughput manner, thus even in the single run they are producing gigantic amounts of data. Practicing the integration of these data from multiomics simultaneously is a very tedious and challenging task.
7. *Incomplete data:* Omics deals with the bio-molecules, research is going on to get inside of each biomolecule and biological process, but still, the available information is far away from the complete picture. The use of these incomplete data to develop any hypothesis or model is prone to provide false positive or negative conclusions.
8. *Variety of data:* Omics data are produced by different experimental methods even in the same type of techniques there are different platforms that produced data in a different manner (e.g., Affymetrix microarray vs. Agilent microarray). Integrating data produced from different experiment platforms is not an easy task for users. In fact, every platform needs its own pipelines to preprocess, normalize, and analyze the data. Integration of these data for a common purpose raises challenges.
9. *Quality of data:* Uses of high-throughput techniques saves time and money, but due to their high speed and large sample size, they compromise the quality of the resulting data. Because of the poor quality or noise in the raw data, irrespective of the kind of "omics," they must have to go through preprocessing and quality enhancement processes before proceeding with any objective. Because of the size of the data, preprocessing is also not an easy task and faces computational challenges (Maturana et al., 2019).
10. *Dynamic nature of data (velocity)*: Except genomics, every other -omics is dynamic in their nature. This dynamic nature of the -omics data and data velocity creates a huge impact on data interpretation and decision-making. As data are increasing day by day and their integration for multiomics-based analysis is anticipated. It is assumed that in the very near future, this field will also face the "concept drift" problem, and every analysis comes with different outcomes.

9. Scope for data science in multi-integrative omics

Observing the features of -omics and multiomics data, it is very much clear that they are very much extreme and their handling is challenging. Using them in an integrative manner to check any holistic hypothesis is not an easy and straightforward task. Near observations of these -omics data conclude that broadly they are facing big data issues which are *volume, variety, velocity*, and *veracity*, sometimes along with value (Tolani et al., 2021). In this information technology era, almost every data-related field (e.g., share market, social networking, etc.) is facing big data issues. Considering the demand, a lot of work is already going on to handle these issues by the data scientists. Data scientists are attempting to use data analytics concepts like artificial intelligence, machine learning, and deep learning to handle all data-related issues and on the basis of these concepts, various tools for data processing are already in practice. Both structured and unstructured data can be handled by these specially designed tools. Various companies are providing their own tools which are mainly designed to handle and process big data in an efficient manner. Cluster, parallel computing, and cloud computing are the main elements that act as the backbone for these data analytics purposes. From the multiomics point of view, the major features which are expected from these data analytics tools are—(i) analytic capabilities, (ii) data import and export, (iii) integration, (iv) visualization, and (v) scalability. Data analytics and big data analytics are now almost established fields and commonly utilized for various domains, the computational biologists also attempt to use big data analytics approaches in the near past with hope (Reel et al., 2021; Tolani et al., 2021). Despite the potential, data analytics is still a new field and its use in computational biology and computational omics are still rare (Katara et al., 2023). The emergence of fields like systems biology, multiomics, and integrative omics raises the demands for the use of big data analytics in computational omics fields (Tolani et al., 2021). It is expected that very soon computational biologists will get bioinformatics-oriented big data analytics tools and will be able to handle various data and big data issues of computational omics in a more user-friendly manner.

10. Conclusion

Traditionally genomics, transcriptomics, proteomics, and other emerging omics are in practice as individuals branch to get inside of living systems with different molecular aspects. The emergence of systems biology redefines the view to understand the systems and raises the need for a holistic approach to view any biological systems for a realistic picture. Integrative omics have the potential to provide a relatively better picture of any biological process and can guide us to connect genome to phenome. Considering the need for multi-integrative omics, databases, software, and data mining software are already in the pipeline to provide integration of diverse omics data and their analysis. In a very short period of time integrative -omics already proved its advantage over any single omics technology, especially in the field of medical sciences. Despite its potential, currently, the wide use of integrative omics is not that straight. Heterogeneity among the omics is reflected in their datasets and acts as a major challenge for data integration and integrative analysis. As the size of omics data is also huge, it causes some additional hurdles in practice. It is assumed that in the near future with the use of advanced integrative data analysis pipelines and data analytics techniques, multi-integrative omics will be frequently utilized by scientists to get a better understanding of living systems.

References

Altelaar, A. F. M., Munoz, J., & Heck, A. J. R. (2013). Next-generation proteomics: Towards an integrative view of proteome dynamics. *Nature Reviews Genetics, 14*(1), 35–48. https://doi.org/10.1038/nrg3356

Balmer, A., Pastor, V., Gamir, J., Flors, V., & Mauch-Mani, B. (2015). The 'prime-ome': Towards a holistic approach to priming. *Trends in Plant Science, 20*(7), 443–452. https://doi.org/10.1016/j.tplants.2015.04.002. http://www.elsevier.com/inca/publications/store/3/0/9/6/0/index.htt

Borrageiro, G., Haylett, W., Seedat, S., Kuivaniemi, H., & Bardien, S. (2018). A review of genome-wide transcriptomics studies in Parkinson's disease. *European Journal of Neuroscience, 47*(1), 1–16. https://doi.org/10.1111/ejn.13760. http://www.blackwell-synergy.com/loi/EJN

Burley, S. K., Berman, H. M., Kleywegt, G. J., Markley, J. L., Nakamura, H., & Velankar, S. (2017). Protein Data Bank (PDB): The single global macromolecular structure archive. *Methods in Molecular Biology, 1607*, 627–641. https://doi.org/10.1007/978-1-4939-7000-1_26. http://www.springer.com/series/7651

Choi, H., & Pavelka, N. (2012). When one and one gives more than two: Challenges and opportunities of integrative omics. *Frontiers in Genetics, 2*. https://doi.org/10.3389/fgene.2011.00105. http://www.frontiersin.org/Journal/DownloadFile.ashx?pdf=1&FileId=1591&articleId=17228&Version=1&ContentTypeId=21&FileName=fgene-02-00105.pdf

Del Giacco, L., & Cattaneo, C. (2012). Introduction to genomics. *Methods in Molecular Biology, 823*, 79–88. https://doi.org/10.1007/978-1-60327-216-2_6

van Dijk, E. L., Auger, H., Jaszczyszyn, Y., & Thermes, C. (2014). Ten years of next-generation sequencing technology. *Trends in Genetics, 30*(9), 418–426. https://doi.org/10.1016/j.tig.2014.07.001

Emwas, A. H., Szczepski, K., Al-Younis, I., Lachowicz, J. I., & Jaremko, M. (2022). Fluxomics—New metabolomics approaches to monitor metabolic pathways. *Frontiers in Pharmacology, 13*. https://doi.org/10.3389/fphar.2022.805782. http://www.frontiersin.org/Pharmacology

Goh, H. H. (2018). Integrative multi-omics through bioinformatics. *Advances in Experimental Medicine and Biology, 1102*, 69–80. https://doi.org/10.1007/978-3-319-98758-3_5. http://www.springer.com/series/5584

Gupta, B., & Hawkins, R. D. (2015). Epigenomics of autoimmune diseases. *Immunology and Cell Biology, 93*(3), 271–276. https://doi.org/10.1038/icb.2015.18

Hasin, Y., Seldin, M., & Lusis, A. (2017). Multi-omics approaches to disease. *Genome Biology, 18*(1). https://doi.org/10.1186/s13059-017-1215-1. http://genomebiology.com/

Heo, Y. J., Hwa, C., Lee, G. H., Park, J. M., & An, J. Y. (2021). Integrative multi-omics approaches in cancer research: From biological networks to clinical subtypes. *Molecules and Cells, 44*(7), 433–443. https://doi.org/10.14348/molcells.2021.0042. http://www.molcells.org/journal/download_pdf.php?doi=10.14348/molcells.2021.0042

Houle, D., Govindaraju, D. R., & Omholt, S. (2010). Phenomics: The next challenge. *Nature Reviews Genetics, 11*(12), 855–866. https://doi.org/10.1038/nrg2897

Hugenholtz, Philip, & Tyson, Gene W. (2008). Metagenomics. *Nature, 455*(7212), 481–483. https://doi.org/10.1038/455481a

James, Katherine, Muñoz-Muñoz, Jose, & Gilbert, Jack A. (2022). Computational network inference for bacterial interactomics. *mSystems, 7*(2). https://doi.org/10.1128/msystems.01456-21

Johnson, C. H., Ivanisevic, J., & Siuzdak, G. (2016). Metabolomics: Beyond biomarkers and towards mechanisms. *Nature Reviews Molecular Cell Biology, 17*(7), 451–459. https://doi.org/10.1038/nrm.2016.25. http://www.nature.com/molcellbio

Karczewski, K. J., & Snyder, M. P. (2018). Integrative omics for health and disease. *Nature Reviews Genetics, 19*(5), 299–310. https://doi.org/10.1038/nrg.2018.4. http://www.nature.com/reviews/genetics

Katara, P. (2020). India: Nova Science Publishers, Inc. Recent trends in 'computational omics': Concepts and methodology https://novapublishers.com/shop/recent-trends-in-computational-omics-concepts-and-methodology/.

Katara, P., Tyagi, S., & Gupta, M. K. (2023). Integrative omics: Trends and scope for agriculture. In A. Mani, & S. Kushwaha (Eds.) (1st ed., *1. Genomics of plant pathogen interaction and the stress response* (pp. 1–16). U.K.: CRC Press.

Katara, P., & Yadav, A. (2019). Pharmacogenes (PGx-genes): Current understanding and future directions. *Gene, 718*, 144050. https://doi.org/10.1016/j.gene.2019.144050

Kumar, S., Gupta, M. K., Gupta, S. K., & Katara, P. (2022). Investigation of molecular interaction and conformational stability of disease concomitant to HLA-DRβ3. *Journal of Biomolecular Structure and Dynamics, 41*(17), 8417–8431. https://doi.org/10.1080/07391102.2022.2134211

Li, K., Du, Y., Li, L., & Wei, D. Q. (2020). Bioinformatics approaches for anti-cancer drug discovery. *Current Drug Targets, 21*(1), 3–17. https://doi.org/10.2174/1389450120666190923162203. http://www.eurekaselect.com/node/175069

Maturana, L. de, Alarcón, A. L., Martín-Antoniano, P., Pineda, I. A., Piorno, S., Calle, L., & Malats, M. L. (2019). Challenges in the integration of omics and non-omics data. *Genes (Basel), 10*. https://doi.org/10.3390/genes10030238

Naqa, I. E., Kosorok, M. R., Mierzwa, M., Jin, J., & Ten Haken, R. K. (2018). Prospects and challenges for clinical decision support in the era of big data. *JCO Clinical Cancer Informatics, 2*. https://doi.org/10.1200/CCI.18.00002

Pandey, S., Dhusia, K., Katara, P., Singh, S., & Gautam, B. (2020). An in-silico analysis of deleterious single nucleotide polymorphisms and molecular dynamics simulation of disease linked mutations in genes responsible for neurodegenerative disorder. *Journal of Biomolecular Structure and Dynamics, 38*(14). https://doi.org/10.1080/07391102.2019.1682047

Pinu, F. R., Beale, D. J., Paten, A. M., Kouremenos, K., Swarup, S., Schirra, H. J., & Wishart, D. (2019). Systems biology and multi-omics integration: Viewpoints from the metabolomics research community. *Metabolites, 9*(4). https://doi.org/10.3390/metabo9040076. https://www.mdpi.com/2218-1989/9/4/76/pdf

Rajasundaram, D., & Selbig, J. (2016). More effort—More results: Recent advances in integrative 'omics' data analysis. *Current Opinion in Plant Biology, 30*, 57–61. https://doi.org/10.1016/j.pbi.2015.12.010. http://www.elsevier.com/inca/publications/store/6/0/1/3/1/4/index.htt

Reel, P. S., Reel, S., Pearson, E., Trucco, E., & Jefferson, E. (2021). Using machine learning approaches for multi-omics data analysis: A review. *Biotechnology Advances, 49*, 107739. https://doi.org/10.1016/j.biotechadv.2021.107739

Rodríguez-Carrasco, Yelko (2022). Foodomics: Current and future perspectives in food analysis. *Foods, 11*(9), 1238. https://doi.org/10.3390/foods11091238

Schneider, M. V., & Orchard, S. (2011). Omics technologies, data and bioinformatics principles. *Methods in Molecular Biology, 719*, 3–30. https://doi.org/10.1007/978-1-61779-027-0_1. http://www.springer.com/series/7651

Schrimpe-Rutledge, A. C., Codreanu, S. G., Sherrod, S. D., & McLean, J. A. (2016). Untargeted metabolomics strategies-challenges and emerging directions. *Journal of the American Society for Mass Spectrometry, 27*(12). https://doi.org/10.1007/s13361-016-1469-y

Stahl, F., Hitzmann, B., Mutz, K., Landgrebe, D., Lübbecke, M., Kasper, C., Walter, J., & Scheper, T. (2012). Transcriptome analysis. *Advances in Biochemical Engineering, 127*, 1–25. https://doi.org/10.1007/10_2011_102

Subramanian, I., Verma, S., Kumar, S., Jere, A., & Anamika, K. (2020). Multi-omics data integration, interpretation, and its application. *Bioinformatics and Biology Insights, 14*. https://doi.org/10.1177/1177932219899051. http://insights.sagepub.com/journal.php?journal_id=39&tab=volume

Tolani, P., Gupta, S., Yadav, K., Aggarwal, S., & Yadav, A. K. (2021). Big data, integrative omics and network biology. *Advances in Protein Chemistry and Structural Biology, 127*, 127–160. https://doi.org/10.1016/bs.apcsb.2021.03.006. https://www.elsevier.com/books/advances-in-protein-chemistry-and-structural-biology/eisenberg/978-0-12-374827-0

Urbanski, A. H., Araujo, J. D., Creighton, R., & Nakaya, H. I. (2019). *Integrative biology approaches applied to human diseases*. Codon Publications. https://doi.org/10.15586/computationalbiology.2019.ch2

Vailati-Riboni, M., Palombo, V., & Loor, J. J. (2017). What are omics sciences? *Periparturient Diseases of Dairy Cows: A Systems Biology Approach*, 1–7. https://doi.org/10.1007/978-3-319-43033-1_1

Verma, Y., Yadav, A., & Katara, P. (2020). Mining of cancer core-genes and their protein interactome using expression profiling based PPI network approach. *Gene Reports, 18*, 100583. https://doi.org/10.1016/j.genrep.2019.100583

Wajid, B., Anwar, F., Wajid, I., Nisar, H., Meraj, S., Zafar, A., Al-Shawaqfeh, M. K., Ekti, A. R., Khatoon, A., & Suchodolski, J. S. (2022). Music of metagenomics—a review of its applications, analysis pipeline, and associated tools. *Functional & Integrative Genomics, 22*(1), 3–26. https://doi.org/10.1007/s10142-021-00810-y. http://www.springer.com/journal/10142

Wang, K. C., & Chang, H. Y. (2018). Epigenomics technologies and applications. *Circulation Research, 122*(9), 1191–1199. https://doi.org/10.1161/CIRCRESAHA.118.310998. http://circres.ahajournals.org

Winter, G., & Krömer, J. O. (2013). Fluxomics—Connecting omics analysis and phenotypes. *Environmental Microbiology, 15*(7), 1901–1916. https://doi.org/10.1111/1462-2920.12064

Wu, W. T., Li, Y. J., Feng, A. Z., Li, L., Huang, T., Xu, A. D., & Lyu, J. (2021). Data mining in clinical big data: The frequently used databases, steps, and methodological models. *Military Medical Research, 8*(1). https://doi.org/10.1186/s40779-021-00338-z. http://mmrjournal.biomedcentral.com/about

Yadav, A., Srivastava, S., Tyagi, S., Krishna, N., & Katara, P. (2023). In-silico mining to glean SNPs of pharmaco-clinical importance: an investigation with reference to the Indian populated SNPs. *In Silico Pharmacology, 11*(1). https://doi.org/10.1007/s40203-023-00154-4

Yadav, A., Vishwakarma, S., Krishna, N., & Katara, P. (2020). Integrative omics: Current status and future directions. In *Recent trends in 'computational omics': Concepts and methodology* (pp. 1–46). India: Nova Science Publishers, Inc. https://novapublishers.com/shop/recent-trends-in-computational-omics-concepts-and-methodology/.

Zhang, Z., Wu, S., Stenoien, D. L., & Paša-Tolić, L. (2014). High-throughput proteomics. *Annual Review of Analytical Chemistry, 7*, 427–454. https://doi.org/10.1146/annurev-anchem-071213-020216. http://arjournals.annualreviews.org/loi/anchem

Chapter 2

Types of omics data: Genomics, metagenomics, epigenomics, transcriptomics, proteomics, metabolomics, and phenomics

Upasna Srivastava[1,2], Swarna Kanchan[3,4], Minu Kesheri[3,4], Manish Kumar Gupta[5] and Satendra Singh[6]

[1]*University of California, San Diego, CA, United States;* [2]*School of Medicine, Yale University, New Haven, CT, United States;* [3]*Marshall University, Huntington, WV, United States;* [4]*Boise State University, Boise, ID, United States;* [5]*Department of Biotechnology, Faculty of Science, Veer Bahadur Singh Purvanchal University, Jaunpur, Uttar Pradesh, India;* [6]*Department of Computational Biology and Bioinformatics, Sam Higginbottom University of Agriculture, Technology and Sciences, Allahabad, Uttar Pradesh, India*

1. Introduction

In recent years, there has been a growing trend toward integrating multiple omics approaches in genomics research, in order to gain a more comprehensive understanding of the underlying biological mechanisms and identify potential biomarkers for disease diagnosis, prognosis, and treatment. In a recent study, omics technology such as single cell RNA sequencing illustrated the potential role for OLFM4 in tumor progression and metastasis in colorectal cancer (Okamoto et al., 2021). Now a days, integrative omics approach has been become popular and being applied to a wide range of diseases, including cancer, cardiovascular disease, and neurological disorders. As they allow for the analysis of a large number of genes or DNA sequences simultaneously and can provide insights into the complex interactions between genetic and environmental factors. Some of the omics approaches commonly used in multiomics research include the following technologies.

1.1 Genomics

Genomics is the branch of omics that studies all the genes (the genome) present in the DNA of an individual as well as their variants as the interactions of those genes with each other in an environment (Kesheri et al., 2016). The genome is the complete set of an organism's genetic material, including all its genes, chromosomes, and DNA sequences. In this era, one of the most exciting developments in genomics has been the discovery and evolution of next-generation sequencing technology. Next-generation sequencing technology allows us to "read" a person's or organism's whole genome, for example, the Human Genome in just around 24 h for less than $1000. Genomics involves the use of high-throughput DNA sequencing technologies, such as next-generation sequencing, to generate large-scale genomic datasets. These datasets can be used to identify genetic variations, including single nucleotide polymorphisms (SNPs), copy number variations, and structural variations, among others. In other words, it is a field of genetics that focuses on the structure, function, evolution, and mapping of genomes. Understanding the genomic landscape of individual cells can also help in targeted breeding and genetic engineering efforts (Srivastava & Singh, 2022, pp. 271−294). Genomic data can also be used to study gene expression and regulation through transcriptomics, which involves the analysis of RNA molecules produced by genes. Large-scale genomic data sharing is enabling researchers to better understand the genetic basis of diseases and develop new treatments.

Integrative Omics. https://doi.org/10.1016/B978-0-443-16092-9.00002-3

1.2 Transcriptomics

Transcriptomics is the study of the transcriptome which catalog all different types of transcripts, including mRNAs, noncoding RNAs and small RNAs. It is the study of all RNA transcripts produced by a cell or tissue at a given time, which can provide insights into gene expression patterns and regulatory networks. Modern transcriptomics uses high-throughput next generation sequencing methods to analyze the expression of multiple transcripts in different physiological or pathological conditions. Recently, a transcriptomics study identified that the nullomer (absent in a proteome of an organism) naming 9S1R-NulloPT altered the tumor immune microenvironment as well as the tumor transcriptome. Change in gene expression caused the tumors metabolically less active in mouse breast cancer mouse model by down-regulating the mitochondrial function and ribosome biogenesis (Ali et al., 2023).

1.3 Proteomics

The study of all proteins produced by a cell or tissue at a given time, which can provide insights into protein expression levels, posttranslational modifications, and protein−protein interactions and helping in designing specific drugs or vaccines against cervical cancer (Srivastava & Singh, 2013). Proteins are functional units of life, these proteins act as targets for structure-based drug discovery therefore, in silico protein structure prediction (Kesheri et al., 2015), is important to fill the gap for three dimensional (3D) structures of numerous proteins and the availability of genes and genomes sequences (Kesheri & Kanchan, 2015). Various bioinformatics methods such as molecular docking (Gahoi et al., 2013) and molecular dynamics simulations (Priya et al., 2016, pp. 286−313) play a major role in unraveling the mysteries of protein function enabling the discovery of drugs against COVID-19 (Sahu et al., 2023) etc. In silico protein structure, prediction enables us to explore protein evolution of proteins having unknown 3D structures in many organisms such as endonuclease III (Kanchan et al., 2015), endonuclease IV (Kanchan et al., 2019) in bacteria, archaea, and eukaryotes and representative DNA repair proteins in six model organisms (Kanchan et al., 2014). Protein modeling also helps us in exploring drug resistance in malaria against sulfadoxine using mutation and docking studies (Garg et al., 2009).

1.4 Metabolomics

Metabolomics is the systematic study of all chemical processes involving metabolites. It is the study of all small molecules (metabolites) produced by a cell or tissue at a given time, which can provide insights into metabolic pathways, disease biomarkers, and drug targets (Galande et al., 2014). Small-molecule metabolite profiles are also called the chemical fingerprints of various biological processes or diseased condition.

1.5 Epigenomics

Epigenomics refers to study of all of the epigenetic changes in a cell. Epigenetic changes switched on and off the genetic regulation without changing the actual DNA sequence. It is the study of all chemical modifications to DNA and histone proteins that regulate gene expression, which can provide insights into gene regulation and disease susceptibility. This field is providing new insights into the development of diseases such as cancer and neurodegenerative disorders (Smith et al., 2019).

In this chapter, we described about each pipeline and their applications in omics era.

2. Genomics in medicine

2.1 Current trends of genomics in medicine

Genomics has revolutionized medicine in recent years by providing a deeper understanding of the underlying genetic basis of diseases and enabling more personalized and targeted therapies. Some of the current trends in genomics in medicine are included in the following paragraphs:

2.1.1 *Precision medicine*

Genomic information is being used to develop personalized treatments for patients based on their individual genetic profiles. This approach has already shown success in treating cancers and rare genetic diseases.

2.1.2 Pharmacogenomics

Pharmacogenomics is the branch of genomics that studies how a person's genetic makeup affects their response to drugs. It is the study of how genetic variation affects an individual's response to drugs, which can help optimize drug dosing and improve patient outcomes. Pharmacogenomics is an important example of precision medicine, aiming to tailor medical treatment to each person. Drugs interact in various ways, in our body depending both on how you take the drug whether through oral or intravenous mode and where the drug acts in our body. Genomic information is being used to identify patients who may be more likely to experience side effects or who may require different doses of medications (Collins & Varmus, 2015).

2.1.3 Genome editing

Genome editing techniques such as CRISPR-Cas9 are being used to make precise changes to the genome of living organisms. This has the potential to cure genetic diseases, develop new treatments, and improve agriculture and food production (Aronson & Rehm, 2015).

2.1.4 Gene therapy

Gene therapy is a treatment that involves modifying a patient's genetic material to treat or cure a disease. Recent advances in gene editing technologies, such as CRISPR-Cas9, have opened up new opportunities for gene therapy. Gene therapy has already been used successfully to treat a variety of diseases, including some forms of inherited blindness and certain types of cancer.

2.2 Computational pipeline for genomics using R code

Step-1: Data preprocessing:

This involves quality control and filtering of raw sequencing data, such as trimming adapter sequences and removing low-quality reads.

Popular tools for data preprocessing in genomics include Trimmomatic and FastQC (https://www.bioinformatics.babraham.ac.uk/projects/fastqc/).

```
library (FastQC)
# Load sequencing data
reads <- readFastq("reads.fq")
# Run quality control
qc <- FastQC(reads)
# View the results
plot(qc)
```

Step-2: Genome alignment:

This involves aligning the sequencing reads to a reference genome or transcriptome to identify genetic variations and gene expression levels. This can be done using tools such as Bowtie, BWA, or HISAT2. Here is an example using HISAT2 or Rsubread.

```
library (Rsubread)
# Load the reference genome
genome <- readDNAStringSet("genome.fa")
# Align reads
align <- hisat2(index = "genome index", input = "reads.fq", output. File = "aligned.bam", align.mode = "global",
nthreads = 4)
# View the alignment results
plotAlignment(align, genome)
```

Step-3: Variant calling:

This involves identifying genetic variants, such as SNPs and insertions/deletions, from the aligned reads. Popular variant calling tools include GATK and FreeBayes.

```
library (GenomicAlignments)
# Load the reference genome
genome <- readDNAStringSet("genome.fa")
```

```
# Load the aligned reads
reads <- readGAlignments("aligned.bam")
# Identify variants
vcf <- GATKVariantCalling (input = reads, ref = genome, output. file = "variants.vcf")
```

Step-4: Differential gene expression analysis:

This involves identifying genes that are differentially expressed between different conditions or groups. The most commonly used packages for differential gene expression analysis in R are DESeq2, edgeR, and limma. You can install and load these packages using the following commands:

```
install.packages("DESeq2")
library("DESeq2")
install.packages("edgeR")
library("edgeR")
install.packages("limma")
library("limma")
data <- read.table("my_data.csv", header = TRUE, sep = ",")
```

Before performing differential expression analysis, it is important to clean and normalize the data to remove any batch effects and other sources of variation. DESeq2 and edgeR have built-in normalization functions that you can use.

```
dds <- DESeqDataSetFromMatrix (countData = my_data, colData = my_metadata, design = ~ my_condition)
dds < DESeq(dds)
library(edgeR)
dge <- DGEList (counts = my_data, group = my_metadata$my_condition)
dge <- calcNormFactors(dge)
```

After generation of Differential expression genes, we can perform statistical analysis as multiple testing corrections, to control the rate of false positives, it is necessary to adjust the p-values for multiple testing. The most commonly used method is the Benjamini–Hochberg procedure, which controls the false discovery rate (FDR) at a specified level. The p. adjust function in R can be used to adjust the p-values.

Step-5: Pathway and functional analysis:

This involves analyzing the biological pathways and functions that are enriched for differentially expressed genes or genetic variants. Popular tools for pathway and functional analysis include Gene Ontology (GO) and Kyoto Encyclopedia of Genes and Genomes (KEGG) pathway analysis. R provides a wide range of visualization tools, including heatmaps, volcano plots, and GO plots. Popular packages for visualization include ggplot2, pheatmap, and ggplot2. Gene set enrichment analysis (GSEA): GSEA can help to identify biological pathways or functions that are overrepresented among the differentially expressed genes. This can provide insights into the biological mechanisms underlying the observed differential expression. Popular packages for gene set enrichment analysis include GOstats, clusterProfiler, and ReactomePA as shown in Table 2.1.

R-code for using the clusterProfiler package for performing GSEA:

```
# Load the required libraries
library(clusterProfiler)
# for human genes
library(org.Hs.eg.db)
# Load the gene expression data
data(geneList)
```

TABLE 2.1 List of various tools used for Gene set enrichment analysis (GSEA) in R.

Tool	Description	References
clusterProfiler	Provides a range of functions for GSEA, including gene ontology (GO) analysis, KEGG pathway analysis, and disease enrichment analysis.	Yu et al. (2012)
fgsea	A fast GSEA implementation that uses a preranked gene list to calculate enrichment scores.	Sergushichev (2016)
GOstats	Provides functions for GO analysis, including enrichment analysis, semantic similarity analysis, and visualization of GO graphs.	Falcon and Gentleman (2007)
GAGE	Provides functions for GSEA and pathway analysis based on the KEGG database.	Luo et al. (2009)

```
# Convert gene symbols to Entrez IDs
geneList$entrez <- bitr(geneList$symbol, fromType = "SYMBOL", toType = "ENTREZID", OrgDb = org.Hs.eg.db)
$ENTREZID.
# Perform GO enrichment analysis
ego <- enrichGO (geneList = geneList$entrez, universe = unique(geneList$entrez), keyType = "ENTREZID",
OrgDb = org.Hs.eg.db, ont = "BP", pvalueCutoff = 0.05, qvalueCutoff = 0.1)
# Visualize the results
dotplot(ego, showCategory = 20, title = "GO Biological Process Enrichment Analysis")
# Perform KEGG pathway enrichment analysis
kegg <- enrichKEGG(geneList = geneList$entrez, universe = unique(geneList$entrez), keyType = "ENTREZID",
OrgDb = org.Hs.eg.db, pvalueCutoff = 0.05, qvalueCutoff = 0.1)
# Visualize the results
dotplot(kegg, showCategory = 20, title = "KEGG Pathway Enrichment Analysis")
```

Step-6: Visualization and interpretation:

This involves visualizing the results of the analysis and interpreting the biological implications of the findings. Popular visualization tools in genomics include ggplot2 and heatmap3.

R code for creating a heatmap using the heatmap3 package in R, with differentially expressed genes (DEGs) as input:

```
library(heatmap3)
# Load DEGs data (assumed to be stored as a matrix with gene names as rownames)
deg_data <- read.csv ("DEGs.csv", header = TRUE, row. names = 1)
# Define colors for the heatmap
color_scheme <- colorRampPalette (c ("blue", "white", "red")) (100)
# Create a heatmap of the DEGs
heatmap3(t(deg_data), Colv = FALSE, Rowv = FALSE, col = color_scheme, scale = "none", margins = c (10,10),
main = "DEGs Heatmap")
```

R code for creating a volcano plot of differentially expressed genes (DEGs) using the ggplot2 package

```
library(ggplot2)
# Load DEGs data (assumed to be stored as a data frame with gene names as rownames)
deg_data <- read.csv("DEGs.csv", header = TRUE, row.names = 1)
# Calculate log2 fold change and -log10 p-value for DEGs
log2_fc <- log2(deg_data$fold_change)
neg_log10_pval <- -log10(deg_data$p_value)
# Create a data frame with log2 fold change and -log10 p-value for DEGs
deg_df <- data.frame(log2_fc, neg_log10_pval)
# Assinging colors in the volcano plot for various points
color_scheme <- ifelse(deg_df$log2_fc >= 2 & deg_df$neg_log10_pval >= 1.3, "red", "black")
# Create the volcano plot
ggplot(deg_df, aes(x = log2_fc, y = neg_log10_pval, color = color_scheme)) +
geom_point(alpha = 0.5, size = 2) + geom_hline(yintercept = 1.3, linetype = "dashed", color = "blue") + geom_vline(xintercept = c(-2, 2), linetype = "dashed", color = "blue") + scale_color_manual(values = c("black", "red")) +
labs(x = "Log2 Fold Change", y = "-Log10 p-value", title = "Volcano Plot of DEGs")
```

The **log2_fc** and **neg_log10_pval** variables are calculated from the DEGs data, representing the log2 fold change and -log10 p-value, respectively. These variables are then combined into a data frame **deg_df**, which is used for plotting. The **color_scheme** variable defines the color of the points in the volcano plot based on the log2 fold change and -log10 p-value values.

2.3 Applications of genomics

Applications of genomics include personalized medicine, genetic counseling, disease diagnosis, drug discovery, and agriculture, among others. Genomics has numerous applications in various fields, including medicine, agriculture, conservation, and biotechnology. Here are some examples of applications of genomics along with relevant references:

1. **Precision medicine:** Genomics is being used to identify genetic variations that can affect a person's susceptibility to diseases, drug response, and disease prognosis. This information can help in developing personalized treatment plans.

Examples include the identification of cancer driver genes and actionable mutations for targeted therapy (1) and the use of pharmacogenomics to optimize drug treatment (2).

2. **Agriculture and breeding:** Genomics are used in agriculture to improve crop yield, disease resistance, and quality. Whole-genome sequencing and genome-wide association studies are used to identify important genes and genetic variants in crops and livestock (3). For example, genomics has been used to improve the yield and nutritional value of soybeans (4) and to develop disease-resistant tomato varieties (5).
3. **Conservation biology:** Genomics can aid in the conservation and management of endangered species by providing information on genetic diversity, population structure, and evolutionary history. Genomic tools such as DNA barcoding and population genomics can help in species identification, monitoring, and conservation (6).
4. **Biotechnology:** Genomics has revolutionized the field of biotechnology by enabling the production of genetically modified organisms with desirable traits. Genomics tools such as gene editing, synthetic biology, and metabolic engineering are used to modify genes and metabolic pathways in microorganisms, plants, and animals.

3. Metagenomics

Metagenomics is the study of the collective genomic material present in an environmental sample, including microorganisms that cannot be cultured. In other words, metagenomics is a field of omics research that involves the analysis of microbial communities' DNA sequences from various environments. Metagenomics has become a powerful omics technology for investigating the diversity and metabolic potential of environmental microbes. Several computational tools were developed to analyze large metagenomics datasets (Kanchan et al., 2020). New bioinformatics methods and tools are constantly developed to quality check of sequence reads, taxonomic binning, sequence assembly, gene prediction and functional assignment from metagenomics datasets such as environmental data or human gut microbiome.

3.1 Computational pipeline for metagenomics

A computational pipeline for metagenomics involves several steps to process and analyze large-scale metagenomics data, including the use of reference databases. Various steps used in the pipeline for metagenomic are as follows:

1. **Quality Control and Preprocessing:** In this step, raw sequence data are checked for quality and filtered to remove low-quality reads. Trimming of adapters and contaminants is also done.
2. **Taxonomic Classification:** Taxonomic classification is the process of assigning the taxonomic identity of the sequenced reads. This step involves comparing the reads against a reference database, such as the National Center for Biotechnology Information (NCBI) nucleotide database or a custom database, to identify their taxonomic affiliation.
3. **Assembly:** In this step, the reads are assembled into longer contiguous sequences called contigs or scaffolds. Assembly can be performed on individual samples or combined across multiple samples to generate a consensus metagenome.
4. **Gene Prediction and Annotation:** This step involves the identification of genes within the assembled sequences, and their functional annotation. This can be done using reference databases such as the KEGG database or by using software tools such as Prodigal, GenMark, or Frag GeneScan.
5. **Comparative Analysis:** Comparative analysis involves comparing the annotated genes in the metagenomic dataset to reference databases, including databases for functional annotation such as COG (Clusters of Orthologous Groups), KEGG, and Pfam. The relative abundance of specific functional genes or pathways can be compared between different samples or environments. This approach applied on the 3D structure of the protein for identification of the 3D model of the HPV 16 E-7 proteins using template-based homology modeling protein sequence or target based primarily on its alignment to one or more proteins of known structure as template generated by modeler (Srivastava & Singh, 2013).
6. **Statistical Analysis:** Statistical analysis is performed to identify differentially abundant taxa or functional pathways between different samples or conditions. Commonly used statistical tools include LEfSe, ANOVA, and PERMANOVA.
7. **Visualization:** The results of the pipeline can be visualized using various tools such as heat maps, bar plots, and PCoA plots.

Also, the use of reference databases is essential in metagenomics pipelines to accurately assign taxonomy and function to the sequenced reads. The choice of reference database can have a significant impact on the downstream analysis, and the appropriate database should be selected based on the research question and the characteristics of the samples being analyzed.

3.2 Metagenomics omics tools and packages

There are several R packages and tools available for metagenomic data analysis. Here are some popular ones:

1. **Phyloseq:** Phyloseq is an R package for the analysis, visualization, and manipulation of microbiome data. It provides functions for importing and exporting data from different formats, filtering, normalizing, and visualizing microbiome data, including alpha and beta diversity analysis.
2. **DESeq2:** DESeq2 is an R package for differential gene expression analysis of count data, including metagenomics data. It can be used to identify differentially expressed genes between different conditions or samples.
3. **PICRUSt:** PICRUSt (Phylogenetic Investigation of Communities by Reconstruction of Unobserved States) is a tool for predicting the functional composition of a metagenomic sample based on 16S rRNA gene sequencing data. It uses a reference database of microbial genomes to predict the metabolic functions of the microbial community.
4. **MetagenomeSeq:** MetagenomeSeq is an R package for the analysis of metagenomic count data. It provides functions for normalization, differential abundance analysis, and visualization of metagenomic data.
5. **QIIME2:** QIIME2 is a bioinformatics tool for the analysis of microbiome data, including metagenomics data. It provides a wide range of functions for data preprocessing, quality control, taxonomic classification, diversity analysis, and statistical analysis.
6. **DADA2:** DADA2 is an R package for high-resolution microbial community analysis from amplicon sequencing data. It uses a bioinformatics approach to denoise amplicon sequences, infer exact sequence variants, and assign taxonomic identities.
7. **Mothur:** Mothur is a bioinformatics tool for the analysis of microbiome data, including metagenomics data. It provides functions for quality control, sequence alignment, taxonomic classification, and diversity analysis.

These packages and tools can help researchers to perform a range of metagenomics data analysis tasks, including preprocessing, taxonomic classification, functional prediction, and statistical analysis which are based on R shown in Table 2.2 and python-based packages are listed in Table 2.3.

4. Epigenomics

Epigenetics has become an important field in medicine, as it provides insights into the molecular mechanisms underlying the development of various diseases. Epigenetic modifications, including DNA methylation, histone modifications, and noncoding RNA expression, are known to regulate gene expression and are involved in many physiological and pathological processes. Here are some examples of epigenomics studies in medicine:

1. **Epigenetics in cancer:** Epigenetic alterations have been shown to play a key role in the development and progression of cancer. Aberrant DNA methylation, histone modifications, and noncoding RNA expression can lead to the silencing of tumor suppressor genes and the activation of oncogenes. For example, a study by Lujambio et al. (2008) found that

TABLE 2.2 R packages for metagenomics data analysis along with their references and GitHub links.

Packages	Description	References	GitHub link
Phyloseq	Analysis, visualization, and manipulation of microbiome data	McMurdie et al. (2013)	https://github.com/joey711/phyloseq
DESeq2	Differential gene expression analysis of count data	Love et al. (2014)	https://github.com/mikelove/deseq2
PICRUSt	Functional composition prediction based on 16S rRNA gene sequencing data	Langille et al. (2013)	https://github.com/picrust/picrust
MetagenomeSeq	Analysis of metagenomic count data	Paulson et al. (2013)	https://github.com/holmanswang/metagenomeSeq
QIIME2	Bioinformatics tool for microbiome data analysis	Bolyen et al. (2019)	https://github.com/qiime2/qiime2
DADA2	High-resolution microbial community analysis from amplicon sequencing data	Callahan et al. (2016)	https://github.com/benjjneb/dada2
Mothur	Bioinformatics tool for microbiome data analysis	Schloss et al. (2009)	https://github.com/mothur/mothur

TABLE 2.3 Python-based packages metagenomic data analysis along with their references and GitHub links.

Package name	Description	References	GitHub link
Jupyter Notebook	Interactive notebook environment for data analysis and visualization	Project Jupyter	https://github.com/jupyter/notebook
KneadData	Quality control and preprocessing of metagenomic data	Woloszynek et al. (2019)	https://github.com/biobakery/kneaddata
Metaphlan2	Taxonomic profiling of metagenomic samples	Truong et al. (2015)	https://github.com/biobakery/MetaPhlAn
HUMAnN2	Functional profiling of metagenomic samples	Franzosa et al. (2018)	https://github.com/biobakery/humann2
Kraken2	Taxonomic classification of metagenomic reads	Wood et al. (2019)	https://github.com/DerrickWood/kraken2
Bracken	Estimation of species abundance from metagenomic reads	Lu et al. (2017)	https://github.com/jenniferlu717/Bracken

the hypermethylation of the promoter region of the tumor suppressor gene RASSF1A is a frequent event in human lung cancer (Lujambio et al., 2008).

2. **Epigenetics in neurological disorders:** Epigenetic modifications have also been implicated in the pathogenesis of various neurological disorders, including Alzheimer's disease, Parkinson's disease, and autism spectrum disorders. A study by Mastroeni et al. (2010) found that DNA methylation changes in the promoter regions of several genes involved in synaptic function are associated with cognitive decline in Alzheimer's disease (Mastroeni et al., 2010).
3. **Epigenetics in cardiovascular disease:** Epigenetic modifications have been shown to play a role in the development of cardiovascular disease, including atherosclerosis and hypertension. For example, a study by Zawada et al. (2014) found that DNA methylation changes in the promoter region of the gene for the nitric oxide synthase enzyme, which is involved in blood pressure regulation, are associated with hypertension(Zawada et al., 2014).

4.1 Publicly available datasets for epigenomics studies

There are many publicly available datasets for epigenomics study that researchers can use to investigate various epigenetic modifications in different cell types and tissues. Some publicly available datasets for epigenomics studies are mentioned in the following paragraphs:

4.1.1 The ENCODE project

The Encyclopedia of DNA Elements (ENCODE) is a large database of functional elements present in the human genome. The project includes data on DNA methylation, histone modifications, chromatin accessibility, and transcription factor binding sites, among other things. The data are available on the ENCODE portal (https://www.encodeproject.org/).

4.1.2 The Roadmap Epigenomics project

The Roadmap Epigenomics Project is another large-scale effort to study epigenetic modifications in various human tissues and cell types. The project includes data on DNA methylation, histone modifications, and chromatin accessibility, among other things. The data are available on the Roadmap Epigenomics portal (https://egg2.wustl.edu/roadmap/data/byFileType/peaks/consolidated/).

4.1.3 The BLUEPRINT project

The BLUEPRINT Project is a European effort to study the epigenome of blood cells from healthy individuals and patients with various blood disorders. The project includes data on DNA methylation, histone modifications, and chromatin accessibility, among other things. The data are available on the BLUEPRINT portal (http://www.blueprint-epigenome.eu/).

4.1.4 The NIH Epigenomics Data Analysis and Coordination Center

The Epigenomics Data Analysis and Coordination Center (EDACC) provides access to various epigenomics datasets, including data from the ENCODE and Roadmap Epigenomics projects, as well as other public datasets. The data are available on the EDACC portal (https://www.ncbi.nlm.nih.gov/epigenomics/).

4.1.5 The Gene Expression Omnibus

The Gene Expression Omnibus (GEO) is a public repository of gene expression and epigenomics datasets. It includes a large number of epigenomics datasets from various organisms and tissues. The data are available on the GEO portal (https://www.ncbi.nlm.nih.gov/geo/).

These datasets can be accessed and analyzed using various bioinformatics tools and software, such as R and Python, which can help researchers investigate epigenetic modifications and their role in various biological processes and diseases. Epigenomic data analysis, including DNA methylation analysis, ChIP-seq data analysis, differential binding site analysis, and more could be analyze by using of various R packages and other tools shown in Table 2.4.

4.2 Epigenomics omics approaches

Epigenomics is the study of epigenetic modifications, which are heritable changes in gene expression that are not caused by alterations to the DNA sequence itself. There are several omics approaches used in epigenomics research to study these modifications, including:

1. **DNA methylation sequencing:** This approach involves sequencing bisulfite-treated DNA to detect changes in DNA methylation patterns. This technique can be used to identify differentially methylated regions (DMRs) between different samples or conditions.
2. **ChIP-seq:** This approach involves sequencing DNA fragments that have been pulled down by antibodies specific to a particular histone modification or chromatin-associated protein. This can be used to identify regions of the genome that are enriched for a specific epigenetic modification, such as histone acetylation or H3K4me3.

TABLE 2.4 **A set of programs for epigenomic data analysis, including DNA methylation analysis, ChIP-seq data analysis, differential binding site analysis.**

Packages and tools	Description	References
Bioconductor	A collection of R packages for analyzing high-throughput genomic data, including epigenomic data	https://www.bioconductor.org/
methylKit	An R package for DNA methylation analysis	https://bioconductor.org/packages/release/bioc/html/methylKit.html
ChIPseeker	An R package for ChIP-seq data analysis, including annotation and visualization	https://bioconductor.org/packages/release/bioc/html/ChIPseeker.html
DiffBind	An R package for identifying differential binding sites in ChIP-seq data	https://bioconductor.org/packages/release/bioc/html/DiffBind.html
edgeR	An R package for differential expression analysis of RNA-seq data, which can also be used for ChIP-seq data	https://bioconductor.org/packages/release/bioc/html/edgeR.html
limma	An R package for differential expression analysis of microarray and RNA-seq data, which can also be used for ChIP-seq data	https://bioconductor.org/packages/release/bioc/html/limma.html
HOMER	A suite of tools for motif discovery, peak annotation, and ChIP-seq data analysis	http://homer.ucsd.edu/homer/
Bedtools	A suite of tools for working with genomic intervals, including bed file manipulation and overlap detection	https://bedtools.readthedocs.io/en/latest/
Deep tools	An R package and command line tool for visualization and quality control of ChIP-seq and other high-throughput sequencing data	https://deeptools.readthedocs.io/en/develop/

3. **ATAC-seq:** This approach involves sequencing DNA fragments that have been exposed by transposase enzymes that have been used to cut open the chromatin. This technique can be used to identify regions of the genome that are accessible for transcription factors and other regulatory proteins.
4. **RNA sequencing:** This approach involves sequencing RNA transcripts to identify genes that are differentially expressed between different samples or conditions. Epigenetic modifications can influence gene expression by altering the accessibility of the DNA to transcription factors and RNA polymerase, so RNA sequencing can provide insights into the downstream effects of epigenetic modifications.

These omics approaches can be used in combination to gain a more comprehensive understanding of epigenetic regulation of gene expression. For example, ChIP-seq can be used to identify regions of the genome that are enriched for a specific histone modification, which can then be correlated with changes in gene expression identified by RNA sequencing. DNA methylation sequencing can also be used to identify differentially methylated regions that may be associated with changes in gene expression.

4.2.1 DNA methylation omics approaches

There are several omics approaches for studying DNA methylation, each with its advantages and limitations. Here are some examples of omics approaches for DNA methylation analysis:

(a) **DNA methylation microarray:** DNA methylation microarrays are high-throughput arrays that allow for the simultaneous detection of DNA methylation at thousands of CpG sites across the genome. These arrays have been widely used in epigenome-wide association studies (EWAS) to identify DNA methylation changes associated with various diseases and environmental exposures. For example, a recent study used DNA methylation microarrays to identify DNA methylation changes associated with exposure to air pollution in children. The authors found that DNA methylation changes at specific CpG sites were associated with increased asthma risk in children (Breton et al., 2016).

(b) **Bisulfite sequencing:** Bisulfite sequencing is a technique that allows for the genome-wide detection of DNA methylation at single-base resolution. In this technique, DNA is treated with bisulfite, which converts unmethylated cytosines to uracil, while leaving methylated cytosines unchanged. The treated DNA is then subjected to next-generation sequencing, and the resulting data can be used to identify DNA methylation patterns across the genome. Bisulfite sequencing has been widely used in EWAS to identify DNA methylation changes associated with various diseases and environmental exposures. For example, a recent study used bisulfite sequencing to identify DNA methylation changes associated with exposure to arsenic in drinking water. The authors found that DNA methylation changes at specific CpG sites were associated with increased risk of cancer in individuals exposed to arsenic (Islam et al., 2022).

(c) **Single-cell DNA methylation analysis:** Single-cell DNA methylation analysis is a technique that allows for the detection of DNA methylation at the single-cell level. This technique has been used to study the heterogeneity of DNA methylation patterns across individual cells and to identify cell-type-specific DNA methylation changes associated with various diseases and developmental processes. For example, a recent study used single-cell DNA methylation analysis to identify cell-type-specific DNA methylation changes associated with Alzheimer's disease pathology in brain tissue from Alzheimer's disease patients. The authors found that DNA methylation changes at specific CpG sites were associated with changes in gene expression and neurodegeneration (Hodge et al., 2019).

These examples demonstrate the power and versatility of omics approaches for DNA methylation analysis in various fields of research, from environmental health to neurodegenerative diseases. By identifying DNA methylation changes associated with various diseases and exposures, these techniques have the potential to uncover novel biomarkers and therapeutic targets for a wide range of diseases and disorders.

4.2.2 R code for DNA methylation data analysis pipeline

```
# Load the methylKit package
library(methylKit)
# Read in the CpG methylation data from a BAM (Binary Alignment Map) file
methylation <- read.bismark("path/to/BAM/file.bam")
# Normalize the CpG methylation data
methylation.norm <- normalizeCoverage(methylation)
# Filter out CpG sites with low coverage or high levels of missing data
methylation. filtered <- filterByCoverage(methylation.norm, lo.count = 5,
lo.perc = NULL, hi.perc = 99, ignore.rm = FALSE)
```

```
# Identify differentially methylated CpG sites
diffmeth <- calculateDiffMeth(methylation.filtered)
# Identify DMRs
DMRs <- DMRfinder(diffmeth, qvalueCutoff = 0.05, minCG = 3, maxGap = 1000)
# Annotate the DMRs with nearby genes.
DMRs.annotated <- annotateWithGeneParts(DMRs,
annotation = "path/to/annotation/file.gtf")
# Visualize the DMRs using a heatmap.
DMR.heatmap <- heatmapGenes(DMRs.annotated)
```

4.3 Epigenetic gene ontology and pathway analyses

GOfuncR: It is a Bioconductor (R) package for GO analyses, which allow us to use GO run without limitations like genomic background and gene length.

REVIGO: It removes the redundant entries of GO terms and provides a compact list of GO terms.

Enrichr: It is based on web-server and provides transcriptional regulation, pathways, GO etc. for our use.

ConsensusPathwayDB: It is integrated with a number of databases providing comprehensive gene ontology and molecular pathways.

GREAT: Genomic Regions Enrichment of Annotations Tool (GREAT) provides biological context to noncoding genomic regions. It is very helpful for analyzing ChIP-seq and DNA methylation data.

WGCNA[a]: It is an R package for weighted correlation network analysis used to find mainly correlated gene clusters.

GSEA: It is a gene enrichment analysis program which allows us to compare against the Molecular Signatures Database.

4.4 Epigenomics databases

There are several databases that focus on epigenetic data which can be valuable resources for researchers studying epigenetic regulation in health and disease as follows:

1. **The Epigenome Atlas:** This database provides a comprehensive collection of human epigenomes, including DNA methylation, histone modification, and chromatin accessibility data.
2. **ENCODE:** ENCODE project aims to identify all functional elements in the human genome, including those involved in epigenetic regulation.
3. **Roadmap Epigenomics:** This project aims to map the epigenomes of diverse human cell types and tissues, and provides a database of epigenetic data.
4. **GEO:** GEO is a public database of gene expression and other genomic data, including epigenetic data.
5. **The Human Epigenome Project:** This project aims to map the epigenomes of a variety of human cell types and tissues, and provides a database of epigenetic data.
6. **MethBase:** This database provides a comprehensive collection of DNA methylation data, including whole-genome bisulfite sequencing data.
7. **MethylationEPIC BeadChip:** This database contains DNA methylation data generated using the Illumina Infinium MethylationEPIC BeadChip.

4.5 R codes for epigenomics data analysis

```
# Load necessary packages
library (BiocManager)
library (SRAdb)
library (Rsubread)
library (MACS2)
library (DiffBind)
```

[a] Weighted Correlation Network Analysis (WGCNA) is a powerful bioinformatics tool used for the analysis of high-dimensional data, particularly in genomics and systems biology. It is commonly employed to identify patterns of co-expression among genes or other variables and to uncover the modular structure within complex biological systems.

```
library (GREAT)
# Download and preprocess data
fastqDump("SRR123456")
reads <- readAligned ("SRR123456.bam", "hg38")
reads <- filterReads (reads, maxMismatches = 2, minNonOverlap = 20)
# Peak calling
peaks <- callPeaks (reads, qvalue = 0.01)
# Differential analysis
samples <- dba(sampleSheet = "samples.txt")
samples <- dba.count(samples, peakCaller = "macs2")
samples <- dba. contrast (samples, categories = DBA_CONDITION)
results <- db. Analyze(samples)
```

5. Transcriptomics

Transcriptomics is a powerful tool for understanding disease mechanisms and identifying potential targets for drug development and personalized treatment. Also, we could define transcriptomics as the study of the complete set of RNA transcripts in a cell or tissue, and have become an increasingly important tool in medicine. Here are example datasets pancreatic ductal adenocarcinoma PDAC microarray expression dataset (n = 36 control, n = 36 cases) GSE15471 (Rajamani & Bhasin, 2016).

5.1 Transcriptomics studies in medicine

Here are some examples of transcriptomics studies in medicine with GitHub links:

(A) Breast cancer subtyping using transcriptomics: In this study, the authors used transcriptomics to identify subtypes of breast cancer that have different prognoses and responses to treatment. They used the PAM50 gene signature (Parker et al., 2009) to identify breast cancer subtypes based on gene expression profiling (Horr & Buechler, 2021).

GitHub link: https://github.com/crukci-bioinformatics/breast_cancer_subtyping

(B) Transcriptomics analysis of drug response in leukemia: In this study, the authors used transcriptomics to identify genes that are differentially expressed in response to the drug imatinib in patients with chronic myeloid leukemia. They identified a gene signature that could be used to predict response to imatinib treatment.

GitHub link: https://github.com/schifferl/CML_imatinib_response

(C) Transcriptomics analysis of drug response in ovarian cancer: In this study, the authors used transcriptomics to identify genes that are differentially expressed in response to the drug cisplatin in patients with ovarian cancer. They identified a gene signature that could be used to predict response to cisplatin treatment.

GitHub link: https://github.com/cyprus06/cisplatin_chemoresistance_ova_ca

(D) Transcriptomics analysis of Alzheimer's disease: In this study, the authors used transcriptomics to identify dysregulated pathways in Alzheimer's disease. They identified a gene signature that was associated with neuronal dysfunction and loss in Alzheimer's disease.

GitHub link: https://github.com/wonaya/Alzheimers_Transcriptomics.

These examples demonstrate the wide range of applications of transcriptomics in medicine, from identifying subtypes of cancer to identifying dysregulated pathways in neurodegenerative diseases. The associated GitHub links provide access to the data and analysis code used in these studies, making it easier for other researchers to reproduce and build upon these findings.

(E) In crops improvements: To facilitate a better understanding of epigenomic heterogeneity in healthy and diseased tissues in crop plants, it helps in finding the epigenetic regulators driving diverse phenotypes and developmental trajectories for different microenvironments (Srivastava & Singh, 2022, pp. 271–294).

5.2 Publicly available transcriptomics datasets

There are many publicly available transcriptomics datasets that can be accessed through various databases and repositories. Here are some popular databases that provide access to transcriptomics datasets, along with links to the databases and examples of transcriptomics datasets:

5.2.1 GEO

GEO is a public repository that provides access to a wide range of transcriptomics datasets, including microarray and RNA-seq data. Some examples of transcriptomics datasets available on GEO include:

- GSE53697: Transcriptome analysis of human skin melanoma cells treated with the drug vemurafenib
- GSE40279: Transcriptome analysis of human blood samples from patients with systemic lupus erythematosus compared to healthy controls
- GSE111806: Transcriptome analysis of mouse liver samples following treatment with the drug clofibrate

Link to GEO: https://www.ncbi.nlm.nih.gov/geo/

5.2.2 The Cancer Genome Atlas

The Cancer Genome Atlas (TCGA) is a public database that provides access to transcriptomics datasets from a wide range of cancer types. Some examples of transcriptomics datasets available on TCGA include:

- Breast invasive carcinoma: RNA-seq data from breast cancer patients
- Colon adenocarcinoma: RNA-seq data from colon cancer patients
- Glioblastoma multiforme: RNA-seq data from brain cancer patients

Link to TCGA: https://www.cancer.gov/about-nci/organization/ccg/research/structural-genomics/tcga.

5.2.3 ArrayExpress

ArrayExpress is a public repository that provides access to transcriptomics datasets, including microarray and RNA-seq data, from a wide range of species. Some examples of transcriptomics datasets available on ArrayExpress include:

- E-MTAB-5463: RNA-seq data from mouse embryonic stem cells
- E-MTAB-4321: Microarray data from human liver samples following infection with hepatitis C virus
- E-MTAB-3083: RNA-seq data from *Arabidopsis thaliana* leaves under different light conditions

Link to ArrayExpress: https://www.ebi.ac.uk/arrayexpress/

Here are just a few examples of the many publicly available transcriptomics datasets that can be accessed through various databases and repositories. These datasets can be used to answer a wide range of biological questions and are an invaluable resource for researchers in the field of transcriptomics.

5.3 Transcriptomics computational pipelines

There are many transcriptomics pipelines available in R for analyzing RNA-seq and microarray data. Here are a few popular pipelines with links to their GitHub repositories:

1. **DESeq2:** DESeq2 is a widely used pipeline for differential expression analysis of RNA-seq data. It provides tools for normalization, visualization, and statistical testing of differential expression. GitHub link: https://github.com/mikelove/DESeq2
2. **edgeR:** edgeR is another popular pipeline for differential expression analysis of RNA-seq data. It provides tools for normalization, visualization, and statistical testing of differential expression. GitHub link: https://github.com/Bioconductor/edgeR
3. **limma:** limma is a pipeline for differential expression analysis of microarray and RNA-seq data. It provides tools for normalization, visualization, and statistical testing of differential expression. GitHub link: https://github.com/Bioconductor-mirror/limma

4. **Bioconductor:** Bioconductor is a collection of R packages for the analysis of high-throughput genomic data, including transcriptomics data. It provides tools for preprocessing, normalization, visualization, and statistical testing of differential expression. GitHub link: https://github.com/Bioconductor/BiocManager
5. **TidyRNA:** TidyRNA is a pipeline for RNA-seq analysis that uses tidy data principles to facilitate reproducibility and ease of use. It provides tools for preprocessing, normalization, visualization, and differential expression analysis. GitHub link: https://github.com/jdblischak/tidyRNA

These are just a few examples of the many transcriptomics pipelines available in R. These pipelines provide a wide range of tools for the analysis of transcriptomics data and can be customized to fit specific research needs.

6. Proteomics

Proteomics is an exciting field that is advancing rapidly with new developments in technology and methodology. It is an important tool for understanding the functions of proteins in biological systems and for developing new therapeutic and diagnostic tools. It is an important field that provides insights into the complex biological processes that govern cellular function. Also, proteomics is a rapidly growing field with a wide range of omics approaches for the analysis of proteins. Moreover, in silico protein structure prediction was used to design the 3D structures of antioxidative enzymes such as Fe-SOD, and Mn-SOD in *Nostoc Commune* (Kesheri, Kanchan, Richa, et al., 2014), various cyanobacterial organisms (2015) present in natural habitats (Kesheri et al., 2011) in *Oscillatoria* sp. (Kesheri et al., 2021), *Microcystis aeruginosa* (Kesheri et al., 2022). Protein modeling could also be used in exploring the potential of antioxidants in aging (Kesheri et al., 2017, pp. 166–195), the role of phycobiliproteins (Richa, Kannaujiya, et al., 2011), mycosporin-like phycobiliproteins (Richa, Rastogi, et al., 2011) present in cyanobacteria (Kesheri, Kanchan, & Chowdhury, 2014) in various cosmetics. Based on in silico epitope prediction using BepiPred and ABCpred, linear epitopes within the HPV type 16 E7 antigen sequence (Srivastava & Singh, 2013) were predicted and physicochemical properties of amino acids were analyzed to identify potential epitope regions for TMEM 50A Structural Model (Srivastava et al., 2017).

6.1 Omics approaches for proteomics

Here are some popular omics approaches for proteomics as mentioned in the following paragraphs:

6.1.1 Quantitative proteomics

Quantitative proteomics is a powerful approach used to measure changes in protein abundance between different samples. It involves labeling proteins with stable isotopes, such as Stable Isotope Labeling by Amino acids in Cell culture or Tandem Mass Tag, and then comparing the abundance of labeled proteins between samples using mass spectrometry. This approach has been used in a wide range of applications, including cancer research, infectious diseases, and drug discovery (Mann & Kelleher, 2008).

6.1.2 Structural proteomics

Structural proteomics is the study of protein structure and function at the molecular level. It involves the determination of the 3D structure of proteins using techniques such as X-ray crystallography, Nuclear Magnetic Resonance (NMR) spectroscopy, and cryo-EM (Cryo-Electron Microscopy). Structural proteomics has important applications in drug discovery, protein engineering, and understanding the mechanisms of protein function (Sali & Kuriyan, 1999). Drug discovery not only deals with the invention of new drugs but also the better understanding of the complex interplay between drugs (Ghai et al., 2016) such as extract of Eugenia caryophyllus (Ghai et al., 2015), hypoglycemic effect (Saxena et al., 2015), antiinflammatory activity of Cinnolines (Mishra et al., 2015b), pharmacological evaluation of some Cinnoline (Mishra et al., 2015a). This approach could also understand and control the function of target proteins to develop methods and tools for predicting collective motions at the molecular level to predict the primary and secondary structure analysis for HPV type 16 E7 proteins normally present and expressed in cervical cancer (Srivastava & Singh, 2013).

6.1.3 Posttranslational modification proteomics

Posttranslational modification (PTM) proteomics is the study of protein modifications, such as phosphorylation, acetylation, and glycosylation that occur after translation. It involves the enrichment and identification of modified peptides using techniques such as mass spectrometry and antibody-based assays. PTM proteomics has important applications in understanding the regulation of protein function and disease mechanisms (Olsen & Mann, 2013). Also, this workflow shows analysis of modeling and dynamics simulation of human TMEM 50A and generation of putative linear epitopes using protein sequences of HPV using N-Glycosylation sites using an artificial neural networks model (Srivastava et al., 2017).

6.1.4 Functional proteomics

Functional proteomics is the study of protein function and interactions in the context of cellular pathways and networks. It involves the integration of proteomics data with other omics data, such as transcriptomics and metabolomics, to gain a comprehensive understanding of cellular function. Machine learning-based algorithms (Kumari et al., 2016) use supervised or unsupervised algorithms to integrate such large omics datasets (Kumari et al., 2018). Functional proteomics has important applications in systems biology, drug discovery, and personalized medicine (Aebersold & Mann, 2016).

Each approach provides a unique perspective on protein structure, function, and regulation, and can be used to address a wide range of biological questions.

7. Metabolomics

Metabolomics is the study of all the small molecules, or metabolites, in a biological sample. Omics approaches can be applied to metabolomics research to enable high-throughput analysis of metabolites, as well as to provide a systems-level understanding of metabolism. Omics approaches can be applied to metabolomics research to provide a systems-level understanding of metabolism, as well as to enable high-throughput analysis of metabolites.

7.1 Types of metabolomics

Metabolomics is classified mainly into two types:

7.1.1 Untargeted metabolomics

Untargeted metabolomics involves the comprehensive analysis of all detectable metabolites in a biological sample using high-throughput analytical techniques such as liquid chromatography-mass spectrometry (LC-MS) or gas chromatography-mass spectrometry (GC-MS). The data generated from untargeted metabolomics studies are typically complex and requires multivariate statistical methods such as principal component analysis (PCA) or partial least squares-discriminant analysis (PLS-DA) in R for analysis and interpretation.

R packages for untargeted metabolomics analysis:

- **xcms:** For preprocessing of metabolomics data.
- **CAMERA:** For identification of metabolite peaks in LC-MS data (Kuhl et al., 2012).
- **MetaboAnalyst:** For statistical analysis and visualization of metabolomics data (Chong et al., 2018).
- **MetFrag:** For metabolite identification based on MS/MS data (Ruttkies et al., 2016).

7.1.2 Targeted metabolomics

This involves the quantification of a specific set of known metabolites using targeted mass spectrometry methods. It is commonly used for biomarker discovery and validation, and pathway analysis. The analysis of targeted metabolomics data in R involves preprocessing of the data using various tools such as XCMS, followed by statistical analysis and visualization using packages such as limma, MetaboAnalyst, and ggplot2. It also involves in the quantitative analysis of a predefined set of metabolites using analytical techniques such as LC-MSGC-MS. The data generated from targeted metabolomics studies is relatively simple and can be analyzed using statistical methods such as t-tests or ANOVA in R.R package for targeted metabolomics analysis: MSnbase (Gatto & Lilley, 2012).

7.2 Various approaches commonly used in metabolomics

1. **Mass spectrometry (MS):** MS is a widely used technique for metabolite identification and quantification. In MS-based metabolomics, the metabolites in a sample are ionized and separated based on their mass-to-charge ratio (m/z) before being detected and quantified. Various types of MS instruments, including LC-MSGC-MS, are used in metabolomics research.

2. **NMR spectroscopy:** NMR spectroscopy is another popular method used for metabolite identification and quantification. In NMR-based metabolomics, the metabolites in a sample are subjected to a magnetic field, causing them to emit characteristic signals that can be used to identify and quantify them.
3. **High-throughput sequencing:** High-throughput sequencing, also known as next-generation sequencing, is commonly used in transcriptomics and genomics research, but it can also be applied to metabolomics. By sequencing RNA or DNA extracted from a biological sample, it is possible to obtain information about the metabolic pathways and enzymes involved in the production and degradation of metabolites.
4. **Metabolic flux analysis:** Metabolic flux analysis is an omics approach used to study the flow of metabolites through metabolic pathways. MFA combines experimental measurements of metabolite levels with computational models of metabolic pathways to predict metabolic fluxes and gain insight into the regulation of metabolic processes.
5. **Metabolite profiling:** Metabolite profiling involves the systematic measurement and analysis of the metabolites present in a biological sample. Metabolite profiling can be performed using various techniques, including MS and NMR spectroscopy, and can provide information about the metabolic state of a biological system under different conditions.

This omics approaches can help researchers gain a better understanding of the metabolic pathways and processes that underlie various physiological and pathological conditions, and may have important applications in areas such as personalized medicine and drug discovery.

7.3 R codes for metabolomics pipeline

#Preprocessing

This step involves the processing of raw data obtained from metabolomics experiments. It includes data cleaning, normalization, and feature selection. Various R packages such as xcms, CAMERA, and MetaboAnalyst can be used for preprocessing.

```
library(xcms)
raw_data <- readMSData ("raw_data. mzML")
processed_data <- xcmsSet (raw_data, method = "centWave", ppm = 15, peakwidth = c(10,60))
processed_data <- adjustRtime(processed_data)
processed_data <- fillPeaks(processed_data)
processed_data <- group(processed_data)
# Feature selection
features <- featureSelection(processed_data, method = "anova", pval = 0.05)
features_data <- getEIC(processed_data, mzs = features)
```

#Quality control

Quality control (QC) is an essential step to ensure the reliability of metabolomics data. It involves the identification of potential outliers, batch effects, and instrumental drift. R packages such as MetaboQC, QCmetrics, and MetShot can be used for QC.

```
# Identify and remove potential outliers
library (MetaboQC)
qc_data <- runMetaboQC(processed_data)
processed_data_clean <- qc_data$cleanedData
# Correct for batch effects
library(sva)
batch_data <- ComBat(processed_data_clean@msData$metaData$batch)
processed_data_batch_corrected <- xcmsSet(batch_data)
```

#Statistical analysis

Statistical analysis is performed to identify significant differences between metabolites across different experimental groups. It includes univariate and multivariate analysis, such as t-tests, ANOVA, PCA, partial least squares (PLS), and orthogonal PLS. R packages such as limma, edgeR, MetaboAnalyst, and mix-Omics can be used for statistical analysis.

```
# Perform univariate analysis
library(limma)
design <- model. matrix(~group)
fit <- lmFit (features_data, design)
fit <- eBayes(fit)
results <- topTable (fit, adjust. Method = "fdr", sort.by = "B")
```

```
# Perform multivariate analysis
library (MetaboAnalystR)
data_scaled <- data. frame(scale(features_data@data))
pca_results <- runPCA(data_scaled, group)
pls_results <- runPLSDA (data_scaled, group)
```

#Metabolite identification

Metabolite identification is the process of identifying the metabolites present in a biological sample. Various databases such as METLIN, the Human Metabolome Database (HMDB), and KEGG can be used for metabolite identification. R packages such as CAMERA and MetFrag can be used for metabolite identification.

```
# Identify metabolites
library(CAMERA)
features_data <- camera(features_data, eic = TRUE, iso = TRUE, ppm = 5, mzdiff = 0.01)
features_data <- group(features_data)
# Annotate metabolites
library(MetFragR)
frag_results <- runMetFrag(features_data, ion_mode = "positive", database = "HMDB")
annotated_data <- annotateMetabolites (frag_results, features_data)
```

#Visualization

Visualization is a crucial step in metabolomics pipeline for presenting the results obtained from statistical analysis and metabolite identification. Various R packages such as ggplot2, pheatmap, and shiny can be used for visualization.

```
# Visualize results
library(ggplot2)
ggplot (results, aes (x = logFC, y = -log10(FDR))) + geom_point(aes(color = significant))
ggplot (data_scaled, aes (x = PC1, y = PC2, color = group)) + geom_point ()
ggplot (pls_results, aes (x = PLS1, y = PLS2, color = group)) + geom_point ()
```

7.4 Publicly available metabolomics datasets

1. **Human Serum Metabolome HUSERMET**—A dataset of 20,000 serum metabolites from 200 healthy individuals and 100 patients with various diseases (Psychogios et al., 2011).
2. **Metabolomics Workbench**—A large, public repository of metabolomics data, including data from over 200 studies covering a variety of organisms, tissues, and diseases (Sud et al., 2016).
3. **TCGA**—A collection of multiomics data for various types of cancer, including metabolomics data for some cancer types (Weinstein et al., 2013).
4. **HMDB**—A comprehensive database of human metabolites, including information on their chemical and physical properties, biological functions, and associated diseases (Wishart et al., 2018).
5. **The Women's Health Initiative**—A large-scale longitudinal study of postmenopausal women that investigates the relationship between various factors, including metabolites, and chronic diseases (Wishart et al., 2018).

7.5 Application of omics-based metabolomics tools

Omics-based metabolomics tools have a wide range of applications in various fields such as medicine, environmental sciences, agriculture, and food sciences. Here are some examples of the applications of omics-based metabolomics tools:

1. **Biomarker discovery:** Metabolomics has been used to identify potential biomarkers for a variety of diseases such as cancer, diabetes (Singla et al., 2019), and cardiovascular diseases. For example, a recent study used untargeted metabolomics to identify potential biomarkers for prostate cancer. Another study used targeted metabolomics to identify biomarkers for gestational diabetes mellitus.
2. **Drug discovery:** Metabolomics has been used to identify the targets of drugs and to study the mechanisms of drug action. For example, a study used metabolomics to identify the target of a potential anticancer drug. Another study used metabolomics to study the mechanism of action of an antituberculosis drug.
3. **Environmental monitoring:** Metabolomics has been used to monitor environmental pollution (Singla et al., 2019) and to investigate the various effects of environmental toxins on environmental microbiome. For example, a study used metabolomics to identify the metabolic effects of environmental pollutants on fish. Another study used metabolomics to monitor water quality in the Rhine River.

TABLE 2.5 A list of metabolomics tools.

Tool	Description	Website
XCMS	Preprocessing of metabolomics data	https://xcmsonline.scripps.edu/
MetaboAnalyst	Statistical analysis and visualization of metabolomics data	https://www.metaboanalyst.ca/
MetFrag	Metabolite identification	https://msbi.ipb-halle.de/MetFrag/
CAMERA	Identification of metabolite peaks in LC-MS data	https://www.bioconductor.org/packages/release/bioc/html/CAMERA.html
MetaboQC	Quality control of metabolomics data	https://www.bioconductor.org/packages/release/bioc/html/MetaboQC.html
MS-DIAL	Feature detection and quantification in LC-MS data	http://prime.psc.riken.jp/Metabolomics_Software/MS-DIAL/
MZmine2	Preprocessing and feature detection in LC-MS data	https://mzmine.github.io/2.40.3/
GNPS	Metabolite annotation and spectral network analysis	https://gnps.ucsd.edu/
MetExplore	Interactive visualization of metabolomics data	https://metexplore.toulouse.inra.fr/
Pathway tools	Pathway analysis of metabolomics data	https://bioinformatics.ai.sri.com/ptools/

4. **Agriculture and food sciences:** Metabolomics has been used to study the metabolism of plants and to identify the metabolites responsible for desirable traits such as flavor and nutrition. For example, a study used metabolomics to identify the metabolites responsible for the aroma of green tea. Another study used metabolomics to identify the metabolites responsible for the antioxidant activity of barley grains. Here is a list of few metabolomics-based tools as shown in Table 2.5

8. Phenomics

Phenomics is a branch of biology that studies the physical and biochemical characteristics (or traits) of organisms, known as phenotypes, at the whole-organism level. It involves the comprehensive analysis of all observable traits of an organism, including its behavior, physiology, morphology, and biochemistry. Phenomics aims to understand how the interactions between an organism's genes and its environment give rise to its phenotype. This involves analyzing large amounts of data, including genetic, molecular, and physiological information, to identify the underlying mechanisms that contribute to the observed phenotype. Phenomics has applications in many areas of biology, including agriculture, medicine, and conservation. By understanding the underlying genetic and environmental factors that contribute to an organism's phenotype, researchers can develop more effective strategies for breeding crops with desirable traits, for diagnosing and treating genetic diseases, and for preserving endangered species.

8.1 Phenomics example datasets for using in R

There are several publicly available phenomics datasets that can be used in R for analysis. Here are a few examples:

1. **The Tabula Muris Senis dataset**, which contains single-cell transcriptomic data from aging mice. The dataset can be downloaded from the following link: https://figshare.com/articles/dataset/Tabula_Muris_Senis_single-cell_RNA-seq_data_from_18_month-old_mice/13263209
2. **The Human Cell Atlas dataset**, which contains single cell transcriptomic data from different tissues and organs in the human body. The dataset can be downloaded from the following link: https://www.humancellatlas.org/data
3. **The Genotype-Tissue Expression dataset**, which contains transcriptomic data from different human tissues. The dataset can be downloaded from the following link: https://gtexportal.org/home/datasets
4. **The 1000 Genomes Project dataset**, which contains genomic data from individuals of different ethnicities. The dataset can be downloaded from the following link: https://www.internationalgenome.org/data

5. **The UK Biobank dataset**, which contains genomic and phenotypic data from over 500,000 individuals in the United Kingdom. The dataset can be accessed by applying for access through the following link: https://www.ukbiobank.ac.uk/

8.2 Omics pipeline for phenomics analysis

```
#Data import and preprocessing:
    library (Summarized Experiment)
    library (Biobase)
    library(dplyr)
    # Load the data
    data <- readRDS("path/to/data.rds")
    # Preprocess the data
    data_processed <- data %>%
    # Remove any missing values
    na.omit() %>%
    # Scale the data scale
    () %>%
    # Remove any highly correlated features
    select(-highly_correlated_features)
```

References

Aebersold, R., & Mann, M. (2016). Mass-spectrometric exploration of proteome structure and function. *Nature, 537*(7620), 347–355. https://doi.org/10.1038/nature19949. http://www.nature.com/nature/index.html

Ali, N., Wolf, C., & Kanchan, S. (2023). *Nullomer peptide increases immune cell infiltration and reduces tumor metabolism in triple negative breast cancer mouse model.* https://doi.org/10.21203/rs.3.rs-3097552/v1

Aronson, S. J., & Rehm, H. L. (2015). Building the foundation for genomics in precision medicine. *Nature, 526*(7573), 336–342. https://doi.org/10.1038/nature15816. http://www.nature.com/nature/index.html

Bolyen, E., Rideout, J. R., Dillon, M. R., Bokulich, N. A., Abnet, C. C., Al-Ghalith, G. A., Alexander, H., Alm, E. J., Arumugam, M., Asnicar, F., Bai, Y., Bisanz, J. E., Bittinger, K., Brejnrod, A., Brislawn, C. J., Brown, C. T., Callahan, B. J., Caraballo-Rodríguez, A. M., Chase, J., … Caporaso, J. G. (2019). Reproducible, interactive, scalable and extensible microbiome data science using QIIME 2. *Nature Biotechnology, 37*(8), 852–857. https://doi.org/10.1038/s41587-019-0209-9. http://www.nature.com/nbt/index.html

Breton, C. V., Yao, J., Millstein, J., Gao, L., Siegmund, K. D., Mack, W., Whitfield-Maxwell, L., Lurmann, F., Hodis, H., Avol, E., & Gilliland, F. D. (2016). Prenatal air pollution exposures, DNA methyl transferase genotypes, and associations with newborn line1 and Alu methylation and childhood blood pressure and carotid intima-media thickness in the children's health study. *Environmental Health Perspectives, 124*(12), 1905–1912. https://doi.org/10.1289/EHP181. http://ehp.niehs.nih.gov/wp-content/uploads/124/12/EHP181.alt.pdf

Callahan, B. J., McMurdie, P. J., Rosen, M. J., Han, A. W., Johnson, A. J. A., & Holmes, S. P. (2016). DADA2: High-resolution sample inference from Illumina amplicon data. *Nature Methods, 13*(7), 581–583. https://doi.org/10.1038/nmeth.3869. http://www.nature.com/nmeth/

Chong, J., Soufan, O., Li, C., Caraus, I., Li, S., Bourque, G., Wishart, D. S., & Xia, J. (2018). MetaboAnalyst 4.0: Towards more transparent and integrative metabolomics analysis. *Nucleic Acids Research, 46*(W1), W486–W494. https://doi.org/10.1093/nar/gky310

Collins, F. S., & Varmus, H. (2015). A new initiative on precision medicine. *New England Journal of Medicine, 372*(9), 793–795. https://doi.org/10.1056/NEJMp1500523. http://www.nejm.org/doi/pdf/10.1056/NEJMp1500523

Falcon, S., & Gentleman, R. (2007). Using GOstats to test gene lists for GO term association. *Bioinformatics, 23*(2), 257–258. https://doi.org/10.1093/bioinformatics/btl567. http://bioinformatics.oxfordjournals.org/

Franzosa, E. A., McIver, L. J., Rahnavard, G., Thompson, L. R., Schirmer, M., Weingart, G., Lipson, K. S., Knight, R., Caporaso, J. G., Segata, N., & Huttenhower, C. (2018). Species-level functional profiling of metagenomes and metatranscriptomes. *Nature Methods, 15*(11), 962–968. https://doi.org/10.1038/s41592-018-0176-y. http://www.nature.com/nmeth/

Gahoi, S., Mandal, R. S., Ivanisenko, N., Shrivastava, P., Jain, S., Singh, A. K., Raghunandanan, M. V., Kanchan, S., Taneja, B., Mandal, C., Ivanisenko, V. A., Kumar, A., Kumar, R., Consorti, Open Source Drug Discovery, Ramachandran, S. (2013). Computational screening for new inhibitors of M. tuberculosis mycolyltransferases antigen 85 group of proteins as potential drug targets. *Journal of Biomolecular Structure and Dynamics, 31*(1), 30–43. https://doi.org/10.1080/07391102.2012.691343

Galande, S. H., Kanchan, S., & Kesheri, M. (2014). *Drug discovery and design: A bioinformatics approach.* LAP Lambert Academy Publishing.

Garg, S., Saxena, V., Kanchan, S., Sharma, P., Mahajan, S., Kochar, D., & Das, A. (2009). Novel point mutations in sulfadoxine resistance genes of Plasmodium falciparum from India. *Acta Tropica, 110*(1), 75–79. https://doi.org/10.1016/j.actatropica.2009.01.009

Gatto, L., & Lilley, K. S. (2012). Msnbase-an R/Bioconductor package for isobaric tagged mass spectrometry data visualization, processing and quantitation. *Bioinformatics, 28*(2), 288–289. https://doi.org/10.1093/bioinformatics/btr645

Ghai, R., Nagarajan, K., Kumar, V., Kesheri, M., & Kanchan, S. (2015). Amelioration of lipids by Eugenia caryophyllus extract in atherogenic diet induced hyperlipidemia. *International Bulletin of Drug Research, 5*(8), 90–101.

Ghai, R., Nagarajan, K., Singh, J., Swarup, S., & Keshari, M. (2016). Evaluation of anti-oxidant status in-vitro and in-vivo in hydro-alcoholic extract of Eugenia caryophyllus. *International Journal of Pharmacology and Toxicology, 4*(1), 19. https://doi.org/10.14419/ijpt.v4i1.5880

Hodge, R. D., Bakken, T. E., Miller, J. A., Smith, K. A., Barkan, E. R., Graybuck, L. T., Close, J. L., Long, B., Johansen, N., Penn, O., Yao, Z., Eggermont, J., Höllt, T., Levi, B. P., Shehata, S. I., Aevermann, B., Beller, A., Bertagnolli, D., Brouner, K., ... Lein, E. S. (2019). Conserved cell types with divergent features in human versus mouse cortex. *Nature, 573*(7772), 61–68. https://doi.org/10.1038/s41586-019-1506-7. http://www.nature.com/nature/index.html

Horr, C., & Buechler, S. A. (2021). Breast cancer consensus subtypes: A system for subtyping breast cancer tumors based on gene expression. *Npj Breast Cancer, 7*(1). https://doi.org/10.1038/s41523-021-00345-2. http://www.nature.com/npjbcancer/

Islam, R., Zhao, L., Wang, Y., Lu-Yao, G., & Liu, L.-Z. (2022). Epigenetic Dysregulations in arsenic-induced Carcinogenesis. *Cancers, 14*(18), 4502. https://doi.org/10.3390/cancers14184502

Kanchan, S., Mehrotra, R., & Chowdhury, S. (2014). Evolutionary pattern of four representative DNA repair proteins across six model organisms: An in silico analysis. *Network Modeling Analysis in Health Informatics and Bioinformatics, 3*(1). https://doi.org/10.1007/s13721-014-0070-1. http://www.springer.com/new+%26+forthcoming+titles+%28default%29/journal/13721

Kanchan, S., Mehrotra, R., & Chowdhury, S. (2015). In silico analysis of the endonuclease III protein family identifies key Residues and processes during evolution. *Journal of Molecular Evolution, 81*(1–2), 54–67. https://doi.org/10.1007/s00239-015-9689-5. http://link.springer.de/link/service/journals/00239/index.htm

Kanchan, S., Sharma, P., & Chowdhury, S. (2019). Evolution of endonuclease IV protein family: An in silico analysis. *3 Biotech, 9*(5). https://doi.org/10.1007/s13205-019-1696-6. http://www.springerlink.com/content/2190-572x/

Kanchan, S., Sinha, R. P., Chaudière, J., & Kesheri, M. (2020). Computational metagenomics: current status and challenges. In *Recent trends in 'computational omics': Concepts and methodology* (pp. 371–395). France: Nova Science Publishers, Inc., https://novapublishers.com/shop/recent-trends-in-computational-omics-concepts-and-methodology/

Kesheri, M., Kanchan, S., Chowdhury, S., & Sinha, R. P. (2015). Secondary and tertiary structure prediction of proteins: A bioinformatic approach. *Studies in Fuzziness and Soft Computing, 319*, 541–569. https://doi.org/10.1007/978-3-319-12883-2_19

Kesheri, M., Kanchan, S., & Sinha, R. P. (2017). *Exploring the potentials of antioxidants in retarding ageing* (pp. 166–195). IGI Global. https://doi.org/10.4018/978-1-5225-0607-2.ch008

Kesheri, M., Kanchan, S., & Sinha, R. P. (2021). Isolation and in silico analysis of antioxidants in response to temporal variations in the cyanobacterium Oscillatoria sp. *Gene Reports, 23*, 101023. https://doi.org/10.1016/j.genrep.2021.101023

Kesheri, M., & Kanchan. (2015). Oxidative stress: Challenges and its mitigation mechanisms in cyanobacteria. In *Biological sciences: Innovations and dynamics* (pp. 309–324). New India Publishing Agency.

Kesheri, M., Kanchan, S., & Chowdhury, S. (2014). *Cyanobacterial stresses: An ecophysiological, biotechnological and bioinformatic approach.* LAP Lambert Academy Publishing.

Kesheri, M., Kanchan, S., Richa, & Sinha, R. P. (2014). Isolation and in silico analysis of Fe-superoxide dismutase in the cyanobacterium Nostoc commune. *Gene, 553*(2), 117–125. https://doi.org/10.1016/j.gene.2014.10.010. http://www.elsevier.com/locate/gene

Kesheri, M., Kanchan, S., & Sinha, R. P. (2022). Responses of antioxidants for resilience to temporal variations in the cyanobacterium Microcystis aeruginosa. *South African Journal of Botany, 148*, 190–199. https://doi.org/10.1016/j.sajb.2022.04.017. http://www.elsevier.com

Kesheri, M., Richa, & Sinha, R. P. (2011). Antioxidants as natural arsenal against multiple stresses in Cyanobacteria. *International Journal of Pharma and Bio Sciences, 2*(2), 168–187. http://ijpbs.net/volume2/issue2/bio/17.pdf.

Kesheri, M., Sinha, R. P., & Kanchan, S. (2016). Advances in soft computing approaches for gene prediction: A bioinformatics approach. *Studies in Computational Intelligence, 651*, 383–405. https://doi.org/10.1007/978-3-319-33793-7_17. http://www.springer.com/series/7092

Kuhl, C., Tautenhahn, R., Böttcher, C., Larson, T. R., & Neumann, S. (2012). CAMERA: An integrated strategy for compound spectra extraction and annotation of liquid chromatography/mass spectrometry data sets. *Analytical Chemistry, 84*(1), 283–289. https://doi.org/10.1021/ac202450g

Kumari, A., Kanchan, S., Sinha, R. P., & Kesheri, M. (2016). Applications of bio-molecular databases in bioinformatics. *Studies in Computational Intelligence, 651*, 329–351. https://doi.org/10.1007/978-3-319-33793-7_15. http://www.springer.com/series/7092

Kumari, A., Kesheri, M., Sinha, R. P., & Kanchan, S. (2018). Integration of soft computing approach in plant biology and its applications in agriculture. *Soft Computing for Biological Systems*, 265–281. https://doi.org/10.1007/978-981-10-7455-4_16. http://www.springer.com/in/book/9789811074547

Langille, M. G. I., Zaneveld, J., Caporaso, J. G., McDonald, D., Knights, D., Reyes, J. A., Clemente, J. C., Burkepile, D. E., Vega Thurber, R. L., Knight, R., Beiko, R. G., & Huttenhower, C. (2013). Predictive functional profiling of microbial communities using 16S rRNA marker gene sequences. *Nature Biotechnology, 31*(9), 814–821. https://doi.org/10.1038/nbt.2676

Love, M. I., Huber, W., & Anders, S. (2014). Moderated estimation of fold change and dispersion for RNA-seq data with DESeq2. *Genome Biology, 15*(12). https://doi.org/10.1186/s13059-014-0550-8. http://genomebiology.com/

Lu, J., Breitwieser, F. P., Thielen, P., & Salzberg, S. L. (2017). Bracken: Estimating species abundance in metagenomics data. *PeerJ Computer Science, 2017*(1). https://doi.org/10.7717/peerj-cs.104. https://peerj.com/articles/cs-104.pdf

Lujambio, A., Ropero, S., Ballestar, E., Fraga, M. F., Cerrato, C., Setién, F., & Esteller. (2008). Genetic unmasking of an epigenetically silenced microRNA in human cancer cells. *Cancer Research, 68*(9), 2667–2675.

Luo, W., Friedman, M. S., Shedden, K., Hankenson, K. D., & Woolf, P. J. (2009). GAGE: Generally applicable gene set enrichment for pathway analysis. *BMC Bioinformatics, 10*. https://doi.org/10.1186/1471-2105-10-161

Mann, M., & Kelleher, N. L. (2008). Precision proteomics: The case for high resolution and high mass accuracy. *Proceedings of the National Academy of Sciences, 105*(47), 18132–18138. https://doi.org/10.1073/pnas.0800788105

Mastroeni, D., Grover, A., Delvaux, E., Whiteside, C., Coleman, P. D., & Rogers, J. (2010). Epigenetic changes in Alzheimer's disease: Decrements in DNA methylation. *Neurobiology of Aging, 31*(12), 2025–2037. https://doi.org/10.1016/j.neurobiolaging.2008.12.005

McMurdie, P. J., Holmes, S., & Watson, M. (2013). phyloseq: An R package for reproducible interactive analysis and graphics of microbiome census data. *PLoS One, 8*(4), Article e61217. https://doi.org/10.1371/journal.pone.0061217

Mishra, P., Saxena, V., Kesheri, M., & Saxena, A. (2015a). Synthesis, characterization and pharmacological evaluation of cinnoline (thiophene) derivatives. *The Pharma Innovation Journal, 4*(10), 68–73.

Mishra, P., Saxena, V., Kesheri, M., & Saxena, A. (2015b). Synthesis, characterization and antiinflammatory activity of cinnolines (pyrazole) derivatives. *IOSR Journal of Pharmacy and Biological Sciences, 10*(6), 77–82.

Okamoto, T., duVerle, D., Yaginuma, K., Natsume, Y., Yamanaka, H., Kusama, D., Fukuda, M., Yamamoto, M., Perraudeau, F., Srivastava, U., Kashima, Y., Suzuki, A., Kuze, Y., Takahashi, Y., Ueno, M., Sakai, Y., Noda, T., Tsuda, K., Suzuki, Y., Nagayama, S., & Yao, R. (2021). Comparative analysis of patient-matched PDOs revealed a reduction in OLFM4-associated clusters in metastatic lesions in colorectal cancer. *Stem Cell Reports, 16*(4), 954–967. https://doi.org/10.1016/j.stemcr.2021.02.012. http://www.elsevier.com/journals/stem-cell-reports/2213-6711

Olsen, J. V., & Mann, M. (2013). Status of large-scale analysis of posttranslational modifications by mass spectrometry. *Molecular and Cellular Proteomics, 12*(12), 3444–3452. https://doi.org/10.1074/mcp.O113.034181. http://www.mcponline.org/content/12/12/3444.full.pdf+html

Parker, J. S., Mullins, M., Cheang, M. C. U., Leung, S., Voduc, D., Vickery, T., Davies, S., Fauron, C., He, X., Hu, Z., Quackenbush, J. F., Stijleman, I. J., Palazzo, J., Marron, J. S., Nobel, A. B., Mardis, E., Nielsen, T. O., Ellis, M. J., Perou, C. M., & Bernard, P. S. (2009). Supervised risk predictor of breast cancer based on intrinsic subtypes. *Journal of Clinical Oncology, 27*(8), 1160–1167. https://doi.org/10.1200/jco.2008.18.1370

Paulson, J. N., Stine, O. C., Bravo, H. C., & Pop, M. (2013). Differential abundance analysis for microbial marker-gene surveys. *Nature Methods, 10*(12), 1200–1202. https://doi.org/10.1038/nmeth.2658

Priya, P., Kesheri, M., Sinha, R. P., & Kanchan, S. (2016). *Molecular dynamics simulations for biological systems*. IGI Global. https://doi.org/10.4018/978-1-4666-8811-7.ch014

Psychogios, N., Hau, D. D., Peng, J., Guo, A. C., Mandal, R., Bouatra, S., Sinelnikov, I., Krishnamurthy, R., Eisner, R., Gautam, B., Young, N., Xia, J., Knox, C., Dong, E., Huang, P., Hollander, Z., Pedersen, T. L., Smith, S. R., Bamforth, F., … Flower, D. (2011). The human serum metabolome. *PLoS One, 6*(2), Article e16957. https://doi.org/10.1371/journal.pone.0016957

Rajamani, D., & Bhasin, M. K. (2016). Identification of key regulators of pancreatic cancer progression through multidimensional systems-level analysis. *Genome Medicine, 8*(1). https://doi.org/10.1186/s13073-016-0282-3. http://www.genomemedicine.com/

Richa, Kannaujiya, V. K., Kesheri, M., Singh, G., & Sinha, R. P. (2011). Biotechnological potentials of phycobiliproteins. *International Journal of Pharma and Bio Sciences, 2*(4), 446–454. http://www.ijpbs.net/vol-2_issue-4/bio_science/50.pdf.

Richa, Rastogi, R. P., Kumari, S., Singh, K. L., Kannaujiya, V. K., Singh, G., Kesheri, M., & Sinha. (2011). Biotechnological potential of mycosporine-like amino acids and phycobiliproteins of cyanobacterial origin. *Biotechnology, Bioinformatics and Bioengineering, 1*(2), 159–171.

Ruttkies, C., Schymanski, E. L., Wolf, S., Hollender, J., & Neumann, S. (2016). MetFrag relaunched: Incorporating strategies beyond in silico fragmentation. *Journal of Cheminformatics, 8*(1). https://doi.org/10.1186/s13321-016-0115-9. http://www.jcheminf.com/

Sahu, N., Mishra, S., Kesheri, M., Kanchan, S., & Sinha, R. P. (2023). Identification of cyanobacteria-based natural inhibitors against SARS-CoV-2 druggable target ACE2 using molecular docking study, ADME and toxicity analysis. *Indian Journal of Clinical Biochemistry, 38*(3), 361–373. https://doi.org/10.1007/s12291-022-01056-6. https://www.springer.com/journal/12291

Sali, A., & Kuriyan, J. (1999). Challenges in protein structure prediction. *Current Opinion in Structural Biology, 9*(3), 397–402.

Saxena, A., Saxena, V., Kesheri, M., & Mishra. (2015). Comparative hypoglycemic effects of different extract of Clitoria ternatea leaves on rats. *IOSR Journal of Pharmacy and Biological Sciences, 10*(2), 60–65.

Schloss, P. D., Westcott, S. L., Ryabin, T., Hall, J. R., Hartmann, M., Hollister, E. B., Lesniewski, R. A., Oakley, B. B., Parks, D. H., Robinson, C. J., Sahl, J. W., Stres, B., Thallinger, G. G., Van Horn, D. J., & Weber, C. F. (2009). Introducing mothur: Open-source, platform-independent, community-supported software for describing and comparing microbial communities. *Applied and Environmental Microbiology, 75*(23), 7537–7541. https://doi.org/10.1128/AEM.01541-09. http://aem.asm.org/cgi/reprint/75/23/7537

Sergushichev, A. A. (2016). An algorithm for fast preranked gene set enrichment analysis using cumulative statistic calculation. *bioRxiv*. https://doi.org/10.1101/060012

Singla, S., Kesheri, M., Kanchan, S., & Aswath, S. (2019). Current status and data analysis of diabetes in India. *International Journal of Innovative Technology and Exploring Engineering, 8*(9), 1920–1934. https://doi.org/10.35940/ijitee.i8403.078919. https://www.ijitee.org/wp-content/uploads/papers/v8i9/I8403078919.pdf

Singla, S., Kesheri, M., Kanchan, S., & Mishra, A. (2019). Data analysis of air pollution in India and its effects on health. *International Journal of Pharma and Bio Sciences, 10*(2), 155–169.

Smith, C. C., McMahon, K. W., & Pedersen, B. S. (2019). APOBEC3A and APOBEC3B promote haploinsufficiency-driven genome instability and tumorigenesis. *Cell, 176*(1–2), 432–447.

Srivastava, U., & Singh, G. (2013). Comparative homology modelling for HPV type 16 E 7 proteins by using MODELLER and its validations with SAVS and ProSA web server. *Journal of Computational Intelligence in Bioinformatics, 6*(1), 27. https://doi.org/10.37622/jcib/6.1.2013.27-33

Srivastava, U., & Singh, S. (2022). *Approaches of single-cell analysis in crop improvement*. Springer Science and Business Media LLC. https://doi.org/10.1007/978-1-0716-2533-0_14

Srivastava, U., Singh, S., Gautam, B., Yadav, P., Yadav, M., Thomas, G., & Singh, G. (2017). Linear epitope prediction in HPV type 16 E7 antigen and their docked interaction with human TMEM 50A structural model. *Bioinformation, 13*(05), 122–130. https://doi.org/10.6026/97320630013122

Sud, M., Fahy, E., Cotter, D., Azam, K., Vadivelu, I., Burant, C., Edison, A., Fiehn, O., Higashi, R., Nair, K. S., Sumner, S., & Subramaniam, S. (2016). Metabolomics Workbench: An international repository for metabolomics data and metadata, metabolite standards, protocols, tutorials and training, and analysis tools. *Nucleic Acids Research, 44*(D1), D463–D470. https://doi.org/10.1093/nar/gkv1042

Truong, D. T., Franzosa, E. A., Tickle, T. L., Scholz, M., Weingart, G., Pasolli, E., Tett, A., Huttenhower, C., & Segata, N. (2015). MetaPhlAn2 for enhanced metagenomic taxonomic profiling. *Nature Methods, 12*(10), 902–903. https://doi.org/10.1038/nmeth.3589. http://www.nature.com/nmeth/

Weinstein, J. N., Collisson, E. A., Mills, G. B., Shaw, K. R. M., Ozenberger, B. A., Ellrott, K., Sander, C., Stuart, J. M., Chang, K., Creighton, C. J., Davis, C., Donehower, L., Drummond, J., Wheeler, D., Ally, A., Balasundaram, M., Birol, I., Butterfield, Y. S. N., Chu, A., … Kling, T. (2013). The cancer genome atlas pan-cancer analysis project. *Nature Genetics, 45*(10), 1113–1120. https://doi.org/10.1038/ng.2764. http://www.nature.com/ng/index.html

Wishart, D. S., Feunang, Y. D., Marcu, A., Guo, A. C., Liang, K., Vázquez-Fresno, R., Sajed, T., Johnson, D., Li, C., Karu, N., Sayeeda, Z., Lo, E., Assempour, N., Berjanskii, M., Singhal, S., Arndt, D., Liang, Y., Badran, H., Grant, J., … Scalbert, A. (2018). Hmdb 4.0: The human metabolome database for 2018. *Nucleic Acids Research, 46*(D1), D608–D617. https://doi.org/10.1093/nar/gkx1089

Woloszynek, S., Mell, J. C., Zhao, Z., Simpson, G., O'Connor, M. P., Rosen, G. L., & Loor, J. J. (2019). Exploring thematic structure and predicted functionality of 16S rRNA amplicon data. *PLoS One, 14*(12), e0219235. https://doi.org/10.1371/journal.pone.0219235

Wood, D. E., Lu, J., & Langmead, B. (2019). Improved metagenomic analysis with Kraken 2. *Genome Biology, 20*(1). https://doi.org/10.1186/s13059-019-1891-0. http://genomebiology.com/

Yu, G., Wang, L. G., Han, Y., & He, Q. Y. (2012). ClusterProfiler: An R package for comparing biological themes among gene clusters. *OMICS: A Journal of Integrative Biology, 16*(5), 284–287. https://doi.org/10.1089/omi.2011.0118

Zawada, A. M., Rogacev, K. S., Müller, S., Rotter, B., Winter, P., Fliser, D., & Heine, G. H. (2014). Massive analysis of cDNA Ends (MACE) and miRNA expression profiling identifies proatherogenic pathways in chronic kidney disease. *Epigenetics, 9*(1), 161–172. https://doi.org/10.4161/epi.26931. https://www.landesbioscience.com/journals/epigenetics/2013EPI0349R.pdf

Chapter 3

Biological omics databases and tools

Atifa Hafeez, Archana Gupta and Manish Kumar Gupta
Department of Biotechnology, Faculty of Science, Veer Bahadur Singh Purvanchal University, Jaunpur, Uttar Pradesh, India

1. Introduction

In Molecular Research, Omics has become a new catchword. Rapid advancements of high-throughput omics (genomics, transcriptomics, proteomics, and metabolomics) by high-throughput technologies have permit to the rapid aggregation of patient data. The "omic" technologies are high-throughput technologies and they rise the number of genes or proteins that can be identified concomitantly to associate intricate mixtures to complex effects in the protein or gene expression profiles (Debnath et al., 2005). The biomedical research has plentiful data but still suffer for knowledge. Computational biology refers to bioinformatics, a helping hand in knowledge discovery by handling with storage, retrieval and optimal use of omics data (Kaur et al., 2021). At the beginning, omics experiments provide *single-omics* data; apart from these data, researchers merged different assays from same set of samples and generated an integrated omics data known as *multiomics* data (Conesa & Beck, 2019). Different types of molecules are available with different functions in a cell. These molecules are categorized into numerous categories based on their functions and types. The molecules are categorized into DNA, RNA, proteins, and metabolites. To examine and analyze the extinction and mechanism of the molecule of a cell, the bioinformatics tools are specific. For well rudimentary, these molecules are divided into different omics types as shown in Fig. 3.1, such as genomics (structural, functional analysis of genes and identification of genetic variants), transcriptomics (gene expression profiling, functional regulations, and differential expression), proteomics (protein identification, quantification, and posttranslational modification), and metabolomics (profiling of metabolites, hormones, and signaling molecules). The studies of these categories are being analyzed by various bioinformatics tools and their respective biological databases which are used to extract their biological information (Shankar et al., 2021). Technologies for omics analysis have advanced rapidly, ever since the DNA microarray was established as the first high-throughput technology. In accordance with central dogma, omics technologies have been applied to record the static genomic changes, temporal transcriptome perturbations, alternative splicing, spatio-temporal proteome dynamics, and posttranslational modifications. Additionally, omics technologies have been developed to analyze different omics at the epi-level (including epigenome, epitranscriptome, and epiproteome) which are characterized as the library of all modifications of the referred omics beyond information it surrounded in a single cell, molecular interactions (different level of interactome), as well as disease-associated defining characteristics like metabolome and immunome (Dai & Shen, 2022). The potential to integrate biological characteristics in order to examine a system as a whole is becoming more important with the development of Next-Generation Sequencing (NGS) and Mass Spectrometry (MS). The host's reaction to many diseases and cancers is influenced by features like the transcriptome, proteome, histone post-translational changes, and microbiome (Graw et al., 2021). Single-cell sequencing provides extra resolving power that permits research at the single-cell level, and multiomics integration has become a common practice for building a thorough causal link between molecular signatures and phenotypic symptoms of a specific disease (Dai & Shen, 2022).

2. Omics datasets study based on different technology

For analyzing the omics of a particular biological system, sequencing and MS are the two fundamental experimental tools. Techniques based on MS can be utilized to examine the proteome and metabolome, while the study of genome, transcriptome, and their epitomes can be conducted using sequencing-based techniques (Dai & Shen, 2022). The key characteristics and output of popular sequencing technologies have been carefully examined (Kulski, 2016).

Integrative Omics. https://doi.org/10.1016/B978-0-443-16092-9.00003-5

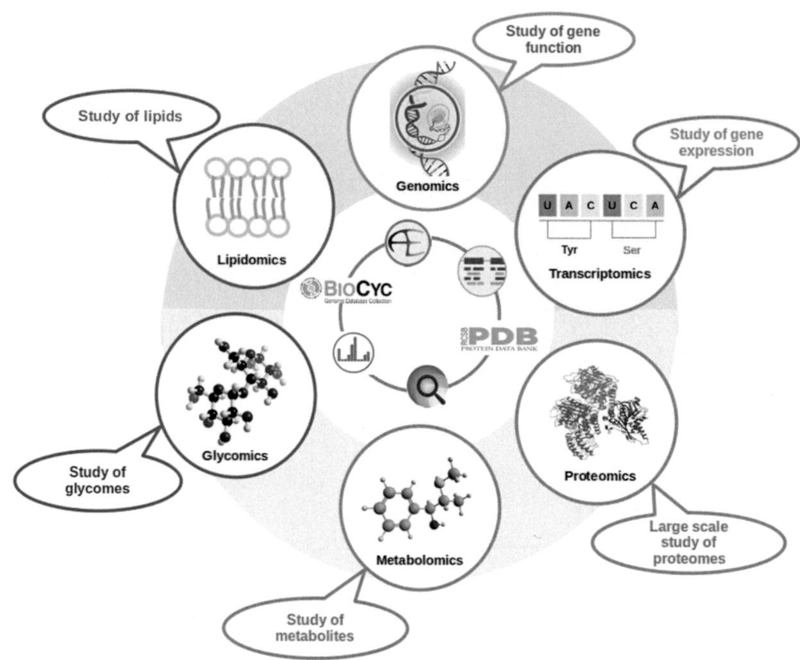

FIGURE 3.1 Overview of omics datasets. Omics approaches seek to determine the complete composition of a certain biochemical cluster: genomics (study of gene function), transcriptomics (study of gene expression), proteomics (large-scale study of proteomes), metabolomics (study of metabolites), glycomics (study of glycomes), and lipidomics (study of lipids).

2.1 Next-generation sequencing

The deep, high-throughput, parallel DNA sequencing technology known as NGS was created a few decades after the Sanger DNA sequencing method (Kulski, 2016). The area of NGS has achieved impressive advancements including lower expenses, unparalleled sequencing speed, high resolution, and precision in genetic analysis (Lee et al., 2013). Nowadays, this technology is developing into a molecular microscope that is permeating almost all sectors of biological research (Buermans & den Dunnen, 2014). NGS has transformed genome sequencing and personal medicine, but it has also changed how genome research was conducted in the past. Sequencing is the use of a technology that aids in identifying the quantity and arrangement of nucleotides in a certain organism's DNA (Thakur et al., 2018). NGS pipeline has been shown in Fig. 3.2.

The utilization of NGS technology is expanding across numerous studies. Their strength is in the ability to gather vast amounts of data and reveal new as well as crucial knowledge about the human genome. This characteristic unlocked a wide range of productive application contexts (Del Vecchio et al., 2017). Application of NGS technologies includes whole genome sequencing, resequencing, de novo assembly sequencing as well as transcriptome sequencing at the DNA or RNA level. It has also been extensively utilized to investigate small RNAs, such as the prediction of new miRNAs (microRNAs) and detection of miRNAs that are differentially expressed and an annotation of other small noncoding RNAs (Lee et al., 2013).The different sequencing methods shown in Fig. 3.3.

2.2 Microarray

Microarray and RNA sequencing (RNA-Seq), which are currently the most popular methods, were established in the mid-1990s and early 2000s (Lowe et al., 2017). They can be used to study the transcriptome of the cell. In comparison to RNA-Seq, microarray has a lower throughput and a number of constraints (Shankar et al., 2021). Microarray technology is a hybridization based approach (Fig. 3.4) that incorporates an application of fluorescent dye for labeling and miniaturization. It enables the concurrent study of the RNA expression of thousands of genes (Solanke & Tribhuvan, 2015). The genes are quite particular to them (Shankar et al., 2021). The gene expression profiling is the main application of microarray (Solanke & Tribhuvan, 2015). One of the earliest applications of microarray was in the investigations of the mode of action of antibiotic isoniazid, which is useful in the treatment of tuberculosis. Microarray could be a benefit to investigators as it offers a platform for evaluating a large number of genetic samples at once (Debnath et al., 2005).

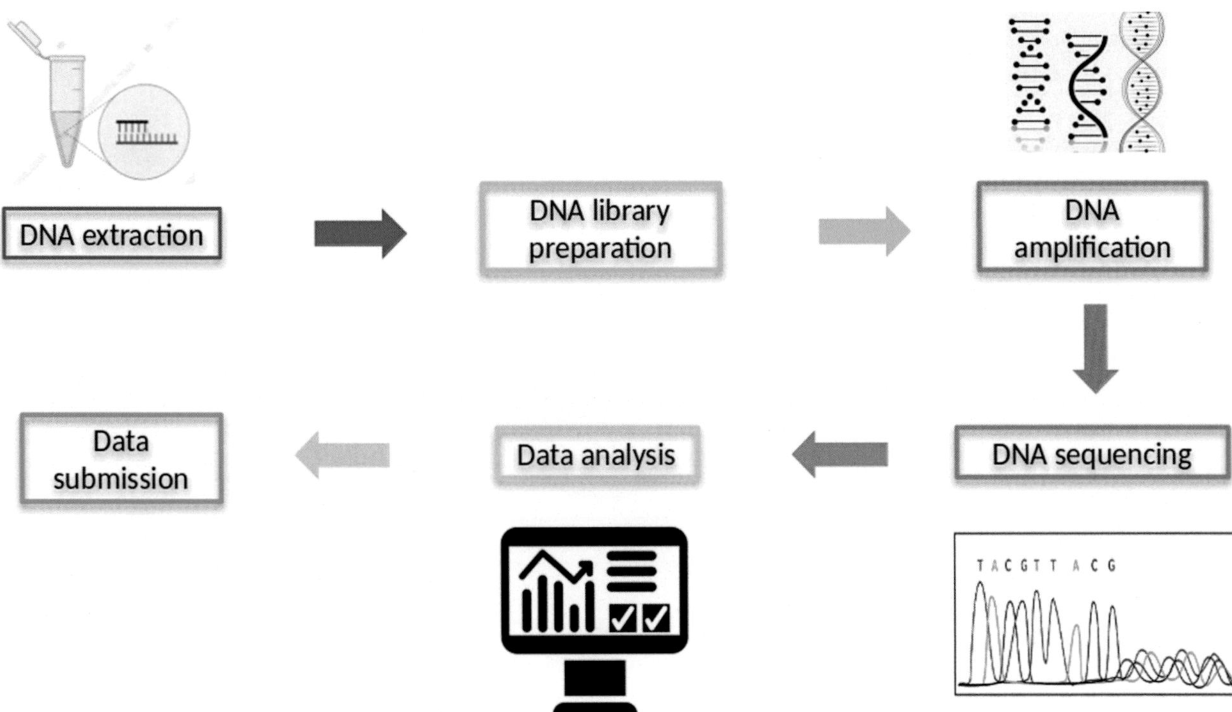

FIGURE 3.2 Next-generation sequencing techniques pipeline. NGS is an innovative tool that promises to significantly expand the knowledge of how nucleic acid functions.

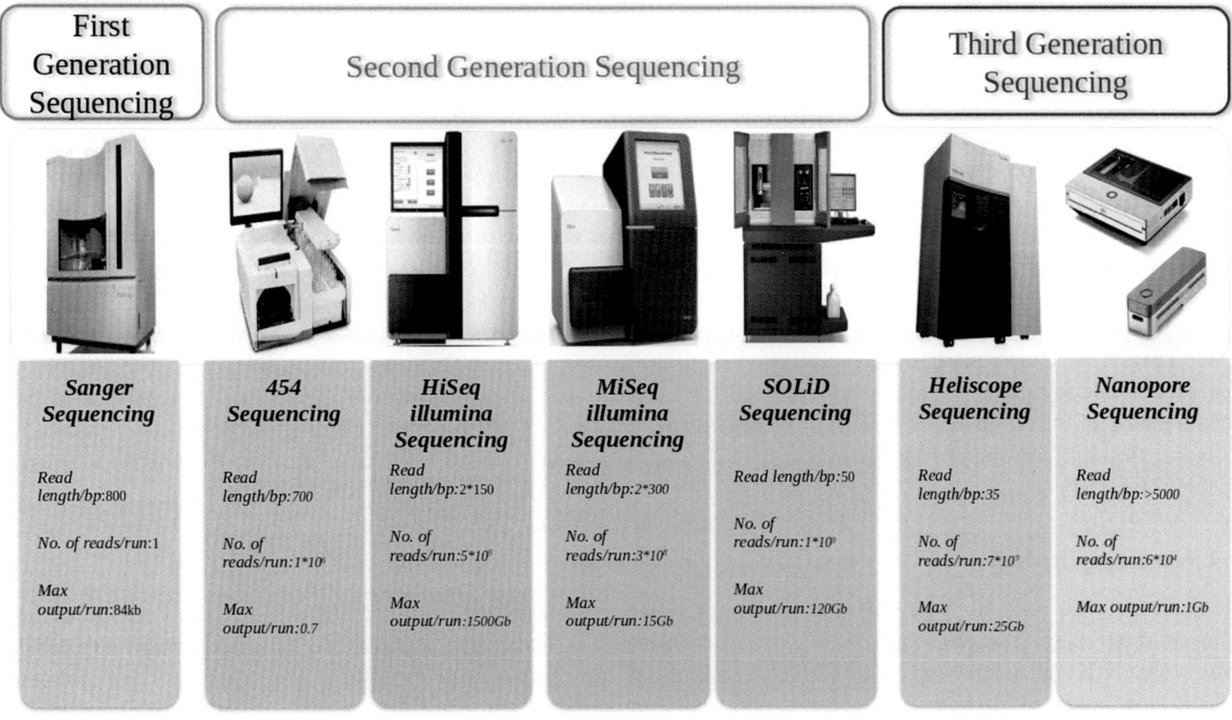

FIGURE 3.3 The different sequencing methods. Depending on the number of reads and read length, various organizations uses multiple platforms with distinctive properties and benefits to safeguard assembly accuracy as well as quality.

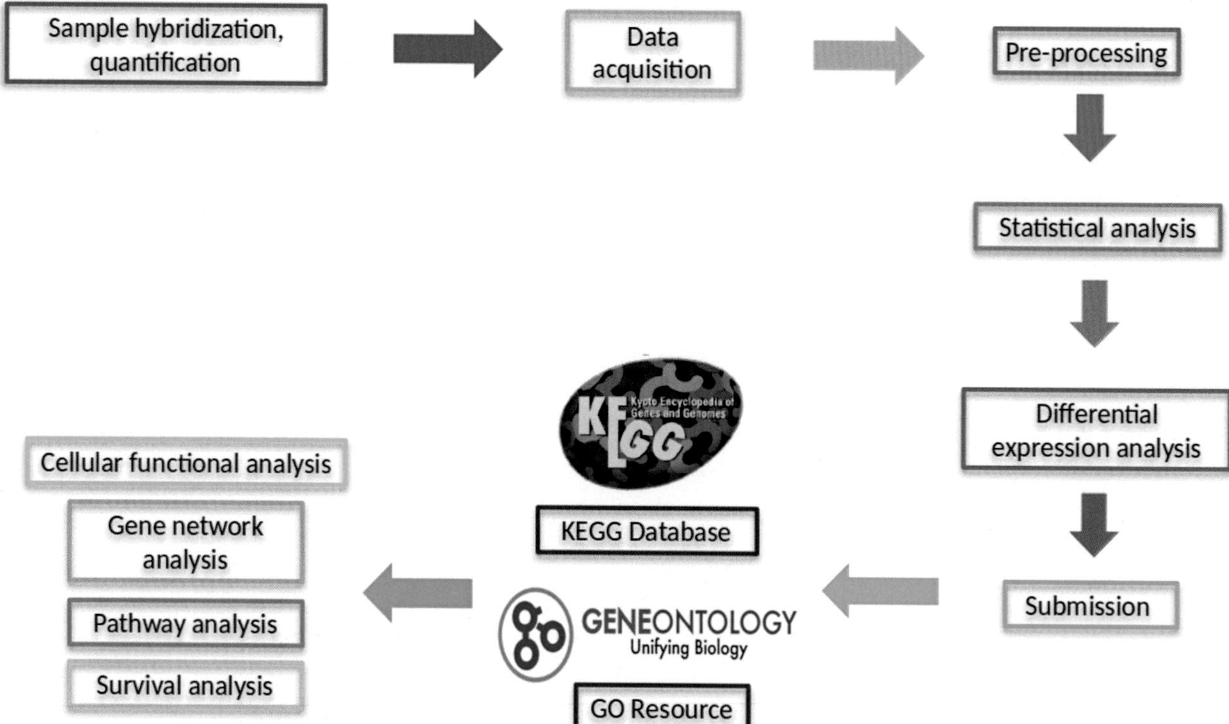

FIGURE 3.4 Microarray pipeline. Microarray: the approach for examining gene expression data. Numerous types of research, which involves gene expression profiling, translation profiling, genotyping, and epigenetics, can be carried out using microarrays.

2.3 RNA-Seq

Recent developments in deep-sequencing technology have led to the development of the transcriptome profiling method known as RNA-Seq (Wang et al., 2009). RNA-Seq data generation workflow has been shown in Fig. 3.5. Mammalian cells and yeast served as the first applications for high-throughput RNA sequencing utilizing cDNA fragments and currently a variety of species use it. For comprehending the complexities of genome function in biology, the genome sequence solely is of little relevance without transcriptome data (Kulski, 2016). The introduction of modern high-throughput DNA sequencing techniques has provided a novel technology (i.e., RNA-Seq) for both mapping as well as measuring transcriptomes. This technology offers definite benefits over existing methods and is predicted to transform the manner in which eukaryotic transcriptomes are examined (Wang et al., 2009). After 2008, when new Solexa/Illumina technology made it possible to capture 10^9 transcript sequences, RNA-Seq started to gain popularity. This yield is adequate for precise quantification of the complete human transcriptome (Lowe et al., 2017). It provides a number of benefits over microarray including the ability to analyze alternate splicing, polyadenylation, and the identification of new genes/transcripts (Shankar et al., 2021).

2.4 Mass spectrometry

MS is an analytical method that divides ionized particles including atoms, molecules and clusters based on variations in the ratios of their charges to mass (m/z; mass/charge) and it could be utilized to estimate the molecular weight of particles (Murayama & Kimura, 2009, pp. 131–139). There are three steps in the entire procedure. In the first step, the molecules must be converted to gas-phase ions, which is difficult for biomolecules in a liquid or solid phase. In the second step, ions are separated based on their m/z values while being exposed to magnetic or electric fields in a space known as mass analyzer. In the end, separated ions are calculated, together with the quantity of each species having a certain m/z value (Aslam et al., 2017).

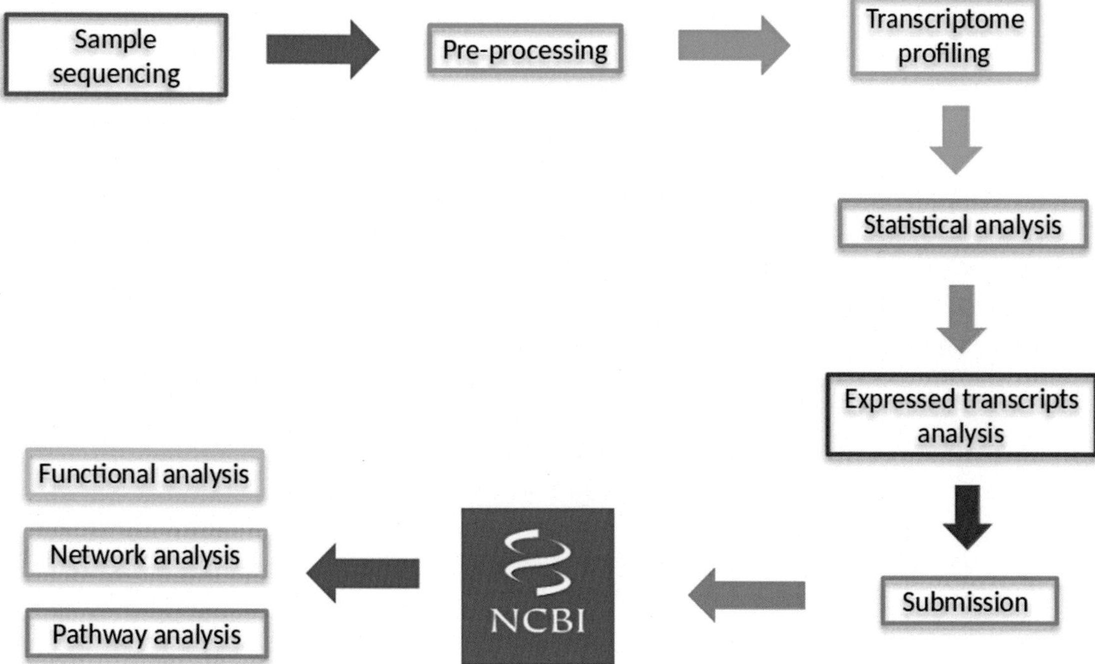

FIGURE 3.5 RNA-Seq data generation workflow. A high-throughput approach offering distinct insights regarding the transcriptome.

2.5 GC-MS (gas chromatography MS), LC-MS (liquid chromatography MS), and CE-MS (capillary electrophoresis MS)

For measuring volatile substances like fatty acids and organic acids, GC-MS is a frequently used technology in metabolomics. Analytes must be volatile and thermally stable. GC separation takes place in oven at high temperatures. It is often essential to derivatize samples before analysis. Derivatization is a supplementary sample processing step that is essential but also has the potential to cause metabolite loss, which is one of the biggest issues with this technology. One of the most advantageous features of this technology is that, molecules will have distinctive spectrum and huge libraries are accessible online (Wang et al., 2011). Similar to proteomics, MS based metabolomics often uses separation (e.g., LC, GC, or CE) before MS detection. By using the whole scan and chosen ion monitoring modes of the GC-MS, it is possible to perform untargeted and targeted metabolomics respectively. MALDI-MS, on the other hand, performs high-throughput screening without separation. High-throughput metabolic fingerprinting is accomplished by nuclear magnetic resonance (NMR), which also gives more accurate metabolites structure (Amer & Baidoo, 2021).

3. Specific omics datasets based on sequencing approach

3.1 Genomics

The study of an organism's genome or genetic material is termed as genomics. In order to comprehend the intricate, biological function of genome, the genomics was established as a new branch of molecular biology (Debnath et al., 2005). In 1986, a geneticist T.H. Roderick created the word "Genomics". This study examines the arrangements of genes and genetic data inside the genome and how this arrangement determines their role. The list of various databases that are available for genomics study have been discussed in Table 3.1. Additionally, it helps to comprehend the variability among genomes derives from various sources such as duplication events, transposons (a small DNA fragment which can transfer from one section of the genome to another), and microsatellites (repetitive DNA region which may dramatically enlarge or reduce in size) (Solanke & Tribhuvan, 2015). Rapid advancements in the understanding of the molecular characteristics of cell and tissue activity are achieved through the molecular biology revolution and the development of genomic technology. These developments have the ability to change toxicological and clinical practice, and are likely to augment or displace

TABLE 3.1 Genomics databases.

Database	Availability	Function	File format	URL
Array Express (Function Genomics Data Collection)	Public access	Saves data from high-throughput functional genomic analysis and offers it to the research community for reuse	MAGE-TAB (MicroArray Gene Expression Tabular)	https://www.ebi.ac.uk/biostudies/arrayexpress
COSMIC (Catalogue Of Somatic Mutation In Cancer)	Private access	Provides genomic data exploration with an emphasis on mutational types and frequency statistics for user-defined gene or cancer type	CSV (Comma Separated Values), TSV (Tab-Separated Values)	https://cancer.sanger.ac.uk/cosmic
dbVar (Database of Genomic structural variation)	Public access	Human genomic structural variation database from NCBI	Excel, XML (Extensible Markup Language), Tab-delimited	https://www.ncbi.nlm.nih.gov/dbvar
dbGaP (Database of Genotypes and Phenotypes)	Public access	To store and disseminate information from research that looked at how genotype and phenotype interacted in humans	SRF (Short Read Archive), VCF (Variant Call Format)	https://www.ncbi.nlm.nih.gov/gap/
dbSNP (Database of Single Nucleotide Polymorphisms)	Public access	Comprises small scale insertions and deletions, microsatellites, and human single nucleotide variants	XML, FASTA	https://www.ncbi.nlm.nih.gov/snp/
EGA (European Genome Phenome Archive)	Public access	Datasets from genomics research	VCF, FASTAQ, CRAM (Compressed Reference-oriented Alignment Map), BAM (Binary Alignment Map)	https://ega-archive.org/
Genbank	Public access	The NIH genetic sequence database, a compendium of all publicly accessible DNA sequences that has been annotated	XML, GenBank, ASN.1	https://www.ncbi.nlm.nih.gov/genbank/
euGenes (Genomic Information of Eukaryotic Organisms)	Public access	A standard overview of genomic data can be obtained from databases of eukaryotic organism	FASTA	http://eugenes.org/
GEO Datasets	Public access	In addition to the original series and platform records from the GEO repository, also contains curated gene expression datasets	Spreadsheet, plain text, XML	https://www.ncbi.nlm.nih.gov/gds
GWAS Atlas (Genome-Wide SNP trait Association Atlas)	Public access	A database documented genome-wide SNP trait association for plants and animals that has been carefully curated	Excel file, plain text file	http://bigd.big.ac.cn/gwas
RefSeq (Reference Sequence)	Public access	DNA, transcripts and proteins are all included in this thorough, incorporated, unique collection of sequences	FASTA	https://www.ncbi.nlm.nih.gov/refseq/about/
SGD (*Saccharomyces* Genome Database)	Public access	A comprehensive, integrated biological resource for *Saccharomyces cerevisiae*	BLAST	https://www.yeastgenome.org/
WormBase	Public access	Offering precise, up-to-date, easily available data to the research community on the genetics, genomics, and biology of *Caenorhabditis elegans* and related nematodes	BAM, BED, BedGraph	https://wormbase.org//#012-34-5

current biomarkers of cellular integrity, cell and tissue homeostasis as well as morphological changes resulting from cell injury or death. The area of functional genomics which is primarily concerned with patterns of gene expression throughout varied situations, has been made possible by the knowledge of entire genome (Debnath et al., 2005).

3.2 Epigenomics

A key regulatory mechanism for gene transcription is explained by epigenomics, which describes changes in the control of gene activity that work without changing genetic sequences. It entails the analysis of the higher order chromatin structure, which also makes the DNA–DNA interactome, RNA/DNA modifications including RNA/DNA methylation. Despite becoming practical for identifying epigenetic changes in large genomic areas, illumina short-read sequencing has been combined with immunoprecipitation for DNA modification detection, however, it is not able to achieve resolution at a single base level or distinguish reads from individual cells. In order to study the epigenome, long-read sequencing methods like PacBio & Oxford nanopore sequencing have been modified (Dai & Shen, 2022).

3.3 Metagenomics

In order to explore the genetic makeup of all populations, the term "metagenomic" was initially put out in 1998 as the "collective genomes of soil microflora" (Kumar Awasthi et al., 2020). The study of the structure and function of genetic material in complex samples of multiple organisms along with whole microbial communities without the use of cultivation techniques is known as metagenomics. It helps in the identification of new genes, enzymes as well as metabolic pathways. Sequence-based and function-based screenings are the two types of metagenomics techniques that are used to find and identify novel natural genes and chemicals from environmental materials, respectively (Amer & Baidoo, 2021). The greatest thing about this approach is that it provides environmental microbiologists more freedom to quickly discover the enormous genetic diversity of microbial populations (Kumar Awasthi et al., 2020).

3.4 Transcriptomics

In 1996, Charles Auffray came up with the term "Transcriptome", and in 1997, it was used in a scholarly publication for the first time (McGettigan, 2013). Transcriptome refers to the collection of all RNA molecules generated in single or in population of cells, which include mRNA, rRNA, tRNA and other noncoding RNA. It fluctuates in various tissues and in response to environmental factors. Several novel functional RNA species have been discovered, such as micro RNAs on the genome, which regulates the expression of numerous mRNAs and are involved in a variety of cellular activities, including development. An analysis of the transcriptome from a cell, tissue, or organism as well as its dynamics known as transcriptomics (Solanke & Tribhuvan, 2015). The list of transcriptomics databases has been shown in Table 3.2.

3.5 Epitranscriptomics

The goal of epitranscriptomics is to clarify how RNA structure and alterations affect the regulation of gene expression, with RNA modification concentrating on altered nucleotides in mRNA. Enzyme-based and chemical-based in vitro methods as well as chemical-based in vivo approaches all could be used to map RNA structures using sequencing-based technologies. To obtain a more thorough understanding of RNA structure information, it is deemed preferable to combine data from both chemical-based and enzyme-based sequencing technologies. The existence of chemical modifications on structural RNAs like rRNA, tRNA as well as IncRNAs and short regulatory RNAs (srRNA) has invigorated interest in defining these molecules functions in the regulation of gene expression and the onset or development of diseases. The development of tools to explore the topography of RNA alterations over the whole transcriptome has been made possible through the rapid advancement of RNA sequencing technologies (Dai & Shen, 2022).

4. Specific omics datasets based on MS approach

4.1 Proteomics

The total protein composition of a cell that is evaluated in terms of its interactions, localization, posttranslational changes, and turnover at a given period is known as proteome (Aslam et al., 2017). In 1995, Marc Wilkins coined the term "proteomics" for the first time. Proteomics is the comprehensive analysis of the structure and function of proteins to

TABLE 3.2 Transcriptomics databases.

Database	Availability	Function	File format	URL
Expression Atlas	Public access	Research community on the abundance and distribution of RNA in various cell types, tissues, phases of development across species	UniProt/Gene ID	https://www.ebi.ac.uk/gxa/home
GEO (Gene Expression Omnibus)	Public access	Stores and unrestrictedly distributes NGS, microarray and other forms of high-throughput genomics data	Spreadsheet, TXT, XML	https://www.ncbi.nlm.nih.gov/geo/
Genevestigator	Public access	Makes possible to analyze highly curated bulk tissue and single cell transcriptomic data from public sources	PDF, JPG, GIF, PNG	https://genevestigator.com/
MTD (Mammalian Transcriptomic Database)	Public access	The primary focus on mammalian transcriptomes	SRA	http://mtd.cbi.ac.cn/
NONCODE	Public access	Integrated database dedicated to noncoding RNAs	BED, GTF	http://www.noncode.org/
Stomics DB (Spatial Transcript Omics Database)	Public access	A thorough database of materials and information about spatial transcriptomics	METADATA	https://db.cngb.org/stomics
TWAS Atlas (Transcriptome-Wide Association Studies)	Public access	Integrates trait-associated transcriptome signals from research that looked at the whole transcriptome	Gene symbol, Publication/ PubMed ID	https://ngdc.cncb.ac.cn/twas/

comprehend the biology of an organism. Table 3.3 summarizes the list of proteomics databases. It may examine a protein's expression at various levels, enabling the evaluation of certain quantitative and qualitative biological responses connected to that protein. At the posttranslational, transcriptomics, and geneomics levels, proteomes are evaluated both quantitatively and qualitatively. The analysis of proteomics has multiple applications in numerous fields including medicine, food microbiology, agriculture, and oncology (Al-Amrani et al., 2021).

4.2 Epiproteomics

The term "epiproteome" refers to the posttranslational modifications including protein acetylome, methylome, phosphorylome, ubiquitinome, and SUMOylation, these occur in histone and nonhistone proteins and are also recently found in lactylome, succinome, and others. The first method for epiproteomics studies to employ Edman degradation to identify protein sequence id microsequencing which takes a lot of time and also a number of highly purified materials. The three types of MS-based technologies to epiproteome analysis can be distinguished as follows:

Bottom—up: a method in which a target protein is digested into short peptides (i.e., 5—20 aa) before MS analysis.
Top—down: describes the proteome contained in a sample by examining intact proteins.
Middle—down: a method which are intended for evaluating histone posttranslational modifications because histone N-terminal tails could be cleaved off through particular proteases to produce polypeptides with accurate size for MS identification (Dai & Shen, 2022).

4.3 Metabolomics

Metabolites are frequently the byproduct of intricate biochemical cascades that can connect the transcriptome, proteome, and genome to phenotype, making them a crucial tool for identifying the genetic basis of metabolic diversity (Misra et al., 2019). The study of an organism's whole metabolic response to an environmental stimulation or a genetic alteration is known as metabolomics (Kulski, 2016). It has the ability to negotiate both the relative and absolute quantities of sugars, amino acids, lipids, nucleotides, organic acids, steroids, drugs, and environmental constituents from a variety of sample types such as primary cells, cells lines, tissues, biofluids, entire organisms, and various geo-climatic environments. The

TABLE 3.3 Proteomics databases.

Database	Availability	Description	File format	URL
CATH/Gene3D	Public access	Give details on how protein domains have changed over time	PDB code, GEO terms	https://www.cathdb.info/
GPMDB (Generalised Proteoform Meta-Analysis)	Public access	Take use of the knowledge gathered by GPM servers to help with the challenging task of verifying peptide MS/MS spectra as well as protein coverage patterns	DTA, PKL, MGF	https://gpmdb.thegpm.org/
Human Protein Atlas	Public access	Purpose to integrated several omics technologies to map all human proteins in cells, tissues, and organs	XML, RDF, TSV	https://www.proteinatlas.org/
HPRD (Human Protein Reference Database)	Public access	A repository for carefully curated proteomic data on human proteins	XML	http://www.hprd.org/
ModBase	Public access	Database of comparative protein structural models with annotations	FASTA	https://modbase.compbio.ucsf.edu/
neXtProt	Public access	For better understanding of what all proteins perform in human bodies	XML, FASTA	https://www.nextprot.org/
PAXdb	Public access	Vast absolute protein abundance database	Tab-separated format	https://pax-db.org/
Peptide Atlas	Public access	Collection of peptides found in several tandem MS proteomic experiment	FASTA	http://www.peptideatlas.org/#
PDB (Protein Data Bank)	Public access	Archive 3D structure data for larger molecules such as proteins DNA, RNA	PDB file	https://www.rcsb.org/
PRIDE (Proteomics identification)	Public access	Unified,open-access database for MS proteomic data	XML, QSTAR, QTRAP	https://www.ebi.ac.uk/pride/archive/
SCOP (Structural Classification of Proteins)	Public access	Give a thorough explanation of structural and evolutionary links between proteins whose 3D structure is known and stored in PDB	SCOP ID, PDB ID, UniProt ID	https://scop.mrc-lmb.cam.ac.uk/
STRING (Search Tool for the Retreival of Interacting Genes/Proteins)	Public access	Web resource and biological database of known and predicted protein–protein interactions	TSV, CSV, TXT	https://string-db.org/
TopFIND (Termini oriented protein Function Inferred Database)	Public access	First open knowledge-base and analytical framework for protein termini and protease processing	SQL, XML, CSV	http://clipserve.clip.ubc.ca/topfind
UniParc (uniprot Archive)	Public access	A vast and nonredundant database that has the majority of the world's publicly accessible protein sequences	XML, FASTA, TXT	https://www.uniprot.org/help/uniparc
UniProt	Public access	World's top source of comprehensive, high-quality and publicly available information on protein sequences and functions	XML, FASTA, TXT	https://www.uniprot.org/

TABLE 3.4 Metabolomics databases.

Database	Availability	Function	File format	URL
BiGG	Public access	Database of genome-scale metabolomic network reconstruction	JSON	http://bigg.ucsd.edu/
BioCyc	Public access	An authoritative source on genes, pathways of metabolism, and (in certain situations) networks of regulations	XML, JSON	https://biocyc.org/intro.shtml
ChEBI (Chemical Entities of Biological Interest)	Public access	An open glossary of molecular entities especially prioritizes on minor chemical substances	SDF	https://www.ebi.ac.uk/chebi/
ECMDB (Enterprise Configuration Management Database)	Public access	Well manages database with substantial metabolic information and metabolic pathway diagrams regarding *Escherichia coli*	JSON, SDF, FASTA	https://ecmdb.ca/
HMDB (Human Metabolome Database)	Public access	Publicly accessible electronic database has comprehensive data on small molecule metabolites present in human body	FASTA, SDF, XML	https://hmdb.ca/
MassBank	Public access	For the purpose of identifying tiny chemical compounds with significance to metabolomics	NIST msp	https://massbank.eu/
Metagene	Public access	This disease database creates a proficient system for the diagnosis of genetic metabolic diseases	BAM	https://www.metagene.de/
METLIN	Public access	Technology platform for the identification of known and unknown metabolites and chemical entities	TXT, TSV	https://metlin.scripps.edu/
MetaboLights	Public access	Database for metabolomics study and related data	MzML, nmrML, CDF, ABF, JPF, ISA-TAB	https://www.ebi.ac.uk/metabolights/
Metabolomics workbench	Public access	Has information on the structure and annotations of physiologically significant metabolites	mzML, mzXML, CDF	https://www.metabolomicsworkbench.org/

various metabolomics databases have been shown in Table 3.4. Metabolomics utilizes spectroscopy (i.e., NMR; Nuclear Magnetic Resonance) and MS (i.e., LC/GC-MS or tandem MS) depending on the instrumentation and application, to collect small molecule details in solid (i.e., solid state NMR) liquid (LC-MS), gas phase (GC-MS) or capillary electrophoresis MS (Misra et al., 2019).

4.4 Glycomics

The field of research known as "glycomics", which aims to describe and analyze carbohydrates, is one that is only starting to develop in informatics. It needs for quite complex computational techniques. The glycomics is described as the approach to identify carbohydrate sequences through the utilization of mass spectrum data (Aoki-Kinoshita, 2008). Even though MS has the potential to be a potent tool for glycomics, there are still a number of problems that need to be overcome. Because the glycome must be streamlined and purified before the evaluation, mass spectrometric methods frequently give an overview of a particular region of the glycome (Krishnamoorthy & Mahal, 2009).

4.5 Lipidomics

The term "lipidomics" describes the comprehensive examination of the lipidome of collection of all lipids in any biological system. MS is significantly utilized in this approach. The study of lipids and their metabolic pathways holds considerable possibilities for the identification of novel biomarkers and the creation of innovative therapeutic approaches since lipids have a variety of physiological functions and are impacted by a broad range of diseases (Sinha & Mann, 2020). Lipid solutions are studied during MS either by shotgun lipidomics or by chromatography lipidomics, notably LC-based lipidomics (Yang Kui, 2017). Using soft ionization methods like matrix-assisted laser desorption ionization and electron spray ionization, the first mass spectrometric analyses of complicated lipid mixtures were published in the 1990s (Wenk, 2010).

5. Specific omics datasets based on knowledge

The definition of omics has evolved from a collection of computational and scientific techniques, in addition to a specific layer of molecular data queried using these well-established tools, to a concoction of knowledge obtained by combining various omics data in a given study topic. This conceptual change has resulted in the development of omics including microbiomics and immunomics (Dai & Shen, 2022).

5.1 Microbiomics

Microbiomics is the area of knowledge to characterise, gather & quantify molecules relevant for structure, function along with dynamics of a microbial community through integrating diverse omics data including genomics, transcriptomics, proteomics, and metabolomics where the entire microorganisms of a certain environment known as microbiome are investigated to explore the possible involvement that such microorganisms have in diseases (Dai & Shen, 2022).

5.2 Immunomics

In 2001, the phrase "immunomics" was first originally used to describe the analysis of immunology using data from genomics, transcriptomics, and proteomics with the goal of transferring molecular immunology into clinics. The group of antigens/epitopes that interact with the host immune system is known as immunome. Immunomics has boosted the detection of disease, diagnosis and prevention also has the ability to transform the understanding about vaccine development and antigen identification (Dai & Shen, 2022).

6. Integration of omics datasets

For the first time ever, newly designed high-throughput experimental technologies commonly to referred as genomics, transcriptomics, proteomics, and metabolomics offer the tools to thoroughly monitor the disease processes at the molecular level. They have the ability to link complicated mixtures with complex effects in the form of gene or protein expression profile by significantly increasing the number of proteins/genes that could be measured concomitantly. The nontargeted identification of all gene products (i.e., transcripts, proteins, and metabolites) present in a given biological sample is the main goal of omic technology. These technologies because of their character disclose unanticipated characteristics of biological systems. The accurate study of quantitative dynamics in biological systems is a second harder concept of omic technology. In order to analyze data at omic level i.e., all DNA sequences, levels of gene expressions, or proteins at once, the high-throughput analysis is required. High-throughput measurements are frequently carried out using the four main approaches (Fig. 3.6): genomic SNP analysis (large scale genotyping of single nucleotide polymorphism), transcriptomic measurements (concurrent measurement of all gene expression values in a cell/tissue type), proteomic measurements (determination of all proteins present in a cell/tissue type), and metabolomics measurements (identification as well as quantification of all metabolites in cell/tissue type) (Debnath et al., 2005).

The genome, epigenome, metagenome, transcriptome, epitranscriptome, proteome, epiproteome, metabolome, glycome, and lipidome are examples of omics data that are available in the big data era. To process, standardize, integrate, and analyze omics data several computational approaches have attracted interest. These includes, machine learning, deep learning, and statistical methods (Kaur et al., 2021). Fig. 3.7 represents the entire hierarchy of omics data analysis. Discussion of recently established techniques, tools, and biological database for the omics data analysis is the primary goal of this study. For the successful analysis of omics data, several techniques such as machine learning, deep learning, and statistical methods are now suggested.

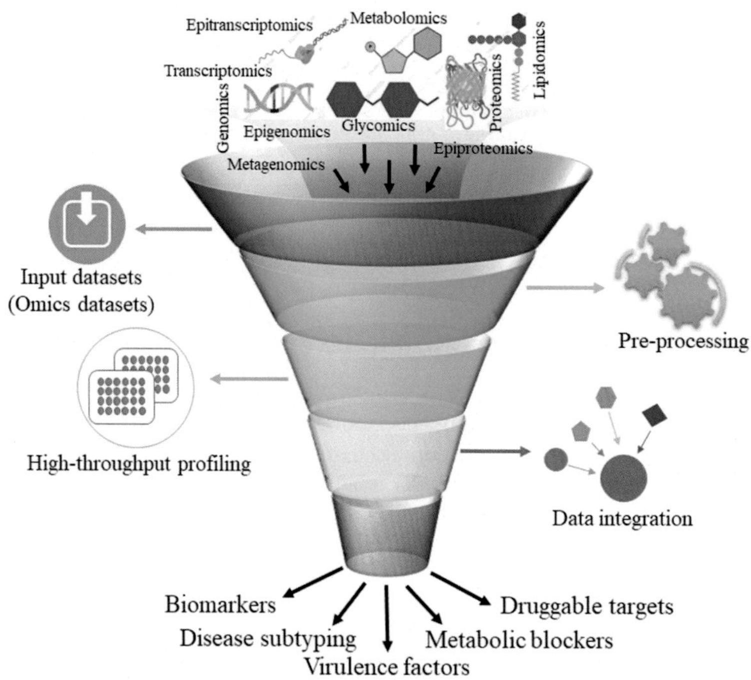

FIGURE 3.6 Data integration of major omics datasets. More detailed comprehension of disease subtyping, metabolic blocker, druggable targets, virulence factors, biomarkers has been made possible by the data integration of multiomics methods.

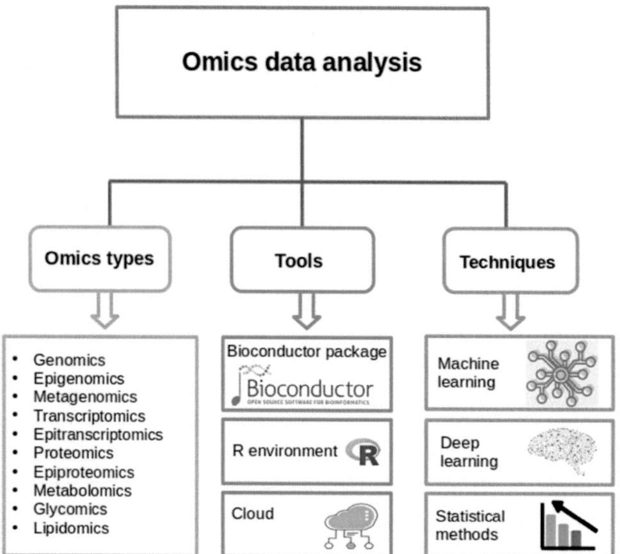

FIGURE 3.7 Categorization of omics data analysis. For the workflow of various omics data analysis, a basic overview has been given in the above figure.

6.1 Methods of integration of omics data

Compared to single datasets, integrated data sources perform better. Data integration methods have been under development in recent years. Numerous methods are proposed to integrate different data sources in an integrative analysis for simultaneous analysis (Kaur et al., 2021). The three primary categories of integration methods are as follows:

Concatenation based integration: By combining several omics data types, this method allows for the study of the combined dataset matrix. Existing analytic techniques for single omics data function properly in this method for the combined matrix.

Transformation based integration: This approach involves first converting these data types into a matrix of graph types. To obtain integrative representation, these intermediate data types are combined. Since many distinct forms of data including categorical, continuous, and sequence data can be integrated, this approach is more reliable than concatenation-based integration.

Model based integration: In this method, the datasets are individually examined, and the conclusions are drawn from the results. This technique is particularly adaptable since several models may be used to analyze various sorts of data. This model-based integration is widely utilized in the field of bioinformatics.

6.2 Techniques used for integration of multiomics data

Precision medicine is aided by the novel application of machine learning for omics data processing in biomedical research (Kaur et al., 2021). Thorough analysis of existing techniques and tools of omics with types of omics data has been studied as mentioned in the following paragraphs:

6.2.1 Machine learning

According to literature, Arthur Samuel is credited for inventing the phrase "Machine Learning" (ML) in 1959 (Arjmand et al., 2022). The omics data are readily available for the medical research community in vast quantities. Medical experts still struggle to identify the course of a disease with accuracy. This technique is gaining popularity as an analytical approach in the medical industry due to its capability to identify important characteristics in large datasets (Kaur et al., 2021). In a basic term, ML is a subset of artificial intelligence (AI) that integrates statistics, mathematics as well as compute science to evaluate input data, discover patterns in data and forecast output data through systematic methods. The primary ML approaches has been identified (Arjmand et al., 2022) as follows:

Supervised learning: The most widely used method for ML technique is supervised learning, which is applied in contexts where input—output data accurately mapped. In other aspects, this approach tends to start with importing training characteristics and target attributes from datasets. The two supervised learning tasks that are chosen based on datasets are classification and regression. Predicting discrete values is the objective of a classification task, to put it simply. Furthermore, the regression task is useful for supervised learning if the dataset contains continuous values.

Unsupervised learning: This approach uses unlabeled data and does not use predefined data. In other terms, no equivalent output variable is provided for the input data. This approach enables use of methods like clustering, association as well as dimensionality reduction.

6.2.2 Deep learning and its application in types of omics data

Deep Learning (DL) is a subcategory of machine learning methods distinguished by the utilization of artificial neural networks. As an effective method, DL is getting importance that could learn from heterogenous and complicated data in both supervised and unsupervised setting. In terms of traditional AI problems such as language processing, speech recognition, and picture identification, DL techniques have made significant advancements. Furthermore, DL algorithms find difficulty in combining many omics layers or even additional sources of data, such as clinical health records or medical imaging, in addition to the problem of studying each type of data independently (Holger, 2019). This approach boosts interpretability and offers further comprehension of biological data structure (Kaur et al., 2021).

In past years, DL techniques have been used to analyze genomics data. In the area of functional genomics, DL algorithms have been used to forecast the sequences of enhancers and regulatory motifs in the genome using variety of data sources (chromatin accessibility and histone modification). DL has been effectively used in a wide variety of transcriptomics applications. For instance, the examination of alternative splicing is one of the primary objectives of gene expression data. Additionally, the use of DL to infer protein secondary structures from their amino acid sequences has been made. Due to their inability to pinpoint specific contributing variables to individual samples, these approaches are particularly difficult to apply metabolomics data (Holger, 2019).

6.3 Biological tools and databases used in omics data analysis

Integrating and analyzing a large number of omics data is a barrier for researchers without bioinformatics expertize. Consequently, a number of projects created tools for the integration and interpretation of complex data generated by omics technologies. List of various biological omics tools and databases for integrated multiomics approach has been summarizes in Tables 3.5 and 3.6 respectively. O-miner is a powerful tool that automatically integrates and analyses data through omics

TABLE 3.5 Biological omics tools and web applications for integrated multiomics datasets.

Tools and web applications	Integrated omics types	Category	Functionality	URL
Cell Illustrator 5.0	Genomics, proteomics, and metabolomics	Unstated	Helps to depict, describe, clarify, and simulate intricate biological processes and systems	http://www.cellillustrator.com/home
CellML (Open source XML Language)	Transcriptomics, proteomics, and metabolomics	Unstated	Storing and exchanging computer-based mathematical models	https://www.cellml.org/
E-Cell	Transcriptomics, proteomics, and metabolomics	Unstated	Modeling, simulation and analysis of complex, heterogeneous and multiscale cellular systems	https://www.e-cell.org/
Escher	Genomics, proteomics, and metabolomics	Unstated	Web application for visualizing data on biological pathways	https://escher.github.io/
3Omics	Transcriptomics, proteomics, and metabolomics	Transcriptomic, proteomics, and metabolomic	An online platform that allows for the quick integration of several inter- or intratranscriptomic, proteomic, and metabolomics human datasets	https://3omics.cmdm.tw/
INMEX (Integrative meta-analysis of expression data)	Transcriptomics and metabolomics	Medical and clinical	Meta and integrative analysis of data; pathway analysis	http://www.inmex.ca/
IMPaLA (Integrated Molecular Pathway Level Analysis)	Transcriptomics, proteomics, and metabolomics	Medical and clinical	Enrichment analysis, pathway analysis	http://impala.molgen.mpg.de/
Ingenuity Pathway Analysis	Metagenomics, transcriptomics, proteomics, and metabolomics	Medical (human) and clinical	Potent search and analysis tool that reveals the importance of omics data	https://digitalinsights.qiagen.com/products-overview/discovery-insights-portfolio/analysis-and-visualization/qiagen-ipa/
IOMA (Integrative Omics-Metabolic Analysis)	Proteomics and metabolomics	Unstated	Integrates proteomics and metabolomics data to predict flux distributions	https://ecoliwiki.org/colipedia/index.php/IOMA
KaPPa-view	Transcriptomics and metabolomics	Plants	Integrates transcriptomics and metabolomics data to map pathways	http://kpv.kazusa.or.jp/kpv4/
MapMan	Metagenomics, transcriptomics, and metabolomics	Plants	Designs enormous datasets into schematics of metabolic pathways or other activities	https://mapman.gabipd.org/mapman#:~:text=MapMan%20is%20a%20user%2Ddriven,the%20group%20of%20Mark%20Stitt.
Marvis-Pathway (Marker Visualization Pathway)	Metagenomics, transcriptomics, and metabolomics	Unstated	An interactive toolset for arranging, narrowing down, integrating, grouping, visualizing, and functional analysis of datasets	http://marvis.gobics.de/
MassTrix	Transcriptomics and metabolomics	Unstated	Integration of data; generation of colored pathway maps KEGG data analysis	https://bio.tools/masstrix

MetaboAnalyst	Genomics, transcriptomics, proteomics, metabolomics, and clinical	Plants, microbial, microbiome, medical, and clinical	Data processing and statistical analysis; pathway analysis; multiomics integration	https://www.metaboanalyst.ca/
mixOmics (R package)	Metagenomics, transcriptomics, proteomics, and metabolomics	Unstated	Integration of data; chemometric analysis	http://mixomics.org/
Omics Integrator	Genomics, epigenomics, and proteomics	Unstated	Program that uses a protein–protein interaction network to combine data from proteomic, gene expression or epigenetic domains	https://github.com/fraenkel-lab/OmicsIntegrator
OMICtools	Genomics, transcriptomics, proteomics, and metabolomics	Medical (human)	Correlation network analysis, coexpression analysis	http://omictools.com/
Omickriging (R package)	Transcriptomics, proteomics, and metabolomics	Unstated	Integration and visualization of omics data	http://www.scandb.org/newinterface/tools/OmicKriging.html
OmicsPLS	Metagenomics, transcriptomics, proteomics, and metabolomics	Unstated	Integration of data; chemometric analysis	https://cran.r-project.org/package=OmicsPLS
Omix visualization tool	Transcriptomics, proteomics, metabolomics, and fluxomics	Unstated	Integration and visualization of omics data	https://www.omix-visualization.com/
PaintOmics 4	Transcriptomics and metabolomics	100 elite species of several biological kingdoms	Integration and visualization of transcriptomics and metabolomics data	https://www.paintomics.org/
Path Visio 3	Transcriptomics, proteomics, and metabolomics	Unstated	Completely public tool that enables for the design, modification and analysis of biological pathways	https://pathvisio.org/
ProMeTra	Transcriptomics and metabolomics	Medical and clinical	Interactive visualizations of metabolite concentrations together with transcript measurements mapped on the pathways and genomemaps	https://www.cebitec.uni-bielefeld.de/brf/software/75-prometra
SIMCA	Metagenomics, transcriptomics, proteomics, and metabolomics	Unstated	Chemometric analysis	https://www.sartorius.com/en/products/process-analytical-technology/data-analytics-software/mvda-software/simca
SimCell	Genomics, transcriptomics, proteomics, and metabolomics	Unstated	Graphical modeling tool	http://wishart.biology.ualberta.ca/SimCell/
VANTED (Visualization and Analysis of Networks with related Experimental Data)	Metagenomics, transcriptomics, proteomics, and metabolomics	Unstated	Comparison of multiomics datasets	https://vanted.sourceforge.net/

TABLE 3.6 Biological omics databases for integrated multiomics datasets.

Database	Integrated omics type	Category	Functionality	URL
Biocyc	Genomics, proteomics and metabolomics	Unstated	Integrates genomic data with a vast array of additional information such as metabolic reconstructions, regulatory networks, and protein characteristics, gene essentiality	https://biocyc.org/
COBRA	Transcriptomics, proteomics and metabolomics	Unstated	Genome scale integrated modeling of cell metabolism and macro molecular expression	https://help.deltek.com/product/cobra/8.4/ga/Cobra%20Database.html
FlyBase	Genomics and transcriptomics	*Drosophila*	Genes and RNA-Seq data of different *Drosophila* spp	https://flybase.org/
GIM3E (Gene Inactivation Moderated by Metabolism, Metabolomics and Expression)	Transcriptomics and metabolomics	Unstated	Establishes metabolite use requirement with metabolomics data	https://github.com/brianjamesschmidt/gim3e
KEGG	Genomics, transcriptomics, proteomics, and metabolomics	Plants, animals and microbes	Comprehensive resource for comprehending the fundamental purposes and applications of the biological system	https://www.genome.jp/kegg/
MADMAX	Metagenomics, transcriptomics, and metabolomics	Plants, medical, clinical	Statistical analysis and pathway mapping	http://madmax2.bioinformatics.nl/
MetScape 2	Transcriptomics and metabolomics	Medical and clinical	Integrates data from KEGG and EHMN database	https://metscape.ncibi.org/metscape2/index.html
PMN (Plant Metabolics Network)	Genomics, proteomics, and metabolomics	Plants	Plant-specific database containing pathways, enzymes, reactions and compounds	https://plantcyc.org/
Reactome	Genomics, transcriptomics, proteomics, and metabolomics	Unstated	Multiomics data visualization; metabolic map of known biological processes and pathways	https://reactome.org/
Recon3D	Genomics, proteomics, and metabolomics	Human	Computation resource and comprehensive human metabolic network model	http://bigg.ucsd.edu/models/Recon3D
VitisNet	Metagenomics, transcriptomics, proteomics, and metabolomics	Grapes	Visualization of connectivity	http://www.grapegenomics.com/pages/VitisNet/

technologies. From datasets containing genome and transcriptome data coupled with clinical and biological information, the tools helps in the discovery of important pathways as well as the prioritization of biomarkers. The pipelines created for the tool make advantage of statistical methods of Bioconductor packages and run in R environment. Molecular modeling, omics data analysis (including RNA sequencing, proteomics, or metabolomics datasets), and the integration, analysis, and interpretation of phenotypic data are all quickly finding usage of cloud computing. Data scientists benefit from cloud computing because it gives them access to computing frameworks. Even though cloud computing is still in its early stages of research and eventual implementation in real-world applications, it has the potential to tackle the large data analysis problem of efficiency in terms of time, memory utilization and storage (Koppad et al., 2021). Omics Pipe (a structure for analyzing both already existing and freshly created data) is a cloud-based tool for automating multiomics data pipeline processing. For the study of genomics data, the cloud platform KnowEnG (Knowledge Engine for Genomics) has been suggested (Kaur et al., 2021).

6.4 Challenges of integration of omics data

Numerous progresses have been made in the era of omics research. But there are a number of issues that still need to be addressed. The continuous analysis of omics data is complicated because it is necessary to integrate heterogenous and massive omics data, which is both a conceptual and practical challenge. With the development of innovative omics technologies and large-scale consortia initiatives, biological systems are being studied at a level never before achieved, resulting in diverse and frequently huge datasets. Researchers are encouraged by these datasets to invent new integration techniques (Gomez-cabrero et al., 2014). Proteomics and metabolomics research are still in their infancy, and new advances in large-scale proteomics data represent a significant challenge to the ability of current bioinformatics tools to confirm these findings. In addition to sample preparation, there were other constraints that needed to be overcome throughout the proteomic analysis, including data assembling and database searches for functional annotation. Only proteins whose information is already in the database could be annotated, although it can be quite difficult to locate new proteins (Shankar et al., 2021). Fig. 3.8 outlines the challenges that integrated omics approaches are now addressing.

Sample preparation, study design, reproducibility, steady state assumption, and statistical power are the basic experimental challenges. Each omics platform has particular restrictions. Due to variances in the data included in a dataset normalization, transformation and scaling procedures in the three primary omics areas of transcriptomics, proteomics, and metabolomics are quite diverse. For integration and subsequent analysis, there are no tools for scaling datasets and eliminating false positives from three or more different systems. The challenges of integration are many. It becomes challenging to account for false positives in the combined datasets when data from many sources is integrated. Results are significantly impacted by the choice of how to handle false positives in specific omics datasets. Making sense of the data is still the most difficult challenge for every omics dataset. Biomarker discovery is one of the main goals of multidimensional omics approaches; regardless of the omics layer from which the vital molecules are obtained, the sensitivity and specificity of molecular biomarkers are crucial for their applicability in biomedical research and the clinical translation of findings. It is challenging, time-consuming, and computationally costly to interpret and curate vast multilayered networks and it also demands in-depth understanding of the biological system being investigated (Misra et al., 2019).

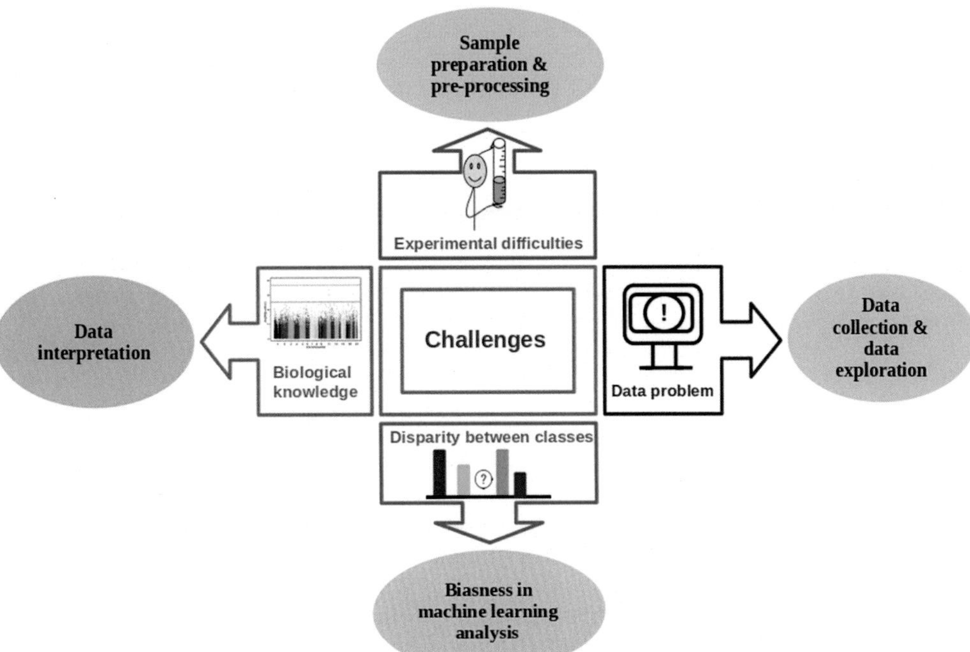

FIGURE 3.8 Challenges associated with integration of omics data. The challenges of omics data integration include experimental challenges, data problem, disparity between classes, and biological knowledge.

6.5 Application of integrated omics data

To comprehend how molecules interact, integrated techniques incorporate various omics data in a sequential or simultaneous manner. They assist in determining how information is transferred from one omics level to the next, which helps in bridging the genotype and phenotype gap. Integrative approaches have the potential to enhance prognostics and prediction accuracy of disease phenotypes, which may ultimately help in effective treatment and prevention. This is due to their ability to examine the biological phenomenon holistically. Research has shown that merging omics datasets results in greater knowledge of the system being analyzed. Integrating genomic and transcriptomic data with proteomics data to highlight driver genes in colon and rectal cancer while integrating metabolomics and transcriptomics led to the discovery of prostate cancer-related molecular alterations. The application of integrative analysis is expanding and new investigations have shown further applications that might bring about a revolutionary transformation in the area of diseases, diagnosis, prognosis as well as therapy. Personalized medicines, clinical assessment predictions as well as risk prediction and clinical outcome are few integrative approach applications that help to enhance the treatment (Subramanian et al., 2020).

References

Al-Amrani, S., Al-Jabri, Z., Al-Zaabi, A., Alshekaili, J., & Al-Khabori, M. (2021). Proteomics: Concepts and applications in human medicine. *World Journal of Biological Chemistry, 12*(5), 57–70. https://doi.org/10.4331/wjbc.v12.i5.57

Amer, B., & Baidoo, E. E. K. (2021). Omics-driven biotechnology for industrial applications. *Frontiers in Bioengineering and Biotechnology, 9*(February), 1–19. https://doi.org/10.3389/fbioe.2021.613307

Aoki-Kinoshita, K. F. (2008). An introduction to bioinformatics for glycomics research. *PLoS Computational Biology, 4*(5). https://doi.org/10.1371/journal.pcbi.1000075

Arjmand, B., Hamidpour, S. K., & Tayanloo-beik, A. (2022). Machine learning: A new prospect in multi-omics data analysis of cancer. *Frontiers in Genetics, 13*(January), 1–18. https://doi.org/10.3389/fgene.2022.824451

Aslam, B., Basit, M., Nisar, M. A., Khurshid, M., & Rasool, M. H. (2017). Proteomics: Technologies and their applications. *Journal of Chromatographic Science, 55*(2), 182–196. https://doi.org/10.1093/chromsci/bmw167

Buermans, H. P. J., & den Dunnen, J. T. (2014). Next generation sequencing technology: Advances and applications. *Biochimica et Biophysica Acta—Molecular Basis of Disease, 1842*(10), 1932–1941. https://doi.org/10.1016/j.bbadis.2014.06.015. Elsevier B.V.

Conesa, A., & Beck, S. (2019). Making multi-omics data accessible to researchers. *Scientific Data, 6*(1). https://doi.org/10.1038/s41597-019-0258-4. Nature Research.

Dai, X., & Shen, L. (2022). Advances and trends in omics technology development. *Frontiers of Medicine, 9*(July). https://doi.org/10.3389/fmed.2022.911861

Debnath, M., Prasad, G. B. K. S., & Bisen, P. S. (2005). Molecular diagnostics: Promises and possibilities. In *Molecular diagnostics: Promises and possibilities*. Springer Netherlands. https://doi.org/10.1007/978-90-481-3261-4

Del Vecchio, F., Mastroiaco, V., Di Marco, A., Compagnoni, C., Capece, D., Zazzeroni, F., Capalbo, C., Alesse, E., & Tessitore, A. (2017). Next-generation sequencing: Recent applications to the analysis of colorectal cancer. *Journal of Translational Medicine, 15*(1). https://doi.org/10.1186/s12967-017-1353-y. BioMed Central Ltd.

Gomez-Cabrero, D., Abugessaisa, I., Maier, D., Teschendorff, A., Merkenschlager, M., Gisel, A., Ballestar, E., Bongcam-Rudloff, E., Conesa, A., & Tegnér, J. (2014). Data integration in the era of omics: Current and future challenges. *BMC Systems Biology, 8*(2), 1–10.

Graw, S., Chappell, K., Washam, C. L., Gies, A., Bird, J., Robeson, M. S., & Byrum, S. D. (2021). Multi-omics data integration considerations and study design for biological systems and disease. *Molecular Omics, 17*(2), 170–185. https://doi.org/10.1039/d0mo00041h. Royal Society of Chemistry.

Holger, H. (2019). *Computational biology*.

Kaur, P., Singh, A., & Chana, I. (2021). Computational techniques and tools for omics data analysis: State-of-the-art, challenges, and future directions. *Archives of Computational Methods in Engineering, 28*(7), 4595–4631. https://doi.org/10.1007/s11831-021-09547-0

Koppad, S., Gkoutos, G. V., & Acharjee, A. (2021). Cloud computing enabled big multi-omics data analytics. *Bioinformatics and Biology Insights, 15*(2). https://doi.org/10.1177/11779322211035921

Krishnamoorthy, L., & Mahal, L. K. (2009). Glycomic analysis: An array of technologies. *ACS Chemical Biology, 4*(9), 715–732. https://doi.org/10.1021/cb900103n

Kulski, J. K. (2016). Next-generation sequencing—an overview of the history, tools, and "omic" applications. In *Next generation sequencing—advances, applications and challenges*. InTech. https://doi.org/10.5772/61964

Kumar Awasthi, M., Ravindran, B., Sarsaiya, S., Chen, H., Wainaina, S., Singh, E., Liu, T., Kumar, S., Pandey, A., Singh, L., & Zhang, Z. (2020). Metagenomics for taxonomy profiling: Tools and approaches. *Bioengineered, 11*(1), 356–374. https://doi.org/10.1080/21655979.2020.1736238

Lee, C. Y., Chiu, Y. C., Wang, L. B., Kuo, Y. L., Chuang, E. Y., Lai, L. C., & Tsai, M. H. (2013). Common applications of next-generation sequencing technologies in genomic research. *Translational Cancer Research, 2*(1), 33–45. https://doi.org/10.3978/j.issn.2218-676X.2013.02.09. AME Publishing Company.

Lowe, R., Shirley, N., Bleackley, M., Dolan, S., & Shafee, T. (2017). Transcriptomics technologies. *PLoS Computational Biology, 13*(5). https://doi.org/10.1371/journal.pcbi.1005457

McGettigan, P. A. (2013). Transcriptomics in the RNA-seq era. *Current Opinion in Chemical Biology, 17*(1), 4–11. https://doi.org/10.1016/j.cbpa.2012.12.008

Misra, B. B., Langefeld, C., Olivier, M., & Cox, L. A. (2019). Integrated omics: Tools, advances and future approaches. *Journal of Molecular Endocrinology, 2016.*

Murayama, C., & Kimura, Y. (2009). Imaging mass spectrometry: Principle and application. *Biophysical Reviews.* https://doi.org/10.1007/s12551-009-0015-6

Shankar, R., Dwivedi, V., & Arya, G. C. (2021). Relevance of bioinformatics and database in omics study. *Omics Technologies for Sustainable Agriculture and Global Food Security, 1*, 19–39. https://doi.org/10.1007/978-981-16-0831-5_2. Springer Singapore.

Sinha, A., & Mann, M. (2020). A beginner's guide to RT-beginner's guide. *0*(June), 1–6 https://doi.org/10.1042/BIO20200034.

Solanke, A., & Tribhuvan, K. (2015). *Genomics: An integrative approach for molecular biology genomics of cotton boll and fibre development view project regulation of quality of tomato fruits view project.* https://www.researchgate.net/publication/283084372.

Subramanian, I., Verma, S., Kumar, S., Jere, A., & Anamika, K. (2020). Multi-omics data integration, interpretation, and its application. *Bioinformatics and Biology Insights, 14.* https://doi.org/10.1177/1177932219899051. SAGE Publications Inc.

Thakur, N., Shirkot, P., & Pandey, H. (2018). Next generation sequencing-techniques and its applications. Food security, nutrition and sustainable agriculture-emerging technologies view project production, partial purification and characterization of laccase from rhizospheric bacteria Pseudomonas putida strain LUA15.1 View project http://www.illumina.com/systems/sequencing.html.

Wang, J. H., Byun, J., & Pennathur, S. (2011). Analytical approaches to metabolomics and applications to systems biology. *Seminars in Nephrology, 30*(5), 500–511. https://doi.org/10.1016/j.semnephrol.2010.07.007

Wang, Z., Gerstein, M., & Snyder, M. (2009). RNA-Seq: A revolutionary tool for transcriptomics. *Nature Reviews Genetics, 10*(1), 57–63. https://doi.org/10.1038/nrg2484

Wenk, M. R. (2010). Lipidomics: New tools and applications. *Cell, 143*(6), 888–895. https://doi.org/10.1016/j.cell.2010.11.033

Yang Kui, H. X. (2017). Lipidomics: Techniques, applications, and outcomes related to biomedical sciences. *Physiology & Behavior, 176*(3), 139–148. https://doi.org/10.1016/j.tibs.2016.08.010.Lipidomics

Chapter 4

Systematic benchmarking of omics computational tools

Sanjay Kumar[1], Manjusa Singh[3], Rajesh Sharma[2] and Manish Kumar Gupta[3]

[1]*Bioinformatics Center, Biotech Park, Lucknow, Uttar Pradesh, India;* [2]*Department of Biotechnology, Veer Bahadur Singh Purvanchal University, Jaunpur, Uttar Pradesh, India;* [3]*Department of Biotechnology, Faculty of Science, Veer Bahadur Singh Purvanchal University, Jaunpur, Uttar Pradesh, India*

1. Introduction

Systematic benchmarking studies are crucial in Omics data analysis to inform the biomedical research community about the strengths and weaknesses of existing analytical tools (Wren, 2016). However, researchers often lack contextual knowledge to assess new standardized datasets, metrics, and algorithms against proven ones systematically (Hackl et al., 2016; Nagarajan & Pop, 2013). This creates a communication gap between tool developers and biomedical researchers. To overcome this, benchmarking studies should complement simulated data with experimental data generated by previous studies (Mangul et al., 2019). Gold standard datasets are obtained using highly accurate, cost-prohibitive experimental procedures in routine biomedical research (Zook et al., 2014). Researchers have little consensus on what constitutes a gold standard experimental dataset (Sczyrba et al., 2017). However, systematic benchmarking studies can provide the data and tools to support the informed dialogue necessary to explore these inquiries (Zheng, 2017). A gold standard is a standard against which the performance of a method can be evaluated and can be either real or simulated data (Ewing et al., 2015). Measures used to identify the best-performing methods for a particular analytical task include precision and sensitivity. Benchmarking studies evaluate various tools, some excluding those with complicated installation processes or those lacking comprehensive documentation (Alberti et al., 2016; Costello et al., 2014). Sharing a tool's supporting documentation through an easy-to-use interface is more effective than through paper and/or supplementary materials.

Omics refers to the study of various biological molecules, like genes, proteins, and metabolites, and the computational tools used in Omics exploration are designed to reuse, dissect, and interpret large quantities of data generated by these molecules. Several enterprises have been established to totally standardize Omics tools, similar to the Critical Assessment of Metagenome Interpretation for metagenomics tools (Meyer et al., 2022), the Critical Assessment of Protein Structure vaticination for protein structure vaticination tools (https://predictioncenter.org/index.cgi), and the Automated Function Vaticination for functional reflection tools (Friedberg, 2006). These enterprises generally involve the development of standardized datasets and evaluation criteria, as well as the participation of multiple exploration groups in testing and assessing the tools. The results of these benchmarking exercises are frequently published in peer-reviewed journals and can give precious guidance to experimenters in opting for and using Omics tools. Omics technologies, like genomics, transcriptomics, proteomics, metabolomics, and epigenomics, have revolutionized the field of natural exploration by enabling comprehensive and high-outturn analysis of different molecular factors within cells and organisms (Dai & Shen, 2022). These technologies induce massive quantities of data, furnishing unknown openings to unravel complex natural processes, and understand the underpinning mechanisms. Still, the effective analysis and interpretation of Omics data pose significant computational challenges due to their sheer volume, complexity, and diversity. These data generated by Omics technologies are vast and complex. To make sense of these data, scientists need to use computational tools by following points; the essential part of this is to get rid of any mistakes or discrepancies in the data by cleaning and filtering data. Recognize the patterns by locating genes or proteins linked to a specific phenotype or illness and create a model to forecast biological system behavior or to find fresh therapeutic targets.

Integrative Omics. https://doi.org/10.1016/B978-0-443-16092-9.00004-7

The adding vacuity of computational tools for Omics data analysis presents a challenge for experimenters in opting for the most applicable tool for their specific exploration requirements (Berger et al., 2013). While some tools have established reports and wide operations, others may be new or less well-known. Assessing the performance and trust ability of Omics tools becomes pivotal to ensure accurate and reproducible results (Cole et al., 2019; Conesa et al., 2016; Corchete et al., 2020; Fröhlich et al., 2022; Välikangas et al., 2018). This is where methodical benchmarking comes into play. Methodical benchmarking involves a comprehensive evaluation of Omics tools using standardized criteria and standard datasets (Duan et al., 2021). By subjugating the tools to rigorous assessment, experimenters can make informed opinions regarding their felicity for different exploration questions and datasets. Researchers can assess the performance of several tools, get insight into their advantages and disadvantages, and choose the tool that best meets their unique analytical needs through systematic benchmarking (Ettorchi-Tardy et al., 2012). Additionally, benchmarking offers a chance to identify areas that still require improvement, spurring the creation of more reliable and effective computational tools (Zheng, 2017).

This chapter explores the wide range of computational tool benchmarks and their effects on upstream user instruction. While finding methods and software/tools with enhanced efficacy is the focus, it's important to realize that benchmarking covers a wider variety of goals. To make it easier for users to choose the most effective tools, we collected the most frequently used software from Omics fields along with their most recent URLs and case studies.

2. Methodology for setting up the benchmarking study

The methodology for setting up a benchmarking for Omics data analysis is different for each dataset (Table 4.1). The Omics data systematic benchmarking necessitates the use of topic applications such as RNA-Seq analysis, error correction in big data sequences, large genome assembly, variation analysis, and microbiome analysis. Additionally, essential computational tools like SeqWare (O'Connor et al., 2010) and RSEM (Li & Dewey, 2011) are needed for Omics data analysis. To set up the benchmarking study, it is necessary to examine previously published benchmarking studies for optimized parameter values, algorithm features, and computational costs. Furthermore, evaluating the results involves sharing the benchmarking results using appropriate sharing methods.

In this section, we outline the process of establishing a pipeline for the comprehensive benchmarking study of omics data.

2.1 Selection of reference model

Most benchmarking studies in genomics focus on the competitive model, while fewer studies utilize the individual model to collect data and conduct systematic analyses (Table 4.2). Some researchers have adopted a hybrid strategy, incorporating features from both benchmarking techniques. The first step in conducting a benchmarking study of omics data is to carefully select an appropriate reference model. Monagul et al. have proposed a descriptive model for this selection process, emphasizing the consideration of key features such as the model type, input data type, method of gold standard data preparation, and optimization parameters (Mangul et al., 2019) (Table 4.2).

2.2 Construction of gold standard dataset

The most crucial step in establishing a benchmarking study's gold standard data involves selecting appropriate methods. Initially, users have the option to adopt an alternative approach by examining previously published benchmarking work and utilizing their gold standard parameters for data analysis. Another popular method for generating gold standards is through computational simulation, like ProBAMsuite, which is used for benchmarking proteomics data (Wang et al., 2016). This approach includes tools for creating synthetic data, simulating proteomics experiments, and evaluating identification and quantification algorithms. However, computationally simulated data are best used as a supplement to actual gold standard data since it may not accurately capture true experimental variability.

The third method for creating gold standards is through expert manual evaluation (Aghaeepour et al., 2013). Additionally, curated databases and literature references-based methods are considered alternative approaches to establishing gold standard parameters (Altenhoff et al., 2016; Jiang et al., 2016; Thompson et al., 2011; Łabaj et al., 2011). Fig. 4.1 illustrates the process of creating a gold standard protocol using existing technologies accessible for raw data analysis.

2.3 Default parameters selection and optimization

It takes a lot of work to evaluate software tools since a researcher needs to select parameter settings and input preliminary processing methods (Mangul et al., 2019). However, compared to using default parameter values, parameter tuning can lead to a significant increase in result accuracy. This process of parameter optimization is computationally intensive,

TABLE 4.1 List of benchmarking studies applied to omics data analysis.

#	Omics dataset	Author
1.	Genome Sequencing	Weisweiler and Stich (2023)
2.	Genome, epigenome, transcriptome, proteome, metabolome, and microbiome	Wang et al. (2023)
3.	Whole genome bisulfite sequencing	Lin et al. (2023)
4.	Spatial transcriptomics technologies	Li et al. (2023)
5.	Spatially resolved transcriptomics	Cheng et al. (2023)
6.	Proteomics	Dimitsaki et al. (2023)
7.	Analyzing proteins from single cells	Gatto et al. (2023)
8.	Proteomics	Koopmans et al. (2023)
9.	Metabolomics analyses	Wandy et al. (2023)
10.	Metabolomics	Bongaerts et al. (2023)
11.	DIA proteomics and phospho-proteomics data analysis	Lou et al. (2023)
12.	Spatial transcriptomics	Yan and Sun (2023)
13.	Metagenomic sequencing data	Zhang et al. (2023)
14.	Single cell genomics	Luecken et al. (2022)
15.	Whole-genome (WGS) and whole-exome (WES) sequencing datasets	Barbitoff et al. (2022)
16.	Genome sequencing	Rajkumari et al. (2022)
17.	Genomic, transcriptomic, and epigenetic	Wissel et al. (2022)
18.	Single-cell RNA-sequencing	Yu et al. (2022)
19.	Proteomics	Fröhlich et al. (2022)
20.	Proteomics analysis	Wooller et al. (2022)
21.	Proteomics	Välikangas et al. (2022)
22.	Metabolomics analyses	de Jonge et al. (2022)
23.	Proteomics	Van Puyvelde et al. (2022)
24.	Proteomics	Canto et al. (2022)
25.	Label-free proteomics	Dowell et al. (2021)
26.	Metabolomics analyses	Rampler et al. (2021)
27.	Proteomics analysis	Sánchez et al. (2021)
28.	Single-cell genomics	Garcia-Heredia et al. (2021)
29.	Whole genome or whole exome sequencing	Pei et al. (2021)
30.	Single cell RNA sequencing	Cao et al. (2021)
31.	Single-cell RNA-sequencing	You et al. (2021)
32.	Single-cell RNA sequencing	Xi and Li (2021)
33.	Deconvolution tools on RNA-seq data.	Jin and Liu (2021)
34.	Protein sequencing and databases	Awan et al. (2021)
35.	Whole genome sequencing	Zhao et al. (2020)
36.	Single-cell RNA sequencing	Tsuyuzaki et al. (2020)
37.	Transcriptomics	Avila Cobos et al. (2020)
38.	Single-cell RNA sequencing	Zhang et al. (2020)
39.	Single-cell transcriptomic	Tran et al. (2020)
40.	Single-cell RNA sequencing	Mereu et al. (2020)

Continued

TABLE 4.1 List of benchmarking studies applied to omics data analysis.—cont'd

#	Omics dataset	Author
41.	RNA sequencing	Baik et al. (2020)
42.	Metagenomic sequencing	Ye et al. (2019)
43.	Genomic sequencing (Global alliance for genomics and health)	Krusche et al. (2019)
44.	Proteomics analysis	Donnelly et al. (2019)
45.	RNA sequencing	Quinn et al. (2018)
46.	Mass spectrometry (MS) based proteomics	Awan and Saeed (2018)

requiring the repeated use of the same tool with various parameter configurations. For instance, in the context of RNA-Seq aligners, researchers have observed that fine-tuning the parameter settings consistently improves the number of correctly mapped reads by an average of 10% across all 14 cutting-edge aligners (Baruzzo et al., 2017). In Table 4.2, many researchers applied parameter optimization methods.

2.4 Selection of benchmarking datasets

Benchmark datasets or databases are crucial for evaluating the effectiveness of computational techniques. These databases need to be carefully curated, including a range of biological circumstances and complexities pertinent to the Omics sector (Table 4.3). Benchmark datasets may be obtained from controlled data sources, simulated data, or publicly accessible datasets. Equally crucial is choosing the right metrics for performance. A big database such as The International Genome Sample Resource, often known as the 1000 Genomes Project, has WGS from a sizable number of people representing various populations (https://www.internationalgenome.org/) and The Human Atlas (https://www.proteinatlas.org/) contains RNA-Seq data for several tissues and cell lines, revealing information about protein-protein interactions and tissue-specific gene expression. Similarly, The Cancer Genome Atlas Program (https://www.cancer.gov/ccg/research/genome-sequencing/tcga), also offers many types of cancer data, such as data on somatic mutations, DNA methylation, and gene expression. The Human Metabolome Database (https://hmdb.ca/) has human tissues, biofluids, and cell lines.

2.5 Selection of computational tools

Currently, there are hundreds of software tools available and new applications are being developed monthly (see Table 4.4 and Section 4: Case studies). For instance, more than 200 computational tools have been developed for the variant analysis of data from next-generation genome sequencing (Huttenhower et al., 2009; Warwick Vesztrocy & Dessimoz, 2020).The first crucial step in the benchmarking process is carefully selecting representative Omics computational tools. These tools should encompass a wide range of functionalities and enjoy significant popularity and usage within the research community. The selection process considers factors like relevance to the specific Omics domain, availability, and diversity of features. It is essential to ensure that the chosen tools provide a fair and comprehensive representation of the available options, including both established and emerging ones.

For example, the R language-supported Bioconductor package currently offers 2230 free and open-source software tools that can be utilized for omics data processing (https://www.bioconductor.org/). Additionally, GitHub provides a platform for hosting numerous tools and repositories dedicated to data analysis and visualization. Users can easily search for specific tools using topic options (https://github.com/topics/), such as Genome Analysis (https://github.com/topics/genome-analysis), which yields over 100 relevant tools like Winnowmap (https://github.com/marbl/Winnowmap). Winnowmap is a program that follows a long-read mapping algorithm optimized for mapping ONT and PacBio sequences, making it suitable for genome alignment. By thoughtfully selecting a diverse set of representative tools, users can lay a solid foundation for an effective benchmarking study that provides valuable insights into the performance of various Omics computational tools.

2.6 Computational efficiency and scalability

It is important to consider the computing efficiency and scalability of Omics tools, especially considering the massive and highly dimensional nature of Omics datasets. This criterion assesses how effectively the analytical tools make use of

TABLE 4.2 A gold standard method adopted by benchmarking study performed by various research groups.

Benchmarking study	Application	No. of tools	Model of study	Raw input data type	Gold standard data preparation method	Parameter optimization
Aghaeepour et al. (2013)	Flow cytometry analysis	14	C	R	EXPERT	N
Bradnam et al. (2013)	Genome assembly	21	C	R	ALTECH	n/a
Lindgreen et al. (2016)	Microbiome analysis	14	I	S	SIMUL	No
McIntyre et al. (2017)	Microbiome analysis	11	I	R, S	MOCK	N
Altenhoff et al. (2016)	Ortholog prediction	15	I	DB	DB	Y
Baruzzo et al. (2017)	Read alignment	14	I	S	SIMUL	Y
Łabaj and Kreil (2016)	RNA-Seq analysis	7	I	R	ALTECH	N
Thompson et al. (2011)	Sequence alignment	8	I	DB	DB	N
Bohnert et al. (2017)	Variant analysis	19	I	R, S	I&A	Y
Ewing et al. (2015)	Variant analysis	14	C	S	SIMUL	n/a
Zhuo et al. (2023)	Functional enrichment analysis	6	C	DB	CIBERSORT	Y

Data from Mangul, S., Martin, L. S., Hill, B. L., Lam, A. K. M., Distler, M. G., Zelikovsky, A., Eskin, E., & Flint, J. (2019). Systematic benchmarking of omics computational tools. *Nature Communications*, *10*(1). https://doi.org/10.1038/s41467-019-09406-4.

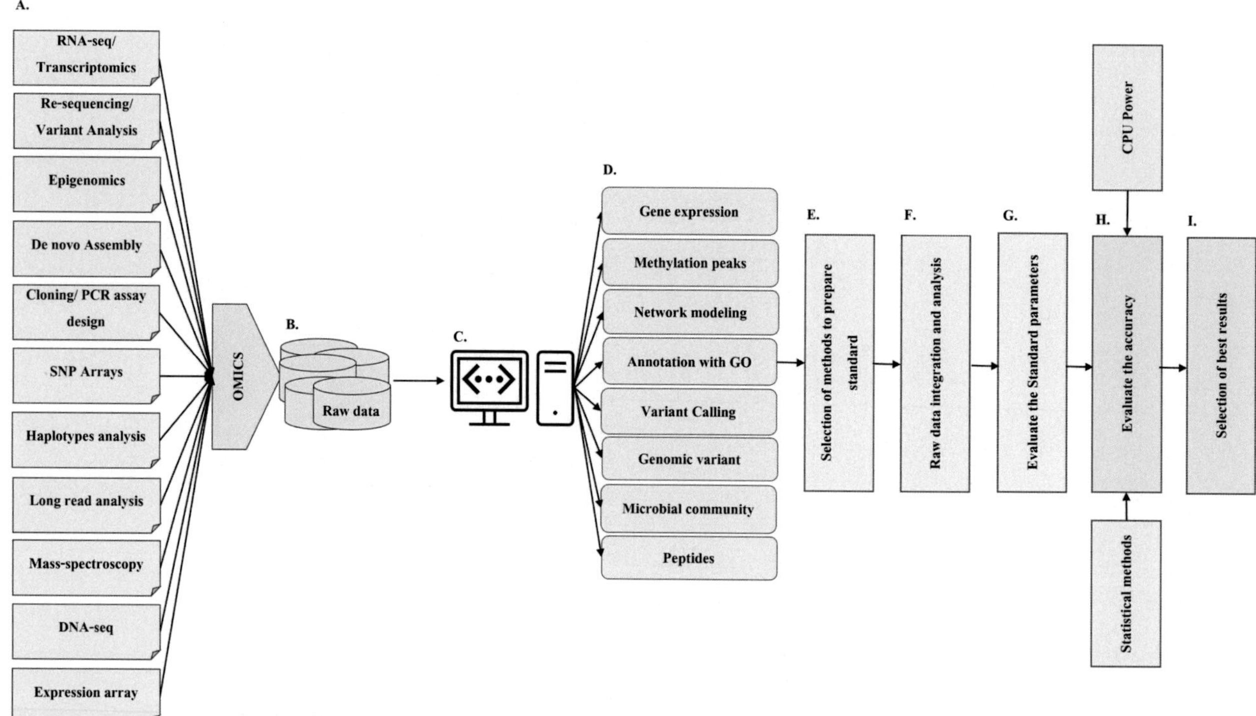

FIGURE 4.1 Flowchart of systematic benchmarking to process omics data. Flowchart of systematic benchmarking including number of steps involved in omics data analysis. (A) Omics data from different technologies, (B) raw data from experiments or public databases, (C) computational tools/packages, (D) result after omics data analysis, (E) gold standard preparation using selected methods, (F) integration of raw data into other tools and compare the results, (G) evaluate the gold standard selection parameters, (H) estimate the computational power, statistical methods, and evaluate the accuracy of omics data, (I) finalize the data and selection of results.

computational resources, such as CPU, memory, and storage (Table 4.5). On benchmark datasets of various sizes and complexity levels, the runtime or execution time of the tools is measured. The tools' ability to manage growing dataset sizes and computing needs is also evaluated as part of the scalability assessment, which shows whether they are appropriate for analyzing large-scale Omics data.

The following features should be considered during the measuring the computational efficiency:

- User-friendliness and ease of implementation are important criteria for benchmarking, as they influence their adoption and usage by researchers with varying levels of computational expertize. This criterion assesses the tools' usability, intuitiveness of their interfaces or command-line interfaces, availability of graphical user interfaces, and ease of installation and setup. Tools with clear and comprehensive documentation, tutorials, and examples are considered more user-friendly, facilitating their implementation and utilization in diverse research settings. Look for intuitive interfaces, drag-and-drop functionality, and clear workflows that simplify analysis processes. Software that doesn't require extensive scripting or programming skills may be more accessible to biologists without a strong bioinformatics background. For example, base pair and NextGenE software is easy to use, and simple click-and-point navigation is used to perform the work.
- Comprehensive and up-to-date documentation enables users to understand the tools' functionalities, parameters, and usage guidelines. Adequate support channels, such as user forums, mailing lists, or dedicated help desks, allow users to seek assistance, clarify doubts, troubleshoot issues, address user queries, and the frequency of tool updates or bug fixes.
- Omics datasets encompass a wide range of data types, such as DNA sequencing data, gene expression data, protein–protein interaction data, and metabolomic profiles. Check if the software is compatible with the sequencing platforms and data formats commonly used in your research. Ensure that it supports the data types (e.g., DNA-seq, RNA-seq) and file formats (e.g., FASTQ, BAM) relevant to your work.
- Accuracy and reliability of results means assessing the tools perform in producing outputs that are consistent with the ground truth or gold standard. It involves evaluating the tools' ability to correctly identify and quantify Omics features,

TABLE 4.3 A list of the most popular databases for omics data access and submission that may be utilized to benchmark datasets.

Database name	Data type	URLs
NCBI- dbVar	Insertions, deletions, duplications, inversions, mobile element insertions, translocations, and complicated chromosomal rearrangements	https://www.ncbi.nlm.nih.gov/dbvar
NCBI- dbGaP	Archive and distribute the data from studies of the interaction of genotype and phenotype in Humans	https://www.ncbi.nlm.nih.gov/gap/
NCBI- dbSNP	SNP and multiple small-scale variations that include insertions/deletions, microsatellites, and nonpolymorphic variants	https://www.ncbi.nlm.nih.gov/snp/
NCBI- GenBank	All publicly available DNA sequences	https://www.ncbi.nlm.nih.gov/genbank/
NCBI- gene	Reference sequences	https://www.ncbi.nlm.nih.gov/gene
NCBI- GEO	Curated gene expression DataSets	https://www.ncbi.nlm.nih.gov/geo/
NCBI- RefSeq	Genomic DNA, transcripts, and proteins	https://www.ncbi.nlm.nih.gov/refseq/about/
Ensembl	Genome annotations	https://useast.ensembl.org/index.html
ExAC	Exome sequencing data	http://exac.broadinstitute.org/
GTEx	Gene expression data from multiple human tissues	https://gtexportal.org/home/
ArrayExpress	Functional genomics data	https://www.ebi.ac.uk/biostudies/arrayexpress
Pfam/InterPro	Protein families and domains	https://www.ebi.ac.uk/interpro/
COSMIC	Somatic mutations in various cancer types	https://cancer.sanger.ac.uk/cosmic
Mouse genome informatics	Genomic information for the laboratory mouse	https://wormbase.org/
WormBase	Model organism genomics data	https://www.informatics.jax.org/

TABLE 4.4 List of available tools that can be used for systematic benchmarking processing of omics data analysis.

Toole name	Features	File format	OS	License	URLs
BWA	Mapping DNA sequences against a large reference genome		Linux	Free	https://github.com/lh3/bwa
SAMtools	Manipulating and analyzing DNA sequence data.	FastQ to SAM and BAM	Linux	Free	https://www.htslib.org/
GATK	Variant discovery, genotyping, and genomic annotations,	FASTQ	Window	License	https://gatk.broadinstitute.org/hc/en-us
PiCard	High-throughput sequencing of the genome	SAM/BAM/CRAM and VCF	Linux	License	https://broadinstitute.github.io/picard/
BEDTools	Henomics analysis	BAM, BED, GFF/GTF, VCF	Unix	License	https://bedtools.readthedocs.io/
VarScan	Massively parallel sequencing technologies, exome, and whole genome resequencing data	SAM/BAM	Linux, Mac OS, Windows	License	http://varscan.sourceforge.net/
ANNOVAR	Annotate genetic variants, including human genome hg18, hg19, hg38, as well as mouse, worm, fly, yeast and many others.	CADD, Clinvar	Dos based	Free	https://annovar.openbioinformatics.org/
IGV	De novo Genome assembly	FastQ	Windows	License	http://www.broadinstitute.org/igv/
RStudio	Integrated development environment	PNG, JPEG, TIFF, SVG, BMP, Meta-file, and Postscript	Windows, Mac, and Linux	Free and open source	https://rstudio.com/
Bioconductor	High-throughput genomic analysis, e.g., sequencing, expression and other microarrays, flow cytometry, mass spectrometry, image analysis	BAM, FASTA, BigWig	Linux, Mac OS X, and Windows	Free open source	https://bioconductor.org/
Galaxy	Genome Diversity, RNA seq analysis	SAM/BAM,	UNIX		https://usegalaxy.org/
Splice center	Evaluating the impact of gene-splicing variation			License	https://doi.org/10.1186/1471-2105-9-313
PhRed	DNA sequence data	SCF, ABI, and ESD	Windows/Linux/macOS	License	www.phrap.org/phredphrapconsed.html
Dialign	Multiple sequence alignment of nucleic acid or protein sequencing	FASTA		Free	http://dialign.gobics.de/
BioGPS	Gene expression in organisms, genomic screening technologies	JSON, PNG	Windows	Free	http://biogps.org/
Genome browser	Most recent assembly of each genome, and visualize genomic data	HTML, Javascript	Linux	License	https://genome.ucsc.edu/
MaGplotR	Analyzes multiple CRISPR screen datasets	FastQ or read	Linux, MacOS or Windows	License	https://github.com/alematia/MaGplotR

ASCAT	Infer tumor purity, ploidy and allele-specific copy number profiles, and high-throughput sequencing (HTS) data	BAM or CRAM			https://github.com/VanLoo-lab/ascat
MaxQuant	Analyzing mass spectrometry (MS) quantitative proteomics data	Perseus and proteus	Windows	License	https://www.biorxiv.org/content/10.1101/416511v2.full
Proteome discoverer	Both labeled and label-free quantification to crosslink analysis, glycoproteomics and even top-down proteomics	Mascot Generic Format (MGF), mzML, mzXML and mzData, FASTA	Windows	License	https://www.mdpi.com/2227-7382/9/1/15
Skyline	Large-scale quantitative mass spectrometry	pepXML and mzXML and a background proteome from a FASTA	Windows	Free	http://skyline.maccosslab.org/
Scaffold	Wrapper scripts to run genome assembly	FASTA, FASTQ	Linux, macOS, Windows	License	https://github.com/martinghunt/Scaffolder-evaluation/pulls
MASCOT	Mass spectrometry data to identify proteins from primary sequence databases	DTA, ASC, PKL, PKS, XML,Fastq	Linux, Windows	Free, License	www.matrixscience.com/mascot_support.html
OpenNMS	LC-MS data management and analyses, analysis and visualization of high-throughput mass spectrometry data	PSI, .tsv, .mzML, .msalign, and, feature	Linux, macOS, Windows	License	https://openms.de
Trans-proteomic	MS/MS-based shotgun proteomics analysis	mzXML	Linux and Windows	Free	http://tools.proteomecenter.org/wiki/index.php?title=Software:TPP
ProteoWizard	Proteomics data analysis	Wine/Docker, mzML, mzXML, and mzData	Linux, macOS, Windows	License	https://proteowizard.sourceforge.io

TABLE 4.5 Computational configuration to process omics data.

Omics data	RNA-seq data	ChIP-seq data	MassSpec data
Computation tools	STAR, HISAT2, TopHat, Cufflinks, and DESeq2	Bowtie, BWA, SAMtools, MACS2	MaxQuant, XCMS, OpenNMS
Processor	>32 cores	12–24 cores	6–12 cores
GPU	4 GB, Choose the GeForce series	The 2G Quadro series is enough	Not necessary
Storage	1–6 TB	1 TB	1 TB
RAM	16–32 GB	4–16 GB	16–32 GB

such as genes, proteins, or metabolites, as well as their performance in distinguishing true positives from false positives and true negatives from false negatives. Benchmarking should consider metrics such as sensitivity, specificity, precision, recall, and F1 score to assess the accuracy and reliability of the tools' results.

- Simulated datasets can be generated to mimic specific genomic scenarios, allowing researchers to evaluate tool performance under controlled conditions. Simulated data can help assess how well a tool performs in terms of sensitivity, specificity, false positive rate, and false discovery rate.
- Cross-validation is a technique where the available data is divided into multiple subsets for training and testing purposes. It can be used to assess tool performance by evaluating its ability to generalize well to unseen data. Common cross-validation methods include k-fold cross-validation and leave-one-out cross-validation.
- A Receiver Operator Characteristic (ROC) curve is a graphical representation of the true positive rate (sensitivity) against the false positive rate (1−specificity) for different threshold values. It provides a measure of the discrimination ability of a tool and can be used to compare the performance of different tools.
- Precision-recall curve is another graphical representation that shows the trade-off between precision and recall (sensitivity) for different threshold values. It is particularly useful when dealing with imbalanced datasets and can provide insights into a tool's performance at different decision thresholds.
- Evaluating the computational time required by a tool is also important, especially for large-scale genomics data analysis. Tools that provide faster analysis without compromising accuracy and quality can be preferred for time-sensitive applications.

- Assess the range of analysis modules and tools provided by the software. Look for features that align with your specific research requirements, such as variant calling, genome assembly, transcriptome analysis, or metagenomics analysis. For example, which software can handle the terabyte of genomics sequences; for this type of query, we use BWA (https://github.com/lh3/bwa), MiniMap2 (https://github.com/lh3/minimap2), and LoReTTA (https://github.com/salvocamiolo/LoReTTA/releases/tag/v0.1) programs while for short reads we use SPAdes (https://github.com/ablab/spades) and BWA.

2.7 Integration and interoperability

Consider the software's ability to integrate with other bioinformatics tools or databases. Look for compatibility with commonly used file formats, databases, and analysis pipelines. Integration with public databases, such as NCBI (https://www.ncbi.nlm.nih.gov/) or Ensembl (https://www.ensembl.org/index.html), can be beneficial for data retrieval and annotation.

3. Experimental setup

A well-defined experimental setup is necessary for any kind of work, and this needs good design of protocols (https://www.scribbr.com/methodology/experimental-design/). Protocols include providing detailed information on the hardware and software environment used for testing, specifying the versions of the benchmarked tools, and documenting any preprocessing steps applied to the benchmark datasets (Beyer et al., 2019). It is essential to conduct the benchmarking experiments in a controlled and consistent manner, considering factors such as parallelization settings, parameter optimization, and resource allocation (Hoffmann et al., 2019). The benchmarking process should be well-documented that enabling other researchers to reproduce the experiments and validate the results independently (Weber et al., 2019). Reproducibility considerations also involve documenting any potential limitations or biases in the benchmarking study. These may include assumptions made during data generation or preprocessing steps, potential conflicts of interest, and disclosure of any affiliations with the tool developers. Transparent reporting of the methodology and results is crucial for the scientific community to evaluate the benchmarking study and its implications accurately as in Fig. 4.1, we connect each step of Omics data analysis with the systematic benchmarking protocol.

The following steps can be selected for setting up an experimental design for systematic benchmarking of Omics data analysis using the computational tools:

- In the benchmarking study, create a comprehensive list of tools to be evaluated by identifying software tools that are best suited for the analytical tasks and data types involved. To do this, perform a PubMed search (https://www.ncbi.nlm.nih.gov/pubmed/) for relevant articles and examine the software tools mentioned in the references of each identified publication. Carefully consider the suitability of each tool for the project's requirements. During this process, be aware that some tools may be challenging to install and run within a reasonable amount of time. In such cases, it is advisable to document these challenges in log files to save the efforts of other researchers who may encounter similar difficulties. By curating a well-rounded and appropriate list of tools, the benchmarking study can provide valuable insights into the performance of various software options for specific analytical tasks and data types. Alternatively, researchers could examine all available tools and select which algorithms to use in the benchmarking study depending on the level of citations or the reputation of the article. Choosing the best tool just based on tool popularity, journal impact, and citation volume, on the other hand, is problematic because these indications do not always indicate that an algorithm is the best (Gardner et al., 2022; Wren, 2016).
- Prepare and describe the benchmarking data by creating a spreadsheet to summarize all relevant information. Clearly outline the protocols used for preparing both the raw data and the gold standard datasets. Additionally, document any potential limitations of the data, including the possibility of biasing the performance of specific types of algorithms. To ensure transparency and reproducibility, record all methods employed for benchmarking data preparation, maintaining complete provenance of data sources, and any necessary code used for data gathering and cleaning.
- Thoroughly describing the benchmarking data, along with its preparation protocols and potential limitations, establishes a robust foundation for accurately and transparently evaluating the performance of different algorithms and tools. This comprehensive approach ensures that the study's results are reliable and provide valuable insights into the effectiveness of the analyzed methods.
- Parameter optimization is typically better understood by the developers of a particular method, as they are familiar with how the method should be applied to a given dataset, including the selection of specific parameters and input preprocessing. In a competition-based model, participants are responsible for determining the optimal parameters for each tool, allowing for a fair comparison of the methods' performance.
- Provide a concise overview of algorithm features and share instructions for installing and running the tools. For instance, it is essential to mention if a tool requires a substantial number of dependencies to run. Having a centralized source of information on such issues, including dependencies, would be highly beneficial to the research community. Complex computational tasks can pose significant barriers for potential users, and a comprehensive resource would help alleviate these challenges and encourage wider tool adoption (e.g., virtual machine images, containers, Docker, and Cloud machines).
- In cases where each tool produces different output formats, it is essential to create and share a script that can generate a universal format. Given the rapidly evolving nature of data types and formats in computational biology, software developers and benchmarking studies can play a crucial role in spearheading standardization efforts. By adopting standardized data types and formats, we can improve compatibility and facilitate seamless data exchange and comparison among different tools and studies.
- Maximizing data reusability can be achieved by providing an accessible interface for downloading both the input raw data and the gold standard data. To enhance user flexibility and evaluation capabilities, it is advisable to share the raw output data generated by each benchmarked tool. This approach allows end users to apply their evaluation metrics and facilitates a more comprehensive understanding of the tools' performance.
- Benchmarking provides an opportunity to identify the strengths and limitations of each Omics computational tool (Krassowski et al., 2020). By analyzing the benchmarking results, researchers can pinpoint the specific areas where each tool excels or underperforms. These strengths may include superior accuracy, computational efficiency, or compatibility with specific data types (Zhang et al., 2019). On the other hand, limitations may arise from suboptimal performance, computational bottlenecks, or inadequate support for certain analysis tasks (Flores et al., 2023).

- The benchmarking results can guide tool developers in refining algorithms, optimizing computational performance, enhancing user-friendliness, and addressing specific limitations highlighted during the benchmarking process. This iterative improvement cycle fosters the advancement of Omics data analysis tools and drives the field forward. By conducting a thorough comparative evaluation, performance assessment, and identification of strengths and limitations, researchers can provide valuable insights to the scientific community about the capabilities and suitability of different Omics computational tools.

The next Sections 3.1—3.9 discussed more about widely used publicly and commercially available software in the Genomics, Metagenomics, and Transcriptomics field in the expect of their strength.

3.1 QIAGEN Genomics Workbench

The QIAGEN Genomics Workbench is a powerful software tool designed to support systematic genomics analysis (https://digitalinsights.qiagen.com/). It includes a complete set of tools and procedures such as resequencing, operations, read mapping, de novo assembly, variant detection, RNA-Seq, ChIP-Seq, and Genome Browser to help researchers handle, analyze, and interpret genomics data quickly. This allows users to import genomics data from GenBank (https://www.ncbi.nlm.nih.gov/genbank/), Blast (https://blast.ncbi.nlm.nih.gov/Blast.cgi), and PubMed (https://pubmed.ncbi.nlm.nih.gov), and raw sequencing data or alignment files in FASTA, GFF/GTF/GVF, BED, Wiggle, Cosmic, UCSC variant database, complete genomics master var file format. It provides a user-friendly interface to organize and manage large datasets efficiently. Users can organize their data into projects, folders, and subfolders, enabling systematic storage and retrieval of genomics data. Data preprocessing and quality control are applied to ensure the reliability of downstream analyses. Users can perform tasks such as adapter trimming, read mapping, and filtering of low-quality reads. This systematic preprocessing step helps improve the accuracy and integrity of the subsequent analyses. The workbench provides predefined and standardized workflows with analysis pipelines for common genomics analysis tasks. Users can easily navigate through the workflow steps, customize parameters, and track the progress of their analyses. It also supports variant analysis, including single nucleotide variants, insertions and deletions, and structural variants. It provides tools for variant calling, annotation, and filtering. Systematic variant analysis allows researchers to identify and prioritize genomic variants of interest, aiding in understanding the genetic basis of diseases or phenotypes. Users can also explore biological pathways and perform functional enrichment analysis to gain insights into the biological processes underlying their data. The systematic examination of pathways and functional annotations enhances the understanding of the biological context and potential implications of the analyzed genomic variants. Users can generate interactive genome browsers, heat maps, scatter plots, and other visual representations. These visualizations help researchers identify patterns, trends, and potential relationships within the data. The workbench also supports the generation of comprehensive reports summarizing the analysis results, facilitating systematic documentation and communication of findings.

3.2 Partek Genomics Suite

The suite is designed to provide researchers and scientists with comprehensive tools for analyzing and interpreting high-throughput genomic data (https://www.partek.com/partek-genomics-suite/). It offers a range of features and algorithms including machine learning methods to facilitate the exploration, visualization, and statistical analysis of various types of genomic data, such as microarray, qPCR, miRNA-Seq, gene expression, single-cell RNA sequencing, DNA methylation, ChIP-seq. It preprocesses raw genomic data in text, CEL (Cell intensity file), IDAT (BeadArray data file), or BAM (Binary Alignment Map files) format, including quality control, normalization, filtering steps, and identifying the differentially expressed genes across experimental conditions or groups using advanced statistical algorithms such as logistic regression, Chi-square, false discovery rate, batch effect removal, fisher effect, multiple test correction, predictive modeling, power analysis, and flexible scaling. The alignment stage uses multiple programs such as Bowtie, BWA (https://bowtie-bio.sourceforge.net/index.shtml), Isaac (https://github.com/sequencing/isaac_variant_caller), TopHat (https://cole-trapnell-lab.github.io/projects/tophat/), TMAP (https://github.com/GPZ-Bioinfo/tmap), HISAT2 (http://daehwankimlab.github.io/hisat2/), STAR (https://github.com/alexdobin/STAR), and GSNAP (http://research-pub.gene.com/gmap/). The quality control and analysis were performed by pre and postalignment, ERCC spike-in, and single-cell quality. Single-cell RNA sequencing allows users to identify cell types, perform clustering either hierarchical, K-means or Graph based, and explore gene expression patterns at the single-cell level. The variant calling analysis is performed by Samtools (http://www.htslib.org/), FreeBayes (https://github.com/freebayes/freebayes), LoFreg (https://csb5.github.io/lofreq/), Strelka (https://github.com/Illumina/strelka), CNVkit (https://cnvkit.readthedocs.io/en/stable/), and GATK (https://gatk.broadinstitute.org/hc/en-us)

software. It also integrates pathway analysis tools to identify enriched biological pathways and functional gene sets related to the analyzed data. It also provides integration with various public databases, such as RefSeq (https://www.ncbi.nlm.nih.gov/refseq/), dbSNP (https://www.ncbi.nlm.nih.gov/snp/), miRbase (https://www.mirbase.org/), JASPAR (https://jaspar.genereg.net/), Ensemble (https://www.ensembl.org/index.html?redirect=no), GENCODE (https://www.gencodegenes.org/human/), TargetScan (https://www.targetscan.org/vert_80/), Gene Ontology (http://geneontology.org/), and KEGG (https://www.genome.jp/kegg/), allowing users to access additional functional annotations and biological information. The outcomes of genomic data can be visualized using a Venn diagram, dot plot, Karyogram, Valcon plot, Violin plot, heatmaps, 2D and 3D scatterplots, box plots, and Box and Whishers. The data also can be explored using partial least squares, Principal component analysis, Multidimensional Scaling, Self-Organizing Maps, portioning and hierarchical clustering, correspondence analysis and pattern recognition.

3.3 Golden Helix

It worked on SNP and Variation Suite data that are used for managing, analyzing, and visualizing genotypic and phenotypic features (https://www.goldenhelix.com/). It performs genome-wide association studies, genomic prediction, copy number analysis, small sample DNA-Seq workflows, large sample DNA-seq analysis, and RNA-seq analysis. It allows files in any of the following formats as inputs such as txt, excel XLS & XLSX, CEL, CHP, CNTs, Illumina, Plink PED, TPED, BED, Agilent files, NimbleGen data summary files, VCF files, Impute2 GWAS files, HapMap format, MACH output, and 50 others. The data visualization can be done by using plots, charts, and heatmaps to visualize genetic variants, linkage disequilibrium, genotype concordance, and other relevant genomic information.

3.4 Genomatix

It is an integrated solution for comprehensive second-level analysis of next-generation sequencing (NGS) data from ChIP-Seq, RNASeq or genotyping experiments, DNA methylation; enable personalized medicine and Mining Stations (http://mygga.genomatix.de/). It supports all established NGS sequencing platforms i.e., Supported Oligonucleotide Ligation and Detection (SOLiD: https://apollo-institute.org/solid-sequencing/), 454 Life Sciences (Patrick, 2007), Illumina's Genome Analyzer (Minoche et al., 2011), HiSeq (https://www.illumina.com/systems/sequencing-platforms/hiseq-x.html), MiSeq (https://www.illumina.com/systems/sequencing-platforms/miseq.html), and IonTorrent (https://www.thermofisher.com/us/en/home/brands/ion-torrent.html). It supports text file formats of sequences in BED (Browser Extensible Data), VCF, and BAM format files. Generally, its module can be accessed through command-line such as MACS, MatInspector, FrameWorker, ModelInspector, CoreSearch, PromoterInspector, DiAlign, Variant analysis, and Replicate analysis programs. It supports two databases; ElDorado (http://mygga.genomatix.de/online_help/help_eldorado/introduction.html) and MatBase (http://mygga.genomatix.de/online_help/help_matbase/matbase_help.html#search) to integrate and search for additional information.

3.5 Biodatomics

This is an open-source SaaS (Software-as-a-Service) platform that specializes in analysis and genome sequencing tools. It seamlessly integrates over 200 open-source tools and pipelines for genomic analysis (https://www.biodatomics.com/bioinformatician/). It offers both private and public cloud versions, catering to different deployment preferences such as BioDT in pro and SaaS. Notable features of the platform include intuitive genomic data visualization, a user-friendly drag-and-drop interface, accelerated analysis capabilities, and real-time collaboration functionalities. Moreover, the platform has incorporated components of the Sequence Read Archive (SRA) Toolkit (https://hpc.nih.gov/apps/sratoolkit.html), enabling users to access and process data from the SRA database (https://www.ncbi.nlm.nih.gov/sra) more efficiently.

3.6 Basepair

It is a web-based workflow, no need to install it in the local system (https://www.basepairtech.com/). It provides workflows for all common NGS applications (RNA-Seq, ChIP-Seq, DNA-Seq, and Metagenomics) and is very fast—gets all results in 1–2 h. It also provides cloud-based platform with no storage or computing limits also an user-friendly that complete an analysis in under a minute. REST is a way of accessing web services in a simple and flexible way without having any processing (https://www.geeksforgeeks.org/rest-api-introduction/) and Python Application programming interface (API) (https://docs.python.org/3/c-api/index.html) to manage large projects.

3.7 DNAnexus

It is a cloud-based genomics platform that provides tools and infrastructure for the analysis, management, and sharing of multiomics and biomedical data (https://www.dnanexus.com/). It offers a secure and scalable environment that enables researchers and organizations to efficiently process and analyze large-scale genomics datasets. It allows users to upload, store, and manage genomic data securely in the cloud. It provides tools for organizing data into projects, applying metadata, and controlling access permissions. The platform offers prebuilt and customizable analysis pipelines for various genomics applications, such as DNA sequencing, RNA sequencing, variant calling, and more. Users can select and run these pipelines on their data, leveraging optimized algorithms and workflows. It enables collaboration among researchers by providing features for sharing data, analyses, and results within and across research teams. It allows for controlled access to data and collaboration with external partners. With its cloud-based infrastructure, DNAnexus offers high scalability and computational power to handle large-scale genomics datasets. It can efficiently process and analyze data from projects of varying sizes, accommodating both small research studies and large-scale genomics initiatives.

3.8 Lasergene Genomics Suite

It is a comprehensive software suite developed by DNASTAR (https://www.dnastar.com/software/lasergene/genomics/). It offers a wide range of tools and functionalities to support various genomic analysis tasks such as reference-guided and de novo genome and transcriptome assembly and analysis, metagenomics sample assembly, targeted resequencing, exome alignment, gene panels with validation control, variant analysis, and RNA-Seq, ChIP-Seq and miRNA alignment and analysis enabling researchers to effectively analyze and interpret genomic data. The suite also includes modules for RNA-seq analysis, ChIP-seq analysis, and metagenomics analysis, allowing users to explore gene expression, identify regulatory elements, and study microbial communities, respectively. It has a user-friendly interface, which allows researchers to navigate through different analysis workflows with ease. The software provides interactive visualizations and intuitive tools for data exploration and interpretation. It also supports seamless integration with other Lasergene modules, enabling users to perform multiomics analyses and leverage a comprehensive suite of bioinformatics tools. It supports all major NGS technologies (Illumina, Ion Torrent, Pac Bio and Roche 454) and is available on Windows, Mac OS X, Linux, and the Amazon Cloud.

3.9 NextGENe

It employs unique platform-specific technologies in one free-standing multiapplication package and supports sequencing data generated by various Illumina systems like iSeq, Miniseq, MiSeq, NextSeq, HiSeq, and NovaSeq (https://softgenetics.com/products/nextgene/). It is also compatible with Ion Torrent Ion GeneStudio S5, PGM, proton systems, and other sequencing platforms. It is designed to run on the Windows Operating System, offering a biology-friendly interface that allows users to navigate through analysis workflows using simple "point & click" interactions.

Furthermore, open-source software is free and provides scripts that can be customized, making it a viable alternative to commercial software (Tables 4.4 and 4.6). These programs allow for customization based on specific study requirements. Researchers can tweak and enhance the software's functionality to suit their experimental designs, data formats, or analysis methods. This versatility enables the introduction of various data formats as well as the incorporation of fresh methods or approaches.

4. Case studies

Case studies provide practical examples of how benchmarking results can be applied in specific Omics domains. The case studies focus on applying benchmarking findings to real-world scenarios and research questions within genomics, proteomics, metabolomics, or other relevant Omics disciplines. By showcasing the application of benchmarking results in these domains, we can demonstrate the relevance and utility of the benchmarking study to address specific research challenges and enhance data analysis in various biological contexts.

Here, we listed the specific strengths and unique features of individual tools and software used for the Omics study and it will help to select the tools for designing a systematic benchmarking-based study.

4.1 Case study 1: Hybrid assembly strategy on genomics data analysis

In their study, Gavrielatos et al. discussed the potential of third generation sequencing technologies for de novo genome assembly (Gavrielatos et al., 2021). They explored the emergence of innovative strategies that utilize the distinct features of

TABLE 4.6 List of tools with their function and types of omics data.

Name	Data domain	Type of analysis	Omics integrated	URLs	References
SAM	Vertebrates, invertebrates, pathogens, plants, and viruses	Processing and analysis of sequencing data, whole-chromosome aberrations	Genomics	https://github.com/samtools/samtools	Danecek et al. (2021)
BCF	Vertebrates, invertebrates, pathogens, plants, and viruses	Copy-number variation, homozygosity, and single-nucleotide polymorphism array data	Genomics	https://github.com/samtools/bcftools	Danecek et al. (2021)
HTSeq	Human	Single-cell omics, gene expression, chromatin accessibility, transcription factor binding affinities, and 3D chromatin conformation	Genomics, Transcriptomics	https://pypi.python.org/pypi/HTSeq	Putri et al. (2022)
Pyrpipe	Human, and virus	Analyzes RNA-Seq data	Transcriptomics	https://github.com/urmi-21/pyrpipe	Singh et al. (2021)
Pygenprop	Animals, plants, fungi, bacteria, virus	Genome analysis	Genomics	https://anaconda.org/lbergstrand/pygenprop	Bergstrand et al. (2019)
Gosling	Human	Visualizations of genomics and epigenomics data	Genomics	https://gosling.js.org/	L'Yi et al. (2022)
GenoREC	Mammals	Genomics analysis	Genomics	https://osf.io/y73pt/	Pandey et al. (2023)
PyHMMER	Microbial species	Protein sequences	Proteomics	https://pypi.org/project/pyhmmer/	Larralde and Zeller (2023)
Htseq-clip	NA	Next-generation sequencing (NGS), transcriptome-wide detection of RBP binding sites	Transcriptomics	https://github.com/EMBL-Hentze-group/htseq-clip	Sahadevan et al. (2023)
eCLIP	NA	RNA binding proteins, RNA processing, RNA ImmunoPrecipitation, UV Cross-Linking and ImmunoPrecipitation	Transcriptomics	https://www.encodeproject.org/	Van Nostrand et al. (2016)
SPARTA	Bacteria	Next-generation sequencing (NGS), RNA-seq analysis	Transcriptomics	http://sparta.readthedocs.org/	Johnson et al. (2016)
EDGAR	Microbial genomes	Pan-genome, core genome, or singleton genes.	Genomics	http://edgar3.computational.bio/	Dieckmann et al. (2021)
CloVR	Microbial genomes	Genome sequence analysis	Genomics	http://www.clovr.org/	Agrawal et al. (2017)
PRAWNS	Bacteria, viruses and fungi	Genomic analysis	Genomics	https://github.com/KiranJavkar/PRAWNS.git	Javkar et al. (2023)
RSAT	Animals, bacteria, plants	De novo motif, genomic sequences scanning, motif analysis, analysis of regulatory variations, and comparative genomics	Genomics, Transcriptomics	https://rsa-tools.github.io/managing-RSAT	Santana-Garcia et al. (2022)

Continued

TABLE 4.6 List of tools with their function and types of omics data.—cont'd

Name	Data domain	Type of analysis	Omics integrated	URLs	References
BamBam	NA	Next-generation sequencing (NGS), Genome sequence analysis, gene expression analyses, Single Nucleotide Polymorphisms, and Copy Number Variants	Genomics, Transcriptomics	http://sourceforge.net/projects/bambam/	Page et al. (2014)
HSRA	Human	Next-Generation Sequencing (NGS), genome Sequencing, RNA data sequencing analysis	Genomics, Transcriptomics	http://hsra.dec.udc.es/	Expósito et al. (2018)
Picard	Plants, animals, and fungus	Next-generation Sequencing (NGS), gene expression, genetic instability, and genome evolution	Genomics	http://broadinstitute.github.io/picard	Sohrab et al. (2021)
WebQUAST	Bacterial, Plant, animal, human, and fungus	Genome assembly	Genomics	https://www.ccb.uni-saarland.de/quast/	Mikheenko et al. (2023)
GenoVi	Microbes	Genome visualization, and Sequencing	Genomics	https://github.com/robotoD/GenoVi	Cumsille et al. (2023)
digIS	Bacteria	Prokaryotics genomic and metagenomic data	Genomics	https://github.com/janka2012/digIS	Puterová and Martínek (2021)
ISEScan	Bacteria	Prokaryotic genome sequences analysis	Genomics	https://github.com/xiezhq/ISEScan	Xie et al. (2017)
panISa	Bacteria	Next-Generation Sequencing (NGS), bacterial genomes to analyze genomic evolution	Genomics	https://github.com/bvalot/panISa	Treepong et al. (2018)
ChemGAPP	Microorganism	Chemical genomics analysis	Genomics	https://github.com/HannahMDoherty/ChemGAPP	Doherty et al. (2023)
BATCH-GE	Animals	NGS generated genome editing data, evaluation of CRISPR/Cas9	Genomics	https://github.com/WouterSteyaert/BATCH-GE.git	Boel et al. (2016)
CRISPRMatch	Plant	NGS, genome-editing data of CRISPR nuclease transformed protoplasts	Genomics, Transcriptomics	https://github.com/zhangtaolab/CRISPRMatch	You et al. (2018)
QUAST-LG	Eukaryotes	Next-Generation Sequencing (NGS), DNA sequencing,	Genomics	http://cab.spbu.ru/software/quast-lg	Mikheenko et al. (2018)
Ensembl	Prokaryotes, Eukaryotes	Genome annotations, including genes, variants, regulatory regions and comparative genomics resources	Genomics	https://www.ensembl.org/	Martin et al. (2023)
PyQuant	Bacteria	MS-based quantitative analysis	Proteomics	https://pandeylab.github.io/pyquant/	Mitchell et al. (2016)
StPeter	Mammals, Bacteria	Trans-proteomic pipeline, mass spectrometer, and label-free quantification of LC–MS/MS data.	Proteomics	http://tools.proteomecenter.org/wiki/index.php?title=Software:StPeter	Hoopmann et al. (2018)
PANDA	Human	Proteomics data quantification, label-free and labeled quantitative proteomics data	Proteomics	https://sourceforge.net/projects/panda-tools/	Chang et al. (2019)
pGlycoQuant	Human	Mass spectrometry-based intact glycopeptide quantitation.	Proteomics	https://github.com/Power-Quant/pGlycoQuant/releases	Kong et al. (2022)

each sequencing platform to address the limitations inherent in individual sequencing types, ultimately aiming to create comprehensive and accurate genome maps. The research involved benchmarking two assembly strategies: the hybrid strategy, combining Illumina short, paired-end reads with Nanopore long reads (https://nanoporetech.com/), using MaSuRCA and Wengan assemblers, and the long-read assembly strategy, relying solely on Nanopore or PacBio HiFi reads (https://www.pacb.com/), benchmarked with Hifiasm and HiCanu. It is important to note that typical Nanopore and PacBio Sequel I long-reads have an average accuracy of approximately 90%, whereas typical Illumina short-reads exhibit a much higher accuracy of around 99.9% (Kent & Haussler, 2001; Lin et al., 2016). Assemblies solely based on long reads demonstrated greater continuity but also exhibited more errors, making subsequent tasks like genome annotation, variant calling, and other genome analyses challenging (Chaisson et al., 2015; Jain et al., 2018). By adopting the hybrid assembly strategy, the researchers combined the advantages of both generations, effectively incorporating information from both read types and overcoming their respective drawbacks (Nowak et al., 2019; Tan et al., 2018).

4.1.1 Data collection

In the first step, they collected the data from the three organisms, *Drosophila virilis*, *Drosophila melanogaster*, and *Homo sapiens* sequence data downloaded in FASTQ format, in which some FASTQ files were subsampled using Reformat tool from BBtools (https://sourceforge.net/projects/bbmap/). The low complexity genome of *D. virilis* and the high complexity genome of *H. sapiens* were constructed using the hybrid assembly strategy, using short paired-end Illumina reads in combination with long Nanopore reads, and read data were downloaded from the European Nucleotide Archive (https://www.ebi.ac.uk/ena/) and the T2T Consortium (https://www.genome.gov/about-genomics/telomere-to-telomere).

4.1.2 Genome assembly

They chose two pipelines to evaluate, MaSuRCA (https://github.com/alekseyzimin/masurca) stands for (Maryland Super Read Cabog Assembler) and Wengan (https://github.com/adigenova/wengan) which is an accurate and ultra-fast hybrid genome assembler. MaSuRCA workflow offers three different assemblers, Celera Assembler with the Best Overlap Graph (CABOG: http://wgs-assembler.sf.net/), which is no longer used for PacBio data processing while SOAPdenovo (https://github.com/aquaskyline/SOAPdenovo2) also called MEGAHIT and Flye (https://github.com/fenderglass/Flye) assembler still used for single-cell and metagenomics data from PacBio or Oxford nanopore. Wengan pipeline is based on the DISCOVAR de novo assembler (https://github.com/broadinstitute/discovar_de_novo) which is a whole genome shotgun assembler. The PacBio RS II/Sequel or Oxford Nanopore MinION data can be processed through Canu (https://github.com/marbl/canu) which is a fork of the Celera Assembler or long-read assembler, designed to use long high-noise single-molecule sequencing data. Hifiasm (https://github.com/chhylp123/hifiasm) and HiCanu (https://github.com/marbl/canu) are fast haplotype-resolved de novo assemblers initially designed for PacBio HiFi reads. The most recent version, which makes use of extremely long Oxford Nanopore readings, might promote the telomere-to-telomere assembly. Hifiasm is one of the finest haplotype-resolved assemblers for the trio-binning assembly given parental short reads, and it creates perhaps the best single-sample telomere-to-telomere assemblies using HiFi, ultralong, and Hi-C reads.

4.1.3 Scaffolding

To verify the need for scaffolding, a scaffolder was utilized to improve assembly consistency and accuracy, as seen below: Arima mapping pipeline (https://arimagenomics.com/) maps Hi-C data to the primary assembly to produce a BAM file, which is then translated to a BED file. SALSA (https://github.com/marbl/SALSA) scaffolds the primary assembly using this BED file, which contains the mapping information for Hi-C reads on the assembly. Scaffolds are also converted to .hic format for visualization.

4.1.4 Metrics for quality control

QUAST v5 (https://github.com/ablab/quast), stands for QUality ASsessment Tool. It evaluates genome/metagenome assemblies by computing various metrics using a reference genome. QUAST makes use of BUSCO (https://busco.ezlab.org/), to assess genome assembly and annotation completeness, based on evolutionarily informed expectations of the gene content of near-universal single-copy orthologs, BUSCO metric is complementary to technical metrics like N50.

4.1.5 Genome consistency plots

JupiterPlot v1 (https://github.com/JustinChu/JupiterPlot) is used for generating a Circos-based genome assembly consistency plot given a set of contigs relative to the reference genome. Intended to visualize large-scale translocations or misassembles in draft assemblies, but it can also be useful when trying to show synteny between WGS.

4.2 Case study 2: Computational cost analysis by assessing Oxford nanopore read assemblers

This study was done by Sun et al. (2021). In this they used an Oxford Nanopore Technology (ONT) long-read sequencing method which has become a popular platform for microbial researchers due to the accessibility and affordability of its devices. Nanopore readings can be used to assemble high-quality bacterial genomes easily and automatically, however this task is still difficult. Guppy 3.6.0's "fast" and "high-accuracy" modes were used to rebase call the raw ONT reads from the Scaly-foot Snail genome sequencing experiment. The "high accuracy" option was only assessed using GPU servers because running it on CPUs would have resulted in an unrealistic time to completion. The quick mode was evaluated using both GPU and CPU servers. The Illumina and ONT read from the *M. coruscus* genome sequencing project was downloaded from the NCBI SRA database (https://www.ncbi.nlm.nih.gov/sra:ERR3415816 and ERR3431204). Illumina reads were cleaned to remove bacterial contamination using Kraken 2 (https://ccb.jhu.edu/software/kraken2/), and the genome size and heterozygosity were calculated by Jellyfish (https://github.com/gmarcais/Jellyfish) with the k-mer size of 17, 19 and 21 and GenomeScope (http://qb.cshl.edu/genomescope/genomescope2.0/) which is an Estimate genome heterozygosity, repeat content, and size from sequencing reads using a kmer-based statistical approach. To attain high accuracy and quick classification speeds, Kraken 2 is a taxonomic classification method that uses exact k-mer matching. This classifier compares each k-mer in a query sequence to the lowest common ancestor (LCA) of all genomes that contain that particular k-mer. The classification algorithm is informed by the k-mer assignments. By adopting an efficient hash table encoding and utilizing the "compare-and-swap" CPU instruction to boost parallelism, the Jellyfish can count k-mers with an order of magnitude less memory and an order of magnitude faster than existing k-mer counting packages. They used the following assemblers for the benchmarking, including the long noisy read-only assemblers: Canu (https://canu.readthedocs.io/en/latest/), Flye (https://github.com/fenderglass/Flye), Wtdbg2 (https://github.com/ruanjue/wtdbg2), Miniasm (https://github.com/lh3/miniasm), NextDenovo (https://github.com/Nextomics/NextDenovo), NECAT (https://github.com/xiaochuanle/NECAT), Raven (https://github.com/lbcb-sci/raven) used for long uncorrected reads and Shasta (https://github.com/paoloshasta/shasta) and hybrid assemblers: MaSuRCA (https://github.com/alekseyzimin/masurca) and QuickMerge (https://github.com/mahulchak/quickmerge) which is a simple and fast metassembler and assembly gap filler designed for long molecule based assemblies. The only difference between these two assemblers is if you already have assembled contigs/scaffolds that you want to merge (Quickmerge) or if you need to perform the entire genome assembly process (MaSuRCA). The assembled contigs were polished at least three times with Flye, and heterozygous contigs were removed with the purge_dup pipeline (https://github.com/dfguan/purge_dups). The resultant genomes were polished twice using Pilon (https://github.com/broadinstitute/pilon/releases/) with Illumina reads. The genome completeness of each assembly was thoroughly monitored at each step using BUSCO (https://busco.ezlab.org/) with the odb10 metazoan dataset (https://busco.ezlab.org/frames/euka.htm). QUAST calculates genome assembly characteristics such as N50 and total size but also assesses misassemblies with minimap2 (https://github.com/lh3/minimap2) which is a versatile pairwise aligner for genomic and spliced nucleotide sequences. Repeat content was initially predicted with RepeatModeler (https://www.repeatmasker.org/RepeatModeler/) and Genomes were hard-masked using RepeatMasker (https://www.repeatmasker.org/) with a species-specific repeat library generated by RepeatModeler and all the known repeat content in the RepeatMasker repeat database. Augustus (https://bioinf.uni-greifswald.de/augustus/), an ab initio gene predictor in eukaryotic genomic sequences, was trained using Braker (https://github.com/Gaius-Augustus/BRAKER) with the hard-masked genome and the transcriptome assemblies. Genome annotation was then performed using Maker (https://www.yandell-lab.org/software/maker.html) with the trained Augustus predictor plus each species' transcriptome assembly and molluscan protein sequences downloaded from the NCBI protein database and each gene was annotated by InterProScan (http://www.ebi.ac.uk/interpro/).

4.3 Case study 3: Genome assembly methods on metagenomic sequencing data

Metagenome assembling is a fast method for reassembling microbial genomes from metagenomic sequencing datasets. Although short-read sequencing has been widely used for metagenome assembly, linked- and long-read sequencing has shown their advancements in assembly by providing long-range DNA connectedness. To resolve the repeats in microbial genomes and simplify the assembly graphs, numerous metagenome assembly algorithms were created. However, there is yet to be a thorough assessment of metagenomic sequencing methods, and there is a paucity of information on how to choose the best metagenome assembly tools. In this study performed by Zhang et al. (2023) used simulated 10X linked reads and ONT long-reads using LRTK-SIM (https://github.com/zhanglu295/LRTK-SIM) and CAMISIM (https://github.com/CAMI-challenge/CAMISIM) given the taxonomic composition in Critical Assessment of Metagenome Annotation datasets.

4.3.1 Metagenome assembly

The MEGAHIT (https://github.com/voutcn/megahit) and metaSPAdes (https://cab.spbu.ru/software/meta-spades/) used for short-read assembly; cloudSPAdes (https://cab.spbu.ru/software/cloudspades/) and Athena (https://github.com/abishara/athena_meta) used for linked-read assembly; Shasta (https://github.com/chanzuckerberg/shasta), wtdbg2 (https://github.com/ruanjue/wtdbg2), MECAT2 (https://github.com/xiaochuanle/MECAT2), NECAT (https://github.com/xiaochuanle/NECAT), metaFlye (https://github.com/fenderglass/Flye), Canu (https://canu.readthedocs.io/en/latest/), and Lathe (https://github.com/bhattlab/lathe) for long-read assembly; and DBG2OLC (https://github.com/yechengxi/DBG2OLC), and metaFlye, OPERA-LG and OPERA-MS (Gao et al., 2016) used for hybrid assembly. All the assembly tools were run on Linux machines with a 64-core AMD processor.

4.3.2 Contig statistics

Contig N50, and MAG N50 generated by QUAST (https://quast.sourceforge.net/) after removing the contigs shorter than 1 kb. The MetaQUAST mode (Mikheenko et al., 2016) enabled to obtain contig NA50 and NGA50 for each species from the datasets for which reference genomes were available.

4.3.3 Contig binning and MAG qualities

To prepare the inputs of MetaBat2 (https://bitbucket.org/berkeleylab/metabat/src/master/) for contig binning, BWA (https://github.com/lh3/bwa) and minimap2 (https://github.com/lh3/minimap2) were used to align short-reads (or linked-reads) and long-reads to the contigs, respectively. The alignment file was sorted by coordinates using SAMtools (http://www.htslib.org/), and the contig coverage was extracted by the "jgi_summarize_bam_contig_depths" program in MetaBat2. The single-copy gene completeness and contamination of each MAG performed using CheckM (https://github.com/Ecogenomics/CheckM). The transfer RNAs (tRNAs) and ribosomal RNAs (5S, 16S, and 23S rRNAs) were detected by ARAGORN (http://www.ansikte.se/ARAGORN/) and barrnap (https://github.com/tseemann/barrnap), respectively.

4.3.4 Annotate MAGs into species

The poorly assembled MAGs removed with contig N50s <50 kbp, completeness <75%, or contamination >25%, and annotated the contigs in MAGs with Kraken2 (https://ccb.jhu.edu/software/kraken2/).

4.4 Case study 4: Data-independent acquisition mass spectrometry for immunopeptidomics

Shahbazy M et al. conducted this investigation in which peptide-spectrum matches (PSMs) were utilized to create spectral libraries to explore MS/MS spectra from data-dependent acquisition matched with Human leukocyte antigen (HLA)-bound peptide sequences (Shahbazy et al., 2023). This approach provides a sizable peptide repertoire containing pHLA targets for spectral library-based DIA data searches. PEAKS Studio Xpro ver (https://www.bioinfor.com/peaks-xpro/) was utilized to process and search LC−MS/MS Data-dependent acquisition (DDA) approach against the human proteome database (UniProtKB/SwissProt v.26072021 UP000005640; 20,375 entries) with a contaminant database of iRT peptide sequences. The false discovery rate (FDR) was adjusted by 1% to identify peptides confidently by deploying a target-decoy algorithm. The decoy sequences (the same size as the protein database) were generated using the PEAKS decoy-fusion method (Ivanov et al., 2016).

4.4.1 Library search-based DIA data analysis

They used four "peptide-centric" DIA software tools (https://biognosys.com/software/spectronaut/), DIA-NN (https://github.com/vdemichev/DiaNN), and PEAKS Xpro to analyze DIA datasets at 1% FDR (at peptide level). Each tool's immunopeptidome coverages were compared using a two-way Analysis of Variance (ANOVA) and Tukey's multiple pairwise tests.

4.4.2 Software tools for peptide sequence and statistical data analysis

GraphPad Prism (https://www.graphpad.com) and MATLAB computer code routines (https://www.mathworks.com/products/matlab.html) were used to conduct all statistical analyses. Allelic specificity was calculated using NetMHCpan (https://services.healthtech.dtu.dk/services/NetMHCpan-4.1/) based on peptide binding rank. With the aid of GibbsCluster (https://services.healthtech.dtu.dk/services/GibbsCluster-2.0/), the peptides were additionally divided based on sequence characteristics. The discovered HLA-bound peptides were sequence motif analyzed using Seq2Logo (https://services.

healthtech.dtu.dk/services/Seq2Logo-2.0/). Venn overlap graphs were produced using BioVenn (https://www.biovenn.nl/) and InteractiVenn (http://www.interactivenn.net/). UpSet plots were created using the UpSetR Shiny App (https://gehlenborglab.shinyapps.io/upsetr/). Microsoft PowerPoint and BioRender (https://www.biorender.com) were used to design and produce schematic diagrams and experimental processes.

4.5 Case study 5: Long-read RNA-sequencing analysis tools using in silico mixtures

This study done by Dong et al. (2022) in which ONT cDNA libraries were constructed with SQK-PCS109 cDNA-PCR sequencing and SQK-PBK004 PCR Barcoding kits using the supplied protocol. A pooled barcoded libraries were sequenced on five PromethION R9 flow cells and sequenced using the ONT PromethION platform (https://nanoporetech.com/products/promethion). Guppy module which is available to ONT customers via the community site (https://community.nanoporetech.com/), was used to base-call the fast5 files. The NEBNext Ultra II Directional RNA Library Prep Kit (https://www.neb.com/) was used to create the Illumina mRNA libraries. bcl2fastq (Illumina) was used to produce and demultiplex the fastq files.

4.5.1 Long-read isoform detection

A six different tools to perform isoform detection and quantification of the long-read data: Bambu (https://github.com/GoekeLab/bambu), Cupcake (https://github.com/Magdoll/cDNA_Cupcake), FLAIR (https://github.com/BrooksLabUCSC/flair), FLAMES (https://github.com/LuyiTian/FLAMES), StringTie2 (https://github.com/skovaka/stringtie2), and TALON (https://github.com/mortazavilab/TALON).

On a sorted BAM file, the default parameters of Cupcake's script to collapse redundant isoforms using a reference genome were applied to produce a collapsed GFF file containing the unique transcripts in our ONT sequences. Before running TALON, TranscriptClean (https://github.com/mortazavilab/TranscriptClean) ran using default parameters to perform reference-based error correction. To compare the isoforms detected by all six tools, the R/Bioconductor package Repitools (https://www.bioconductor.org/packages/release/bioc/html/Repitools.html) was used to first create two-kilobase bins along the genome, and Isoform search by using IRanges (https://bioconductor.org/packages/release/bioc/html/IRanges.html) To calculate short-read splicing junction coverage and classify transcripts found by each tool into structural categories, the SQANTI3 (https://github.com/ConesaLab/SQANTI3) quality control protocol was additionally run with parameter—short reads specifying Illumina FASTQ files. A down-sampled dataset containing 10% of reads was generated with seqtk (https://github.com/lh3/seqtk). The isoform detection tools were first run on this dataset with the same parameters and annotation GTF. Isoforms identified in the full and the down-sampled datasets were compared using GffCompare (https://github.com/gpertea/gffcompare) against the reference annotation file provided with the -r argument.

4.5.2 Transcript-level quantification of detected isoforms

The downstream analysis used transcript-level counts produced by bambu, FLAIR, FLAMES, and TALON. Cupcake and StringTie2 required some extra steps: transcript assemblies were created using annotation GTF files by Cufflinks (http://cole-trapnell-lab.github.io/cufflinks/), and Salmon (https://combine-lab.github.io/salmon/) was used to assign long reads to individual isoforms using its "mapping-based mode".

4.5.3 Read count processing

Transcripts were removed from both long- and short-read datasets if they had fewer than 10 counts in the long-read data. Counts were transformed into log2CPM values using the count-per-million (CPM) function from edgeR (https://bioconductor.org/packages/release/bioc/html/edgeR.html). Counts were transformed into log2TPM values (transcripts per million) using a prior count of 0.5 and transcript lengths generated by Salmon from the counting process.

4.5.4 Transcriptomic alignment

Using minimap2 (https://github.com/lh3/minimap2), the long Reads were mapped to the RNA sequin transcript sequences and the GENCODE Human Release 33 transcript sequences. Using Samtools (http://www.htslib.org/), secondary alignments were taken out of BAM files then, using Salmon (https://combine-lab.github.io/salmon/), "alignment-based mode" to achieve transcript-level quantification. Sort reads transcript-level quantification was carried out by Salmon in its "mapping-based mode". For quality control purposes, reads were also mapped to the GENCODE Human Release 33 genome and RNA sequin decoy chromosome using STAR (https://github.com/alexdobin/STAR). Coverage fractions of transcripts by individual reads were calculated using the R/Bioconductor package GenomicAlignments (https://github.com/Bioconductor/GenomicAlignments). Gene body read coverage was calculated using the Python package RSeQC (https://github.com/MonashBioinformaticsPlatform/RSeQC). An annotated transcripts and gene biotypes using the R/Bioconductor package AnnotationHub (https://bioconductor.org/packages/release/bioc/html/AnnotationHub.html) to retrieve Ensembl-based human annotation. The function catchSalmon from R/Bioconductor package edgeR was used to read transcript-level counts and estimate transcriptome mapping ambiguity (overdispersion coefficient) for each transcript using the bootstrap samples generated by Salmon. The linear relationship of sequin transcript quantification with annotated abundance was calculated by fitting a linear model using the R function lm, and the 95% confidence intervals of R-squared were calculated using the function chi-squared from the R package confintr (https://github.com/mayer79/confintr). The estimated number of transcript counts was loaded into R using the R package tximport (https://github.com/mikelove/tximport) for differential transcript expression and differential transcript usage analysis.

4.5.5 Differential transcript expression and usage analysis

Five different tools used to perform differential transcript expression analysis and each tool to perform low-count gene filtering and data normalization and chose a representative pipeline from each tool for the DGE analysis as described in their vignettes:

- limma (https://kasperdanielhansen.github.io/genbioconductor/html/limma.html),
- edgeR (https://bioconductor.org/packages/release/bioc/html/edgeR.html),
- DESeq2 (https://lashlock.github.io/compbio/R_presentation.html),
- EBSeq (https://bioconductor.org/packages/release/bioc/html/EBSeq.html) and
- NOISeq (https://bioconductor.org/packages/release/bioc/html/NOISeq.html).

Similarly, differential transcript analysis, five different methods selected for differential transcript usage analysis:

- DRIMSeq (https://github.com/gosianow/DRIMSeq),
- DEXSeq (https://bioconductor.org/packages/release/bioc/html/DEXSeq.html),
- diffSplice function in limma (https://kasperdanielhansen.github.io/genbioconductor/html/limma.html),
- the diffSpliceDGE function in edgeR (https://bioconductor.org/packages/release/bioc/html/edgeR.html),
- satuRn (https://www.bioconductor.org/packages/release/bioc/html/satuRn.html).

DRIMSeq was designed for differential transcript usage (DTU) analysis with transcript-level counts and gives adjusted *P*-values at both the gene- and transcript level. DEXSeq, limma-diffSplice and edgeR-diffSplice were designed to take exon-level short-read counts to analyze differential exon usage.

5. The future of benchmarking challenges

Benchmarking Omics tools have become a standard practice and a vital part of the computational methods' development cycle. However, the current benchmarking standard still has significant room for improvement. One issue is that benchmarks can quickly become outdated, especially in rapidly evolving subfields like genomic data analysis. To establish a better benchmarking standard, it is crucial to continuously update and run benchmarks as new methods emerge, and as more effective metrics become available. In the computational biology community, making code available has been common, and data sharing mirrors this trend in the genomics community (Byrd et al., 2020; Deshpande et al., 2021). However, the current benchmarking standard often involves little more than a mere code dump—a minimal record of the procedures followed. Ideally, methodologists should have access not only to code and input data but also to complete software environments and modular workflow systems that facilitate running methods on datasets and evaluating results against a ground truth. In an ideal scenario, benchmark components such as datasets, methods, and evaluation metrics could be repurposed for different computational tasks. However, the lack of general interoperability between components is a significant obstacle. Currently, most benchmarks do not provide a mechanism to add new methods, datasets, or metrics, which hinders the ability to update and advance the state-of-the-art for developers, researchers, and users of these methods. Establishing mechanisms for incorporating new elements into benchmarks is an essential consideration for promoting progress in the field. Establishing and maintaining seamless and enduring extensibility or reproducibility can be challenging, especially for individual research groups, given the constantly evolving nature of software tool chains in computational biology. The absence of provenance tracking further underscores that intermediate results from benchmarks are not being directly reused. For instance, developers evaluating their methods within existing benchmarks often have access to only the bare minimum—usually reference datasets and some code. Consequently, they are expected to rerun all previous methods, essentially reconstructing the entire benchmark from scratch. This presents an opportunity to make intermediate results systematically available, utilizing provenance tracking to record method results on datasets and metric calculations comparing them to the ground truth. Method developers can then reexecute relevant new components of a workflow, enhancing the reusability and extensibility of benchmarks and increasing their visibility beyond a static and dated perspective. Software environments also pose a challenge, as onboarding new methods across diverse computing systems requires addressing different library requirements of each software tool, such as python versions or bioconductor releases. Despite the considerable time and effort required to set up comprehensive benchmarks, the common practice is to build new benchmarks from scratch instead of extending existing ones. This pattern exists even though the single-cell community has developed numerous methods for specific tasks. Establishing systems for the orchestration of neutral benchmarks could alleviate this burden, as many benchmarks share a similar structure involving datasets, methods, and metrics, which could be effectively managed by a general benchmarking orchestration system. A systematic approach also benefits method developers, as using established and vetted benchmarks to evaluate their methods offers considerable time savings, avoids some of the pitfalls of self-assessment (Buchka et al., 2021), and enhances transparency for end users.

To successfully organize benchmarks, the community must populate the system with high-quality tests of methods and implement strategies for constant reevaluation of their suitability, in a systematic and community-informed manner. Several questions remain, such as determining the responsibility for the cost of computation, defining the composition of high-quality method tests, and finding optimal ways to combine performance metrics for method ranking. Addressing these questions collectively will strengthen benchmarking practices and foster ongoing advancements in computational biology.

6. Conclusion

Systematic benchmarking of Omics computational tools plays a crucial role in enabling informed tool selection. By conducting rigorous evaluations and comparisons, benchmarking studies empower researchers to make evidence-based decisions when choosing tools for their specific research needs. The study helps researchers navigate through the plethora of available tools, understanding their strengths, limitations, and suitability for different analysis tasks. Systematic benchmarking ensures that researchers can maximize the effectiveness and efficiency of their Omics data analysis, leading to more reliable and insightful research outcomes.

7. Tools for omics data analysis

There are several tools, packages, software, and web servers available in public domains in which we collected the open-source programs and made a list that can help to designing in systematic benchmarking, see Tables 4.4 and 4.6.

References

Aghaeepour, N., Finak, G., Hoos, H., Mosmann, T. R., Brinkman, R., Gottardo, R., Scheuermann, R. H., Dougall, D., Khodabakhshi, A. H., Mah, P., Obermoser, G., Spidlen, J., Taylor, I., Wuensch, S. A., Bramson, J., Eaves, C., Weng, A. P., Fortuno, E. S., Ho, K., ... Vilar, J. M. G. (2013). Critical assessment of automated flow cytometry data analysis techniques. *Nature Methods.* ISSN: 15487105, *10*(3), 228–238. https://doi.org/10.1038/NMETH.2365

Agrawal, S., Arze, C., Adkins, R. S., Crabtree, J., Riley, D., Vangala, M., Galens, K., Fraser, C. M., Tettelin, H., White, O., Angiuoli, S. V., Mahurkar, A., & Fricke, W. F. (2017). CloVR-Comparative: Automated, cloud-enabled comparative microbial genome sequence analysis pipeline. *BMC Genomics.* ISSN: 14712164, *18*(1). https://doi.org/10.1186/s12864-017-3717-3. http://www.biomedcentral.com/bmcgenomics

Alberti, C., Daniels, N., Hernaez, M., Voges, J., Goldfeder, R. L., Hernandez-Lopez, A. A., Mattavelli, M., & Berger, B. (2016). An evaluation framework for lossy compression of genome sequencing quality values. *Data Compression Conference Proceedings*, 221–230. https://doi.org/10.1109/DCC.2016.39, 9781509018536 http://ieeexplore.ieee.org/xpl/conhome.jsp?punumber=1000177.

Altenhoff, A. M., Boeckmann, B., Capella-Gutierrez, S., Dalquen, D. A., DeLuca, T., Forslund, K., Huerta-Cepas, J., Linard, B., Pereira, C., Pryszcz, L. P., Schreiber, F., Da Silva, A. S., Szklarczyk, D., Train, C. M., Bork, P., Lecompte, O., Von Mering, C., Xenarios, I., Sjölander, K., ... Dessimoz, C. (2016). Standardized benchmarking in the quest for orthologs. *Nature Methods.* ISSN: 15487105, *13*(5), 425–430. https://doi.org/10.1038/nmeth.3830. http://www.nature.com/nmeth/

Avila Cobos, F., Alquicira-Hernandez, J., Powell, J. E., Mestdagh, P., & De Preter, K. (2020). Benchmarking of cell type deconvolution pipelines for transcriptomics data. *Nature Communications.* ISSN: 20411723, *11*(1). https://doi.org/10.1038/s41467-020-19015-1. http://www.nature.com/ncomms/index.html

Awan, M. G., Awan, A. G., & Saeed, F. (2021). Benchmarking mass spectrometry based proteomics algorithms using a simulated database. *Network Modeling Analysis in Health Informatics and Bioinformatics.* ISSN: 21926670, *10*(1). https://doi.org/10.1007/s13721-021-00298-3. https://www.springer.com/journal/13721

Awan, M. G., & Saeed, F. (2018). MaSS-simulator: A highly configurable simulator for generating MS/MS datasets for benchmarking of proteomics algorithms. *Proteomics*. ISSN: 16159861, *18*(20). https://doi.org/10.1002/pmic.201800206. http://onlinelibrary.wiley.com/journal/10.1002/(ISSN)1615-9861

Baik, B., Yoon, S., & Nam, D. (2020). Benchmarking RNA-seq differential expression analysis methods using spike-in and simulation data. *PLoS One*. ISSN: 19326203, *15*(4). https://doi.org/10.1371/journal.pone.0232271. https://journals.plos.org/plosone/article/file?id=10.1371/journal.pone.0232271&type=printable

Barbitoff, Y. A., Abasov, R., Tvorogova, V. E., Glotov, A. S., & Predeus, A. V. (2022). Systematic benchmark of state-of-the-art variant calling pipelines identifies major factors affecting accuracy of coding sequence variant discovery. *BMC Genomics*. ISSN: 14712164, *23*(1). https://doi.org/10.1186/s12864-022-08365-3. http://www.biomedcentral.com/bmcgenomics

Baruzzo, G., Hayer, K. E., Kim, E. J., DI Camillo, B., Fitzgerald, G. A., & Grant, G. R. (2017). Simulation-based comprehensive benchmarking of RNA-seq aligners. *Nature Methods*. ISSN: 15487105, *14*(2), 135−139. https://doi.org/10.1038/nmeth.4106. http://www.nature.com/nmeth/

Berger, B., Peng, J., & Singh, M. (2013). Computational solutions for omics data. *Nature Reviews Genetics*. ISSN: 14710064, *14*(5), 333−346. https://doi.org/10.1038/nrg3433

Bergstrand, L. H., Neufeld, J. D., & Doxey, A. C. (2019). Pygenprop: A Python library for programmatic exploration and comparison of organism genome properties. *Bioinformatics*. ISSN: 14602059, *35*(23), 5063−5065. https://doi.org/10.1093/bioinformatics/btz522. http://bioinformatics.oxfordjournals.org/

Beyer, D., Löwe, S., & Wendler, P. (2019). Reliable benchmarking: Requirements and solutions. *International Journal on Software Tools for Technology Transfer*. ISSN: 14332787, *21*(1), 1−29. https://doi.org/10.1007/s10009-017-0469-y. http://springerlink.metapress.com/app/home/journal.asp?wasp=e2ggqmluwmc226vrfl1y&referrer=parent&backto=linkingpublicationresults,1:101563,1

Boel, A., Steyaert, W., De Rocker, N., Menten, B., Callewaert, B., De Paepe, A., Coucke, P., & Willaert, A. (2016). BATCH-GE: Batch analysis of Next-Generation Sequencing data for genome editing assessment. *Scientific Reports*. ISSN: 20452322, *6*. https://doi.org/10.1038/srep30330. http://www.nature.com/srep/index.html

Bohnert, Regina, Vivas, Sonia, Jansen, Gunther, & Galli, Alvaro (2017). Comprehensive benchmarking of SNV callers for highly admixed tumor data. *PLoS One, 12*(10), e0186175. https://doi.org/10.1371/journal.pone.0186175, 1932-6203.

Bongaerts, M., Kulkarni, P., Zammit, A., Bonte, R., Kluijtmans, L. A. J., Blom, H. J., Engelke, U. F. H., Tax, D. M. J., Ruijter, G. J. G., & Reinders, M. J. T. (2023). Benchmarking outlier detection methods for detecting IEM patients in untargeted metabolomics data. *Metabolites*. ISSN: 22181989, *13*(1). https://doi.org/10.3390/metabo13010097. http://www.mdpi.com/journal/metabolites

Bradnam, K. R., Fass, J. N., Alexandrov, A., Baranay, P., Bechner, M., Birol, I., Boisvert, S., Chapman, J. A., Chapuis, G., Chikhi, R., Chitsaz, H., Chou, W. C., Corbeil, J., Fabbro, C. D., Docking, R. R., Durbin, R., Earl, D., Emrich, S., Fedotov, P., … Korf, I. F. (2013). Assemblathon 2: Evaluating de novo methods of genome assembly in three vertebrate species. *GigaScience*. ISSN: 2047217X, *2*(1). https://doi.org/10.1186/2047-217X-2-10. http://www.gigasciencejournal.com/

Buchka, S., Hapfelmeier, A., Gardner, P. P., Wilson, R., & Boulesteix, A. L. (2021). On the optimistic performance evaluation of newly introduced bioinformatic methods. *Genome Biology*. ISSN: 1474760X, *22*(1). https://doi.org/10.1186/s13059-021-02365-4. http://genomebiology.com/

Byrd, J. B., Greene, A. C., Prasad, D. V., Jiang, X., & Greene, C. S. (2020). Responsible, practical genomic data sharing that accelerates research. *Nature Reviews Genetics*. ISSN: 14710064, *21*(10), 615−629. https://doi.org/10.1038/s41576-020-0257-5. http://www.nature.com/reviews/genetics

Canto, A. M., Godoi, A. B., Matos, A. H. B., Geraldis, J. C., Rogerio, F., Alvim, M. K. M., Yasuda, C. L., Ghizoni, E., Tedeschi, H., Veiga, D. F. T., Henning, B., Souza, W., Rocha, C. S., Vieira, A. S., Dias, E. V., Carvalho, B. S., Gilioli, R., Arul, A. B., Robinson, R. A. S., Cendes, F., & Lopes-Cendes, I. (2022). Benchmarking the proteomic profile of animal models of mesial temporal epilepsy. *Annals of Clinical and Translational Neurology*. ISSN: 23289503, *9*(4), 454−467. https://doi.org/10.1002/acn3.51533. http://onlinelibrary.wiley.com/journal/10.1002/(ISSN)2328-9503

Cao, Y., Yang, P., & Yang, J. Y. H. (2021). A benchmark study of simulation methods for single-cell RNA sequencing data. *Nature Communications*. ISSN: 20411723, *12*(1). https://doi.org/10.1038/s41467-021-27130-w. http://www.nature.com/ncomms/index.html

Chaisson, M. J. P., Wilson, R. K., & Eichler, E. E. (2015). Genetic variation and the de novo assembly of human genomes. *Nature Reviews Genetics*. ISSN: 14710064, *16*(11), 627−640. https://doi.org/10.1038/nrg3933. http://www.nature.com/reviews/genetics

Chang, C., Li, M., Guo, C., Ding, Y., Xu, K., Han, M., He, F., & Zhu, Y. (2019). Panda: A comprehensive and flexible tool for quantitative proteomics data analysis. *Bioinformatics*. ISSN: 14602059, *35*(5), 898−900. https://doi.org/10.1093/bioinformatics/bty727. http://bioinformatics.oxfordjournals.org/

Cheng, A., Hu, G., & Li, W. V. (2023). Benchmarking cell-type clustering methods for spatially resolved transcriptomics data. *Briefings in Bioinformatics*. ISSN: 14774054, *24*(1). https://doi.org/10.1093/bib/bbac475. http://bib.oxfordjournals.org

Cole, M. B., Risso, D., Wagner, A., DeTomaso, D., Ngai, J., Purdom, E., Dudoit, S., & Yosef, N. (2019). Performance assessment and selection of normalization procedures for single-cell RNA-seq. *Cell Systems*. ISSN: 24054720, *8*(4), 315−328.e8. https://doi.org/10.1016/j.cels.2019.03.010. http://www.journals.elsevier.com/cell-systems/

Conesa, A., Madrigal, P., Tarazona, S., Gomez-Cabrero, D., Cervera, A., McPherson, A., Szcześniak, M. W., Gaffney, D. J., Elo, L. L., Zhang, X., & Mortazavi, A. (2016). A survey of best practices for RNA-seq data analysis. *Genome Biology*. ISSN: 1474760X, *17*(1). https://doi.org/10.1186/s13059-016-0881-8. http://genomebiology.com/

Corchete, L. A., Rojas, E. A., Alonso-López, D., De Las Rivas, J., Gutiérrez, N. C., & Burguillo, F. J. (2020). Systematic comparison and assessment of RNA-seq procedures for gene expression quantitative analysis. *Scientific Reports*. ISSN: 20452322, *10*(1). https://doi.org/10.1038/s41598-020-76881-x. http://www.nature.com/srep/index.html

Costello, J. C., Heiser, L. M., Georgii, E., Gönen, M., Menden, M. P., Wang, N. J., Bansal, M., Ammad-Ud-Din, M., Hintsanen, P., Khan, S. A., Mpindi, J. P., Kallioniemi, O., Honkela, A., Aittokallio, T., Wennerberg, K., Collins, J. J., Gallahan, D., Singer, D., Saez-Rodriguez, J., … Van Westen, G. J. P. (2014). A community effort to assess and improve drug sensitivity prediction algorithms. *Nature Biotechnology*. ISSN: 15461696, *32*(12), 1202–1212. https://doi.org/10.1038/nbt.2877. http://www.nature.com/nbt/index.html

Cumsille, A., Durán, R. E., Rodríguez-Delherbe, A., Saona-Urmeneta, V., Cámara, B., Seeger, M., Araya, M., Jara, N., Buil-Aranda, C., & GenoVi. (2023). An open-source automated circular genome visualizer for bacteria and archaea. *PLoS Computational Biology, 19*(4). https://doi.org/10.1371/journal.pcbi.1010998. PMID: 37014908; PMCID: PMC10104344.

Dai, X., & Shen, L. (2022). Advances and trends in omics Technology development. *Frontiers of Medicine*. ISSN: 2296858X, *9*. https://doi.org/10.3389/fmed.2022.911861. http://journal.frontiersin.org/journal/medicine

Danecek, P., Bonfield, J. K., Liddle, J., Marshall, J., Ohan, V., Pollard, M. O., Whitwham, A., Keane, T., McCarthy, S. A., & Davies, R. M. (2021). Twelve years of SAMtools and BCFtools. *GigaScience*. ISSN: 2047217X, *10*(2). https://doi.org/10.1093/gigascience/giab008. https://academic.oup.com/gigascience/issue/8/7

Deshpande, Sarkar, A., Guo, Moore, A., Darci-Maher, N., & Mangul, S. (2021). *A comprehensive analysis of code and data availability in biomedical research*. https://doi.org/10.31219/osf.io/uz7m5

Dieckmann, M. A., Beyvers, S., Nkouamedjo-Fankep, R. C., Hanel, P. H. G., Jelonek, L., Blom, J., & Goesmann, A. E. D. G. A. R.3 (2021). 0: Comparative genomics and phylogenomics on a scalable infrastructure. *Nucleic Acids Research, 49*(W1). https://doi.org/10.1093/nar/gkab341. PMID: 33988716; PMCID: PMC8262741.

Dimitsaki, S., Gavriilidis, G. I., Dimitriadis, V. K., & Natsiavas, P. (2023). Benchmarking of Machine Learning classifiers on plasma proteomic for COVID-19 severity prediction through interpretable artificial intelligence. *Artificial Intelligence in Medicine*. ISSN: 18732860, *137*. https://doi.org/10.1016/j.artmed.2023.102490. http://www.elsevier.com/locate/artmed

Doherty, H. M., Kritikos, G., Galardini, M., Banzhaf, M., & Moradigaravand, D. (2023). ChemGAPP: A tool for chemical genomics analysis and phenotypic profiling. *Bioinformatics*. ISSN: 13674811, *39*(4). https://doi.org/10.1093/bioinformatics/btad171. http://bioinformatics.oxfordjournals.org/

Dong, X., Du, M. R. M., Gouil, Q., Tian, L., Jabbari, J. S., Bowden, R., Baldoni, P. L., Chen, Y., Smyth, G. K., Amarasinghe, S. L., Law, C. W., & Ritchie, M. E. (2022). Benchmarking long-read RNA-sequencing analysis tools using in silico mixtures. *bioRxiv*. ISSN: 26928205 https://doi.org/10.1101/2022.07.22.501076. https://www.biorxiv.org

Donnelly, D. P., Rawlins, C. M., DeHart, C. J., Fornelli, L., Schachner, L. F., Lin, Z., Lippens, J. L., Aluri, K. C., Sarin, R., Chen, B., Lantz, C., Jung, W., Johnson, K. R., Koller, A., Wolff, J. J., Campuzano, I. D. G., Auclair, J. R., Ivanov, A. R., Whitelegge, J. P., … Agar, J. N. (2019). Best practices and benchmarks for intact protein analysis for top-down mass spectrometry. *Nature Methods*. ISSN: 15487105, *16*(7), 587–594. https://doi.org/10.1038/s41592-019-0457-0. http://www.nature.com/nmeth/

Dowell, J. A., Wright, L. J., Armstrong, E. A., & Denu, J. M. (2021). Benchmarking quantitative performance in label-free proteomics. *ACS Omega*. ISSN: 24701343, *6*(4), 2494–2504. https://doi.org/10.1021/acsomega.0c04030. http://pubs.acs.org/journal/acsodf

Duan, Ran, Gao, Lin, Gao, Yong, Hu, Yuxuan, Xu, Han, Huang, Mingfeng, Song, Kuo, Wang, Hongda, Dong, Yongqiang, Jiang, Chaoqun, Zhang, Chenxing, Jia, Songwei, & Wang, Edwin (2021). Evaluation and comparison of multi-omics data integration methods for cancer subtyping. *PLoS Computational Biology, 17*(8), e1009224. https://doi.org/10.1371/journal.pcbi.1009224, 1553-7358.

Ettorchi-Tardy, A., Levif, M., & Michel, P. (2012). Benchmarking: A method for continuous quality improvement in health. *Healthcare Policy*. ISSN: 17156572, *7*(4), e101–e119. http://www.longwoods.com/product/download/code/22872.

Ewing, A. D., Houlahan, K. E., Hu, Y., Ellrott, K., Caloian, C., Yamaguchi, T. N., Bare, J. C., P'Ng, C., Waggott, D., Sabelnykova, V. Y., Kellen, M. R., Norman, T. C., Haussler, D., Friend, S. H., Stolovitzky, G., Margolin, A. A., Stuart, J. M., & Boutros, P. C. (2015). Combining tumor genome simulation with crowdsourcing to benchmark somatic single-nucleotide-variant detection. *Nature Methods*. ISSN: 15487105, *12*(7), 623–630. https://doi.org/10.1038/nmeth.3407. http://www.nature.com/nmeth/

Expósito, González-Domínguez, & Hsra, T. J. (2018). Hadoop-based spliced read aligner for RNA sequencing data. *PLoS One, 13*(7). https://doi.org/10.1371/journal.pone.0201483. PMID: 30063721; PMCID: PMC6067734.

Flores, J. E., Claborne, D. M., Weller, Z. D., Webb-Robertson, Waters, K. M., & Bramer, L. M. (2023). Missing data in multi-omics integration: Recent advances through artificial intelligence. *Front Artif Intell, 6*. https://doi.org/10.3389/frai.2023.1098308. PMID: 36844425; PMCID: PMC9949722.

Friedberg, I. (2006). Automated protein function prediction - the genomic challenge. *Briefings in Bioinformatics*. ISSN: 14774054, *7*(3), 225–242. https://doi.org/10.1093/bib/bbl004

Fröhlich, K., Brombacher, E., Fahrner, M., Vogele, D., Kook, L., Pinter, N., Bronsert, P., Timme-Bronsert, S., Schmidt, A., Bärenfaller, K., Kreutz, C., & Schilling, O. (2022). Benchmarking of analysis strategies for data-independent acquisition proteomics using a large-scale dataset comprising inter-patient heterogeneity. *Nature Communications*. ISSN: 20411723, *13*(1). https://doi.org/10.1038/s41467-022-30094-0. http://www.nature.com/ncomms/index.html

Gao, S., Bertrand, D., Chia, B. K. H., & Nagarajan, N. (2016). OPERA-LG: Efficient and exact scaffolding of large, repeat-rich eukaryotic genomes with performance guarantees. *Genome Biology*. ISSN: 1474760X, *17*(1). https://doi.org/10.1186/s13059-016-0951-y. http://genomebiology.com/

Garcia-Heredia, I., Bhattacharjee, A. S., Fornas, O., Gomez, M. L., Martínez, J. M., & Martinez-Garcia, M. (2021). Benchmarking of single-virus genomics: A new tool for uncovering the virosphere. *Environmental Microbiology*. ISSN: 14622920, *23*(3), 1584–1593. https://doi.org/10.1111/1462-2920.15375. http://onlinelibrary.wiley.com/journal/10.1111/(ISSN)1462-2920

Gardner, P. P., Paterson, J. M., McGimpsey, S., Ashari-Ghomi, F., Umu, S. U., Pawlik, A., Gavryushkin, A., & Black, M. A. (2022). Sustained software development, not number of citations or journal choice, is indicative of accurate bioinformatic software. *Genome Biology*. ISSN: 1474760X, *23*(1). https://doi.org/10.1186/s13059-022-02625-x. http://genomebiology.com/

Gatto, L., Aebersold, R., Cox, J., Demichev, V., Derks, J., Emmott, E., Franks, A. M., Ivanov, A. R., Kelly, R. T., Khoury, L., Leduc, A., MacCoss, M. J., Nemes, P., Perlman, D. H., Petelski, A. A., Rose, C. M., Schoof, E. M., Van Eyk, J., Vanderaa, C., Yates, J. R., & Slavov, N. (2023). Initial recommendations for performing, benchmarking and reporting single-cell proteomics experiments. *Nature Methods*. ISSN: 15487105, *20*(3), 375–386. https://doi.org/10.1038/s41592-023-01785-3. https://www.nature.com/nmeth/

Gavrielatos, M., Kyriakidis, K., Spandidos, D. A., & Michalopoulos, I. (2021). Benchmarking of next and third generation sequencing technologies and their associated algorithms for de novo genome assembly. *Molecular Medicine Reports*. ISSN: 17913004, *23*(4). https://doi.org/10.3892/mmr.2021.11890. http://spandidos-publications.com//10.3892/mmr.2021.11890

Hackl, H., Charoentong, P., Finotello, F., & Trajanoski, Z. (2016). Computational genomics tools for dissecting tumour-immune cell interactions. *Nature Reviews Genetics*. ISSN: 14710064, *17*(8), 441–458. https://doi.org/10.1038/nrg.2016.67. http://www.nature.com/reviews/genetics

Hoffmann, Frank, Bertram, Torsten, Mikut, Ralf, Reischl, Markus, & Nelles, Oliver (2019). Benchmarking in classification and regression. *WIREs Data Mining and Knowledge Discovery, 9*(5). https://doi.org/10.1002/widm.1318, 1942-4787.

Hoopmann, M. R., Winget, J. M., Mendoza, L., & Moritz, R. L. (2018). StPeter: Seamless label-free quantification with the trans-proteomic pipeline. *Journal of Proteome Research*. ISSN: 15353907, *17*(3), 1314–1320. https://doi.org/10.1021/acs.jproteome.7b00786. http://pubs.acs.org/journal/jprobs

Huttenhower, C., Hibbs, M. A., Myers, C. L., Caudy, A. A., Hess, D. C., & Troyanskaya, O. G. (2009). The impact of incomplete knowledge on evaluation: An experimental benchmark for protein function prediction. *Bioinformatics*. ISSN: 14602059, *25*(18), 2404–2410. https://doi.org/10.1093/bioinformatics/btp397

Ivanov, M. V., Levitsky, L. I., & Gorshkov, M. V. (2016). Adaptation of decoy fusion strategy for existing multi-stage search workflows. *Journal of the American Society for Mass Spectrometry*. ISSN: 18791123, *27*(9), 1579–1582. https://doi.org/10.1007/s13361-016-1436-7. http://www.springerlink.com/content/1044-0305/

Jain, M., Koren, S., Miga, K. H., Quick, J., Rand, A. C., Sasani, T. A., Tyson, J. R., Beggs, A. D., Dilthey, A. T., Fiddes, I. T., Malla, S., Marriott, H., Nieto, T., O'Grady, J., Olsen, H. E., Pedersen, B. S., Rhie, A., Richardson, H., Quinlan, A. R., … Loose, M. (2018). Nanopore sequencing and assembly of a human genome with ultra-long reads. *Nature Biotechnology*. ISSN: 15461696, *36*(4), 338–345. https://doi.org/10.1038/nbt.4060. http://www.nature.com/nbt/index.html

Javkar, Kiran, Rand, Hugh, Strain, Errol, Pop, Mihai, & Marschall, Tobias (2023). Prawns: Compact pan-genomic features for whole-genome population genomics. *Bioinformatics, 39*(1). https://doi.org/10.1093/bioinformatics/btac844, 1367-4811.

Jiang, Y., Oron, T. R., Clark, W. T., Bankapur, A. R., D'Andrea, D., Lepore, R., Funk, C. S., Kahanda, I., Verspoor, K. M., Ben-Hur, A., Koo, D. C. E., Penfold-Brown, D., Shasha, D., Youngs, N., Bonneau, R., Lin, A., Sahraeian, S. M. E., Martelli, P. L., Profiti, G., … Radivojac, P. (2016). An expanded evaluation of protein function prediction methods shows an improvement in accuracy. *Genome Biology*. ISSN: 1474760X, *17*(1). https://doi.org/10.1186/s13059-016-1037-6. http://genomebiology.com/

Jin, H., & Liu, Z. (2021). A benchmark for RNA-seq deconvolution analysis under dynamic testing environments. *Genome Biology*. ISSN: 1474760X, *22*(1). https://doi.org/10.1186/s13059-021-02290-6. http://genomebiology.com/

Johnson, B. K., Scholz, M. B., Teal, T. K., & Abramovitch, R. B. (2016). Sparta: Simple program for automated reference-based bacterial RNA-seq transcriptome analysis. *BMC Bioinformatics*. ISSN: 14712105, *17*(1). https://doi.org/10.1186/s12859-016-0923-y. http://www.biomedcentral.com/bmcbioinformatics/

de Jonge, N. F., Mildau, K., Meijer, D., Louwen, J. J. R., Bueschl, C., Huber, F., & van der Hooft, J. J. J. (2022). Good practices and recommendations for using and benchmarking computational metabolomics metabolite annotation tools. *Metabolomics*. ISSN: 15733890, *18*(12). https://doi.org/10.1007/s11306-022-01963-y. https://www.springer.com/journal/11306

Kent, W. J., & Haussler, D. (2001). Assembly of the working draft of the human genome with GigAssembler. *Genome Research*. ISSN: 10889051, *11*(9), 1541–1548. https://doi.org/10.1101/gr.183201

Kong, S., Gong, P., Zeng, W. F., Jiang, B., Hou, X., Zhang, Y., Zhao, H., Liu, M., Yan, G., Zhou, X., Qiao, X., Wu, M., Yang, P., Liu, C., & Cao, W. (2022). pGlycoQuant with a deep residual network for quantitative glycoproteomics at intact glycopeptide level. *Nature Communications*. ISSN: 20411723, *13*(1). https://doi.org/10.1038/s41467-022-35172-x. https://www.nature.com/ncomms/

Koopmans, F., Li, K. W., Klaassen, R. V., & Smit, A. B. (2023). MS-DAP platform for downstream data analysis of label-free proteomics uncovers optimal workflows in benchmark data sets and increased sensitivity in analysis of Alzheimer's biomarker data. *Journal of Proteome Research*. ISSN: 15353907, *22*(2), 374–386. https://doi.org/10.1021/acs.jproteome.2c00513. http://pubs.acs.org/journal/jprobs

Krassowski, M., Das, V., Sahu, S. K., & Misra, B. B. (2020). State of the field in multi-omics research: From computational needs to data mining and sharing. *Frontiers in Genetics*. ISSN: 16648021, *11*. https://doi.org/10.3389/fgene.2020.610798. https://www.frontiersin.org/journals/genetics#

Krusche, P., Trigg, L., Boutros, P. C., Mason, C. E., De La Vega, F. M., Moore, B. L., Gonzalez-Porta, M., Eberle, M. A., Tezak, Z., Lababidi, S., Truty, R., Asimenos, G., Funke, B., Fleharty, M., Chapman, B. A., Salit, M., & Zook, J. M. (2019). Best practices for benchmarking germline small-variant calls in human genomes. *Nature Biotechnology*. ISSN: 15461696, *37*(5), 555–560. https://doi.org/10.1038/s41587-019-0054-x. http://www.nature.com/nbt/index.html

L'Yi, S., Wang, Q., Lekschas, F., & Gehlenborg, N. (2022). Gosling: A grammar-based Toolkit for scalable and interactive genomics data visualization. *IEEE Transactions on Visualization and Computer Graphics*. ISSN: 19410506, *28*(1), 140–150. https://doi.org/10.1109/TVCG.2021.3114876. http://ieeexplore.ieee.org/xpl/RecentIssue.jsp?punumber=2945

Larralde, M., & Zeller, G. (2023). PyHMMER: A Python library binding to HMMER for efficient sequence analysis. *Bioinformatics*. ISSN: 13674811, *39*(5). https://doi.org/10.1093/bioinformatics/btad214. http://bioinformatics.oxfordjournals.org/

Li, B., & Dewey, C. N. (2011). Rsem: Accurate transcript quantification from RNA-Seq data with or without a reference genome. *BMC Bioinformatics*. ISSN: 14712105, *12*. https://doi.org/10.1186/1471-2105-12-323. http://www.biomedcentral.com/1471-2105/12/323

Li, H., Zhou, J., Li, Z., Chen, S., Liao, X., Zhang, B., Zhang, R., Wang, Y., Sun, S., & Gao, X. (2023). A comprehensive benchmarking with practical guidelines for cellular deconvolution of spatial transcriptomics. *Nature Communications*. ISSN: 20411723, *14*(1). https://doi.org/10.1038/s41467-023-37168-7. https://www.nature.com/ncomms/

Lin, Q.t., Yang, W., Zhang, X., Li, Q.g., Liu, Y.f., Yan, Q., & Sun, L. (2023). Systematic and benchmarking studies of pipelines for mammal WGBS data in the novel NGS platform. *BMC Bioinformatics*. ISSN: 14712105, *24*(1). https://doi.org/10.1186/s12859-023-05163-w. https://bmcbioinformatics.biomedcentral.com/

Lin, Y., Yuan, J., Kolmogorov, M., Shen, M. W., Chaisson, M., & Pevzner, P. A. (2016). Assembly of long error-prone reads using de Bruijn graphs. *Proceedings of the National Academy of Sciences of the United States of America*. ISSN: 10916490, *113*(52), E8396–E8405. https://doi.org/10.1073/pnas.1604560113. http://www.pnas.org/content/113/52/E8396.full.pdf

Lindgreen, S., Adair, K. L., & Gardner, P. P. (2016). An evaluation of the accuracy and speed of metagenome analysis tools. *Scientific Reports*. ISSN: 20452322, *6*. https://doi.org/10.1038/srep19233. http://www.nature.com/srep/index.html

Lou, R., Cao, Y., Li, S., Lang, X., Li, Y., Zhang, Y., & Shui, W. (2023). Benchmarking commonly used software suites and analysis workflows for DIA proteomics and phosphoproteomics. *Nature Communications*. ISSN: 20411723, *14*(1). https://doi.org/10.1038/s41467-022-35740-1. https://www.nature.com/ncomms/

Łabaj, P. P., & Kreil, D. P. (2016). Sensitivity, specificity, and reproducibility of RNA-Seq differential expression calls. *Biology Direct*. ISSN: 17456150, *11*(1). https://doi.org/10.1186/s13062-016-0169-7. http://www.biology-direct.com/

Łabaj, P. P., Leparc, G. G., Linggi, B. E., Markillie, L. M., Wiley, H. S., & Kreil, D. P. (2011). Characterization and improvement of RNA-seq precision in quantitative transcript expression profiling. *Bioinformatics*. ISSN: 14602059, *27*(13), i383–i391. https://doi.org/10.1093/bioinformatics/btr247

Luecken, M. D., Büttner, M., Chaichoompu, K., Danese, A., Interlandi, M., Mueller, M. F., Strobl, D. C., Zappia, L., Dugas, M., Colomé-Tatché, M., & Theis, F. J. (2022). Benchmarking atlas-level data integration in single-cell genomics. *Nature Methods*. ISSN: 15487105, *19*(1), 41–50. https://doi.org/10.1038/s41592-021-01336-8. http://www.nature.com/nmeth/

Mangul, S., Martin, L. S., Hill, B. L., Lam, A. K. M., Distler, M. G., Zelikovsky, A., Eskin, E., & Flint, J. (2019). Systematic benchmarking of omics computational tools. *Nature Communications*. ISSN: 20411723, *10*(1). https://doi.org/10.1038/s41467-019-09406-4. http://www.nature.com/ncomms/index.html

Martin, F. J., Amode, M. R., & Aneja, a (2023). *D1. PMCIDNucleic acids res*. https://doi.org/10.1093/nar/gkac958. PMID: 36318249.

McIntyre, A. B. R., Ounit, R., Afshinnekoo, E., Prill, R. J., Hénaff, E., Alexander, N., Minot, S. S., Danko, D., Foox, J., Ahsanuddin, S., Tighe, S., Hasan, N. A., Subramanian, P., Moffat, K., Levy, S., Lonardi, S., Greenfield, N., Colwell, R. R., Rosen, G. L., & Mason, C. E. (2017). Comprehensive benchmarking and ensemble approaches for metagenomic classifiers. *Genome Biology*. ISSN: 1474760X, *18*(1). https://doi.org/10.1186/s13059-017-1299-7. http://genomebiology.com/

Mereu, E., Lafzi, A., Moutinho, C., Ziegenhain, C., McCarthy, D. J., Álvarez-Varela, A., Batlle, E., Sagar, Grün, D., Lau, J. K., Boutet, S. C., Sanada, C., Ooi, A., Jones, R. C., Kaihara, K., Brampton, C., Talaga, Y., Sasagawa, Y., Tanaka, K., ... Heyn, H. (2020). Benchmarking single-cell RNA-sequencing protocols for cell atlas projects. *Nature Biotechnology*. ISSN: 15461696, *38*(6), 747–755. https://doi.org/10.1038/s41587-020-0469-4. http://www.nature.com/nbt/index.html

Meyer, F., Fritz, A., Deng, Z. L., Koslicki, D., Lesker, T. R., Gurevich, A., Robertson, G., Alser, M., Antipov, D., Beghini, F., Bertrand, D., Brito, J. J., Brown, C. T., Buchmann, J., Buluç, A., Chen, B., Chikhi, R., Clausen, P. T. L. C., Cristian, A., ... McHardy, A. C. (2022). Critical assessment of metagenome interpretation: The second round of challenges. *Nature Methods*. ISSN: 15487105, *19*(4), 429–440. https://doi.org/10.1038/s41592-022-01431-4. http://www.nature.com/nmeth/

Mikheenko, A., Prjibelski, A., Saveliev, V., Antipov, D., & Gurevich, A. (2018). Versatile genome assembly evaluation with QUAST-LG. *Bioinformatics*. ISSN: 14602059, *34*(13), i142–i150. https://doi.org/10.1093/bioinformatics/bty266. http://bioinformatics.oxfordjournals.org/

Mikheenko, A., Saveliev, V., & Gurevich, A. (2016). MetaQUAST: Evaluation of metagenome assemblies. *Bioinformatics*. ISSN: 14602059, *32*(7), 1088–1090. https://doi.org/10.1093/bioinformatics/btv697. http://bioinformatics.oxfordjournals.org/

Mikheenko, A., Saveliev, V., Hirsch, P., & Gurevich, A. (2023). WebQUAST: Online evaluation of genome assemblies. *Nucleic Acids Research*. ISSN: 13624962, *51*(1), W601–W606. https://doi.org/10.1093/nar/gkad406

Minoche, A. E., Dohm, J. C., & Himmelbauer, H. (2011). Evaluation of genomic high-throughput sequencing data generated on Illumina HiSeq and Genome Analyzer systems. *Genome Biology*. ISSN: 1474760X, *12*(11). https://doi.org/10.1186/gb-2011-12-11-r112. http://genomebiology.com/2011/12/11/R112

Mitchell, C. J., Kim, M. S., Na, C. H., & Pandey, A. (2016). PyQuant: A versatile framework for analysis of quantitative mass spectrometry data. *Molecular and Cellular Proteomics*. ISSN: 15359484, *15*(8), 2829–2838. https://doi.org/10.1074/mcp.O115.056879. http://www.mcponline.org/content/15/8/2829.full.pdf

Nagarajan, N., & Pop, M. (2013). Sequence assembly demystified. *Nature Reviews Genetics*. ISSN: 14710064, *14*(3), 157–167. https://doi.org/10.1038/nrg3367

Nowak, R. M., Jastrzebski, J. P., Kuśmirek, W., Sałamatin, R., Rydzanicz, M., Sobczyk-Kopcioł, A., Sulima-Celińska, A., Paukszto, Ł., Makowczenko, K. G., Płoski, R., Tkach, V. V., Basałaj, K., & Młocicki, D. (2019). Hybrid de novo whole-genome assembly and annotation of the model tapeworm Hymenolepis diminuta. *Scientific Data*. ISSN: 20524463, *6*(1). https://doi.org/10.1038/s41597-019-0311-3. http://www.nature.com/sdata/

O'Connor, B. D., Merriman, B., & Nelson, S. F. (2010). SeqWare Query Engine: Storing and searching sequence data in the cloud. *BMC Bioinformatics*. ISSN: 14712105, *11*(12). https://doi.org/10.1186/1471-2105-11-S12-S2. http://www.biomedcentral.com/1471-2105/11/S12/S2

Page, Justin T., Liechty, Zachary S., Huynh, Mark D., & Udall, Joshua A. (2014). BamBam: Genome sequence analysis tools for biologists. *BMC Research Notes, 7*(1), 829. https://doi.org/10.1186/1756-0500-7-829, 1756-0500.

Pandey, A., L'Yi, S., Wang, Q., Borkin, M. A., & Gehlenborg, N. (2023). GenoREC: A recommendation system for interactive genomics data visualization. *IEEE Transactions on Visualization and Computer Graphics*. ISSN: 19410506, *29*(1), 570–580. https://doi.org/10.1109/TVCG.2022.3209407. http://ieeexplore.ieee.org/xpl/RecentIssue.jsp?punumber=2945

Patrick, K. L. (2007). 454 Life Sciences: Illuminating the future of genome sequencing and personalized medicine. *Yale Journal of Biology and Medicine*. ISSN: 00440086, *80*(4), 191–194. http://www.pubmedcentral.nih.gov/picrender.fcgi?artid=2347365&blobtype=pdf.

Pei, S., Liu, T., Ren, X., Li, W., Chen, C., & Xie, Z. (2021). Benchmarking variant callers in next-generation and third-generation sequencing analysis. *Briefings in Bioinformatics*. ISSN: 14774054, *22*(3). https://doi.org/10.1093/bib/bbaa148. http://bib.oxfordjournals.org

Puterová, J., & Martínek, T. (2021). digIS: towards detecting distant and putative novel insertion sequence elements in prokaryotic genomes. *BMC Bioinformatics*. ISSN: 14712105, *22*(1). https://doi.org/10.1186/s12859-021-04177-6. http://www.biomedcentral.com/bmcbioinformatics/

Putri, G. H., Anders, S., Pyl, P. T., Pimanda, J. E., & Zanini, F. (2022). Analysing high-throughput sequencing data in Python with HTSeq 2.0. *Bioinformatics*. ISSN: 14602059, *38*(10), 2943–2945. https://doi.org/10.1093/bioinformatics/btac166. http://bioinformatics.oxfordjournals.org/

Quinn, T. P., Crowley, T. M., & Richardson, M. F. (2018). Benchmarking differential expression analysis tools for RNA-Seq: Normalization-based vs. log-ratio transformation-based methods. *BMC Bioinformatics*. ISSN: 14712105, *19*(1). https://doi.org/10.1186/s12859-018-2261-8. http://www.biomedcentral.com/bmcbioinformatics/

Rajkumari, J., Katiyar, P., Dheeman, S., Pandey, P., & Maheshwari, D. K. (2022). The changing paradigm of rhizobial taxonomy and its systematic growth upto postgenomic technologies. *World Journal of Microbiology and Biotechnology*. ISSN: 15730972, *38*(11). https://doi.org/10.1007/s11274-022-03370-w. http://www.wkap.nl/journalhome.htm/0959-3993

Rampler, E., Hermann, G., Grabmann, G., El Abiead, Y., Schoeny, H., Baumgartinger, C., Köcher, T., & Koellensperger, G. (2021). Benchmarking non-targeted metabolomics using yeast- derived libraries. *Metabolites*. ISSN: 22181989, *11*(3). https://doi.org/10.3390/metabo11030160. https://www.mdpi.com/2218-1989/11/3/160/pdf

Sahadevan, Sudeep, Sekaran, Thileepan, Ashaf, Nadia, Fritz, Marko, Hentze, Matthias W., Huber, Wolfgang, Schwarzl, Thomas, & Alkan, Can (2023). htseq-clip: a toolset for the preprocessing of eCLIP/iCLIP datasets. *Bioinformatics, 39*(1). https://doi.org/10.1093/bioinformatics/btac747, 1367-4811.

Santana-Garcia, W., Castro-Mondragon, J. A., Padilla-Gálvez, M., Nguyen, N. T. T., Elizondo-Salas, A., Ksouri, N., Gerbes, F., Thieffry, D., Vincens, P., Contreras-Moreira, B., Van Helden, J., Thomas-Chollier, M., & Medina-Rivera, A. (2022). Rsat 2022: Regulatory sequence analysis tools. *Nucleic Acids Research*. ISSN: 13624962, *50*(1), W670–W676. https://doi.org/10.1093/nar/gkac312. https://academic.oup.com/nar/issue

Sánchez, B. J., Lahtvee, P. J., Campbell, K., Kasvandik, S., Yu, R., Domenzain, I., Zelezniak, A., & Nielsen, J. (2021). Benchmarking accuracy and precision of intensity-based absolute quantification of protein abundances in *Saccharomyces cerevisiae*. *Proteomics*. ISSN: 16159861, *21*(6). https://doi.org/10.1002/pmic.202000093. http://onlinelibrary.wiley.com/journal/10.1002/(ISSN)1615-9861

Sczyrba, A., Hofmann, P., Belmann, P., Koslicki, D., Janssen, S., Dröge, J., Gregor, I., Majda, S., Fiedler, J., Dahms, E., Bremges, A., Fritz, A., Garrido-Oter, R., Jørgensen, T. S., Shapiro, N., Blood, P. D., Gurevich, A., Bai, Y., Turaev, D., … McHardy, A. C. (2017). Critical assessment of metagenome interpretation - a benchmark of metagenomics software. *Nature Methods*. ISSN: 15487105, *14*(11), 1063–1071. https://doi.org/10.1038/nmeth.4458. http://www.nature.com/nmeth/

Shahbazy, M., Ramarathinam, S. H., Illing, P. T., Jappe, E. C., Faridi, P., Croft, N. P., & Purcell, A. W. (2023). Benchmarking bioinformatics pipelines in data-independent acquisition mass spectrometry for immunopeptidomics. *Molecular and Cellular Proteomics*. ISSN: 15359484, *22*(4). https://doi.org/10.1016/j.mcpro.2023.100515. https://www.mcponline.org/article/S1535-9476(23)00024-5/pdf

Singh, U., Li, J., Seetharam, A., & Wurtele, E. S. (2021). Pyrpipe: A Python package for RNA-seq workflows. *NAR Genomics and Bioinformatics*. ISSN: 26319268, *3*(2), 1–8. https://doi.org/10.1093/nargab/lqab049. https://academic.oup.com/nargab

Sohrab, V., López-Díaz, C., Di Pietro, A., Ma, L. J., & Ayhan, D. H. (2021). Tefinder: A bioinformatics pipeline for detecting new transposable element insertion events in next-generation sequencing data. *Genes*. ISSN: 20734425, *12*(2), 1–11. https://doi.org/10.3390/genes12020224. https://www.mdpi.com/2073-4425/12/2/224/pdf

Sun, Jin, Li, Runsheng, Chen, Chong, Sigwart, Julia D., & Kocot, Kevin M. (2021). Benchmarking Oxford Nanopore read assemblers for high-quality molluscan genomes. *Philosophical transactions of the Royal Society of London. Series B, Biological sciences, 376*(1825), 20200160. https://doi.org/10.1098/rstb.2020.0160

Tan, M. H., Austin, C. M., Hammer, M. P., Lee, Y. P., Croft, L. J., & Gan, H. M. (2018). Finding Nemo: Hybrid assembly with Oxford Nanopore and Illumina reads greatly improves the clownfish (*Amphiprion ocellaris*) genome assembly. *GigaScience*. ISSN: 2047217X, *7*(3), 1–6. https://doi.org/10.1093/gigascience/gix137. http://www.gigasciencejournal.com/

Thompson, Julie D., Linard, Benjamin, Lecompte, Odile, Poch, Olivier, & Badger, Jonathan (2011). A comprehensive benchmark study of multiple sequence alignment methods: Current challenges and future perspectives. *PLoS One, 6*(3), e18093. https://doi.org/10.1371/journal.pone.0018093, 1932-6203.

Tran, H. T. N., Ang, K. S., Chevrier, M., Zhang, X., Lee, N. Y. S., Goh, M., & Chen, J. (2020). A benchmark of batch-effect correction methods for single-cell RNA sequencing data. *Genome Biology*. ISSN: 1474760X, *21*(1). https://doi.org/10.1186/s13059-019-1850-9. http://genomebiology.com/

Treepong, P., Guyeux, C., Meunier, A., Couchoud, C., Hocquet, D., & Valot, B. (2018). PanISa: Ab initio detection of insertion sequences in bacterial genomes from short read sequence data. *Bioinformatics*. ISSN: 14602059, *34*(22), 3795–3800. https://doi.org/10.1093/bioinformatics/bty479. http://bioinformatics.oxfordjournals.org/

Tsuyuzaki, K., Sato, H., Sato, K., & Nikaido, I. (2020). Benchmarking principal component analysis for large-scale single-cell RNA-sequencing. *Genome Biology*. ISSN: 1474760X, *21*(1). https://doi.org/10.1186/s13059-019-1900-3. http://genomebiology.com/

Van Nostrand, E. L., Pratt, G. A., Shishkin, A. A., Gelboin-Burkhart, C., Fang, M. Y., Sundararaman, B., Blue, S. M., Nguyen, T. B., Surka, C., Elkins, K., Stanton, R., Rigo, F., Guttman, M., & Yeo, G. W. (2016). Robust transcriptome-wide discovery of RNA-binding protein binding sites with enhanced CLIP (eCLIP). *Nature Methods*. ISSN: 15487105, *13*(6), 508–514. https://doi.org/10.1038/nmeth.3810. http://www.nature.com/nmeth/

Van Puyvelde, B., Daled, S., Willems, S., Gabriels, R., Gonzalez de Peredo, A., Chaoui, K., Mouton-Barbosa, E., Bouyssié, D., Boonen, K., Hughes, C. J., Gethings, L. A., Perez-Riverol, Y., Bloomfield, N., Tate, S., Schiltz, O., Martens, L., Deforce, D., & Dhaenens, M. (2022). A comprehensive LFQ benchmark dataset on modern day acquisition strategies in proteomics. *Scientific Data*. ISSN: 20524463, *9*(1). https://doi.org/10.1038/s41597-022-01216-6. http://www.nature.com/sdata/

Välikangas, T., Suomi, T., Chandler, C. E., Scott, A. J., Tran, B. Q., Ernst, R. K., Goodlett, D. R., & Elo, L. L. (2022). Benchmarking tools for detecting longitudinal differential expression in proteomics data allows establishing a robust reproducibility optimization regression approach. *Nature Communications*. ISSN: 20411723, *13*(1). https://doi.org/10.1038/s41467-022-35564-z. https://www.nature.com/ncomms/

Välikangas, T., Suomi, T., & Elo, L. L. (2018). A systematic evaluation of normalization methods in quantitative label-free proteomics. *Briefings in Bioinformatics*. ISSN: 14774054, *19*(1), 1–11. https://doi.org/10.1093/bib/bbw095. http://bib.oxfordjournals.org

Wandy, J., McBride, R., Rogers, S., Terzis, N., Weidt, S., van der Hooft, J. J. J., Bryson, K., Daly, R., & Davies, V. (2023). Simulated-to-real benchmarking of acquisition methods in untargeted metabolomics. *Frontiers in Molecular Biosciences*. ISSN: 2296889X, *10*. https://doi.org/10.3389/fmolb.2023.1130781. http://journal.frontiersin.org/journal/molecular-biosciences

Wang, X., Slebos, R. J. C., Chambers, M. C., Tabb, D. L., Liebler, D. C., & Zhang, B. (2016). ProBAMsuite, a bioinformatics framework for genome-based representation and analysis of proteomics data. *Molecular and Cellular Proteomics*. ISSN: 15359484, *15*(3), 1164–1175. https://doi.org/10.1074/mcp.M115.052860. http://www.mcponline.org/content/15/3/1164.full.pdf+html

Wang, X. W., Wang, T., Schaub, D. P., Chen, C., Sun, Z., Ke, S., Hecker, J., Maaser-Hecker, A., Zeleznik, O. A., Zeleznik, R., Litonjua, A. A., DeMeo, D. L., Lasky-Su, J., Silverman, E. K., Liu, Y. Y., & Weiss, S. T. (2023). Benchmarking omics-based prediction of asthma development in children. *Respiratory Research*. ISSN: 1465993X, *24*(1). https://doi.org/10.1186/s12931-023-02368-8. https://respiratory-research.biomedcentral.com/

Warwick Vesztrocy, Alex, & Dessimoz, Christophe (2020). Benchmarking gene ontology function predictions using negative annotations. *Bioinformatics, 36*(Suppl. 1), i210–i218. https://doi.org/10.1093/bioinformatics/btaa466, 1367-4803.

Weber, L. M., Saelens, W., Cannoodt, R., Soneson, C., Hapfelmeier, A., Gardner, P. P., Boulesteix, A. L., Saeys, Y., & Robinson, M. D. (2019). Essential guidelines for computational method benchmarking. *Genome Biology*. ISSN: 1474760X, *20*(1). https://doi.org/10.1186/s13059-019-1738-8. http://genomebiology.com/

Weisweiler, M., & Stich, B. (2023). Benchmarking of structural variant detection in the tetraploid potato genome using linked-read sequencing. *Genomics*. ISSN: 10898646, *115*(2). https://doi.org/10.1016/j.ygeno.2023.110568. http://www.elsevier.com/inca/publications/store/6/2/2/8/3/8/index.htt

Wissel, D., Toniato, E., Janakarajan, N., Martínez, M. R., Grover, A., & Boeva, V. (2022). SurvBoard: Standardised benchmarking for multi-omics cancer survival models. *bioRxiv*. ISSN: 26928205 https://doi.org/10.1101/2022.11.18.517043. https://www.biorxiv.org

Wooller, S., Anagnostopoulou, A., Kuropka, B., Crossley, M., Benjamin, P. R., Pearl, F., Kemenes, I., Kemenes, G., & Eravci, M. (2022). A combined bioinformatics and LC-MS-based approach for the development and benchmarking of a comprehensive database of Lymnaea CNS proteins. *Journal of Experimental Biology*. ISSN: 14779145, *225*(7). https://doi.org/10.1242/jeb.243753. https://journals.biologists.com/jeb/article/225/7/jeb243753/275025/A-combined-bioinformatics-and-LC-MS-based-approach

Wren, J. D. (2016). Bioinformatics programs are 31-fold over-represented among the highest impact scientific papers of the past two decades. *Bioinformatics*. ISSN: 14602059, *32*(17), 2686–2691. https://doi.org/10.1093/bioinformatics/btw284. http://bioinformatics.oxfordjournals.org/

Xi, N. M., & Li, J. J. (2021). Benchmarking computational doublet-detection methods for single-cell RNA sequencing data. *Cell Systems*. ISSN: 24054720, *12*(2), 176–194.e6. https://doi.org/10.1016/j.cels.2020.11.008. http://www.journals.elsevier.com/cell-systems/

Xie, Zhiqun, Tang, Haixu, & Hancock, John (2017). ISEScan: Automated identification of insertion sequence elements in prokaryotic genomes. *Bioinformatics, 33*(21), 3340–3347. https://doi.org/10.1093/bioinformatics/btx433, 1367-4803.

Yan, L., & Sun, X. (2023). Benchmarking and integration of methods for deconvoluting spatial transcriptomic data. *Bioinformatics*. ISSN: 13674811, *39*(1). https://doi.org/10.1093/bioinformatics/btac805

Ye, S. H., Siddle, K. J., Park, D. J., & Sabeti, P. C. (2019). Benchmarking metagenomics tools for taxonomic classification. *Cell*. ISSN: 10974172, *178*(4), 779–794. https://doi.org/10.1016/j.cell.2019.07.010. https://www.sciencedirect.com/journal/cell

You, Q., Zhong, Z., Ren, Q., Hassan, F., Zhang, Y., & Zhang, T. (2018). CRISPRMatch: An automatic calculation and visualization tool for high-throughput CRISPR genome-editing data analysis. *International Journal of Biological Sciences*. ISSN: 14492288, *14*(8), 858–862. https://doi.org/10.7150/ijbs.24581. http://www.ijbs.com/v14p0858.pdf

You, Y., Tian, L., Su, S., Dong, X., Jabbari, J. S., Hickey, P. F., & Ritchie, M. E. (2021). Benchmarking UMI-based single-cell RNA-seq preprocessing workflows. *Genome Biology*. ISSN: 1474760X, *22*(1). https://doi.org/10.1186/s13059-021-02552-3. http://genomebiology.com/

Yu, L., Cao, Y., Yang, J. Y. H., & Yang, P. (2022). Benchmarking clustering algorithms on estimating the number of cell types from single-cell RNA-sequencing data. *Genome Biology*. ISSN: 1474760X, *23*(1). https://doi.org/10.1186/s13059-022-02622-0. http://genomebiology.com/

Zhang, Y., Ma, Y., Huang, Y., Zhang, Y., Jiang, Q., Zhou, M., & Su, J. (2020). Benchmarking algorithms for pathway activity transformation of single-cell RNA-seq data. *Computational and Structural Biotechnology Journal*. ISSN: 20010370, *18*, 2953–2961. https://doi.org/10.1016/j.csbj.2020.10.007. http://www.csbj.org

Zhang, Z., Yang, C., Veldsman, W. P., Fang, X., & Zhang, L. (2023). Benchmarking genome assembly methods on metagenomic sequencing data. *Briefings in Bioinformatics*. ISSN: 14774054, *24*(2). https://doi.org/10.1093/bib/bbad087. http://bib.oxfordjournals.org

Zhang, Z., Zhao, Y., Liao, X., Shi, W., Li, K., Zou, Q., & Peng, S. (2019). Deep learning in omics: A survey and guideline. *Briefings in Functional Genomics*. ISSN: 20412657, *18*(1), 41–57. https://doi.org/10.1093/bfgp/ely030. https://academic.oup.com/bfg/issue

Zhao, S., Agafonov, O., Azab, A., Stokowy, T., & Hovig, E. (2020). Accuracy and efficiency of germline variant calling pipelines for human genome data. *Scientific Reports*. ISSN: 20452322, *10*(1). https://doi.org/10.1038/s41598-020-77218-4. http://www.nature.com/srep/index.html

Zheng, S. (2017). Benchmarking: Contexts and details matter. *Genome Biology*. ISSN: 1474760X, *18*(1). https://doi.org/10.1186/s13059-017-1258-3. http://genomebiology.com/

Zhuo, J., Wang, K., Shi, Z., & Yuan, C. (2023). Immunogenic cell death-led discovery of COVID-19 biomarkers and inflammatory infiltrates. *Frontiers in Microbiology, 14*. https://doi.org/10.3389/fmicb.2023.1191004. PMID: 37228369; PMCID: PMC10203236.

Zook, J. M., Chapman, B., Wang, J., Mittelman, D., Hofmann, O., Hide, W., & Salit, M. (2014). Integrating human sequence data sets provides a resource of benchmark SNP and indel genotype calls. *Nature Biotechnology*. ISSN: 15461696, *32*(3), 246–251. https://doi.org/10.1038/nbt.2835. http://www.nature.com/nbt/index.html

Chapter 5

Pharmacogenomics, nutrigenomics, and microbial omics

Nutan Prakash Vishwakarma and Sahista Zulfikar Keshavani
Department of Biotechnology, Atmiya University, Rajkot, Gujarat, India

1. Introduction

Since the inception of the Human Genome Project, new technologies have made it possible to obtain a huge amount of molecular data within a tissue or cell. The scientific disciplines that are concerned with the high-throughput measurement of biological molecules are collectively referred to as "omics". The concept of "integrated omics" refers to the integration and analysis of multiple types of omics data, such as proteomics, metabolomics, genomics, and transcriptomics to gain a more comprehensive understanding of a biological system. For example, genomics provides information about an organism's genetic makeup, while transcriptomics provides information about which genes are being expressed and to what extent. Proteomics provides information about the proteins that are being produced, and metabolomics provides information about the small molecules or metabolites present in an organism. By integrating all of these data, researchers can gain a more complete understanding of the different molecular pathways and interactions that occur within a biological system. This can help identify new targets for drug development or to understand the underlying causes of disease. Additionally, considering all of these omics data together can provide a holistic view of the organism and its interaction with the environment.

Pharmacogenomics, nutrigenomics, and microbial genomics are three fields of growing importance in today's healthcare landscape. For example, pharmacogenomics is being used to help doctors prescribe the most effective and safe medications for their patients by identifying genetic variations that may influence a patient's response to a drug. Similarly, nutrigenomics is allowing for personalized nutrition plans that take into account an individual's genetic predispositions to certain health conditions. The field of microbial genomics is providing us with an enhanced understanding of how the gut microbiome influences both general well-being and the development of various diseases.

An example of how these fields can be applied to routine lifestyle is discussing how a person's genetic variations can influence their response to a specific diet or exercise routine. A person with a certain genetic variation may respond differently to a low-carbohydrate diet than someone without that variation, and nutrigenomic testing can help identify these differences. Similarly, pharmacogenomic testing can help identify genetic variations that may cause a person to have a different response to a certain medication, allowing for more effective and safer treatment. Understanding an individual's genetic makeup can help create a more effective and safer lifestyle routine.

2. Pharmacogenomics

Pharmacogenomics investigates how an individual's genetic composition can impact their reaction to medications. It is a rapidly developing field that combines pharmacology (the study of drugs) and the study of genes and their functions to tailor medical treatment to an individual's unique genetic makeup (Ma & Lu, 2011). The goal of pharmacogenomics is to use genetic information to optimize drug therapy, reduce the risk of adverse drug reactions, and improve the overall effectiveness of treatment. This is achieved by using genetic testing to determine an individual's likelihood of responding to a particular medication and to predict the likelihood of experiencing side effects.

Integrative Omics. https://doi.org/10.1016/B978-0-443-16092-9.00005-9

2.1 Why pharmacogenomics?

Pharmacogenomics is a scientific discipline focused on investigating how an individual's distinct genetic composition can impact their reactions and outcomes when exposed to medications (Motulsky, 2002). Just like how you would plan for a trip and customize your itinerary based on your specific needs and preferences, pharmacogenomics aims to tailor medical treatment to an individual's genetic profile to optimize the effectiveness of the treatment and minimize the risk of adverse reactions. The body metabolizes drugs through enzymes that are produced by specific genes, and variations in these genes can cause people to metabolize drugs differently. This can affect the way the body responds to the medication and can lead to different outcomes in terms of effectiveness and side effects (Kalow, 2006). By understanding an individual's unique genetic makeup, pharmacogenomics can help doctors predict how a patient will respond to a particular medication and tailor the treatment accordingly. Warfarin, a medication used to prevent blood clots, is one example of how pharmacogenomics can optimize treatment. The mechanism of action of warfarin involves the inhibition of vitamin K, an essential factor in the synthesis of specific proteins that play a crucial role in the process of blood clotting (McLeod & Evans, 2001). Genetic variations in a specific gene called CYP2C9 can affect how the body metabolizes warfarin, leading to either too much or too little of the medication and increasing the risk of adverse reactions. By identifying these genetic variations, pharmacogenomics can help doctors tailor the dose of warfarin to an individual patient's unique genetic profile to optimize the effectiveness of the treatment and minimize the risk of adverse reactions (Duster, 2005).

2.2 Discovery of pharmacogenomics

One of the key early discoveries in the field was the identification of a genetic variation that affects the way the body metabolizes a drug called primaquine. In the 1950s, it was observed that some people who took primaquine for the treatment of malaria experienced severe side effects, while others did not. It was later discovered that this variation in the way the body responded to the drug was due to a genetic variation that affected the activity of an enzyme that metabolizes the drug (Marcsisin et al., 2016). Since this early discovery, pharmacogenomics has continued to evolve and grow as a field of study. Today, advances in genetic testing and genomic technologies have made it possible to identify a wide range of genetic variations that can affect the way the body responds to medications. This has led to a greater understanding of the role that genetics plays in drug therapy and has opened up new opportunities for personalized medicine. Some examples of treatments for Alzheimer's disease that have been studied in the context of pharmacogenomics are included in the following paragraphs.

2.2.1 Cholinesterase inhibitors

These drugs, such as donepezil, rivastigmine, and galantamine, are used to increase levels of the neurotransmitter acetylcholine in the brain. Genetic variations in the CHRM2 gene have been found to affect an individual's response to cholinesterase inhibitors, with certain genetic variations associated with a greater risk of treatment failure (Sharma, 2019).

2.2.2 Memantine

These drugs are used to moderate the symptoms of moderate to severe Alzheimer's disease by regulating the activity of the neurotransmitter glutamate. Genetic variations in the SLC1A2 gene have been found to affect an individual's response to memantine, with certain genetic variations associated with a greater risk of treatment failure (Cacabelos et al., 2022).

2.2.3 BACE inhibitors

These drugs are used to reduce the production of the protein beta-amyloid, which is believed to play a role in the development of Alzheimer's disease. Genetic variations in the BACE1 gene have been found to affect an individual's response to BACE inhibitors, with certain genetic variations associated with a greater risk of treatment failure. It's important to note that the field of pharmacogenomics in Alzheimer's disease is still in its early stages and more research is needed to fully understand the role of genetic variations in treatment response. Personalized medicine using pharmacogenomics is still in the research phase and more studies are needed to validate the findings and for it to be widely used in the clinic.

2.3 The promise of pharmacogenomics

One of the main promises of pharmacogenomics is the ability to tailor treatment to an individual's specific genetic profile. By understanding how genetic variations can affect an individual's response to medications, doctors can prescribe the most appropriate and effective medications for a particular patient. This personalized approach to treatment has the potential to improve patient outcomes and reduce the risk of adverse reactions to medications. In addition to improving patient care, pharmacogenomics has the potential to reduce health-care costs by helping to identify patients who are likely to respond poorly to certain medications, potentially avoiding the use of expensive and ineffective therapies.

Overall, the promise of pharmacogenomics lies in its ability to improve patient outcomes and reduce health-care costs through the use of personalized medicine. One example of the promise of pharmacogenomics is in the treatment of cancer. Many chemotherapy medications are metabolized by the enzyme CYP3A4, and variations in the gene that codes for this enzyme can affect an individual's response to these medications. For example, individuals with a certain genetic variation in the CYP3A4 gene may metabolize certain chemotherapy medications more slowly than others, leading to higher levels of the medication in the body and an increased risk of side effects. On the other hand, individuals with a different variation may metabolize the medication more quickly, leading to lower levels of the medication in the body and a reduced therapeutic effect.

By understanding these genetic variations, doctors can tailor treatment to an individual's specific genetic profile, potentially leading to more effective and safer chemotherapy regimens. This personalized approach to treatment has the potential to improve patient outcomes and reduce the risk of adverse reactions to chemotherapy medications. Pharmacogenomics plays a crucial role in optimizing drug treatments and improving patient outcomes. Here are some key roles of pharmacogenomics in drug response mentioned in the following paragraphs.

2.3.1 Personalized medicine

Pharmacogenomics allows health-care providers to personalize drug treatments based on an individual's genetic profile. By identifying specific genetic variations that impact drug response, doctors can predict how patients will metabolize and respond to medications. This enables the selection of the most effective drugs and optimal dosages for each patient, maximizing treatment efficacy and minimizing the risk of adverse effects (Evans & Relling, 1999).

2.3.2 Drug selection and dosing

Genetic variations can significantly affect an individual's ability to metabolize and respond to specific drugs. Pharmacogenomics helps in selecting the most appropriate medications for each patient based on their genetic profile. Additionally, it assists in determining the optimal dosage that achieves the desired therapeutic effect while avoiding toxicity or suboptimal response (Caudle et al., 2017).

2.3.3 Adverse drug reactions

Some individuals are more prone to experiencing adverse drug reactions due to their genetic variations. Pharmacogenomics can identify genetic markers associated with increased risk of adverse reactions, allowing health-care providers to avoid certain medications or adjust dosages to minimize these risks. This promotes patient safety and reduces the occurrence of severe side effects.

2.3.4 Drug development and clinical trials

Pharmacogenomics plays a crucial role in drug development and clinical trials. By considering genetic factors in early stages of drug development, researchers can design more targeted and effective medications. In clinical trials, pharmacogenomics information helps identify subpopulations of patients who are most likely to respond positively to a specific drug, allowing for more precise evaluation of drug efficacy and safety (Relling & Klein, 2011).

2.3.5 Cost-effectiveness

Pharmacogenomics has the potential to improve cost-effectiveness in health care. By tailoring drug treatments to individual patients based on their genetic profiles, health-care providers can reduce the risk of adverse reactions, ineffective treatments, and hospitalizations. This leads to better patient outcomes and potentially reduces health-care costs in the long run.

Overall, pharmacogenomics provides valuable insights into how genetic variations influence drug response. By incorporating this knowledge into clinical practice, health-care providers can optimize drug selection, dosing, and overall treatment strategies, leading to more personalized and effective patient care.

2.4 Pharmacogenomics as a powerful tool

Pharmacogenomics has the potential to be used as a powerful public health tool in several ways. Here are a few examples mentioned in the following paragraphs.

2.4.1 Improving the safety and effectiveness of medications

By using genetic information to understand how an individual is likely to respond to a particular medication, pharmacogenomics can help to optimize treatment decisions and reduce the risk of adverse reactions. This can improve the safety and effectiveness of medications for the general population.

2.4.2 Reducing health-care costs

By tailoring treatment to each individual based on their genetic makeup, pharmacogenomics can help to ensure that patients receive the most appropriate and effective medications for their needs. This can help to reduce the risk of ineffective treatment and the need for multiple trial-and-errors prescribing, which can save money and resources.

2.4.3 Improving the accuracy of clinical guidelines and treatment recommendations

By incorporating genetic information into clinical guidelines and treatment recommendations, pharmacogenomics can help to ensure that these recommendations are based on the best available evidence and are tailored to the individual needs of patients.

2.4.4 Enhancing the understanding of the underlying causes of diseases

By studying the genetic factors that influence drug response, pharmacogenomics can also help to enhance our understanding of the underlying causes of diseases and identify new targets for treatment.

Overall, pharmacogenomics has the potential to be a powerful public health tool that can improve the safety and effectiveness of medications, reduce health-care costs, enhance the accuracy of treatment recommendations, and deepen our understanding of the underlying causes of disease.

2.5 Pharmacogenomics—A solution

Just as a computer uses algorithms and software to process and analyze information, pharmacogenomics uses genetic testing and other tools to analyze a person's genetic makeup and understand how it can affect their response to medications. In the same way that a computer can use information and algorithms to make predictions and solve problems, pharmacogenomics can use genetic information to predict and understand how different medications will be metabolized and respond in different individuals.

Just as computers have revolutionized many aspects of our lives, pharmacogenomics has the potential to revolutionize the way that we approach drug treatment and improve the way that medications are prescribed and administered. By tailoring treatment to each individual based on their genetic makeup, pharmacogenomics can help to optimize the therapeutic benefit and minimize the risk of adverse reactions.

One example of how pharmacogenomics can be used as a solution is in the treatment of depression. Depression is a common and debilitating mental health disorder that is often treated with medications called selective serotonin reuptake inhibitors (SSRIs). However, not all people respond equally well to these medications, and some people may experience side effects. Pharmacogenomics can be used to understand how an individual's genetic makeup may affect their response to SSRIs. For example, a person may have a variation in a gene that codes for an enzyme called CYP2D6, which is involved in the metabolism of many medications, including SSRIs. If this person has a variation in the CYP2D6 gene that leads to reduced enzyme activity, they may metabolize SSRIs more slowly, which can affect their therapeutic effectiveness and increase the risk of side effects. By using pharmacogenomics to understand how an individual's genetic makeup may affect their response to SSRIs, health-care providers can tailor treatment decisions to optimize the therapeutic benefit and minimize the risk of adverse reactions. This can help to improve the effectiveness of treatment for depression and improve the quality of life for individuals with this disorder.

2.6 Role of pharmacogenomics in drug development

Pharmacogenomics is a major emerging trend in medicinal biology that enhances the discovery of drugs and its therapeutics. In the current scenario, though pharmacogenetic studies (Fig. 5.1) development of medicines needs to get started on a huge measure. By this approach, drug development is prompt as per the safety and efficacy. Pharmacogenetic studies are employing in various phases of drug development. Effect of drug targeting polymorphisms on response of drug, utilization of stratification for patients as per their genotypes, metabolizing capacity, prevention of adverse drug reactions and supports in better consequence of clinical trials, can be measured and acknowledged (Surendiran et al., 2008) (Fig. 5.2).

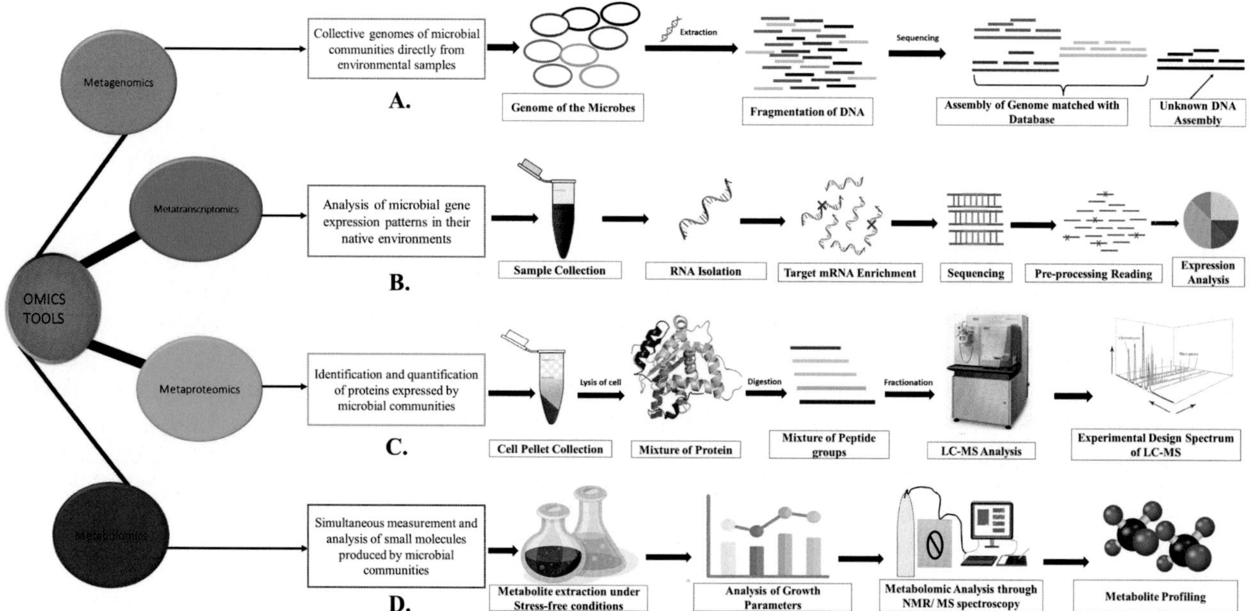

FIGURE 5.1 Implementation of the process involved in the OMICS tools, (A) metagenomics; showcasing the workflow of collective genome analysis directly from the environment samples through DNA fragments by shotgun determination, while (B) metatranscriptomics; showcasing the process preparation and analysis of gene expression from mRNA samples, (C) metaproteomics; showing the quantification of proteins from the microbial communities and lastly, (D) Metabolomics; representing the analysis of metabolite profiling. Exploring the microbial world: Insights into OMICS tools for metagenomics, metatranscriptomics, metaproteomics, and metabolite profiling.

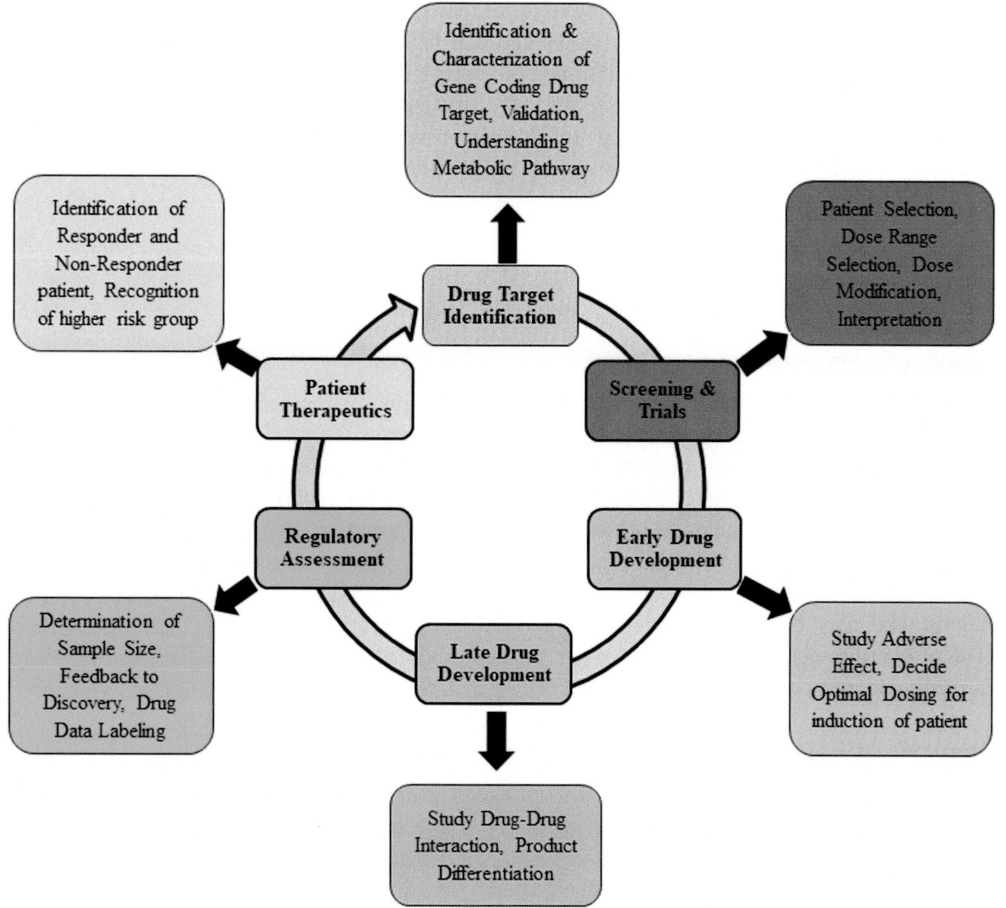

FIGURE 5.2 Unraveling the future of medicine: phases and applications involved in pharmacogenomics drug development.

2.7 Research scopes in pharmacogenomics

There are many areas of research in pharmacogenomics that are currently being actively pursued by scientists and clinicians around the world. Some examples of research areas that could be explored further include treatment strategies: Scientists are currently engaged in the development of personalized therapies that are customized according to an individual's genetic profile. This approach aims to enhance treatment efficacy and minimize the potential for adverse reactions to medications.

2.7.1 Identifying genetic predictors of drug response

Scientists are studying the genetic basis of individual differences in drug response to identify genetic markers that can predict how well a person will respond to a particular medication. These genetic predictors can provide valuable insights into understanding why certain individuals may experience adverse effects or lack therapeutic efficacy with a particular drug, while others may respond positively. Through advanced genomic analysis techniques and association studies, researchers can uncover genetic variants that play a significant role in drug metabolism, drug targets, or drug transporters. This knowledge can pave the way for personalized medicine, allowing health-care professionals to tailor drug therapies based on an individual's genetic profile, thereby optimizing treatment outcomes and minimizing potential adverse reactions. Identifying genetic predictors of drug response is a promising field that holds immense potential for improving patient care and advancing precision medicine approaches.

2.7.2 Investigating the role of genetic variation in adverse drug reactions

Researchers are studying how genetic variations can affect the risk of adverse reactions to medications, to develop strategies to prevent or mitigate these reactions. Investigating the role of genetic variation in adverse drug reactions is a critical area of research within pharmacogenomics. Adverse drug reactions, also known as adverse drug events, can range from mild to severe and can have significant implications for patient safety and treatment outcomes. Genetic variations play a crucial role in determining an individual's susceptibility to adverse reactions to specific medications. By studying the genetic makeup of patients who have experienced adverse drug reactions, researchers can identify specific genetic variants that may be associated with increased risk. This knowledge can help health-care providers in several ways. Firstly, it enables them to identify patients who may be at higher risk of adverse reactions based on their genetic profile, allowing for personalized treatment plans and dosing adjustments. Secondly, understanding the genetic mechanisms underlying adverse drug reactions can contribute to the development of safer medications and improved prescribing guidelines. Ultimately, investigating the role of genetic variation in adverse drug reactions can significantly enhance patient safety and enable more precise and tailored approaches to medication use.

2.7.3 Exploring the use of pharmacogenomics in personalized medicine

The utilization of pharmacogenomics is under investigation by scientists for the purpose of creating personalized treatment strategies that are specifically designed to align with an individual's distinctive genetic profile. Exploring the use of pharmacogenomics in personalized medicine is revolutionizing the field of health care. Personalized medicine aims to tailor medical treatments to an individual's unique genetic profile, and pharmacogenomics plays a crucial role in achieving this goal. By studying how an individual's genetic variations influence their response to specific drugs, pharmacogenomics allows health-care providers to make informed decisions about drug selection, dosage, and treatment plans. This approach helps optimize therapeutic outcomes while minimizing adverse effects. By identifying genetic markers associated with drug response, pharmacogenomics enables health-care professionals to predict an individual's likelihood of responding positively or negatively to certain medications. This information can guide treatment decisions, preventing ineffective treatments and reducing the risk of adverse reactions. Furthermore, pharmacogenomics can enhance medication safety by identifying patients who may be at increased risk of adverse drug reactions based on their genetic profile. Overall, exploring the use of pharmacogenomics in personalized medicine holds tremendous potential for improving patient care, optimizing treatment outcomes, and advancing the field of health care toward more precise and individualized approaches.

2.7.4 Improving drug development and regulatory processes

Researchers are working on developing strategies to incorporate pharmacogenomic information into the drug development and regulatory process to improve the safety and effectiveness of medications. Improving drug development and regulatory processes is crucial for ensuring the safety and efficacy of pharmaceutical products. In recent years, there has been a growing recognition of the need to streamline and enhance these processes to accelerate the availability of new therapies to patients while maintaining rigorous standards. Advances in technology and the availability of large-scale data have provided opportunities for more efficient drug development. For instance, the use of computational modeling, in silico testing, and high-throughput screening techniques has enabled researchers to identify potential drug candidates and predict their safety and effectiveness in a more cost-effective and time-efficient manner. Additionally, regulatory agencies have been adopting innovative approaches such as adaptive clinical trial designs and accelerated approval pathways, allowing for faster evaluation and access to promising therapies. Moreover, collaborations between academia, industry, and regulatory bodies have facilitated the exchange of knowledge, data sharing, and harmonization of standards, promoting transparency and efficiency in the drug development process. These efforts aim to bring safe and effective treatments to patients in a timely manner, addressing unmet medical needs and improving public health outcomes. By continuously improving drug development and regulatory processes, we can ensure that patients have access to innovative therapies while maintaining the highest standards of safety and efficacy.

3. Nutrigenomics

Nutrigenomics is a field of study that examines how nutrients in the diet interact with an individual's genetic makeup to affect their health. It aims to understand how an individual's genes can influence their response to different nutrients, and how nutrients can affect the expression of an individual's genes. This understanding can be used to develop personalized nutrition recommendations that are based on an individual's specific genetic makeup. Just like how a car's performance is influenced by factors such as the quality of its fuel, the condition of its engine, and the efficiency of its components, an individual's health is influenced by the nutrients they consume and how those nutrients interact with their genetic makeup. Just as a mechanic can use diagnostic tools to understand how a car's various systems are functioning and make adjustments to improve its performance, a health-care professional can use nutrigenomic testing to understand how an individual's genes are influencing their response to nutrients and make recommendations for a personalized nutrition plan to support their health. Just as a mechanic can use different types of fuel and oil to optimize a car's performance, a nutrigenomic approach can help an individual identify the specific nutrients that their body needs to function optimally and make recommendations for dietary changes that can support their health.

3.1 Why nutrigenomics?

Nutrigenomics is an emerging field of study that explores the intricate relationship between an individual's genetic makeup and their response to nutrients and dietary components. It aims to understand how genetic variations can influence an individual's nutritional requirements, metabolism, and overall health outcomes. By examining the interactions between genes, diet, and health, nutrigenomics provides valuable insights into personalized nutrition recommendations and interventions.

According to a study conducted by Ferguson et al. (2016), nutrigenomics has the potential to revolutionize the field of nutrition by enabling tailored dietary recommendations based on an individual's genetic profile. The study highlights the role of nutrigenomics in identifying genetic variations that affect nutrient metabolism and the development of chronic diseases. It emphasizes the importance of incorporating genetic information into dietary interventions for improved health outcomes.

3.2 Discovery of nutrigenomics

Nutrigenomics emerged as a scientific discipline in the early 21st century, driven by advancements in genetic research and technology. The field aimed to unravel the complex interplay between an individual's genes and their dietary choices, shedding light on the varying responses to nutrients observed among individuals. Researchers began investigating genetic variations that influenced nutrient metabolism, food preferences, and the risk of developing chronic diseases (Kohlmeier et al., 2010).

3.3 The promise of nutrigenomics

Nutrigenomics offers a novel and promising approach to understanding the intricate interplay between an individual's genetic variations and their response to diet and nutrition. Through in-depth research on the relationship between genes, nutrients, and health outcomes, nutrigenomics enables the identification of specific genetic markers that play a crucial role in determining an individual's unique nutritional needs. This knowledge empowers researchers and health-care professionals to develop personalized nutrition interventions that provide tailored dietary recommendations based on an individual's genetic profile, thus optimizing their nutritional status and overall health outcomes.

By analyzing an individual's genetic variations related to nutrient metabolism, absorption, and utilization, nutrigenomics has the potential to revolutionize the field of personalized nutrition. This approach allows for a deeper understanding of how specific genetic markers can influence an individual's response to different dietary components. For example, certain genetic variations may affect the way individuals metabolize certain nutrients, such as carbohydrates, fats, or vitamins. By identifying these variations, personalized nutrition interventions can be designed to optimize nutrient intake and metabolism, leading to improved health outcomes and reduced risk of chronic diseases.

3.4 Nutrigenomics as a powerful tool

Nutrigenomics, a field that combines nutrition and genomics, has emerged as a powerful tool in unraveling the intricate interplay between nutrition and genetics. By studying how nutrients interact with an individual's genetic makeup, nutrigenomics aims to understand how specific dietary components influence gene expression and ultimately impact health outcomes. This article explores the significance of nutrigenomics in providing personalized dietary recommendations and advancing our understanding of the complex relationship between nutrition and genetics.

3.4.1 Personalized dietary recommendations

A notable advantage of nutrigenomics is its capacity to offer personalized dietary recommendations that are specifically tailored to an individual's genetic profile. By analyzing genetic variations that affect nutrient metabolism, researchers can identify specific dietary requirements tailored to an individual's unique genetic makeup (Meiliana & Wijaya, 2020). For example, certain genetic variants may influence the efficiency of nutrient absorption, metabolism, or utilization. Nutrigenomic research enables the identification of these genetic markers and helps in designing personalized dietary interventions to optimize health outcomes (Ferguson et al., 2016).

3.4.2 Gene–nutrient interactions

Nutrigenomics investigates how dietary components interact with genes to influence health and disease. It explores the molecular mechanisms underlying gene–nutrient interactions, such as how specific nutrients activate or suppress certain genes (Afman & Müller, 2006). This knowledge enhances our understanding of the biological pathways through which nutrients exert their effects and provides insights into the development of targeted nutritional interventions. For instance, nutrigenomic studies have elucidated the interactions between dietary factors, such as polyunsaturated fatty acids, and gene variants associated with cardiovascular health (Ferguson et al., 2016).

3.4.3 Chronic disease prevention and management

The integration of nutrigenomics into research and clinical practice has the potential to transform chronic disease prevention and management. By understanding how genetic variations affect an individual's response to specific nutrients, researchers can develop tailored dietary recommendations to reduce the risk of chronic diseases (Marcum, 2020). Nutrigenomic approaches have shown promise in identifying genetic variants associated with obesity, diabetes, cardiovascular diseases, and certain types of cancer (Afman & Müller, 2006). This knowledge can guide the development of targeted nutritional interventions to mitigate disease risk and improve therapeutic outcomes.

3.4.4 Ethical considerations and challenges

As with any emerging field, nutrigenomics also presents ethical considerations and challenges. Issues such as privacy, data security, and potential stigmatization due to genetic information need to be addressed (Stewart-Knox et al., 2013). Additionally, translating nutrigenomic research findings into practical dietary recommendations for the general population requires careful consideration and validation in diverse populations. Collaboration between researchers, clinicians, policymakers, and regulatory bodies is essential to ensure the responsible integration of nutrigenomics into health-care practices.

3.5 Nutrigenomics as a solution

Nutrigenomics, the study of how genetic variations affect individual responses to nutrients, has emerged as a promising solution to personalize nutrition and optimize health outcomes.

Genetic variations and nutrient metabolism: Genetic variations can influence nutrient metabolism by affecting enzyme activities, transporters, receptors, and other molecular targets involved in nutrient absorption, metabolism, and utilization. An example of this is the influence of single nucleotide polymorphisms in genes responsible for carbohydrate metabolism, which can affect how an individual responds to dietary carbohydrates. Understanding these genetic variations can provide insights into personalized dietary recommendations tailored to an individual's genetic profile.

Nutrigenomics and disease prevention: Nutrigenomics has the potential to revolutionize disease prevention by identifying individuals who are genetically susceptible to certain diseases and tailoring their diets accordingly. For example, individuals with specific genetic variations associated with increased risk of cardiovascular disease may benefit from a diet rich in omega-3 fatty acids. Nutrigenomic approaches can also help in the prevention and management of conditions such as obesity, diabetes, and cancer by optimizing nutrient intake based on an individual's genetic makeup (Ferguson et al., 2016).

Challenges and Future Prospects: Despite the promise of nutrigenomics, several challenges need to be addressed for its widespread implementation. These challenges include the cost of genetic testing, the need for large-scale studies to establish robust associations between genetic variations and nutrient response, and the integration of nutrigenomic data into health-care systems. However, advancements in technology and decreasing costs of genetic sequencing are making nutrigenomics more accessible (Nielsen et al., 2014).

4. Microbial omics

Microbial omics refers to the study of the collective genomes, transcriptomes, proteomes, and metabolomes of a group of microorganisms. It is a field within the broader discipline of genomics that involves the analysis and interpretation of the genetic and functional data of microorganisms at a large scale. The goal of microbial omics is to understand the genetic basis of microbial functions and to identify the genetic elements and pathways that underlie their roles in various ecological and biotechnological processes. It involves the use of various high-throughput technologies such as DNA sequencing, microarray analysis, and mass spectrometry to generate large datasets that can be used to study the genetic and functional properties of microbes.

One of the profound discoveries in microbial omics is the identification of the genetic basis for the production of a natural product called artemisinin. Artemisinin is a potent antimalarial drug that is produced by a type of plant called *Artemisia annua*. However, the yield of artemisinin from plant sources is low, making it difficult to meet the demand for this drug. In the early 2000s, researchers used microbial omics techniques to identify the genetic basis for the production of artemisinin in *A. annua* and to identify the genes responsible for its synthesis. This discovery led to the development of a microbial fermentation process for the production of artemisinin, which has significantly increased the supply of this drug and made it more affordable.

A notable breakthrough in the field of microbial omics is the advancement of metagenomics, which involves studying the genetic material obtained directly from environmental samples. This approach enables a comprehensive exploration of the genetic diversity and functional potential of microbial communities in their natural habitats. Before the development of metagenomics, researchers were limited to studying individual microbial species in isolation, which meant that they were unable to study the complex communities of microorganisms that exist in natural environments. With the development of metagenomics, researchers can now analyze the genomic content of entire microbial communities, providing insights into the diversity, distribution, and functions of microorganisms in different environments.

An example of the impact of metagenomics is the discovery of the human microbiome, which refers to the collection of microbes that colonize the human body. The human microbiome has been found to play important roles in various aspects of human health, including the immune system, metabolism, and digestion. The use of metagenomic techniques has allowed researchers to identify the microbial communities present in different parts of the human body and to study their roles in health and disease. This has led to the development of new therapies and interventions that target the human microbiome, such as probiotics and fecal microbiota transplantation, which are being used to treat a variety of conditions.

Overall, the development of metagenomics has revolutionized the field of microbial omics, enabling researchers to study the complex communities of microorganisms that exist in natural environments and to understand their roles in various ecological and biotechnological processes.

4.1 Why microbial omics?

Microorganisms play a vital role in numerous biological processes, including human health, environmental sustainability, and industrial applications. Traditional microbiological techniques have significantly contributed to our understanding of microbial diversity and function. However, these methods often suffer from limitations in terms of scalability, sensitivity, and the ability to capture the complexity of microbial communities. In recent years, the emergence of high-throughput techniques collectively known as "microbial omics" has revolutionized the field of microbiology. The most advantages of Microbial Omics are mentioned in the following sections.

4.1.1 Metagenomics

Metagenomics allows researchers to study the collective genomes of microbial communities directly from environmental samples, providing a comprehensive understanding of microbial diversity and functional potential. By bypassing the need for cultivation, metagenomics can capture the full scope of microbial communities, including uncultivable organisms, thereby unlocking the previously untapped reservoir of microbial genetic information (Smith et al., 2019).

4.1.2 Metatranscriptomics

Metatranscriptomics enables the analysis of microbial gene expression patterns in their native environments. By examining the transcribed RNA molecules, researchers gain insights into the active metabolic pathways, response mechanisms, and community interactions occurring within microbial communities. This approach has proven invaluable in deciphering the functional dynamics of complex microbial ecosystems.

4.1.3 Metaproteomics

Metaproteomics involves the identification and quantification of proteins expressed by microbial communities. By characterizing the proteome, researchers can elucidate the functional profiles, metabolic activities, and cellular processes of individual microorganisms within a community (Wilmes & Bond, 2004). This information provides a more comprehensive understanding of microbial interactions and the roles played by specific species or functional groups.

4.1.4 Metabolomics

Metabolomics allows the simultaneous measurement and analysis of small molecules produced by microbial communities. By profiling the metabolites, researchers can decipher the metabolic pathways and interactions within microbial communities, as well as gain insights into the overall functional state and health of the system. Metabolomics data can also serve as a valuable resource for identifying novel bioactive compounds and understanding the impact of microbial communities on their environment. Fig. 5.1.

4.2 Discovery of microbial omics

The field of microbial omics has transformed our understanding of microorganisms and their impact on diverse ecosystems. This article explores the discovery of microbial omics techniques and their applications in unraveling the hidden world of microorganisms. It discusses the various omics approaches, including genomics, metagenomics, transcriptomics, proteomics, and metabolomics, and their role in elucidating microbial diversity, functional potential, and interactions. Through a comprehensive review of scientific literature and empirical evidence, this article highlights the significance of microbial omics in advancing our knowledge of microbial communities and their ecological significance. Microorganisms are ubiquitous and play vital roles in various ecosystems, yet their true diversity and functional potential remained largely unexplored until the advent of microbial omics techniques (Handelsman, 2004).

4.2.1 Genomics

Decoding microbial DNA: genomics, the study of complete sets of genes in organisms, has revolutionized our understanding of microbial diversity and functional capabilities. Through genome sequencing and analysis, researchers can explore the genetic content of microorganisms, revealing insights into their evolutionary relationships, metabolic pathways, and potential roles in ecological processes (Handelsman, 2004).

4.2.2 Metagenomics Fig. 5.2

Unveiling community genomes: Metagenomics allows for the study of microbial communities without the need for cultivation. By extracting and sequencing the collective genetic material from an environmental sample, researchers can identify and characterize the diversity of microorganisms within a community, uncover novel genes, and assess functional potential (Wooley et al., 2010).

4.2.3 Transcriptomics

Unraveling gene expression: Transcriptomics provides a snapshot of gene expression within microbial communities, shedding light on their activity and responses to environmental stimuli. By analyzing the RNA transcripts, researchers can identify the genes that are actively expressed, gaining insights into cellular processes, metabolic pathways, and ecological interactions (Wooley et al., 2010).

4.2.4 Proteomics

Investigating protein expression: Proteomics enables the study of the complete set of proteins produced by microorganisms, providing insights into their functional repertoire and responses to environmental conditions. By analyzing protein profiles, researchers can identify key enzymes, signaling molecules, and metabolic pathways, further enhancing our understanding of microbial physiology and ecological interactions (Wilmes & Bond, 2004).

4.2.5 Metabolomics

Profiling small molecules: Metabolomics focuses on the comprehensive analysis of small molecules produced by microorganisms, including metabolites and signaling compounds. By studying the metabolic profiles, researchers can gain insights into the metabolic pathways, nutrient utilization, and secondary metabolite production, unveiling the chemical intricacies of microbial communities (Wooley et al., 2010).

Applications and Future Perspectives Microbial omics techniques have widespread applications, from understanding the ecological roles of microorganisms in diverse environments to improving biotechnology, human health, and environmental management. Furthermore, advancements in high-throughput sequencing technologies and bioinformatics tools continue to enhance the resolution and accessibility of microbial omics data, further propelling discoveries in this field (Handelsman, 2004).

4.3 Promise of microbial omics

Microbial omics, an interdisciplinary field combining genomics, metagenomics, transcriptomics, proteomics, and metabolomics, has revolutionized our understanding of microbial communities and their functional potential.

Microorganisms constitute a significant fraction of earth's biodiversity and play pivotal roles in various ecosystems. Understanding the diversity, functional potential, and interactions of microorganisms has long been a challenge. However, the advent of microbial omics approaches has opened new avenues for unraveling the secrets of the microbial world (Handelsman, 2004).

4.3.1 Genomics

Illuminating microbial genes: Genomics involves the study of complete sets of genes present in microorganisms. By sequencing and analyzing microbial genomes, researchers can identify key functional genes, uncover novel metabolic pathways, and gain insights into the genetic diversity and evolutionary relationships within microbial communities.

4.3.2 Metagenomics

Decoding environmental microbes: Metagenomics enables the analysis of genetic material directly extracted from environmental samples, bypassing the need for cultivation. By studying the collective genomes of microbial communities, metagenomics allows for the identification of novel species, functional gene clusters, and potential contributions of microbes to biogeochemical cycles and ecosystem functioning.

4.3.3 Transcriptomics

Revealing gene expression: Transcriptomics focuses on the study of RNA transcripts produced by microorganisms in a particular environment. By analyzing gene expression patterns, transcriptomics provides insights into the functional activity, metabolic capabilities, and responses of microbial communities to changing environmental conditions.

4.3.4 Proteomics

Investigating microbial proteins: Proteomics involves the comprehensive study of proteins produced by microorganisms. By identifying and quantifying microbial proteins, proteomics sheds light on their functional repertoire, enzymatic activities, and cellular processes, providing valuable information about microbial physiology and ecological interactions (Wilmes & Bond, 2004).

4.3.5 Metabolomics

Profiling microbial metabolites: Metabolomics focuses on the study of small molecules, such as metabolites and signaling compounds, produced by microorganisms. By analyzing the metabolic profiles, metabolomics provides insights into microbial nutrient utilization, secondary metabolite production, and potential ecological roles, contributing to a deeper understanding of microbial communities and their interactions (Fiehn, 2002).

Applications and future directions: Microbial omics approaches have a wide range of applications, including studying microbial diversity, functional genomics, and ecological interactions in various ecosystems. They have contributed to fields such as environmental microbiology, biotechnology, human health, and agriculture. Furthermore, advancements in high-throughput sequencing technologies, data analysis tools, and bioinformatics are enhancing the resolution and applicability of microbial omics data, promising further breakthroughs.

5. Correlation between nutrigenomics, pharmacogenomics, and microbial omics

Yes, there is a correlation between nutrigenomics, pharmacogenomics, and microbial omics. Nutrigenomics is the study of how nutrition affects the expression of an individual's genes. It involves the identification of genetic variations that can affect an individual's response to specific nutrients, and the use of this information to personalize nutrition recommendations. Pharmacogenomics is a scientific discipline focused on examining how an individual's genetic composition can influence their response to medications. It involves the identification of genetic variations that can alter an individual's response to a particular medication, and the use of this information to tailor treatment recommendations. Microbial omics is the study of the collective genomes of microbes in a particular environment, such as the human gut microbiome. It involves the analysis of the genes and other genetic material of microbes to understand their role in health and disease.

There is a relationship between these fields because the nutrients an individual consumes, the medications they take, and the microbes they are exposed to can all affect the expression of their genes. For example, certain nutrients or medications may alter the composition of the gut microbiome, which in turn can affect the expression of an individual's genes.

References

Afman, L., & Müller, M. (2006). Nutrigenomics: From molecular nutrition to prevention of disease. *Journal of the American Dietetic Association*. ISSN: 00028223, *106*(4), 569–576. https://doi.org/10.1016/j.jada.2006.01.001

Cacabelos, R., Naidoo, V., Martínez-Iglesias, O., Corzo, L., Cacabelos, N., Pego, R., & Carril, J. C. (2022). Pharmacogenomics of Alzheimer's disease: Novel strategies for drug utilization and development. *Methods in Molecular Biology, 2547*, 275–387. https://doi.org/10.1007/978-1-0716-2573-6_13, 19406029 http://www.springer.com/series/7651.

Caudle, K. E., Dunnenberger, H. M., Freimuth, R. R., Peterson, J. F., Burlison, J. D., Whirl-Carrillo, M., Scott, S. A., Rehm, H. L., Williams, M. S., Klein, T. E., Relling, M. V., & Hoffman, J. M. (2017). Standardizing terms for clinical pharmacogenetic test results: Consensus terms from the Clinical Pharmacogenetics Implementation Consortium (CPIC). *Genetics in Medicine*. ISSN: 15300366, *19*(2), 215–223. https://doi.org/10.1038/gim.2016.87. http://www.nature.com/gim/index.html

Duster, T. (2005). Race and reification in science. *Science, 307*(5712), 1050–1051. https://doi.org/10.1126/science.1110303

Evans, W. E., & Relling, M. V. (1999). Pharmacogenomics: Translating functional genomics into rational therapeutics. *Science*. ISSN: 00368075, *286*(5439), 487–491. https://doi.org/10.1126/science.286.5439.487

Ferguson, L. R., De Caterina, R., Görman, U., Allayee, H., Kohlmeier, M., Prasad, C., Choi, M. S., Curi, R., De Luis, D. A., Gil, Á., Kang, J. X., Martin, R. L., Milagro, F. I., Nicoletti, C. F., Nonino, C. B., Ordovas, J. M., Parslow, V. R., Portillo, M. P., Santos, J. L., … Martinez, J. A. (2016). Guide and position of the international society of nutrigenetics/nutrigenomics on personalised nutrition: Part 1—Fields of precision nutrition. *Journal of Nutrigenetics and Nutrigenomics*. ISSN: 16616758, *9*(1), 12–27. https://doi.org/10.1159/000445350. http://content.karger.com/ProdukteDB/produkte.asp?Aktion=JournalHome&ProduktNr=232009

Fiehn, O. (2002). Metabolomics—The link between genotypes and phenotypes. *Plant Molecular Biology, 48*(1–2), 155–171. https://doi.org/10.1023/A:1013713905833

Handelsman, J. (2004). Metagenomics: Application of genomics to uncultured microorganisms. *Microbiology and Molecular Biology Reviews, 68*(4), 669–685. https://doi.org/10.1128/MMBR.68.4.669-685.2004

Kalow, W. (2006). Pharmacogenetics and pharmacogenomics: Origin, status, and the hope for personalized medicine. *The Pharmacogenomics Journal*. ISSN: 14731150, *6*(3), 162–165. https://doi.org/10.1038/sj.tpj.6500361

Kohlmeier, M., Caterina, Ferguson, L. R., Görman, U., Allayee, H., Prasad, C., & Roche, H. M. (2010). Guide and position of the international society of nutrigenetics/nutrigenomics on personalized nutrition: Part 1—Fields of precision nutrition. *Journal of Nutrigenetics and Nutrigenomics, 3*(4–6), 209–219. https://doi.org/10.1159/000334958

Ma, Q., & Lu, A. Y. H. (2011). Pharmacogenetics, pharmacogenomics, and individualized medicine. *Pharmacological Reviews*. ISSN: 15210081, *63*(2), 437–459. https://doi.org/10.1124/pr.110.003533. http://pharmrev.aspetjournals.org/content/63/2/437.full.pdf+html

Marcsisin, S. R., Reichard, G., & Pybus, B. S. (2016). Primaquine pharmacology in the context of CYP 2D6 pharmacogenomics: Current state of the art. *Pharmacology and Therapeutics*. ISSN: 01637258, *161*, 1–10. https://doi.org/10.1016/j.pharmthera.2016.03.011

Marcum, J. A. (2020). Nutrigenetics/nutrigenomics, personalized nutrition, and precision healthcare. *Current Nutrition Reports*. ISSN: 21613311, *9*(4), 338–345. https://doi.org/10.1007/s13668-020-00327-z. http://www.springer.com/new+%26+forthcoming+titles+%28default%29/journal/13668?changeHeader

McLeod, H. L., & Evans, W. E. (2001). Pharmacogenomics: Unlocking the human genome for better drug therapy. *Annual Review of Pharmacology and Toxicology*. ISSN: 03621642, *41*, 101–121. https://doi.org/10.1146/annurev.pharmtox.41.1.101

Meiliana, A., & Wijaya, A. (2020). Nutrigenetics, nutrigenomics and precision nutrition. *The Indonesian Biomedical Journal, 12*(3), 189–200. https://doi.org/10.18585/inabj.v12i3.1158, 2355-9179.

Motulsky, A. G. (2002). From pharmacogenetics and ecogenetics to pharmacogenomics. *Medicina Nei Secoli*. ISSN: 03949001, *14*(3), 683–705.

Nielsen, D. E., El-Sohemy, A., & DeAngelis, M. M. (2014). Disclosure of genetic information and change in dietary intake: A randomized controlled trial. *PLoS One, 9*(11), e112665. https://doi.org/10.1371/journal.pone.0112665, 1932-6203.

Relling, M. V., & Klein, T. E. (2011). CPIC: Clinical pharmacogenetics implementation consortium of the pharmacogenomics research network. *Clinical Pharmacology and Therapeutics (St. Louis)*. ISSN: 15326535, *89*(3), 464–467. https://doi.org/10.1038/clpt.2010.279

Sharma, K. (2019). Cholinesterase inhibitors as Alzheimer's therapeutics (Review). *Molecular Medicine Reports, 20*(2), 1479–1487. https://doi.org/10.3892/mmr.2019.10374, 17913004 https://www.spandidos-publications.com/mmr/.

Smith, A. M., Archer, M. J., Cowan, D. A., & Green, S. J. (2019). Molecular tools for deciphering microbial community structure and diversity in environmental samples. *FEMS Microbiology Ecology, 95*(11).

Stewart-Knox, B., Bunting, B., Gilpin, S., Parr, H., Pinhão, S., Strain, J. J., & Kuznesof. (2013). Attitudes toward genetic testing and personalized nutrition in a representative sample of European consumers. *British Journal of Nutrition, 110*(11), 2068–2077. https://doi.org/10.1017/S000711451300148X

Surendiran, A., Pradhan, S., & Adithan, C. (2008). Role of pharmacogenomics in drug discovery and development. *Indian Journal of Pharmacology*. ISSN: 19983751, *40*(4), 137–143. https://doi.org/10.4103/0253-7613.43158

Wilmes, P., & Bond, P. L. (2004). Metaproteomics: Studying functional gene expression in microbial ecosystems. *Trends in Microbiology, 12*, 499–504.

Wooley, J. C., Godzik, A., Friedberg, I., & Bourne, P. E. (2010). A primer on metagenomics. *PLoS Computational Biology, 6*(2), 1553–7358. https://doi.org/10.1371/journal.pcbi.1000667

Chapter 6

Proteomics: Present and future prospective

Tejaswini Hipparagi, Shivaleela Biradar, Srushti S.C. and Babu R.L.
Laboratory of Natural Compounds and Drug Discovery, Department of Bioinformatics, Karnataka State Akkamahadevi Women University, Vijayapura, Karnataka, India

1. Introduction

The objective of this young and promising area of proteomics is to examine the complete proteome, which is the totality of all proteins from an organism, tissue, cell, biofluid, or a subfraction thereof. As a result, a detailed topography of expressed proteins and how they are affected by certain conditions is created. Up until now, the majority of proteomic research has been devoted to the study of cancer, the discovery of pharmacological and therapeutic targets, and the study of biomarkers.

The study of the many kinds, quantities, purposes, and dynamics of proteins in a cell, tissue, or organism is known as proteomics. The amino acid sequences that make up proteins are assembled from DNA and RNA templates. Proteins are either structural or functional components of cells. Their sequences dictate their structure, which in turn dictates how cells operate. Most proteins undergo posttranscriptional changes. The processes of protein synthesis and breakdown, which are tightly regulated in healthy physiology, produce protein dynamics. Concurrent expressions, localization, and physical interactions can give information on their cellular activities because they also function in groups. Proteins have the ability to perform in extracellular space, traveling via the bloodstream to perform distant from the site of their production. Because of this, numerous proteins found in serum or urine are used as clinical biomarkers. There are still many areas of clinical importance that need to be learned. Here, we examine the fundamental ideas, useful computational techniques, and future possibilities in proteomics research.

The focus of the study is moving as the human genome project approaches completion to the monumental challenge of figuring out the structure, function, and relationships of the proteins made by specific genes, as well as their roles in certain disease processes. There is presently a plethora of knowledge available on the sequencing of individual genes, thanks to the human genome project. Proteomics, a recently coined word for the comprehensive examination of biological proteins, is a crucial field of expanding study in the postgenome age. Proteins are resolved thoroughly, quantified, and characterized using proteomics, which combines a variety of advanced methods including two-dimensional (2D) gel electrophoresis (2D GE), image analysis, mass spectrometry (MS), amino acid sequencing, and bioinformatics. It makes sense to concentrate on protein analysis for a variety of bases, including the fact that post-translational modifications, which can be important for protein activity and function, are not described by gene sequences and that the study of the genome falls short of adequately capturing dynamic cellular processes. The approach of proteomics is projected to possess a substantial effect due to its capacity to offer a combined perspective of various disease activities at the protein level. Proteomics has been made possible by recent advancements in protein analysis methods, particularly the creation of sophisticated bioinformatics databases and processing tools. Today's methods are trustworthy and robust enough to answer precise queries about how proteins are expressed in certain disorders. Proteomics in particular gives pathologists the opportunity to find protein markers linked with illness to aid in diagnosis or prognosis and to choose prospective targets for targeted medication treatment.

Fig. 6.1.

Integrative Omics. https://doi.org/10.1016/B978-0-443-16092-9.00006-0

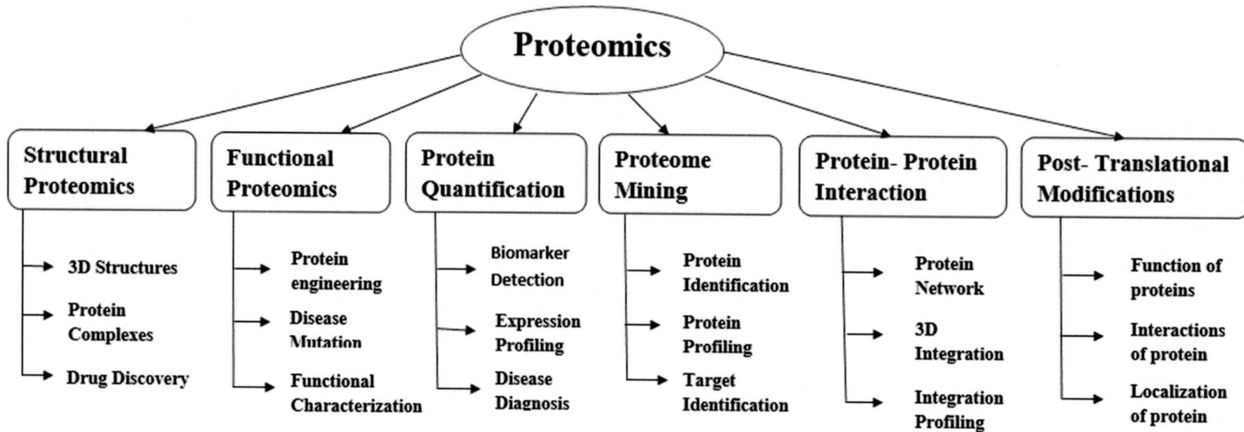

FIGURE 6.1 Proteomics evolution of proteomics.

Proteomics is like other "discovery science" technologies (Aebersold et al., 2000), like microarray analysis of proteins, profiling of metabolites, and sequencing of genomes, is a direct result of both the discoveries made by determined projects to map and sequence the entire genomes of numerous genus and the alterations to our copy that these projects have sparked. The core of this newly developed "systems biology" concept is that the space of potential biological molecules and their groups into pathways and activities is vast but restricted for any particular species. Therefore, in principle, the biological systems that are active in the species may be fully defined if an abundant density of consideration for every component of the system can be attained. Since proteins are implicated in practically all living or biological processes and have a variety of unique features, proteomics is a peculiarly prosperous source of biological living knowledge that helps us better understand how biological systems work. Although technically difficult, genome sequencing is theoretically straightforward and has a clear goal: to definitively identify the whole genome amino acid sequence of the targeted genus. Discovery research initiatives with a focus on testing the management and operation of biological systems are barely well-defined technologically and lack definite objectives. Proteomics originally aimed to rapidly point out or know every protein exhibited by a tissue or cell; however, no one species has still succeeded in achieving this objective. The systematic determination of various protein characteristics is one of the more diversified and current objectives of proteome research. Sequence, volume, modification status, interactivity with other proteins, performance of proteins, subcellular distribution, and structure of proteins are a few of these. To gather the knowledge contained in the characteristics of proteins, several different methods have been created and are continuously being worked on. These proteomic technologies have three characteristics that are exactly evident: first, none of the single technology platforms can facilitate all requested proteomic quantification; second, the near quantification is to a protein activity, less advanced the technology is; and third, there is immature "true" proteomic technology as of right now.

2. Label-free quantitative proteomics

Whichever label-free quantitative proteomics technique is employed, that always incorporates the successive essential steps: (i) Proteins are extracted, degraded, alkylated, and digested as part of sample preparation; (ii) Separation of biomolecules by liquid chromatography (LC) and analysis by MS/MS; and (iii) Analysis of data or results, which includes the quantification and statistical analysis of peptides and proteins. When protein samples are labeled, they are mixed into pooled mixes, which are put through the preparation of the sample process before being analyzed by a single liquid chromatography-mass spectrometry (LCMS)/MS or LC/LCMS/MS experiment. In contrast, each sample is individually prepared and then put across a separate LCMS/MS or LC/LC-MS/MS run in label-free quantitative procedures. Two types of measurements are often used to quantify proteins. Ion intensity change measurements, such as peptide peak lengths or chromatography peak heights, are made in the first. The second method is based on the spectral of protein counting that has been recognized using MS/MS analysis. For each LCMS/MS or LC/LCMS/MS run, monitoring is carried out for the intensity of the peptide peak or spectral count, and the results of several types of research are directly compared to estimate changes in protein abundance.

2.1 Relative quantification by peak intensity of LCMS

In LCMS, the ion with a specific mass-to-charge (m/z) ratio is characterized and noted at a specific time and intensity. Electrospray ionization (ESI) signal strength has been shown to correlate with ion concentration (Voyksner & Lee, 1999). The first research on label-free protein/peptide quantification by peak intensity in LCMS was achieved by loading 10 fmol−100 pmol of myoglobin digests into Nano LC and analyzing the results with LC/MS/MS (Chelius & Bondarenko, 2002). When the chromatographic peak fields of the recognized peptides were brought out and calculated, it was noted that the peak fields improved with larger peptide concentrations. The peak fields of all detected myoglobin peptides were aggregated and plotted as opposed to the protein concentration, and it was discovered that this relationship was linear. The significant association between chromatographic peak positions and the protein or peptide content maintained when myoglobin was added to a complex or huge mixture (human serum) and its digests was identified by LCMS/MS. The computed peak regions were normalized to further enhance the findings of quantitative profiling (Bondarenko et al., 2002; Chelius & Bondarenko, 2002). Although this preliminary research showed that relative peptide quantification could be done by comparing the peak intensities of one and all peptide ions in different LCMS datasets, this approach had significant practical limitations when used to investigate alters in protein in huge amounts in complicated biological materials. Firstly, variations in the peak intensities of the peptides from one run to the next might occur with even the same sample. These variances are the result of experimental flaws, such as inconsistencies in sample injection and preparation. To take into consideration this sort of fluctuation, normalization is necessary. Second, any expletory changes to maintenance time and m/z will make it very difficult to compare different LCMS datasets directly and accurately. Collective sample injection into the same reverse-phase LC column may assist in chromatographic changes. Peak comparisons that are not aligned will have a high degree of variability and inaccurate quantification. Therefore, a fundamental component of this comparison technique is rigorous chromatographic peak alignment and highly repeatable LCMS. Last but not least, automated data processing of these spectra is necessary due to the greater capacity of data acquired in the time of LCMS/MS analysis of complicated protein complex mixtures. During resolving those problems and naturally comparing the data of peak intensity between LCMS samples on a broad scale, competent computer algorithms were created in the subsequent investigations. These label-free quantifications involved a number of related data processing processes. First, nearby peaks and background noise were used to separate peptide peaks (peak detection). Deconvolution was used to assign the isotope patterns. Careful adjustments were made to the LCMS retention periods in order to accurately contest the relevant mass peaks between several LCMS runs (matching of peaks). For more precise matching and quantification, the chromatographic peak intensity (peak area or peak height) was computed and standardized. The significance of variations across several samples was finally determined using statistical analysis, such as the Student's t-test (Higgs et al., 2005; Wang et al., 2003; Wiener et al., 2004). Clinical biomarker discovery, which often necessitates high sample throughput, is ideally suited for automatic comparison of peak intensity from several LCMS datasets. The label-free quantitative method was used in all of the experiments that followed. The differentiation of control and radio-exposed cells of human colon cancer demonstrated the dependability of this label-free methodology (Lengqvist et al., 2009); profiling of the serum proteomic of familial adenomatous polyposis (FAP) patients exhibited various novel celecoxib modulated proteins (Fatima et al., 2009); proteins crucially related with development were recognized by study paraffin-embedded archival melanomas (Huang, Darfler, & Nicholl, 2009); and the study of 55 clinical serum samples from schizophrenia patients and healthy volunteers recognized proteins significantly associated with metastasis (Levin et al., 2007); diagnostic labels and protein signatures were identified from the serum of gaucher patients (Vissers et al., 2007) and the cerebrospinal fluid of schizophrenia patients (Huang et al., 2007).

2.2 Relative quantification by spectral count

In order to identify the relative protein quantification, the spectral counting approach compares the number of identified MS/MS spectra from the same protein in all of the different LCMS/MS or LC/LC-MS/MS datasets. This is feasible because a rise in the quantity of a protein proteolytic peptides, and inversely, usually follows an increase in the abundance of that protein. The number of recognized unique peptides, total MS/MS spectra (spectral count), and protein sequence coverage are all often raised as a result of the increased number of (tryptic) digests (Washburn et al., 2001). The relationship between sequence coverage, number of peptides, spectral calculations, and relative protein affluence was investigated by Liu et al. One element of identification, spectral count, was shown to have the strongest linear association with relative protein affluence (r2 = 0.9997) and the widest dynamic range (more than two orders of magnitude) (Liu et al., 2004). Spectral calculation can therefore be employed as a straightforward yet accurate metric for relative protein measurement. A fascinating research examined the relative quantification of protein complexes using ion chromatographic methods that were 15N/14N isotope-tagged and spectral counting methods (Zybailov et al., 2005). MudPIT was used to evaluate and quantify the primitive membrane proteins isolated from *S. cerevisiae* cultured in a rich and minimal medium. When the peptides with the highest signal-to-noise ratio in the extricate ion chromatogram were chosen for differentiation, it was discovered that the two quantitative approaches had a substantial association. Additionally, compared to peptide ion chromatogram-based quantification, spectral-calculating-based quantification is shown to be more repeatable and to have a wider active range (Zybailov et al., 2005). Since spectral calculating is so simple to use, no particular tools or computer algorithms have been created for it, unlike the chromatographic peak intensity technique, which necessitates complex computer algorithms for automated LCMS peak alignment and comparison. However, for precise and dependable identification of protein changes in normalization, complex mixtures and statistical analysis of spectral counting records are required. To account for the fluctuation from run to run, a straightforward normalizing approach based on total spectral calculations has been published (Dong et al., 2007). The effects of length of protein on spectral calculation were taken into consideration by the establishment of a normalization spectral abundance factor (NSAF), as big proteins often give more peptide/spectra than short proteins (Florens et al., 2006; Zybailov et al., 2006). NSAF is determined as the total of the SpC/L values for each of the proteins in the demonstration is divided by the number of spectral counts (SpC) that uniquely identify a protein. NSAF permits the comparison of protein abundance across several independent samples and has been used to measure expression changes within various complexes (Florens et al., 2006; Paoletti et al., 2006). To evaluate the importance of SpC-based comparative quantification, Zhang et al. investigated five different statistical techniques (Zhang et al., 2006). On the spectral count data gathered by the MudPIT investigation of yeast digests, the Fisher's exact test, goodness of fit test (G test), AC test, Student's t-test, and Local Pooled Error test were run. When three or more replicates are available, it was discovered that the Student's t-test is more effective. When there are just one or two replications, there are three techniques that may be used: The Fisher's exact test, the G test, and the AC test, although the G test has the advantage owing to its estimation simplicity.

Various biological complexes were examined through relative quantification by spectral counting, including urine samples from healthy donors and patients with acute inflammation (Pang et al., 2002), the identification of indicators for type 2 diabetes in the human saliva proteome (Rao et al., 2009), the study of protein expression in yeast and mammalian cells under various conditions of growth level (Florens et al., 2006; Zybailov et al., 2005), the contrast between normal and lung cancer (Pan et al., 2008), and the examination of mammalian phosphotyrosine binding proteins (Asara et al., 2008; Seyfried et al., 2008).

2.3 Absolute label-free quantification

Label-free proteomics techniques can be used to determine the absolute abundance of proteins in addition to their relative quantification. The protein abundance index (PAI), which is calculated as the ratio of the number of detected peptides to the amount of potentially observable tryptic peptides for each protein, was applied to estimate the protein abundance of the human spliceosome complexes (Rappsilber et al., 2002). Later, this index was transformed into exponentially modified PAI (emPAI, which is PAI minus one in exponential form) (Ishihama et al., 2005). In a mouse whole-cell lysate that had been examined using synthetic peptides, the emPAI demonstrated its efficiency by estimating the absolute abundance of 46 proteins. In protein identification investigations, the values of emPAI may be determined quickly with the use of a straightforward script without the need for extra testing. In a large-scale study, it can be regularly utilized to report approximative absolute protein abundance. In the recently evolved absolute protein expression (APEX) profiling approach, a modified version of spectral counting, the link between protein abundance and the quantity of identified peptides can be used to determine the absolute protein concentration per cell (Lu et al., 2007). The secret to APEX is to introduce the proper correction factors, which will make the percentage of peptides that are predicted to be present and the proportion of peptides that are really present proportionate to one another. The absolute abundance of a protein is demonstrated by an APEX score, which is produced from the proportion of observed peptide mass spectra associated with one protein and corrected by the comparison of expectations of the number of distinct peptides anticipated from a particular protein during a MudPIT experiment. The important correction factor for each protein (known as the Oi value) is obtained using a machine-learning classification algorithm to predict the observed tryptic peptides from a given protein based on peptide length and amino acid composition. APEX successfully identified 10 proteins that were introduced in a yeast cell extract with a known amount being abundant. When compared to data from other absolute expression assessments, such as high-throughput analysis of fusion proteins by western blotting or flow cytometry, the absolute protein abundance of the yeast and *E. coli* proteomes examined by APEX corresponded well. APEX Quantitative Proteomics Tool (Braisted et al., 2008), a free open-source Java implementation for the absolute quantification of proteins, was now created, and it may be found at http://pfgrc.jcvi.org.

2.4 Commercially available software for label-free quantitative proteomics

The creation of advanced bioinformatics applications that provide automated label-free analysis for comparative LCMS has accelerated recently. Data simplification, temporal alignment, peak identification, peak calculation, peak matching, and statistical analysis are typically included in the data processing pipelines. There are now several free-source and paid programs available. MSight, MZmine, OpenMs, MapQuant, SuperHirn, MsInspect, and PEPPeR are some of the open-source applications (Mueller et al., 2008; Rajcevic et al., 2009). Table 6.1 is a list of the software applications that are presently for sale. DeCyder 2D Differential Analysis Software is the foundation of Decyder MS. It has two key analytical features: Run-to-run complementing with the PepMatch module and peptide reorganization with the PepDetect module. Utilizing imaging methods, the PepDetect module contributes to horizon transport and isotopes and alters state deconvolution and peak-volume estimations. Additionally, this module offers the choice of submitting all peptides or a chosen subset for protein detection through database screening. The PepMatch application aligns peptides from many LCMS runs and confidently and statistically finds tiny quantitative variations between peptides across numerous runs. Different normalization methods can be used to enhance outcomes even further (Lengqvist et al., 2009). To discover statistically significant differences, SIEVE software uses chromatographic alignment using the ChromAlign method. For each differential peaks expression ratio, the program may calculate a P-value, adding another level of assurance. To identify peptides and proteins, protein databases may be searched against peptides that exhibit statistically significant differences. Its profiteering feature speeds up complex biomarker discovery operations by lowering the number of spectra that must be examined, cutting down on identification time, and increasing throughput. The Rosetta Elucidator system is a data management platform for storing and managing massive amounts of MS data, in addition to being label-free quantitative software. Additionally, it backs labeling analyses like stable isotope labeling by amino acids in cell culture (SILAC) and ICAT. For peak reorganization, extraction, and quantification or calculation from the MS result, the Elucidator employs the Peak Teller algorithm. It makes use of Protein Teller and Peptide Teller to confirm that all feature peptide/protein assignments are accurate. The system is compatible with a large number of MS instruments, database scavenging techniques, and extensive visualization and analysis tools (Neubert et al., 2008). Additionally, it offers label-free evaluation by spectral calculation. Protein Lynx Global Server companion for label-free quantification by peak potency. It is also a database-penetrating engine for peptide or protein spotting (Huang et al., 2007; Levin et al., 2007).

3. Techniques used in proteomics

3.1 Isotope coded affinity tag (ICAT) labeling

The isotope-coded affinity tag (ICAT) methodology quantitative analysis of protein profiles permits pairwise comparisons of the expression of proteins in biosamples like cells, tissue extracts, and biological fluids. ICATs are chemical labeling reagents used for in-vitro isotopic labeling in quantitative proteomics by MS. This method provides simultaneous recognition of the proteins in a composite combination as well as accurate assessment of the abundance differences for one and all proteins present in two or many protein samples (Gygi et al., 1999). Three functioning parts make up the ICAT reagents. The first one is a disulfide reactive group that is selective for the sulfhydryl groups on the side chain of reduced cysteines when this is given to peptides. The ethylene glycol linker group, the second component, can be observed in nondeuterated (isotopically normal) or deuterated (isotopically heavy) forms and thus serves as the foundation for precise quantification. The third component, using avidin affinity chromatography, is the biotin segment, which provides an affinity tag for the selective extraction of the tagged peptides from heterogeneous peptide mixtures. During an experiment,

TABLE 6.1 Commercially available application software's for label free analysis.

Sl. no.	Software	Producer	Quantification	Website
1	Decyder MS	GE Healthcare	Peak intensity	http://www5.gelifesciences.com/
2	SIEVE	Thermo Electron	Peak intensity	http://www.thermo.com/
3	Elucidator	Rosetta	Peak intensity, spectral count	http://www.rosettabio.com/
4	ProteinLynx	Waters	Peak intensity	http://www.waters.com/

From Neubert, H., Bonnert, T. P., Rumpel, K., Hunt, B. T., Henle, E. S., & Lames, I. T. (2008). Label-Free detection of differential protein expression by LC/MALDI mass spectrometry. *Journal of Proteome Research, 7*(6), 2270–2279. https://doi.org/10.1021/pr700705u

the proteins in the two samples being examined have their reduced cysteine residues tagged with either the isotopically heavy or regular reagent. Following the mixing of the two samples, trypsin digestion, avidin affinity chromatography separation of the tagged peptides, and MS/MS analysis of the separated tagged peptides, the amino acid composition and relative abundance of each protein in the two samples are then determined. Special software tools, such as the Sequent application, are used to determine the sequence of a peptide by comparing the CID spectrum generated from a tandem mass spectrometry with a sequence database (Ducret et al., 1998). The samples were mixed, separated using chromatography, and then subsequently analyzed using a mass spectrometry to discover the protein mass-to-charge ratio. Analysis is at most possible with peptides that include cysteine. The posttranslational change is typically disregarded since only peptides containing cysteine are analyzed (Colangelo & Williams, 2006).

3.2 Stable isotope labeling with amino acids in cell culture (SILAC)

The technique known as SILAC (stable isotope labeling with amino acids in cell culture), which expects the incorporation of amino acid-accommodated substituted stable isotopic nuclei into proteins in living cells, is quick, easy, direct, and accurate for embedding a tag into proteins for mass spectroscopic analysis and relative calculation. By using nonradioactive, stable isotopes containing amino acids in thousands of proteins, SILAC marks cellular proteomes during normal metabolic functions. In place of natural or light amino acids, "heavy" SILAC amino acids are used to make a growth medium. The major amino acids are assimilated by cell cultures in this medium after 5 cell divisions, and SILAC amino acids have no effect on the cell structure or development rate. When light and heavy cell populations are mixed, protein abundances can still be determined by comparing the relative MS signal intensities since MS can still tell one type of cell from the other (Ong & Mann, 2005). Without the requirement for chemical treatment or chemical modification, SILAC allows for the development of intricate functional proteomics experiments and precise relative measurements. Workflow for quantitative proteomics studies: Stable isotope labels are applied to the different research procedures at numerous locations to allow protein populations to merge. The analysis of up to four sample state isobaric tags for relative and absolute quantitation (iTRAQ) or three different states (triple encoding SILAC) in a single MS analysis is facilitated by the latest methods that permit sample combining. Despite being careful to handle samples in parallel and employing chemical modification techniques to identify isolated proteins or peptides, this could lead to quantitative inaccuracy. When using the sample processing method for metabolic labeling, samples can be mixed as a whole, as well as cells and related diseases. The quantitative accuracy is therefore unaffected by sample losses at a particular phase (Ong & Mann, 2006).

3.3 Isobaric tag for relative and absolute quantitation (iTRAQ)

Proteomics uses the iTRAQ technique or isobaric tags for relative and absolute quantification to study quantitative proteome alterations using tandem MS. Isobaric reagents are used to label the main amines of peptides and proteins. Protein quantification, which involves mass spectrometric analysis and stable isotopes, is a powerful method in modern proteomics research for systematically and quantitatively evaluating the variations in protein abilities. It is especially simpler to compare peptides and proteins using iTRAQ-based quantization, including comparisons of normal or treated states, by labeling peptides that can be detected and estimated by the study of corresponding groups produced on fragmentation in the mass spectrometry (MS). The iTRAQ approach is now one of the primary quantification technologies used in differential plant proteome research (Vélez-Bermúdez et al., 2016). In order to use iTRAQ, peptides formed from protein digestion "N" terminus and side chain amines must be covalently tagged with tags of varying mass that comprise three regions that are a reporter region, a peptide reactive area, and a balancing region (Ross et al., 2004). Separate samples are extracted, digested, and chemically labeled with one of the iTRAQ reagents prior to being used for assessment. A tandem MS exploration is then executed on the combined, 1 or 2-D HPLC-separated, and merged protein sample pools. Tandem MS capacity to fragment the isolated precursor peptide ions and provide for a quantitative method of the intense reporter ions in the tandem mass spectrometry. The signals of the reporter ions in each MS/MS spectrum can be used to determine the relative abundance of each peptide identified by this spectrum at the peptide level (Gafken & Lampe, 2006). Then, using the collected data and a database search, the tagged peptides and the corresponding proteins are discovered. Environments that prioritize quality control and are user-friendly can process iTRAQ data (Vaudel et al., 2012).

3.4 2D gel electrophoresis

In the history and advancement of proteomics, 2D GE has been a key factor. In the field of proteomics, however, it is no longer the only separation method used. The basic idea behind 2D electrophoresis is that although the name implies that it is a two-step process, it typically involves five steps: The first step starts with sample preparation preceding the first separation; the second step is followed by the first separation; the third step is interfacing with the second separation; the fourth step is the second separation; and finally, the fifth step is protein detection. Iso-electric focusing is frequently carried out first, followed by SDS electrophoresis, in terms of the order of separation. However, the classical order is actually used, both for commercial (a large, second separation, which is less expensive, is preferred, that is SDS type electrophoresis) and technical reasons (SDS electrophoresis gels interface with downstream protein analytical techniques like protein blotting and MS considerably greater easily than isoelectric ones, and they are also quicker to stain than iso-electric ones.). Both are highly sensitive to ampholytes and nonionic detergents, two chemical groups that are actually required in denaturing isoelectric focusing (IEF) but are limited to the buffer front when the usual sequence of separations is being used. Detailed processes for sample preparation for 2D electrophoresis, sample runs on 2D gels, and protein detection are much beyond the scope of an instructional paper (Rabilloud & Lelong, 2011). A popular method in proteomics is 2D electrophoresis. Since its creation in 1975, 2D GE has operated depending on the same basic ideas (Klose, 1975). In addition, the orthogonal sodium dodecyl sulfate polyacrylamide GE separation and the first iso-electric focusing (one dimension, IEF) separation of proteins (SDS-PAGE) (Görg et al., 2000). High-resolution proteomic research is made possible by the fact that one gel can successfully distinguish up to 10,000 protein spots (Klose & Kobalz, 1995). IEF separates proteins based on their isoelectric point (pI). Amphoteric molecules, including proteins, may carry on the functions of an acid or a base. The side chains of the amino acids in proteins contain systems to work with or deprotonate acidic or basic buffering groups, depending on the pH of the solutions in which the protein is present. At a specific pH, the total of all charges generated by the amino acids in a protein tends to zero. This property is used in IEF, which includes placing proteins in a pH gradient and implementing an electric potential. The protein travels either toward the anode or the cathode, depending on its net charge. The protein migratory process will terminate once it reaches its pI. The pH gradients essential for IEF were first incredibly challenging and unpredictable to produce and just use. To generate these pH gradients, tube gels with ampholytes as carriers were often used. Chemicals are added to a gel matrix to moderate the pH gradient (Bjellqvist et al., 1982). The establishment of the gel on a strong backing was a significant step toward improving the use of IEF for researchers and delivering more accurate results (Görg et al., 1995).

The IEF gel or block is adjusted in SDS before being placed on top of the SDS PAGE gel. This balancing step is crucial for allowing the SDS molecules to interact with the proteins and form the anionic complexes with a net negative (-ve) charge. Many proteins migrate from the IEF gel into the SDS gel as a result of the electric force, where they are then separated by their molecular weights. Similarly, strategies for protein separation under nondenaturing conditions have been developed. The majority of applications employ denaturing SDS-PAGE. The proteins in the gel must be recognized after electrophoresis. Traditionally, a noticeable stain is used to achieve this. Silver nitrate (Winkler et al., 2007) is a very prevalent visual stain that is quite sensitive. Silver staining is problematic due to undesirable background generation, a lack of quantitative factors, inconsistent results, and incompatibility with MS. Other noticeable stains have been created, including Coomassie blue (Neuhoff et al., 1988) and zinc imidazole (Fernandez-Patron et al., 1998), with the latter finding virtually ubiquitous use. Both Coomassie staining and zinc imidazole are more compatible with MS while having lower sensitivity than silver staining and having less capacity for quantification, respectively. Certain luminous stains, including ruthenium bathophenanthroline disulfonate (Rabilloud et al., 2001) and Deep Purple (Mackintosh et al., 2003), attempt to incorporate MS compatibility, sensitivity, and accessibility of use.

3.5 2D differential gel electrophoresis

Two-dimensional differential gel electrophoresis (2D DIGE), which Unlu et al. developed, gracefully overcomes the issues with traditional 2D GE. Quantification and consistency problems have made it challenging to implement conventional 2D GE for applications like protein expression profiling in brain research. A new progressive incremental approach would stimulate the scientific proteomics community and advance the subject, just the same as significant research was done on the constraints of 2D GE around the turn of the century (Ünlü et al., 1997). Before EF, direct protein tagging with fluorescent dyes was utilized in different GEs (known as CyDyes: Cy5, Cy3, and Cy2) (Alban et al., 2003). CyDyes are spectrally resolvable cyanine dyes with only an N-hydroxysuccinimidyl ester reactive group that interacts covalently with proteins lysine amino groups. One dye molecule is able to be added to each protein at low dye concentrations. The potential of DIGE technology to separate two or even more samples on the same gel by labeling them with separate dyes and minimizing gel-to-gel variability is its main advantage. As an outcome, spot matching and quantification accuracy are significantly improved and made much easier. The normalization pool used is called Cy2, and it combines every sample from the experiment. This Cy2-labeled pool is used to run every gel, allowing for spot matching and standardizing signals from various gels (Diez et al., 2010, pp. 51–69). For researchers, the DIGE technology has a lot of potential. The dyes are compatible with MS, can provide good quantification of any method now in use, and have sensitivity levels that are on par with silver staining procedures.

In 2D DIGE for differential display proteomics, sample duplication and the use of an internal standard allow the analysis of repeated samples from various experimental circumstances with the highest possible level of statistical confidence (Minden, 2007). The significant interpersonal variation anticipated from biological samples can be easily controlled in DIGE experiments by using enough independent (biological) replication samples. Then, by focusing on the fundamental distinctions that might describe diverse disorders or by removing the noise from technological variation, one may describe biological states and naturally occurring biological variation (Diez et al., 2010, pp. 51–69). A sophisticated collection of independent samples can be reviewed through multivariate statistical analysis. Following protein visualization, image analysis will be required (Goldfarb, 2007). The great sensitivity and linearity of the dyes used, the simple method, and the significant decrease in intergel variability are the great features of DIGE. These aspects decrease experimental bias and improve the capability of recognizing biological variability clearly. Furthermore, a pooled internal standard that is loaded sequentially with the control and experimental samples is used in order to increase statistical confidence and quantification reliability (Alban et al., 2003).

3.6 Protein microarray

High-throughput scientific approaches, as distinct from this conventional method, have been created in the last 10 years to enhance the research of numerous molecules, such as proteins, DNA, and metabolites. In particular, in genomic research, DNA microarrays have proven effective (Schena et al., 1995). They are frequently used to evaluate the patterns of gene expression, identify the binding sites for transcription factors, and spot mutations and deletions in patterns. However, DNA microarrays mostly reveal details about the genes themselves and nothing about the activities of the proteins that certain genes encode. High-throughput approaches to analyzing proteins, such as protein microarrays, tagging and subcellular localization, and MS protein profiling, have recently been developed (Hall et al., 2007).

Protein microarrays, commonly referred to as protein chips, are microscopic parallel assay devices that contain high-density arrays of small amounts of pure proteins. They make it simpler for several molecules to be identified sequentially from modest amounts of materials in a single experiment (Tao et al., 2007). These techniques have been employed in clinical diagnosis, environmental food safety analysis, biomarker discovery, cell surface marker glycosylation profiling, and protein expression profiling. Protein microarrays exist in three varieties: Functional, analytical, and reverse phase, and they are all used to investigate the biological functions of proteins. In order to profile a complex protein mixture and define the binding affinities, specificities, and levels of protein expression, analytical microarrays are frequently used. Using this approach, a number of antibodies, aptamers, or affibodies are shown on a microscope slide. Afterward, a protein solution is used to probe the array (Bertone & Snyder, 2005). The most prevalent kind of analytical microarray is an antibody microarray. These microarrays can be used for patient assessment and differential expression profile analysis. Examining how healthy and sick tissues respond to environmental stress is one example (Sreekumar et al., 2001). Analytical arrays differentiate themselves from functional protein microarrays in that they are constructed of arrays that accurately represent functional proteins or protein domains. These protein chips are exploited in a single experiment to explore the crucial function that the whole proteome performs in biochemistry. Numerous protein interactions, among them those involving protein-RNA, protein-DNA, protein-protein, protein-phospholipids, and protein-small molecule interactions, are being investigated using proteomics (Hall et al., 2004; Zhu et al., 2001).

A third stage of protein microarrays connected to analytical microarrays is known as reverse-phase protein microarrays (RPA). Cells are extracted from various target tissues and lyzed in RPA. On a nitrocellulose slide, the lysate is organized using a contact pin microarrayer. To probe the target protein of interest into the slides, antibodies are usually used together with fluorescent, chemiluminescent, or colorimetric techniques. Reference peptides are positioned on the slides to provide protein quantification of the sample lysates. RPAs allow the identification and quantification of changes that might have been brought on by disease. RPAs can be used specifically to identify posttranslational modifications, which are typically modified by diseases (Speer et al., 2005). Once the protein pathway that may be malfunctioning in the cell has already been discovered, a chemically treated drug may be developed to target the malfunctioning protein pathway and treat the targeted diseases.

3.7 Reverse phased protein microarray

An antibody-based microarray technology called the reverse phase protein array (RPPA) runs immunoassays on a large number of samples at once, including serum, cell lysates, tissue, other bodily fluids, or plasma. RPPA is a specialized proteomic platform that can evaluate posttranslational modifications, like phosphorylation, which is a marker of protein functions, in addition to MS profiling. Additionally, relative levels of expression of all proteins may be evaluated (Creighton & Huang, 2015). A RPPA may be used to identify protein pathways at any length in different disorders and cell biology processes. It was first created for the purpose of identifying modifications in cancer protein signaling pathways. The evaluation of protein pathways in common diseases and cell biology processes can be performed relatively simply using reverse-phase protein arrays. It was initially developed to investigate alterations in cancer protein signaling pathways. In a microarray format, concordant probing of several protein signaling pathways and their functional status is possible with the help of the reliable, highly repeatable proteomic technology known as RPPA. By using control activities and assay conditions for each distinct antibody, the test can be improved by arraying protein targets in reverse and incubating repeat arrays, each with a single well-characterized antibody. Relative to MS proteomics, RPPA provides a greater throughput and cheaper cost, making it more suitable for quick analysis of several experimental samples (Coarfa et al., 2021). Furthermore, it is desirable to use serum or cell/tissue lysates without prefractionation operations. Proteins across a wide dynamic range can be identified by RPPA using small sample volumes and protein concentrations (5 g). However, it is extremely sensitive and successful in recognizing low levels of regulatory proteins, which can occasionally be problematic to evaluate using MS profiling (Grubb et al., 2009). These characteristics make RPPA suitable for validating possible protein biomarkers or protein pathways identified by gene-expression profiling, or MS. The potential of RPPA as a method for identifying protein signal transduction associated with the progression of cancer, as well as pathways accessible to therapy or implicated in resistance, has also been demonstrated (Coarfa et al., 2021).

3.8 Mass spectrometry

To estimate the m/z ratio of charged molecules, use a mass spectrometry. (MS) in analytical chemistry has undergone a series of modifications to optimize its performance. There have been several techniques proposed to focus, ionize, fragment, scan, and detect chemical structures (Griffiths, 2008). In order to identify different precursor product ion pairs, early MS techniques relied on the ability to recognize various structures by fragmenting larger molecules into smaller components. Today, high-resolution MS may be used to get extremely exact masses that can be used to identify analytes (HRMS). More sensitive, selective, reliable, and repeatable analyses have been made possible by recent innovations. The latest generation of mass spectrometry, with improved accuracy and repeatability, enables the use of these tools not only for the identification but also for the quantification of analytes. The use of these tools for both analyte identification and quantification is facilitated by the most recent generation of mass spectrometry improved precision and reproducibility. Current developments have enabled more sensitive, focused, trustworthy, and repeatable analyses (Loos et al., 2016). It is a common misconception that a mass spectrometry is only a replacement for flame ionization, UV/VIS, electrochemical, and fluorescence detection. A mass spectrometry may be used as a distinctive separation method because of its detection approach, which is based on evaluating certain mass-to-charge ratios (Zhou et al., 2010). The variety of samples used in emerging domains, including environmental, toxicological, and biomarker discovery research, which required extremely sensitive and selective readings to reduce interferences, makes it exceptionally difficult to apply an MS alone (Verplaetse & Henion, 2016; Verplaetse et al., 2012).

To get oversensitive or keep away from injecting samples that are significantly altered, it is suggested to carry out a sample preparation step before the analysis. Solid-phase extraction, protein precipitation, liquid-liquid extraction, and all of its derivations are a few examples of regularly used strategies for preparing samples. The procedure frequently includes an additional separation step in place of or in addition to sample preparation. To make sure that not all molecules enter the ionization source at once, the most often employed procedures are GC (gas chromatography), CE (capillary electrophoresis), and LC. As an effect, the signal experiences less ME (matrix effects), increasing its sensitivity (Stahnke et al., 2012).

3.9 LCMS/MS

LCMS is currently commonly used in referral laboratories and clinical reference centers around the globe, and it is also beginning to expand to large and medium-sized hospitals and district clinical labs. There are already hundreds of different assays that employ it in medical labs, ranging from large-volume testing in drug or toxicology, newborn screening, and endocrinology to rare and extremely esoteric analytes (Want et al., 2005). MS evolved rapidly in terms of technology and extended to the physical and chemical sciences, where it is now a standard analytical method. However, the ionization approaches that were available were only suited for relatively low molecular weight chemicals (200 Da or less than 200 Da), and because there were no appropriate ways to quickly introduce biospecimens into the high vacuum of the mass spectrometry, application for biological samples remained limited. When John Fenn developed soft ionization of huge macromolecules using ESI, which created some of the crucial methodologies that permitted for simple sample input into the mass spectrometry, the situation drastically altered in the 1980s (Fenn et al., 1989). The m/z ratio (abbreviated as m/q, m/Q, m/z, or m/Z) of charged particlesis determined using a mass spectrometry. The tandem mass spectrometer key parts are: The source ionizes the sample, the first mass filter (Q1) which then travels through the collision cell (Q2), the second mass filter (Q3), and finally the identified. And illustratively represent the two primary ionization source types currently utilized in clinical LCMS/MS devices: Atmospheric pressure chemical ionization (APCI, C) and ESI, B. Through a positively charged, exceptionally small capillary, the solvent analyte flow from the LC is positively charged solvent analyte droplets and injected in ESI as small as possible. Droplets follow the direction of the -ve charged faceplate, volatizing solvent as they go, until the revolting charge of their ionized components reaches the interfacial tension of the faceplate, creating a Coulomb explosion. The faceplate intake hole is then used to introduce certain ionized analyte molecules into the mass spectrometry. In the APCI, a high-current Corona needle releases ion that ionizes the solvent polar components, and heated nebulizer gas vaporizes the solvent analyte stream from the LC. The ionizable analyte molecules then get the charge from the solvent molecules, while those pass through the mass spectrometry faceplate entrance hole (Grebe & Singh, 2011). Globally, MS is now a unique area of laboratory medicine, and it is extremely probable that it will remain that way in the future. Valid statistics on the scope and dynamics of an international application are not yet established, despite the fact that information from expertize testing and the literature indicate that the LCMS approach has expanded dramatically in the previous 10 years. Furthermore, there is no instrumentation registry and no regular data from healthcare or insurance providers. Nonetheless, with less than 1% of analyses done in worldwide laboratory diagnostics based on MS, LCMS use is generally limited. In other application sectors, like toxicology or therapeutic-drug monitoring (TDM), comparable numbers are substantially larger because of the lack of substitute approaches available. Up to 70% of laboratories taking part in ability testing used LCMS/MS to provide outcomes when testing for the immunosuppressive medication TDM (ISD-TDM), as an example (Greaves, 2016; Seger et al., 2016). The limitations of protein and peptide immunoassays are the following: Objectives and dispute for LCMS/MS. Our expertise in fundamental protein science has already been significantly improved by MS technologies with a variety of front ends. It has been challenging to use this information in clinical situations. There have been numerous translational strives in the well-known sectors of cardiovascular and cancer diseases. As an outcome of this effort, numerous marker profiles with qualitative patterns were developed. Furthermore, incorrect signals that were driven about by preanalytical problems or different sample preparations between sites were found during the validation of these profiles (McLerran et al., 2008). Contrarily, targeted proteomics, or the deployment of LCMS/MS to known, validated peptide and protein biomarkers, has proven effective. LCMS/MS approaches for analyzing peptides or proteins can be beneficial in a variety of clinical settings, including: The analyte has no immunoassay available. Some important clinical problems cannot be answered by an existing immunoassay. A commonly used immunoassay is susceptible to interference. The analyte exists in a number of isoforms. The results of various tests for the same analyte demonstrate considerable diversity. Workflows for assays today are extremely demanding (Grebe & Singh, 2011). There are currently many broadcasted modals of focused clinical LCMS/MS assays of peptides and proteins that deal with such circumstances, and we will present some instances of such LCMS/MS assays that have been profitably validated, designed, and may be used in research laboratories.

3.10 MALDI TOF/TOF

MALDI TOF/MS, or matrix-assisted laser desorption ionization, or "time of flight" MS, is a faster, more accurate, and more economical approach for identifying and identifying microorganisms. This method can produce different mass spectral fingerprints for each bacterium, which makes it perfect for precise microbiological identification of species and genus levels. It also possesses the potential to be developed for strain type and reorganization (Anhalt & Fenselau, 1975). A mass spectrometry contains three types of functional units: (1) A source of ions sufficient for ionizing molecules in the sample and transmitting their ions into the gas phase; (2) equipment for mass analysis that differentiates ions depending on their m/z; and (3) a tool for observing ion separation. FAB, PD, atmospheric pressure chemical ionization (CI), MALDI, electrospray (ESI), and LD are just a couple of the ionization techniques that have been used. The kind of aim and the sample of the MS analysis decide the ionization technique; however, MALDI and ESI are soft ionization processes that allow the ionization and vaporization of immense, nonvolatile biological molecules like complete proteins (Emonet et al., 2010). MALDI yields mostly singly charged ions as compared to ESI; therefore, MALDI-derived spectra may contain more proteins. In order to characterize microorganisms, LD has been profitably associated with a diversity of mass analyzers, comprising TOF, Fourier-transform ion cyclotron resonance, quadrupole TOF, and quadrupole ion trap. As assessed by mass validity, mass scale, resolution, scan speed, susceptivity, and cost, the various working modes of mass analyzers have strengths and limitations. Mass analyzer performance can be increased by combining and/or summing the benefits of one kind of analyzer, tandem MS, or different analyzers through the development of multistage instruments like hybrid quadrupole TOF (Q-Q-TOF), triple quadrupole TOF (hybrid MS), and tandem TOF (TOF-TOF). Overall, the variety of material to be evaluated (complex or simple mixtures, proteins, peptides, lipids, and polysaccharides) and the purpose of the analysis define the mass analyzer predicted performance (protein identification, quantification, biotyping, and microorganism identification). According to their compatibility with pulse laser ionization, ability for quick analysis, and ability to be miniaturized, TOF mass analyzers have been used to identify intact microbes for a long time (De Carolis et al., 2014). When samples are prepared for MALDI analysis, they are mixed with a matrix, which causes the material to crystallize inside the matrix. Small acid molecules that have a high optical absorption in the laser wavelength range make the matrix. Depending on the biological molecule being analyzed and the type of laser being applied, the matrix properties change (Fenselau & Demirev, 2001).

3.11 NMR spectroscopy

NMR is a fundamental and frequently employed analytical method in both academic and commercial research. It provides a method for measuring complete molecule structures even in mixtures and allows for a singular and, in principle, quantitative measurement of the relative quantity of chemical groups. The first quantitative measurements (qNMR) were written about in the literature in 1963 by Jungnickel, Forbes, and Hollis. In proteomics, sophisticated technologies are required. In the first instance, the intramolecular proton ratios of 26 pure organic compounds were evaluated; in contrast, Hollis looked at the quantity fractions of three analytes, phenacetin, aspirin, and caffeine, in their respective combinations. However, due to certain features, qNMR has found broad applications despite the drawback of expensive NMR instrument costs, such as (i) the capability of determining molecular structures, (ii) when determining ratios, intensity calibrations are also not required (the nuclei count and signal area are directly correlated), (iii) relatively fast measurement periods, (iv) due to its lack of destructiveness, and (v) no isolation of the analyte beforehand. Quantitative 1H and 13C NMR in liquids is employed in a range of applications despite its poor precision, including agriculture, pharmacy, military applications, and material science, where determining a substance purity or content is the primary concern. This progress has been accelerated by important advancements in the sensitivity and homogeneity of high-field NMR spectrometers, as well as by trimming software tools that allow precise and accurate data processing and interpretation (Malz & Jancke, 2005). NMR spectroscopy examines the atomic-level organization and activity of molecules in three dimensions (3D). Nuclear magnetic resonance is the rhythmic response of nonzero-spin nuclei in a magnetic field to resonant excitation by radiofrequency (RF) radiation. Atoms with nuclear spins higher than zero lift their degeneracy when exposed to an external magnetic field, with an energy difference of ΔE given by the equation $\Delta E = \gamma\hbar(1\text{-}\sigma)B_0$. In this formula of the equation, γ is the gyromagnetic ratio; it is an essential characteristic of each isotope; B_0 is extremely strong, as the static magnetic field is; and σ is the chemical coating that protects a nucleus. The transitions between these nuclear-spin states can then be generated by electromagnetic radiation (Levitt, 2013). The transition frequencies are in the RF region of the electromagnetic spectrum with the normal 5−28 T magnetic fields utilized in NMR nowadays (213−1200 MHz 1H Larmor frequencies) NMR transition frequencies are sensitive to the nucleus electron dispersion, which protects it from the applied magnetic field. The shielding constant, σ, resulting in slightly different frequencies, varies for different nuclei of the same isotope in a molecule. Thus, NMR frequencies give clear details on the sample chemical composition (Facelli, 2011; Schmidt-Rohr & Spiess, 2012). The fractional difference between a given nucleus frequency and the frequency of a common substance like trimethylsilane is known as the chemical shift. Chemical shift differences among isotopes can be as low as 10 parts per million (ppm) for ^{1}H, as high as 200 ppm for ^{13}C, and as high as 1000 ppm for ^{17}O. Numerous couplings affect the NMR frequencies, in addition to chemical shifts: spin and spin scalar couplings, which are often in the 0−1 kHz range and depend on covalent bonding; spin and spin dipolar couplings, which are often in the 0−20 kHz range and depend on internuclear distances; and quadrupolar couplings between the electric field gradient at the nucleus and when the nuclear spin is greater than 1/2. Although all of these NMR interactions are anisotropic, sample orientation with respect to the magnetic field direction impacts them; it's indeed significant to mention. Due to quadrupolar couplings, internuclear couplings, and orientation-dependent chemical shifts, NMR spectra can encode 3D structural information. As a result of molecular rotations partially averaging these anisotropic interactions, studies of motionally averaged NMR spectra and the largest continuous-induced nuclear spin relaxation provide data on the geometries and rates of motion (Reif et al., 2021).

4. Present prospective of proteomics

Cancer is a very complicated illness with a wide range of molecular abnormalities that affect how it manifests clinically. Therefore, conventional studies of only one or even a few biological characteristics have shown that they are insufficient for precise illness outcome prediction. Clinical data and genome-wide expression can now be correlated, thanks to developments in gene expression profiling (Weinstein et al., 1997). Recently, proteomic profiles were used to establish similar associations (Voss et al., 2001). Importantly, it is anticipated that novel therapeutic targets and chemotherapy regimens tailored to the needs of particular patients will emerge in the near future as a result of the identification of functional components, or proteins, as accurate prognostic markers. The 2D-GE protein expression patterns of 24 B-cell chronic lymphocytic leukemia patients (B-CLL) were correlated with clinical staging, allowing the identification of 20 proteins with distinctive expression in the three patient distributions under investigation. Among them, patients with shorter life periods had higher levels of the gene HSP27. Survival periods were also linked with down-regulation of protein disulfide isomerase and thioredoxin peroxidase 2 (Voss et al., 2001). These proteins identification is particularly valuable for predicting outcomes in B-CLL patients, but with more research, it's possible that they could also serve as useful therapeutic targets. There are currently studies that are similar in various cancer typess. Other proteomics techniques can potentially be used to predict outcomes. As previously mentioned, identifying protein activity in particular pathways that are known to be active over the course of a disease or in response to treatment is a potent application of proteomics, and phosphorylation-specific arrays enable the simultaneous detection of pathway activation. Tissue microarrays are a different strategy that, as previously mentioned, can be very helpful for analyzing known sets of protein markers and require considerably less sample than 2D-gels (Voss et al., 2001). Large-scale proteome analysis of clinical samples can occasionally be prevented by a lack of access to relevant clinical data. The use of animal models for proteomics research can be a potent way to get around this during the discovery phase. Additionally, studies on in vivo drug response and drug resistance can be carried out under controlled experimental circumstances utilizing such models (Jacquemier et al., 2005).

Blood or urine proteomes are also being studied by several researchers due to the ease with which these samples may be obtained. In fact, studies on the effectiveness of the cyclooxygenase-2 inhibitor celecoxib for the treatment and prevention of cancer have revealed that serum profiling can predict response (Xiao et al., 2004). FAP, an inherited autosomal dominant disorder, has been observed to respond to celecoxib with varying degrees of success in preventing cancer (Steinbach et al., 2000). Xiao et al. discovered expression variations of numerous indicators that were altered following treatment with celecoxib using SELDI TOF MS serum proteome profiles from patients on this FAP/celecoxib clinical trial, and they identified one marker as a powerful discriminator between response and nonresponse (Poon et al., 2005). Additionally, SELDI-TOF MS serum profiling has been utilized to predict the development of liver fibrosis and cirrhosis in chronic hepatitis B infection (Steinbach et al., 2000) and, as previously mentioned, to forecast the success of SARS therapy (R. T. K. Pang et al., 2006).

Genes responsible for illness can also be found using proteomics, which is a very effective method. A pepstatin-insensitive lysosomal peptidase known as CLN2 was discovered to be absent in late-infantile neuronal ceroid lipofuscinosis (LINCL) patients compared to normal in a seminal study using 2D-GE and affinity chromatography of autopsy brain tissue from LINCL patients. LINCL is a fatal neurodegenerative disease whose defective gene has remained elusive. The sequencing study of this gene revealed important mutations in LINCL patients (Sleat et al., 1997). Future research into the plethora of proteomic markers seen in various disorders will undoubtedly uncover additional disease-causing genes.

5. Future prospective of proteomics

Proteomic approaches give information with clinical value since they track the expression of chemicals that directly affect cell phenotypic. These methods may be used to study protein expression in vitro or in vivo and can be applied to a wide range of species and biological materials. Thousands of proteins and their PTMs may be analyzed using proteomic methods, which also enable objective assessments of the molecular biology of disease states by emphasizing components that could otherwise be missed in hypothesis-driven situations. Proteomic research is a relatively new field of study, as was previously mentioned. Proteomics is now under fire for a number of issues, most notably inadequate reproducibility as shown by gel-to-gel and interlaboratory variance. Despite this, it's probable that when protocols and reporting requirements grow standardized and are used everywhere, these problems will become less of a concern. Although methods like western blotting may be utilized to get over this obstacle, the identification of compounds that are poorly expressed using MS remains a hurdle.

Proteomic investigations just offer lists of protein data with limited practical utility without following downstream analysis, and the application of proteomics to vascular disease is still in its infancy. Big datasets produced by proteomic studies need specialized processing, particularly when large or complicated experiments are involved. Proteomics ongoing application has stimulated advancements in data visualization, helping to develop bioinformatics as a specialized subject. Proteomic data are increasingly being presented as interaction networks, emphasizing relationships (such as synchronous up/down regulation) between identified molecules (Huang et al., 2009; Krycer et al., 2008). When examining disease processes or treatment routes, this is very useful. The most common use of proteomic methods in vascular medicine is to find biomarkers with potential for diagnostic, therapeutic, and prognostic purposes. Proteomic protein identifications, however, are still speculative and must be confirmed using more dependable antibodies and established methods like ELISA, western blotting, and immunohistochemistry. Additionally, high-throughput approaches must be used to verify molecules of interest in large patient populations in order to evaluate the performance of biomarkers, which takes time and is constrained by the availability of large patient cohorts.

Combining proteomics with other postgenomic methods considerably increases the exploratory capability of proteomics. The impact of protein expression on biological activities may be demonstrably shown by combined proteomic and metabolomic investigations. Perlman et al. (2009) explored the cardioprotective pathways after nitrate exposure using this method. Nitrate delivery led to significant increases in cardiac ascorbate oxidation, but only a transient rise in cardiac nitrate levels. After preconditioning with low (0.1 mg/kg) or high (10 mg/kg) nitrate dosages, this was accompanied by considerable improvements in heart contractile recovery following ischemia-reperfusion. A proteomic study of cardiac mitochondria demonstrated dose-dependent PTM of three protein isoforms implicated in cell metabolism, antioxidant defense, and serine/threonine kinase signaling. The scientists concluded as a result that a similar mechanism could be responsible for the cardioprotective benefits of exercise and a diet high in nitrite-/nitrate-rich foods (Perlman et al., 2009). Although very few vascular investigations have adopted this multiomics strategy, such systemic methodologies have been effectively used in other biological contexts. Therefore, it is likely that the future application of proteomics in a "multiomics" environment would considerably enhance vascular medicine.

6. Advance tools used for proteomics

Bioinformatics is a prerequisite for proteomics; therefore, its implications are increasing in step with the advancement of high-throughput techniques that depend on accurate information processing. These new and resulting exposures provide novel methods to handle enormous and different proteome data and improve the discovery process (Vihinen, 2001).

Endolysins are a class of antibacterial enzymes that are beneficial for preventing the spread of bacteria with multiple resistances. Domain switching, mutagenesis, or gene switching can change or enhance the antibacterial activity. Insilico examination of the protein domains found in prophage and phage endolysins has demonstrated the difficulty of creating particular endolysins. Multiple sequence alignments have been used to identify the sequence type, domain layout, and conserved amino acids after studying the combination of domains. Researchers have also looked at the kinds, numbers, and presence of binding domains in the endolysin sequence (Vidová et al., 2014). The distribution of plant food allergens into protein families was calculated using an in silico analytic technique, and the conserved surface needed for IgE cross-reactivity was identified. Four families of plant food allergen sequences were identified, demonstrating how conserved biological processes and structural features contribute to the induction of allergic characteristics (Jenkins et al., 2005).

Human Factor Xa (FXa), a blood coagulation enzyme, catalyzes the activation of prothrombin to thrombin and is crucial for thrombosis and hemostasis. Blood diseases are brought on by an imbalance in the activation of enzymes, which interferes with hemostasis. Direct inhibition of FXa may be used to create safe and efficient anticoagulants without affecting the thrombin activity required for proper hemostasis. Through Discovery Studio, a study assisted in the creation of more effective ligands. Sulfonamide compounds were identified as FXa inhibitors by docking investigations and binding confirmations (Abubacker et al., 2013).

The application of bioinformatics to proteomics has developed significantly during the last few years. The establishment of a new algorithm helps in the identification and quantification of proteins, allowing for the analysis of larger amounts of data with more specificity and accuracy, making it possible to get comprehensive information on the expression of proteins. The key issue with these kinds of analyses is how to manage such large amounts of data. It has become challenging to establish connections between proteomic data and other omics technologies, such as genomics and metabolomics. However, the effective instruments that might help to get within these constraints include database technology and new semantic statistical approaches.

The sample of proteins is collected, and one or more proteases are used to break down them in order to produce a specific set of peptides that may be used in MS (Wiśniewski & Mann, 2012). Additional techniques, such as enrichment and fractionation, can be used at the protein or peptide level to reduce the complexity of the sample or when the analysis of a specific protein subset is desired (Altelaar & Heck, 2012; Lee et al., 2010; Schmidt et al., 2014). The peptides are then examined using LC-MS. Common strategies include either quantitative inquiry for a specific collection of proteins by targeted MS or the study of deep proteome coverage by shotgun MS (Picotti & Aebersold, 2012; Schmidt et al., 2014). The sequence information in the produced spectra is essential for identifying proteins. With the m/z ratio, retention time (RT), peptide intensities, and fragmentation spectra collectively, the information obtained can be shown as a 3D map. The intensity of the mass-to-charge ratio for a particular peptide is plotted along the RTs to obtain the chromatographic peak. Proteins may be recognized by their fragmentation spectra, while peptides can be quantified using the region under this curve. The repositories where the proteomic data may be uploaded and used for database searches (Riffle & Eng, 2009). The most comprehensive proteome archives, including Proteome Commons, the PeptideAtlas project, the PRIDE proteomics identification database, and others, provide direct access to a significant amount of stored data and are helpful tools for data mining (Desiere et al., 2005; Vizcaino et al., 2013).

The protein pathways are a set of cellular processes that each have a specific biological consequence. There are many tools and databases available for the protein pathways since pathway databases integrate the proteins that are directly engaged in reactions with those that govern the pathways. The Knowledge Base for KEGG, Ingenuity, and Pathways Among the route databases with extensive information about metabolism, signaling, and interactions are Reactome and BioCarta (Croft et al., 2011; Kanehisa et al., 2012). These extensive databases have been supplemented by the development of specialized databases for signal transduction pathways like GenMAPP or PANTHER (Mi et al., 2007; Salomonis et al., 2007; Schaefer et al., 2009). Additionally, databases like NetPath have been created, which include cancer-related pathways useful for identifying proteins pertinent to a particular cancer type (Kandasamy et al., 2010).

New protein discoveries are made possible by the enhanced connectivity of these open databases. Most of the time, proteins interact with other proteins to create fleeting or persistent complexes rather than acting alone. For a thorough knowledge of a biological system, it is crucial to examine protein complexes and the circumstances that lead to their formation or dissociation. Proteins can be complexes with varying compositions and be rather complicated. Information on protein interactions in complexes may be found in databases like BioGRID, IntAct, MINT, and HRPD (Chatr-aryamontri et al., 2007; Kerrien et al., 2012; Schmidt et al., 2014). STRING connects to several other resources for literature mining in addition to being a popular database for information on protein interactions. Furthermore, by utilizing the STRING database and the specified list of genes and interactions, protein networks may be created (Franceschini et al., 2013; Glaab et al., 2012; Schmidt et al., 2014) Table 6.2 (Aslam et al., 2017).

TABLE 6.2 Variety of tools used for proteomics studies such as alignment tools, general protein sequence databases, structural analysis and prediction servers, and sequence similarity searches.

Sl. No.	Database	Type	Web links
1	GenBank	Database	http://www.ncbi.nih.gov/entrez/query.fcgi?db=protein
2	RefSeq	Database	https://www.ncbi.nlm.nih.gov/refseq/
3	Nr	Database	http://www.ncbi.nlm.nih.gov/BLAST/
4	UniProt	Database	http://www.pir.uniprot.org/
5	UniRef	Database	http://www.pir.uniprot.org/database/nref.shtml
6	UniParc	Database	http://www.pir.uniprot.org/database/archive.shtml
7	TrEMBL	Database	http://kr.expasy.org/sprot/
8	SwissProt	Database	http://kr.expasy.org/sprot/
9	PIR	Database	http://pir.georgetown.edu/
10	OWL	Database	http://www.bioinf.man.ac.uk/dbbrowser/OWL/
11	BLASTP	BLAST	http://blast.ncbi.nlm.nih.gov
12	TBLASTN	BLAST	https://blast.ncbi.nlm.nih.gov
13	PSI-BLAST	Position Specific Iterated BLAST	https://blast.ncbi.nlm.nih.gov
14	PHI-BLAST	Pattern Hit Initiated BLAST	https://blast.ncbi.nlm.nih.gov
15	DELTA-BLAST	Domain Enhanced Lookup Time Accelerated BLAST	https://blast.ncbi.nlm.nih.gov
16	InterProScan	Protein domain servers	http://www.ebi.ac.uk/InterProScan/

TABLE 6.2 Variety of tools used for proteomics studies such as alignment tools, general protein sequence databases, structural analysis and prediction servers, and sequence similarity searches.—cont'd

Sl. No.	Database	Type	Web links
17	CD server	Protein domain servers	http://www.ncbi.nlm.nih.gov/Structure/cdd/cdd.shtml
18	ProWleScan	Protein domain servers	http://hits.isb-sib.ch/cgi-bin/PFSCAN
19	ScanProsite	Protein domain servers	http://us.expasy.org/tools/scanprosite/
20	PATTINPROT	Protein motif search tools	http://pbil.ibcp.fr/html/pbiljndex.html
21	SIRW	Protein motif search tools	http://sirw.embl.de/index.html
22	Match Box	Motif based alignment server	http://www.sciences.fundp.ac.be/biologie/bms/help.html
23	MEME	Motif based alignment server	http://meme.sdsc.edu/meme/website/meme.html
24	Gibbs	Motif based alignment server	http://bayesweb.wadsworth.org/gibbs/gibbs.html
25	Dialign	Motif based alignment server	http://bibiserv.techfak.uni-bielefeld.de/dialign/
26	BlockMakei	Motif based alignment server	http://blocks.fhcrc.org/make_blocks.html
27	PDB	Protein structure databases	http://www.rcsb.org/pdb/
28	SwissModel	Protein structure databases	http://swissmodel.expasy.org/repository/
29	SCOP	Protein structure databases	http://scop.mrc-lmb.cam.ac.uk/scop/
30	ModBase	Protein structure databases	http://alto.compbio.ucsf.edu/modbase-cgi/index.cgi
31	CATH	Protein structure databases	http://www.biochem.ucl.ac.uk/bsm/cath/
32	MMDD	Protein structure databases	http://www.ncbi.nlm.nih.gov/Structure/
33	ConSurf	Protein structure analysis server	http://consurf.tau.ac.il/
34	CASTp	Protein structure analysis server	http://sts.bioe.uic.edu/castp/index.php
35	ProtSkin	Protein structure analysis server	http://www.mcgnmr.ca/ProtSkin/intro/
36	LigandProtein	Protein structure analysis server	http://bip.weizmann.ac.il/oca-bin/lpccsu
37	PredictProtein	Protein structure prediction sever	http://www.embl-heidelberg.de/predictprotein
38	O-GlycoBase	Protein structure prediction sever	http://www.cbs.dtu.dk/services/NetOGlyc/
39	PhosphoBase	Protein structure prediction sever	http://www.cbs.dtu.dk/services/NetPhos
40	SwissModel	Protein structure modeling server	http://www.expasy.org/swissmod
41	WhatIf	Protein structure modeling server	http://www.cmbi.kun.nl/gv/servers/WIWWWI
42	ESyPred3D	Protein structure modeling server	http://www.fundp.ac.be/urbm/bioinfo/esypred
43	EBI	Protein structure modeling server	http://biotech.ebi.ac.uk:8400/

From Aslam, B., Basit, M., Nisar, M. A., Khurshid, M., & Rasool, M. H. (2017). Proteomics: Technologies and their applications. *Journal of Chromatographic Science, 55*(2), 182–196. https://doi.org/10.1093/chromsci/bmw167

7. Conclusion

Proteomics technologies are based on bottom-up display as well as bottom-up identification exploiting peptides that have significantly and incredibly usefully advanced in the past years. It is commonly asserted, both explicitly and implicitly, that these approaches will ultimately replace more conventional top-down analytical techniques, particularly those based on 2D gels, as they provide higher sensitivity, greater speed, and broader proteome coverage. A whole protein collection may be probed using this superb high-throughput approach for a particular function or biochemistry. It is a fantastic new method for identifying novel functions for proteins that have been extensively investigated, as well as previously unknown multifunctional proteins. Results from large-scale experiments require a methodical and effective review: (1) Automated information extraction from user-defined sources to build a customized, large database; (2) a user-friendly graphical and query platform for displaying and analyzing experimental data inside the framework of the custom database; (3) effective use of web-based bioinformatics tools for modeling, function prediction, and data interpretation; and (4) reconfigurability and scalability. It is challenging to continue to create software solutions with the four crucial components listed above, especially in light of the growing body of publicly accessible knowledge and bioinformatics tools. The complete toolkit may be combined into a single linear pipeline to eliminate the bottleneck in data processing and analysis. The good news is that all of the various proteomics approaches are undergoing rapid advancement in technology, and in the near future, throughput, considerable improvements in sensitivity, and proteome coverage may all be anticipated for all.

References

Abubacker, S. M., Pavanchand, A., Basheer, S. B., Sriveena, K., Paul, R., & Enaganti, S. (2013). In silico assessment of factor Xa inhibitors by docking studies. *Vedic Research International Bioinformatics and Proteomics, 1*(1), 9. https://doi.org/10.14259/bp.v1i1.43

Aebersold, R., Hood, L. E., & Watts, J. D. (2000). Equipping scientists for the new biology. *Nature Biotechnology, 18*(4), 359. https://doi.org/10.1038/74325

Alban, A., David, S. O., Bjorkesten, L., Andersson, C., Sloge, E., Lewis, S., & Currie, I. (2003). A novel experimental design for comparative two-dimensional gel analysis: Two-dimensional difference gel electrophoresis incorporating a pooled internal standard. *Proteomics, 3*(1), 36–44. https://doi.org/10.1002/pmic.200390006

Altelaar, A. F. M., & Heck, A. J. R. (2012). Trends in ultrasensitive proteomics. *Current Opinion in Chemical Biology, 16*(1–2), 206–213. https://doi.org/10.1016/j.cbpa.2011.12.011

Anhalt, J. P., & Fenselau, C. (1975). Identification of bacteria using mass spectrometry. *Analytical Chemistry, 47*(2), 219–225. https://doi.org/10.1021/ac60352a007

Asara, J. M., Christofk, H. R., Freimark, L. M., & Cantley, L. C. (2008). A label-free quantification method by MS/MS TIC compared to SILAC and spectral counting in a proteomics screen. *Proteomics, 8*(5), 994–999. https://doi.org/10.1002/pmic.200700426

Aslam, B., Basit, M., Nisar, M. A., Khurshid, M., & Rasool, M. H. (2017). Proteomics: Technologies and their applications. *Journal of Chromatographic Science, 55*(2), 182–196. https://doi.org/10.1093/chromsci/bmw167. http://chromsci.oxfordjournals.org/content

Bertone, P., & Snyder, M. (2005). Advances in functional protein microarray technology. *FEBS Journal, 272*(21), 5400–5411. https://doi.org/10.1111/j.1742-4658.2005.04970.x

Bjellqvist, B., Ek, K., Giorgio Righetti, P., Gianazza, E., Görg, A., Westermeier, R., & Postel, W. (1982). Isoelectric focusing in immobilized pH gradients: Principle, methodology and some applications. *Journal of Biochemical and Biophysical Methods, 6*(4), 317–339. https://doi.org/10.1016/0165-022X(82)90013-6

Bondarenko, P. V., Chelius, D., & Shaler, T. A. (2002). Identification and relative quantitation of protein mixtures by enzymatic digestion followed by capillary reversed-phase liquid chromatography—Tandem mass spectrometry. *Analytical Chemistry, 74*(18), 4741–4749. https://doi.org/10.1021/ac0256991

Braisted, J. C., Kuntumalla, S., Vogel, C., Marcotte, E. M., Rodrigues, A. R., Wang, R., Huang, S. T., Ferlanti, E. S., Saeed, A. I., Fleischmann, R. D., Peterson, S. N., & Pieper, R. (2008). The APEX quantitative proteomics tool: Generating protein quantitation estimates from LC-MS/MS proteomics results. *BMC Bioinformatics, 9*. https://doi.org/10.1186/1471-2105-9-529

Chatr-aryamontri, A., Ceol, A., Palazzi, L. M., Nardelli, G., Schneider, M. V., Castagnoli, L., & Cesareni, G. (2007). MINT: The Molecular INTeraction database. *Nucleic Acids Research, 35*(1), D572–D574. https://doi.org/10.1093/nar/gkl950

Chelius, D., & Bondarenko, P. V. (2002). Quantitative profiling of proteins in complex mixtures using liquid chromatography and mass spectrometry. *Journal of Proteome Research, 1*(4), 317–323. https://doi.org/10.1021/pr025517j

Coarfa, C., Grimm, S. L., Rajapakshe, K., Perera, D., Lu, H. Y., Wang, X., Christensen, K. R., Mo, Q., Edwards, D. P., & Huang, S. (2021). Reverse-phase protein array: Technology, application, data processing, and integration. *Journal of Biomolecular Techniques, 32*(1), 15–29. https://doi.org/10.7171/jbt.21-3202-001. https://www.ncbi.nlm.nih.gov/pmc/articles/PMC8121080/pdf/jbt.32-15.pdf

Colangelo, C. M., & Williams, K. R. (2006). Isotope-coded affinity tags for protein quantification. *Methods in Molecular Biology, 328*, 151–158. https://doi.org/10.1385/1-59745-026-x:151

Creighton, C. J., & Huang, S. (2015). Reverse phase protein arrays in signaling pathways: A data integration perspective. *Drug Design, Development and Therapy, 9*, 3519–3527. https://doi.org/10.2147/DDDT.S38375. http://www.dovepress.com/getfile.php?fileID=25837

Croft, D., O'Kelly, G., Wu, G., Haw, R., Gillespie, M., Matthews, L., Caudy, M., Garapati, P., Gopinath, G., Jassal, B., Jupe, S., Kalatskaya, I., MayMahajan, S., May, B., Ndegwa, N., Schmidt, E., Shamovsky, V., Yung, C., Birney, E., & Stein, L. (2011). Reactome: A database of reactions, pathways and biological processes. *Nucleic Acids Research, 39*(1), D691–D697. https://doi.org/10.1093/nar/gkq1018

De Carolis, E., Vella, A., Vaccaro, L., Torelli, R., Spanu, T., Fiori, B., Posteraro, B., & Sanguinetti, M. (2014). Application of MALDI-TOF mass spectrometry in clinical diagnostic microbiology. *Journal of Infection in Developing Countries, 8*(9), 1081–1088. https://doi.org/10.3855/jidc.3623. http://www.jidc.org/index.php/journal/article/view/4451/1145

Desiere, F., Deutsch, E. W., Nesvizhskii, A. I., Mallick, P., King, N. L., Eng, J. K., Aderem, A., Boyle, R., Brunner, E., Donohoe, S., Fausto, N., Hafen, E., Hood, L., Katze, M. G., Kennedy, K. A., Kregenow, F., Lee, H., Lin, B., Martin, D., & Aebersold, R. (2005). Integration with the human genome of peptide sequences obtained by high-throughput mass spectrometry. *Genome Biology, 6*(1), R9.

Diez, R., Osorio, C., Alzate, O., & Herbstreith, M. (2010). *2-D fluorescence difference gel electrophoresis (DIGE) in neuroproteomics*. Informa UK Limited. https://doi.org/10.1201/9781420076264.ch4

Dong, M. Q., Venable, J. D., Au, N., Xu, T., Sung, K. P., Cociorva, D., Johnson, J. R., Dillin, A., & Yates, J. R. (2007). Quantitative mass spectrometry identifies insulin signaling targets in C. elegans. *Science, 317*(5838), 660–663. https://doi.org/10.1126/science.1139952

Ducret, A., Van Oostveen, I., Eng, J. K., Yates, J. R., & Aebersold, R. (1998). High throughput protein characterization by automated reverse-phase chromatography/electrospray tandem mass spectrometry. *Protein Science, 7*(3), 706–719. https://doi.org/10.1002/pro.5560070320. http://onlinelibrary.wiley.com/journal/10.1002/(ISSN)1469-896X

Emonet, S., Shah, H. N., Cherkaoui, A., & Schrenzel, J. (2010). Application and use of various mass spectrometry methods in clinical microbiology. *Clinical Microbiology and Infection, 16*(11), 1604–1613. https://doi.org/10.1111/j.1469-0691.2010.03368.x. http://onlinelibrary.wiley.com/journal/10.1111/(ISSN)1469-0691

Facelli, J. C. (2011). Chemical shift tensors: Theory and application to molecular structural problems. *Progress in Nuclear Magnetic Resonance Spectroscopy, 58*(3–4), 176–201. https://doi.org/10.1016/j.pnmrs.2010.10.003. http://www.sciencedirect.com/science/journal/00796565

Fatima, N., Chelius, D., Luke, B. T., Yi, M., Zhang, T., Stauffer, S., Stephens, R., Lynch, P., Miller, K., Guszczynski, T., Boring, D., Greenwald, P., & Ali, I. U. (2009). Label-free global serum proteomic profiling reveals novel celecoxib-modulated proteins in familial adenomatous polyposis patients. *Cancer Genomics & Proteomics, 6*(1), 41–50. http://cgp.iiarjournals.org.

Fenn, J. B., Mann, M., Meng, C. K., Wong, S. F., & Whitehouse, C. M. (1989). Electrospray ionization for mass spectrometry of large biomolecules. *Science, 246*(4926), 64–71. https://doi.org/10.1126/science.2675315

Fenselau, C., & Demirev, F. A. (2001). Characterization of intact microorganisms by MALDI mass spectrometry. *Mass Spectrometry Reviews, 20*(4), 157–171. https://doi.org/10.1002/mas.10004

Fernandez-Patron, C., Castellanos-Serra, L., Hardy, E., Guerra, M., Estevez, E., Mehl, E., & Frank, R. W. (1998). Understanding the mechanism of the zinc-ion stains of biomacromolecules in electrophoresis gels: Generalization of the reverse-staining technique. *Electrophoresis, 19*(14), 2398–2406. https://doi.org/10.1002/elps.1150191407

Florens, L., Carozza, M. J., Swanson, S. K., Fournier, M., Coleman, M. K., Workman, J. L., & Washburn, M. P. (2006). Analyzing chromatin remodeling complexes using shotgun proteomics and normalized spectral abundance factors. *Methods, 40*(4), 303–311. https://doi.org/10.1016/j.ymeth.2006.07.028

Franceschini, A., Szklarczyk, D., Frankild, S., Kuhn, M., Simonovic, M., Roth, A., Lin, J., Minguez, P., Bork, P., Von Mering, C., & Jensen, L. J. (2013). STRING v9.1: Protein-protein interaction networks, with increased coverage and integration. *Nucleic Acids Research, 41*(1), D808–D815. https://doi.org/10.1093/nar/gks1094

Görg, A., Boguth, G., Obermaier, C., Posch, A., & Weiss, W. (1995). Two-dimensional polyacrylamide gel electrophoresis with immobilized pH gradients in the first dimension (IPG-Dalt): The state of the art and the controversy of vertical versus horizontal systems. *Electrophoresis, 16*(1), 1079–1086. https://doi.org/10.1002/elps.11501601183

Görg, A., Obermaier, C., Boguth, G., Harder, A., Scheibe, B., Wildgruber, R., & Weiss, W. (2000). The current state of two-dimensional electrophoresis with immobilized pH gradients. *Electrophoresis, 21*(6), 1037–1053. https://doi.org/10.1002/(SICI)1522-2683(20000401)21:6<1037::AID-ELPS1037>3.0.CO;2-V

Gafken, P. R., & Lampe, P. D. (2006). Methodologies for characterizing phosphoproteins by mass spectrometry. *Cell Communication and Adhesion, 13*(5–6), 249–262. https://doi.org/10.1080/15419060601077917

Glaab, E., Baudot, A., Krasnogor, N., Schneider, R., & Valencia, A. (2012). EnrichNet: Network-based gene set enrichment analysis. *Bioinformatics, 28*(18), i451–i457. https://doi.org/10.1093/bioinformatics/bts389

Goldfarb, M. (2007). Computer analysis of two-dimensional gels. *Journal of Biomolecular Techniques, 18*(3), 143–146. http://www.ncbi.nlm.nih.gov/pmc/articles/PMC2062544/pdf/jbt-18-143.pdf.

Greaves, R. F. (2016). Recent advances in the clinical application of mass spectrometry. *The Journal of the International Federation of Clinical Chemistry and Laboratory Medicine, 27*(4), 264–271. https://www.ncbi.nlm.nih.gov/pmc/articles/PMC5282912/.

Grebe, S. K. G., & Singh, R. J. (2011). LC-MS/MS in the clinical laboratory—Where to from here? *Clinical Biochemist Reviews, 32*(1), 5–31. http://www.ncbi.nlm.nih.gov/pmc/articles/PMC3052391/pdf/cbr_32_1_005.pdf.

Griffiths, J. (2008). A brief history of mass spectrometry. *Analytical Chemistry, 80*(15), 5678–5683. https://doi.org/10.1021/ac8013065

Grubb, R. L., Deng, J., Pinto, P. A., Mohler, J. L., Chinnaiyan, A., Rubin, M., Linehan, W. M., Liotta, L. A., Petricoin, E. F., & Wulfkuhle, J. D. (2009). Pathway biomarker profiling of localized and metastatic human prostate cancer reveal metastatic and prognostic signatures. *Journal of Proteome Research, 8*(6), 3044–3054. https://doi.org/10.1021/pr8009337. http://pubs.acs.org/doi/pdfplus/10.1021/pr8009337

Gygi, S. P., Rist, B., Gerber, S. A., Turecek, F., Gelb, M. H., & Aebersold, R. (1999). Quantitative analysis of complex protein mixtures using isotope-coded affinity tags. *Nature Biotechnology, 17*(10), 994–999. https://doi.org/10.1038/13690

Hall, D. A., Ptacek, J., & Snyder, M. (2007). Protein microarray technology. *Mechanisms of Ageing and Development, 128*(1), 161–167. https://doi.org/10.1016/j.mad.2006.11.021

Hall, D. A., Zhu, H., Zhu, X., Royce, T., Gerstein, M., & Snyder, M. (2004). Regulation of gene expression by a metabolic enzyme. *Science, 306*(5695), 482–484. https://doi.org/10.1126/science.1096773

Higgs, R. E., Knierman, M. D., Gelfanova, V., Butler, J. P., & Hale, J. E. (2005). Comprehensive label-free method for the relative quantification of proteins from biological samples. *Journal of Proteome Research, 4*(4), 1442–1450. https://doi.org/10.1021/pr050109b

Huang, J. T. J., McKenna, T., Hughes, C., Leweke, F. M., Schwarz, E., & Bahn, S. (2007). CSF biomarker discovery using label-free nano-LC-MS based proteomic profiling: Technical aspects. *Journal of Separation Science, 30*(2), 214–225. https://doi.org/10.1002/jssc.200600350

Huang, S. K., Darfler, M. M., & Nicholl, M. B. (2009). LC/MSbased quantitative proteomic analysis of paraffin-embedded archival melanomas reveals potential proteomic biomarkers associated with metastasis. *PLoS One, 4*.

Ishihama, Y., Oda, Y., Tabata, T., Sato, T., Nagasu, T., Rappsilber, J., & Mann, M. (2005). Exponentially modified protein abundance index (emPAI) for estimation of absolute protein amount in proteomics by the number of sequenced peptides per protein. *Molecular and Cellular Proteomics, 4*(9), 1265–1272. https://doi.org/10.1074/mcp.M500061-MCP200

Jacquemier, J., Ginestier, C., Rougemont, J., Bardou, V. J., Charafe-Jauffret, E., Geneix, J., Adélaïde, J., Koki, A., Houvenaeghel, G., Hassoun, J., Maraninchi, D., Viens, P., Birnbaum, D., & Bertucci, F. (2005). Protein expression profiling identifies subclasses of breast cancer and predicts prognosis. *Cancer Research, 65*(3), 767–779.

Jenkins, J. A., Griffiths-Jones, S., Shewry, P. R., Breiteneder, H., & Mills, E. N. C. (2005). Structural relatedness of plant food allergens with specific reference to cross-reactive allergens: An in silico analysis. *Journal of Allergy and Clinical Immunology, 115*(1), 163–170. https://doi.org/10.1016/j.jaci.2004.10.026

Kandasamy, K., Mohan, S., Raju, R., Keerthikumar, S., Kumar, G. S. S., Venugopal, A. K., Telikicherla, D., Navarro, D. J., Mathivanan, S., Pecquet, C., Gollapudi, S. K., Tattikota, S. G., Mohan, S., Padhukasahasram, H., Subbannayya, Y., Goel, R., Jacob, H. K. C., Zhong, J., Sekhar, R., & Pandey, A. (2010). NetPath: A public resource of curated signal transduction pathways. *Genome Biology, 11*(1), R3. https://doi.org/10.1186/gb-2010-11-1-r3

Kanehisa, M., Goto, S., Sato, Y., Furumichi, M., & Tanabe, M. (2012). KEGG for integration and interpretation of large-scale molecular data sets. *Nucleic Acids Research, 40*(1), D109–D114. https://doi.org/10.1093/nar/gkr988

Kerrien, S., Aranda, B., Breuza, L., Bridge, A., Broackes-Carter, F., Chen, C., Duesbury, M., Dumousseau, M., Feuermann, M., Hinz, U., Jandrasits, C., Jimenez, R. C., Khadake, J., Mahadevan, U., Masson, P., Pedruzzi, I., Pfeiffenberger, E., Porras, P., Raghunath, A., & Hermjakob, H. (2012). The IntAct molecular interaction database in 2012. *Nucleic Acids Research, 40*(1), D841–D846. https://doi.org/10.1093/nar/gkr1088

Klose, J., & Kobalz, U. (1995). Two-dimensional electrophoresis of proteins: An updated protocol and implications for a functional analysis of the genome. *Electrophoresis, 16*(1), 1034–1059. https://doi.org/10.1002/elps.11501601175

Klose, J. (1975). Protein mapping by combined isoelectric focusing and electrophoresis of mouse tissues—A novel approach to testing for induced point mutations in mammals. *Human Genetics, 26*(3), 231–243. https://doi.org/10.1007/BF00281458

Krycer, J. R., Pang, C. N., & Wilkins, M. R. (2008). High throughput protein protein interaction data: Clues for the architecture of protein complexes. *Proteome Science, 26*(6).

Lee, Y. H., Tan, H. T., & Chung, M. C. M. (2010). Subcellular fractionation methods and strategies for proteomics. *Proteomics, 10*(22), 3935–3956. https://doi.org/10.1002/pmic.201000289

Lengqvist, J., Andrade, J., Yang, Y., Alvelius, G., Lewensohn, R., & Lehtiö, J. (2009). Robustness and accuracy of high speed LC–MS separations for global peptide quantitation and biomarker discovery. *Journal of Chromatography B, 877*(13), 1306–1316. https://doi.org/10.1016/j.jchromb.2009.02.052

Levin, Y., Schwarz, E., Wang, L., Leweke, F. M., & Bahn, S. (2007). Label-free LC-MS/MS quantitative proteomics for large-scale biomarker discovery in complex samples. *Journal of Separation Science, 30*(14), 2198–2203. https://doi.org/10.1002/jssc.200700189

Levitt, M. H. (2013). *Spin dynamics: Basics of nuclear magnetic resonance*. John Wiley & Sons.

Liu, H., Sadygov, R. G., & Yates, J. R. (2004). A model for random sampling and estimation of relative protein abundance in shotgun proteomics. *Analytical Chemistry, 76*(14), 4193–4201. https://doi.org/10.1021/ac0498563

Loos, G., Van Schepdael, A., & Cabooter, D. (2016). Quantitative mass spectrometry methods for pharmaceutical analysis. *Philosophical Transactions of the Royal Society A: Mathematical, Physical & Engineering Sciences, 374*(2079), 20150366. https://doi.org/10.1098/rsta.2015.0366

Lu, P., Vogel, C., Wang, R., Yao, X., & Marcotte, E. M. (2007). Absolute protein expression profiling estimates the relative contributions of transcriptional and translational regulation. *Nature Biotechnology, 25*(1), 117–124. https://doi.org/10.1038/nbt1270

Mackintosh, J. A., Choi, H. Y., Bae, S. H., Veal, D. A., Bell, P. J., Ferrari, B. C., Van Dyk, D. D., Verrills, N. M., Paik, Y. K., & Karuso, P. (2003). A fluorescent natural product for ultra sensitive detection of proteins in one-dimensional and two-dimensional gel electrophoresis. *Proteomics, 3*(12), 2273–2288. https://doi.org/10.1002/pmic.200300578

Malz, F., & Jancke, H. (2005). Validation of quantitative NMR. *Journal of Pharmaceutical and Biomedical Analysis, 38*(5), 813–823. https://doi.org/10.1016/j.jpba.2005.01.043. http://www.elsevier.com/locate/jpba

McLerran, D., Grizzle, W. E., Feng, Z., Thompson, I. M., Bigbee, W. L., Cazares, L. H., Chan, D. W., Dahlgren, J., Diaz, J., Kagan, J., Lin, D. W., Malik, G., Oelschlager, D., Partin, A., Randolph, T. W., Sokoll, L., Srivastava, S., Srivastava, S., Thornquist, M., & Semmes, O. J. (2008). SELDI-TOF MS whole serum proteomic profiling with IMAC surface does not reliably detect prostate cancer. *Clinical Chemistry, 54*(1), 53–60. https://doi.org/10.1373/clinchem.2007.091496. http://www.clinchem.org/cgi/reprint/54/1/53

Mi, H., Guo, N., Kejariwal, A., & Thomas, P. D. (2007). PANTHER version 6: Protein sequence and function evolution data with expanded representation of biological pathways. *Nucleic Acids Research, 35*(1), D247–D252. https://doi.org/10.1093/nar/gkl869

Minden, J. (2007). Comparative proteomics and difference gel electrophoresis. *Biotechniques, 43*(6), 739–745. https://doi.org/10.2144/000112653

Mueller, L. N., Brusniak, M. Y., Mani, D. R., & Aebersold, R. (2008). An assessment of software solutions for the analysis of mass spectrometry based quantitative proteomics data. *Journal of Proteome Research, 7*(1), 51–61. https://doi.org/10.1021/pr700758r

Neubert, H., Bonnert, T. P., Rumpel, K., Hunt, B. T., Henle, E. S., & Lames, I. T. (2008). Label-Free detection of differential protein expression by LC/MALDI mass spectrometry. *Journal of Proteome Research, 7*(6), 2270–2279. https://doi.org/10.1021/pr700705u

Neuhoff, V., Arold, N., Taube, D., & Ehrhardt, W. (1988). Improved staining of proteins in polyacrylamide gels including isoelectric focusing gels with clear background at nanogram sensitivity using Coomassie Brilliant Blue G-250 and R-250. *Electrophoresis, 9*(6), 255–262. https://doi.org/10.1002/elps.1150090603

Ong, S. E., & Mann, M. (2005). Mass spectrometry–based proteomics turns quantitative. *Nature Chemical Biology, 1*(5), 252–262. https://doi.org/10.1038/nchembio736

Ong, S.-E., & Mann, M. (2006). A practical recipe for stable isotope labeling by amino acids in cell culture (SILAC). *Nature Protocols, 1*(6), 2650–2660. https://doi.org/10.1038/nprot.2006.427

Pan, J., Chen, H. Q., Sun, Y. H., Zhang, J. H., & Luo, X. Y. (2008). Comparative proteomic analysis of non-small-cell lung cancer and normal controls using serum label-free quantitative shotgun technology. *Lung, 186*(4), 255–261. https://doi.org/10.1007/s00408-008-9093-7

Pang, J. X., Ginanni, N., Dongre, A. R., Hefta, S. A., & Opiteck, G. J. (2002). Biomarker discovery in urine by proteomics. *Journal of Proteome Research, 1*(2), 161–169. https://doi.org/10.1021/pr015518w

Pang, R. T. K., Poon, T. C. W., Chan, K. C. A., Lee, N. L. S., Chiu, R. W. K., Tong, Y. K., Wong, R. M. Y., Chim, S. S. C., Ngai, S. M., Sung, J. J. Y., & Lo, Y. M. D. (2006). Serum proteomic fingerprints of adult patients with severe acute respiratory syndrome. *Clinical Chemistry, 52*(3), 421–429. https://doi.org/10.1373/clinchem.2005.061689

Paoletti, A. C., Parmely, T. J., Tomomori-Sato, C., Sato, S., Zhu, D., Conaway, R. C., Conaway, J. W., Florens, L., & Washburn, M. P. (2006). Quantitative proteomic analysis of distinct mammalian Mediator complexes using normalized spectral abundance factors. *Proceedings of the National Academy of Sciences of the United States of America, 103*(50), 18928–18933. https://doi.org/10.1073/pnas.0606379103. http://www.pnas.org/cgi/reprint/103/50/18928

Perlman, D. H., Bauer, S. M., Ashrafian, H., Bryan, N. S., Garcia-Saura, M. F., Lim, C. C., Fernandez, B. O., Infusini, G., McComb, M. E., Costello, C. E., & Feelisch, M. (2009). Mechanistic insights into nitrite-induced cardioprotection using an integrated metabolomic/proteomic approach. *Circulation Research, 104*(6), 796–804. https://doi.org/10.1161/CIRCRESAHA.108.187005

Picotti, P., & Aebersold, R. (2012). Selected reaction monitoring-based proteomics: Workflows, potential, pitfalls and future directions. *Nature Methods, 9*(6), 555–566. https://doi.org/10.1038/nmeth.2015

Poon, T. C. W., Hui, A. Y., Chan, H. L. Y., Ang, I. L., Chow, S. M., Wong, N., & Sung, J. J. Y. (2005). Prediction of liver fibrosis and cirrhosis in chronic hepatitis B infection by serum proteomic fingerprinting: A pilot study. *Clinical Chemistry, 51*(2), 328–335. https://doi.org/10.1373/clinchem.2004.041764

Rabilloud, T., & Lelong, C. (2011). Two-dimensional gel electrophoresis in proteomics: A tutorial. *Journal of Proteomics, 74*(10), 1829–1841. https://doi.org/10.1016/j.jprot.2011.05.040

Rabilloud, T., Strub, J. M., Luche, S., Van Dorsselaer, A., & Lunardi, J. (2001). A comparison between Sypro Ruby and ruthenium ii tris (bathophenanthroline disulfonate) as fluorescent stains for protein detection in gels. *Proteomics, 1*(5), 699–704. https://doi.org/10.1002/1615-9861(200104)1:5<699::AID-PROT699>3.0.CO;2-C. http://onlinelibrary.wiley.com/journal/10.1002/(ISSN)1615-9861

Rajcevic, U., Niclou, S. P., & Jimenez, C. R. (2009). Proteomics strategies for target identification and biomarker discovery in cancer. *Frontiers in Bioscience, 14*(9), 3292–3303. https://doi.org/10.2735/3452. https://www.fbscience.com/archives

Rao, P. V., Reddy, A. P., Lu, X., Dasari, S., Krishnaprasad, A., Biggs, E., Roberts, C. T., & Nagalla, S. R. (2009). Proteomic identification of salivary biomarkers of type-2 diabetes. *Journal of Proteome Research, 8*(1), 239–245. https://doi.org/10.1021/pr8003776

Rappsilber, J., Ryder, U., Lamond, A. I., & Mann, M. (2002). Large-scale proteomic analysis of the human spliceosome. *Genome Research, 12*(8), 1231–1245. https://doi.org/10.1101/gr.473902. http://www.genome.org/cgi/reprint/12/8/1231

Reif, B., Ashbrook, S. E., Emsley, L., & Hong, M. (2021). Solid-state NMR spectroscopy. *Nature Reviews Methods Primers, 1*(1). https://doi.org/10.1038/s43586-020-00002-1. https://www.nature.com/nrmp/journal-information

Riffle, M., & Eng, J. K. (2009). Proteomics data repositories. *Proteomics, 9*(20), 4653–4663. https://doi.org/10.1002/pmic.200900216. http://www3.interscience.wiley.com/cgi-bin/fulltext/122613662/PDFSTART

Ross, P. L., Huang, Y. N., Marchese, J. N., Williamson, B., Parker, K., Hattan, S., Khainovski, N., Pillai, S., Dey, S., Daniels, S., Purkayastha, S., Juhasz, P., Martin, S., Bartlet-Jones, M., He, F., Jacobson, A., & Pappin, D. J. (2004). Multiplexed protein quantitation in Saccharomyces cerevisiae using amine-reactive isobaric tagging reagents. *Molecular and Cellular Proteomics, 3*(12), 1154–1169. https://doi.org/10.1074/mcp.M400129-MCP200. http://www.mcponline.org/content/by/year

Salomonis, N., Hanspers, K., Zambon, A. C., Vranizan, K., Lawlor, S. C., Dahlquist, K. D., Doniger, S. W., Stuart, J., Conklin, B. R., & Pico, A. R. (2007). GenMAPP 2: New features and resources for pathway analysis. *BMC Bioinformatics, 8*. https://doi.org/10.1186/1471-2105-8-217

Schaefer, C. F., Anthony, K., Krupa, S., Buchoff, J., Day, M., Hannay, T., & Buetow, K. H. (2009). PID: The pathway interaction database. *Nucleic Acids Research, 37*(1), D674–D679. https://doi.org/10.1093/nar/gkn653

Schena, M., Shalon, D., Davis, R. W., & Brown, P. O. (1995). Quantitative monitoring of gene expression patterns with a complementary DNA microarray. *Science, 270*(5235), 467–470. https://doi.org/10.1126/science.270.5235.467

Schmidt, A., Forne, I., & Imhof, A. (2014). Bioinformatic analysis of proteomics data. *BMC Systems Biology, 8*, S3. https://doi.org/10.1186/1752-0509-8-S2-S3

Schmidt-Rohr, K., & Spiess, H. W. (2012). *Multidimensional solid-state NMR and polymers* (pp. 1–478). Germany: Elsevier Ltd. https://doi.org/10.1016/C2009-0-21335-3. http://www.sciencedirect.com/science/book/9780080925622

Seger, C., Shipkova, M., Christians, U., Billaud, E. M., Wang, P., Holt, D. W., Brunet, M., Kunicki, P. K., Pawinski, T., Langman, L. J., Marquet, P., Oellerich, M., Wieland, E., & Wallemacq, P. (2016). Assuring the proper analytical performance of measurement procedures for immunosuppressive drug concentrations in clinical practice: Recommendations of the international association of therapeutic drug monitoring and clinical toxicology immunosuppressive drug scientific committee. *Therapeutic Drug Monitoring, 38*(2), 170–189. https://doi.org/10.1097/FTD.0000000000000269. http://journals.lww.com/drug-monitoring

Seyfried, N. T., Huysentruyt, L. C., Atwood, J. A., Xia, Q., Seyfried, T. N., & Orlando, R. (2008). Up-regulation of NG2 proteoglycan and interferon-induced transmembrane proteins 1 and 3 in mouse astrocytoma: A membrane proteomics approach. *Cancer Letters, 263*(2), 243–252. https://doi.org/10.1016/j.canlet.2008.01.007

Sleat, D. E., Donnelly, R. J., Lackland, H., Liu, C. G., Sohar, I., Pullarkat, R. K., & Lobel, P. (1997). Association of mutations in a lysosomal protein with classical late- infantile neuronal ceroid lipofuscinosis. *Science, 277*(5333), 1802–1805. https://doi.org/10.1126/science.277.5333.1802

Speer, R., Wulfkuhle, J. D., Liotta, L. A., & Petricoin, E. F. (2005). Reverse-phase protein microarrays for tissue-based analysis. *Current Opinion in Molecular Therapeutics, 7*(3), 240–245.

Sreekumar, A., Nyati, M. K., Varambally, S., Barrette, T. R., Ghosh, D., Lawrence, T. S., & Chinnaiyan, A. M. (2001). Profiling of cancer cells using protein microarrays: Discovery of novel radiation-regulated proteins. *Cancer Research, 61*(20), 7585–7593.

Stahnke, H., Kittlaus, S., Kempe, G., & Alder, L. (2012). Reduction of matrix effects in liquid chromatography-electrospray ionization-mass spectrometry by dilution of the sample extracts: How much dilution is needed? *Analytical Chemistry, 84*(3), 1474–1482. https://doi.org/10.1021/ac202661j

Steinbach, G., Lynch, P. M., Phillips, R. K. S., Wallace, M. H., Hawk, E., Gordon, G. B., Wakabayashi, N., Saunders, B., Shen, Y., Fujimura, T., Su, L. K., Levin, B., Godio, L., Patterson, S., Rodriguez-Bigas, M. A., Jester, S. L., King, K. L., Schumacher, M., Abbruzzese, J., & Kelloff, G. (2000). The effect of celecoxib, a cyclooxygenase-2 inhibitor, in familial adenomatous polyposis. *New England Journal of Medicine, 342*(26), 1946–1952. https://doi.org/10.1056/NEJM200006293422603

Tao, S. C., Chen, C. S., & Zhu, H. (2007). Applications of protein microarray technology. *Combinatorial Chemistry and High Throughput Screening, 10*(8), 706–718. https://doi.org/10.2174/138620707782507386. http://docserver.ingentaconnect.com/deliver/connect/ben/13862073/v10n8/s8.pdf?expires=1194862519&id=40586010&titleid=3879&accname=Elsevier&checksum=CE6AFBF3E600C9F84142F21051349447

Ünlü, M., Morgan, M. E., & Minden, J. S. (1997). Difference gel electrophoresis: A single gel method for detecting changes in protein extracts. *Electrophoresis, 18*(11), 2071–2077. https://doi.org/10.1002/elps.1150181133. http://onlinelibrary.wiley.com/doi/10.1002/elps.v34.9-10/issuetoc

Vélez-Bermúdez, I. C., Wen, T. N., Lan, P., & Schmidt, W. (2016). Isobaric tag for relative and absolute quantitation (iTRAQ)-based protein profiling in plants. *Methods in Molecular Biology, 1450*, 213–221. https://doi.org/10.1007/978-1-4939-3759-2_17. http://www.springer.com/series/7651

Vaudel, M., Burkhart, J. M., Zahedi, R. P., Martens, L., & Sickmann, A. (2012). ITRAQ data interpretation. *Methods in Molecular Biology, 893*, 501–509. https://doi.org/10.1007/978-1-61779-885-6_30

Verplaetse, R., Cuypers, E., & Tytgat, J. (2012). The evaluation of the applicability of a high pH mobile phase in ultrahigh performance liquid chromatography tandem mass spectrometry analysis of benzodiazepines and benzodiazepine-like hypnotics in urine and blood. *Journal of Chromatography A, 1249*, 147–154. https://doi.org/10.1016/j.chroma.2012.06.023

Verplaetse, R., & Henion, J. (2016). Quantitative determination of opioids in whole blood using fully automated dried blood spot desorption coupled to on-line SPE-LC-MS/MS. *Drug Testing and Analysis, 8*(1), 30–38. https://doi.org/10.1002/dta.1927. http://onlinelibrary.wiley.com/journal/10.1002/(ISSN)1942-7611

Vidová, B., Šramková, Z., Tišáková, L., Oravkinová, M., & Godány, A. (2014). Bioinformatics analysis of bacteriophage and prophage endolysin domains. *Biologia, 69*(5), 541–556. https://doi.org/10.2478/s11756-014-0358-8

Vihinen, M. (2001). Bioinformatics in proteomics. *Biomolecular Engineering, 18*(5), 241–248. https://doi.org/10.1016/S1389-0344(01)00099-5. http://www.elsevier.com/inca/publications/store/5/0/5/7/6/2

Vissers, J. P. C., Langridge, J. I., & Aerts, J. M. F. G. (2007). Analysis and quantification of diagnostic serum markers and protein signatures for Gaucher disease. *Molecular and Cellular Proteomics, 6*(5), 755–766. https://doi.org/10.1074/mcp.M600303-MCP200

Vizcaino, J. A., Cote, R. G., Csordas, A., Dianes, J. A., Fabregat, A., Foster, J. M., & et al., (2013). The PRoteomics IDEntifications (PRIDE) database and associated tools: Status in 2013. *Nucleic Acids Research.*

Voss, T., Ahorn, H., Haberl, P., Dhner, H., & Wilgenbus, K. (2001). Correlation of clinical data with proteomics profiles in 24 patients with B-cell chronic lymphocytic leukemia. *International Journal of Cancer, 91*(2), 180–186. https://doi.org/10.1002/1097-0215(200002)9999:9999<::AID-IJC1037>3.0.CO;2-J

Voyksner, R. D., & Lee, H. (1999). Investigating the use of an octupole ion guide for ion storage and high-pass mass filtering to improve the quantitative performance of electrospray ion trap mass spectrometry. *Rapid Communications in Mass Spectrometry, 13*(14), 1427–1437. https://doi.org/10.1002/(SICI)1097-0231(19990730)13:14<1427::AID-RCM662>3.0.CO;2-5. http://onlinelibrary.wiley.com/journal/10.1002/(ISSN)1097-0231

Wang, W., Becker, C. H., Zhou, H., Lin, H., Roy, S., Shaler, T. A., Hill, L. R., Norton, S., Kumar, P., & Anderle, M. (2003). Quantification of proteins and metabolites by mass spectrometry without isotopic labeling or spiked standards. *Analytical Chemistry, 75*(18), 4818–4826. https://doi.org/10.1021/ac026468x

Want, E. J., Cravatt, B. F., & Siuzdak, G. (2005). The expanding role of mass spectrometry in metabolite profiling and characterization. *ChemBioChem, 6*(11), 1941–1951. https://doi.org/10.1002/cbic.200500151

Washburn, M. P., Wolters, D., & Yates, J. R. (2001). Large-scale analysis of the yeast proteome by multidimensional protein identification technology. *Nature Biotechnology, 19*(3), 242–247. https://doi.org/10.1038/85686

Weinstein, J. N., Myers, T. G., O'Connor, P. M., Friend, S. H., Fornace, A. J., Kohn, K. W., Fojo, T., Bates, S. E., Rubinstein, L. V., Anderson, N. L., Buolamwini, J. K., Van Osdol, W. W., Monks, A. P., Scudiero, D. A., Sausville, E. A., Zaharevitz, D. W., Bunow, B., Viswanadhan, V. N., Johnson, G. S., Wittes, R. E., & Paull, K. D. (1997). An information-intensive approach to the molecular pharmacology of cancer. *Science, 275*(5298), 343–349. https://doi.org/10.1126/science.275.5298.343

Wiśniewski, J. R., & Mann, M. (2012). Consecutive proteolytic digestion in an enzyme reactor increases depth of proteomic and phosphoproteomic analysis. *Analytical Chemistry, 84*(6), 2631–2637. https://doi.org/10.1021/ac300006b

Wiener, M. C., Sachs, J. R., Deyanova, E. G., & Yates, N. A. (2004). Differential mass spectrometry: A label-free LC-MS method for finding significant differences in complex peptide and protein mixtures. *Analytical Chemistry, 76*(20), 6085–6096. https://doi.org/10.1021/ac0493875

Winkler, C., Denker, K., Wortelkamp, S., & Sickmann, A. (2007). Silver-and Coomassie-staining protocols: Detection limits and compatibility with ESI MS. *Electrophoresis, 28*(12), 2095–2099. https://doi.org/10.1002/elps.200600670. http://onlinelibrary.wiley.com/journal/10.1002/(ISSN)1522-2683

Xiao, Z., Luke, B. T., Izmirlian, G., Umar, A., Lynch, P. M., Phillips, R. K. S., Patterson, S., Conrads, T. P., Veenstra, T. D., Greenwald, P., Hawk, E. T., & Ali, I. U. (2004). Serum proteomic profiles suggest celecoxib-modulated targets and response predictors. *Cancer Research, 64*(8), 2904–2909. https://doi.org/10.1158/0008-5472.CAN-03-3754

Zhang, B., VerBerkmoes, N. C., Langston, M. A., Uberbacher, E., Hettich, R. L., & Samatova, N. F. (2006). Detecting differential and correlated protein expression in label-free shotgun proteomics. *Journal of Proteome Research, 5*(11), 2909–2918. https://doi.org/10.1021/pr0600273

Zhou, M., McDonald, J. F., & Fernández, F. M. (2010). Optimization of a direct analysis in real time/time-of-flight mass spectrometry method for rapid serum metabolomic fingerprinting. *Journal of the American Society for Mass Spectrometry, 21*(1), 68–75. https://doi.org/10.1016/j.jasms.2009.09.004

Zhu, H., Bilgin, M., Bangham, R., Hall, D., Casamayor, A., Bertone, P., Lan, N., Jansen, R., Bidlingmaier, S., Houfek, T., Mitchell, T., Miller, P., Dean, R. A., Gerstein, M., & Snyder, M. (2001). Global analysis of protein activities using proteome chips. *Science, 293*(5537), 2101–2105. https://doi.org/10.1126/science.1062191

Zybailov, B., Coleman, M. K., Florens, L., & Washburn, M. P. (2005). Correlation of relative abundance ratios derived from peptide ion chromatograms and spectrum counting for quantitative proteomic analysis using stable isotope labeling. *Analytical Chemistry, 77*(19), 6218–6224. https://doi.org/10.1021/ac050846r

Zybailov, B., Mosley, A. L., Sardiu, M. E., Coleman, M. K., Florens, L., & Washburn, M. P. (2006). Statistical analysis of membrane proteome expression changes in Saccharomyces cerevisiae. *Journal of Proteome Research, 5*(9), 2339–2347. https://doi.org/10.1021/pr060161n

Chapter 7

Foodomics: Integrated omics for the food and nutrition science

Smriti Mall and Apoorva Srivastava
Molecular Plant Pathology Laboratory, Department of Botany, Deen Dayal Upadhayay Gorakhpur University, Gorakhpur, Uttar Pradesh, India

1. Introduction

In present scenario, we are unable to justify the numerous health, nutritional, and environmental targets associated to food. Foodomics can be simply defined as the integration of advanced omics technologies in the study of food and nutrition domains, with the aim of improving knowledge and health (Cifuentes, 2009). It encompasses various omics technologies, including proteomics (proteins), genomics (gene detection), nutrigenomics (nutrients), metabolomics (metabolites), and transcriptomics (mRNA) (Bordoni et al., 2014; Cifuentes, 2017). The primary objective of food characterization is to ensure food safety, which has prompted researchers in food technology to transition from old techniques to new analytical methods to meet global demands. Foodomics provides researchers and scientists in the field of food and nutrition science with enhanced access to data that can be used to evaluate the effects of food on human health. Commonly used techniques for food analysis can be categorized as follows: Immunological techniques, biosensors, PCR; mass spectrometry, fluorescence; separation techniques such as gas chromatography, supercritical fluid chromatography, HPLC, capillary electrophoresis; electrochemical methods; sample preparation techniques including fluid extraction, solid extraction, flow injection analysis, purge and trap, liquid pressurized extraction, microwave-assisted extraction, thermal desorption; and hyphenated techniques, among others. On the basis of the following techniques, researchers are able to better understand the development and application of technology in food science and food safety. However, the development of foodomics also leads to the development of other sub-disciplines. For example, the most interesting sub-discipline of foodomics is Nutrigenomics. Nutrigenomics combines omics with nutrition and gene (Fig. 7.1).

2. Major four areas of omics that are involved in foodomics are

1. Genomics which deals with genome.
2. Proteomics deals with the study of proteins its functions, structure, and interactions.
3. Transcriptomics explores a combination of gene and identifies the difference among various organisms.
4. Metabolomics deals with the study of chemical behavior of the cell (Figs. 7.2 and 7.3).

Proteomics involves the examination of proteins, including their functions, structure, and interactions. Metabolomics focuses on studying the chemical behavior of cells. Fig. 7.1 illustrates the various subdisciplines within foodomics that contribute to improving nutrition and food safety. In recent times, advanced omics technologies, as mentioned above, have been predominantly utilized to uncover diverse aspects of food, such as quality, composition, food allergens, toxins, and the genetic-level functioning of food proteins. The acquired knowledge plays a crucial role in enhancing our understanding of different facets of food for the well-being of individuals (Schasteen, 2016; Andjelković & Josić, 2018).

2.1 Genomics

The application of high-throughput genomics and functional genomic technologies for the study of nutritional sciences and food technology is an encouraging scientific field that comes under nutritional genomics. Nutritional Sciences encompasses

Integrative Omics. https://doi.org/10.1016/B978-0-443-16092-9.00007-2

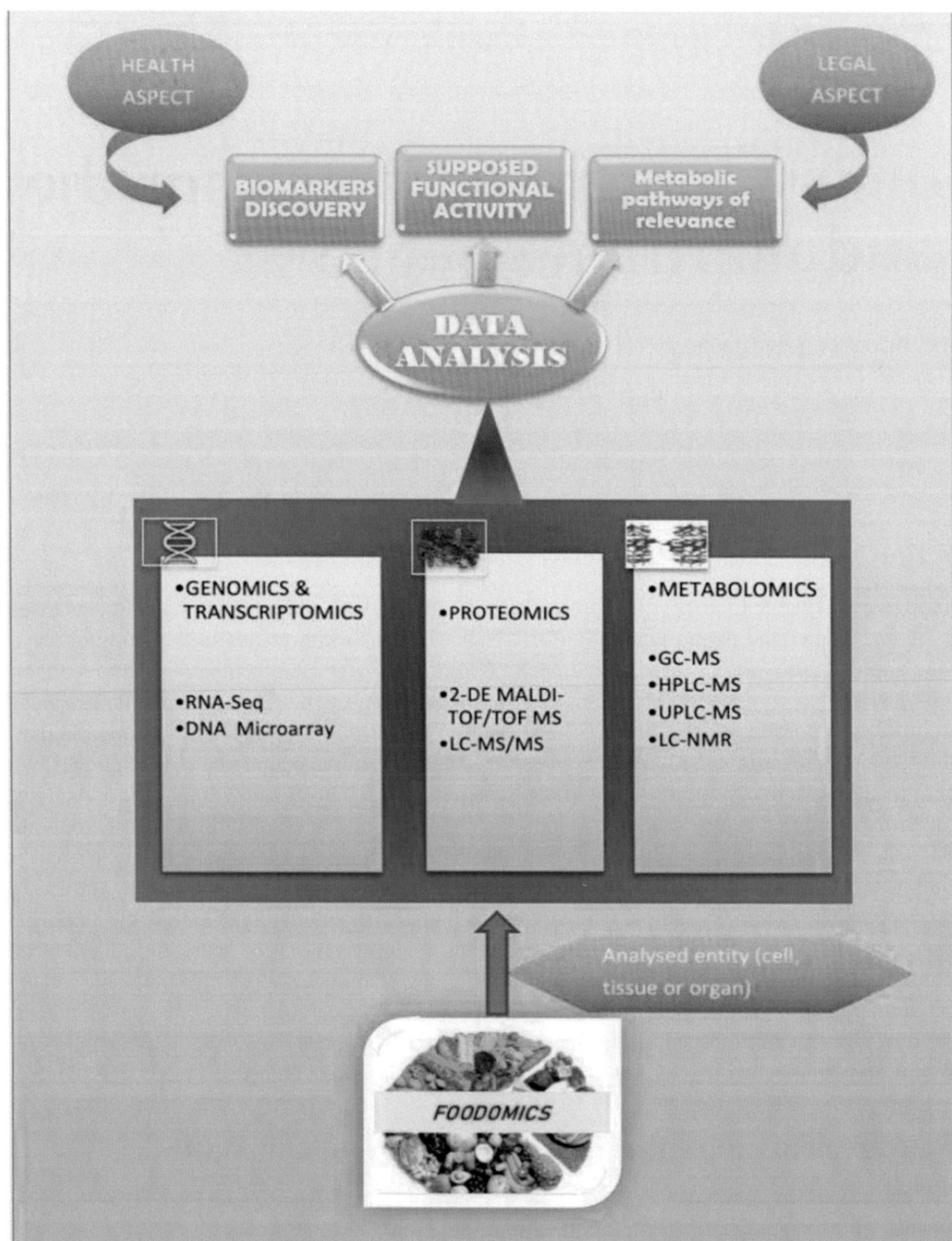

FIGURE 7.1 Foodomics different subdisciplines utilized and techniques involved for better nutrition and safety of food.

the study of nutrients, food, and related substances, examining their impact on health and disease, the biochemical processes involved in food digestion and utilization, and the utilization of this knowledge for policy and program development. A rising field known as foodomics applies modern "-omics" technologies to explore nutrition and food, aiming to enhance knowledge for the sake of people's health and safety. By leveraging tools from transcriptomics, genomics, epigenomics, metabolomics, and proteomics, foodomics provides researchers with an opportunity to address current challenges in nutrition and food science. The integration of metabolism and intermediary metabolism has shaped the concept of nutritional sciences, uncovering interconnections between various metabolic pathways and transformations and paving the way for a new era in nutrition research.

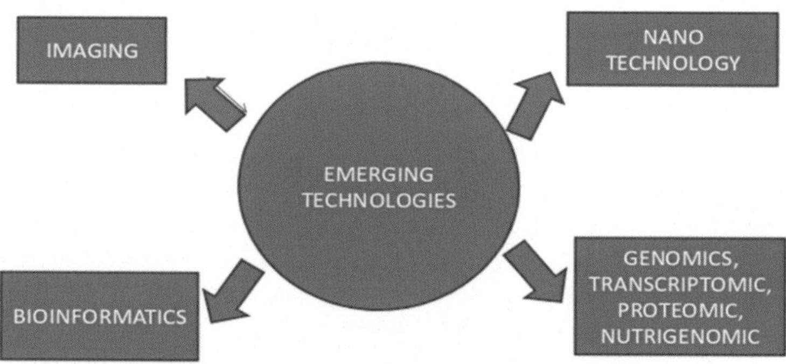

FIGURE 7.2 Emerging technologies utilized in foodomics.

2.1.1 Nutritional genomics

Molecular assays, successfully employed in fields like pharmacology and physiology, have contributed to understanding the functions of signaling cascades, drug actions, and membrane trafficking. Nutrition scientists have progressively adopted these techniques to explore the potential of nutrients as modulators of metabolic processes, and distinct signaling pathways (Cifuentes, 2017). Afman and Müller (2006) highlighted nutrient interactions with signaling pathways, demonstrating how they coordinate gene expression and influence biological functions. Ligand-activated transcription factors, including nuclear receptors, have been identified as key players in binding and activating fatty acids (Nagy & Schwabe, 2004). For instance, peroxisome proliferator-activated receptors (PPARs), activated by docosahexanoate, regulate numerous genes involved in fatty acid metabolism (Valdés et al., 2017). Another member of the nuclear receptor family, liver X receptor (LXR), closely interacts with nuclear receptors such as FXR, RXR, and PPARs. LXRs play a crucial role in controlling fatty acid, glucose homeostasis, and cholesterol metabolism (Mazzeo et al., 2008). Nuclear receptors are known to regulate the expression of a wide range of genes involved in metabolic pathways. Given the importance of fatty acids in metabolic regulation, their presence in the diet is essential.

2.1.2 Applications of genomics in food science

Genomic technologies is well utilized in the agriculture sector for examples approaches for developing new, nontransgenic plant varieties; detection of new genomic markers for animal and plant breeding programmes; study diet–gene interactions for enhancing product quality and health. According to UN Report data, 2004 It is projected that the world population will reach 9 billion by 2050 thus it becomes a social concern and responsibility to meet the Food Security of the world population. Now food productivity is increases with a broadening of agricultural base and adoption of new technology for plant improvement. In coming days, we have to see some other options of crops like millets, jowar, bajra, etc. instead of dependency on main crops like rice, wheat, soyabean etc. With the upgradation in genomic technologies, improvement in techniques of food processing, quality assurance, food safety, and also development of functional food products would be successfully implemented. Due to advancement in technologies new concepts are emerging and one of them is personalized nutrition that is diet of a person based on genomic information of that person to minimize health issues.

Food scientists have shown significant interest in metabolomics technology over the past decade as a means to gain insights into the physiological aspects of food. This technology has been employed for various purposes, including food quality assessment, analysis of food components, investigation of the effects of food processing, shelf life determination, and monitoring of food consumption. Metabolomics offers a wide range of applications in the fields of food and nutrition science, providing opportunities to explore the science of food for the betterment of human health. However, it is worth noting that the analysis of food can be complex due to the presence of a large number of metabolites within food materials (Onuh & Aluko, 2019). While traditional analysis techniques are commonly used for the analysis of macronutrients such as sugars, proteins, and lipids in food (Liang et al., 2022), are of key importance to ensure food safety and quality. "Foodomics" is the combination of modern omics technologies, such as proteomics, metabolomics and transcriptomics, together with chemometrics, bioinformatics and biostatistics, to allow the identification of the mechanisms of biologically active compounds present in food and evaluation of complex biological systems. The application of these analytical methods has permitted a dramatic change in the field of food science. In foodomics, the emerging technologies (Fig. 7.2), are also used for the identification of toxic compounds associated with food that has a harmful effect on human health and food safety (Yan et al., 2022).

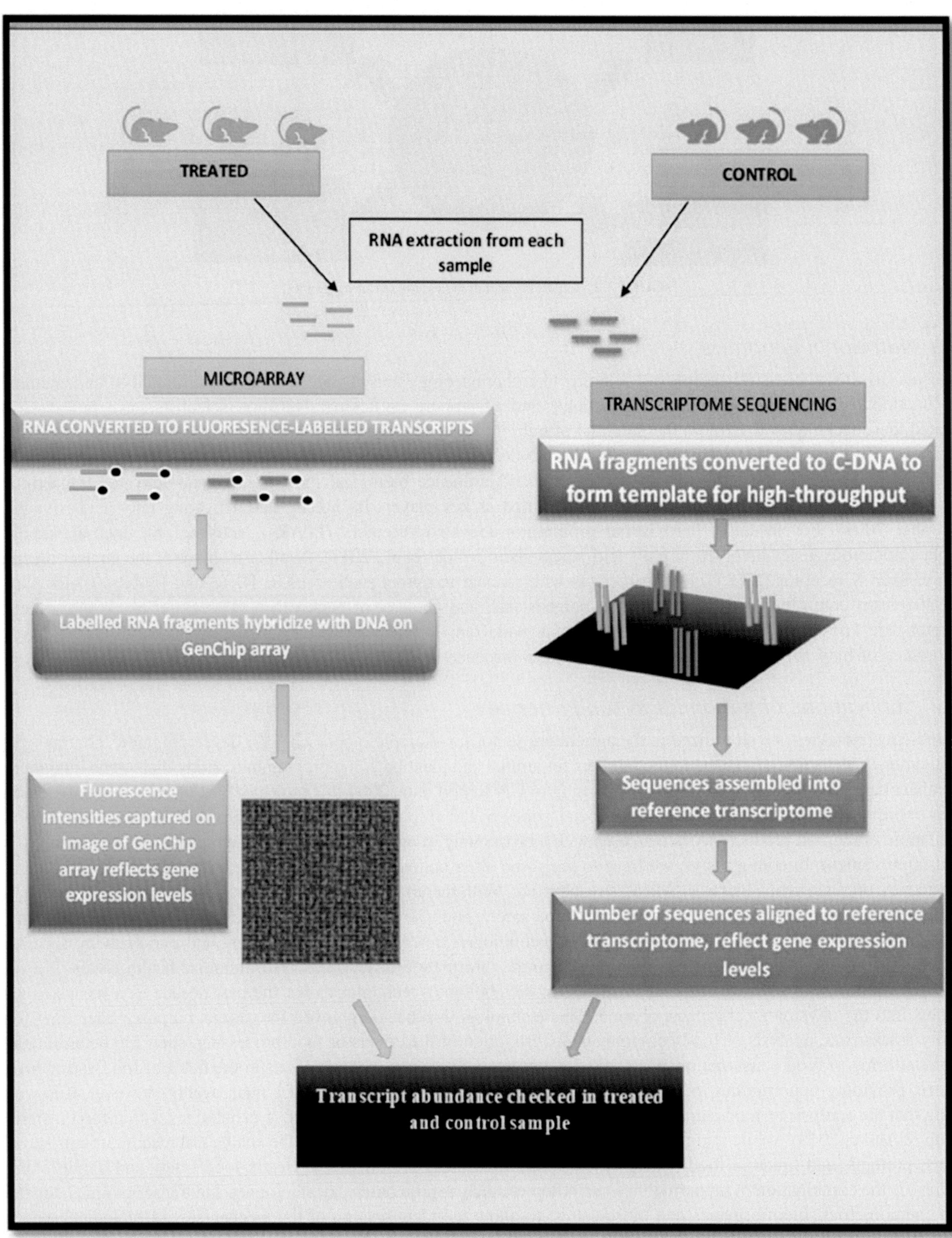

FIGURE 7.3 A detailed schematic presentation of transcriptomics procedures involved in foodomics technology.

2.1.3 Green foodomics

Various scientists also talked about the adaption of Green technologies in food science. Green foodomics related to the use of technologies in the advancement of food in a sustainable manner, in other words utilization of green techniques in sample preparation, analyzing, extraction and development of new foods (Castro-Puyana et al., 2013). According to Gilbert-López et al. (2017), concept of Green technologies would be successful only when there is reduction in the use of chemicals and minimize the use of energy and waste. And on that basis, analytical green chemistry employs the following principles:

(a) reduce the use of chemicals in the analytical methods.
(b) toxic reagents are replaced by safer counterparts.
(c) minimizing the waste volume.
(d) the reduction of the energy used.

To achieve green and eco-friendly environment in our society, concepts like green analytical chemistry and green chemistry would be extensively promoted. The analytical approaches based on Green Analytical Chemistry have contributed to quality assessment as well as to food biological activity studies and food safety (Ballesteros-Vivas et al., 2021).

2.2 Proteomics in food science

In food omics proteomics is most widely used technology concerned with protein-coding section of a genome. Proteins are essential to life, with structural, immune, metabolic, transport, signaling and regulatory functions. Peptidomics helps to investigate the characteristics and interactions of peptide sequences and is the subresearch field of proteomics. Various scientists utilize mass spectrometry with chromatography methods for the identification of proteins from food samples and are very much useful for the detection of adulterations in food. In recent term, metaproteomics can be described as proteomics at the microbial level and it is utilized to identify the total protein role in spoilage, and beneficial or pathogenic activities of microorganisms in food (Wilmes et al., 2015).

In food science profiling of proteins offers wide range of applications that can be grouped into three categories:

(i) food safety and quality manage.
(ii) studies related to food processing, and
(iii) characterization of new food ingredients and nutrition that shows beneficial effects on health.

Electrospray-ion trap mass spectrometry and matrix-assisted laser ionization—time of flight are known as an essential tool for proteomic studies (Aebersold & Mann, 2003).

2.2.1 Application of proteomic approaches in food science

One of the major challenges is to authenticate the ingredients present in food and analysis of food quality and the adulteration of food products means the nondeclared introduction of harmful food ingredients, such as toxic compounds, allergenic in the food items that represents a safety risk (Sotelo et al., 1993). Some traditional methods are extensively used for authentication and detection of microbes present in food is immunological protein-based methods and electrophoresis (Rehbein, 1990). For identification of species in food, the authenticated method is the separation of proteins on a polyacrylamide gel using a pH gradient and isoelectric focusing (IEF). Limitations of these old methods, such as the cross-reaction between closely related species, some proteins shows lack of stability during food processing, and labor intensiveness, were resolved with the introduction of molecular techniques and PCR analysis.

Mass spectrophotometry proteomic have recently developed as a sensitive, fast and high-throughput approach for the evaluation of the authenticity and traceability of species in food products. For the development of species—specific peptide markers in reference samples and for the identification of the diagnostic peptides in real samples, Mass Spectrometry tool is frequently used (Carrera et al., 2007).

2.2.1.1 Marine fish

In Proteomic approaches we are able to differentiate at very minor difference like 2-DE has recognized as a valuable tool for differentiation between five different hake species and among flat fish species, and different gadoid fish species (Piñeiro et al., 1999). Two-dimensional electrophoresis (2-DE) can even differentiate between species that show identical *IEF* protein patterns. The differential categorization has been made on the basis of specific 2-DE profiles of the parvalbumin fractions, which are heat-resistant and low-molecular weight proteins. MALDI—TOF and 2-DE analysis of parvalbumin were used for the categorization of two populations of grenadier and 10 hake species. MALDI—TOF MS profiles from 25 different fish species, presenting the most complete fish authentication study in terms of the numbers of species included (Mazzeo et al., 2008).

2.2.1.2 Shellfish

Marine mussels *Mytilus galloprovincialis* and *Mytilus edulis* are two related species shows changes in protein expression when observed by 2-DE (López et al., 2002). Three European marine mussel species namely *M. galloprovincialis*, *M. edulis*, and *M. trossulus* tropomyosin peptides were detected by MALDI—TOF PMF and sequenced by nESI-IT MS/MS methods. With the use of HPLC—ESI-IT MS species—specific markers were validated using the (SIM) selected ion monitoring configuration. Fishery products especially Decapoda crustaceans are of major commercial value, especially those within the superfamily Penaeoidea represent significant resources for both aquafarming and fisheries facilities worldwide and account for more than 20%—17% of the world consumption of seafood products. Penaeid species are phenotypically similar, their differentiation is difficult even when there is no processing and the external features remain intact. But due to sensitivity of MALDI—TOFMSPMF technique able to differentiate between two very close *Fennero penaeus indicus* and *Penaeus monodon* penaeid species (Ortea et al., 2009). Sarcoplasmic protein (AK) arginine kinase-specic to *P. monodon,* which may be useful for the identification of these species as markers. Several peptides were reported which are species-specific, together with their MS spectra. SCPSarcoplasmic calcium-binding proteins (SCP) and AK protein, both have been reported as potential species—specific biomarkers in a study that are used IEF of sarcoplasmic proteins for the identification of 14 prawn and shrimp species of commercial value (Ortea et al., 2010). But according to Shiomi et al. (2008), AK and SCPs have been identified as allergens and, thus, an analysis targeting this protein focuses on two important aspects i.e., food safety and identification of species.

2.2.1.3 Meat and food products

According to Taylor et al. (1993), the potential of mass spectrometry as an analytical method for meat species identification is very much highlighted. With the use of electrospray ionization mass spectrometry, the scientists were able to identify the origin of purified myoglobin and hemoglobin from different sources (beef, pig, and sheep). Addition of 5% bovine gelling agents to different food matrices was reported by Grundy et al. (2007), this method was based on the MS-based detection of the species-specific fibrino peptides, released from fibrinogen during gelling. With the help of sensitive LC—MS/MS methodology researchers are able to detect chicken meat in a mixture of more than two to three types of meat. For calculation of the amount of biomarker peptides in the samples, the stable isotope-labeled peptides were used. In the processed meat products we are able to detect soybean proteins added in processed meat with the help of MS/MS-based proteomics tool (Leitner et al., 2006) and also used for the identification of pea and soy proteins in milk powder (Cordewener et al., 2009). MALDI—TOF protein profiling was used to identify the geographical origin of honey obtained from Hawaiian bees (Wang et al., 2009). Guarino et al. (2010) reported electrospray ionization mass spectrometry method based on the detection of a sheep-specific peptide from the digestion of casein, to detect upto 1%—2% of sheep milk in cow and goat cheeses.

2.2.2 Proteomics in food microbial industry

For the accurate identification of microorganisms, mass spectrometry is become an important tool. Anhalt and Fenselau (1975) first applied MS for the characterization of small compounds from lyophilized bacteria. The introduction of soft ionization techniques such as MALDI (Karas & Hillenkamp, 1988) and ESI has made the identification of proteins possible. Few MS method based studies have been performed for the identification of microorganisms responsible for food spoilage and foodborne pathogens. Alterations or food spoilage are mainly due to unspecific and specific microflora, growth conditions for microorganisms related to extrinsic and intrinsic factors (pH and temperature), and contamination during processing. Significant economic losses in the food industries and serious food borne diseases are reported due to different types of food spoilages. Certain bacteria such as *Clostridium botulinum* and *Staphylococcus aureus* causes food intoxication from the intake of toxins present in spoiled food. The important food poisoning bacteria are *E. coli* O157:H7, *E. coli* O104:H4, *Listeria monocytogenes*, *Salmonella* spp., *S. aureus*, *B. cereus*, *Shigella* spp., *Vibrio parahaemolyticus*, *Vibrio cholera* etc. Challenges related to food safety have been increased due to the development of market globalization, new products and new food preservation processes. Mass spectrometry-based proteomic techniques, are currently complementing traditional and genetics-based identification techniques to overcome or meet the challenges of food safety. MALDI−TOFMS of whole bacterial cells has been used for the identification and detection of 25 different food spoilage bacteria and foodborne pathogens, such as *Yersinia*, *Proteus*, *Escherichia*, *Morganella*, *Staphylococcus*, *Salmonella*, *Lactococcus*, *Micrococcus*, *Pseudomonas*, *Listeria* and *Leuconostoc*, (Mazzeo et al., 2006). Even with such proteomic tools, scientists are able to identify the nonpathogenic and pathogenic microorganisms that are present in food. MALDI is used to identify 146 strains of *Listeria* spp. in poultry, meat, dairy, and vegetables, and same methodology was used for bacterial identification at the species level in yogurts and probiotic foods (Angelakis et al., 2011).

2.3 Metabolomics in food science

Food scientist have been interested in metabolomics technology from a decade to understand the physiological aspects of food alongside food quality assessment, food component analysis, effects of food processing, shelf life and monitoring on food consumption. Thus metabolomics shows broad range of applications in food and nutrition science and scope to explore food science for betterment of human health. Although complexities are there in food as large number of metabolites are present in food materials (Scalbert et al., 2014). Macronutrients that are present in food like sugar, proteins, lipids etc are usually analyzed by using traditional analysis techniques (Liang et al., 2022). With the development of advance techniques like imaging techniques MSI can provide a good insight into the distribution of metabolites in food. This helps us to analyze mechanisms and principles for determining the functionality of the food and also the nutritional profile of different parts of the food. According to Yan et al. (2022) metabolomics techniques allows rapid localization of certain types of substances in foods and quantification which is a very convenient method in food composition studies not only this but also similar principle will be utilized to identified different substances in food such as hormones, toxins, drug residues, pesticides in food for the food safety control.

2.3.1 Applications of metabolomics in food science

Traditional analysis techniques usually only analyze the macronutrients of food such as sugars, lipids, or proteins. The components within food products are complex and highly susceptible to transformation. The analysis of macronutrients is no longer sufficient to meet the needs of the current food industry. Metabolomics techniques help in the analysis of changes in composition during food processing and storage, this is highly beneficial for quality control in the processing or storage of food products. In addition, MSI-like imaging techniques can provide a good insight into the spatial distribution of metabolites in food. This facilitates the analysis of the nutritional profile of different parts of the food and provides principles and mechanisms for determining the functionality of the food. The use of metabolomics techniques allows for the rapid quantification or localization of certain types of substances in foods, which is a very convenient method in food composition studies. Based on this same principle, toxins, hormones, pesticides or drug residues in food can be identified for the purpose of food safety control. Flavor components of food products are also a popular research topic. To resolve the volatiles, present behind unique food aroma by using GC/MS techniques. As Gas chromatography−mass spectrometry is a powerful technique mainly applies during processes such as fermentation in dairy products and roasting in coffee. Some more analytical techniques, including NMR and LC/MS appears suitable for the profiling of nonvolatile polar phytonutrients such as glycosides, phenolics to more contribute to food health benefits.

2.4 Transcriptomics in food science

Recent advances in high-throughput technologies have proposed groundbreaking methods that enable the detailed characterization of transcriptomes. The most relevant application of transcriptomic technologies in modern food science includes the transcriptome characterization and gene expression analysis in individual organisms, such as food crops, food animals, foodborne pathogens, fermenting microorganisms, and probiotics.

2.4.1 Major techniques that are involved in transcriptomics are as follow

2.4.1.1 High-Throughput Analytical Technologies in transcriptomics

High-Throughput Analytical Technologies are now become very significant because with the help of such advanced throughput techniques, scientists are able to analyze large number of samples in a single laboratory per day. This technique involves ultra-chromatographic analysis, and other most widely used technologies are: Microarray and RNA or Transcriptome sequencing.

2.4.1.2 Microarray

Gene expression microarrays are collections of short oligonucleotide probes, representing thousands of genes, attached to a substrate, which can be microspheres randomly distributed on the surface of fiber-optic bundles, or more frequently, a glass slide, at predefined locations within a grid pattern. Regardless of the microarray format, this technique is based on complementary probe hybridization and it can be used to simultaneously measure the relative abundance of transcripts among two or more samples for thousands of genes.

2.4.1.3 RNA-seq or transcriptome sequencing

The development of groundbreaking DNA-sequencing strategies known as next-generation sequencing (NGS) technologies is revolutionizing the way we study biological systems. Steps involved in NGS are- (1) library preparation involving fragmentation of RNA molecules, cDNA synthesis and ligation to specific adaptors at both ends; (2) clonal amplification of each template; (3) attachment of the amplified DNA templates to a solid support in a flow cell or a reaction chamber; and (4) iterative and synchronized flowing and washing off the reagents for DNA strand extension while signals are acquired by the detection system.

RNA-seq offers good opportunities to discover new transcripts, identify fusion transcripts and unknown splice variants since this technique has the potential to truly cover the whole transcriptome. Compared to microarray technology, because detection of sequences does not rely upon the availability of an annotated genome, RNA-Seq is particularly suitable for the investigation of organisms for which their genome has not been totally sequenced. In addition, RNA-Seq provides better sensitivity and wider dynamic range (spanning over five orders of magnitude) to measure RNA abundance with high reproducibility and in a nonrelative quantitative mode.

2.4.2 Applications of transcriptomics in food science

2.4.2.1 Food microbiology

With the development of advance techniques, food scientists are able to better understand the wide aspects of fermentation, food spoilage, starter culture, and probiotics (Walsh et al., 2017). Fig. 7.3, Shows one such example with the help of new transcriptomics tool elucidation of the molecular mechanisms behind metabolic transformations in fermented food improves quality of food (De Filippis et al., 2017). Publications of whole genome of food microorganisms has develop the species-specific microarrays that help us to identify alterations in transcriptome under different conditions provides new insight into metabolic processes. Some examples of microarray applications are gene expression study during different fermentation processes in natural substrate or artificial media; stress factors involved in fermentation on the transcriptomal response for industrial or laboratory wine and yeast and transcriptional differences between mutants or diverse strains (Musher et al., 2006; Komatsuzaki et al., 2005; Bartra et al., 2010; Penacho et al., 2012).

RNA-seq has been utilized in studying individual microorganisms, aiming to uncover new insights into the transcriptomes of fermentative microorganisms (Solieri et al., 2013). This method has been employed to elucidate the transcriptome structures of *Aspergillus oryzae* and *Saccharomyces cerevisiae* (Nagalakshmi et al., 2008; Wang et al., 2009). NGS has become a critical technique in food microbiology, particularly in whole genome sequencing for rapid and precise subtyping, allowing the identification of harmful pathogens and their toxins in food with greater efficiency (Riedmaier et al., 2012). Metatranscriptomics has been applied to identify specific genes being transcribed and translated within microbial populations present in foods.

2.4.2.2 Crop improvement

Traditional breeding and recent markers assisted breeding programs have brought essential developments that helped us to meet the food demands. In the near future, the demand for food increases, by considering the availability of limited natural resources and the threats of population explosion and environmental changes. But with urgent needs for food and other farm related produce, there is an extreme need to use genomic resources for precise crop breeding fully. By applying transcriptome, studies in plant breeding can benefit in improving crops for several biotic and abiotic stresses. Irrespective of genome availability and (or) genome complexity of the crops, transcriptome studies can efficiently bridge the phenotype with the genotypes of the traits under study, thus helping us in a practical selection of superior genotypes.

2.4.2.3 Production of food crops: Production of food animals and food products

Understanding the relationship between gene activity and agricultural traits in crops has become crucial due to climate change and the increasing demand for food from a growing population. Transcriptomic techniques have provided valuable insights into how genomes respond to cellular disturbances and have been employed to detect genes controlling traits such as tolerance to adverse environmental conditions and yield. Several food crops, including citrus, soybean, rice, maize, wheat, melon, and grape, have had their genomes sequenced, and advancements in microarray techniques have facilitated the generation of gene expression datasets specific to cells, tissues, and developmental stages (Valdés et al., 2017). Gene expression microarray technology has been extensively used to study various responses, such as salinity, cold, drought, heat, and diseases. This technology has greatly improved postharvest treatments and storage methods, contributing to the maintenance of food quality. To comprehend the molecular processes involved in the production of metabolites and proteins relevant to food science, RNA-seq data analysis has been extensively utilized. Several studies have been published in the public domain, focusing on the RNA-seq analysis of various food-related plants such as bayberry (Feng et al., 2012), pomegranate (Ono et al., 2011), zucchini (Carvajal et al., 2018), sweet potato (Xie et al., 2019), apple (Qi et al., 2017), Guava (Mittal et al., 2020), Summer squash (Xanthopoulou et al., 2021) and strawberry (Gaete-Eastman et al., 2022).

In the realm of food production from animals and related products, RNA-seq combined with biostatistical tools for pattern recognition has been widely employed. For instance, in the study of liver gene expression in Nguni heifers (Riedmaier et al., 2012). RNA-seq analysis facilitated the investigation of the effects of specific chemicals, enabling the identification of gene expression biomarkers independent of species or breed. This approach allowed differentiation between nontreated and treated animals Now a days it becomes possible to identify varying quality of forages on milk production for feeding dairy cows with the use of RNA-seq analysis technique definitely improve the strategies for feeding dairy cows (Cifuentes, 2009). Furthermore, the utilization of RNA-seq analysis has made it possible to identify variations in forage quality and its impact on milk production in dairy cows. By studying the differences in biological processes and molecular pathways in various tissues of dairy cows fed different forages, RNA-seq has provided insights into the strategies for feeding dairy cows and their subsequent milk production.

MS-based proteomics is applied for the authentication of seafoods and meats. By using ESI MS, the scientists were able to identify the origin of purified myoglobin and hemoglobin from various sources (sheep, pig, beef, and horse), able to detect soybean proteins in the processed meat products, MALDI−TOF protein profiling help to detect milk and mozzarella cheese adulteration (Cozzolino & Murray, 2004). With the help of LC−MS/MS a sensitive methodology we are able to detect chicken meat in a mixture of more than one meat.

2.4.2.4 Foodomics: In treatment of deadly diseases

The Foodomics approach has been employed to investigate the diverse chemical compounds present in Rosemary. An intriguing finding through RNA sequencing analysis is the identification of Carnosic acid, obtained from the rosemary plant, as a potential treatment for colon cancer. This discovery was made by conducting RNA sequencing analysis on mice that had been administered rosemary extract, both in controlled and treated groups. The analysis helped in identifying the effects of rosemary extract on gene expression patterns and specifically it's potential anticancer properties in the context of colon cancer. Thus, foodomics techniques present a significant example and also explaining the treatment of deadly colon cancer disease, with rosemary extract.

3. Challenges of foodomics

- In the transcriptomics field, the RNA-seq technology has been applied to the characterization of transcriptomes of different foods, and its wider application in the study of the effects of bioactive food compounds is expected.
- Other tools, such as molecular engineering of microorganisms through CRISPR-Cas9 (clustered regularly interspaced short palindromic repeats), together with synthetic biology applications pose a great potential to modify microbial communities in food, improving processes such as fermentation or generating enhanced probiotic strains.
- In the proteomics field, the combination of more sensitive, faster, and higher-resolution MS instruments coupled to different separations systems and fractionation techniques will increase the coverage of proteomes, subproteomes, and peptidomes.
- However, when the time aspect is considered, there are still some limitations making it difficult to understand the metabolic and physiological changes occurring during molecular and cellular processes.
- All these technologies that we are using in foodomics are very costly and require a sophisticated lab with technically-sound persons. In the case of metabolomics, great advances in extraction, separation, and detection techniques have been performed but the main limitations are still the identification and accurate quantification of metabolites.
- Another major challenge is the integration of the different omics approaches, because of the lack of adequate bioinformatics tools and our limited understanding of the biological and chemical process occurring inside any biological system, what makes especially demanding the study about the effect of food components on health.

- The achievement of all these goals also requires a collaborative work within the scientific community to compare and share data. Therefore, more harmonized and standardized sampling methods, improvements in computational techniques and biological databases (i.e., with functional annotations), and further developments in the analytical technologies used on each specific omics field are essential.
- Overcoming the above-mentioned challenges will allow scientists to gain a more comprehensive foodomics insight about the relationship between food and health.

4. Conclusions

Overall, we can say that foodomics is a new science moving around food, and with the use of modern transcriptomic technologies, it provides an impressive analytical "toolbox" for decoding transcriptomes in the context of food science for the well-being of humans and to overcome the global scarcity of food by developing more nutritious and healthy food. In many gene–expression studies, advanced tools especially like RNA sequencing have immense potential instead of microarray techniques. However, in food science such powerful evolving technology will have to overcome several challenges to realize its full potentiality. Although RNAseq is still expensive cost per sequenced base compared with conventional sequencing methods. Thus, in the near future it would be expected to develop faster and cheaper library for transcriptome sequencing in food science. But with the development of omics and integration of omic techniques like proteomics, nutrigenomics, transcriptomics, and metabolomics in food science, researchers and scientists are now able to modify traditional food for e.g., rice, wheat, pulses etc. into nutritionally more rich and genetically modified food for betterment of health and food safety. Scientist working on different aspects of food like quality, quantity, chemical constituents of food, tolerance toward changing environment etc. would be overcome in coming days.

References

Aebersold, R., & Mann, M. (2003). Mass spectrometry-based proteomics. *Nature, 422*(6928), 198–207. https://doi.org/10.1038/nature01511

Afman, L., & Müller, M. (2006). Nutrigenomics: From molecular nutrition to prevention of disease. *Journal of the American Dietetic Association*. ISSN: 00028223, *106*(4), 569–576. https://doi.org/10.1016/j.jada.2006.01.001. http://www.elsevier.com/inca/publications/store/6/6/2/1/7/3/index.htt

Andjelković, U., & Josić, D. (2018). Mass spectrometry based proteomics as foodomics tool in research and assurance of food quality and safety. *Trends in Food Science and Technology*. ISSN: 09242244, *77*, 100–119. https://doi.org/10.1016/j.tifs.2018.04.008. http://www.elsevier.com/wps/find/journaldescription.cws_home/601278/description#description

Angelakis, E., Million, M., Henry, M., & Raoult, D. (2011). Rapid and accurate bacterial identification in probiotics and yoghurts by MALDI-TOF mass spectrometry. *Journal of Food Science*. ISSN: 17503841, *76*(8), M568–M572. https://doi.org/10.1111/j.1750-3841.2011.02369.x

Anhalt, J. P., & Fenselau, C. (1975). Identification of bacteria using mass spectrometry. *Analytical Chemistry*. ISSN: 15206882, *47*(2), 219–225. https://doi.org/10.1021/ac60352a007

Ballesteros-Vivas, D., Socas-Rodríguez, B., Mendiola, J. A., Ibáñez, E., & Cifuentes, A. (2021). Green food analysis: Current trends and perspectives. *Current Opinion in Green and Sustainable Chemistry*. ISSN: 24522236, *31*, 100522. https://doi.org/10.1016/j.cogsc.2021.100522

Bartra, E., Casado, M., Carro, D., Campamà, C., & Piña, B. (2010). Differential expression of thiamine biosynthetic genes in yeast strains with high and low production of hydrogen sulfide during wine fermentation. *Journal of Applied Microbiology*. ISSN: 13652672, *109*(1), 272–281. https://doi.org/10.1111/j.1365-2672.2009.04652.x

Bordoni, A., & Francesco, C. (2014). Foodomics for healthy nutrition. *Current Opinion in Clinical Nutrition & Metabolic Care, 17*(5), 418–424.

Carrera, M., Cañas, B., Piñeiro, C., Vázquez, J., & Gallardo, J. M. (2007). De novo mass spectrometry sequencing and characterization of species-specific peptides from nucleoside diphosphate kinase B for the classification of commercial fish species belonging to the family merlucciidae. *Journal of Proteome Research*. ISSN: 15353893, *6*(8), 3070–3080. https://doi.org/10.1021/pr0701963

Carvajal, F., Rosales, R., Palma, F., Manzano, S., Cañizares, J., Jamilena, M., & Garrido, D. (2018). Transcriptomic changes in Cucurbita pepo fruit after cold storage: Differential response between two cultivars contrasting in chilling sensitivity. *BMC Genomics*. ISSN: 14712164, *19*(1). https://doi.org/10.1186/s12864-018-4500-9. http://www.biomedcentral.com/bmcgenomics

Castro-Puyana, M., Mendiola, J. A., & Ibañez, E. (2013). Strategies for a cleaner new scientific discipline of green foodomics. *TrAC, Trends in Analytical Chemistry*. ISSN: 18793142, *52*, 23–35. https://doi.org/10.1016/j.trac.2013.06.013. http://www.elsevier.com/locate/trac

Cifuentes, A. (2009). Food analysis and foodomics. *Journal of Chromatography A*. ISSN: 00219673, *1216*(43), 7109. https://doi.org/10.1016/j.chroma.2009.09.018. http://www.sciencedirect.com

Cifuentes, A. (2017). Foodomics, foodome and modern food analysis. *TrAC, Trends in Analytical Chemistry*. ISSN: 18793142, *96*, 1. https://doi.org/10.1016/j.trac.2017.09.001. http://www.elsevier.com/locate/trac

Cordewener, J. H. G., Luykx, D. M. A. M., Frankhuizen, R., Bremer, M. G. E. G., Hooijerink, H., & America, A. H. P. (2009). Untargeted LC-Q-TOF mass spectrometry method for the detection of adulterations in skimmed-milk powder. *Journal of Separation Science*. ISSN: 16159314, *32*(8), 1216–1223. https://doi.org/10.1002/jssc.200800568. http://www3.interscience.wiley.com/cgi-bin/fulltext/122267695/PDFSTART

Cozzolino, D., & Murray, I. (2004). Identification of animal meat muscles by visible and near infrared reflectance spectroscopy. *LWT—Food Science and Technology*. ISSN: 00236438, *37*(4), 447−452. https://doi.org/10.1016/j.lwt.2003.10.013

De Filippis, F., Parente, E., & Ercolini, D. (2017). Metagenomics insights into food fermentations. *Microbial Biotechnology*. ISSN: 17517915, *10*(1), 91−102. https://doi.org/10.1111/1751-7915.12421. http://onlinelibrary.wiley.com/journal/10.1111/(ISSN)1751-7915

Feng, C., Chen, M., Xu, C. J., Bai, L., Yin, X. R., Li, X., Allan, A. C., Ferguson, I. B., & Chen, K. S. (2012). Transcriptomic analysis of Chinese bayberry (*Myrica rubra*) fruit development and ripening using RNA-Seq. *BMC Genomics*. ISSN: 14712164, *13*(1). https://doi.org/10.1186/1471-2164-13-19. http://www.biomedcentral.com/1471-2164/13/19

Gaete-Eastman, C., Stappung, Y., Molinett, S., Urbina, D., Moya-Leon, M. A., & Herrera, R. (2022). RNAseq, transcriptome analysis and identification of DEGs involved in development and ripening of Fragaria chiloensis fruit. *Frontiers in Plant Science*. ISSN: 1664462X, *13*. https://doi.org/10.3389/fpls.2022.976901. https://www.frontiersin.org/journals/plant-science

Gilbert-López, B., Mendiola, J. A., & Ibáñez, E. (2017). Green foodomics. Towards a cleaner scientific discipline. *TrAC, Trends in Analytical Chemistry*. ISSN: 18793142, *96*, 31−41. https://doi.org/10.1016/j.trac.2017.06.013. http://www.elsevier.com/locate/trac

Grundy, H. H., Reece, P., Sykes, M. D., Clough, J. A., Audsley, N., & Stones, R. (2007). Screening method for the addition of bovine blood-based binding agents to food using liquid chromatography triple quadrupole mass spectrometry. *Rapid Communications in Mass Spectrometry*. ISSN: 10970231, *21*(18), 2919−2925. https://doi.org/10.1002/rcm.3160

Guarino, Carmine, De Simone, Luciana, Santoro, Simona, Caira, Simonetta, Lilla, Sergio, Calabrese, Maria Grazia, Chianese, Lina, & Addeo, Francesco (2010). The proteomic changes in *Cynara Cardunculus* L. Var. altilis DC following the Etiolation Phenomena using De novo sequence analysis. *Journal of Botany, 2010*, 1−16. https://doi.org/10.1155/2010/496893, 2090-0120.

Karas, M., & Hillenkamp, F. (1988). Laser desorption ionization of proteins with molecular masses exceeding 10 000 daltons. *Analytical Chemistry*. ISSN: 15206882, *60*(20), 2299−2301. https://doi.org/10.1021/ac00171a028

Komatsuzaki, N., Shima, J., Kawamoto, S., Momose, H., & Kimura, T. (2005). Production of γ-aminobutyric acid (GABA) by Lactobacillus paracasei isolated from traditional fermented foods. *Food Microbiology*. ISSN: 07400020, *22*(6), 497−504. https://doi.org/10.1016/j.fm.2005.01.002. http://www.elsevier.com/inca/publications/store/6/2/2/8/3/3/index.htt

Leitner, A., Castro-Rubio, F., Marina, M. L., & Lindner, W. (2006). Identification of marker proteins for the adulteration of meat products with soybean proteins by multidimensional liquid chromatography-tandem mass spectrometry. *Journal of Proteome Research*. ISSN: 15353893, *5*(9), 2424−2430. https://doi.org/10.1021/pr060145q

Liang, L., Duan, W., Zhao, C., Zhang, Y., & Sun, B. (2022). Recent development of two-dimensional liquid chromatography in food analysis. *Food Analytical Methods*. ISSN: 1936976X, *15*(5), 1214−1225. https://doi.org/10.1007/s12161-021-02190-2. http://www.springer.com/life+sci/food+science/journal/12161

López, J. L., Marina, A., Vázquez, J., & Alvarez, G. (2002). A proteomic approach to the study of the marine mussels *Mytilus edulis* and *M. galloprovincialis*. *Marine Biology, 141*(2), 217−223. https://doi.org/10.1007/s00227-002-0827-4

Mazzeo, M. F., De Giulio, B., Guerriero, G., Ciarcia, G., Malorni, A., Russo, G. L., & Siciliano, R. A. (2008). Fish authentication by MALDI-TOF mass spectrometry. *Journal of Agricultural and Food Chemistry*. ISSN: 00218561, *56*(23), 11071−11076. https://doi.org/10.1021/jf8021783. http://pubs.acs.org/doi/pdfplus/10.1021/jf8021783

Mazzeo, M. F., Sorrentino, A., Gaita, M., Cacace, G., Di Stasio, M., Facchiano, A., Comi, G., Malorni, A., & Siciliano, R. A. (2006). Matrix-assisted laser desorption ionization-time of flight mass spectrometiy for the discrimination of food-borne microorganisms. *Applied and Environmental Microbiology, 72*(2), 1180−1189. https://doi.org/10.1128/AEM.72.2.1180-1189.2006

Mittal, A., Yadav, I. S., Arora, N. K., Boora, R. S., Mittal, M., Kaur, P., Erskine, W., Chhuneja, P., Gill, M. I. S., & Singh, K. (2020). RNA-sequencing based gene expression landscape of guava cv. Allahabad Safeda and comparative analysis to colored cultivars. *BMC Genomics*. ISSN: 14712164, *21*(1). https://doi.org/10.1186/s12864-020-06883-6. http://www.biomedcentral.com/bmcgenomics

Musher, D. M., Logan, N., Hamill, R. J., DuPont, H. L., Lentnek, A., Gupta, A., & Rossignol, J. F. (2006). Nitazoxanide for the treatment of *Clostridium difficile* colitis. *Clinical Infectious Diseases*. ISSN: 10584838, *43*(4), 421−427. https://doi.org/10.1086/506351

Nagalakshmi, U., Wang, Z., Waern, K., Shou, C., Raha, D., Gerstein, M., & Snyder, M. (2008). The transcriptional landscape of the yeast genome defined by RNA sequencing. *Science*. ISSN: 10959203, *320*(5881), 1344−1349. https://doi.org/10.1126/science.1158441

Nagy, L., & Schwabe, J. W. R. (2004). Mechanism of the nuclear receptor molecular switch. *Trends in Biochemical Sciences, 29*(6), 317−324. https://doi.org/10.1016/j.tibs.2004.04.006

Ono, N. N., Britton, M. T., Fass, J. N., Nicolet, C. M., Lin, D., & Tian, L. (2011). Exploring the transcriptome landscape of pomegranate fruit peel for natural product biosynthetic gene and SSR marker discovery. *Journal of Integrative Plant Biology*. ISSN: 17447909, *53*(10), 800−813. https://doi.org/10.1111/j.1744-7909.2011.01073.x

Onuh, J. O., & Aluko, R. E. (2019). Metabolomics as a tool to study the mechanism of action of bioactive protein hydrolysates and peptides: A review of current literature. *Trends in Food Science and Technology*. ISSN: 09242244, *91*, 625−633. https://doi.org/10.1016/j.tifs.2019.08.002. http://www.elsevier.com/wps/find/journaldescription.cws_home/601278/description#description

Ortea, I., Barros, L., Cañas, B., Calo-Mata, P., Barros-Velázquez, J., & Gallardo, J. M. (2009). A method to compare MALDI-TOF MS PMF spectra and its application in phyloproteomics. *Lecture Notes in Computer Science*. ISSN: 16113349, *5518*(2), 1147−1153. https://doi.org/10.1007/978-3-642-02481-8_174

Ortea, I., Cañas, B., Calo-Mata, P., Barros-Velázquez, J., & Gallardo, J. M. (2010). Identification of commercial prawn and shrimp species of food interest by native isoelectric focusing. *Food Chemistry*. ISSN: 03088146, *121*(2), 569−574. https://doi.org/10.1016/j.foodchem.2009.12.049

Penacho, V., Valero, E., & Gonzalez, R. (2012). Transcription profiling of sparkling wine second fermentation. *International Journal of Food Microbiology*. ISSN: 18793460, *153*(1–2), 176–182. https://doi.org/10.1016/j.ijfoodmicro.2011.11.005

Piñeiro, C., Barros-Velázquez, J., Sotelo, C. G., & Gallardo, J. M. (1999). The use of two-dimensionalelectrophoresis in the characterization of the wáter-soluble protein fraction of comercialflat fish species. *ZeitschriftfurLebensmittel-Untersuchung und-Forschung A, 208*, 342–348.

Qi, Y., Lei, Q., Zhang, Y., Liu, X., Zhou, B., Liu, C., & Ren, X. (2017). Comparative transcripome data for commercial maturity and physiological maturity of 'Royal Gala' apple fruit under room temperature storage condition. *Scientia Horticulturae*. ISSN: 03044238, *225*, 386–393. https://doi.org/10.1016/j.scienta.2017.07.024. http://www.elsevier.com/inca/publications/store/5/0/3/3/1/6

Rehbein, H. (1990). Electrophoretic techniques for species identification of fishery products. *Zeitschrift für Lebensmittel-Untersuchung und -Forschung*. ISSN: 14382385, *191*(1), 1–10. https://doi.org/10.1007/BF01202356

Riedmaier, I., Benes, V., Blake, J., Bretschneider, N., Zinser, C., Becker, C., Meyer, H. H. D., & Pfaffl, M. W. (2012). RNA-sequencing as useful screening tool in the combat against the misuse of anabolic agents. *Analytical Chemistry*. ISSN: 15206882, *84*(15), 6863–6868. https://doi.org/10.1021/ac301433d

Scalbert, A., Brennan, L., Manach, C., Andres-Lacueva, C., Dragsted, L. O., Draper, J., Rappaport, S. M., Van Der Hooft, J. J. J., & Wishart, D. S. (2014). The food metabolome: A window over dietary exposure. *American Journal of Clinical Nutrition*. ISSN: 19383207, *99*(6), 1286–1308. https://doi.org/10.3945/ajcn.113.076133. http://ajcn.nutrition.org/content/99/6/1286.full.pdf+html

Schasteen, C. (2016). *Food omics*. https://doi.org/10.1016/B978-0-08-100596-5.03446-6

Shiomi, K., Sato, Y., Hamamoto, S., Mita, H., & Shimakura, K. (2008). Sarcoplasmic calcium-binding protein: Identification as a new allergen of the black tiger shrimp *Penaeus monodon*. *International Archives of Allergy and Immunology*. ISSN: 10182438, *146*(2), 91–98. https://doi.org/10.1159/000113512

Solieri, L., Dakal, T. C., & Giudici, P. (2013). Next-generation sequencing and its potential impact on food microbial genomics. *Annals of Microbiology*. ISSN: 18692044, *63*(1), 21–37. https://doi.org/10.1007/s13213-012-0478-8. https://rd.springer.com/journal/13213

Sotelo, C. G., Piñeiro, C., Gallardo, J. M., & Pérez-Martin, R. I. (1993). Fish species identification in seafood products. *Trends in Food Science and Technology*. ISSN: 09242244, *4*(12), 395–401. https://doi.org/10.1016/0924-2244(93)90043-A

Taylor, A. J., Linforth, R., Weir, O., Hutton, T., & Green, B. (1993). Potential of electrospray mass spectrometry for meat pigment identification. *Meat Science*. ISSN: 03091740, *33*(1), 75–83. https://doi.org/10.1016/0309-1740(93)90095-Y

Valdés, Alberto, Cifuentes, Alejandro, & León, Carlos (2017). Foodomics evaluation of bioactive compounds in foods. *TrAC, Trends in Analytical Chemistry*. ISSN: 01659936, *96*, 2–13. https://doi.org/10.1016/j.trac.2017.06.004

Walsh, Alexandra M., Duncan, Susan E., Bell, Martha Ann, O'Keefe, Sean F., & Gallagher, Daniel L. (2017). Breakfast meals and emotions: Implicit and explicit assessment of the visual experience. *Journal of Sensory Studies*. ISSN: 08878250, *32*(3), e12265. https://doi.org/10.1111/joss.12265

Wang, J., Kliks, M. M., Qu, W., Jun, S., Shi, G., & Li, Q. X. (2009). Rapid determination of the geographical origin of honey based on protein fingerprinting and barcoding using MALDI TOF MS. *Journal of Agricultural and Food Chemistry*. ISSN: 00218561, *57*(21), 10081–10088. https://doi.org/10.1021/jf902286p. http://pubs.acs.org/doi/pdfplus/10.1021/jf902286p

Wilmes, P., Heintz-Buschart, A., & Bond, P. L. (2015). A decade of metaproteomics: Where we stand and what the future holds. *Proteomics*. ISSN: 16159861, *15*(20), 3409–3417. https://doi.org/10.1002/pmic.201500183. http://onlinelibrary.wiley.com/journal/10.1002/(ISSN)1615-9861

Xanthopoulou, A., Montero-Pau, J., Picó, B., Boumpas, P., Tsaliki, E., Paris, H. S., Tsaftaris, A., Kalivas, A., Mellidou, I., & Ganopoulos, I. (2021). A comprehensive RNA-Seq-based gene expression atlas of the summer squash (Cucurbita pepo) provides insights into fruit morphology and ripening mechanisms. *BMC Genomics*. ISSN: 14712164, *22*(1). https://doi.org/10.1186/s12864-021-07683-2. http://www.biomedcentral.com/bmcgenomics

Xie, Z., Zhou, Z., Li, H., Yu, J., Jiang, J., Tang, Z., Ma, D., Zhang, B., Han, Y., & Li, Z. (2019). High throughput sequencing identifies chilling responsive genes in sweetpotato (*Ipomoea batatas* Lam.) during storage. *Genomics*. ISSN: 10898646, *111*(5), 1006–1017. https://doi.org/10.1016/j.ygeno.2018.05.014. http://www.elsevier.com/inca/publications/store/6/2/2/8/3/8/index.htt

Yan, X., Chen, H., Du, G., Guo, Q., Yuan, Y., & Yue, T. (2022). Recent trends in fluorescent aptasensors for mycotoxin detection in food: Principles, constituted elements, types, and applications. *Food Frontiers*. ISSN: 26438429, *3*(3), 428–452. https://doi.org/10.1002/fft2.144. http://onlinelibrary.wiley.com/journal/26438429

Chapter 8

Vaccinomics: Structure based drug designing computational approaches for the designing of novel vaccines

Madhulika Jha[1], Nidhi Yadav[2], Swasti Rawal[3], Payal Gupta[1], Navin Kumar[1], Ravi Kumar Yadav[4] and Tara Chand Yadav[5,6]

[1]*Department of Biotechnology, Graphic Era University, Dehradun, Uttarakhand, India;* [2]*Department of Chemistry, SS Khanna Girls Degree College, University of Allahabad, Allahabad, Uttar Pradesh, India;* [3]*SRC Division of Biophysics, Medical University of Graz, Graz, Austria;* [4]*Department of Botany, Kashi Naresh Government Post Graduate College, Gyanpur, Uttar Pradesh, India;* [5]*Department of Computer Science and Biosciences, Marwadi University, Rajkot, Gujarat, India;* [6]*Department of Electronics, Electric, and Automatic Engineering, Rovira I Virgili University (URV), Tarragona, Spain*

1. Introduction

Vaccinomics advocated for an innovative strategy to vaccine development defined as a “identify—evaluate—classify—execute” paradigm predicated on vaccinomics and tailored vaccine development. Vaccine-preventable diseases are a continuous issue to global health that may be prevented by protective and long-term immunization coverage. Vaccines now provide protection against important infections, save three million lives annually. However, present coverage rates are not ideal, particularly for the so-called “vulnerable groups,” who include newborn, premature babies, pregnant women, senior citizens, and patients with chronic and immune-compromising illnesses (Doherty et al., 2016). Several factors contribute to this underimmunization, including a lack of knowledge about illnesses that may be prevented by vaccination and skepticism or false beliefs about the effectiveness and safety of vaccination among at-risk health-care professionals, parents, and patients. Therefore, in many diseases that are preventable by vaccination, the immune responses produced by the presently available vaccinations and schedules may not be sufficient, providing less protection than in healthy people (Doherty et al., 2016; O’Shea et al., 2014). This condition places a significant financial and health burden on society, and resolving it will be challenging in areas with limited public resources. Due to the many obstacles, public health authorities must overcome in order to increase the effectiveness of vaccination programs, greater focus and creative approaches are needed. To improve current vaccination methods, two strategies are required: (1) address vaccine hesitancy through education and management, and (2) create novel tools that allow for an explanation of the mechanisms underlying low or no responsiveness to existing vaccination regimens in these organizations and the development of targeted interventions (Fig. 8.1).

A look back at the study of vaccinology demonstrates medical science’s victory against infectious illnesses that were previously fatal.

In order to overcome the difficulties presented by these obstacles, novel approaches have been developed, including vaccinomics, reverse vaccinology, and structure-based vaccine design. These approaches make use of high-dimensional techniques and tools as well as generate novel information that can be used to produce new vaccines (Poland et al., 2018; Rino Rappuoli, 2000). Such genomics-based techniques have been used in the last 10 years to design and produce novel vaccines, such as the approved meningococcus B vaccine. Genomic research is undergoing rapid development of new vaccines in the 21st century, closely paralleling the application of genomics to other parts of human medicine, including such individualized medicine, with the rising sophistication and falling cost of next-generation sequencing and gene-based assays technologies (Catalanotto et al., 2016; Stewart-Morgan et al., 2020). The basic nucleic acid sequence of

Integrative Omics. https://doi.org/10.1016/B978-0-443-16092-9.00008-4

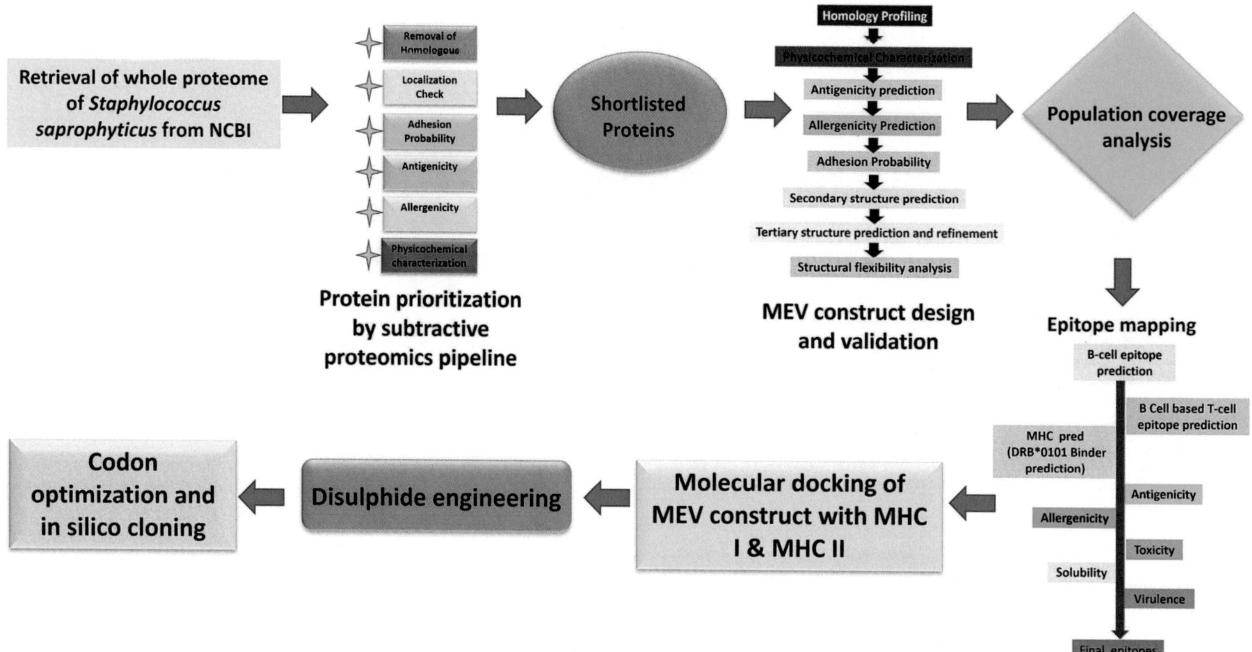

FIGURE 8.1 Steps involved in the optimization of vaccine constructs.

a particular organism is now just a small part of genetics. Although it mostly focuses on particular genes, it also encompasses the many regulatory systems that regulate gene expression (Giani et al., 2020). In a similar vein, the field of genomics has broadened to include the thorough characterization of gene editing, pre- and posttranscriptional alterations, gene regulation, epistasis, complementarity, pleiotropy, and other complicated relationships (Brodie & Tosevski, 2018; Reeves et al., 2018). In terms of the technology and platforms that may be utilized to develop, manufacture, and research vaccines, genomics is not the only field that has lately seen a dramatic revolution. Several examples are as follows: proteomics, mass spectrometry, and metabolomics, which have been strongly connected to immunologic activity and vaccination response, mass cytometry, which enables very sophisticated immunophenotyping (Johnson et al., 2005; Sarkizova et al., 2020). The inventive use of these techniques as well as the biological insights they provide have the potential to completely transform how we create, produce, analyze, and use vaccines.

2. Tools for vaccinomics

The massive amount of genome data obtained through sequencing projects has opened the way for multiple in silico screenings and computational analysis. Bioinformatics is a field that utilizes new software tools and mathematical elements to organize and analyze biological data (Gibas et al., 2001). Vaccine design approaches are seeking computer-aided assistance to save time and money. It is an ever-evolving area that gives fruitful results related to biological science that will be further verified by in vitro techniques. The main goal of making and using computational algorithms and software tools is to help the agriculture and pharmaceutical industries, health care, crop improvement, drug discovery, forensic analysis, food analysis, and biodiversity management, understand how biological processes work (Singh, 2016).

2.1 ANTIGENpro

ANTIGENpro is a sequence-based predictor that does not need alignment and is independent of pathogens. It uses a two-stage architecture, several representations of the primary sequence, and five machine-learning methods for predicting protein antigenicity (Rasheed et al., 2021). ANTIGENpro is the first protein-predicting tool that utilizes data from protein microarray analysis and then evaluates data from protein microarray analysis and then evaluates total proteins' antigenicity (Magnan et al., 2010).

2.2 AllergenFP

The AllergenFP is created using a dataset defined by five E-descriptors with texts that have been auto-cross covariance converted into uniform vectors (Rasheed et al., 2021). Based on specificity and sensitivity static analyses, the overall accuracy reveals that AllerTOP and AllergenFP are the top allergen identification tools for sequencing compared to other analytic tools and servers (Dimitrov et al., 2014).

2.3 Ellipro

The three-dimensional (3D) structure of a protein antigen is used as the basis for ElliPro's prediction of linear and discontinuous antibody epitopes. Protein Data Bank structures may be uploaded into ElliPro. Methods for modeling, protein 3D structure, and antibiotic docking had more details on these techniques, given a protein sequence as input. Antibody epitopes may be predicted and seen in a protein structure and sequence using Thornton's technique, which is implemented as a web platform that employs a residue clustering algorithm, the MODELER tool, and the Jmol viewer. ElliPro is based on the geometrical properties of protein structures and needs no training. It can anticipate a wide variety of protein–protein interactions. ElliPro employs epitope properties such as amino acid propensities, residue solvent accessibility, intermolecule, and epitope spatial distribution to enhance prediction performance, in contrast to DiscoTope's reliance on training datasets (Ponomarenko et al., 2008).

2.4 Epipred

EpiPred is a computer software used to identify structural epitopes unique to a specific antibody. The accuracy of antibody–antigen docking may be enhanced by using EpiPred's predicted epitopes. As input, an antibody homology model may be used in this approach. Patches on antigen structure prioritized based on their propensity to contain the epitope. Other methods, such as DiscoTope or PEPITO, annotate vast immunogenic/epitope-like areas on the antigen without requiring any input about the antibody, in contrast to EpiPred (Krawczyk et al., 2014).

2.5 BCPred

A continuous B-cell epitope prediction method uses support vector machine algorithms. These algorithms were taught to distinguish between 701 linear B-cell epitopes obtained from Bcipep database, and 701 nonepitopes obtained randomly from Swiss-Prot sequences using a homology-reduced dataset. During the training phase, five distinct kernel techniques and five-fold cross-validations were utilized (El-Manzalawy et al., 2008).

One of the benefits of using this tool is that it includes experimentally determined samples collected from both the training and the test dataset. To create predictions and enable the use of a large number of datasets that may enhance statistical analysis and characteristics of B-cell epitopes, deep learning algorithms were put into operation (Shi & Liu, 2020, pp. 11–13).

2.6 LBtope

It is constructed based on the non-B cell epitopes and B cell epitopes that have been experimentally confirmed from the Immune Epitope Database (IEDB). The LBtope variable dataset contains 14,876 B-cell epitope and 23,321 nonepitope datasets of varying lengths, whereas the LBtope fixed size dataset has 12,063 B-cell epitope, and 20,589 nonepitope datasets of consistent lengths, both kinds of datasets were produced from one another. In addition, the epitopes that were quite similar to one another were eliminated so the performance could be enhanced (Harinder Singh et al., 2013).

2.7 BepiPred

BepiPred is an antigen–antibody structure prediction system that uses a random forest method for training. These epitopes are taken from antibody–antigen protein structures. It is a novel approach that outperforms the previous tools since it is based on known 3D structures and a massive number of linear epitopes accessible from the IEDB. It provides the findings in a format that is approachable and helpful for users with varying degrees of expertise and experience with technology (Jespersen et al., 2017).

2.8 Immune Epitope Database

IEDB and Analysis Resource is a resource that may be used without any restrictions. It includes a wide variety of potentially protective epitopes and a collection of tools for identifying and analyzing epitopes (Vita et al., 2019). The IEDB incorporates counteracting agents and T cell epitopes for immune system illnesses, nonhuman primates, irresistible sickness, and allergens relocated to alloantigen concentrated mice and other species. Professionals in the life sciences may use the IEDB to develop novel diagnostics, treatments, and antibodies. The dataset is made up of data that was caught or gathered from information that was peer-reviewed and put together by scientists. There are over 1,200,000 B cell, T cell, MHC restriction, MHC ligand elution assays, and over 260,000 epitopes in the curated collection as of December 2016 (Andreatta & Nielsen, 2016).

2.9 DiscoTope

The DiscoTope service predicts discontinuous B-cell epitopes using 3D protein structures. Surface accessibility (measured in terms of contact counts) and an individual amino acid score representing the epitope's tendency to be exposed are considered. The scores are derived by adding the probabilities of nearby residues and the total number of contacts (Kringelum et al., 2012).

DiscoTope identifies 15.5% of residues inside discontinuous epitopes with a specificity of 95%. The assumptions may direct experimental epitope mapping for rational vaccine design and diagnostic tool development, which may improve epitope identification (Haste Andersen et al., 2006).

2.10 AllerTop

Auto cross-covariance is a protein sequence mining technology invented by Wold et al. (1993) that transforms a protein sequence into vectors of uniform length. The Quantitative structure-activity relationships of peptides of varying lengths were investigated using this method. Five E descriptors were used to describe the most essential things about amino acids. The information shows that amino acids range in molecular size, forming properties, Helix, hydrophobicity, the relative abundance of amino acids, and the capacity to create strands. A k-nearest neighbor method (kNN, $k = 1$) is used to categorize proteins based on a training set comprised of 2427 allergens from various species and 2427 nonallergens (Shen & Chou, 2005). AllerTOP v.2 is the most precise, user-friendly, and robust allergy prediction tool.

2.11 ABCpred

For antigen sequence prediction of linear B-cell epitope areas, the service ABCpred uses artificial neural networks. Through this database, scientists may more easily locate potential epitope regions to select synthetic vaccine candidates, diagnose diseases, and conduct allergy studies. This is the first server built on a recurrent neural network that uses pre-defined pattern lengths. The testing and training databases include 2100 peptides, 700 of which are B-cell epitopes and 700 of which are non-B-cell epitopes, all of which are no longer than 20 amino acids. Using a recurrent neural network, around 65.93% accuracy was attained. Using epitopes learned from antibody—antigen protein structures, BepiPred is a random forest method. It is a novel approach that outperforms the previous tools since it is based on known 3D structures and a vast number of linear epitopes accessible from the IEDB database. It provides the findings in an approachable and helpful format for users with different extents of expertise and familiarity with computers (Saha & Raghava, 2006).

3. T- and B-cell epitope identification

Epitope accessibility is crucial for the development of epitope-based antibodies. It is recommended that the idea of epitopes contained in an antigen be considered while designing such vaccinations. Differentiating among both T and B cell epitope recognition is possible (Sanchez-Trincado et al., 2017). The binding of B-cell receptors to antigen epitopes on soluble structures or the surface of microorganisms is a particle-independent process. B-cell epitopes are reliable, conformationally stable, and localized to the target protein (Adhikari et al., 2019). The amino acids that make up consistent epitopes sometimes referred to as consecutive or straight epitopes are the ones that appear in the order that they do in the protein. B-cell epitopes are generally surface visible antigen regions that are hydrophilic, polar, and capable of rapidly binding to individual particles of the counteracting drug. T-cell epitopes, in contrast to B-cell epitopes, which may be directly recognized, call for the presentation of the epitope in conjunction with MHC atoms (Jespersen et al., 2019). Only direct or sequential connections exist between lymphocyte epitopes, and antigens need to be processed before their receptors can recognize them (Alberts et al., 2002). The protein is first cleaved into specific peptides, which then attach to MHC particles and combine with T-cell receptors to create a trimolecular complex. Cytotoxic T cells, also known as Tc cells, contain protein particles known as CD8 on their surface, but T-helper cells and Th cells have a protein known as CD4 on their surface. In the MHC system, Tc-cell epitopes are denoted by Class I particles, while Class II atoms denote Th-cell epitopes. Two kinds of T lymphocytes have different epitope delivery and preparation mechanisms (Fleri et al., 2017).

4. The procedure of vaccine antigen designing

In the field of immunology, the process of producing vaccine-induced immunity might be considered to be fairly complex. Conventional vaccines were made by trial and error when scientists knew little or nothing about how vaccines activate the immune system. Several studies have been undertaken in an attempt to gain a deeper understanding of this issue. However, owing to its complexity, a new approach is required (Goh et al., 2019; Kim et al., 2019). Immunoinformatics aims to establish a method that takes into account a variety of elements that have an impact on vaccine production, such as pathogen antigenic diversity, the onset of infectious illness, and human genetic variation. Induction of the immunological memory is one of the several steps involved in immune system activation. The potency of a vaccination is dependent on the intensity of this induction. Therefore, immunological memory stimulating factors, persistent antibodies, and the type and number of immune memory cells generated have an impact on vaccination effectiveness over time. The primary vaccine-mediated immunological effectors are mainly the antibodies (from B lymphocytes/cell) and sometimes $CD8^+$ and $CD4^+$ T cells (Giacomet et al., 2018). These antibodies selectively bind to specific pathogens or toxins. The majority of antigens and vaccines elicit both humoral and cell-mediated immune responses. Vaccines that induce these types of immunological responses, namely B and T cell responses, are believed to be more effective. T cells create immunological memory cells and high-affinity antibodies, while B cells are often thought of being the major vaccine immune effectors. After the identification of new vaccination targets using EpiMatrix, studies in reverse vaccinology and immunomics also demonstrated T cells as the primary immune effectors (Munang'andu et al., 2015). Successful improvements in vaccine design have resulted from this shift in immunological focus.

Immune-related research is currently concentrated on certain susceptible groups, like the young, old, or immunocompromised, even in the face of prospective advancements in vaccine design. These worries have sparked a deeper knowledge of the effectiveness of the present vaccines on this susceptible group and have opened the door for the use of novel strategies that can take into account demographic variations and improved targets that may induce the best immune response (Rappuoli et al., 2016) of the type II T-cell-independent (TI-2) antigens being the only exception. The germinal centers are activated by antigens that might trigger both B and T lymphocyte responses. This results in antigen-specific, highly effective B-cell proliferation and ultimate differentiation into memory B cells and plasma cells that produce antibodies. All known protein and DNA antigens, as opposed to type II T-cell-independent (TI-2) antigens, produce immunological memory B cells. However, even in the absence of recall reactions, these polysaccharide antigens may still promote long-lasting humoral immunity (Simon et al., 1821). It's possible that vaccinations that just stimulate B cells won't provide long-term protection. Although the development of vaccines with weak immunogenicity, needing heavy adjuvantation, is a common outcome of the conventional method, additional vaccine design strategies, like as cloning and expressing important surface antigens, have been included to combat infectious disease threats (Berical et al., 2016). More complicated infections or those with a high rate of mutation are likely to escape this method's sensitivity. The virulence of these diseases' pathogenesis in humans does not rely on a single pathway, thus improving the specificity of the vaccine should be the goal rather than simply its efficacy as is the case with the present conventional vaccinations in order to change this process (Burton, 2017). Rather than focusing just on effectiveness, it is important to fund research that focuses on the specificity of vaccines against disease antigens. Thankfully, genomes for a number of pathogens causing neglected tropical illnesses and a number of new pathogens are now becoming accessible thanks to international research efforts (Rauch et al., 1963). These genomes may now be screened for potential vaccine targets using computational vaccinology. The most important gene of interest may be modeled for a possible vaccine candidate specific for that disease and several proteins of virulence interest can be sequenced using these technologies. Immunoinformatics is the next step in the process of identifying potential vaccination candidates for infections that produce a variety of antigens and in developing customized treatments for diseases.

5. Immunoinformatics in COVID-19 vaccine development

It has been determined that vaccination is the most successful method for protecting against infectious diseases (Bozzola et al., 2013). In general, there are two stages involved in the vaccination process: (1) the presentation of an alien antigen, such as a weakened virus, and (2) the induction of an immunological response by the body's immune system (Strugnell et al., 2011). According to the principles of classical immunology, a contact must first occur between the host receptors and the foreign antigen which needs to be preceded by activating both the innate and adaptive immune systems (Chaplin, 2010). Even though vaccines have come a long way and are used by many people, not much is known about how they stimulate the immune system. The role of adjuvants only makes what little is known about how these fundamental molecular interactions work even more confusing. Several bioinformatics tools have been made to find and learn more about how the immune system and the antigen work together at the molecular level. Immunoinformatics is the name of the new field that has grown out of this and the experimental data (Buonaguro et al., 2011). Immunoinformatics uses a variety of computer programs, databases, and tools to make it easier to analyze immunologic data (Patronov & Doytchinova, 2013). Immunoinformatics is predicted to play a significant role in the progress of immunological investigations, such as those involving bioinformatics tools in vaccine creation. The development of immunology and molecular biology has also given birth to reverse vaccinology, which, using in silico methods, can enhance vaccine manufacture and vaccination regimens. Due to their specificity, affinity, and stability, antibodies have been given more and more thought as possible treatments in recent years (Lu et al., 2020). In addition, specialized computational methods and online databases have been created to anticipate the epitopes of these immune cells since T and B cells are crucial for finding epitopes (Raoufi et al., 2020). The development of antibody-based therapeutics is directly aided by the ability of antibodies to the molecules on the surface of these cells capable of mounting an adaptive immune response. The study of these interactions may probably benefit from computational developments in peptide modeling and design advances (Fig. 8.2). In reality, antibodies have evolved throughout time into the primary components of protein-based treatments, and studies of the link between their function and structure provide a foundation for protein engineering that may lead to effective therapeutic strategies.

Additionally, the current SARS-CoV-2 epidemic has highlighted the significance of antibodies in biomedical research, both in diagnostic and therapeutic methods. Consequently, there is a rising need to comprehend antibody features and

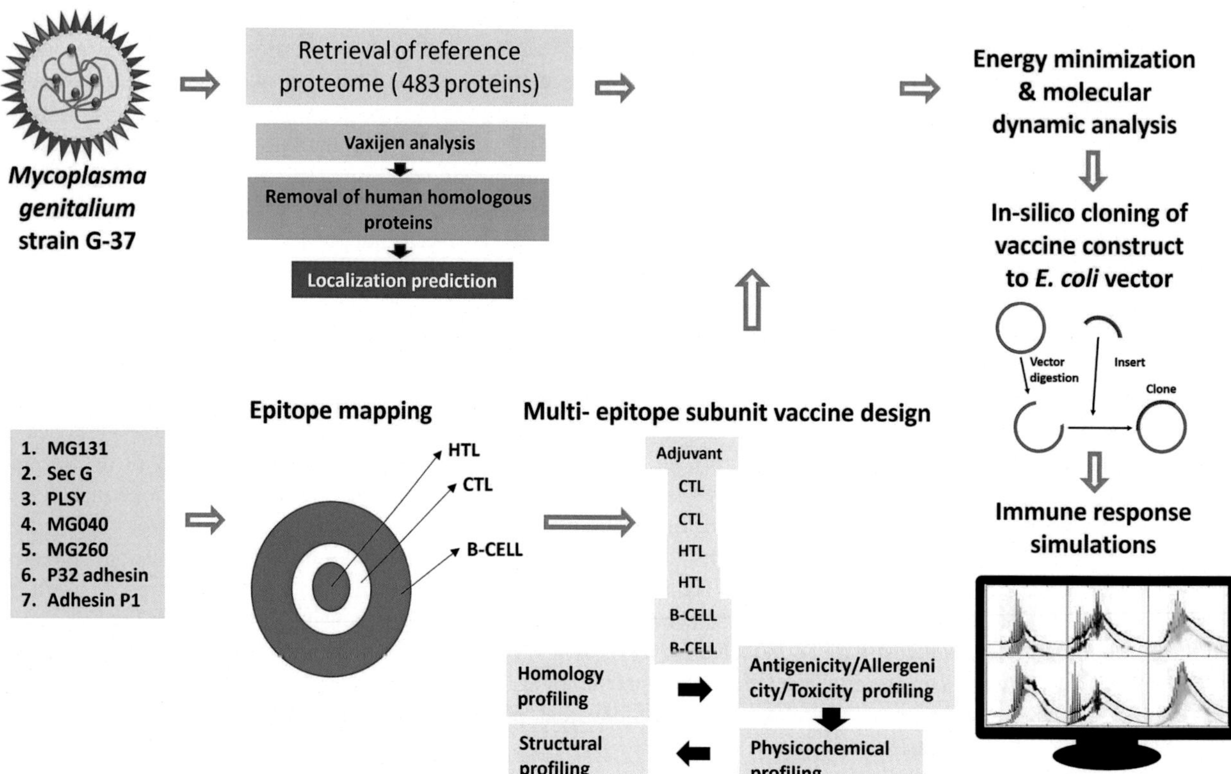

FIGURE 8.2 Steps employed in the designing of vaccines using structure-based drug designing approach.

structure, which has led to the development of several databases and tools for structure and sequence analysis (Chiu et al., 2019). The use of vaccinomics as a multidisciplinary strategy in SARS-CoV-2 antigen discovery based on structural proteins and logical vaccine design, as well as the use of open databases and servers for analyzing the structure and sequence of antigens and antibodies.

Various in silico algorithms anticipate epitopes for vaccine creation thanks to new informatics methods for assessing the vast quantities of accessible data. Computational methods, in particular, might enhance in silico identification of SARS-CoV-2 epitopes and antigenic proteins for vaccine development (De Sousa & Doolan, 2016). Various immunoinformatics approaches have been used to create a multiepitope vaccine polypeptide with the most significant potential for stimulating the human immune system against SARS-CoV-2 (Behmard et al., 2020). A variety of potential peptides for a T-cell epitope-based peptide vaccine were made using the SARS-CoV-2 proteins as a target, comparative genomics as the technical method, and immunoinformatics as a tool (Abdelmageed et al., 2020).

6. Vaccinomics limitations for vaccine design

Although vaccinomics tools are great for making vaccines, they have some drawbacks when it comes to vaccine development and design:

- Vaccinomics depends on wet laboratory work to provide raw data for vaccine production (Kumar & Hasija, 2022).
- The vaccinomics technique does not technically proof concepts and thus cannot replace conventional experimental research methods that include the actual hypothesis testing.
- In designing vaccines, the accuracy of the output data depends on how complicated the immunoinformatics tool used is. The next conclusions of the analysis will also be wrong.
- An immunoinformatics method to vaccine creation may be used to produce vaccines from proteins, linear and discontinuous epitopes, but not other macromolecules like polysaccharides.
- Antigen residues have been proven in some cases to create epitopes, influencing prediction algorithms. Improved prediction approaches in immunoinformatics tools should be required (Kumar & Hasija, 2022).

7. Conclusion

Vaccinomics is a revolutionary approach in the development of newer, safe, and effective personalized vaccination. This paradigm shifts the domain away from the traditional pragmatic approach of "isolate, inactivate/attenuate, inject," and direct toward "identify—evaluate—classify—execute" paradigm. The next phase in vaccinomics advancement is to comprehend immune "signature profiles" from a systems biology viewpoint to formulate vaccine responsiveness in terms of "markers" to assist individually tailored vaccine development as well as to make informed vaccine development. This new paradigm is underpinned by a comprehensive systems biology framework that explores the genetic and immunologic processes as well as the determinants of adaptive immune responses and antigen-mediated innate responses.

References

Abdelmageed, M. I., Abdelmoneim, A. H., Mustafa, M. I., Elfadol, N. M., Murshed, N. S., Shantier, S. W., & Makhawi, A. M. (2020). Design of a multiepitope-based peptide vaccine against the E protein of human COVID-19: An immunoinformatics approach. *BioMed Research International*. ISSN: 23146141, *2020*. https://doi.org/10.1155/2020/2683286. http://www.hindawi.com/journals/biomed/

Adhikari, A., Simha, M. V., Singh, V., Jha, R. K., & Upadhyay, H. (2019). A review on immunosuppressive drugs of organ transplantation. *Think India Journal, 22*(14), 1657—1671.

Alberts, B., Johnson, A., Lewis, J., Raff, M., Roberts, K., & Walter, P. (2002). Genesis, modulation, and regeneration of skeletal muscle. *Molecular Biology of the Cell, 18*(3), 273—289.

Andreatta, Massimo, & Nielsen, Morten (2016). Gapped sequence alignment using artificial neural networks: Application to the MHC class I system. *Bioinformatics, 32*(4), 511—517. https://doi.org/10.1093/bioinformatics/btv639, 1367-4811.

Behmard, Esmaeil, Soleymani, Bijan, Najafi, Ali, & Barzegari, Ebrahim (2020). Immunoinformatic design of a COVID-19 subunit vaccine using entire structural immunogenic epitopes of SARS-CoV-2. *Scientific Reports, 10*(1). https://doi.org/10.1038/s41598-020-77547-4, 2045-2322.

Berical, A. C., Harris, D., Dela Cruz, C. S., & Possick, J. D. (2016). Pneumococcal vaccination strategies: An update and perspective. *Annals of the American Thoracic Society*. ISSN: 23256621, *13*(6), 933—944. https://doi.org/10.1513/AnnalsATS.201511-778FR. http://www.atsjournals.org/doi/pdf/10.1513/AnnalsATS.201511-778FR

Bozzola, Elena, Bozzola, Mauro, Calcaterra, Valeria, Barberi, Salvatore, & Villani, Alberto (2013). Infectious diseases and vaccination strategies: How to protect the "unprotectable"? *ISRN Preventive Medicine, 2013*, 1—5. https://doi.org/10.5402/2013/765354, 2090-8784.

Brodie, T. M., & Tosevski, V. (2018). Broad immune monitoring and profiling of T cell subsets with mass cytometry. *Methods in Molecular Biology*. ISSN: 10643745, *1745*, 67–82. https://doi.org/10.1007/978-1-4939-7680-5_4. http://www.springer.com/series/7651

Buonaguro, L., Wang, E., Tornesello, M. L., Buonaguro, F. M., & Marincola, F. M. (2011). Systems biology applied to vaccine and immunotherapy development. *BMC Systems Biology*. ISSN: 17520509, *5*. https://doi.org/10.1186/1752-0509-5-146. http://www.biomedcentral.com/1752-0509/5/146

Burton, D. R. (2017). What are the most powerful immunogen design vaccine strategies?: Reverse vaccinology 2.0 shows great promise. *Cold Spring Harbor Perspectives in Biology*. ISSN: 19430264, *9*(11). https://doi.org/10.1101/cshperspect.a030262. http://cshperspectives.cshlp.org/content/9/11/a030262.full.pdf

Catalanotto, Caterina, Cogoni, Carlo, & Zardo, Giuseppe (2016). MicroRNA in control of gene expression: An overview of nuclear functions. *International Journal of Molecular Sciences, 17*(10). https://doi.org/10.3390/ijms17101712, 1422-0067.

Chaplin, D. D. (2010). Overview of the immune response. *Journal of Allergy and Clinical Immunology*. ISSN: 00916749, *125*(2), S3–S23. https://doi.org/10.1016/j.jaci.2009.12.980

Chiu, M. L., Goulet, D. R., Teplyakov, A., & Gilliland, G. L. (2019). Antibody structure and function: The basis for engineering therapeutics. *Antibodies*. ISSN: 20734468, *8*(4). https://doi.org/10.3390/antib8040055. https://www.mdpi.com/2073-4468/8/4/55/pdf

De Sousa, K. P., & Doolan, D. L. (2016). Immunomics: A 21st century approach to vaccine development for complex pathogens. *Parasitology*. ISSN: 14698161, *143*(2), 236–244. https://doi.org/10.1017/S0031182015001079. http://journals.cambridge.org/action/displayJournal?jid=PAR

Dimitrov, I., Bangov, I., Flower, D. R., & Doytchinova, I. (2014). AllerTOP v. 2—A server for in silico prediction of allergens. *Journal of Molecular Modeling, 20*(6), 1–6.

Doherty, M., Schmidt-Ott, R., Santos, J. I., Stanberry, L. R., Hofstetter, A. M., Rosenthal, S. L., & Cunningham, A. L. (2016). Vaccination of special populations: Protecting the vulnerable. *Vaccine*. ISSN: 18732518, *34*(52), 6681–6690. https://doi.org/10.1016/j.vaccine.2016.11.015. www.elsevier.com/locate/vaccine

El-Manzalawy, Yasser, Dobbs, Drena, & Honavar, Vasant (2008). Predicting linear B-cell epitopes using string kernels. *Journal of Molecular Recognition*. ISSN: 09523499, *21*(4), 243–255. https://doi.org/10.1002/jmr.893

Fleri, W., Paul, S., Dhanda, S. K., Mahajan, S., Xu, X., Peters, B., & Sette, A. (2017). The immune epitope database and analysis resource in epitope discovery and synthetic vaccine design. *Frontiers in Immunology*. ISSN: 16643224, *8*. https://doi.org/10.3389/fimmu.2017.00278. http://journal.frontiersin.org/article/10.3389/fimmu.2017.00278/full

Giacomet, Vania, Masetti, Michela, Nannini, Pilar, Forlanini, Federica, Clerici, Mario, Zuccotti, Gian Vincenzo, Trabattoni, Daria, & Jhaveri, Ravi (2018). Humoral and cell-mediated immune responses after a booster dose of HBV vaccine in HIV-infected children, adolescents and young adults. *PLoS One, 13*(2). https://doi.org/10.1371/journal.pone.0192638, 1932-6203.

Giani, A. M., Gallo, G. R., Gianfranceschi, L., & Formenti, G. (2020). Long walk to genomics: History and current approaches to genome sequencing and assembly. *Computational and Structural Biotechnology Journal*. ISSN: 20010370, *18*, 9–19. https://doi.org/10.1016/j.csbj.2019.11.002. www.csbj.org

Gibas, Jambeck, P., & Fenton, J. (2001). *Developing bioinformatics computer skills*, 2001.

Goh, Y. S., McGuire, D., & Rénia, L. (2019). Vaccination with sporozoites: Models and correlates of protection. *Frontiers in Immunology*. ISSN: 16643224, *10*. https://doi.org/10.3389/fimmu.2019.01227. https://www.frontiersin.org/journals/immunology#

Haste Andersen, Pernille, Nielsen, Morten, & Lund, Ole (2006). Prediction of residues in discontinuous B-cell epitopes using protein 3D structures. *Protein Science*. ISSN: 09618368, *15*(11), 2558–2567. https://doi.org/10.1110/ps.062405906

Jespersen, M. C., Mahajan, S., Peters, B., Nielsen, M., & Marcatili, P. (2019). Antibody specific B-cell epitope predictions: Leveraging information from antibody-antigen protein complexes. *Frontiers in Immunology*. ISSN: 16643224, *10*. https://doi.org/10.3389/fimmu.2019.00298. https://www.frontiersin.org/journals/immunology#

Jespersen, M. C., Peters, B., Nielsen, M., & Marcatili, P. (2017). BepiPred-2.0: Improving sequence-based B-cell epitope prediction using conformational epitopes. *Nucleic Acids Research*. ISSN: 13624962, *45*(1), W24–W29. https://doi.org/10.1093/nar/gkx346. http://nar.oxfordjournals.org/

Johnson, K. L., Ovsyannikova, I. G., Poland, G. A., & Muddiman, D. C. (2005). Identification of class II HLA-DRB1*03-bound measles virus peptides by 2D-liquid chromatography tandem mass spectrometry. *Journal of Proteome Research*. ISSN: 15353893, *4*(6), 2243–2249. https://doi.org/10.1021/pr0501416

Kim, H. I., Ha, N. Y., Kim, G., Min, C. K., Kim, Y., Yen, N. T. H., Choi, M. S., & Cho, N. H. (2019). Immunization with a recombinant antigen composed of conserved blocks from TSA56 provides broad genotype protection against scrub typhus. *Emerging Microbes and Infections*. ISSN: 22221751, *8*(1), 946–958. https://doi.org/10.1080/22221751.2019.1632676. https://www.tandfonline.com/loi/temi20

Krawczyk, K., Liu, X., Baker, T., Shi, J., & Deane, C. M. (2014). Improving B-cell epitope prediction and its application to global antibody-antigen docking. *Bioinformatics*. ISSN: 14602059, *30*(16), 2288–2294. https://doi.org/10.1093/bioinformatics/btu190. http://bioinformatics.oxfordjournals.org/

Kringelum, Jens Vindahl, Lundegaard, Claus, Lund, Ole, Nielsen, Morten, & Peters, Bjoern (2012). Reliable B cell epitope predictions: Impacts of method development and improved benchmarking. *PLoS Computational Biology, 8*(12). https://doi.org/10.1371/journal.pcbi.1002829, 1553-7358.

Kumar, S., & Hasija, Y. (2022). Immunoinformatics tools: A boon in vaccine development against Covid-19. In *2022 IEEE Delhi section conference, DELCON 2022*. India: Institute of Electrical and Electronics Engineers Inc.. https://doi.org/10.1109/DELCON54057.2022.9753152, 9781665458832 http://ieeexplore.ieee.org/xpl/mostRecentIssue.jsp?punumber=9752769.

Lu, R. M., Hwang, Y. C., Liu, I. J., Lee, C. C., Tsai, H. Z., Li, H. J., & Wu, H. C. (2020). Development of therapeutic antibodies for the treatment of diseases. *Journal of Biomedical Science*. ISSN: 14230127, *27*(1). https://doi.org/10.1186/s12929-019-0592-z. http://www.jbiomedsci.com

Magnan, C. N., Zeller, M., Kayala, M. A., Vigil, A., Randall, A., Felgner, P. L., & Baldi, P. (2010). High-throughput prediction of protein antigenicity using protein microarray data. *Bioinformatics*. ISSN: 14602059, *26*(23), 2936–2943. https://doi.org/10.1093/bioinformatics/btq551

Munang'andu, H. M., Mutoloki, S., & Evensen, O. (2015). A review of the immunological mechanisms following mucosal vaccination of finfish. *Frontiers in Immunology*. ISSN: 16643224, *6*. https://doi.org/10.3389/fimmu.2015.00427. http://journal.frontiersin.org/article/10.3389/fimmu.2015.00427/full

O'Shea, D., Widmer, L. A., Stelling, J., & Egli, A. (2014). Changing face of vaccination in immunocompromised hosts. *Current Infectious Disease Reports*. ISSN: 15343146, *16*(9). https://doi.org/10.1007/s11908-014-0420-2. http://www.springerlink.com/content/1523-3847/

Patronov, Atanas, & Doytchinova, Irini (2013). T-cell epitope vaccine design by immunoinformatics. *Open Biology, 3*(1). https://doi.org/10.1098/rsob.120139, 2046-2441.

Poland, G. A., Ovsyannikova, I. G., & Kennedy, R. B. (2018). Personalized vaccinology: A review. *Vaccine*. ISSN: 0264410X, *36*(36), 5350–5357. https://doi.org/10.1016/j.vaccine.2017.07.062

Ponomarenko, J., Bui, H. H., Li, W., Fusseder, N., Bourne, P. E., Sette, A., & Peters, B. (2008). ElliPro: A new structure-based tool for the prediction of antibody epitopes. *BMC Bioinformatics*. ISSN: 14712105, *9*. https://doi.org/10.1186/1471-2105-9-514

Raoufi, E., Hemmati, M., Eftekhari, S., Khaksaran, K., Mahmodi, Z., Farajollahi, M. M., & Mohsenzadegan, M. (2020). Epitope prediction by novel immunoinformatics approach: A state-of-the-art review. *International Journal of Peptide Research and Therapeutics*. ISSN: 15733904, *26*(2), 1155–1163. https://doi.org/10.1007/s10989-019-09918-z. http://www.springeronline.com/sgw/cda/frontpage/0,11855,4-40109-70-35677487-0,00.html

Rappuoli, R., Bottomley, M. J., D'Oro, U., Finco, O., & De Gregorio, E. (2016). Reverse vaccinology 2.0: Human immunology instructs vaccine antigen design. *Journal of Experimental Medicine*. ISSN: 15409538, *213*(4), 469–481. https://doi.org/10.1084/JEM.20151960. http://jem.rupress.org/content/jem/213/4/469.full.pdf

Rappuoli, Rino (2000). Reverse vaccinology. *Current Opinion in Microbiology*. ISSN: 13695274, *3*(5), 445–450. https://doi.org/10.1016/s1369-5274(00)00119-3

Rasheed, M. A., Raza, S., Zohaib, A., Riaz, M. I., Amin, A., Awais, M., Khan, S. U., Ijaz Khan, M., & Chu, Y. M. (2021). Immunoinformatics based prediction of recombinant multi-epitope vaccine for the control and prevention of SARS-CoV-2. *Alexandria Engineering Journal*. ISSN: 11100168, *60*(3), 3087–3097. https://doi.org/10.1016/j.aej.2021.01.046. http://www.elsevier.com/wps/find/journaldescription.cws_home/724292/description#description

Rauch, S., Jasny, E., Schmidt, K. E., & Petsch, B. (1963). New vaccine technologies to combat outbreak situations. *Frontiers in Immunology, 9*.

Reeves, P. M., Sluder, A. E., Paul, S. R., Scholzen, A., Kashiwagi, S., & Poznansky, M. C. (2018). Application and utility of mass cytometry in vaccine development. *Federation of American Societies for Experimental Biology Journal*. ISSN: 15306860, *32*(1), 5–15. https://doi.org/10.1096/fj.201700325R. http://www.fasebj.org/content/32/1/5.full.pdf+html

Saha, Sudipto, & Raghava, G. P. S. (2006). Prediction of continuous B-cell epitopes in an antigen using recurrent neural network. *Proteins: Structure, Function, and Bioinformatics*. ISSN: 08873585, *65*(1), 40–48. https://doi.org/10.1002/prot.21078

Sanchez-Trincado, J. L., Gomez-Perosanz, M., & Reche, P. A. (2017). Fundamentals and methods for T-and B-cell epitope prediction. *Journal of Immunology Research*.

Sarkizova, S., Klaeger, S., Le, P. M., Li, L. W., Oliveira, G., Keshishian, H., Hartigan, C. R., Zhang, W., Braun, D. A., Ligon, K. L., Bachireddy, P., Zervantonakis, I. K., Rosenbluth, J. M., Ouspenskaia, T., Law, T., Justesen, S., Stevens, J., Lane, W. J., Eisenhaure, T., ... Keskin, D. B. (2020). A large peptidome dataset improves HLA class I epitope prediction across most of the human population. *Nature Biotechnology*. ISSN: 15461696, *38*(2), 199–209. https://doi.org/10.1038/s41587-019-0322-9. http://www.nature.com/nbt/index.html

Shen, H., & Chou, K. C. (2005). Using optimized evidence-theoretic K-nearest neighbor classifier and pseudo-amino acid composition to predict membrane protein types. *Biochemical and Biophysical Research Communications, 334*(1), 288–292. https://doi.org/10.1016/j.bbrc.2005.06.087

Shi, & Liu. (2020). *Huanjing Kexue/Environmental science* (pp. 11–13).

Simon, A. K., Hollander, G. A., & McMichael, A. (1821). Evolution of the immune system in humans from infancy to old age. *Proceedings of the Royal Society B: Biological Sciences, 282*.

Singh, Harinder, Ansari, Hifzur Rahman, Raghava, Gajendra P. S., & Schönbach, Christian (2013). Improved method for linear B-cell epitope prediction using antigen's primary sequence. *PLoS One, 8*(5). https://doi.org/10.1371/journal.pone.0062216, 1932-6203.

Singh, H. (2016). Bioinformatics: Benefits to mankind. *International Journal of PharmTech Research*. ISSN: 09744304, *9*(4), 242–248. http://www.sphinxsai.com/2016/ph_vol9_no4/1/(242-248)V9N4PT.pdf.

Stewart-Morgan, K. R., Petryk, N., & Groth, A. (2020). Chromatin replication and epigenetic cell memory. *Nature Cell Biology*. ISSN: 14764679, *22*(4), 361–371. https://doi.org/10.1038/s41556-020-0487-y. http://www.nature.com/ncb/index.html

Strugnell, Richard, Zepp, Fred, Cunningham, Anthony, & Tantawichien, Terapong (2011). Vaccine antigens. *Perspectives in Vaccinology*. ISSN: 22107622, *1*(1), 61–88. https://doi.org/10.1016/j.pervac.2011.05.003

Vita, R., Mahajan, S., Overton, J. A., Dhanda, S. K., Martini, S., Cantrell, J. R., Wheeler, D. K., Sette, A., & Peters, B. (2019). The immune epitope database (IEDB): 2018 update. *Nucleic Acids Research*. ISSN: 13624962, *47*(1), D339–D343. https://doi.org/10.1093/nar/gky1006. https://academic.oup.com/nar/issue

Wold, S., Jonsson, J., Sjöström, M., Sandberg, M., & Rännar, S. (1993). DNA and peptide sequences and chemical processes multivariately modelled by principal component analysis and partial least squares projections to latent structures. *Analytica Chimica Acta, 277*, 239–253. https://doi.org/10.1016/0003-2670(93)80437-P

Chapter 9

Integrative omics approach for identification of genes associated with disease

Keerti Kumar Yadav and Ajay Kumar Singh
Department of Bioinformatics, Central University of South Bihar, Gaya, Bihar, India

1. Introduction

The omics terms are mainly referred to a molecular term, which suggests a global evaluation of a set of biological molecules. The omics-based study mainly focused on advanced technology that has enabled cost-effective, high-throughput of biological molecules (Hasin et al., 2017). The genomic research mainly offered an extremely valuable framework for identifying and analyzing specific genetic variants that involved in inherited and complex diseases. In recent decade, high-throughput genotyping, in addition with the advancement of high-quality reference map and large number of clinical obtained patient data, has helpful for the localization and mapping of thousands of genetic variants in human genome (Feero et al., 2010; LaFramboise, 2009). The second level of data is transcriptomics data, which deals the RNA data and their transcripts. The RNA transcripts are the linking of DNA data and protein data. The proteomic approach of omics data mainly deals the protein-level information of the biological molecules. They mainly considered the protein structure and their functional aspect of biological molecules. Metabolomic data considered the metabolic activity and its reaction by the help of obtained omics analysis. These all omics-level data are helpful to understanding of the biological process.

The omics data mainly provides the complete analysis report related to the diseases which are helpful for the understanding of the occurrence of the disease. These data also useful to understand the disease markers as well as the specific biological pathways or processes differ between disease and control groups. The integration and deep analysis of multiomics data is mainly used to explain potential causative abnormalities that lead to disease, or therapeutic targets, which may then be evaluated in various molecular research (Hasin et al., 2017). High-throughput omics technologies based on next-generation sequencing (NGS) and mass spectrometry have evolved significantly in recent years, enabling for in-depth genetic and functional analyses of diseases development and progression (Khan et al., 2020). In the past decade, technological improvements including ease of sample preparation workflows (Cui et al., 2013), developed well instrument sensitivity and accuracy (Palmfeldt & Bross, 2017), and advanced computational analysis methods also help in the development of new diagnosis techniques for disease treatment (Huang et al., 2017).

Furthermore, the utilization of genomes, transcriptomics, proteomics, and metabolomics data has provided valuable insights into pathogenic pathways involved in pathologies, such as molecular stress responses and substantial metabolome and proteome modifications (Buzkova et al., 2018). The integration of omics data at various levels (multiomics) allows for more robust insights into complex molecular functional mechanisms by strengthening complementary evidence from multiple levels (Hasin et al., 2017). Fig. 9.1 depicts the omics data analysis methodologies, which are generally used for the development of advance disease treatment strategy.

Integrative Omics. https://doi.org/10.1016/B978-0-443-16092-9.00009-6

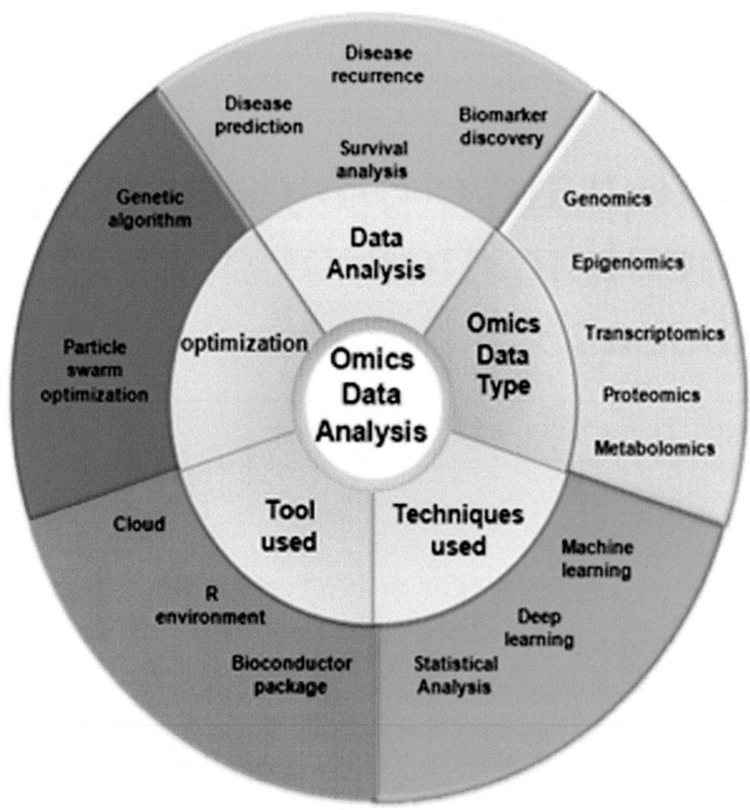

FIGURE 9.1 Methodology useful for the omics data analysis.

2. Various types of omics data

2.1 Genomics

Genomics are the most advanced omics analysis approach for biological data analysis. Genomic researches of biomolecules are mainly focused on identifying genetic variants linked to response of patient disease prognosis and treatment. Genome-wide association study (GWAS) is an effective method useful to find large numbers of genetic variations related with complex diseases in large human population's data. The large numbers of people are genotyped for over a million genetic markers in this type of research and based on the statistically significant changes in minor allele frequencies between cases and controls. The GWASs contribute significantly to our understanding of complex traits of disease. Genotype arrays (Ragoussis, 2009), NGS for whole-genome sequencing (Koboldt et al., 2013), and exome sequencing (Ng et al., 2009) are all associated technologies helpful for handling and understanding of disease-associated data.

2.2 Epigenomics

The epigenomics study is mainly concerned with the genome-wide characterization of reversible DNA or DNA-linked protein changes such as DNA methylation or histone acetylation. The covalent modifications to DNA and histones are important regulators of gene transcription and helpful to define the cellular fate (Piunti & Shilatifard, 2016). These changes can be influenced by both genetic and environmental influences, can continue a long time, and are sometimes transferred to parents to offspring (Keating & El-Osta, 2015). While the function of epigenetic changes as mediators of transgenerational environmental impacts has still been controversial (Barrès & Zierath, 2016), many epigenome-wide association studies have been published, demonstrating their importance in biological processes and disease development. Differentially methylated DNA regions can be employed as disease markers for metabolic syndrome (Horvath, 2013), cardiovascular disease (Kim et al., 2010), cancer (Baylin et al., 2001), and various other pathophysiologic conditions (Raghuraman et al., 2016). Epigenetic signatures are frequently tissue-specific (Zhu et al., 2013), and the several large consortia are working to create comprehensive epigenomic maps. The various human tissues databases such as International Human Epigenome Consortium (http://ihec-epigenomes.org/) and Roadmap Epigenomics (http://www.roadmapepigenomics.org/) are also helpful in epigenetic research. In addition to providing insight into the epigenetic modifications associated with diseases, these research' results have the potential to improve the functional interpretation of disease-specific genetic variations existing in those locations, as well as epigenetic markers linked with disease independent of genetic variation. The NGS analysis of DNA alterations analysis is also an example of epigenetic-based associated technology (Consortium et al., 2015).

2.3 Transcriptomics

The transcriptomics studies are also important for the biological data analysis. The transcriptomics studies are performed both qualitatively and quantitatively. Qualitative transcriptomics analysis approach identifies the transcripts present in the genome and identifies the novel splicing sites and RNA edition site, which are helpful for the research purpose. In the quantitative analysis, how much the each transcript is expressed in the disease and normal condition is analyzed. The central dogma of biology process depicts that RNA as a biochemical bridge between DNA and proteins that mainly represent the primary function of DNA. RNA functions are mainly classified based on structural (e.g., ribosomal complexes) and regulatory (e.g., Xist in chromosome X inactivation) function. The large transcriptome-based investigations have revealed that just 3% of the genome encodes proteins, up to 80% of the transcribed genome (Roadmap Epigenomics Consortium, 2015). The RNA-Seq experiments discovered thousands of new isoforms and revealed that the protein-coding transcriptome is more complicated than previously identified by different biological methods (Trapnell et al., 2010). However, the establishment of the noncoding RNA analysis was an even more significant contribution of these RNA-Seq investigations. At present, we all know that thousands of long noncoding RNAs (lncRNAs) (http://www.gencodegenes.org/) transcribed in mammalian cells play important roles in various physiological processes, such as brown adipose differentiation (Alvarez-Dominguez et al., 2015), endocrine regulation (Knoll et al., 2015), and neuron development in brain tissue (Yao et al., 2016). Dysregulation of lncRNAs had been participated in various diseases, e.g., myocardial infarction (Ishii et al., 2006), diabetes (Morán et al., 2012), cancer (Gupta et al., 2010), and others diseases (Schmitz et al., 2016). The NGS-based approach is also helpful in finding the probing of short RNAs (microRNAs, small nuclear RNAs, and Piwi-interacting RNAs) as well as the detection of circular RNAs (Barrett & Salzman, 2016). An increasing amount of evidence suggests that short and circular RNAs, such as long noncoding RNAs, are involved in the dysregulation in illness (Chen et al., 2016), and it's also a potential indicator of therapeutic targets. The probe-based arrays and RNA-Seq analysis approaches are also two principal technologies used for the transcriptomics analysis of biological data (Wang et al., 2009).

2.4 Proteomics

The proteomics-based approach is mainly used to quantify peptide richness, modification, and their interaction. At present, the quantification of proteins has been more developed by the help of MS-based approach, and in recent decades, these are also adapted for high-throughput analyses of a large number of proteins in cells or biological fluids. The phage display method and yeast two-hybrid assays are helpful in the analysis of interaction between the proteins. Affinity purification methods can also be utilized, in which one molecule is isolated by the help of an antibody or a specific genetic marker. Then, using MS, any related proteins are identified. These affinity approaches, are sometimes helpful to combined with chemical cross-linking and develop to identification of global interactions between proteins and nucleic acids (e.g., ChIP-Seq). Nowadays posttranslational changes such as proteolysis, glycosylation, phosphorylation, nitrosylation, and ubiquitination mediate the various fraction of proteins (Mann & Jensen, 2003). These proteomics-based alterations also plays an important role in enzyme activity control, protein turnover, intracellular level cell signaling, and cell structure maintenance (Wu et al., 2011). MS-based approach can directly measure covalent changes by the help of the corresponding shift in protein mass. There are attempts underway to produce genome-level studies of biological changes (Choudhary & Mann, 2010). These protein data are helpful in the analysis of structure and functional mechanism of the disease.

2.5 Metabolomics

The metabolomics are also advanced method for the analysis of the biological data. Metabolomics are also helpful to analyze multiple small biological molecules, such as carbohydrates, amino acids, fatty acids, and other biomolecules, which are involved in cellular metabolic functions. Metabolite levels and their relative ratios are helpful to understand the metabolic function, and deviations from the normal range are helpful to understand the of disease stages. Quantitative measurements of metabolite levels have been helpful in the identification of new genetic loci that are involved in regulation of small biomolecules or their relative ratios in plasma fluid and body tissues (Kettunen et al., 2012). Furthermore, metabolomics in conjunction with biological modeling has been helpful in the deep understanding of metabolite flux of the

TABLE 9.1 Some important multiomics data resources that are obtains for research.

S. no.	Name of database	Web link	Disease	Types of multiomics data available
1	The Cancer Genome Atlas (TCGA)	https://www.cancer.gov/about-nci/organization/ccg/research/structural-genomics/tcga	Cancer	RNA-Seq, DNA-Seq, miRNA-Seq, SNV, CNV, DNA methylation, and RPPA
2	Clinical Proteomic Tumor Analysis Consortium (CPTAC)	https://proteomics.cancer.gov/programs/cptac	Cancer	Collection of proteomics data corresponding to TCGA
3	International Cancer Genomics Consortium (ICGC)	https://dcc.icgc.org/	Cancer	Whole-genome sequencing, genomic variations data (somatic and germline mutation)
4	Cancer Cell Line Encyclopedia (CCLE)	https://portals.broadinstitute.org/ccle	Cancer cell line	Gene expression, copy number, and sequencing data; pharmacological profiles of 24 anticancer drugs
5	Molecular Taxonomy of Breast Cancer International Consortium (METABRIC)	https://www.mercuriolab.umassmed.edu/metabric	Breast cancer	Clinical traits, gene expression, SNP, and CNV
6	TARGET	https://ocg.cancer.gov/programs/target	Pediatric cancers	Gene expression, miRNA expression, copy number, and sequencing data
7	Omics Discovery Index	https://www.omicsdi.org	Consolidated data sets from 11 repositories in a uniform framework	Collection of genomics, transcriptomics, proteomics, and metabolomics
8	Gene bank	https://www.ncbi.nlm.nih.gov/genbank/	All types of diseases	Collection of nucleotide database
9	KEGG	https://www.genome.jp/kegg/pathway.html	All types of diseases	Collection of pathway information

biological process. The MS-based approaches are also helpful in quantifying both relative and targeted small molecule abundances among the associated technologies (Steuer, 2006). The other metabolic-based approaches are also helpful to integrate biological data across multiple omics layers depending on the study design for the analysis of biological process. Two frequently used approaches involve simple correlation and comapping, which are mainly used for the multiomics data analysis. Some important multiomics data platforms are shown in Table 9.1.

3. Methods of omics data analysis

3.1 Genome-wide association study

The GWAS is mainly used to combine the SNP-based association summary data into a single gene-based statistics using the rapid set-based association analysis implemented in the GCTA 1.26.0 software package (Yang et al., 2011). To examine complicated disorders, GWAS has created an approach for integrating association results from multiomics data. The results of the multiomics integration were then merged with gene interaction networks constructed from different tissues to further understand their tissue-specific relationships. To the contrary, it is critical to select the appropriate tissue type when constructing gene interaction networks. Otherwise, there will be little or no improvement in statistical power (Uffelmann et al., 2021). As more omics data becomes available, multiomics data integration is predicted to reveal more disease-related genes beyond GWAS. The epigenome-wide association study (EWAS) and transcriptome-wide association study (TWAS) methods are also used for whole-blood gene expression and DNA methylation measurements. GWAS analysis tools and their functions are shown in Table 9.2, and detailed methodology of GWAS analysis is shown in Fig. 9.2.

TABLE 9.2 Some important software and tools used for the GWAS analysis.

S. no.	Tools name	Function
1	PLINK	Most commonly used tool for GWAS study
2	SUGEN	Useful for complicated survey design, weighted estimation to account for differential sampling probability
3	SNPtest	Useful for analysis of single SNP data and Bayesian association tests
4	GenABEL Project suite	GWAS analyses and statistical "omics" applications, including mixed model-based GWAS analysis
5	GWASTools	An R/bioconductor package for quality control and association analyses of GWAS data
6	EMMAX, SOLAR, GEMMA	Mixed models for association analysis accounting for sample structure (e.g., family-based studies)
7	GMMAT	Generalized linear mixed model test for GWAS of binary traits accounting for population structure and relatedness
8	BigTop	Visualization of Manhattan plot in 3Dof GWAS data
9	Pascal	Score and analysis of GWAS data
10	Metal	Metaanalysis of GWAS data

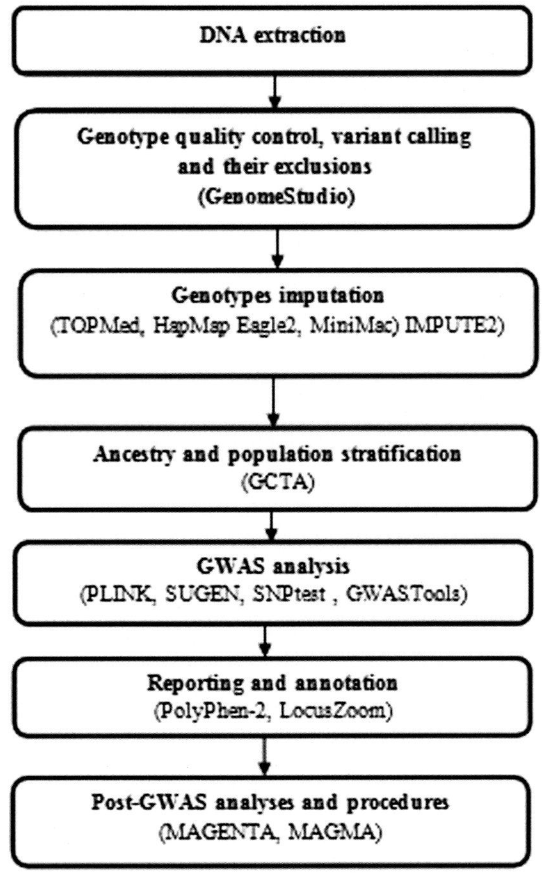

FIGURE 9.2 Flowchart of methodology of GWAS analysis. Steps used in GWAS analysis and their related tools are shown in the flowchart.

3.2 Epigenome-wide association study

The EWAS is a systematic strategy to identify epigenetic variations that explain common diseases/phenotypes. The EWASs are derived from the developing subject of epigenetic epidemiology, with both attempting to understand the molecular basis for illness risk. While genetic risk of disease is unmodifiable till date, epigenetic risk may be reversible or changeable. In early EWAS using the 27k Illumina methylation bead arrays and at present, most investigators use the 450k Illumina bead arrays, and finally to the future, where next-generation sequencing based methods beckon. EWAS has been used to explore a wide range of diseases, exposures, and lifestyle factors, with several noteworthy findings. However, similar to GWAS research, EWASs are expected to necessitate massive multinational consortium-based techniques to reach the numbers of participants, as well as the statistical and scientific rigor, required for solid conclusions (Flanagan, 2015).

3.3 Transcriptome-wide association study

TWAS method is also important for the multiomics data analysis. The TWAS models were adjusted for age, gender, batch effects, and family relatedness. To find gene—trait connections, TWASs combine GWAS and gene expression data sets. We also use simulations and case studies of literature-identified putative genes for schizophrenia, low-density lipoprotein cholesterol, and Crohn's disease to investigate the features of TWAS as a viable technique to prioritize genes at GWAS sites. This also investigate risk loci where TWAS appropriately prioritizes the likely causal gene as well as loci where TWAS prioritizes many genes, some of which are likely to be noncausal due to expression quantitative trait loci (eQTL) sharing. Due to significant cross-cell-type variance in expression levels and eQTL intensities, TWAS is especially vulnerable to spurious selection using expression data from non—trait-related tissues or cell types. TWAS, on the other hand, ranks putative causative genes more precisely than basic baselines. This approach provides optimal practices for causal gene prioritization using TWAS and highlights potential future improvements. The TWAS approach also highlight the advantages and disadvantages of biological eQTL data sets to identify causative genes at GWAS regions (Wang et al., 2020).

3.4 QTL analysis

To gain insight into the genetic contributions to the regulation of molecular pathways and phenotypes, quantitative trait loci (QTL) analysis is commonly used. The associations of genetic loci or intervals with expression profiles of transcriptome, methylome, proteome, or metabolome are determined to define eQTLs, meQTLs, pQTLs, or mQTLs, respectively. The QTL mainly show the continuous variation in a population. This QTL can be single genes or cluster of linked genes that affect the traits. The QTL have various advantages such as identification of the novel disease-related genes and less laborious and expensive as compared with mutant screening-like approaches.

3.5 NGS analysis

The NGS is an advanced sequencing technology that has been developed and refined for biological data analysis. The NGS has grown in popularity in genomics research over the past decade. NGS has recently begun to be used in clinical oncology to improve cancer therapy through a variety of modalities ranging from the discovery of novel and rare cancer mutations. NGS is helpful in the identification of mutation carriers of cancer and other diseases. NGS is also useful in the development of therapeutic techniques for the creation of personalized medicine (PM) for illness treatment. By moving the existing standard medical method for treating cancer and other diseases to a tailored preventive and predictive approach, PM offers the potential to reduce medical costs. Currently, NGS can aid in the early detection of diseases and the discovery of pharmacogenetic markers that aid in the personalization of therapy. Despite enormous advances in genomic understanding, NGS has the added benefit of providing a more comprehensive picture of the cancer landscape and finding cancer development pathways. NGS offered a comprehensive treatment strategy to diagnosis and treatment prospective in scientific and clinical level in cancer and other diseases, with a special emphasis on pharmacogenomics in the direction of precision medicine treatment choices (Hussen et al., 2022). DNA sequencing tools and platforms are depicted in Table 9.3.

TABLE 9.3 DNA sequencing process tools and their platforms.

S. no.	Method	Process	Tool	Platform
1	DNA-sequencing	DNA alignment/mapping	MAQ	Illumina/ABI
			BWA	Illumina/ABI
			Novo align	Illumina/Roche
			SOAP3	Illumina/Roche/ABI
			BFAST	Illumina/Roche/ABI/Helicos
		De novo assembly construction	Newbler	Roche
			VCAKE	Illumina/Roche
			Velvet	Illumina/Roche/ABI
		SNV detection	GATK	Illumina/Roche/ABI
			SAMtools	Illumina/Roche
			VarScan/VarScan2	Illumina/Roche/ABI
			SomaticSniper	Illumina
			JointSNVMix	Illumina
		Structural variation detection	BreakDancer	Illumina/Roche/ABI
			VariationHunter	Illumina
			SVDetect	Illumina/ABI
			PEMer	Illumina/Roche/ABI

From Inchingolo, F., Santacroce, L., Ballini, A., Topi, S., Dipalma, G., Haxhirexha, K., Bottalico, L. & Charitos, I. A. (2020). Oral Cancer: A Historical Review. *International Journal of Environmental Research and Public Health, 17*(9), 3168. https://doi.org/10.3390/ijerph17093168.

3.6 Chip-Seq data analysis

Chromatin immunoprecipitation sequencing (ChIP-Seq) is a critical method in biological data epigenomic research. This method identifies enriched regions within a genome by using an antibody against a specific DNA-binding protein or a histone modification. In the field of ChIP-Seq analysis, histone modifications are utilized to examine the features and biological activities of epigenetic markers (Nakato & Sakata, 2021). With advances in NGS technology and computational analysis, we can now comprehensively investigate how the epigenomic landscape influences cell identity, development, lineage specification, cancer, and other disorders (Zhao & Shilatifard, 2019).

3.7 RNA-Seq data analysis

RNA-Seq is currently a common method for measuring gene expression. The large number of RNA-based analytic tools has evolved during the previous decade. The steps involved in RNA-Seq analysis are then broken down into three major stages: (1) preprocessing and data preparation, (2) upstream processing, and (3) high-level analyses. These techniques are using freely available for RNA-Seq data analysis (Pavlovich & Cauchy, 2022). The messenger RNA serves as a blueprint for protein construction, although gene transcription differs from cell to cell and over time. Selective gene expression allows a single genome to create many proteins, resulting in variances in cell shape and function. The description of the transcriptome is thus a critical step in the investigation of gene function. RNA-Seq measures transcript and isoform levels significantly more precisely and comprehensively than previous technologies such as microarray and low-throughput sequencing technology (Wang et al., 2009). RNA-Seq has become one of the most widely used technologies in biology, and it has changed our perception of the complexity of transcriptome, providing new insights into transcriptional

TABLE 9.4 RNA sequencing process, tools, and platforms.

S. no.	Method	Process	Tool	Platform
1	RNA-sequencing	De novo transcriptome assembly	Trinity	Illumina/Roche/ABI
			Trans-AbySS	Illumina/Roche/ABI
			Oases	Illumina/Roche/ABI
		Alignment/mapping	Bowtie/Bowtie2	Illumina/Roche/ABI
			TopHat	Illumina/Roche/ABI
		Counting reads per transcript data	HTSeq	Illumina/Roche/ABI
			Cufflinks	Illumina/Roche/ABI
		Normalization, bias correction, and differential expression testing	DESeq	Illumina/Roche/ABI
			baySeq	Illumina/Roche/ABI
			edgeR	Illumina/Roche/ABI
			Cufflinks	Illumina/Roche/ABI
2	Small RNA sequencing	Adapter trimming	cutadapt	Illumina/Roche/ABI
			Flicker	Illumina
			FASTX Clipper	Illumina
			scythe	Illumina
		Quality control	NGS QC Toolkit	Illumina/Roche
			FASTQ Quality Filter	Illumina
		Quality viewer	FastQC	Illumina/Roche
			qrqc	Illumina/Roche/ABI
		Alignment/mapping	Bowtie/Bowtie2	Illumina/Roche/ABI
		miRNA prediction	DSAP	Illumina/Roche/ABI
			miRanalyzer	Illumina/Roche/ABI
			miRDeep/miRDeep2	Illumina/Roche/ABI
			MIReNA	Illumina/Roche/ABI
			mirExplorer	Illumina/Roche/ABI
			miRTRAP	Illumina/Roche/ABI
			miRDeep-P	Illumina/Roche/ABI

Data from From Inchingolo, F., Santacroce, L., Ballini, A., Topi, S., Dipalma, G., Haxhirexha, K., Bottalico, L. & Charitos, I. A. (2020). Oral Cancer: A Historical Review. *International Journal of Environmental Research and Public Health, 17*(9), 3168. https://doi.org/10.3390/ijerph17093168.

and posttranscriptional gene regulation (Marguerat & Bähler, 2010). Furthermore, RNA-Seq has increased our understanding of RNA biology by allowing for the accurate description of transcription as well as the intermolecular interactions that influence RNA function. some RNA sequencing tools and platforms are shown in Table 9.4.

4. Limitations in omics data integration and interpretation

Typically, associations between multiple omics levels of biological data are very complex. These data are obtained from the various methods, so integration of the data is also a huge task. These multiomics platforms are mainly considered the specific types of the data format file, so conversion and integration of the data result is also a problematic task. Some multiomics data analysis approaches only define the genetic aspect of the disease; they are not connected directly to the phenotypic results of the disease so the stage and level of the disease occurrence are not defined. The multiomics data are the large data set so handling of these large amounts of data needs the large computational platform and the dynamic level analysis of these data is also a tedious task. Sometimes these omics data are not well organized, so users could not choose data processing and model selection flexibly for getting the fruitful information.

5. Future research directions

Based on the comprehensive literature-based search, the following are possible future directions in this area of omics data research. The advanced technologies such as deep learning, artificial intelligence, and machine learning methods are helpful for the effective analysis of feature selection and illness prediction of the diseases. The computational-based methodologies developed for omics data analysis can be helpful to multiomics data analysis. The additional complicated biological data obtained from the various levels can be combined for multiomics data analysis. For analysis, machine learning, deep learning, and ensemble models can all be enhanced for biological data analysis in multiomics level. The data set with additional samples will be used to improve the model's capacity. The multiomics-based data analysis and interpretation show the path of identification and development of new strategy for the treatment of various genetics and rare diseases.

6. Conclusion

A multiomics data analysis technique proposed is useful for integrating different omics data to discover significant disease-related genes. These integrated approaches can also be used to improve the creation of novel strategies for the identification and diagnosis of rare features and diseases with small sample sizes (Wang et al., 2020). The integration of omics data is also helpful to provide the large number of data publically, which are helpful to high-quality research. The integrative-based omics data are also helpful on identification of disease prognosis diagnosis classification and identification of biomarkers. Therefore, it can be promising to discover more effective and accurate diagnosis and therapeutic approaches to manage and control disease growth in future generation. The comparative study of multiomics tools is depicted in Table 9.5.

TABLE 9.5 Comparative analysis of existing multiomics tools.

S. no.	Year	Tool	Technology/ platform	Availability link	Limitations	References
1	2008	DAnTE	C#, R language	http://omics.pnl.gov/software/	–	Polpitiya et al. (2008)
2	2013	Rainbow	Cloud computing	http://s3.amazonaws.com/jnj_rainbow/index.html	Failure to move large data sets is a difficult process	Zhao et al. (2013)
3	2014	TCGA-Assembler	R programming language	https://bio.tools/TCGAAssembler	–	Zhu et al. (2014)
4	2015	TCGAbiolinks	R/Bioconductor	https://bioconductor.org/packages/release/bioc/tml/TCGAbiolinks.html	–	Colaprico et al. (2016)
5	2015	Omics Pipe	Python, Cloud	http://sulab.scripps.edu/omicspipe	–	Fisch et al. (2015)
6	2016	Web-TCGA	R language	https://github.com/Mariodeng/web-TCGA	–	Deng et al. (2016)
7	2016	DIABLO	mixOmics R package	http://mixomics.org/mixdiablo/	–	Singh et al. (2019)
8	2016	MONGKIE	Java (NetBeans architecture)	http://yjjang.github.io/mongkie	–	Jang et al. (2016)
9	2017	OASISPRO	Cloud	http://tinyurl.com/oasispro	–	Yu et al. (2018)

Continued

TABLE 9.5 Comparative analysis of existing multiomics tools.—cont'd

S. no.	Year	Tool	Technology/ platform	Availability link	Limitations	References
10	2017	TCGA2BED	Java programming language	http://bioinf.iasi.cnr.it/tcga2bed/	–	Cumbo et al. (2017)
11	2017	MultiDataSet	R programming language	https://bioconductor.org/packages/release/bioc/html/MultiDataSet.html	–	Hernandez-Ferrer et al. (2017)
12	2017	WebMeV	R version repository	http://mev.tm4.org	Limited number of nodes assigned for application	Wang et al. (2017)
13	2017	Omics Database Generator	Java programming language	https://github.com/jguhlin/odg	Data available only in standard file formats such as TSV, GFF3, OBO, FASTA, and so on	Guhlin et al. (2017)
14	2017	CancerDiscover	WEKA, Affy R package	https://github.com/HelikarLab/CancerDiscover	–	Mohammed et al. (2018)
15	2018	TCGAassembler 2	R programming language	http://www.compgenome.org/TCGA-Assembler	–	Wei et al. (2018)
16	2018	MOBCdb	Perl, R, MySQL database	http://bigd.big.ac.cn/MOBCdb/	Exome sequencing is used to obtain SNV data	Xie et al. (2018)
17	2019	O miner	R statistical environment	http://www.o-miner.org	Preprocessing of NGS inputs data sequence alignment required and QC to be conducted	Sangaralingam et al. (2019)
18	2019	OmicsARules	R programming using ARM framework	https//github.com/Bioinformatics-STU/OmicsARules	–	Chen et al. (2019)
19	2019	iOmicsPASS	R programming	https://github.com/cssblab/iOmicsPASS	No functionality to be predict phenotype group	Koh (2019)
20	2020	ShinyOmics	R programming language	https://github.com/dsurujon/ShinyOmics	–	Surujon and van Opijnen (2020)
21	2020	KnowEnG	Cloud platform	https://knoweng.org/analyze/	The platform is not one size that can fit all solutions	Sinha et al. (2015)
22	2021	IOAT		https://github.com/WlSunshine/IOAT-software	Users could not choose data processing and model selection flexibly	Wu et al. (2021)
23	2021	MAINE	–	http://maine.ibemag.pl/	–	Gruca et al. (2022)
24	2022	PaintOmics 4	Java and Python language	https://github.com/ConesaLab/paintomics4	–	Liu et al. (2022)

References

Alvarez-Dominguez, J. R., Bai, Z., Xu, D., Yuan, B., Lo, K., Yoon, M., Lim, Y., Knoll, M., Slavov, N., Chen, S., Chen, P., Lodish, H. F., & Sun, L. (2015). De novo reconstruction of adipose tissue transcriptomes reveals long non-coding RNA regulators of Brown adipocyte development. *Cell Metabolism, 21*(5), 764–776. https://doi.org/10.1016/j.cmet.2015.04.003

Barrès, R., & Zierath, J. R. (2016). The role of diet and exercise in the transgenerational epigenetic landscape of T2DM. *Nature Reviews Endocrinology, 12*(8), 441–451. https://doi.org/10.1038/nrendo.2016.87

Barrett, S. P., & Salzman, J. (2016). Circular RNAs: Analysis, expression and potential functions. *Development, 143*(11), 1838–1847. https://doi.org/10.1242/dev.128074. http://dev.biologists.org/content/develop/143/11/1838.full.pdf

Baylin, S. B., Esteller, M., Rountree, M. R., Bachman, K. E., Schuebel, K., & Herman, J. G. (2001). Abberant patterns of DNA methylation, chromatin formation and gene expression in cancer. *Human Molecular Genetics, 10*(7), 687–692.

Buzkova, J., Nikkanen, J., Ahola, S., Hakonen, A. H., Sevastianova, K., Hovinen, T., Yki-Järvinen, H., Pietiläinen, K. H., Lönnqvist, T., Velagapudi, V., Carroll, C. J., & Suomalainen, A. (2018). Metabolomes of mitochondrial diseases and inclusion body myositis patients: Treatment targets and biomarkers. *EMBO Molecular Medicine, 10*(12). https://doi.org/10.15252/emmm.201809091

Chen, D., Zhang, F., Zhao, Q., & Xu, J. (2019). OmicsARules: a R package for integration of multi-omics datasets via association rules mining. *BMC Bioinformatics, 20*(1). https://doi.org/10.1186/s12859-019-3171-0

Chen, Y., Li, C., Tan, C., & Liu, X. (2016). Circular RNAs: A new frontier in the study of human diseases. *Journal of Medical Genetics, 53*(6), 359–365. https://doi.org/10.1136/jmedgenet-2016-103758

Choudhary, C., & Mann, M. (2010). Decoding signalling networks by mass spectrometry-based proteomics. *Nature Reviews Molecular Cell Biology, 11*(6), 427–439. https://doi.org/10.1038/nrm2900

Colaprico, A., Silva, T. C., Olsen, C., Garofano, L., Cava, C., Garolini, D., Sabedot, T. S., Malta, T. M., Pagnotta, S. M., Castiglioni, I., Ceccarelli, M., Bontempi, G., & Noushmehr, H. (2016). TCGAbiolinks: An R/Bioconductor package for integrative analysis of TCGA data. *Nucleic Acids Research, 44*(8), e71. https://doi.org/10.1093/nar/gkv1507. http://nar.oxfordjournals.org/

Consortium, R., Kundaje, A., Meuleman, W., Ernst, J., Bilenky, M., Yen, A., Heravi-Moussavi, A., Kheradpour, P., Zhang, Z., Wang, J., Ziller, M. J., Amin, V., Whitaker, J. W., Schultz, M. D., Ward, L. D., Sarkar, A., Quon, G., Sandstrom, R. S., Eaton, M. L., … Kellis, M. (2015). Integrative analysis of 111 reference human epigenomes. *Nature, 518*(7539), 317–329. https://doi.org/10.1038/nature14248. http://www.nature.com/nature/index.html

Cui, H., Li, F., Chen, D., Wang, G., Truong, C. K., Enns, G. M., Graham, B., Milone, M., Landsverk, M. L., Wang, J., Zhang, W., & Wong, L. C. (2013). Comprehensive next-generation sequence analyses of the entire mitochondrial genome reveal new insights into the molecular diagnosis of mitochondrial DNA disorders. *Genetics in Medicine, 15*(5), 388–394. https://doi.org/10.1038/gim.2012.144

Cumbo, F., Fiscon, G., Ceri, S., Masseroli, M., & Weitschek, E. (2017). TCGA2BED: Extracting, extending, integrating, and querying the cancer genome atlas. *BMC Bioinformatics, 18*(1). https://doi.org/10.1186/s12859-016-1419-5

Deng, M., Brägelmann, J., Schultze, J. L., & Perner, S. (2016). Web-TCGA: An online platform for integrated analysis of molecular cancer data sets. *BMC Bioinformatics, 17*(1). https://doi.org/10.1186/s12859-016-0917-9

Feero, W. G., Guttmacher, A. E., & Manolio, T. A. (2010). Genomewide association studies and assessment of the risk of disease. *New England Journal of Medicine, 363*(2), 166–176. https://doi.org/10.1056/nejmra0905980

Fisch, K. M., Meißner, T., Gioia, L., Ducom, J., Carland, T. M., Loguercio, S., & Su, A. I. (2015). Omics pipe: A community-based framework for reproducible multi-omics data analysis. *Bioinformatics, 31*(11), 1724–1728. https://doi.org/10.1093/bioinformatics/btv061

Flanagan, J. M. (2015). Epigenome-wide association studies (EWAS): Past, present, and future. *Methods in Molecular Biology, 1238*, 51–63. https://doi.org/10.1007/978-1-4939-1804-1_3. http://www.springer.com/series/7651

Gruca, A., Henzel, J., Kostorz, I., Steclik, T., Wróbel, Ł., Sikora, M., & Wren, J. (2022). Maine: A web tool for multi-omics feature selection and rule-based data exploration. *Bioinformatics, 38*(6), 1773–1775. https://doi.org/10.1093/bioinformatics/btab862

Guhlin, J., Silverstein, K. A. T., Zhou, P., Tiffin, P., & Young, N. D. (2017). Odg: Omics database generator - a tool for generating, querying, and analyzing multi-omics comparative databases to facilitate biological understanding. *BMC Bioinformatics, 18*(1). https://doi.org/10.1186/s12859-017-1777-7

Gupta, R. A., Shah, N., Wang, K. C., Kim, J., Horlings, H. M., Wong, D. J., Tsai, M., Hung, T., Argani, P., Rinn, J. L., Wang, Y., Brzoska, P., Kong, B., Li, R., West, R. B., van de Vijver, M. J., Sukumar, S., & Chang, H. Y. (2010). Long non-coding RNA HOTAIR reprograms chromatin state to promote cancer metastasis. *Nature, 464*(7291), 1071–1076. https://doi.org/10.1038/nature08975

Hasin, Y., Seldin, M., & Lusis, A. (2017). Multi-omics approaches to disease. *Genome Biology, 18*(1). https://doi.org/10.1186/s13059-017-1215-1

Hernandez-Ferrer, C., Ruiz-Arenas, C., Beltran-Gomila, A., & González, J. R. (2017). MultiDataSet: an R package for encapsulating multiple data sets with application to omic data integration. *BMC Bioinformatics, 18*(1). https://doi.org/10.1186/s12859-016-1455-1

Horvath, S. (2013). DNA methylation age of human tissues and cell types. *Genome Biology, 14*(10), R115. https://doi.org/10.1186/gb-2013-14-10-r115

Huang, S., Chaudhary, K., & Garmire, L. X. (2017). More is better: Recent progress in multi-omics data integration methods. *Frontiers in Genetics, 8*. https://doi.org/10.3389/fgene.2017.00084

Hussen, B. M., Abdullah, S. T., Salihi, A., Sabir, D. K., Sidiq, K. R., Rasul, M. F., Hidayat, H. J., Ghafouri-Fard, S., Taheri, M., & Jamali, E. (2022). The emerging roles of NGS in clinical oncology and personalized medicine. *Pathology, Research and Practice, 230*, 153760. https://doi.org/10.1016/j.prp.2022.153760

Ishii, N., Ozaki, K., Sato, H., Mizuno, H., Saito, S., Takahashi, A., Miyamoto, Y., Ikegawa, S., Kamatani, N., Hori, M., Saito, S., Nakamura, Y., & Tanaka, T. (2006). Identification of a novel non-coding RNA, MIAT, that confers risk of myocardial infarction. *Journal of Human Genetics, 51*(12), 1087–1099. https://doi.org/10.1007/s10038-006-0070-9

Jang, Y., Yu, N., Seo, J., Kim, S., & Lee, S. (2016). Mongkie: An integrated tool for network analysis and visualization for multi-omics data. *Biology Direct, 11*(1). https://doi.org/10.1186/s13062-016-0112-y

Keating, S. T., & El-Osta, A. (2015). Epigenetics and metabolism. *Circulation Research, 116*(4), 715–736. https://doi.org/10.1161/circresaha.116.303936

Kettunen, J., Tukiainen, T., Sarin, A., Ortega-Alonso, A., Tikkanen, E., Lyytikäinen, L., Kangas, A. J., Soininen, P., Würtz, P., Silander, K., Dick, D. M., Rose, R. J., Savolainen, M. J., Viikari, J., Kähönen, M., Lehtimäki, T., Pietiläinen, K. H., Inouye, M., McCarthy, M. I., … Ripatti, S. (2012). Genome-wide association study identifies multiple loci influencing human serum metabolite levels. *Nature Genetics, 44*(3), 269–276. https://doi.org/10.1038/ng.1073

Khan, S., Ince-Dunn, G., Suomalainen, A., & Elo, L. L. (2020). Integrative omics approaches provide biological and clinical insights: Examples from mitochondrial diseases. *Journal of Clinical Investigation, 130*(1), 20–28. https://doi.org/10.1172/jci129202

Kim, M., Long, T. I., Arakawa, K., Wang, R., Yu, M. C., Laird, P. W., & Bader, J. S. (2010). DNA methylation as a biomarker for cardiovascular disease risk. *PLoS One, 5*(3), e9692. https://doi.org/10.1371/journal.pone.0009692

Knoll, M., Lodish, H. F., & Sun, L. (2015). Long non-coding RNAs as regulators of the endocrine system. *Nature Reviews Endocrinology, 11*(3), 151–160. https://doi.org/10.1038/nrendo.2014.229

Koboldt, D. C., Steinberg, K. M., Larson, D. E., Wilson, R. K., & Mardis, E. R. (2013). The next-generation sequencing revolution and its impact on genomics. *Cell, 155*(1), 27–38. https://doi.org/10.1016/j.cell.2013.09.006

Koh, H. L. (2019). "iOmicsPASS: Network-based integration of multiomics data for predictive subnetwork discovery. *Systems Biology and Applications, 5*(1).

LaFramboise, T. (2009). Single nucleotide polymorphism arrays: A decade of biological, computational and technological advances. *Nucleic Acids Research, 37*(13), 4181–4193. https://doi.org/10.1093/nar/gkp552

Liu, T., Salguero, P., Petek, M., Martinez-Mira, C., Balzano-Nogueira, L., Ramšak, Ž., McIntyre, L., Gruden, K., Tarazona, S., & Conesa, A., 2022. PaintOmics 4: New tools for the integrative analysis of multi-omics datasets supported by multiple pathway databases. *Nucleic Acids Research, 50*(W1), W551–W559. https://doi.org/10.1093/nar/gkac352.

Mann, M., & Jensen, O. N. (2003). Proteomic analysis of post-translational modifications. *Nature Biotechnology, 21*(3), 255–261. https://doi.org/10.1038/nbt0303-255

Marguerat, S., & Bähler, J. (2010). RNA-Seq: From technology to biology. *Cellular and Molecular Life Sciences, 67*(4), 569–579. https://doi.org/10.1007/s00018-009-0180-6

Mohammed, A., Biegert, G., Adamec, J., & Helikar, T. (2018). CancerDiscover: An integrative pipeline for cancer biomarker and cancer class prediction from high-throughput sequencing data. *Oncotarget, 9*(2), 2565–2573. https://doi.org/10.18632/oncotarget.23511

Morán, I., Akerman, İ., van de Bunt, M., Xie, R., Benazra, M., Nammo, T., Arnes, L., Nakić, N., García-Hurtado, J., Rodríguez-Seguí, S., Pasquali, L., Sauty-Colace, C., Beucher, A., Scharfmann, R., van Arensbergen, J., Johnson, P., Berry, A., Lee, C., Harkins, T., ... Ferrer, J. (2012). Human β cell transcriptome analysis uncovers lncRNAs that are tissue-specific, dynamically regulated, and abnormally expressed in type 2 diabetes. *Cell Metabolism, 16*(4), 435–448. https://doi.org/10.1016/j.cmet.2012.08.010

Nakato, R., & Sakata, T. (2021). Methods for ChIP-seq analysis: A practical workflow and advanced applications. *Methods, 187*, 44–53. https://doi.org/10.1016/j.ymeth.2020.03.005

Ng, S. B., Turner, E. H., Robertson, P. D., Flygare, S. D., Bigham, A. W., Lee, C., Shaffer, T., Wong, M., Bhattacharjee, A., Eichler, E. E., Bamshad, M., Nickerson, D. A., & Shendure, J. (2009). Targeted capture and massively parallel sequencing of 12 human exomes. *Nature, 461*(7261), 272–276. https://doi.org/10.1038/nature08250

Palmfeldt, J., & Bross, P. (2017). Proteomics of human mitochondria. *Mitochondrion, 33*, 2–14. https://doi.org/10.1016/j.mito.2016.07.006

Pavlovich, P. V., & Cauchy, P. (2022). Sequences to differences in gene expression: Analysis of RNA-seq data. *Methods in Molecular Biology, 2508*, 279–318. https://doi.org/10.1007/978-1-0716-2376-3_20. http://www.springer.com/series/7651

Piunti, A., & Shilatifard, A. (2016). Epigenetic balance of gene expression by Polycomb and COMPASS families. *Science, 352*(6290). https://doi.org/10.1126/science.aad9780

Polpitiya, A. D., Qian, W., Jaitly, N., Petyuk, V. A., Adkins, J. N., Camp, D. G., Anderson, G. A., & Smith, R. D. (2008). DAnTE: A statistical tool for quantitative analysis of -omics data. *Bioinformatics, 24*(13), 1556–1558. https://doi.org/10.1093/bioinformatics/btn217

Raghuraman, S., Donkin, I., Versteyhe, S., Barrès, R., & Simar, D. (2016). The emerging role of epigenetics in inflammation and immunometabolism. *Trends in Endocrinology and Metabolism, 27*(11), 782–795. https://doi.org/10.1016/j.tem.2016.06.008

Ragoussis, J. (2009). Genotyping technologies for genetic research. *Annual Review of Genomics and Human Genetics, 10*(1), 117–133. https://doi.org/10.1146/annurev-genom-082908-150116

Roadmap Epigenomics Consortium, Kundaje, A., Meuleman, W., et al. (2015). Integrative analysis of 111 reference human epigenomes. *Nature, 518*, 317–330. https://doi.org/10.1038/nature14248.

Sangaralingam, A., Dayem Ullah, A. Z., Marzec, J., Gadaleta, E., Nagano, A., Ross-Adams, H., Wang, J., Lemoine, N. R., & Chelala, C. (2019). 'Multi-omic' data analysis using O-miner. *Briefings in Bioinformatics, 20*(1), 130–143. https://doi.org/10.1093/bib/bbx080

Schmitz, S. U., Grote, P., & Herrmann, B. G. (2016). Mechanisms of long noncoding RNA function in development and disease. *Cellular and Molecular Life Sciences, 73*(13), 2491–2509. https://doi.org/10.1007/s00018-016-2174-5

Singh, A., Shannon, C. P., Gautier, B., Rohart, F., Vacher, M., Tebbutt, S. J., Lê Cao, K. A., & Birol, I. (2019). Diablo: An integrative approach for identifying key molecular drivers from multi-omics assays. *Bioinformatics, 35*(17), 3055–3062. https://doi.org/10.1093/bioinformatics/bty1054

Sinha, S., Song, J., Weinshilboum, R., Jongeneel, V., & Han, J. (2015). KnowEnG: A knowledge engine for genomics. *Journal of the American Medical Informatics Association, 22*(6), 1115–1119. https://doi.org/10.1093/jamia/ocv090

Steuer, R. (2006). Review: On the analysis and interpretation of correlations in metabolomic data. *Briefings in Bioinformatics, 7*(2), 151–158. https://doi.org/10.1093/bib/bbl009

Surujon, D., & van Opijnen, T. (2020). ShinyOmics: Collaborative exploration of omics-data. *BMC Bioinformatics, 21*(1). https://doi.org/10.1186/s12859-020-3360-x

Trapnell, C., Williams, B. A., Pertea, G., Mortazavi, A., Kwan, G., van Baren, M. J., Salzberg, S. L., Wold, B. J., & Pachter, L. (2010). Transcript assembly and quantification by RNA-Seq reveals unannotated transcripts and isoform switching during cell differentiation. *Nature Biotechnology, 28*(5), 511–515. https://doi.org/10.1038/nbt.1621

Uffelmann, E., Huang, Q., Munung, N., de Vries, J., Okada, Y., Martin, A. R., Martin, H. C., Lappalainen, T., & Posthuma, D. (2021). Genome-wide association studies. *Nature Reviews Methods Primers, 1*(1). https://doi.org/10.1038/s43586-021-00056-9

Wang, Y.E., Kutnetsov, L., Partensky, A., Farid, J., & Quackenbush, J. (2017). WebMeV: A cloud platform for analyzing and visualizing cancer genomic data. *Cancer Research, 77*(21), e11-e14. https://doi.org/10.1158/0008-5472.CAN-17-0802. PMID: 29092929; PMCID: PMC5679251.

Wang, B., Lunetta, K. L., Dupuis, J., Lubitz, S. A., Trinquart, L., Yao, L., Ellinor, P. T., Benjamin, E. J., & Lin, H. (2020). Integrative omics approach to identifying genes associated with atrial fibrillation. *Circulation Research*, 350–360. https://doi.org/10.1161/CIRCRESAHA.119.315179. http://circres.ahajournals.org

Wang, Z., Gerstein, M., & Snyder, M. (2009). RNA-seq: A revolutionary tool for transcriptomics. *Nature Reviews Genetics, 10*(1), 57–63. https://doi.org/10.1038/nrg2484

Wei, L., Jin, Z., Yang, S., Xu, Y., Zhu, Y., Ji, Y., & Kelso, J. (2018). TCGA-Assembler 2: Software pipeline for retrieval and processing of TCGA/CPTAC data. *Bioinformatics, 34*(9), 1615–1617. https://doi.org/10.1093/bioinformatics/btx812

Wu, L., Liu, F., & Cai, H. (2021). Ioat: An interactive tool for statistical analysis of omics data and clinical data. *BMC Bioinformatics, 22*(1). https://doi.org/10.1186/s12859-021-04253-x

Wu, R., Haas, W., Dephoure, N., Huttlin, E. L., Zhai, B., Sowa, M. E., & Gygi, S. P. (2011). A large-scale method to measure absolute protein phosphorylation stoichiometries. *Nature Methods, 8*(8), 677–683. https://doi.org/10.1038/nmeth.1636

Xie, B., Yuan, Z., Yang, Y., Sun, Z., Zhou, S., & Fang, X. (2018). MOBCdb: A comprehensive database integrating multi-omics data on breast cancer for precision medicine. *Breast Cancer Research and Treatment, 169*(3), 625–632. https://doi.org/10.1007/s10549-018-4708-z

Yang, J., Lee, S. H., Goddard, M. E., & Visscher, P. M. (2011). Gcta: A tool for genome-wide complex trait analysis. *The American Journal of Human Genetics, 88*(1), 76–82. https://doi.org/10.1016/j.ajhg.2010.11.011

Yao, B., Christian, K. M., He, C., Jin, P., Ming, G., & Song, H. (2016). Epigenetic mechanisms in neurogenesis. *Nature Reviews Neuroscience, 17*(9), 537–549. https://doi.org/10.1038/nrn.2016.70

Yu, K., Fitzpatrick, M. R., Pappas, L., Chan, W., Kung, J., Snyder, M., & Kelso, J. (2018). Omics Analysis system for Precision oncology (OASISPRO): A web-based omics analysis tool for clinical phenotype prediction. *Bioinformatics, 34*(2), 319–320. https://doi.org/10.1093/bioinformatics/btx572

Zhao, S., Prenger, K., Smith, L., Messina, T., Fan, H., Jaeger, E., & Stephens, S. (2013). Rainbow: A tool for large-scale whole-genome sequencing data analysis using cloud computing. *BMC Genomics, 14*(1), 425. https://doi.org/10.1186/1471-2164-14-425

Zhao, Z., & Shilatifard, A. (2019). Epigenetic modifications of histones in cancer. *Genome Biology, 20*(1). https://doi.org/10.1186/s13059-019-1870-5

Zhu, J., Adli, M., Zou, J. Y., Verstappen, G., Coyne, M., Zhang, X., Durham, T., Miri, M., Deshpande, V., De Jager, P. L., Bennett, D. A., Houmard, J. A., Muoio, D. M., Onder, T. T., Camahort, R., Cowan, C. A., Meissner, A., Epstein, C. B., Shoresh, N., & Bernstein, B. E. (2013). Genome-wide chromatin state transitions associated with developmental and environmental cues. *Cell, 152*(3), 642–654. https://doi.org/10.1016/j.cell.2012.12.033

Zhu, Y., Qiu, P., & Ji, Y. (2014). TCGA-assembler: Open-source software for retrieving and processing TCGA data. *Nature Methods, 11*(6), 599–600. https://doi.org/10.1038/nmeth.2956

Chapter 10

Integrative omics approaches for identification of biomarkers

Upasna Srivastava[1], Swarna Kanchan[2,3], Minu Kesheri[2,3], Manish Kumar Gupta[4] and Satendra Singh[5]

[1]University of California, San Diego, CA, United States; [2]Marshall University, Huntington, WV, United States; [3]Boise State University, Boise, ID, United States; [4]Department of Biotechnology, Faculty of Science, Veer Bahadur Singh Purvanchal University, Jaunpur, Uttar Pradesh, India; [5]Department of Computational Biology and Bioinformatics, Sam Higginbottom University of Agriculture, Technology and Sciences, Allahabad, Uttar Pradesh, India

1. Introduction

In this chapter, we'll go through a variety of procedures and equipment for generating differential gene expression (DGE) and carrying out pathway analyses. The identification of genes that exhibit differential expression in various situations or populations via DEG analysis is essential for gaining an understanding of the molecular processes behind complicated disorders. In the following, we will also go through several statistical tools and machine learning techniques that were used to study the relationship between DEGs and their annotated pathways, improving our comprehension of biological mechanisms and possible treatment targets.

1.1 Differential gene expression analysis

1.1.1 RNA-Seq

We will explore the RNA-Seq technology, which enables the quantification of gene expression levels and identification of DEGs using statistical methods such as edgeR (Robinson et al., 2009) and DESeq2 (Love et al., 2014).

1.1.2 Microarray analysis

We will discuss the microarray-based approaches, including the popular limma package (Ritchie et al., 2015), which offers robust statistical methods for DEG analysis.

1.1.3 Single-cell RNA-Seq

We will highlight the unique considerations and specialized tools for DEG analysis in single-cell RNA-Seq data, such as Seurat (Stuart et al., 2019) and Monocle (Qiu et al., 2017).

1.2 Pathway analysis

1.2.1 Gene set enrichment analysis

We will describe the gene set enrichment analysis (GSEA) method, which assesses whether a predefined set of genes shows statistically significant enrichment in DEGs, revealing potential biological pathways associated with the condition of interest.

1.2.2 Overrepresentation analysis

We will discuss overrepresentation analysis (ORA) methods, including hypergeometric tests and Fisher's exact tests, which identify the overrepresentation of DEGs within predefined gene sets or biological pathways.

Integrative Omics. https://doi.org/10.1016/B978-0-443-16092-9.00010-2

1.3 Statistical tools for correlation analysis

Correlation analysis plays a vital role in exploring the relationships between differentially expressed genes (DEGs) and their associated pathways. Here are some commonly used statistical tools for correlation analysis in DEG analysis and pathway enrichment.

1.3.1 Pearson correlation

The Pearson correlation coefficient is a measurement of the linear relationship between two variables. The relationship between the DEGs' pathway activity and their gene expression levels is widely studied using this method. A positive connection shows that DEG expression and pathway activity grow in tandem, whereas a negative correlation suggests the opposite. The Pearson's correlation coefficient has a range of values between −1 and 1, with 1 denoting a perfect positive correlation, −1 a perfect negative correlation, and 0 denoting no association. We will cover the widely used Pearson correlation coefficient to measure the linear relationship between DEGs and their pathway activity.

1.3.2 Spearman correlation

Using Spearman's rank correlation coefficient, one can evaluate the monotonic relationship between two variables. It is especially helpful when there is a nonlinear link between DEGs and route activity. The rankings of the data are used to construct Spearman's rank correlation coefficient rather than the actual values, which makes it more resistant to outliers. It has a similar interpretation to Pearson's correlation coefficient and has a range of −1 to 1 (Spearman, 1904). We will explore the Spearman correlation coefficient, which assesses the monotonic relationship between DEGs and pathway activity, suitable for nonlinear associations.

1.4 Machine learning methods

1.4.1 Support vector machines

We will discuss how support vector machine (SVM) can be employed to classify DEGs based on their pathway information, aiding in the identification of relevant pathways in disease classification.

1.4.2 Random forest

We will explore the application of random forest algorithms in feature selection and classification of DEGs, leveraging pathway information for improved accuracy.

By reviewing these protocols, tools, and methods, we aim to provide researchers with a comprehensive understanding of the available options for DEG analysis and pathway analysis. Furthermore, the inclusion of statistical tools and machine learning methods for analyzing DEG−pathway correlations will facilitate the extraction of meaningful insights from gene expression data, aiding in the identification of potential therapeutic targets and advancing our knowledge of complex diseases. In addition to traditional statistical methods and pathway analysis, machine learning methods can also be used to analyze the correlation between DEGs and their annotated pathways. For example, SVMs, random forests, and neural networks can be used to classify samples based on their gene expression profiles and identify the key genes and pathways that contribute to the classification.

1.5 Biomarkers, their importance in disease diagnosis, prognosis, and treatment

A biomarker is a measurable indicator of a biological state or condition. According to the National Institutes of Health (NIH), "a biomarker is a characteristic that is objectively measured and evaluated as an indicator of normal biological processes, pathogenic processes, or pharmacologic responses to a therapeutic intervention." Biomarkers can be used in a variety of ways, including disease diagnosis, prognosis, and treatment. For example, biomarkers can be used to identify individuals who are at risk for developing a disease, to monitor disease progression, and to evaluate the effectiveness of treatment. Biomarkers can be measured using a variety of techniques, including imaging, genomics, proteomics, metab olomics, and other omics technologies. The identification and validation of biomarkers is an important area of research, with the potential to improve disease diagnosis and treatment, and to advance personalized medicine approaches (Diamandis et al., 2010). It plays an important role in disease diagnosis, prognosis, and treatment. In diagnosis, biomarkers can be used to identify individuals who are at risk for developing a disease, as well as to detect the presence of a disease before symptoms appear. For example, prostate-specific antigen (PSA) is a biomarker that is commonly used to screen for prostate cancer. In prognosis, biomarkers can be used to predict the likely course of a disease and to estimate the risk of disease progression. For example, the presence of certain biomarkers in breast cancer patients can be used to predict the likelihood of disease recurrence and to guide treatment decisions. Also in treatment, they can be used to personalize treatment approaches and to monitor the effectiveness of treatment, for example, to identify patients who are likely to respond to a particular treatment and to adjust treatment dosage as needed.

Overall, biomarkers are essential tools in disease diagnosis, prognosis, and treatment. They can improve patient outcomes by enabling early detection and personalized treatment approaches. Research in biomarker discovery and validation is ongoing, with the goal of improving disease management and advancing personalized medicine.

1.6 Overview of omics technologies and their potential for identifying biomarkers

Computational-based multiomics methods are powerful approaches for analyzing large data sets generated by omics technologies, such as genomics, transcriptomics, proteomics, and metabolomics, to identify potential biomarkers. These methods involve the use of statistical and machine learning algorithms to integrate and analyze data from multiple omics technologies, enabling the identification of key biological pathways and molecular interactions that may be associated with disease. One common approach is network analysis, which involves the construction and analysis of complex molecular interaction networks using data from multiple omics technologies. These networks can help identify key pathways and molecular interactions that are disrupted in disease and can help identify potential biomarkers. For example, a study using a network-based approach to analyze multiomics data from Alzheimer's disease patients identified several key biomarkers associated with the disease, including altered expression levels of several genes and proteins involved in inflammation and immune response pathways. Besides that, another common approach is machine learning, which involves the development and application of algorithms that can learn from and make predictions based on large data sets. Machine learning algorithms can be trained on multiomics data from both healthy and diseased individuals to identify patterns and biomarkers that are associated with disease. For example, a study using a machine learning approach to analyze multiomics data from liver cancer patients identified several potential biomarkers that were associated with disease progression and patient survival (Cheng, 2019). Also, the field of biomarker discovery and validation is rapidly advancing, with recent developments including the use of artificial intelligence and machine learning to analyze large data sets and identify complex patterns that traditional statistical methods may not detect. For example, a study used machine learning to identify a panel of protein biomarkers that predict the risk of breast cancer recurrence. Single-cell omics technologies are also being employed to identify cell type−specific biomarkers, with a recent study by Eichner et al. (2021) using single-cell RNA sequencing to identify biomarkers for early detection of acute kidney injury.

Advancements in technology have also expanded the scope of biomarker discovery, with the development of new omics technologies such as spatial omics and glycomics. Spatial omics allows for the analysis of molecules within their tissue microenvironment, providing a more detailed understanding of disease biology. Glycomics focuses on the study of complex sugars, which have potential as biomarkers for diseases such as cancer. Liquid biopsies, which involve the analysis of circulating biomarkers in blood, are also an emerging area of biomarker research. Liquid biopsies offer noninvasive and real-time monitoring of disease progression and treatment response, with ongoing research on their use in monitoring treatment response in patients with non−small-cell lung cancer.

Overall, the continued advancement of technology and computational methods in biomarker discovery and validation hold the promise of improving disease diagnosis, prognosis, and treatment through the identification of increasingly complex and precise biomarkers.

1.7 Biomarker identification in cohort studies and its importance

Biomarker identification in cohort studies can be performed using various omics approaches, including genomics, transcriptomics, proteomics, and metabolomics. Here are some examples of each.

1.7.1 Genomics

Genomics is the study of the entire genome, including coding regions called genes (Kesheri et al., 2016) present in DNA sequences, variants, and epigenetic modifications. In cohort studies, genomics can be used to identify genetic variants associated with disease risk or progression. For example, a study using genomics identified a genetic variant associated with increased risk of developing Alzheimer's disease in a large cohort of individuals (Kunkle et al., 2019).

1.7.2 Transcriptomics

Transcriptomics focuses on the study of gene expression, including mRNA and noncoding RNA. In cohort studies, transcriptomics can be used to identify differentially expressed genes associated with disease. For example, a study using transcriptomics identified a set of genes associated with the development of type 2 diabetes in a large cohort of individuals (Mohlke et al., 2018). Another transcriptomics study identified that the nullomer 9S1R-NulloPT altered the tumor immune microenvironment as well as the tumor transcriptome, rendering tumors metabolically less active in cancer mouse model by downregulating the mitochondrial function and ribosome biogenesis (Ali et al., 2023).

1.7.3 Proteomics

Proteomics is the study of the entire set of proteins expressed in a cell or tissue. In cohort studies, proteomics can be used to identify differentially expressed proteins associated with disease. For example, a study using proteomics identified a set of proteins associated with the risk of developing cardiovascular disease in a large cohort of individuals (Ganz et al., 2016). Proteins being functional molecules may serve as a crucial bioinformatics tool for structure prediction filling the gap for 3D structures of numerous proteins (Kesheri et al., 2015), molecular docking (Gahoi et al., 2013), and molecular dynamics simulations (Priya et al., 2016, p. 286−313), thereby playing a major role in unraveling the mysteries of protein function used for the discovery of drugs such as Covid-19 (Sahu et al., 2023). Protein modeling also helps in exploring protein evolution such as endonuclease III (Kanchan et al., 2015) and endonuclease IV (Kanchan et al., 2019) in all three domains of life and representative DNA repair proteins in six model organisms (Kanchan et al., 2014), exploring drug resistance in malaria against sulfadoxine (Garg et al., 2009). Moreover, protein modeling has been instrumental in unveiling the 3D protein models for antioxidative enzymes such as Fe-SOD, and Mn-SOD in *Nostoc Commune* (Kesheri et al., 2014), various cyanobacterial organisms present in natural habitats (Kesheri et al., 2011) in *Oscillatoria* sp. (Minu Kesheri et al., 2021), helping in exploring the potential of antioxidants in aging (Kesheri et al., 2017, p. 166−195), *Microcystis aeruginosa* (Kesheri et al., 2022), phycobiliproteins (Richa, Kannaujiya, et al., 2011), and mycosporin-like phycobiliproteins (Richa, Rastogi, et al., 2011). Based on bioinformatics tools and algorithms in proteomics such as available methods for linear epitope prediction such as BepiPred, ABCpred, and SVMTriP, it predicted linear epitopes within the HPV type 16 E7 antigen sequence (Srivastava & Singh, 2013) and analyzed physicochemical properties of amino acids to identify potential epitope regions for TMEM 50A Structural Model (Srivastava et al., 2017).

1.7.4 Metabolomics

Metabolomics is the study of small molecules, including metabolites and lipids, involved in cellular metabolism. In cohort studies, metabolomics can be used to identify differentially expressed metabolites associated with disease. For example, a study using metabolomics identified a set of metabolites associated with the risk of developing colorectal cancer in a large cohort of individuals (Guertin et al., 2014).

Overall, these omics approaches can provide a comprehensive view of biological systems and enable the identification of potential biomarkers for disease diagnosis, prognosis, and treatment in cohort studies (Fig. 10.1).

2. Multimodal omics methods for a single cell

The term "omics" refers to the study of biological molecules where genomic research focuses on genes, Transcriptomics is the field of study of RNA transcripts, and proteomics is the study of proteins. While single-cell omics directly addresses cellular heterogeneity by isolating and analyzing molecular components from individual cells and offering in-depth knowledge on each cell's resolution, metabolomics focuses on metabolites. Here are computational pipeline steps as shown in Fig. 10.2.

2.1 Study design

Bioinformatics approaches play a critical role in the design and analysis of integrative omics studies for biomarker identification. Computational tools and algorithms are used for data preprocessing, integration, and analysis. Examples of such tools include R, Python, Bioconductor, and various machine learning algorithms. A well-designed study and analysis can lead to the identification of robust and clinically useful biomarkers. It shows several steps, including study planning, sample collection, data acquisition, data preprocessing, data integration, and biomarker discovery. The flowchart in Fig. 10.3 illustrates the basic steps involved in the study design process.

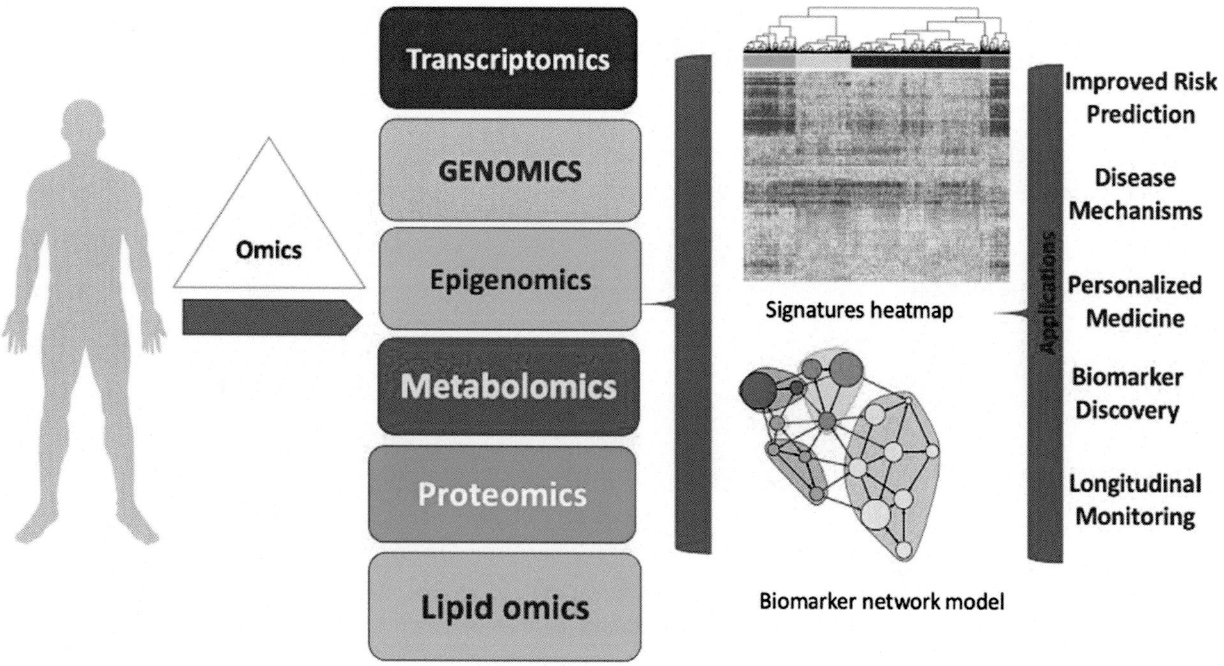

FIGURE 10.1 Omics approaches for biomarker identification. Biomarker identification in cohort studies based on omics approaches.

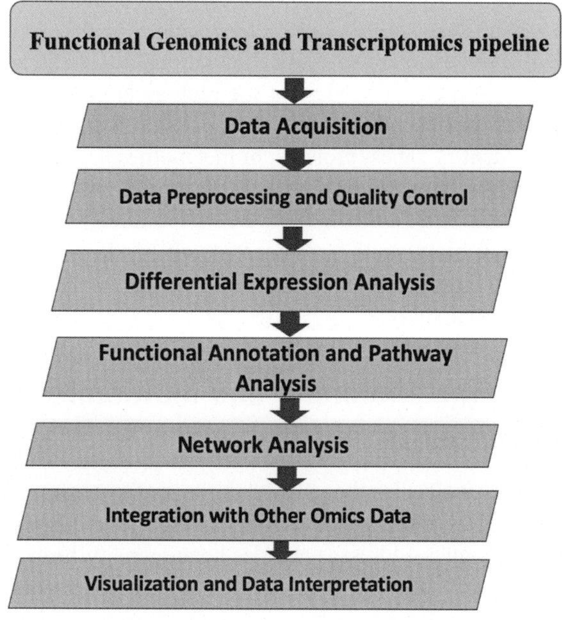

FIGURE 10.2 Single-cell omics methods.

1. **Study planning:** The first step in the study design process is to plan the study. This involves defining the research question, selecting the study population, and designing the experimental protocols. The study design should also take into account the requirements of different omics technologies to ensure that appropriate samples are collected and processed.
2. **Sample collection:** The next step is to collect samples from the study population. The sample collection should be performed according to the experimental protocols to ensure that the samples are of high quality and can be used for multiple omics analyses.

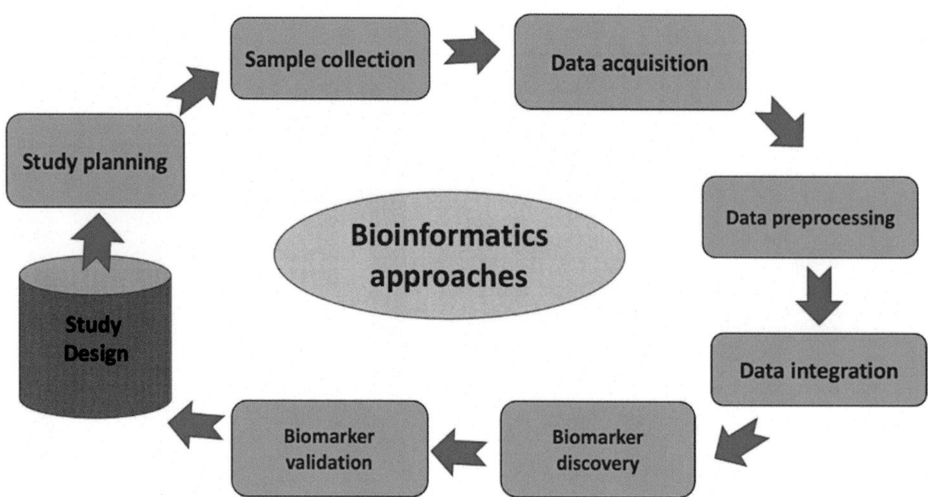

FIGURE 10.3 Bioinformatics approaches for biomarker identification.

3. **Data acquisition:** The samples are then subjected to different omics analyses, such as genomics, transcriptomics, proteomics, and metabolomics. Each of these analyses generates large amounts of data that need to be carefully processed and curated.
4. **Data preprocessing:** The next step is to preprocess the data to remove noise, correct errors, and normalize the data. This step is critical to ensure that the data is of high quality and can be used for downstream analyses.
5. **Data integration:** The preprocessed data from different omics analyses is then integrated to generate a more comprehensive view of the biological system. This can be challenging because the data may be in different formats and have different levels of complexity. However, there are several computational tools and approaches that can be used to integrate the data and generate a more comprehensive view of the biological system.
6. **Biomarker discovery:** Once the data has been integrated, the next step is to identify potential biomarkers. This may involve using statistical methods to identify genes, proteins, or metabolites that are differentially expressed or altered in the disease state. Machine learning algorithms can also be used to identify patterns in the data that may be indicative of a disease state.
7. **Biomarker validation:** After potential biomarkers have been identified, they need to be validated using independent data sets. This may involve testing the biomarkers in a larger cohort of patients or experimental models, and comparing the results to those obtained in the initial study.

3. Approaches to integrating metabolomics and multiomics data

3.1 Single-cell multimodal omics strategies

Omics is known as the study of biological molecules that ends with the suffix -omics: Genomics is the study of the genes. Transcriptomics refers to the analysis of RNA transcripts. Proteomics describes the investigation of proteins. Metabolomics focuses on metabolites where single-cell omics addresses cellular heterogeneity head on by isolating and analyzing molecular components from individual cells and provide deep insight on each cell resolution. Single-cell chromatin accessibility profiling (scATAC-seq) provides information about genetic variations and epigenetic states at the single-cell level. These approaches can help identify genetic mutations, structural variations, and regulatory elements that influence crop traits and phenotypes. Understanding the genomic landscape of individual cells can aid in targeted breeding and genetic engineering efforts (Srivastava & Singh, 2022, p. 271−294).

The acquisition of single-cell multimodal omics strategies from molecular data types such as genome, transcriptome, methylome, or proteome from single cells is represented as "integrative multiomics approaches," which shows the unification of different data modalities to discover the complex biological mechanisms on the cellular level such as the reconstruction of gene regulatory and signaling networks pathways; however, many challenges occur, which shows the batch effects problem that is mainly caused due to data sparsity and technical noise, which is often high due to low sequencing coverage and missing values. So therefore, many machine learning approaches are existed through which we can remove the batch effect problems. For preanalysis, we can prepare the single-cell RNA-Seq libraries using the Chromium Single Cell 3′ Reagent kit from 10× Genomics USA, where single-cell suspensions were loaded on the chromium single-cell controller Instrument 10× Genomics to generate single-cell gel beads in emulsions. Then, reverse transcription reactions used barcoded full-length cDNA followed by the disruption of emulsions using the recovery agent, and for final step to clean up cDNA, we can use the Dyna Beads and Myone Silane Beads. Besides this for handling of 10× multiplex data, Cell Ranger software pipeline version 3.0.0 is good to fit as to demultiplex cellular barcodes and map reads to the genome. STAR aligner was applied to generate normalized aggregate data across samples and obtain a matrix of gene counts versus cells.

The use of integrated multiomics can reveal novel molecular networks relevant to complex phenotypes. This approach also allows the identification of biomarkers and of previously unknown relationships between the data sets. In case of cancer, research has established that tumorigenesis and tumor progression typically result from the accumulation of somatic genomic alterations over the life span. It is also likely that at least in some cases, somatic mutations cooperate with inherited germline variants that predispose certain individuals to have a higher risk of disease. At the same time, however, science has been unable to consistently identify specific somatic mutations that are conserved across all individuals who present with the same type of cancer. In luminal A breast cancer, for example, fewer than half of tumors have two or more somatic mutations that have been identified as being associated with the disease.

3.2 Overview of computational tools and approaches for integrating omics data

The integration of multiple omics data types, such as genomics, transcriptomics, proteomics, and metabolomics, has become increasingly popular in recent years. Integrating omics data can provide a more comprehensive understanding of biological systems and help identify underlying biological mechanisms. However, integrating omics data is a complex task that requires specialized computational tools and approaches. Here is an overview of some of the commonly used tools and approaches for integrating omics data.

3.2.1 Multiomics factor analysis

Multiomics factor analysis (MOFA) is a probabilistic framework for integrative analysis of multiple omics data sets. It is based on the idea of factor analysis, which seeks to identify underlying patterns in the data. MOFA models each omics dataset as a combination of shared and dataset-specific factors and estimates the relationships between the factors across different omics data sets. MOFA can handle missing data and can integrate data sets of different sizes and types.

3.2.2 Canonical correlation analysis

Canonical correlation analysis (CCA) is a widely used method for integrating two omics data sets. It seeks to identify linear combinations of variables from each data set that are maximally correlated with each other. CCA can be extended to multiple omics data sets using a multiview CCA approach. One limitation of CCA is that it assumes linear relationships between the variables in the data sets.

3.2.3 Joint and individual variation explained

Joint and individual variation explained (JIVE) is a matrix factorization method that separates the data into joint and individual components. The joint component captures the shared information across all data sets, while the individual components capture the data set-specific information. JIVE can be used to identify shared and unique patterns of variation across multiple omics datasets.

3.2.4 Network-based integration

Network-based integration techniques establish biological networks and examine the connections between the network's variables in an effort to integrate omics data sets. These techniques can reveal groups of genes or proteins that are coexpressed or coregulated across many data sets. Network-based stratification (NBS), weighted gene coexpression network analysis (WGCNA), and coexpression network (CoNet) analysis are a few examples of network-based integration techniques shown in Table 10.1.

3.3 Machine learning–based integration

Machine learning–based methods (Kumari et al., 2016) use supervised or unsupervized algorithms to integrate omics data sets (Kumari et al., 2018). For example, classification algorithms can be used to identify features that distinguish between

TABLE 10.1 Network-based integration tools.

Tool	Language	Description
Cytoscape	R/Java	A powerful software platform for network visualization and analysis. It provides numerous plugins for data integration and functional analysis.
Network Integration Analysis (NIA)	R	An R package that utilizes network-guided regularization to integrate diverse omics data and identify molecular signatures.
NetGestalt	R/Python	A web-based platform for integrative analysis of omics data. It incorporates network-based methods for data integration and exploration.
pyBioTools	Python	A Python library that offers network-based integration methods for diverse omics data, including gene expression and protein-protein interaction networks.
OmicsIntegrator	Python	A Python package that integrates multiomics data by incorporating biological network information. It offers various algorithms for data integration and network-based analysis.

different phenotypes, while clustering algorithms can be used to group samples based on their omics profiles. Machine learning—based integration methods can handle large and complex data sets and can identify nonlinear relationships between the variables. Machine learning algorithms have been widely used in biomarker discovery and disease diagnosis by identifying patterns in high-dimensional data sets that are indicative of a disease state. Here are some commonly used machines learning algorithms for biomarker discovery, which are also shown in Fig. 10.4.

1. **Random forest**: This technique employs decision trees to create a model that can forecast the result based on input factors. It is renowned for its accuracy and robustness and can handle both continuous and categorical information.
2. **Support vector machine:** This algorithm is used for classification problems and can handle nonlinearly separable data. It works by finding the hyperplane that maximally separates the classes.
3. **Artificial neural networks:** This algorithm mimics the functioning of the human brain to identify patterns in data. It can be used for both classification and regression problems and is particularly useful for handling noisy data.
4. **K-nearest neighbors:** This algorithm is used for classification and regression problems and works by finding the k-nearest neighbors of a data point to predict the outcome.
5. **Deep learning:** This is a subset of machine learning that uses neural networks with multiple layers to identify patterns in data. It has been used successfully in image and speech recognition and is gaining popularity in biomarker discovery

3.4 R packages for 10× scRNA-Seq data analysis

A list of R packages is listed in Table 10.2.

The real scRNA-Seq data sets can be obtained from the following websites:

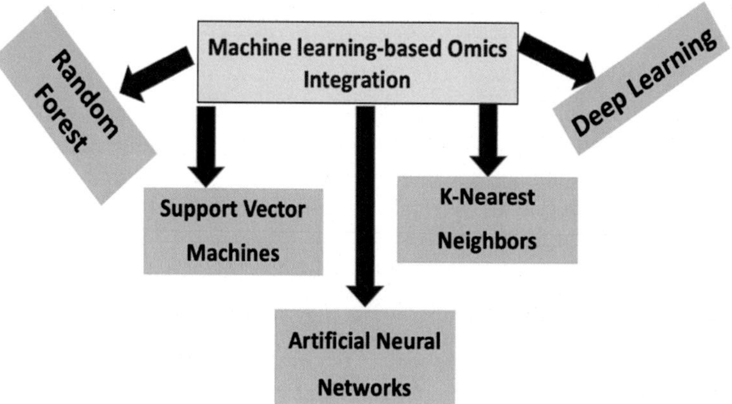

FIGURE 10.4 Machine learning methods for biomarker discovery.

TABLE 10.2 R packages for 10× scRNA-Seq data analysis.

Package name	Description
Seurat	Comprehensive toolkit for single-cell RNA-Seq data analysis
Scater	Quality control, normalization, and visualization of scRNA-Seq data
SingleR	Cell type identification based on reference gene expression profiles
Monocle	Trajectory analysis and differential expression analysis
Scran	Normalization and batch effect removal for scRNA-Seq data
MAST	Differential expression analysis for scRNA-Seq data
SCENIC	Gene regulatory network inference for scRNA-Seq data
scRNAseq	Preprocessing, visualization, and clustering of scRNA-Seq data
SeuratWrappers	Additional functions and wrappers for Seurat package
Harmony	Batch effect correction and integration of scRNA-Seq data sets

- https://hemberg-lab.github.io/scRNA.seq.datasets
- https://support.10xgenomics.com/single-cell-gene-expression/datasets
- https://figshare.com/articles/MCA_DGE_Data/5435866
- https://oncoscape.v3.sttrcancer.org/atlas.gs.washington.edu.mouse.rna/downloads

3.4.1 Gene set testing in limma

Sometimes after differential expression testing, we have a long list of 1000's of genes for which we need to do enrichment analysis, so various methods are available for this process as: goana, DAVID, ToppFun, GOstats are the most web-based tools, whereas goana has the ability to take into account gene length using the "covariate" argument. The GOseq bioconductor package (Young et al., 2010) contains the original method while in case of CAMERA, an "overlap" analysis assumes the genes are independent. CAMERA tests the ranking of the gene set relative to the other genes in the experiment, while taking into account intergene correlations and also taking into account strength of evidence of DE by using the moderated t-statistics (Kuhl et al., 2012).

3.4.2 Identification of downstream effects in signaling pathways

Gene expression analysis reveals correlations between drug sensitivity and elevated or decreased expression levels without prior knowledge on the affected biological processes. Functional enrichment analysis of the differentially expressed genes (DEGs), Gene ontology (GO) enrichment analysis was performed with the DAVID database available at https://david.ncifcrf.gov/ (Sherman et al., 2022), and for Kyoto Encyclopedia of Genes and Genomes (KEGG) analysis, the differentially expressed genes or transcripts were analyzed on the KEGG online website available at http://www.genome.jp/kegg/. After that, we could analyze the gene–pathway interaction networks for the DEGs that were visualized with Cytoscape 3.4.0 available at http://www.cytoscape.org/ (Otasek et al., 2019).

4. Integration of multiomics data

The integration of different types of sequencing data is useful to construct a model that can be used to predict complex traits and phenotypes and shows that main advantageous role is to reduce the gap between different omics platform data generation and the ability to analyze and understand the biological mechanisms against a host response to their diseases or environments. The use of a multiomics approach (Aho, 2013) will also help reduce the incidence of false positives generated from single source data sets (Ritchie et al., 2015). Performing genome sequencing on tumors in tens of thousands cancer patients could provide the statistical power to produce additional actionable information, but this would come at an enormous cost. Moreover, identifying large numbers of mutations associated with disease does not necessarily make cancer more treatable, as differentiating between disease-associated and disease-driving alterations remains a challenge, and the time and financial costs of developing specific therapeutics targeted to each of those individual mutations would be astronomical. In the end, the polygenic nature of the problem will prevent us from identifying and characterizing all but the most recurrent alteration patterns, leaving most tumors untreatable. This could also promote the understanding of the molecular mechanism of various cancer mechanisms and provide new insight into clinical research and treatment. It also provides advancement in cell—cell communication analysis and biomarkers prediction for identification of differential essential genes based on hierarchical clustering methods.

4.1 Pipeline for cell—cell communication analysis

To preprocess expression data for cell—cell communication analysis, biomarker prediction, different clustering methods, hierarchical visualization, and cell type annotation in R, you can follow the steps outlined in the following:

1 Load required packages:

Install and load the necessary R packages for data preprocessing, analysis, and visualization. Commonly used packages include dplyr, tidyverse, Seurat, SingleR, edgeR, limma, ggplot2, and ComplexHeatmap. You can install packages using the install. packages () function and load them using the library () function.

```
install. packages (c ("dplyr", "tidyverse", "Seurat", "SingleR", "edgeR", "limma", "ggplot2", "ComplexHeatmap"))
library(dplyr)
library(tidyverse)
library (Seurat)
library (SingleR)
library(edgeR)
library(limma)
library(ggplot2)
library (ComplexHeatmap)
```

2 Data preprocessing:

Read and preprocess the expression data, which typically includes normalization, log transformation, and quality control. Depending on the data format, you can use functions like read.csv (), read. Table (), or Read10X () for 10x Genomics data.

```
# Read expression data
expression_data <- read.csv ("expression_data.csv", header = TRUE, row. names = 1)
# Normalize and transform the expression data
normalized_data <- log2(normalize (expression data +1, method = "log"))
# Perform quality control and filtering
filtered_data <- normalized_data # Apply your specific quality control criteria.
# Perform dimensionality reduction if required (e.g., PCA or t-SNE)
reduced_data <- RunPCA(filtered_data) # Example using Seurat package for PCA
```

3 Cell—cell ccommunication analysis

Use expression data to identify potential cell—cell communication interactions. This can be done using various methods, such as ligand—receptor analysis or coexpression analysis. For ligand—receptor analysis, you can utilize tools such as CellPhoneDB or scRNAseqNet.

```
# Perform ligand—receptor analysis using CellPhoneDB.
# Install the CellPhoneDB package
devtools: install_github("Teichlab/CellPhoneDB")
library (CellPhoneDB)
# Prepare the expression matrix for CellPhoneDB
cpdb_data <- cpdb_prepare(expression_data)
# Run CellPhoneDB analysis
cpdb_results <- cpdb_analysis(cpdb_data)
# Explore and interpret the CellPhoneDB results
```

4 Biomarker prediction:

Identify differential essential genes or biomarkers that distinguish between different conditions or cell types. Commonly used methods include differential expression analysis using packages like edgeR, limma, or Seurat.

```
# Perform differential expression analysis using edgeR.
# Install the edgeR package
if (!requireNamespace("BiocManager", quietly = TRUE))
install.packages("BiocManager")
BiocManagerinstall("edgeR")
library(edgeR)
# Prepare the data for differential expression analysis
group_labels <- c("GroupA", "GroupB") # Labels for different groups or conditions
design_matrix <- model.matrix(~group_labels)
# Perform differential expression analysis
edgeR_object <- DGEList(counts = expression_data, group = group_labels)
edgeR_object <- calcNormFactors(edgeR_object)
edgeR_object <- estimateDisp.
```

5. Advantages of multiomics integration

Enabling a more comprehensive and holistic understanding of biological systems and diseases. Here are some examples:

1 **Enhanced molecular characterization:** Integration of multiomics data allows a more comprehensive molecular characterization of diseases, capturing multiple layers of biological information simultaneously by combining genomic, transcriptomic, epigenomic, proteomic, metabolomic, metagenomics, and other omics data (Kanchan et al., 2020). Researchers gain a more complete view of the underlying molecular mechanisms and pathways involved in diseases (Sun & Hu, 2016).
2 **Identification of disease subtypes and biomarkers:** Integration of multiomics data enables the identification of disease subtypes or patient stratification based on shared molecular profiles. It facilitates the discovery of novel disease biomarkers that can aid in early diagnosis, prognosis, and personalized treatment (Wang et al., 2014).
3 **Improved understanding of disease mechanisms**: Integration of multiomics data helps uncover the intricate interactions and cross-talk between different molecular layers, shedding light on disease mechanisms. It facilitates the identification of key driver genes, regulatory networks, and signaling pathways underlying diseases.
4 **Discovery of therapeutic targets and drug repurposing:** Integration of multiomics data helps identify potential therapeutic targets and enables drug repurposing by uncovering shared molecular features across different diseases. It provides a more comprehensive understanding of the complex interplay between drugs (Ghai et al., 2016) such as extract of *Eugenia caryophyllus* (Ghai et al., 2015), antiinflammatory activity of cinnolines (Mishra et al., 2015a,b), pharmacological evaluation of some cinnolines (Mishra et al., 2015a,b), hypoglycemic effects of different extract of *Clitoria ternatea* leaves (Saxena et al., 2015), molecular pathways, and disease-associated alterations (Zeng et al., 2020).
5 **Predictive modeling and precision medicine**: Integration of multiomics data allows the development of predictive models that incorporate diverse molecular features to improve disease prediction and treatment response prediction. It enables the application of precision medicine approaches by considering individual molecular profiles and tailoring therapies accordingly (Zhang et al., 2020).

These advantages highlight the potential of integrating multiomics data to advance our understanding of diseases, discover new biomarkers and therapeutic targets, and facilitate personalized medicine approaches.

6. Application of omics approaches in cohort studies

Omics-based cohort studies have become effective methods for examining the connection between molecular traits and numerous aspects of human health and disease. Our understanding of the causes of disease, risk assessment, therapeutic response, and customized medicine have all been fundamentally altered by these investigations. The following is a review of cohort studies using omics methods:

1 **Improved risk prediction:** The ability to predict illness risk has considerably improved because of omics-based cohort studies. Researchers can find biomarkers linked to greater vulnerability to particular diseases by examining genetic variations, gene expression profiles, or metabolic signatures in large cohorts. By including these biomarkers in risk prediction models, early identification and focused preventative measures are made possible.

2 **Identification of disease mechanisms:** Cohort studies using omics methods have provided insight into the underlying biological causes of disease. Researchers are able to understand the intricate relationships between genes, proteins, and metabolites that contribute to the onset and progression of disease by investigating genomic, transcriptomic, proteomic, or metabolomic changes. The creation of precision medicine strategies and the identification of new therapeutic targets are made easier by this knowledge. In proteomics studies of Yadav et al. (2013) it could help in identification of active site in the target protein is the starting point for virtual screening using the metaPocket 2.0 server, ligand binding site as pocket which located in the 3D structure of the fructose biphosphate aldolase (FBA) of CA-MRSA.
3 **Subtyping and personalized medicine:** The identification of molecular subtypes within diseases is possible through cohort studies using omic data. The combination of genomic and clinical data enables the identification of various disease subgroups with varying prognoses, therapeutic responses, or molecular signatures. Based on the molecular profiles of the patients, this data can direct the creation of targeted medicines and individualized treatment programs.
4 **Biomarker discovery:** Finding biomarkers for many diseases has been significantly sped up using omics techniques. Researchers can identify specific molecules or molecular patterns related to the presence, progression, or responsiveness to treatment of a certain disease by comparing molecular profiles between healthy people and people with that condition. These biomarkers may enhance early diagnosis, track the development of the illness, and forecast how well a treatment will work.
5 **Longitudinal monitoring:** Cohort studies allow for the collection of longitudinal data, enabling the assessment of molecular changes over time. Omics approaches can be applied to measure dynamic molecular alterations, such as changes in gene expression or metabolite levels, as individuals progress from healthy states to disease onset and progression. This longitudinal molecular monitoring provides valuable insights into disease trajectories and can guide the development of intervention strategies.
6 **Investigation of cancer metastasis:** Patient-derived organoids (PDOs) are three-dimensional cell culture models derived from patient tumor samples, providing a more representative and personalized model system for studying cancer biology. By comparing PDOs derived from primary colorectal tumors with PDOs derived from matched metastatic lesions, researchers can gain insights into the molecular changes associated with tumor progression and metastasis. The comparative analysis of patient-matched PDOs has uncovered a reduction in OLFM4-associated clusters in metastatic lesions of colorectal cancer. This finding suggested a potential role for OLFM4 in tumor progression and metastasis in colorectal cancer (Okamoto et al., 2021).

7. Computational methods for biomarker identification in complex disease

Here, we described the types of computational methods for biomarker identification in complex disease as shown in Fig. 10.5.

(a) **Genomics in biomarker discovery:** Genomic studies enable the identification of genetic variations, such as single-nucleotide polymorphisms (SNPs), copy number variations (CNVs), and structural variants, which contribute to disease

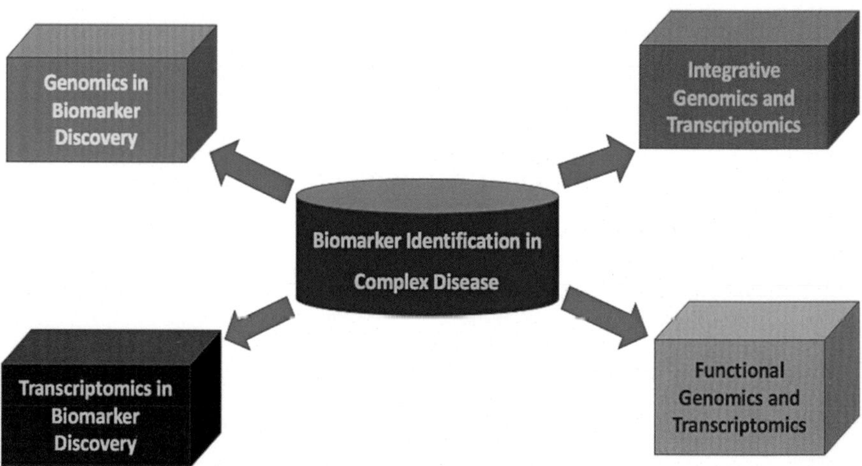

FIGURE 10.5 Methods for biomarker identification in complex diseases.

susceptibility and progression. These genetic markers (Visscher et al., 2017) can serve as potential biomarkers for disease diagnosis, prognosis, and treatment response (Manolio et al., 2009).

(b) Transcriptomics in biomarker discovery: Transcriptomics involves the study of gene expression patterns and the quantification of RNA molecules, providing valuable information about the functional state of cells and tissues. Transcriptomic analyses, such as microarray and RNA sequencing (RNA-Seq), enable the identification of differentially expressed genes (Huang, 2015) between disease and control samples, as well as the characterization of disease-specific gene expression signatures (Wang et al., 2009).

(c) Integrative genomics and transcriptomics: Integrating genomics and transcriptomics data allows for a comprehensive understanding of the molecular mechanisms underlying disease. Integration approaches, such as expression quantitative trait loci (eQTL) mapping (Li et al., 2018), help identify genetic variants that influence gene expression, linking genetic variations to disease-related molecular phenotypes (Battle et al., 2014).

(d) Functional genomics and transcriptomics: Functional genomics and transcriptomics approaches, such as gene ontology analysis, pathway analysis, and functional enrichment analysis, help uncover the biological processes, molecular functions, and pathways associated with disease phenotypes. This knowledge aids in identifying relevant biomarkers (Kanehisa et al., 2017) and potential therapeutic targets (Subramanian et al., 2005).

Through their insightful analyses of the molecular traits and functional mechanisms underpinning complex diseases, genomics and transcriptomics play a crucial role in the search for biomarkers. In addition to some important references, the following is a succinct review of the roles that genomics and transcriptomics play in the discovery of biomarkers. The role of genomics in the discovery of biomarkers. Genomic analyses make it possible to identify genetic variations that affect disease susceptibility and development, such as SNPs, CNVs, and structural variants. These genetic markers have the potential to be used as biomarkers to diagnose, predict, and assess the effectiveness of diseases and their treatments. However, transcriptomics is the study of gene expression patterns and the quantification of RNA molecules providing important knowledge about the functional condition of cells and tissues, and microarray transcriptomic analyses.

8. Network analysis and visualization with igraph and cystoscope

We need to acquire the necessary interaction data, such as protein–protein interactions, gene regulatory interactions, or coexpression correlations, for the network's construction. After that, preprocess the interaction data by removing low-confidence interactions or by using statistical criteria to filter out noise and to generate a graph object that represents the network structure in R, use the igraph package. Depending on the nature of the interactions, the graph might be either directed or undirected. In biology, network structures are everywhere because many biological systems depend on intricate interactions between the parts that make them up. The most frequent kind of species interaction in ecosystems such as predator–prey relationships that are essential to maintaining biodiversity. Synapses allow neurons in the human brain to exchange electrical and chemical impulses. At the cellular level, DNA, RNA, proteins, and other components take part in a number of biochemical processes that control how a cell functions and "sets of items, which we will call vertices or sometimes nodes, with connections between them, called edges," networks provide a concise mathematical model of these systems. R Code to generate igraph network analysis is as follows:

```
install. packages (c ("igraph", "graphlayouts", "ggraph", "ggforce"))
devtools: install_github("schochastics/networkdata")
library(igraph)
library(ggraph)
library(graphlayouts)
library(ggforce)
autograph(data)
ggraph (data, layout = "stress") +
geom_edge_link0(aes (edge_linewidth = weight), edge_colour = "gray66") +
geom_node_point (aes (fill = clu, size = size), shape = 21) +
geom_node_text (aes (filter = size ≥ 26, label = name), family = "serif") +
scale_fill_manual (values = got_palette) +
scale_edge_width (range = c (0.2, 3)) +
scale_size (range = c (1, 6)) +
thermograph () +
theme (legend. position = "none")
```

```
ggraph (data, layout = "phenotype")
##########
ggraph (greys, "phenotype", bbox = 15) +
geom_edge_link0(edge_colour = "gray66", edge_linewidth = 0.5) +
geom_node_point (aes (fill = sex), shape = 21, size = 3) +
geom_node_text (aes (label = name, size = degree(greys))),
family = "serif", repel = TRUE.
() +
scale_fill_manual (values = c ("gray66", "#EEB422", "#424242")) +
scale_size (range = c (2, 5), guide = "none") +
theme_graph () +
theme (legend. position = "bottom")
```

8.1 Weighted gene co-expression analysis

Weighted gene coexpression network analysis is also known as "WGCNA." It is used to create coexpression networks, grouping genes with similar expression pattern in clusters and relating these cluster with phenotypic characteristics. The accuracy and dependability of biomarker discoveries are increased through the integration of coexpression networks and machine learning techniques, which enable systems-level knowledge of the disease. It helps in making the framework for the discovery of reliable prognostic biomarkers.

R code for WGCNA analysis.

```
#Reading the raw data (columns represent the genes and rows represent the sample).
COUNT = read.csv ("COUNT.csv", sep = "; ")
#Create a new format expression data - remove gene name column.
COUNT = as.data. frame (COUNT [, -c (1)])
COUNT = t(COUNT)
#Column one - gene names
colnames (COUNT) = COUNT $EST
rownames (COUNT) = names (COUNT) [-c (1)]
#Group data in a dendogram to check outliers.
Count.Tree = hclust (dist (Count), method = "average")
#Plot Count.Tree.
#dev.off ()
pdf (file = "ClusterTreeDendrogram.pdf", width = 12, height = 9);
par (cex = 0.6)
par (mar = c (0,4,2,0))
plot (Count.Tree, main = "Sample clustering to detect outliers", sub = "", xlab = "", cex.lab = 1.5, cex.axis = 1.5, cex.main = 2)
#Plot a line showing the cut-off
abline (h = 31,000, col = "blue")
```

9. Conclusion

Integrative omics approaches for identification of biomarkers have transformed our understanding of human health and disease. These studies have elucidated disease mechanisms, improved risk prediction, facilitated personalized medicine, and accelerated biomarker discovery. The integration of multiomics data has further enhanced the power of these studies, enabling a more comprehensive and holistic view of molecular characteristics and their association with various health outcomes.

References

Aho, K. A. (2013). *Foundational and applied statistics for Biologists using R. Illustrated.* CRC Press.

Ali, N., Wolf, C., & Kanchan, S. (2023). *Nullomer peptide increases immune cell infiltration and reduces tumor metabolism in triple negative breast cancer mouse model.* https://doi.org/10.21203/rs.3.rs-3097552/v1

Battle, A., Mostafavi, S., Zhu, X., Potash, J. B., Weissman, M. M., McCormick, C., Haudenschild, C. D., Beckman, K. B., Shi, J., Mei, R., Urban, A. E., Montgomery, S. B., Levinson, D. F., & Koller, D. (2014). Characterizing the genetic basis of transcriptome diversity through RNA-sequencing of 922 individuals. *Genome Research, 24*(1), 14–24. https://doi.org/10.1101/gr.155192.113

Cheng, S. (2019). Multi-omics analysis reveals potential biomarkers and therapeutic targets of hepatocellular carcinoma. *Frontiers in Oncology, 9.*

Diamandis, E. P., Allard, W. J., & Shaw, J. L. (2010). Biomarkers in cancer diagnosis, prognosis, and treatment. *Biomarkers in Cancer*, 1–13.

Eichner, F., Kolleritsch, S., Huber, S., Stelzer, G., Thallinger, G. G., Mann, M., & Parapatics. (2021). Identification of early detection biomarkers for acute kidney injury using single-cell RNA sequencing. *Communications Biology, 4*(1), 1–11.

Gahoi, S., Mandal, R., Ivanisenko, N., Shrivastava, P., Jain, S., Singh, A., Raghunandanan, M., Kanchan, S., Taneja, B., Mandal, C., Ivanisenko, V., Kumar, A., Kumar, R., & Ramachandran, S. (2013). Computational screening for new inhibitors of *M. tuberculosis* mycolyltransferases antigen 85 group of proteins as potential drug targets. *Journal of Biomolecular Structure and Dynamics, 31*(1), 30–43. https://doi.org/10.1080/07391102.2012.691343

Ganz, P., Heidecker, B., Hveem, K., Jonasson, C., Kato, S., Segal, M. R., Sterling, D. G., & Williams, S. A. (2016). Development and validation of a protein-based risk score for cardiovascular outcomes among patients with stable coronary heart disease. *JAMA, the Journal of the American Medical Association, 315*(23), 2532–2541. https://doi.org/10.1001/jama.2016.5951

Garg, S., Saxena, V., Kanchan, S., Sharma, P., Mahajan, S., Kochar, D., & Das, A. (2009). Novel point mutations in sulfadoxine resistance genes of *Plasmodium falciparum* from India. *Acta Tropica, 110*(1), 75–79. https://doi.org/10.1016/j.actatropica.2009.01.009

Ghai, R., Nagarajan, K., Kumar, V., Kesheri, M., & Kanchan, S. (2015). Amelioration of lipids by *Eugenia caryophyllus* extract in atherogenic diet induced hyperlipidemia. *International Bulletin of Drug Research, 5*(8), 90–101.

Ghai, R., Nagarajan, K., Singh, J., Swarup, S., & Kesheri, M. (2016). Evaluation of antioxidant status in-vitro and in-vivo in hydro-alcoholic extract of Eugenia caryophyllus. *International Journal of Pharmacology and Toxicology, 4*(1), 19–24.

Guertin, K. A., Moore, S. C., Sampson, J. N., Huang, W. Y., Xiao, Q., Stolzenberg-Solomon, R. Z., Sinha, R., & Cross, A. J. (2014). Metabolomics in nutritional epidemiology: Identifying metabolites associated with diet and quantifying their potential to uncover diet-disease relations in populations. *American Journal of Clinical Nutrition, 100*(1), 208–217. https://doi.org/10.3945/ajcn.113.078758

Huang, J. K. (2015). Systematic evaluation of molecular networks for discovery of disease genes. *Cell Systems, 1*(4), 253–265.

Kanchan, S., Mehrotra, R., & Chowdhury, S. (2014). Evolutionary pattern of four representative DNA repair proteins across six model organisms: An in silico analysis. *Network Modeling Analysis in Health Informatics and Bioinformatics, 3*(1). https://doi.org/10.1007/s13721-014-0070-1

Kanchan, S., Mehrotra, R., & Chowdhury, S. (2015). In silico analysis of the endonuclease III protein family identifies key residues and processes during evolution. *Journal of Molecular Evolution, 81*(1–2), 54–67. https://doi.org/10.1007/s00239-015-9689-5

Kanchan, S., Sharma, P., & Chowdhury, S. (2019). Evolution of endonuclease IV protein family: An in silico analysis. *3 Biotech, 9*(5). https://doi.org/10.1007/s13205-019-1696-6

Kanchan, S., Sinha, R. P., Chaudière, J., & Kesheri, M. (2020). Computational metagenomics: Current status and challenges. *Recent Trends in Computational Omics: Concepts and Methodology*, 371–395. https://novapublishers.com/shop/recent-trends-in-computational-omics-concepts-and-methodology/.

Kanehisa, M., Furumichi, M., Tanabe, M., Sato, Y., & Morishima, K. (2017). Kegg: New perspectives on genomes, pathways, diseases and drugs. *Nucleic Acids Research, 45*(D1), D353–D361. https://doi.org/10.1093/nar/gkw1092

Kesheri, M., Kanchan, S., & Sinha, R. P. (2021). Isolation and in silico analysis of antioxidants in response to temporal variations in the cyanobacterium oscillatoria sp. *Gene Reports, 23*, 101023. https://doi.org/10.1016/j.genrep.2021.101023

Kesheri, M., Kanchan, S., & Sinha, R. P. (2017). *Exploring the potentials of antioxidants in retarding ageing* (pp. 166–195). https://doi.org/10.4018/978-1-5225-0607-2.ch008

Kesheri, M., Kanchan, S., Chowdhury, S., & Sinha, R. P. (2015). Secondary and tertiary structure prediction of proteins: A bioinformatic approach. *Studies in Fuzziness and Soft Computing, 319*, 541–569. https://doi.org/10.1007/978-3-319-12883-2_19

Kesheri, M., Kanchan, S., Richa, & Sinha, R. P. (2014). Isolation and in silico analysis of Fe-superoxide dismutase in the cyanobacterium nostoc commune. *Gene, 553*(2), 117–125. https://doi.org/10.1016/j.gene.2014.10.010. http://www.elsevier.com/locate/gene

Kesheri, M., Kanchan, S., & Sinha, R. P. (2022). Responses of antioxidants for resilience to temporal variations in the cyanobacterium *Microcystis aeruginosa. South African Journal of Botany, 148*, 190–199. https://doi.org/10.1016/j.sajb.2022.04.017

Kesheri, M., Richa, & Sinha, R. P. (2011). Antioxidants as natural arsenal against multiple stresses in cyanobacteria. *International Journal of Pharma and Bio Sciences, 2*(2), 168–187. http://ijpbs.net/volume2/issue2/bio/17.pdf.

Kesheri, M., Sinha, R. P., & Kanchan, S. (2016). Advances in soft computing approaches for gene prediction: A bioinformatics approach. *Studies in Computational Intelligence, 651*, 383–405. https://doi.org/10.1007/978-3-319-33793-7_17

Kuhl, C., Tautenhahn, R., Böttcher, C., Larson, T. R., & Neumann, S. (2012). Camera: An integrated strategy for compound spectra extraction and annotation of liquid chromatography/mass spectrometry data sets. *Analytical Chemistry, 84*(1), 283–289. https://doi.org/10.1021/ac202450g

Kumari, A., Kanchan, S., Sinha, R. P., & Kesheri, M. (2016). Applications of bio-molecular databases in bioinformatics. *Studies in Computational Intelligence, 651*, 329–351. https://doi.org/10.1007/978-3-319-33793-7_15

Kumari, A., Kesheri, M., Sinha, R. P., & Kanchan, S. (2018). Integration of soft computing approach in plant biology and its applications in agriculture. *Soft Computing for Biological Systems*, 265–281. https://doi.org/10.1007/978-981-10-7455-4_16

Kunkle, B. W., Grenier-Boley, B., Sims, R., Bis, J. C., Damotte, V., Naj, A. C., Boland, A., Vronskaya, M., van der Lee, S. J., Amlie-Wolf, A., Bellenguez, C., Frizatti, A., Chouraki, V., Martin, E. R., Sleegers, K., Badarinarayan, N., Jakobsdottir, J., Hamilton-Nelson, K. L., Moreno-Grau, S., … Pericak-Vance, M. A. (2019). Genetic meta-analysis of diagnosed Alzheimer's disease identifies new risk loci and implicates Aβ, tau, immunity and lipid processing. *Nature Genetics, 51*(3), 414−430. https://doi.org/10.1038/s41588-019-0358-2

Li, Y. I., Knowles, D. A., Humphrey, J., Barbeira, A. N., Dickinson, S. P., Im, H. K., & Pritchard, J. K. (2018). Annotation-free quantification of RNA splicing using LeafCutter. *Nature Genetics, 50*(1), 151−158. https://doi.org/10.1038/s41588-017-0004-9

Love, M. I., Huber, W., & Anders, S. (2014). Moderated estimation of fold change and dispersion for RNA-seq data with DESeq2. *Genome Biology, 15*(12). https://doi.org/10.1186/s13059-014-0550-8

Manolio, T. A., Collins, F. S., Cox, N. J., Goldstein, D. B., Hindorff, L. A., Hunter, D. J., McCarthy, M. I., Ramos, E. M., Cardon, L. R., Chakravarti, A., Cho, J. H., Guttmacher, A. E., Kong, A., Kruglyak, L., Mardis, E., Rotimi, C. N., Slatkin, M., Valle, D., Whittemore, A. S., … Visscher, P. M. (2009). Finding the missing heritability of complex diseases. *Nature, 461*(7265), 747−753. https://doi.org/10.1038/nature08494

Mishra, P., Saxena, V., Kesheri, M., & Saxena, A. (2015a). Synthesis, characterization and antiinflammatory activity of Cinnolines (pyrazole) derivatives. *IOSR Journal of Pharmacy and Biological Sciences, 10*(6), 77−82.

Mishra, P., Saxena, V., Kesheri, M., & Saxena, A. (2015b). Synthesis, characterization and pharmacological evaluation of cinnoline (thiophene) derivatives. *The Pharma Innovation Journal, 4*(10), 68−73.

Mohlke, K. L., Boehnke, M., & Abecasis, G. R. (2018). Metabolic and cardiovascular traits: An abundance of discovery, a paucity of replication. *Circulation Research, 122*(9), 1200−1210.

Okamoto, T., duVerle, D., Yaginuma, K., Natsume, Y., Yamanaka, H., Kusama, D., Fukuda, M., Yamamoto, M., Perraudeau, F., Srivastava, U., Kashima, Y., Suzuki, A., Kuze, Y., Takahashi, Y., Ueno, M., Sakai, Y., Noda, T., Tsuda, K., Suzuki, Y., Nagayama, S., & Yao, R. (2021). Comparative analysis of patient-matched PDOs revealed a reduction in OLFM4-associated clusters in metastatic lesions in colorectal cancer. *Stem Cell Reports, 16*(4), 954−967. https://doi.org/10.1016/j.stemcr.2021.02.012

Otasek, D., Morris, J. H., Bouças, J., Pico, A. R., & Demchak, B. (2019). Cytoscape Automation: Empowering workflow-based network analysis. *Genome Biology, 20*(1). https://doi.org/10.1186/s13059-019-1758-4

Priya, P., Kesheri, M., Sinha, R. P., & Kanchan, S. (2016). *Molecular dynamics simulations for biological systems*. IGI Global. https://doi.org/10.4018/978-1-4666-8811-7.ch014

Qiu, X., Mao, Q., Tang, Y., Wang, L., Chawla, R., Pliner, H. A., & Trapnell, C. (2017). Reversed graph embedding resolves complex single-cell trajectories. *Nature Methods, 14*(10), 979−982. https://doi.org/10.1038/nmeth.4402

Richa, Kannaujiya, V. K., Kesheri, M., Singh, G., & Sinha, R. P. (2011). Biotechnological potentials of phycobiliproteins. *International Journal of Pharma and Bio Sciences, 2*(4), 446−454. http://www.ijpbs.net/vol-2_issue-4/bio_science/50.pdf.

Richa, Rastogi, R. P., Kumari, S., Singh, K. L., Kannaujiya, V. K., Singh, G., Kesheri, M., & Sinha, R. P. (2011). Biotechnological potential of mycosporine-like amino acids and phycobiliproteins of cyanobacterial origin. *Biotechnology, Bioinformatics and Bioengineering, 1*(2), 159−171.

Ritchie, M. D., Holzinger, E. R., Li, R., Pendergrass, S. A., & Kim, D. (2015). Methods of integrating data to uncover genotype-phenotype interactions. *Nature Reviews Genetics, 16*(2), 85−97. https://doi.org/10.1038/nrg3868

Robinson, M. D., McCarthy, D. J., & Smyth, G. K. (2009). edgeR: A Bioconductor package for differential expression analysis of digital gene expression data. *Bioinformatics*. ISSN: 14602059, *26*(1), 139−140. https://doi.org/10.1093/bioinformatics/btp616

Sahu, N., Mishra, S., Kesheri, M., Kanchan, S., & Sinha, R. P. (2023). Identification of cyanobacteria-based natural inhibitors against SARS-CoV-2 druggable target ACE2 using molecular docking study, ADME and toxicity analysis. *Indian Journal of Clinical Biochemistry, 38*(3), 361−373. https://doi.org/10.1007/s12291-022-01056-6

Saxena, A., Saxena, V., Kesheri, M., & Mishra, P. (2015). Comparative hypoglycemic effects of different extract of clitoriaternatea leaves on rats. *IOSR Journal of Pharmacy and Biological Sciences, 10*(2), 60−65.

Sherman, B. T., Hao, M., Qiu, J., Jiao, X., Baseler, M. W., Lane, H. C., Imamichi, T., & Chang, W. (2022). David: A web server for functional enrichment analysis and functional annotation of gene lists (2021 update). *Nucleic Acids Research, 50*(W1), W216−W221. https://doi.org/10.1093/nar/gkac194

Spearman, C. (1904). \General intelligence\objectively determined and measured. *American Journal of Psychology*. ISSN: 00029556, *15*(2), 201. https://doi.org/10.2307/1412107

Srivastava, U., & Singh, G. (2013). Comparative homology modelling for HPV type 16 E 7 proteins by using MODELLER and its validations with SAVS and ProSA web server. *Journal of Computational Intelligence in Bioinformatics, 6*(1), 27. https://doi.org/10.37622/jcib/6.1.2013.27-33

Srivastava, U., & Singh, S. (2022). *Approaches of single-cell analysis in crop Improvement*. Springer Science and Business Media LLC. https://doi.org/10.1007/978-1-0716-2533-0_14

Srivastava, U., Singh, S., Gautam, B., Yadav, P., Yadav, M., Thomas, G., & Singh, G. (2017). Linear epitope prediction in HPV type 16 E7 antigen and their docked interaction with human TMEM 50A structural model. *Bioinformation, 13*(05), 122−130. https://doi.org/10.6026/97320630013122

Stuart, T., Butler, A., Hoffman, P., Hafemeister, C., Papalexi, E., Mauck, W. M., Hao, Y., Stoeckius, M., Smibert, P., & Satija, R. (2019). Comprehensive integration of single-cell data. *Cell, 177*(7), 1888−1902. https://doi.org/10.1016/j.cell.2019.05.031

Subramanian, A., Tamayo, P., Mootha, V. K., Mukherjee, S., Ebert, B. L., Gillette, M. A., Paulovich, A., Pomeroy, S. L., Golub, T. R., Lander, E. S., & Mesirov, J. P. (2005). Gene set enrichment analysis: A knowledge-based approach for interpreting genome-wide expression profiles. *Proceedings of the National Academy of Sciences, 102*(43), 15545−15550. https://doi.org/10.1073/pnas.0506580102

Sun, Y. V., & Hu, Y. J. (2016). Integrative analysis of multi-omics data for discovery and functional studies of complex human diseases. *Advances in Genetics, 93*, 147−190. https://doi.org/10.1016/bs.adgen.2015.11.004

Visscher, P. M., Wray, N. R., Zhang, Q., Sklar, P., McCarthy, M. I., Brown, M. A., & Yang, J. (2017). 10 Years of GWAS discovery: Biology, function, and translation. *The American Journal of Human Genetics, 101*(1), 5–22. https://doi.org/10.1016/j.ajhg.2017.06.005

Wang, B., Mezlini, A. M., Demir, F., Fiume, M., Tu, Z., Brudno, M., Haibe-Kains, B., & Goldenberg, A. (2014). Similarity network fusion for aggregating data types on a genomic scale. *Nature Methods, 11*(3), 333–337. https://doi.org/10.1038/nmeth.2810

Wang, Z., Gerstein, M., & Snyder, M. (2009). RNA-seq: A revolutionary tool for transcriptomics. *Nature Reviews Genetics, 10*(1), 57–63. https://doi.org/10.1038/nrg2484

Yadav, P., Singh, G., Gautam, B., Singh, S., Yadav, M., Srivastav, U., & Singh, B. (2013). Molecular modeling, dynamics studies and virtual screening of fructose 1, 6 biphosphate aldolase-II in community acquired- methicillin resistant *Staphylococcus aureus* (CA-MRSA). *Bioinformation, 9*(3), 158–164. https://doi.org/10.6026/97320630009158

Young, M. D., Wakefield, M. J., Smyth, G. K., & Oshlack, A. (2010). Gene ontology analysis for RNA-seq: Accounting for selection bias. *Genome Biology, 11*(2). https://doi.org/10.1186/gb-2010-11-2-r14

Zeng, X., Ding, N., & Rodríguez-Patón, A. (2020). Integration of multi-omics data for mining of biological insights. *Multidisciplinary Computational Intelligence Techniques*, 205–228. https://doi.org/10.1007/978-3-030-38810-3_8

Zhang, X., Li, J., & Wang, Q. (2020). Integration of multi-omics data for cancer research: From computational strategies to experimental applications. *Frontiers in Genetics, 11*. https://doi.org/10.3389/fgene.2020.607736

Chapter 11

Omics approach for personalized and diagnostics medicine

Deepak Verma[1] and Shruti Kapoor[2]

[1]School of Medicine, Johns Hopkins University, Baltimore, MD, United States; [2]Department of Genetics, University of Alabama, Birmingham, AL, United States

1. Introduction

The omics approach refers to studying large-scale molecular data, including genomics, transcriptomics, proteomics, and metabolomics, to understand biological systems and diseases comprehensively. Genomics focuses on studying an individual's genetic makeup, while transcriptomics involves the study of the transcriptome, which is the set of all RNA molecules in a cell. Proteomics focuses on studying proteins and their interactions, while metabolomics is concerned with analyzing small molecules involved in metabolism (Hasin et al., 2017).

Integrating these omics data types provides a more complete understanding of biological systems and diseases and is a key aspect of personalized and diagnostic medicine. Personalized medicine, also known as precision medicine, utilizes this information to tailor medical treatment to each patient's individual needs and genetic makeup. By combining the information from multiple omics data types, a complete picture of the underlying biology can be obtained, informing the development of new diagnostic tools and treatments (Subramanian et al., 2020).

One of the main advantages of omics approach is that it can provide detailed information about the molecular changes associated with a particular disease. For example, genomic data can identify genetic mutations associated with a specific disease. In contrast, transcriptomic data can provide information about gene expression changes related to disease. Proteomic data can be used to identify changes in protein expression and function, while metabolomic data can be used to identify changes in metabolic pathways associated with disease (Bludau & Aebersold, 2020; Clish, 2015).

The omics approach has already significantly impacted the field of personalized medicine. For example, genomic data have been used to identify genetic mutations associated with specific cancers, leading to the development of new targeted therapies. Transcriptomic data have been used to identify new biomarkers for disease, while proteomic data have been used to develop new diagnostic tests for various diseases. Personalized and diagnostic medicine, also known as precision medicine, utilizes this information to tailor medical treatment to each patient's individual needs and genetic makeup. The omics approach helps to identify the specific molecular changes associated with disease, leading to improved diagnosis, and personalized treatment options. By combining the information from multiple omics data types, a complete picture of the underlying biology can be obtained, verging on the development of new diagnostic tools and treatments (Hasanzad et al., 2022).

In addition to its application in personalized medicine, the omics approach is also being used to develop new diagnostic tests. For example, genomic data helps to identify genetic mutations associated with specific diseases, and this information can be used to develop specific diagnostic tests for those diseases. Similarly, proteomic data can be analyzed to identify protein markers associated with specific diseases, which is useful to develop diagnostic tests particular to those diseases (Ahmed, 2022).

The omics approach is also contributing in the development of new treatments for diseases. For example, information from the genomic data leading to identification of genetic mutations associated with specific diseases, can be used to develop targeted therapies specific to those diseases. Similarly, transcriptomic data can be used to identify changes in gene

Integrative Omics. https://doi.org/10.1016/B978-0-443-16092-9.00011-4

expression that are associated with specific diseases, and this information can be used to develop new treatments for those diseases (Chakravarty & Solit, 2021; Hasin et al., 2017).

Despite the many advances that have been made in the field of personalized medicine, many challenges still need to be addressed. One of the main challenges is the need to integrate the large amounts of data that are generated by the omics approach. Another challenge is the need to develop new methods for analyzing these data, which can be complex and time-consuming. Additionally, there is a need to develop new strategies for validating the results of omics studies and new ways of communicating these results to the broader scientific community.

The omics approach has had a significant impact on personalized medicine and has the potential to revolutionize how we diagnose and treat disease. By providing detailed information about the molecular changes that are associated with specific diseases, the omics approach has the potential to lead to the development of new diagnostic tests and treatments that are tailored to the individual needs and genetic makeup of individuals (Tebani et al., 2016).

2. Multiomics approaches in diseases and medicine

Multiomics approaches in diseases and medicine refer to integrating multiple datasets generated from different "omics" technologies, such as genomics, transcriptomics, proteomics, and metabolomics, to understand diseases' underlying mechanisms and develop new treatments comprehensively.

Integrating multiple datasets provides a more holistic view of the molecular events occurring in diseases, compared to a single "omics" approach. For example, genomics can give information about the genetic basis of disease. At the same time, transcriptomics can provide insight into the changes in gene expression, and proteomics can shed light on changes in protein abundance and function. By combining these datasets, researchers can build a complete picture of the disease processes, identify new biomarkers, and develop targeted therapies (Bludau & Aebersold, 2020; Chakravarty & Solit, 2021; Rusch et al., 2018).

One example of the application of multiomics approaches in diseases is cancer. Cancer is a complex disease with multiple genetic and epigenetic alterations, leading to changes in gene expression, protein abundance, and metabolic pathways. By analyzing various "omics" datasets from cancer patients, researchers can identify new therapeutic targets, develop personalized treatments, and monitor the effectiveness of treatments (Chakravarty & Solit, 2021).

Another example is in the field of neuroscience. Multiomics approaches can be used to study the underlying mechanisms of neurodegenerative diseases, such as Alzheimer's and Parkinson's. By integrating genomics, transcriptomics, proteomics, and metabolomics data, researchers can gain insights into the molecular changes occurring in the brain, identify new biomarkers, and develop new treatments (Guo et al., 2022).

In addition to disease diagnosis and treatment, multiomics approaches also have applications in precision medicine. Precision medicine is an approach that considers individual differences in genes, environment, and lifestyle to develop personalized treatments. Multiomics approaches can provide a comprehensive understanding of an individual's molecular profile, allowing for personalized treatments that target specific molecular changes.

Overall, multiomics approaches in diseases and medicine have the potential to revolutionize our understanding of diseases and the development of new treatments. Integrating multiple datasets from different "omics" technologies provides a more comprehensive understanding of the underlying mechanisms of diseases and allows for the development of targeted and personalized treatments. However, the integration of multiple datasets can be challenging. There is a need for more advanced computational methods and tools to analyze and interpret the large and complex datasets generated by multiomics approaches (Hasin et al., 2017; Olivier et al., 2019).

2.1 Identification of rare diseases through genomics

Rare diseases are a group of conditions that affect a small number of people and often go undiagnosed for long periods. Advances in genomics have led to improved methods for identifying the genetic causes of rare diseases, providing hope for affected individuals and their families.

One approach to identifying the genetic causes of rare diseases is whole genome sequencing (WGS). This process involves determining the entire DNA sequence of an individual, allowing for the identification of genetic variations that may be responsible for a disease. WGS can be performed on a single individual or a small family with multiple affected members, providing insight into the inheritance pattern of the disease and facilitating the identification of disease-causing mutations (Frésard et al., 2019).

Another approach is exome sequencing, which involves sequencing only the genome's exons. Exons are the portions of DNA that encode proteins and are, therefore, the most likely to contain disease-causing mutations. Exome sequencing can be a cost-effective and efficient method for identifying the genetic causes of rare diseases, particularly for conditions that are caused by mutations in a single gene.

Genome-wide association studies (GWAS) is another approach that can be used to identify the genetic causes of rare diseases. This method involves comparing the genomes of large groups of individuals, both with and without the disease, to identify common genetic variations associated with the disease. GWAS have successfully identified the genetic causes of many common diseases, including cardiovascular disease and type 2 diabetes, but have yet to be widely used for rare diseases due to the small number of affected individuals (Tam et al., 2019).

Finally, clinical exome sequencing is a method that combines the power of WGS and exome sequencing with clinical information to identify the genetic causes of rare diseases. This approach involves sequencing the DNA of individuals with undiagnosed or rare diseases and analyzing the data in the context of the individual's medical history, symptoms, and physical examination. Clinical exome sequencing can be beneficial for identifying the genetic causes of complex, multisystem diseases that may not be easily classified into a specific disease category.

Advances in genomics have greatly improved our ability to identify the genetic causes of rare diseases. The use of WGS, exome sequencing, GWAS, and clinical exome sequencing can provide a powerful tool for improving the diagnosis and treatment of rare diseases. As our understanding of the genomic basis of rare diseases continues to grow, we can expect to see further advances in identifying and managing these conditions (Dehghan, 2018; Visscher et al., 2012).

2.2 Major omics approaches in medicine

Omics refers to the large-scale study of biological molecules and their interactions, including genomics (study of genomes), transcriptomics (study of transcripts), proteomics (study of proteins), metabolomics (study of metabolites), and epigenomics (study of epigenetic modifications). These fields have revolutionized how we understand the underlying mechanisms of disease and opened up new avenues for diagnosis and treatment.

Genomics: Genomics is the study of individual genes, their sequence, and any alterations in the genome, including their interaction with other genes or the individual's environment. It extends to the genome's function, evolution, editing, and mapping (total set of genes). The application of genomics in the early detection of diseases has been a revolution where any alteration in the genome can be identified before developing symptoms for the disease. At the same time, it also aids in identifying the causative changes which lead to disease after developing symptoms. Thus, early diagnosis can lead to prevention and effective targeted treatment, which may vary within populations or even from person to person, paving the path for personalized medicine. Pharmacogenomics is another field of genomics involving the effect of genomics on drugs and their effective treatment depending on the person's genome. Thus, depending on an individual's genetic makeup, drugs can be prescribed and designed to treat the diseases. Genomics has provided a revolutionary shift from discovery to treatment and now cure concerning diseases (Rusch et al., 2018).

Transcriptomics: Transcriptomics involves the study of RNA molecules and the transcriptome, the entire set of RNA molecules expressed by a cell or tissue. By analyzing the transcriptome, researchers can determine which genes are expressed and how they are regulated, providing insight into cellular processes and disease mechanisms. For example, transcriptomics can be used to identify disease-specific gene signatures and potential therapeutic targets (Mock et al., 2023; Wang et al., 2009).

Proteomics: Proteomics involves the study of proteins and the proteome, the entire set of proteins expressed by a cell or tissue. This can help to understand the function of proteins and how they interact with each other to carry out cellular processes. Proteomics has been used to identify new biomarkers for disease diagnosis and to understand better the mechanisms of drug action (Bludau & Aebersold, 2020).

Metabolomics: Metabolomics involves the study of metabolites, the small molecules produced as a result of cellular processes. By analyzing the metabolome, researchers can understand how cells respond to stimuli and how disease affects metabolic pathways. Metabolomics has been used to identify new biomarkers for disease and to study the effects of environmental factors, such as diet and toxic exposure, on human health (Clish, 2015).

Epigenomics: Epigenomics refers to the study of epigenetic modifications, changes to the DNA molecule that affect gene expression but do not alter the underlying genetic code. This includes modifications such as DNA methylation and histone modification. By analyzing the epigenome, researchers can better understand how environmental and lifestyle factors can influence gene expression and contribute to disease (Wang & Chang, 2018).

Immunomics: Immunomics is a field of study that combines immunology and genomics to understand the immune system and its interactions with pathogens and diseases. It uses computational tools and techniques to analyze and interpret large-scale genomic and proteomic data related to the immune system. The main goal of immunomics is to gain insights into the immune response and identify potential targets for therapeutic interventions. Immunomics has broad applications in infectious diseases, cancer immunotherapy, autoimmune disorders, and vaccine development. By leveraging the power of genomics and computational tools, researchers aim to enhance our understanding of the immune system and develop more effective strategies for diagnosing, preventing, and treating diseases (Bonaguro et al., 2022; Castle et al., 2014).

2.3 Other omics approaches in medicine

In addition to genomics, proteomics and transcriptomics, several other "omics" approaches are being used in medicine to gain a more complete understanding of biological systems. Some of these include:

Microbiomics: This field is concerned with the study of the genomes of microorganisms, including bacteria, viruses, fungi, and parasites, that interact with each other and with the host (Parello, 2020).

Integrative omics: Integrative omics involves integrating data from multiple omics approaches to provide a more comprehensive understanding of biological processes and disease mechanisms. For example, integrative omics can be used to understand the interplay between genetic, epigenetic, and environmental factors in disease development (Sun & Hu, 2016).

Lipidomics: This field studies lipids, a diverse group of biomolecules including fats, waxes, and phospholipids. Lipidomics helps to understand the role of lipids in health and disease (Han, 2016).

Connectomics: This field focuses on studying neural connections in the brain, including both the wiring diagram of neurons and the molecules that support communication between them (Fornito et al., 2015).

Glycomics: This field studies carbohydrates, also known as sugars, and their role in biological processes (Rudd et al., 2015).

Phosphoproteomics: This field focuses on the study of phosphorylated proteins, which are proteins that the addition of a phosphate group has modified. Phosphorylation plays an essential role in cellular signaling pathways (Savage & Zhang, 2020).

Interactomics: This field studies protein—protein interactions, crucial for many biological processes (Bludau & Aebersold, 2020).

Structural-omics: This field focuses on studying protein structures and how they contribute to the function of proteins (Shi, 2014).

Each "omics" approach provides a unique perspective on biological systems and contributes to a more comprehensive understanding of the underlying mechanisms of health and disease. For example, combining transcriptomics and proteomics can help identify changes in gene expression and protein levels in response to a particular stimulus. Combining metabolomics and lipidomics can help understand the metabolic changes during disease.

The "omics" approaches in medicine are essential for advancing our understanding of biological systems and improving patient outcomes. By integrating information from multiple "omics" domains, scientists, and health-care professionals can gain a more complete picture of the underlying causes of disease and develop more effective therapies. An essential component of multiomics is shown in Fig. 11.1.

3. Integration of clinical data of personnel and group with patient-specific omics datasets

Integrating clinical data, personal information, and group data with patient-specific omics datasets is a complex and evolving field with great promise for advancing personalized medicine. This integration combines various data types to have a better understanding of an individual's health status, disease progression, and treatment response.

Omics datasets, such as genomics, transcriptomics, proteomics, metabolomics, and epigenomics, provide detailed molecular information about an individual's biological processes. These datasets can be generated from various sources, including DNA sequencing, gene expression profiling, mass spectrometry, and other high-throughput technologies.

Integrating these omics datasets with clinical data, personal information, and group data allows for a more comprehensive analysis of an individual's health. Here are some key considerations and approaches for integrating these diverse datasets.

1. **Data preprocessing:** Omics datasets often require extensive preprocessing steps to clean and normalize the data. This may involve removing batch effects, normalizing expression levels, and filtering out low-quality or irrelevant data. Clinical data may also require preprocessing to standardize formats and resolve missing or inconsistent values.

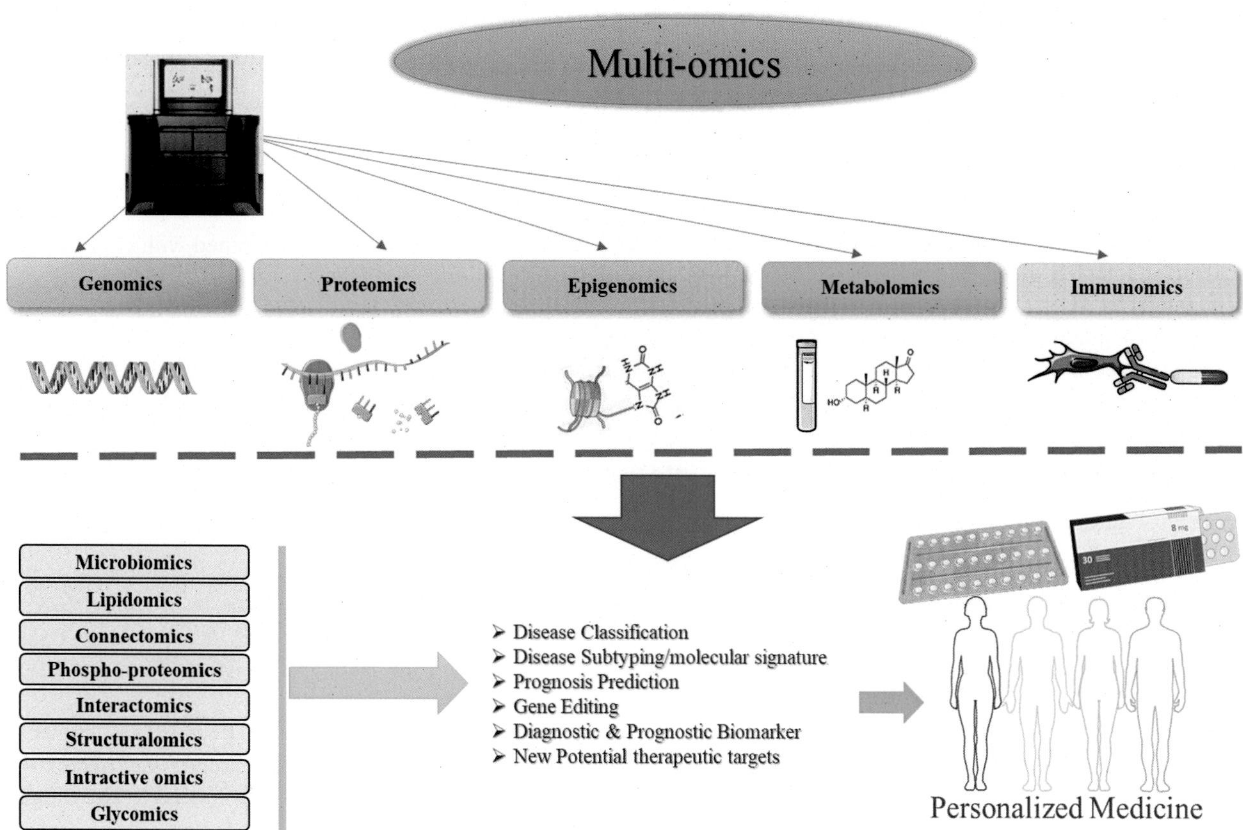

FIGURE 11.1 Different types of multiomics approach. Exploration of multiomics branches.

2. **Data harmonization:** Integrating different data types requires harmonizing the data formats, ensuring compatibility, and establishing common variables for linkage. This step is crucial to ensure that data from various sources can be combined and analyzed effectively.
3. **Data integration methods:** Various computational methods and statistical approaches are available for integrating different datasets. These include correlation analysis, network-based techniques, machine learning algorithms, and statistical modeling. The choice of method depends on the specific research question and the integrated datasets' characteristics.
4. **Privacy and data security:** When integrating personal and clinical data with omics datasets, privacy and data security are of utmost importance. Compliance with data protection regulations and ethical considerations must be ensured to protect patient confidentiality and prevent unauthorized access to sensitive information.
5. **Interpretation and analysis:** Integrated clinical and omics data analysis requires advanced bioinformatics and biostatistics tools. This analysis aims to identify biomarkers, molecular pathways, and potential therapeutic targets associated with specific diseases or patient subgroups. It may involve the identification of genetic variants, gene expression patterns, protein—protein interactions, or metabolite profiles that correlate with clinical outcomes.
6. **Validation and replication:** Findings from integrated analysis should be validated and replicated using independent datasets or experimental validation techniques. Reproducibility and generalizability are crucial to ensure the findings' robustness and applicability to larger patient populations.

Integrating clinical data, personal information, and group data with patient-specific omics datasets holds great potential for advancing precision medicine. It enables a more comprehensive understanding of disease mechanisms, the identification of personalized treatment strategies, and the development of targeted therapies. However, it also presents challenges related to data integration, privacy, and analysis, which need to be carefully addressed to maximize its benefits (Ghosh et al., 2018; Hackl et al., 2010; Holzinger et al., 2019; Karczewski & Snyder, 2018).

Here are a few examples that discuss insights into integrating clinical data with patient-specific omics datasets in different disease contexts, highlighting the potential of this approach in advancing personalized medicine and improving patient outcomes.

Mer et al. demonstrate the integration of patient-derived xenografts with clinical data and omics datasets to identify predictive biomarkers for drug response (Mer et al., 2019).

Cho et al. presented an integrative analysis of proteomic and transcriptomic data and clinical information to identify molecular pathways associated with diabetic retinopathy (Cho et al., 2013).

Song et al. utilize an integrative analysis of genomic, transcriptomic, and proteomic data, combined with clinical data, to identify potential therapeutic targets and biomarkers in ovarian cancer (Song et al., 2019).

Pabinger et al. summarized various tools and methods for analyzing genomic data, including variant analysis, and discussed integrating such data with clinical information (Pabinger et al., 2014).

Xiong et al. proposed an integrative approach that combines genetic and gene expression data, along with clinical information, to perform genome-wide association analysis of gene sets (Xiong et al., 2012).

3.1 Clinical data generation procedures and integration of omics data with clinical data

Several sources of patient-specific omics datasets can be accessed for research and analysis. Here are some familiar sources.

1. **Research consortia and initiatives:** Many large-scale research projects and consortia have been established to generate omics data from patient samples. Examples include The Cancer Genome Atlas (TCGA), Genotype-Tissue Expression (GTEx), International Cancer Genome Consortium (ICGC), and the Human Cell Atlas. These initiatives provide comprehensive omics datasets from various diseases and tissues.
2. **Public databases:** Several public databases have been created to archive and share omics data. The most well-known is the National Center for Biotechnology Information (NCBI), which hosts various databases such as the Gene Expression Omnibus (GEO), Sequence Read Archive (SRA), and Database of Genotypes and Phenotypes (dbGaP). These databases contain many omics data from different diseases and conditions.
3. **Disease-specific databases:** Many diseases and medical conditions have dedicated databases that store patient-specific omics data. For example, the Alzheimer's disease Neuroimaging Initiative (ADNI) database contains omics data related to Alzheimer's disease, and the Parkinson's Progression Markers Initiative (PPMI) database focuses on Parkinson's disease.
4. **Institutional repositories:** Research institutions and hospitals often maintain their own repositories of patient-specific omics data. These repositories may be specific to a particular disease or cover a broader range of conditions. Examples include the European Genome-phenome Archive (EGA) and Genomic Variants (DGV) database.
5. **Data sharing platforms:** Various data sharing platforms have emerged to facilitate the sharing and integration of omics datasets. One notable example is the Global Alliance for Genomics and Health (GA4GH), which provides a framework for data sharing and interoperability across multiple institutions and countries.
6. **Collaborative research networks:** Researchers may establish collaborative networks or partnerships to share and analyze patient-specific omics datasets. These networks often involve multiple research institutions and can provide access to diverse data.

It's important to note that accessing and utilizing patient-specific omics datasets often requires proper ethical approvals and adherence to data protection regulations. Researchers should follow the necessary guidelines and obtain appropriate permissions before using these datasets.

A systematic diagram for clinical data generation procedures and integration of omics data is depicted through a flowchart in Fig. 11.2.

4. Personalized medicine and its importance

Personalized medicine is an approach to health care that involves tailoring medical treatment to each patient's individual needs and characteristics. This approach considers various factors such as a patient's genetic makeup, lifestyle, medical history, and environmental exposures to create a unique care plan.

Personalized medicine aims to improve health outcomes and reduce the risk of adverse effects by using the most appropriate treatments for each individual. It is based on the premise that each person is unique and that a one-size-fits-all health-care approach is ineffective for everyone.

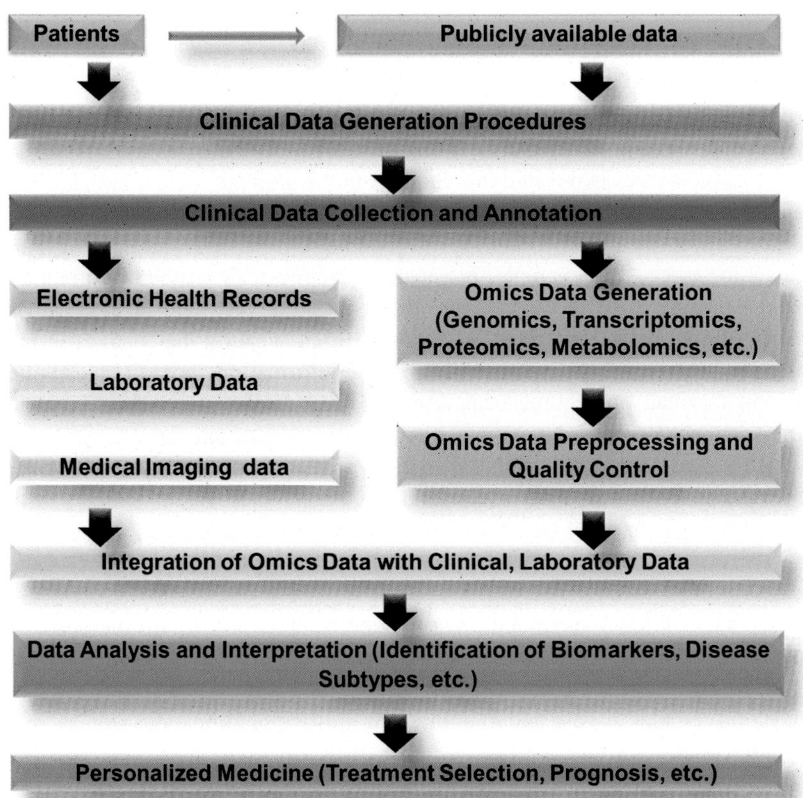

FIGURE 11.2 Generation and integration of clinical data. This flowchart illustrates the sequential steps involved in the generation and integration of clinical data with omics data. It starts with the collection and annotation of clinical data, followed by the generation of omics data (such as genomics, transcriptomics, proteomics, and metabolomics). The omics data then undergo preprocessing and quality control procedures. The next step involves integrating the omics data with clinical data, enabling comprehensive analysis and interpretation. This analysis can include identifying biomarkers, disease subtypes, and other relevant insights. Finally, the integrated data is used to guide personalized medicine approaches, including treatment selection and prognosis.

One of the key benefits of personalized medicine is that it can lead to earlier and more accurate diagnoses. By analyzing an individual's genetic information, physicians can determine their risk for certain diseases and conditions and take proactive steps to prevent or treat them. This can result in earlier detection of conditions such as cancer, allowing for earlier and more effective treatment.

Personalized medicine also has the potential to improve treatment outcomes. For example, genetic testing can help doctors determine which medications are most likely effective for a patient, reducing the risk of ineffective treatments and side effects. This can lead to better health outcomes, increased patient satisfaction, and reduced health-care costs.

Another vital aspect of personalized medicine is that it can help to reduce health-care disparities. This is because the approach considers individual differences such as genetics, lifestyle, and medical history, which can impact health outcomes. By providing personalized care, health-care providers can help reduce these disparities and improve patient outcomes.

The field of personalized medicine is advancing rapidly, driven by technological advances, particularly in genomics and biotechnology. As a result, it is becoming increasingly possible to tailor medical treatment to each patient's unique needs. This has led to the development of new diagnostic tests and treatments that are specifically designed for individual patients.

Personalized medicine is a promising approach to health care that has the potential to improve health outcomes, reduce adverse effects, and reduce health-care disparities. By taking into account individual differences such as genetics, lifestyle, and medical history, health-care providers can create customized plans of care that are more effective and tailored to the needs of each patient. As the field continues to advance, the benefits of personalized medicine are likely to become even more pronounced, providing patients with the best possible care and outcomes (Ahmed, 2022; Collins & Varmus, 2015; Guo et al., 2022; Hasanzad et al., 2022; Subramanian et al., 2020).

4.1 Integrated multiomics approach toward precision and preventive medicine

Integrated multiomics approach toward precision and preventive medicine has become a key focus in modern medical research. It is a combination of different-omics technologies, such as genomics, transcriptomics, proteomics, metabolomics, and epigenomics, that work together to provide a comprehensive understanding of the molecular mechanism of diseases and how to prevent them. This approach aims to develop personalized medicine and reduce the prevalence of noncommunicable diseases such as cancer, cardiovascular disease, and diabetes.

Genomics studies the genetic material in a person's DNA and the variation in different individuals. Transcriptomics involves the analysis of the mRNA transcribed from DNA that is used as a blueprint to build proteins. Proteomics focuses on studying proteins and their interactions with other cellular components, while metabolomics analyses the metabolic pathways and metabolites produced by the body. Epigenomics is the study of changes in the genome that do not alter the DNA sequence but regulate the expression of genes.

Integrating these different-omics technologies allows for a more complete understanding of disease mechanisms and identifying new disease biomarkers and therapeutic targets. For example, in cancer, genomics can provide information about the genetic mutations that drive cancer development. At the same time, transcriptomics can help identify genes expressed differently in cancer cells compared to normal cells. Proteomics can then be used to study the changes in the levels of specific proteins involved in cancer progression, and metabolomics can provide information about the metabolic changes in cancer cells that affect their growth and survival. Epigenomics can also help identify changes in the epigenetic marks that regulate gene expression, which may contribute to the development and progression of cancer.

Precision medicine is a personalized approach to medical treatment that considers a person's genetic, environmental and lifestyle factors. Integrating multiomics data can help identify individualized treatment options for patients with complex diseases such as cancer, where a single therapy may not be effective for all patients. For example, in cancer, multiomics data can be used to predict a patient's responsiveness to specific treatments, allowing for the selection of the most effective therapy for each patient.

Preventive medicine is the use of early screening, risk assessment and lifestyle modifications to prevent the onset of diseases. Multiomics data can provide information about a person's genetic risk for developing certain diseases, allowing for early intervention and lifestyle modifications to prevent disease onset. For example, multiomics data can be used to identify individuals at high risk for developing cardiovascular disease and to implement lifestyle modifications such as diet and exercise, which can reduce the risk of disease onset.

The integration of multiomics technologies provides a comprehensive understanding of the molecular mechanism of diseases and has the potential to revolutionize the way we diagnose and treat complex diseases. The integration of multiomics data has the potential to enable the development of personalized medicine and preventive medicine, reducing the burden of noncommunicable diseases on society (Ahmed, 2022; Bludau & Aebersold, 2020; Guo et al., 2022; Hasanzad et al., 2022; Hasin et al., 2017; Mock et al., 2023; Ota & Fujio, 2021; Subramanian et al., 2020).

5. Challenges, opportunities, and prospects for personalized diagnostics

Personalized diagnostic is an emerging field that aims to tailor medical diagnoses and treatments to individual patients based on their unique characteristics, including genetic makeup, lifestyle factors, and clinical data. This approach can potentially revolutionize health care by improving accuracy, efficacy, and patient outcomes. However, several challenges and opportunities must be considered to implement personalized diagnostics successfully. Here are some key points addressing challenges, opportunities, and future prospects of personalized diagnostics:

5.1 Challenges in personalized diagnostics

a. **Cost:** One of the main limitations of the omics approach is the cost of the technology and the analysis. The cost of sequencing a genome, for example, has decreased dramatically over the past few years, but it is still expensive.
b. **Technical limitations:** Another limitation of the omics approach is the technical limitations of the technology. For example, sequencing technologies are still developing and improving, and some platforms may not be as accurate as others.
c. **Data integration:** Integrating and analyzing heterogeneous data from multiple sources, including genomic data, clinical records, and lifestyle information, is a significant challenge in personalized diagnostics. This requires developing robust computational methods and tools for data integration and interpretation (Hood & Friend, 2011)
d. **Privacy and ethics:** The use of personal data in personalized diagnostics raises concerns about privacy, security, and ethical considerations. Safeguarding patient privacy and obtaining informed consent for data sharing and analysis is crucial (Brothers & Rothstein, 2015).
e. **Validation and clinical utility:** Demonstrating personalized diagnostic tests' clinical utility and validity is essential for their widespread adoption. Robust validation studies, including prospective clinical trials, are necessary to establish the reliability and accuracy of personalized diagnostic approaches (Nabi, 2022; Zhang et al., 2019).

5.2 Opportunities in personalized diagnostics

a. **Precision medicine:** Personalized diagnostics deliver precise and targeted therapies to individual patients, leading to better treatment outcomes. By identifying patient-specific biomarkers and molecular signatures, personalized diagnostics can guide treatment decisions and optimize therapy selection (Collins & Varmus, 2015).
b. **Early disease detection:** Personalized diagnostics have the potential to detect diseases at early stages, even before clinical symptoms appear. This can facilitate timely interventions and improve prognosis (Goetz & Schork, 2018).
c. **Patient engagement:** Personalized diagnostics empowers patients by involving them in decision-making processes and promoting active participation in their health care. Patient engagement can lead to increased adherence to treatment plans and better health outcomes (Siebert et al., 2015).

5.3 Future prospects for personalized diagnostics

Personalized diagnostics, which involves tailoring medical treatments and interventions to individual patients based on their unique characteristics, is a rapidly evolving field with promising future prospects. Here are some key areas that hold potential for personalized diagnostics, along with references for further reading.

a. **Genomics and genetic testing:** Advances in genomic sequencing technologies have identified genetic variations and their association with diseases. Genetic testing can provide valuable insights into an individual's susceptibility to certain conditions, response to specific treatments, and risk of adverse drug reactions. Additionally, pharmacogenomics aims to optimize drug selection and dosing based on an individual's genetic makeup (Manolio, 2013; Relling & Evans, 2015).
b. **Liquid biopsies and circulating biomarkers:** Liquid biopsies involve the analysis of blood or other body fluids to detect and monitor diseases. This noninvasive approach can provide information on genetic mutations, circulating tumor cells, circulating tumor DNA, and other biomarkers, allowing for early detection, monitoring of treatment response, and identification of resistance mechanisms (Diaz & Bardelli, 2014; Pantel & Alix-Panabières, 2019).
c. **Multiomics integration:** Integrating data from multiple omics platforms (genomics, transcriptomics, proteomics, and metabolomics) can provide a more comprehensive view of an individual's health status and disease mechanisms. By combining various omics data, researchers can identify novel biomarkers, molecular pathways, and therapeutic targets for personalized diagnostics and treatment (Olivier et al., 2019; Yu & Zeng, 2018).
d. **Artificial intelligence (AI) and machine learning:** AI and machine learning algorithms can analyze complex datasets, identify patterns, and generate predictive models to support personalized diagnostics. These technologies can aid in disease risk prediction, early diagnosis, treatment response prediction, and outcome forecasting, leading to more precise and tailored patient care (Esteva et al., 2017; Litjens et al., 2017; McKinney et al., 2020).

The omics approach has the potential to revolutionize personalized and diagnostic medicine, but several limitations need to be overcome before it can be widely adopted. The cost of the technology and the analysis, the technical limitations of the platforms, and the challenges of data analysis and interpretation must be addressed. Despite these limitations, the omics approach is a promising field that can improve patient care and advance our understanding of human biology.

This chapter provides a starting point for exploring the omics approaches for personalized and diagnostics medicine and its future prospects. However, given that the field is rapidly evolving, it is advisable to consult more recent literature and stay updated with the latest research advancements.

References

Ahmed, Zeeshan (2022). Multi-omics strategies for personalized and predictive medicine: Past, current, and future translational opportunities. *Emerging Topics in Life Sciences, 6*(2), 215–225. https://doi.org/10.1042/etls20210244, 2397-8554.

Bludau, I., & Aebersold, R. (2020). Proteomic and interactomic insights into the molecular basis of cell functional diversity. *Nature Reviews Molecular Cell Biology*. ISSN: 14710080, *21*(6), 327–340. https://doi.org/10.1038/s41580-020-0231-2. http://www.nature.com/molcellbio

Bonaguro, L., Schulte-Schrepping, J., Ulas, T., Aschenbrenner, A. C., Beyer, M., & Schultze, J. L. (2022). A guide to systems-level immunomics. *Nature Immunology*. ISSN: 15292916, *23*(10), 1412–1423. https://doi.org/10.1038/s41590-022-01309-9. https://www.nature.com/ni

Brothers, K. B., & Rothstein, M. A. (2015). Ethical, legal and social implications of incorporating personalized medicine into healthcare. *Personalized Medicine*. ISSN: 1744828X, *12*(1), 43–51. https://doi.org/10.2217/pme.14.65. http://www.futuremedicine.com/loi/pme;jsessionid=iDlhLxr7W4pgLWINGK

Castle, J. C., Loewer, M., Boegel, S., de Graaf, J., Bender, C., Tadmor, A. D., Boisguerin, V., Bukur, T., Sorn, P., Paret, C., Diken, M., Kreiter, S., Türeci, O., & Sahin, U. (2014). Immunomic, genomic and transcriptomic characterization of CT26 colorectal carcinoma. *BMC Genomics*. ISSN: 14712164, *15*(1). https://doi.org/10.1186/1471-2164-15-190. http://www.biomedcentral.com/1471-2164/15/190

Chakravarty, D., & Solit, D. B. (2021). Clinical cancer genomic profiling. *Nature Reviews Genetics*. ISSN: 14710064, *22*(8), 483–501. https://doi.org/10.1038/s41576-021-00338-8. http://www.nature.com/reviews/genetics

Cho, Y. E., Moon, P. G., Lee, J. E., Singh, T. S. K., Kang, W., Lee, H. C., Lee, M. H., Kim, S. H., & Baek, M. C. (2013). Integrative analysis of proteomic and transcriptomic data for identification of pathways related to simvastatin-induced hepatotoxicity. *Proteomics*. ISSN: 16159861, *13*(8), 1257–1275. https://doi.org/10.1002/pmic.201200368

Clish, Clary B. (2015). Metabolomics: An emerging but powerful tool for precision medicine. *Molecular Case Studies, 1*(1), a000588. https://doi.org/10.1101/mcs.a000588, 2373-2865.

Collins, F. S., & Varmus, H. (2015). A new initiative on precision medicine. *New England Journal of Medicine*. ISSN: 15334406, *372*(9), 793–795. https://doi.org/10.1056/NEJMp1500523. http://www.nejm.org/doi/pdf/10.1056/NEJMp1500523

Dehghan, A. (2018). Genome-wide association studies. *Methods in Molecular Biology*. ISSN: 10643745, *1793*, 37–49. https://doi.org/10.1007/978-1-4939-7868-7_4. http://www.springer.com/series/7651

Diaz, L. A., & Bardelli, A. (2014). Liquid biopsies: Genotyping circulating tumor DNA. *Journal of Clinical Oncology*. ISSN: 15277755, *32*(6), 579–586. https://doi.org/10.1200/JCO.2012.45.2011. http://jco.ascopubs.org/content/32/6/579.full.pdf+html

Esteva, A., Kuprel, B., Novoa, R. A., Ko, J., Swetter, S. M., Blau, H. M., & Thrun, S. (2017). Dermatologist-level classification of skin cancer with deep neural networks. *Nature*. ISSN: 14764687, *542*(7639), 115–118. https://doi.org/10.1038/nature21056. http://www.nature.com/nature/index.html

Fornito, A., Zalesky, A., & Breakspear, M. (2015). The connectomics of brain disorders. *Nature Reviews Neuroscience*. ISSN: 14710048, *16*(3), 159–172. https://doi.org/10.1038/nrn3901. http://www.nature.com/nrn/

Frésard, L., Smail, C., Ferraro, N. M., Teran, N. A., Li, X., Smith, K. S., Bonner, D., Kernohan, K. D., Marwaha, S., Zappala, Z., Balliu, B., Davis, J. R., Liu, B., Prybol, C. J., Kohler, J. N., Zastrow, D. B., Reuter, C. M., Fisk, D. G., Grove, M. E., … Montgomery, S. B. (2019). Identification of rare-disease genes using blood transcriptome sequencing and large control cohorts. *Nature Medicine*. ISSN: 1546170X, *25*(6), 911–919. https://doi.org/10.1038/s41591-019-0457-8. http://www.nature.com/nm/index.html

Ghosh, D., Bernstein, J. A., Hershey, G. K. K., Rothenberg, M. E., & Mersha, T. B. (2018). Leveraging multilayered "omics" data for atopic dermatitis: A road map to precision medicine. *Frontiers in Immunology*. ISSN: 16643224, *9*. https://doi.org/10.3389/fimmu.2018.02727. https://www.frontiersin.org/journals/immunology#

Goetz, L. H., & Schork, N. J. (2018). Personalized medicine: Motivation, challenges, and progress. *Fertility and Sterility*. ISSN: 15565653, *109*(6), 952–963. https://doi.org/10.1016/j.fertnstert.2018.05.006. http://www.elsevier.com/locate/fertnstert

Guo, Y., Wang, S., Chao, X., Li, D., Wang, Y., Guo, Q., & Chen, T. (2022). Multi-omics studies reveal ameliorating effects of physical exercise on neurodegenerative diseases. *Frontiers in Aging Neuroscience*. ISSN: 16634365, *14*. https://doi.org/10.3389/fnagi.2022.1026688. https://www.frontiersin.org/journals/aging-neuroscience#

Hackl, H., Stocker, G., Charoentong, P., Mlecnik, B., Bindea, G., Galon, J., & Trajanoski, Z. (2010). Information technology solutions for integration of biomolecular and clinical data in the identification of new cancer biomarkers and targets for therapy. *Pharmacology and Therapeutics*. ISSN: 01637258, *128*(3), 488–498. https://doi.org/10.1016/j.pharmthera.2010.08.012

Han, X. (2016). Lipidomics for studying metabolism. *Nature Reviews Endocrinology*. ISSN: 17595037, *12*(11), 668–679. https://doi.org/10.1038/nrendo.2016.98. http://www.nature.com/nrendo/index.html

Hasanzad, M., Sarhangi, N., Ehsani Chimeh, S., Ayati, N., Afzali, M., Khatami, F., Nikfar, S., & Aghaei Meybodi, H. R. (2022). Precision medicine journey through omics approach. *Journal of Diabetes and Metabolic Disorders*. ISSN: 22516581, *21*(1), 881–888. https://doi.org/10.1007/s40200-021-00913-0. http://www.jdmdonline.com/

Hasin, Y., Seldin, M., & Lusis, A. (2017). Multi-omics approaches to disease. *Genome Biology*. ISSN: 1474760X, *18*(1). https://doi.org/10.1186/s13059-017-1215-1. http://genomebiology.com/

Holzinger, A., Haibe-Kains, B., & Jurisica, I. (2019). Why imaging data alone is not enough: AI-based integration of imaging, omics, and clinical data. *European Journal of Nuclear Medicine and Molecular Imaging*. ISSN: 16197089, *46*(13), 2722–2730. https://doi.org/10.1007/s00259-019-04382-9. http://link.springer.com/journal/volumesAndIssues/259

Hood, L., & Friend, S. H. (2011). Predictive, personalized, preventive, participatory (P4) cancer medicine. *Nature Reviews Clinical Oncology*. ISSN: 17594782, *8*(3), 184–187. https://doi.org/10.1038/nrclinonc.2010.227

Karczewski, K. J., & Snyder, M. P. (2018). Integrative omics for health and disease. *Nature Reviews Genetics*. ISSN: 14710064, *19*(5), 299–310. https://doi.org/10.1038/nrg.2018.4. http://www.nature.com/reviews/genetics

Litjens, G., Kooi, T., Bejnordi, B. E., Setio, A. A. A., Ciompi, F., Ghafoorian, M., van der Laak, J. A. W. M., van Ginneken, B., & Sánchez, C. I. (2017). A survey on deep learning in medical image analysis. *Medical Image Analysis*. ISSN: 13618423, *42*, 60–88. https://doi.org/10.1016/j.media.2017.07.005. http://www.elsevier.com/inca/publications/store/6/2/0/9/8/3/index.htt

Manolio, T. A. (2013). Bringing genome-wide association findings into clinical use. *Nature Reviews Genetics*. ISSN: 14710064, *14*(8), 549–558. https://doi.org/10.1038/nrg3523

McKinney, S. M., Sieniek, M., Godbole, V., Godwin, J., Antropova, N., Ashrafian, H., Back, T., Chesus, M., Corrado, G. C., Darzi, A., Etemadi, M., Garcia-Vicente, F., Gilbert, F. J., Halling-Brown, M., Hassabis, D., Jansen, S., Karthikesalingam, A., Kelly, C. J., King, D., … Shetty, S. (2020). International evaluation of an AI system for breast cancer screening. *Nature*. ISSN: 14764687, *577*(7788), 89–94. https://doi.org/10.1038/s41586-019-1799-6. http://www.nature.com/nature/index.html

Mer, A. S., Ba-Alawi, W., Smirnov, P., Wang, Y. X., Brew, B., Ortmann, J., Tsao, M. S., Cescon, D. W., Goldenberg, A., & Haibe-Kains, B. (2019). Integrative pharmacogenomics analysis of patient-derived xenografts. *Cancer Research*. ISSN: 15387445, *79*(17), 4539–4550. https://doi.org/10.1158/0008-5472.CAN-19-0349. https://cancerres.aacrjournals.org/content/79/17/4539.full-text.pdf

Mock, A., Braun, M., Scholl, C., Fröhling, S., & Erkut, C. (2023). Transcriptome profiling for precision cancer medicine using shallow nanopore cDNA sequencing. *Scientific Reports*. ISSN: 20452322, *13*(1). https://doi.org/10.1038/s41598-023-29550-8. https://www.nature.com/srep/

Nabi, H. (2022). Personalized approaches for the prevention and treatment of breast cancer. *Journal of Personalized Medicine*. ISSN: 20754426, *12*(8). https://doi.org/10.3390/jpm12081201. http://www.mdpi.com/journal/jpm

Olivier, M., Asmis, R., Hawkins, G. A., Howard, T. D., & Cox, L. A. (2019). The need for multi-omics biomarker signatures in precision medicine. *International Journal of Molecular Sciences*. ISSN: 14220067, *20*(19). https://doi.org/10.3390/ijms20194781. https://www.mdpi.com/1422-0067/20/19/4781/pdf

Ota, M., & Fujio, K. (2021). Multi-omics approach to precision medicine for immune-mediated diseases. *Inflammation and Regeneration*. ISSN: 18808190, *41*(1). https://doi.org/10.1186/s41232-021-00173-8. http://inflammregen.biomedcentral.com/

Pabinger, S., Dander, A., Fischer, M., Snajder, R., Sperk, M., Efremova, M., Krabichler, B., Speicher, M. R., Zschocke, J., & Trajanoski, Z. (2014). A survey of tools for variant analysis of next-generation genome sequencing data. *Briefings in Bioinformatics*. ISSN: 14774054, *15*(2), 256–278. https://doi.org/10.1093/bib/bbs086

Pantel, K., & Alix-Panabières, C. (2019). Liquid biopsy and minimal residual disease—Latest advances and implications for cure. *Nature Reviews Clinical Oncology*. ISSN: 17594782, *16*(7), 409–424. https://doi.org/10.1038/s41571-019-0187-3. http://www.nature.com/nrclinonc/archive/index.html

Parello, Caitlin S. L. (2020). *Microbiomics* (pp. 137–162), ISBN 9780128137628. https://doi.org/10.1016/b978-0-12-813762-8.00006-2

Relling, M. V., & Evans, W. E. (2015). Pharmacogenomics in the clinic. *Nature*. ISSN: 14764687, *526*(7573), 343–350. https://doi.org/10.1038/nature15817. http://www.nature.com/nature/index.html

Rudd, P., Karlsson, N. G., Khoo, K. H., & Packer, N. H. (2015). Essentials of Glycobiology. In La Jolla (Ed.), *Glycomics and Glycoproteomics*. Cold Spring Harbor Laboratory Press. https://doi.org/10.1101/glycobiology.3e.051

Rusch, M., Nakitandwe, J., Shurtleff, S., Newman, S., Zhang, Z., Edmonson, M. N., Parker, M., Jiao, Y., Ma, X., Liu, Y., Gu, J., Walsh, M. F., Becksfort, J., Thrasher, A., Li, Y., McMurry, J., Hedlund, E., Patel, A., Easton, J., … Zhang, J. (2018). Clinical cancer genomic profiling by three-platform sequencing of whole genome, whole exome and transcriptome. *Nature Communications*. ISSN: 20411723, *9*(1). https://doi.org/10.1038/s41467-018-06485-7. http://www.nature.com/ncomms/index.html

Savage, S. R., & Zhang, B. (2020). Using phosphoproteomics data to understand cellular signaling: A comprehensive guide to bioinformatics resources. *Clinical Proteomics*. ISSN: 15590275, *17*(1). https://doi.org/10.1186/s12014-020-09290-x. http://www.clinicalproteomicsjournal.com/

Shi, Y. (2014). A glimpse of structural biology through X-ray crystallography. *Cell*. ISSN: 10974172, *159*(5), 995–1014. https://doi.org/10.1016/j.cell.2014.10.051. https://www.sciencedirect.com/journal/cell

Siebert, U., Jahn, B., Rochau, U., Schnell-Inderst, P., Kisser, A., Hunger, T., Sroczynski, G., Mühlberger, N., Willenbacher, W., Schnaiter, S., Endel, G., Huber, L., & Gastl, G. (2015). Oncotyrol—Center for personalized cancer medicine: Methods and applications of health technology assessment and outcomes research. *Zeitschrift fur Evidenz, Fortbildung und Qualitat im Gesundheitswesen*. ISSN: 18659217, *109*(4–5), 330–340. https://doi.org/10.1016/j.zefq.2015.06.012. http://www.elsevier.com/wps/find/journaldescription.cws_home/715741/description#description

Song, X., Ji, J., Gleason, K. J., Yang, F., Martignetti, J. A., Chen, L. S., & Wang, P. (2019). Insights into impact of DNA copy number alteration and methylation on the proteogenomic landscape of human ovarian cancer via a multi-omics integrative analysis. *Molecular and Cellular Proteomics*. ISSN: 15359484, *18*(8), S52–S65. https://doi.org/10.1074/mcp.RA118.001220. https://www.mcponline.org/content/mcprot/18/8_suppl_1/S52.full.pdf

Subramanian, Indhupriya, Verma, Srikant, Kumar, Shiva, Jere, Abhay, & Anamika, Krishanpal (2020). Multi-omics data integration, interpretation, and its application. *Bioinformatics and Biology Insights, 14*. https://doi.org/10.1177/1177932219899051, 1177-9322, 117793221989905.

Sun, Y. V., & Hu, Y. J. (2016). Integrative analysis of multi-omics data for discovery and functional studies of complex human diseases. *Advances in Genetics*. ISSN: 00652660, *93*, 147–190. https://doi.org/10.1016/bs.adgen.2015.11.004. http://www.elsevier.com/wps/find/bookdescription.cws_home/703716/description#description

Tam, V., Patel, N., Turcotte, M., Bossé, Y., Paré, G., & Meyre, D. (2019). Benefits and limitations of genome-wide association studies. *Nature Reviews Genetics*. ISSN: 14710064, *20*(8), 467–484. https://doi.org/10.1038/s41576-019-0127-1. http://www.nature.com/reviews/genetics

Tebani, A., Afonso, C., Marret, S., & Bekri, S. (2016). Omics-based strategies in precision medicine: Toward a paradigm shift in inborn errors of metabolism investigations. *International Journal of Molecular Sciences*. ISSN: 14220067, *17*(9). https://doi.org/10.3390/ijms17091555. http://www.mdpi.com/1422-0067/17/9/1555/pdf

Visscher, P. M., Brown, M. A., McCarthy, M. I., & Yang, J. (2012). Five years of GWAS discovery. *The American Journal of Human Genetics*. ISSN: 15376605, *90*(1), 7–24. https://doi.org/10.1016/j.ajhg.2011.11.029

Wang, K. C., & Chang, H. Y. (2018). Epigenomics technologies and applications. *Circulation Research*. ISSN: 15244571, *122*(9), 1191–1199. https://doi.org/10.1161/CIRCRESAHA.118.310998. http://circres.ahajournals.org

Wang, Z., Gerstein, M., & Snyder, M. (2009). RNA-seq: A revolutionary tool for transcriptomics. *Nature Reviews Genetics*. ISSN: 14710064, *10*(1), 57–63. https://doi.org/10.1038/nrg2484

Xiong, Q., Ancona, N., Hauser, E. R., Mukherjee, S., & Furey, T. S. (2012). Integrating genetic and gene expression evidence into genome-wide association analysis of gene sets. *Genome Research*. ISSN: 15495469, *22*(2), 386–397. https://doi.org/10.1101/gr.124370.111. http://genome.cshlp.org/content/22/2/386.full.pdf+html

Yu, X. T., & Zeng, T. (2018). Integrative analysis of omics big data. *Methods in Molecular Biology*. ISSN: 10643745, *1754*, 109–135. https://doi.org/10.1007/978-1-4939-7717-8_7. http://www.springer.com/series/7651

Zhang, X., Yang, H., & Zhang, R. (2019). Challenges and future of precision medicine strategies for breast cancer based on a database on drug reactions. *Bioscience Reports*. ISSN: 15734935, *39*(9). https://doi.org/10.1042/BSR20190230. http://www.bioscirep.org/content/39/9/BSR20190230.full-text.pdf

Chapter 12

Role of bioinformatics in genome analysis

Sarika Sahu[1], Puru Supriya[2], Soumya Sharma[1], Aalok Shiv[3] and Dev Bukhsh Singh[4]
[1]ICAR-Indian Agricultural Statistics Research Institute, New Delhi, India; [2]ICAR-National Academy of Agricultural Research Management, Hyderabad, Telangana, India; [3]ICAR-Indian Institute of Sugarcane Research, Lucknow, Uttar Pradesh, India; [4]Department of Biotechnology, Siddharth University, Kapilvastu, Uttar Pradesh, India

1. Applications of bioinformatics

This era of genomics has been witnessing a huge explosion in the quantity of biological information available, mainly due to advances in genome sequencing methodologies. The proteome data available are also expanding at a rapid pace due to the advancement in mass spectrometry and X-ray crystallography techniques. Bioinformatics is a tool to study those biological data, analyze it, and has a huge potential in providing significant inputs for the researchers. Bioinformatics has advanced analytical tools that can be used to search for the agriculturally important genes and proteins within genomes, and further to elucidate their mechanisms and functions. This specific knowledge could be used to develop biotic and abiotic stress tolerant and resistant crops. Now, bioinformatics has spread its wings in different areas such as genomics, proteomics, metabolomics, etc.

1.1 Genotyping by sequencing

Genotyping by sequencing (GBS) is revolutionizing the field of crop genotyping as it allows high-throughput detection of genetic variants for specific agronomic traits. Due to the increased popularity of these approaches, single nucleotide polymorphisms (SNPs) are gaining the status of marker of choice for genotyping. GBS methods are widely applied for crop genotyping, as SNPs serves for practical molecular breeding applications. As there is an increase in the cost of sequencing technology, more and more data will be available for many agricultural crops. This will eventually help researchers in understanding the impact of climate change over crop adaptation (Scheben et al., 2016).

1.2 Genome editing

Genome editing (GE) aids in alteration/replacement of specific nucleotides in the genome of any individual. Thus, many different and new varieties could be developed in a shorter time period compared to traditional breeding approaches. Further, GE is highly beneficial for creating targeted variations (Scheben & Edwards, 2017). The most important fact about GE is that the outcome of GE is not merely a genetically modified organism (GMO) according to the scientific community (Huang et al., 2016). Thus, GE is much more superior and accurate than genetic engineering (Georges & Ray, 2017).

1.3 Transcriptomics analysis

Transcriptome analysis can be studied with or without reference genome (Martin & Wang, 2011). Currently, transcriptomics is revolutionized to enhance plant and animal gene expressions. The expression of genes alters during the developmental stages, biotic and abiotic stress conditions. Transcriptome analysis has a significant role in interpreting the function of gene, protein, and their interactions. It has major role in the identification and generation of tissue-specific transcripts from the genome of plants and animals (mRNAs, small RNAs, and noncoding RNAs) in addition to the

Integrative Omics. https://doi.org/10.1016/B978-0-443-16092-9.00012-6

identification of SNPs. RNA-seq or transcriptome analysis is efficient and cost effective to study expression studies, mutant, and structural alteration.

1.4 Epigenetics

Environmental changes affecting adaptive responses (drought stress, food accessibility, etc.) can trigger physiological changes in plants and animals that affect their viability and reproductive fitness. Wang et al. (2009), Park (2009) help to identify genome modifications such as genome DNA methylation, modifications in chromatin architecture, and small RNA expression studies. Scientists can better comprehend epigenetic variables that contribute controlling changes and other characters in desired species through RNA-seq.

1.5 Targeted resequencing

Targeted resequencing provides information of the particular gene of interest (exome) identified from association mapping studies. Illumina consists of two procedures for target enrichment and amplicon generation (Lo & Chiu, 2009). This strategy is more efficient and economical as genomics regions (SNPs, copy number variations [CNVs], etc.) that show genetic variation only will be sequenced over a big number of samples. These variants may illustrate beneficial mutations that help to notify breeding choice as well as causative mutations for plant or animal disease and susceptibility to different parasites.

1.6 Whole-genome sequencing

Whole-genome sequencing (WGS) is a method for exploring the entire genomes of a cell at a given time of location. Further, the WGS is commonly used to identify the genetic variations among the species. There is various sequencing technology based on the read lengths like short-read sequencing and long-read sequencing. These sequencing technologies have their own pros and cons.

1.6.1 Short reads sequencing

Short reads (less than 300 bp in length) produced by second-generation sequencing are extremely accurate (the sequencing error rate is less than 1%). Compared to Sanger sequencing, short read sequencing methods have made sequencing much simpler, faster, and less expensive. The short-read sequencing technologies are Illumima, 454 pyrosequencing, Ion Torrent, and SOLiD.

1.6.2 Mate-pair sequencing

Mate-pairs refer to methodologies that give information about two reads belonging to a pair. The basic idea involves shearing DNA into random fragments of a selected size, called the insert length, and then sequencing both ends of each fragment. Paired-end sequencing sequences both the forward and reverse template strands of the same DNA molecule. The distance between paired-end reads is limited up to 300 bp (200–600 bp). In mate-pair sequencing, tags that are sequence belong to ends of a much larger molecule, typically between 2 and 10 kb.

1.7 Long reads sequencing

The third-generation sequencing sequences long reads (>400 bp) and is often referred to as long read sequencing (LRS). Single DNA molecules can be sequenced using LRS technology without amplifying them first. It is frequently challenging to produce lengthy continuous consensus sequence using NGS due to the challenge of finding overlaps between NGS short reads, which affects the overall quality of assembly. Having access to long reads is therefore a significant advantage. The short-read sequencing technologies are Pacific Biosciences and Oxford Nanopore.

1.8 Next-generation sequencing

The NGS is a fundamentally unique sequencing methodology that has produced a number of ground-breaking findings and sparked a new revolution in genomic research by providing boundless insight into the genome, transcriptome, and epigenome of any species. As a result, the welfare of human society has undergone a new revolution thanks to NGS technology. NGS expands the concept of capillary electrophoreses (CEs)-based Sanger sequencing in order to carry out massive parallel sequencing, in which millions of DNA fragments from a single sample are precisely sequenced. With the NGS, a long stretch of DNA base pairs can be sequenced, yielding hundreds of gigabases of data in a single sequential run.

1.8.1 First-generation sequencing

Sequencing by synthesis, invented by Sanger et al. (1977), and sequencing by cleavage, invented by Maxam and Gilbert (1977), are examples of first-generation sequencing technologies. The Human Genome Project (HGP), as well as numerous other animal and plant genomes, were completed using the conventional automated Sanger sequencing method, commonly known as first-generation sequencing. However, new technologies (NGS) have been developed to replace the automated Sanger method in the late 20th and early 21st centuries due to its high cost and labor-intensive process. Before 2008, Sanger sequencing was the industry standard for biomedical research. The preferred method for detection by automated CE platforms, commercially available from Applied Biosystems Inc., Life Technologies Inc., and Beckman Coulter Inc., has been standard four-color fluorescent labeling, where each color corresponds to one of the four DNA bases. Craig Venter's whole diploid genome was the first to be sequenced using Sanger's approach in 2007 (Levy et al., 2007). Sanger sequencing-CE platforms will probably continue to be heavily used for targeted sequencing projects (biomarker identification and pathway analysis) and clinical diagnostic applications until small-scale NGS platforms become affordable and quick enough; this is a rapidly developing area of industrial development. Although sequencing tasks in large comprehensive research projects have now moved to NGS platforms. NGS techniques are positioned to become the preeminent genomics technology due to their vastly enhanced cost effectiveness when compared to Sanger sequencing and their wide range of applications (Morozova & Marra, 2008).

1.8.2 Second-generation sequencing

The Roche 454 pyrosequencing method, reversible terminator sequencing by Illumina, sequencing by ligation of ABI/SoLiD, and single-molecule sequencing by Helicos are examples of second-generation sequencing. To achieve clonal amplification of the target sequence, Roche454 employs emulsion polymerase chain reaction (PCR). Numerous picoliter-volume wells, each containing a single bead and sequencing enzymes, are included throughout the sequencing machine. For the purpose of detecting each individual nucleotide integrated into the developing DNA, pyrosequencing uses luciferase to produce light (Margulies et al., 2005). Cluster target sequence amplification is used by Illumina (Solexa) on a solid surface. Four different nucleotide types, each tagged by one of four fluorophores and including a 30 reversible terminator, are added to create a sequence. The Illumina method can only lengthen DNA one nucleotide at a time, in contrast to pyrosequencing. The fluorophore and the 30 reversible terminators are chemically removed from the DNA molecule once a fluorescent image of the integrated nucleotide is captured (Mardis, 2008). This allows the subsequent cycle to begin. The SOLiD technique from Applied Biosystem/Life Technologies uses ligation reaction for sequencing and a library containing all conceivable oligonucleotides of a set length that are tagged in accordance with the position of the sequence. "True single molecule sequencing" technology is used by the HeliScope sequencer (Thompson & Milos, 2011). Attaching DNA fragments and additional polyA tail adapters to the flow cell surface is the first step in extension-based sequencing, which is followed by cyclic washings of the flow cell with fluorescently labeled nucleotides in a manner reminiscent of Sanger sequencing. The throughput, read-length, and operational costs of the second-generation sequencing platforms vary widely. They are typically very expensive devices with high throughput, costing between 0.5 million and 1 million US dollars. Either fluorescent labeling or pyrophosphate chemical conversion, both of which need optical detection, is the signal recording technique. Despite being successful in many research applications, second-generation sequencing systems experience a variety of drawbacks, including high instrument costs, difficult sample preparation processes, instrumentation challenges, and read-length restrictions (Fuller et al., 2009).

1.8.3 Third-generation sequencing

The novel chemistry, shorter operation times, and lower operation costs define the third-generation sequencing systems. PacBioRS is a 1000 bp real-time single molecule, single polymerase-sequencing tool. Each chip contains so-called zero-mode wave guided (ZMW) nanostructures with 100-nm holes inside of which DNA polymerase carries out sequencing by synthesizing phosphor-coupled nucleotides tagged with fluorophores and introducing them sequentially (Eid et al., 2009). Personal genomic machines (PGMs), a benchtop instrument made available to research and clinical laboratories, are perhaps the most adaptable and affordable approach now available. Each time DNA polymerase incorporates a nucleotide, proton release occurs as part of the sequencing chemistry of the Ion Torrent method. DNA polymerase is able to interact with clonally amplified target DNA fragments thanks to the dense microarray of individual microwells. The system does not include optical detection or nucleotide labeling. The sequencing capacity of Ion Torrent's PGM, which costs less than $100K, is sufficient for small-scale research projects or clinical diagnostic laboratories. With its release onto the market, NGS has officially begun to become a commodity for use in biological and clinical applications. It has multiplex barcoding adaptors, a standard component of many other systems, which enable the examination of many samples simultaneously. Despite the labor-intensive nature of the current Ion Torrent operation, automation with one-step library preparation is now possible, which will streamline the procedure. The drawbacks include short read-lengths (100−200 bp) and technical challenges when reading through homopolymers and highly repetitive sequences, both of which have recently undergone improvement.

2. Assembly

With the advent of state-of-art sequencing technologies, the reads short-reads/longs were assembled and arranged at the chromosomes level by using de Bruijn graph algorithm (Fig. 12.1).

Overview of various approach of WGS assembly and meta-analysis of assembled data are also shown in Fig. 12.2.

The following three main steps are required for the de novo assembly:

1. Contig assembly: long consensus sequence also known as contigs, will be formed without gap from the reads.
2. Scaffolding: the contigs are joined together by pair-end reads and form scaffolding with several gap gaps.
3. Gap filling: the gaps in the scaffold are filled by other independent reads.

3. Identification of mutants

Identification of naturally occurring mutant as well as development of novel mutant for desirable traits and their utilization has always been the practice in crop improvement program (Samantara et al., 2021). With the advancement of sophisticated NGS techniques and bioinformatics tools, throughput of identification of mutant as well as their molecular basis has been increased. Several novel methods of mutant identification utilized in different crop species are summarized below.

3.1 Next-generation sequencing approach

DNA sequencing involves determining the exact order of nucleotides (A, C, G, T) in a DNA molecule. By comparing the sequence of a wild-type DNA sample to a sample from a mutated individual, mutations can be identified. NGS is a technology that enables the parallel sequencing of millions of DNA fragments. This high-throughput approach allows for the identification of mutations at a much faster pace and with greater accuracy compared to traditional sequencing methods (Sandhu et al., 2022; Patel et al., 2022). To identify mutations using NGS, the following steps are typically performed:

(a) DNA library preparation: The DNA sample is fragmented into smaller pieces, and the ends are repaired and modified to add adapter sequences (Andrews, 2010).
(b) Sequencing: The DNA library is then amplified and sequenced using NGS technology, producing millions of short reads.

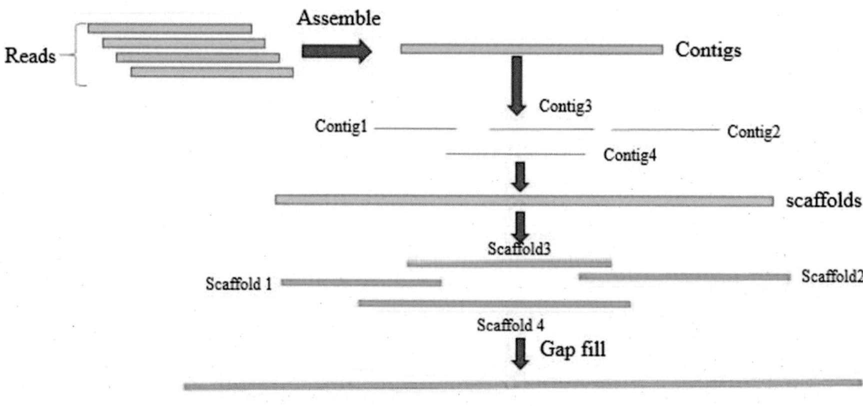

FIGURE 12.1 Workflow of de novo assembly method.

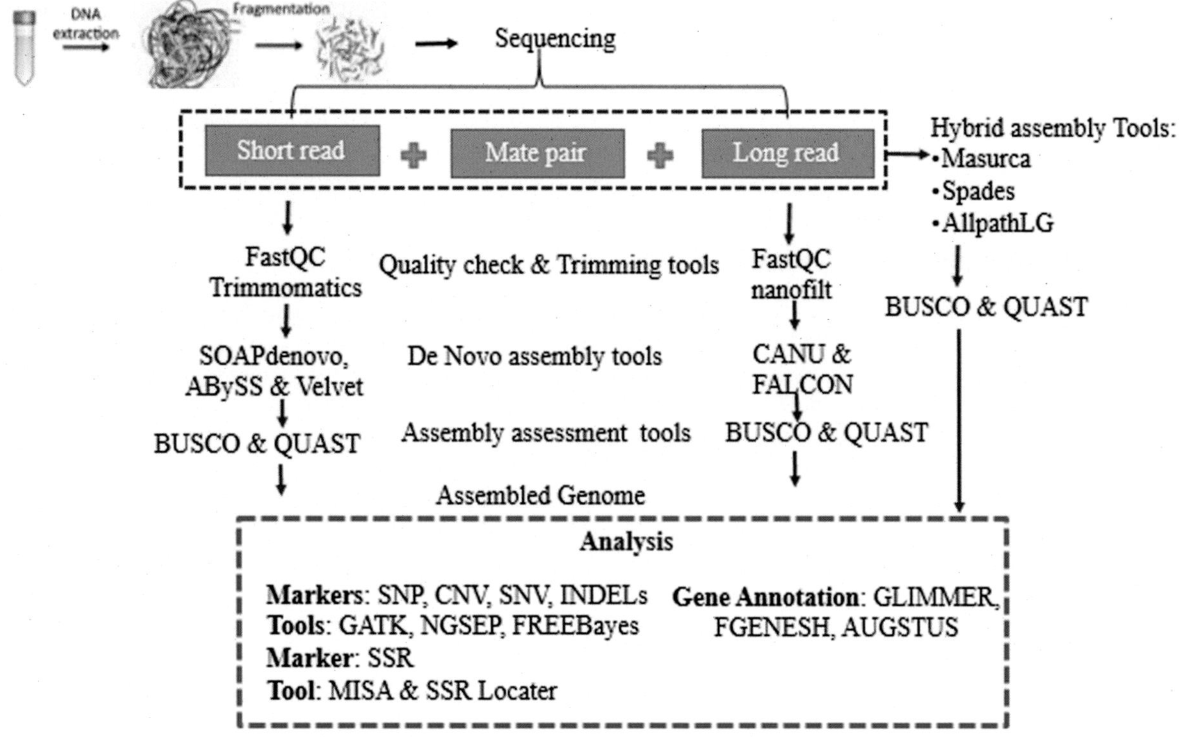

FIGURE 12.2 Overview of various approach of WGS assembly and meta-analysis from the assembled data.

(c) Read mapping: The reads are aligned to a reference genome to determine their location.
(d) Variant calling: The aligned reads are used to identify variations from the reference genome, including SNPs and insertions/deletions (indels).
(e) Annotation and interpretation: The variants identified are annotated and classified based on their potential impact on the gene or protein function. Further analysis is performed to determine the pathogenicity and clinical significance of the mutations.

NGS has been widely used for mutation identification in several crop species for various traits of economic importance (Patel et al., 2022, 2012).

3.2 The MutMap approach

It is a method for identifying mutant loci in plants using a mapping population derived from a cross between a mutant and a wild-type parent. The goal of MutMap is to identify the genomic location of a mutation that causes a phenotypic change of interest, such as changes in growth, development, or stress tolerance (Tran et al., 2020). The MutMap approach involves the following steps:

(a) Generation of mapping population: A mapping population is generated by crossing the mutant with a wild-type parent, and the progeny is self-fertilized to generate a segregating population.
(b) Phenotypic analysis: The progeny is screened for the phenotypic trait of interest, and the plants that display the mutant phenotype are identified.
(c) DNA extraction and genotyping: DNA is extracted from the mutant and wild-type plants and genotyped using a variety of methods, such as amplified fragment length polymorphism (AFLP) or SNP analysis.
(d) Linkage analysis: The genotypic data are analyzed to identify the genomic regions that are linked to the mutant phenotype. This information can be used to generate a linkage map that provides a rough estimate of the position of the mutant locus on the genome.
(e) Fine mapping: The linked regions are further refined using additional methods, such as backcrossing, to generate a more accurate map of the mutant locus.

(f) Identification of causative mutation: The candidate locus is sequenced to identify the exact mutation responsible for the phenotype of interest.

The MutMap approach provides a powerful tool for the identification of mutant loci and has been widely used to study a variety of important traits in crops, such as disease resistance and stress tolerance. The approach can also be combined with other methods, such as TILLING or EcoTILLING, to enhance the accuracy and efficiency of mutation identification (Manchikatla et al., 2021).

4. Visualization tools for genomic data

The genome data visualization is an integrative approach of computational analysis of data. The plethora of sequencing data are available in the public domain and intensive analysis reveals several pattern and signature sequences. The visualization of genome data will help to understand the new trends and diversity of the genomic data. The visualization tool is used to study the SNPs, mutations, location of genes, etc. There are several genomic data visualization tools available like Circular Genome Viewer and Integrative Genomics Viewer.

Integrative Genomics Viewer is an open-source software available at (https://software.broadinstitute.org/software/igv/home). It supports a variety of genomic file formats like GFF, BED, WIG, BAM, and TDF. The GFF, BED, and WIG are simple flat files and easily visualize the small data set. The sequence alignment file like BAM is widely used to visualize the SNPs and expressed genes.

5. Pipeline for the transcriptome data analysis

Transcriptional profiling using RNA-seq technology has emerged as an alluring substitute for conventional microarray platforms. RNA-seq analysis is based on NGS data. The massive amounts of data produced by NGS research cannot be directly or meaningfully evaluated. To extract meaningful information from RNA-seq data and prevent bias or misinterpretation, it is essential to choose the proper analytical strategy and collection of bioinformatics tools. Most data files obtained from sequencing platforms are compressed with ".fastq.gz" extension. These fastq files contain structured information including the called bases, associated quality scores, and a unique identifier for each individual NGS read.

A variety of artifacts that develop throughout the library preparation and sequencing operations can have an adverse effect on NGS and the quality of the raw data used in subsequent studies. Platform-specific error profiles, systematic quality score variation across the sequence read, base composition-driven biases in sequence generation, deviations from ideal library fragment sizes, variations in the proportions of duplicate sequences brought on by PCR amplification bias, and contamination from known and unknown species other than the sequencing target are some of these problems (Schmieder & Edwards, 2011). Therefore, preprocessing of NGS data before stepping to further data analysis is essential (Figs. 12.3 and 12.4).

5.1 Preprocessing

Numerous software program have been released that can identify problems with the quality of NGS data, including poor base quality, adaptor sequence contamination, and base composition biases (Andrews, 2010; Lohse et al., 2012; Patel et al., 2012). In the early stages of the quality control (QC) process, metrics produced by the sequencing platform (such as quality scores) or calculated directly from the raw reads are often used to evaluate the intrinsic quality of the raw reads (e.g., base composition). FastQC (http://www.bioinformatics.babraham.ac.uk/projects/fastqc/) is one of the most used tools for creating these quality indicators. FastQC and other tools of a similar nature are frequently used in NGS data production systems as an initial QC checkpoint and are helpful for evaluating the overall quality of a sequencing run. Mapping the raw data to a known reference and using alignment profiles to derive various metrics are additional standard QC stages. These comprise estimations of the library insert sizes, degrees of fragment or sequence duplication, and the mapping rate to the anticipated target. When a reliable reference is available, these measures are frequently generated for NGS data derived from model organisms and are typically included in QC reports. Other techniques, such as CD-HIT (Fu et al., 2012), FASTX-Toolkit (http://hannonlab.cshl.edu/fastx_toolkit/), and PRINSEQ (Schmieder & Edwards, 2011), can predict the rate of read duplication without a reference, but they have substantial limitations.

After checking the quality of reads, poor-quality regions and adapter sequences should be trimmed from the reads before further analysis. Trimmomatic (Bolger et al., 2014) can be used for trimming the low-quality reads and adapter sequences.

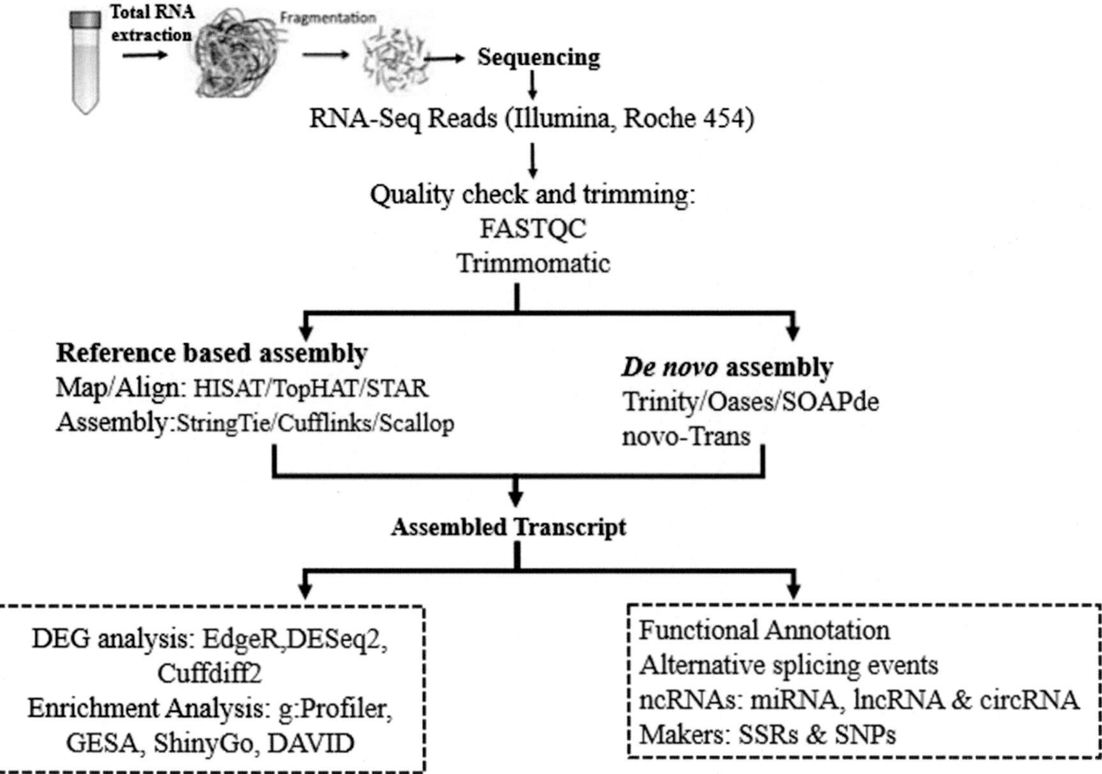

FIGURE 12.3 Schematic pipeline for transcriptome data analysis and meta-analysis.

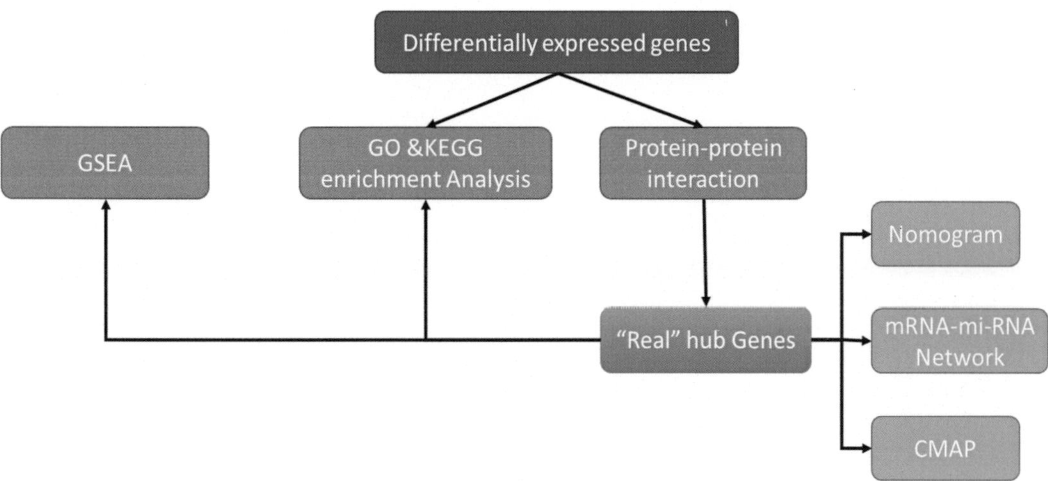

FIGURE 12.4 Enrichment analysis of differentially expressed genes.

5.2 Transcriptome assembly

Next-generation sequencers such as Illumina are capable of producing hundreds of millions of short (less than 100 bp) RNA-seq reads. Except in cases with very short exons, reads of this length often do not cover beyond two exons. To identify gene isoforms and find new genes, it is essential to reconstruct full-length transcripts by putting the short readings together. There are two basic approaches to assemble the RNA-seq reads: De novo and reference-guided assembly. List of tools used for RNA-seq data analysis are given in Table 12.1.

TABLE 12.1 List of popular tools for RNA-Seq data analysis.

Tools	URL	Input data	References
Preprocessing tools			
Quality check			
FASTQC	https://www.bioinformatics.babraham.ac.uk/projects/fastqc/	FASTA/FASTQ	Andrews (2010)
CD-HIT	http://cd-hit.org.	FASTA/FASTQ	Fu et al. (2012)
FASTX-Toolkit	http://hannonlab.cshl.edu/fastx_toolkit/	FASTA/FASTQ	Assaf and Hannon (2010/2010)
PRINSEQ	https://prinseq.sourceforge.net/	FASTA/FASTQ	Schmieder and Edwards (2011)
Trimming			
Trimmomatic	http://www.usadellab.org/cms/?page=trimmomatic	FASTQ	Bolger et al. (2014)
Sickle	https://github.com/najoshi/sickle	FASTQ	Bolger et al. (2014)
RSeQC	https://rseqc.sourceforge.net/	SAM and BAM files	Wang et al. (2012)
Cutadapt	https://github.com/marcelm/cutadapt	FASTA/FASTQ	Martin (2011)
Transcriptome assembly			
References based			
Map/align the reads to reference genome			
TopHat	https://ccb.jhu.edu/software/tophat/index.shtml	FASTA/FASTQ	Trapnell et al. (2013)
HISAT	https://daehwankimlab.github.io/hisat2/	FASTA/FASTQ	Kim et al. (2015)
STAR 9	https://github.com/alexdobin/STAR	FASTA/FASTQ	Dobin et al. (2013)
GSNAP	https://bio.tools/gsnap	FASTQ; FASTA; .txt	Wu and Nacu (2010)
Assemble the mapped reads			
Cufflinks	https://cole-trapnell-lab.github.io/cufflinks/	SAM, BAM	Trapnell et al. (2013)
Bayesembler	https://github.com/bioinformatics-centre/bayesembler	BAM	Trapnell et al. (2013)
TransComb	https://sourceforge.net/projects/transcriptomeassembly/files/	BAM	Liu, Li, et al. (2016); Liu, Yu, et al. (2016)
StringTie	https://ccb.jhu.edu/software/stringtie/	SAM, BAM, or CRAM	Pertea et al. (2015)
Scallop	https://github.com/Kingsford-Group/scallop	BAM	Shao and Kingsford (2017)

De novo assembly			
Trinity	https://github.com/trinityrnaseq	FASTA/FASTQ	Grabherr et al. (2011)
Oases	www.ebi.ac.uk/~zerbino/oases/	FASTA/FASTQ	Schulz et al. (2012)
SPAdes	https://bioinf.spbau.ru/spades	FASTA, FASTQ, BAM	Bankevich et al. (2012)
BinPacker	https://sourceforge.net/projects/transcriptomeassembly/files/BinPacker_1.0.tar.gz/download	FASTA/FASTQ	Liu, Li, et al. (2016); Liu, Yu, et al. (2016)
ABySS	https://www.bcgsc.ca/platform/bioinfo/software/abyss	FASTA/FASTQ	Simpson et al. (2009)
SOAPdenovo-Trans	https://sourceforge.net/projects/soapdenovotrans/	FASTA/FASTQ	Xie et al. (2014)
Differentially expressed gene analysis			
EdgeR	https://bioconductor.org/packages/release/bioc/html/edgeR.html	Count data	Robinson et al. (2013)
DESeq2	https://bioconductor.org/packages/release/bioc/html/DESeq2.html	Count data	Love et al. (2014)
CuffDiff2	https://cole-trapnell-lab.github.io/cufflinks/cuffdiff/	GFF, GTF, BAM, SAM	Trapnell et al. (2013)
Enrichment analysis			
g:Profiler	https://biit.cs.ut.ee/gprofiler/	List of genes/proteins	Raudvere et al. (2019)
Gene Set Enrichment Analysis (GSEA)	https://www.gsea-msigdb.org/gsea/index.jsp	Expression dataset file (res, gct, pcl, or txt) Phenotype labels file (cls) Gene sets file (gmx or gmt) Chip (array) annotation file (chip)	Shi and Walker (2007)
EnrichmentMap	https://apps.cytoscape.org/apps/enrichmentmap	GSEA results	Merico et al. (2010)
ShinyGo	https://bioinformatics.sdstate.edu/go/	A list of gene ids	Sherman et al. (2022)
DAVID	https://david.ncifcrf.gov/	Gene list	Sherman et al. (2022)
agriGO	https://systemsbiology.cpolar.cn/agriGOv2/#:~:text=for%20Agricultural%20Community-,AgriGO%20v2.,the%20realm%20of%20ontology%20analyses	A list of sequence identifiers	Du et al. (2010)

5.2.1 De novo assembly

De novo transcriptome assembly is the process of assembling a transcriptome from scratch without using a reference genome. De novo transcriptome assemblers like Trinity (Grabherr et al., 2011) and Oases (Schulz et al., 2012) look for read overlaps and attempt to chain them together into complete transcripts. The presence of paralogous genes and transcripts with several isoforms that mainly overlap one another complicates this process, and as a result, this method results in severely fragmented data. The workflow in shown in Fig. 12.3.

After transcriptome assembly, transcriptome quantification is done to know the differences in the expression of alternative transcripts under different conditions. The quantification task typically consists of two steps: (1) mapping reads to a reference genome or transcript set and (2) estimating the abundances of the genes and isoforms based on the read mappings.

5.2.2 Reference-based assembly

When a model organism with a sequenced genome for the target transcriptome is available, reference-based transcriptome assembly is frequently used. In order to recreate the transcriptome, reads are mapped to previously known sequences. The reference genome is matched with the short reads, enabling the assembly of transcripts from overlapping areas. Cufflinks (Trapnell et al., 2009), Bayesembler (Maretty et al., 2014), StringTie (Pertea et al., 2015), TransComb (Liu, Yu, et al., 2016), and Scallop (Shao & Kingsford, 2017) are examples of reference-guided assemblers that make use of an existing genome to which the RNA-seq reads are first matched using a spliced aligner like HISAT (Kim et al., 2015) or STAR 9 (Dobin et al., 2013). Based on the alignments, these assemblers can create splice graphs (or other data structures), then use those graphs to create individual transcripts. After transcriptome assembly, the transcripts are quantified resulting in count data to know the differences in the expression of alternative transcripts under different conditions.

5.3 Normalization

Normalization is a crucial step in an RNA-seq analysis, which involves adjusting raw data to take into account the factors that hinder direct comparison of expression values. The raw counts data obtained after transcript quantification are normalized by the total number of reads in the dataset and by the length of individual transcripts. CPM (counts per million reads), RPKM (reads per kilobase per million reads), FPKM (fragments per kilobase per million reads), and TPM (transcripts per million reads) are a few of the normalization techniques that are frequently utilized.

1. RPM or CPM (reads per million mapped reads or counts per million mapped reads)

RPM (also known as CPM) is a basic gene expression unit that normalizes only for sequencing depth (depth-normalized counts). The RPM is biased in some applications where the gene length influences gene expression, such as RNA-seq.

$$\text{RPM} = \frac{\text{number of reads mapped to the gene} * 10^6}{\text{total number of mapped reads}}$$

2. RPKM (reads per kilo base of transcript per million mapped reads)

RPKM is a gene expression unit that measures the expression levels (mRNA abundance) of genes or transcripts. RPKM is a gene length normalized expression unit that is used for identifying the differentially expressed genes by comparing the RPKM values between different experimental conditions.

$$\text{RPKM} = \frac{\text{number of reads mapped to the gene} * 10^3 * 10^6}{\text{total number of mapped reads} * \text{gene length in bp}}$$

3. FPKM (fragments per kilo base of transcript per million mapped fragments)

FPKM is a gene expression unit which is analogous to RPKM. FPKM is used especially for normalizing counts for paired-end RNA-seq data in which two (left and right) reads are sequenced from the same DNA fragment.

4. TPM (transcripts per million)

TPM is suitable for sequencing protocols where reads sequencing depends on gene length. TPM is proposed as an alternative to RPKM because of inaccuracy in RPKM measurement.

$$\text{TPM} = \frac{\text{A} * 10^6}{\sum \text{A}}$$

where $\text{A} = \frac{\text{total reads mapped to the gene} * 10^3}{\text{length of gene in bp}}$

6. Analysis of differential gene expression

A number of computational techniques, such as EdgeR (Robinson et al., 2013), DESeq2 (Love et al., 2014), CuffDiff2 (Trapnell et al., 2013), etc., can be used to find genes that differ in their expression between sample groups. These tools estimate genes or transcripts that exhibit statistically significant differences in gene expression between the samples being compared, which typically represent various biological conditions, by counting NGS reads over individual genes and transcripts across the genome.

7. Gene enrichment analysis

Computational functional analysis based on accumulated biological information has aided in the biological interpretation of DEGs. The result is databases like the Kyoto Encyclopedia of Genes and Genomes (KEGG) (Kanehisa & Goto, 2000) that help put together the functional categories with the highest levels of enrichment, such as pathways. Functional enrichment tools can be divided into two kinds based on the input type. The more conventional approach is to use statistical techniques involving contingency tables to examine the enrichment of each annotated gene set using a list of previously chosen "interesting" genes. The alternative class, which ranks all expressed genes according to the intensity of expression differential and uses Kolmogorov–Smirnov-like tests to determine enrichment significance. Researchers can get mechanistic insight into gene lists produced by transcriptome experiments with the aid of pathway enrichment analysis. Software like g:Profiler (Raudvere et al., 2019), Gene Set Enrichment Analysis (GSEA) (Shi & Walker, 2007), Cytoscape (Shannon et al., 2003), and EnrichmentMap (Daniele Merico et al., 2010) are some publicly available and constantly updated tools for enrichment analysis. The gene enrichment analysis is shown in Fig. 12.4.

8. Conclusion

Recent advances in the field of biotechnology have significantly enhanced the speed of genome sequencing. A vast amount of experimentally validated structural and functional information related to genes and genomes are stored in the databases, which provides the basic foundation for comparison and annotation the genomes of other organism. Several software/servers and tools have been developed for the purpose of sequence data processing, genome assembly, contig assembly and mapping, transcriptome analysis, enrichment analysis, differential expression analysis, genome visualization, comparison, and mutational analysis. These genomic analysis tools have a good accuracy, and it is expected that more automated and integrated tools with high precision may be developed. These biotechnological and computational developments may be helpful in mining and cracking the many hidden facts stored in genomic biodiversity.

References

Andrews, S. (2010). *FastQC: A quality control tool for high throughput sequence data.*

Assaf, G., & Hannon, G. J. (2010). *FASTX-toolkit.*

Bankevich, A., Nurk, S., Antipov, D., Gurevich, A. A., Dvorkin, M., Kulikov, A. S., Lesin, V. M., Nikolenko, S. I., Pham, S., Prjibelski, A. D., Pyshkin, A. V., Sirotkin, A. V., Vyahhi, N., Tesler, G., Alekseyev, M. A., & Pevzner, P. A. (2012). SPAdes: A new genome assembly algorithm and its applications to single-cell sequencing. *Journal of Computational Biology, 19*(5), 455–477. https://doi.org/10.1089/cmb.2012.0021

Bolger, A. M., Lohse, M., & Usadel, B. (2014). Trimmomatic: A flexible trimmer for Illumina sequence data. *Bioinformatics, 30*(15), 2114–2120. https://doi.org/10.1093/bioinformatics/btu170. http://bioinformatics.oxfordjournals.org/

Dobin, A., Davis, C. A., Schlesinger, F., Drenkow, J., Zaleski, C., Jha, S., Batut, P., Chaisson, M., & Gingeras, T. R. (2013). Star: Ultrafast universal RNA-seq aligner. *Bioinformatics, 29*(1), 15–21. https://doi.org/10.1093/bioinformatics/bts635

Du, Z., Zhou, X., Ling, Y., Zhang, Z., & Su, Z. (2010). agriGO: A GO analysis toolkit for the agricultural community. *Nucleic Acids Research, 38*(2), W64–W70. https://doi.org/10.1093/nar/gkq310

Eid, J., Fehr, A., Gray, J., Luong, K., Lyle, J., Otto, G., Peluso, P., Rank, D., Baybayan, P., Bettman, B., Bibillo, A., Bjornson, K., Chaudhuri, B., Christians, F., Cicero, R., Clark, S., Dalal, R., DeWinter, A., Dixon, J., … Turner, S. (2009). Real-time DNA sequencing from single polymerase molecules. *Science, 323*(5910), 133–138. https://doi.org/10.1126/science.1162986

Fu, L., Niu, B., Zhu, Z., Wu, S., & Li, W. (2012). CD-HIT: Accelerated for clustering the next-generation sequencing data. *Bioinformatics, 28*(23), 3150–3152. https://doi.org/10.1093/bioinformatics/bts565

Fuller, C. W., Middendorf, L. R., Benner, S. A., Church, G. M., Harris, T., Huang, X., Jovanovich, S. B., Nelson, J. R., Schloss, J. A., Schwartz, D. C., & Vezenov, D. V. (2009). The challenges of sequencing by synthesis. *Nature Biotechnology, 27*(11), 1013–1023. https://doi.org/10.1038/nbt.1585

Georges, F., & Ray, H. (2017). Genome editing of crops: A renewed opportunity for food security. *GM Crops and Food, 8*(1), 1–12. https://doi.org/10.1080/21645698.2016.1270489. http://www.tandfonline.com/toc/kgmc20/current

Grabherr, M. G., Haas, B. J., Yassour, M., Levin, J. Z., Thompson, D. A., Amit, I., Adiconis, X., Fan, L., Raychowdhury, R., Zeng, Q., Chen, Z., Mauceli, E., Hacohen, N., Gnirke, A., Rhind, N., Di Palma, F., Birren, B. W., Nusbaum, C., Lindblad-Toh, K., Friedman, N., & Regev, A. (2011). Full-length transcriptome assembly from RNA-Seq data without a reference genome. *Nature Biotechnology, 29*(7), 644–652. https://doi.org/10.1038/nbt.1883

Grabherr, M. G., Haas, B. J., Yassour, M., Levin, J. Z., Thompson, D. A., Amit, I., & Regev, A. (2011). Trinity: Reconstructing a full-length transcriptome without a genome from RNA-Seq data. *Nature Biotechnology, 29*(7), 644–652.

Huang, S., Weigel, D., Beachy, R. N., & Li, J. (2016). A proposed regulatory framework for genome-edited crops. *Nature Genetics, 48*(2), 109–111. https://doi.org/10.1038/ng.3484. http://www.nature.com/ng/index.html

Kanehisa, M., & Goto, S. (2000). KEGG: Kyoto encyclopedia of genes and genomes. *Nucleic Acids Research, 28*(1), 27–30. https://doi.org/10.1093/nar/28.1.27. https://academic.oup.com/nar/issue

Kim, D., Langmead, B., & Salzberg, S. L. (2015). HISAT: A fast spliced aligner with low memory requirements. *Nature Methods, 12*(4), 357–360. https://doi.org/10.1038/nmeth.3317. http://www.nature.com/nmeth/

Levy, S., Sutton, G., Ng, P. C., Feuk, L., Halpern, A. L., & Walenz, B. P. (2007). The diploid genome sequence of an individual human. *PLoS Biology, 5*(10), e254.

Liu, J., Li, G., Chang, Z., Yu, T., Liu, B., McMullen, R., Chen, P., Huang, X., & Lengauer, T. (2016). BinPacker: Packing-based de novo transcriptome assembly from RNA-seq data. *PLoS Computational Biology, 12*(2), e1004772. https://doi.org/10.1371/journal.pcbi.1004772

Liu, J., Yu, T., Jiang, T., & Li, G. (2016). TransComb: Genome-guided transcriptome assembly via combing junctions in splicing graphs. *Genome Biology, 17*(1). https://doi.org/10.1186/s13059-016-1074-1. http://genomebiology.com/

Lo, Y. M. D., & Chiu, R. W. K. (2009). Next-generation sequencing of plasma/serum DNA: An emerging research and molecular diagnostic tool. *Clinical Chemistry, 55*(4), 607–608. https://doi.org/10.1373/clinchem.2009.123661

Lohse, M., Bolger, A. M., Nagel, A., Fernie, A. R., Lunn, J. E., Stitt, M., & Usadel, B. (2012). R obi NA: A user-friendly, integrated software solution for RNA-seq-based transcriptomics. *Nucleic Acids Research, 40*(Web server issue), W622–W627.

Love, M. I., Huber, W., & Anders, S. (2014). Moderated estimation of fold change and dispersion for RNA-seq data with DESeq2. *Genome Biology, 15*(12). https://doi.org/10.1186/s13059-014-0550-8. http://genomebiology.com/

Manchikatla, P. K., Kalavikatte, D., Mallikarjuna, B. P., Palakurthi, R., Khan, A. W., Jha, U. C., Bajaj, P., Singam, P., Chitikineni, A., Varshney, R. K., & Thudi, M. (2021). MutMap approach enables rapid identification of candidate genes and development of markers associated with early flowering and enhanced seed size in chickpea (*Cicer arietinum* L.). *Frontiers in Plant Science, 12*. https://doi.org/10.3389/fpls.2021.688694. https://www.frontiersin.org/journals/plant-science

Mardis, E. R. (2008). Next-generation DNA sequencing methods. *Annual Review of Genomics and Human Genetics, 9*, 387–402. https://doi.org/10.1146/annurev.genom.9.081307.164359

Maretty, L., Sibbesen, J. A., & Krogh, A. (2014). Bayesian transcriptome assembly. *Genome Biology, 15*(10). https://doi.org/10.1186/s13059-014-0501-4. http://genomebiology.com/

Margulies, M., Egholm, M., Altman, W. E., Attiya, S., Bader, J. S., Bemben, L. A., Berka, J., Braverman, M. S., Chen, Y. J., Chen, Z., Dewell, S. B., Du, L., Fierro, J. M., Gomes, X. V., Godwin, B. C., He, W., Helgesen, S., Ho, C. H., Irzyk, G. P., … Rothberg, J. M. (2005). Genome sequencing in microfabricated high-density picolitre reactors. *Nature, 437*(7057), 376–380. https://doi.org/10.1038/nature03959

Martin, J. A., & Wang, Z. (2011). Next-generation transcriptome assembly. *Nature Reviews Genetics, 12*(10), 671–682. https://doi.org/10.1038/nrg3068

Martin, M. (2011). Cutadapt removes adapter sequences from high-throughput sequencing reads. *EMBnet Journal, 17*(1), 10. https://doi.org/10.14806/ej.17.1.200

Maxam, A. M., & Gilbert, W. (1977). A new method for sequencing DNA. *Proceedings of the National Academy of Sciences of the United States of America, 74*(2), 560–564. https://doi.org/10.1073/pnas.74.2.560

Merico, D., Isserlin, R., Stueker, O., Emili, A., Bader, G. D., & Ravasi, T. (2010). Enrichment map: A network-based method for gene-set enrichment visualization and interpretation. *PLoS One, 5*(11), e13984. https://doi.org/10.1371/journal.pone.0013984

Morozova, O., & Marra, M. A. (2008). Applications of next-generation sequencing technologies in functional genomics. *Genomics, 92*(5), 255–264. https://doi.org/10.1016/j.ygeno.2008.07.001

Park, P. J. (2009). ChIP-seq: Advantages and challenges of a maturing technology. *Nature Reviews Genetics, 10*(10), 669–680. https://doi.org/10.1038/nrg2641

Patel, M. K., Chaudhary, R., Taak, Y., Pardeshi, P., Nanjundan, J., Vinod, K. K., Saini, N., Vasudev, S., & Yadava, D. K. (2022). Seed coat colour of Indian mustard [*Brassica juncea* (L.) Czern. and Coss.] is associated with Bju.TT8 homologs identifiable by targeted functional markers. *Frontiers in Plant Science, 13*. https://doi.org/10.3389/fpls.2022.1012368. https://www.frontiersin.org/journals/plant-science

Patel, R. K., Jain, M., & Liu, Z. (2012). NGS QC toolkit: A toolkit for quality control of next generation sequencing data. *PLoS One, 7*(2), e30619. https://doi.org/10.1371/journal.pone.0030619

Pertea, M., Pertea, G. M., Antonescu, C. M., Chang, T. C., Mendell, J. T., & Salzberg, S. L. (2015). StringTie enables improved reconstruction of a transcriptome from RNA-seq reads. *Nature Biotechnology, 33*(3), 290−295. https://doi.org/10.1038/nbt.3122. http://www.nature.com/nbt/index.html

Raudvere, U., Kolberg, L., Kuzmin, I., Arak, T., Adler, P., Peterson, H., & Vilo, J. (2019). g: Profiler: A web server for functional enrichment analysis and conversions of gene lists (2019 update). *Nucleic Acids Research, 47*(W1), W191−W198.

Robinson, M., McCarthy, D., Chen, Y., & Smyth, G. K. (2013). *edgeR: Differential expression analysis of digital gene expression data. User's guide.*

Samantara, K., Shiv, A., de Sousa, L. L., Sandhu, K. S., Priyadarshini, P., & Mohapatra, S. R. (2021). A comprehensive review on epigenetic mechanisms and application of epigenetic modifications for crop improvement. *Environmental and Experimental Botany, 188*. https://doi.org/10.1016/j.envexpbot.2021.104479. http://www.elsevier.com/inca/publications/store/2/6/7

Sandhu, K. S., Shiv, A., Kaur, G., Meena, M. R., Raja, A. K., Vengavasi, K., Mall, A. K., Kumar, S., Singh, P. K., Singh, J., Hemaprabha, G., Pathak, A. D., Krishnappa, G., & Kumar, S. (2022). Integrated approach in genomic selection to accelerate genetic gain in sugarcane. *Plants, 11*(16). https://doi.org/10.3390/plants11162139. http://www.mdpi.com/journal/plants

Sanger, F., Nicklen, S., & Coulson, A. R. (1977). DNA sequencing with chain-terminating inhibitors. *Proceedings of the National Academy of Sciences of the United States of America, 74*(12), 5463−5467. https://doi.org/10.1073/pnas.74.12.5463

Scheben, A., & Edwards, D. (2017). Genome editors take on crops. *Science, 355*(6330), 1122−1123. https://doi.org/10.1126/science.aal4680

Scheben, A., Yuan, Y., & Edwards, D. (2016). Advances in genomics for adapting crops to climate change. *Current Plant Biology, 6*, 2−10. https://doi.org/10.1016/j.cpb.2016.09.001

Schmieder, R., & Edwards, R. (2011). Fast identification and removal of sequence contamination from genomic and metagenomic datasets. *PLoS One, 6*(3), e17288.

Schmieder, R., & Edwards, R. (2011). Quality control and preprocessing of metagenomic datasets. *Bioinformatics, 27*(6), 863−864. https://doi.org/10.1093/bioinformatics/btr026

Schulz, M. H., Zerbino, D. R., Vingron, M., & Birney, E. (2012). Oases: Robust de novo RNA-seq assembly across the dynamic range of expression levels. *Bioinformatics, 28*(8), 1086−1092. https://doi.org/10.1093/bioinformatics/bts094

Shannon, P., Markiel, A., Ozier, O., Baliga, N. S., Wang, J. T., Ramage, D., Amin, N., Schwikowski, B., & Ideker, T. (2003). Cytoscape: A software environment for integrated models of biomolecular interaction networks. *Genome Research, 13*(11), 2498−2504. https://doi.org/10.1101/gr.1239303

Shao, M., & Kingsford, C. (2017). Accurate assembly of transcripts through phase-preserving graph decomposition. *Nature Biotechnology, 35*(12), 1167−1169. https://doi.org/10.1038/nbt.4020

Sherman, B. T., Hao, M., Qiu, J., Jiao, X., Baseler, M. W., Lane, H. C., Imamichi, T., & Chang, W. (2022). David: A web server for functional enrichment analysis and functional annotation of gene lists (2021 update). *Nucleic Acids Research, 50*(1), W216−W221. https://doi.org/10.1093/nar/gkac194. https://academic.oup.com/nar/issue

Shi, J., & Walker, M. G. (2007). Gene set enrichment analysis (GSEA) for interpreting gene expression profiles. *Current Bioinformatics, 2*(2), 133−137. https://doi.org/10.2174/157489307780618231. http://www.ingentaconnect.com/content/ben/cbio/2007/00000002/00000002/art00003

Simpson, J. T., Wong, K., Jackman, S. D., Schein, J. E., Jones, S. J. M., & Birol, I. (2009). ABySS: A parallel assembler for short read sequence data. *Genome Research, 19*(6), 1117−1123. https://doi.org/10.1101/gr.089532.108. http://genome.cshlp.org/content/19/6/1117.full.pdf+html

Thompson, J. F., & Milos, P. M. (2011). The properties and applications of single-molecule DNA sequencing. *Genome Biology, 12*(2). https://doi.org/10.1186/gb-2011-12-2-217. http://genomebiology.com/2011/12/2/217

Tran, Q. H., Bui, N. H., Kappel, C., Dau, N. T. N., Nguyen, L. T., Tran, T. T., Khanh, T. D., Trung, K. H., Lenhard, M., & Vi, S. L. (2020). Mapping-by-sequencing via mutmap identifies a mutation in zmcle7 underlying fasciation in a newly developed ems mutant population in an elite tropical maize inbred. *Genes, 11*(3). https://doi.org/10.3390/genes11030281. https://www.mdpi.com/2073-4425/11/3/281/pdf

Trapnell, C., Hendrickson, D. G., Sauvageau, M., Goff, L., Rinn, J. L., & Pachter, L. (2013). Differential analysis of gene regulation at transcript resolution with RNA-seq. *Nature Biotechnology, 31*(1), 46−53. https://doi.org/10.1038/nbt.2450

Trapnell, C., Pachter, L., & Salzberg, S. L. (2009). TopHat: Discovering splice junctions with RNA-Seq. *Bioinformatics, 25*(9), 1105−1111. https://doi.org/10.1093/bioinformatics/btp120

Wang, L., Wang, S., & Li, W. (2012). RSeQC: Quality control of RNA-seq experiments. *Bioinformatics, 28*(16), 2184−2185. https://doi.org/10.1093/bioinformatics/bts356

Wang, Z., Gerstein, M., & Snyder, M. (2009). RNA-seq: A revolutionary tool for transcriptomics. *Nature Reviews Genetics, 10*(1), 57−63. https://doi.org/10.1038/nrg2484

Wu, T. D., & Nacu, S. (2010). Fast and SNP-tolerant detection of complex variants and splicing in short reads. *Bioinformatics, 26*(7), 873−881. https://doi.org/10.1093/bioinformatics/btq057

Xie, Y., Wu, G., Tang, J., Luo, R., Patterson, J., Liu, S., Huang, W., He, G., Gu, S., Li, S., Zhou, X., Lam, T., Li, Y., Xu, X., Wong, G., & Wang, J. (2014). SOAPdenovo-trans: De novo transcriptome assembly with short RNA-seq reads. *Bioinformatics, 30*(12), 1660−1666. https://doi.org/10.1093/bioinformatics/btu077

Chapter 13

Data management in cross-omics

Sanjay Kumar[1] **and Manish Kumar Gupta**[2]
[1]*Bioinformatics Center, Biotech Park, Lucknow, Uttar Pradesh, India;* [2]*Department of Biotechnology, Faculty of Science, Veer Bahadur Singh Purvanchal University, Jaunpur, Uttar Pradesh, India*

1. Introduction

Biological science data management refers to collecting, storing, organizing, maintaining, and using biological data effectively and efficiently. Information governance, advanced analytics, quality of data, digital security, data management, data collection, data cataloging, and visual analytics are all part of it. The primary goal of data management is to ensure that data is accurate, reliable, and easily available for those who require it, while also protecting it from unauthorized users or mishandling. This is accomplished through a combination of processes, policies, and technologies to manage and control data throughout its lifecycle. Effective data management requires a clear understanding of the organization's data needs, as well as the ability to implement and enforce data management best practices. Data management plays a crucial role in the success of many research discoveries and healthcare, including decision-making, analytics, and compliance with regulations. A data management system allows companies to make better use of their data, enhance productivity, and gain an edge over their competitors. A database management system (DBMS) is a software package that collects and processes data by communicating with the database, other applications, and end users. It enables organized data storage, retrieval, and manipulation for users. The DBMS creates and manages a database, a collection of data stored in a specific format. Tables are typically used to organize databases, which are made up of rows (also known as records) and columns (also known as attributes). The 21st century is the era of big data, which refer to large and complex sets of data that are difficult to process and analyze using traditional data processing techniques and tools. This data can come from various sources, such as scientific experiments, and it can be in the form of text, images, videos, or other formats. Similarly, omics data is a term that encompasses large-scale data generated by high-throughput technologies in various fields of biology, including genomics, transcriptomics, proteomics, and metabolomics. These data are used to study the function, structure, and interactions of biological molecules and the genetic and environmental factors that affect them. The goal of omics research is to gain a comprehensive understanding of biological systems at the molecular level, which can lead to new insights into disease mechanisms, drug development, and personalized medicine. Omics technologies such as compound library high-throughput screening, next-generation sequencing (NGS), and system biology are widely used in biomedical research and require advanced methods because they use a large size of gene/protein data, that is, gene expression, gene mutation, and recombination. These techniques focus on disease-level treatment and prevention. The term "cross-omics" represents the combination of different biological datasets to get significant results. In cross-omics data, the computation power is used to analyze the vast data and extract new information. It required new tools to process massive amounts of highly complex biological material with high sensitivity and specificity. The complexity arises when processing and integrating heterogeneous data types; this requires data standardization. The new technology and methods improved the complexity of the data and presented standard data, such as machine learning approaches. The NGS technique generates thousands of gene datasets that are used for clinical diagnostics of autoimmune diseases, mutational analysis, and inborn errors of metabolism. The generated data makes sense only if one is provided with the necessary insightful resources and techniques to understand it fully. As a result, to create efficient cross-omics studies, obtained biological data must be standardized and include heterogeneous data and knowledge in a way understandable by humans and computers. Therefore, the development of laboratory techniques must follow a comparable advancement in diagnostic methods and data handling capabilities. Cross-omics covers various data types from different fields, such as genomics, proteomics, transcriptomics, metabolomics (metabolites and metabolic networks), and system biology (Fig. 13.1).

Integrative Omics. https://doi.org/10.1016/B978-0-443-16092-9.00013-8

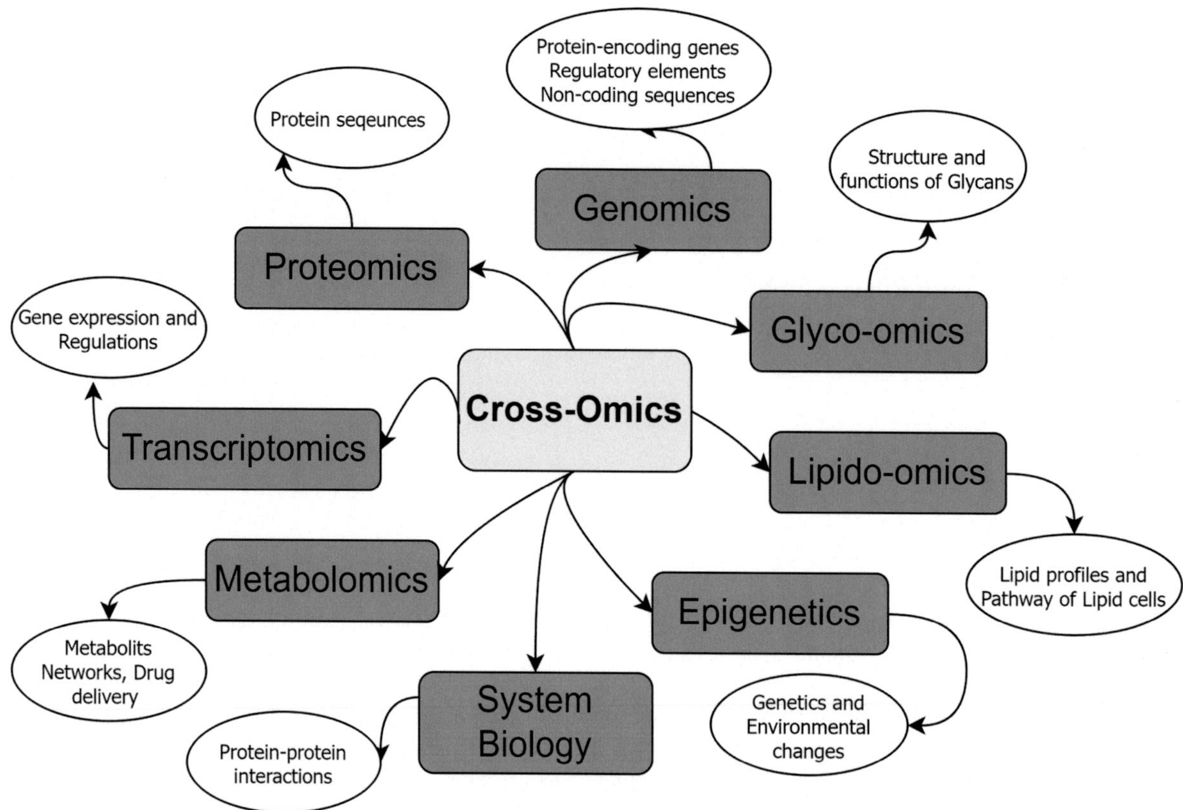

FIGURE 13.1 Mind map of cross-omics domains. This includes different fields that are studied in the area of cross-omics.

Genomics deals with protein-coding genes, regulatory elements, and noncoding sequence data, while proteomics focuses on protein sequences and structural analysis. Similarly, gene expression was used to analyze gene regulation in transcriptomic analysis. Metabolomics deals with biochemical pathways, protein—protein interaction networks, and drug delivery pathway system data. At the same time, similar data is also studied in system biology, where analyses of the involved gene/protein in specific pathways use biochemical pathway data. A vast amount of biological data is already available in the form of text in biological databases. There is a need to address the complexity of these data and understand the regulation of particular components. To tackle the data problem in cross-omics, the computational biology or bioinformatics field can be helpful. It searches the biological databases to compare gene or protein sequences and identify the genes involved in diseases and some specific genes. In this chapter, the new cross-omics technology of databases and its challenges in biological data processing and analysis are discussed.

2. Why cross-omics data management required?

Cross-omics data management plays a critical role in advancing our understanding of biological systems and improving the quality of research in the field of omics. It helps in the integration of multiple data sources and enables the testing of hypotheses across multiple levels of biological data, such as genomics, transcriptomics, proteomics, and metabolomics, to provide a more comprehensive understanding of the biological data. This can also facilitate the discovery of new biological insights by integrating data. It can help to ensure data quality by providing standardized data processing pipelines, reducing the risk of errors, and increasing the accuracy of results. Cross-omics data management involves organizing, storing, and analyzing these data in a way that allows researchers to integrate and interpret the data from different omics disciplines in a meaningful way. This can be a complex task, as different omics data types often have different data structures, formats, and quality standards. To manage cross-omics data effectively, researchers may use specialized software tools and databases and may also need to develop custom analyses and data integration approaches. Several software and tools can assist in cross-omics data management and analysis, such as Galaxy, KNIME, R, and Python. These tools can be used to perform data integration, normalization, visualization, and statistical analysis. Developing common data standards and ontologies across different omics disciplines can also facilitate cross-omics data management and analysis by making it easier to combine and compare data from different sources. It is important to note that proper data management is essential for the reproducibility and reliability of cross-omics research. This includes keeping detailed and accurate records of the data, methods, and analysis steps and making the data available to other researchers through publicly accessible repositories, such as GEO, ArrayExpress, ProteomeXchange, and Metabolomics Workbench. Each subject has been categorized to manage the cross-omics data by its methods, database, and tools. Software for cross-omics data handling must be supported by good hardware from the beginning. This section lists the most recommended software and configurations for cross-omics data management.

3. Hardware setups for managing cross-omics data

The hardware configuration for cross-omics data management will depend on the specific requirements of the database, such as the size of the data, the number of concurrent users, and the performance needs. However, some general considerations for hardware configuration include the following:

1. CPU: A high-end CPU with 32 cores is generally recommended for a large and complex database. Users can attach more to the CPU in their workstations.
2. Memory: The memory size should be at least equal to the size of the data, and ideally, it should be larger to allow for efficient caching of data and indexes. A 128 GB of RAM in our workstation, which is the ideal range of RAM. Generally, workstations support multiple RAM slots; you can set RAM accordingly.
3. Storage: Hard drives are recommended for data storage, but nowadays, solid-state drives (SSDs) are generally recommended for high-performance databases, as they provide faster access times than traditional hard drives.
4. Networking: A fast and reliable network is essential for a data management system, especially if multiple users access the database over a network. A 10 Gigabit Ethernet connection is recommended for high-performance databases.
5. Backup and disaster recovery: A robust backup and disaster recovery strategy is essential for cross-omics data management to ensure data availability and integrity in case of hardware failure or data corruption. This could include a combination of off-site backups, redundant hardware, and failover systems. Approximately 5 TB local hard drive is enough for data storage, and users can prefer cloud storage because nowadays many companies provide cloud storage at cheaper rates, such as Box, Google Drive, Microsoft OneDrive, Amazon, and DropBox.
6. Cloud storage: Many private companies allow individuals and organizations to store and access their data over the Internet. It provides a scalable and flexible solution for data storage and management, which is particularly important for a large amount of omics data. Cloud storage can be cost-effective as it eliminates expensive hardware and IT infrastructure. The following companies provide data storage at cheaper prices; Amazon Web Services (AWS: https://aws.amazon.com/), Google Cloud Storage (https://cloud.google.com/storage), Microsoft Azure (https://azure.microsoft.com/en-us/services/storage/), IBM Cloud Object Storage (https://www.ibm.com/cloud/object-storage), Oracle Cloud Infrastructure Object Storage (https://cloud.oracle.com/object-storage), DropBox (https://www.dropbox.com/), Box (https://www.box.com/), Apple iCloud (https://www.apple.com/icloud/).

It is important to note that the specific hardware configuration will depend on the size and complexity of the database, the number of concurrent users, the performance needs, and the budget available. It is advisable to consult with a database administrator or performance engineer to determine the appropriate hardware configuration for the specific needs of the database.

4. Software required for cross-omics data analysis

Operating system: The cross-omics data management software runs on top of an operating system, such as Windows, Linux, or macOS. The choice of the operating system will depend on the specific requirements of the database, as well as the compatibility of the cross-omics data management software with the chosen operating system. Cross-omics data management typically requires the following software components.

4.1 Database management software

This core software manages the database and provides functionality for creating, modifying, and querying the database. Here is a list of some commonly used database management software.

1. MySQL (https://www.mysql.com/): This is an open-source relational database management system that is widely used for web-based applications and data warehousing.

2. Oracle Database (https://www.oracle.com/database/): A powerful, enterprise-class relational database management system that is widely used for business-critical applications.
3. Microsoft SQL Server (https://www.microsoft.com/en-us/sql-server): A relational database management system that is commonly used for business intelligence and data warehousing.
4. PostgreSQL (https://www.postgresql.org/): An open-source relational database management system that is known for its stability, performance, and compliance with SQL standards.
5. MongoDB (https://www.mongodb.com/): A NoSQL, document-oriented database management system that is designed for high scalability and performance.
6. MariaDB (https://mariadb.org/): An open-source relational database management system that is a fork of MySQL and is designed to be compatible with it.
7. Cassandra (https://cassandra.apache.org/_/index.html): An open-source, distributed NoSQL database management system that is designed for high scalability and performance.
8. Redis (https://redis.io/): An open-source, in-memory data store that is commonly used as a cache, message broker, and database.
9. Firebase (https://firebase.google.com/): A cloud-based, real-time NoSQL database management system that is designed for mobile and web applications.
10. DynamoDB (https://aws.amazon.com/dynamodb/): A NoSQL database management system provided as a service by AWS and designed for high scalability and performance.
11. Solar Winds (https://www.solarwinds.com/): Database performance monitoring and optimization for traditional, open-source, and cloud-native databases.
12. IBM Db2 (https://www.ibm.com/products/db2): It is a cloud-native database built to power low latency transactions and real-time analytics at scale. It provides a single engine for DBAs, enterprise architects, and developers to keep critical applications running, store and query anything, and power faster decision-making and innovation across your organization.
13. Microsoft Access (https://www.microsoft.com/en-us/microsoft-365/access): Create your database apps easily in formats that serve your business best. Runs on basic configuration systems.
14. phpMyAdmin (https://www.phpmyadmin.net/): It is intended to handle the administration of MySQL and MariaDB over the Web.

4.2 Database drivers and connectors

These software components allow application programs to interact with the DBMS. They provide an API (Application Programming Interface) that enables the application to send and receive data to and from the database. Examples: ODBC (Open Database Connectivity), JDBC (Java Database Connectivity), ADO.NET, Ingres, Teradata, and Sybase ASE.

4.3 Backup and recovery software

These programs are used for backing up the database and restoring it in the event of data loss or corruption. Examples: Oracle Recovery Manager (RMAN), MySQL Enterprise Backup, SQL Server Management Studio (SSMS), PostgreSQL pg_dump, MongoDB mongodump, IBM DB2, Informix ON-Bar, Veritas NetBackup, CA ARCserve Backup, EMC Networker.

4.4 Performance monitoring and tuning software

These applications and tools are used to monitor the performance of the DBMS and to optimize the database and operating system settings. Examples include Oracle Enterprise Manager, MySQL Workbench, and Microsoft SQL Server Management Studio.

4.5 Security software

These programs are used to secure and safeguard the database from unwanted access. Examples include firewalls, intrusion detection and prevention systems, and encryption software.

- Oracle Advanced Security
- MySQL Enterprise Firewall

- SQL Server Auditing and Threat Detection
- PostgreSQL
- MongoDB Enterprise Advanced
- IBM DB2 Advanced Security
- Informix Secure Backup
- Veritas NetBackup Secure
- CA ARCserve Backup Security
- EMC Networker Security
- Imperva SecureSphere
- McAfee Database Security
- TrendMicro Deep Security
- SolarWinds Database Performance Analyzer
- IriusRisk
- Redgate SQL Protector
- AWS RDS encryption
- Azure SQL Database TDE
- Google Cloud SQL encryption
- Vormetric Transparent Encryption

5. File formats used in cross-omics data management

Nowadays, bioinformatics analysis requires efficient and quick algorithms due to the enormous amount of data. In principle, the analysis already begins with the sequence analysis. Most of the researchers started their work with sequence analysis approaches and methods. Either they use machine learning or experimental approaches. Generally, the omics data is stored in some specific file extension format that can be read by specific software. Some general formats that are commonly read by any software used in machine learning or bioinformatics applications are discussed here.

1. FASTA: This format is used to store nucleotide or protein sequences. Each sequence is represented by a header line starting with the ">" symbol, followed by the sequence.
2. FASTQ: This format is used to store sequencing reads and their corresponding quality scores. Each read in a metagenomics dataset is typically represented as a single FASTQ record containing four lines of information:
3. SAM/BAM: This is a compressed binary format used to store alignment data from next-generation sequencing experiments. BAM files contain aligned reads and their mapping information, such as the reference genome position and alignment quality scores.
4. VCF: This format is used to store variant call information, such as single-nucleotide polymorphisms (SNPs) and insertions/deletions (indels). Each variant is represented by a line with the genomic coordinates, reference and alternate alleles, and quality scores.
5. BED: This format is used to store genomic regions, such as gene annotations, transcription factor binding sites, and chromatin modifications. Each region is represented by a line with the genomic coordinates and optional metadata.
6. GTF/GFF: These formats are used to store gene annotation data, including exon/intron boundaries, transcript and protein-coding gene information, and other features. Each feature is represented by a line with the genomic coordinates, annotation type, and metadata.
7. SFF format: This is a binary format used to store sequencing data generated by Roche/454 pyrosequencing technology.
8. CSV/TSV: These are spreadsheet file formats that can be used to store the expression values of genes across different samples or conditions.
9. HDF5: This is a hierarchical data format that can be used to store large and complex datasets, such as gene expression matrices and metadata.
10. RDS: This is a file format used by the R programming language to store data objects, including gene expression matrices and statistical models.
11. mzML: This is a standard format for storing raw mass spectrometry data. It is an XML-based format and can be used to store data from a variety of instruments.

12. mzXML: This is another XML-based format for storing mass spectrometry data. It was developed by the Institute for Systems Biology and is widely used in the proteomics community.
13. MGF: This format is used for storing peak lists generated from mass spectrometry data. It is a text-based format and includes information such as m/z ratios, intensities, and charge states.
14. pepXML: This format is used for storing the results of peptide identification from mass spectrometry data. It includes information such as peptide sequence, charge state, and scores.
15. protXML: This format is used for storing protein identification results. It includes information such as protein name, accession number, and the peptides that were used to identify it.
16. SBML (Systems Biology Markup Language): A standard format for representing mathematical models of biochemical reaction networks.
17. CellML (Cellular Markup Language): A format for describing mathematical models of biological processes at the cell level.
18. BioPAX (Biological Pathway Exchange): A format for exchanging biological pathway data, including molecular interactions and cellular processes.
19. GML (Graph Modeling Language): A format for describing graphs, which are commonly used to represent biological networks.
20. PDB (Protein Data Bank): A format for storing three-dimensional structures of biological macromolecules.
21. XML (Extensible Markup Language): A general-purpose format for storing and exchanging data in a structured format.
22. JSON (JavaScript Object Notation): A lightweight format for storing and exchanging data in a structured format.

6. How does the biological database work?

A biological database is a collection of biological data that is organized and stored to allow easy retrieval and analysis. These databases can store various types of information, including DNA and protein sequences, genetic and phenotypic data, and information on organisms and their interactions with each other and their environment. They can be used for various research purposes, such as identifying disease-associated genes, studying evolutionary relationships between different species, and developing new drugs and other biotechnology products (Fig. 13.2). Some common types of biological databases include GenBank, UniProt, and KEGG. These databases can be searched and accessed through web-based interfaces, and many also offer application programming interfaces (APIs) that allow researchers to access and analyze the data programmatically.

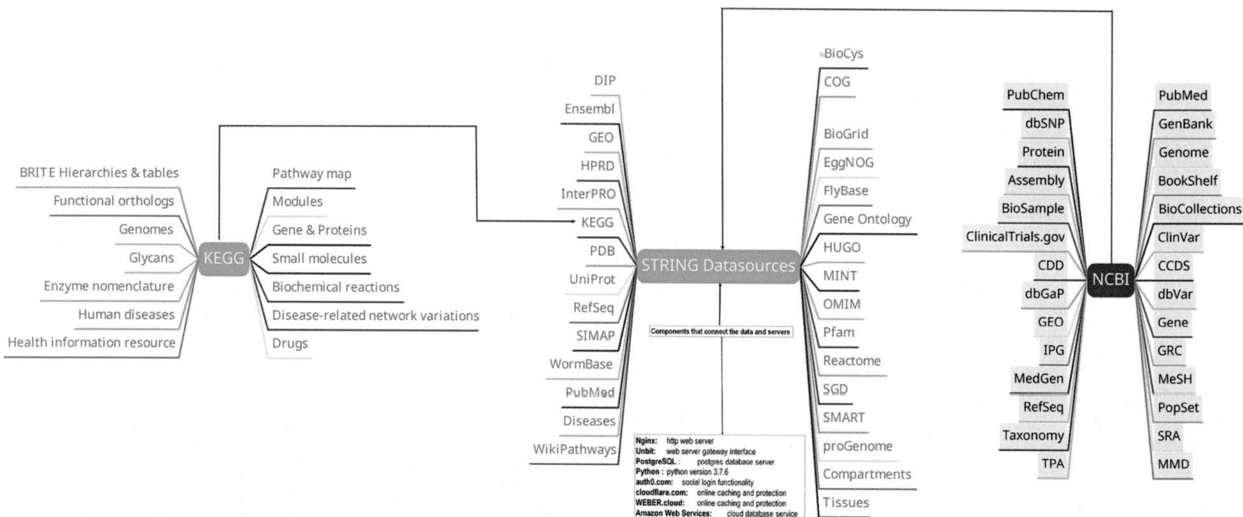

FIGURE 13.2 Biological database is majorly used in omics data analysis. This mind map showed a data integration with two databases and how they work behind the user wall. The showed how databases connect and share the information that can be used in cross-omics data analysis.

7. Cross-omics data management

7.1 Next generation sequencing (genomics)

According to studies, around 93% of human DNA is converted into RNA, but just 2% into pcRNA. The transcriptome is the collection of all messenger RNA (mRNA) molecules, or "transcripts," produced by one or more cells. To obtain expression information at a high-throughput level, several approaches have been devised. Either hybridization using oligonucleotide microarrays or sequence tag counting has been used to analyze global gene expression. Creating genome-wide expression profiles using digital transcriptomics and pyrophosphatase-based ultrahigh-throughput DNA sequencing of ditags is a novel expression analysis method. Several techniques are similar to Genomics data analysis in which few are available for specific transcriptomics data analysis, such as microarrays, RTPCR, RNA-seq, and the latest NGS technologies: ChIP-seq (Park, 2009) and RNA-seq (Wang et al., 2009).

Genomics data management relates to the massive volume of data created by sequencing research being stored, arranged, and analyzed. This can comprise, among other things, clinical information and genetic annotation. A strong and scalable infrastructure that can handle the volume and complexity of genomics data is necessary for the efficient management of this type of data.

NGS was created expressly to be compatible with automation and parallel processing, and they enable the determination of sequence data from amplified single DNA fragments. In a single reaction, current methods can only sequence relatively short (300—1000 nucleotide long) DNA fragments. Short-read sequencing technologies significantly reduce the cost of sequencing. Improvements in accuracy and readability are aimed for as there were initial concerns that the increase in the number would lead to a decline in quality.

To tackle this problem, a second- and third-generation sequencing technologies, such as Roche 454 pyrosequencing (http://www.my454.com), Illumina (Solexa) HiSeq and MiSeq sequencing (http://www.illumina.com), oligonucleotide ligation and detection (SOLiD) (http://www.lifetechnologies.com), DNA nanoball sequencing by BGI Retrovolocity (http://www.completegenomics.com), ion torrent (http://www.iontorrent.com), single-molecule real-time (SMRT) sequencing by pacific biosciences (http://www.pacificbiosciences.com), Helicos sequencing by the genetic analysis system (Shendure & Ji, 2008), Nanopore sequencing by Oxford Nanopore Technologies (MinION and PromethION) (Bayley, 2015), and NGS by electron microscopy (http://omicsmaps.com), can be used. Most of the programs such as DNAscan (Iacoangeli et al., 2019) and SEQprocess (Joo et al., 2019) are implemented in R packages for NGS data processing. It easily runs each step in the R console by giving simple commands that can be customizable with interactive mode.

NGS data management involves several steps, starting from sample collection and preparation to data analysis and interpretation (Table 13.1).

7.1.1 File nomenclature

Creating a file naming convention can help groups by giving them a simple way to identify and organize their files.

TABLE 13.1 Type of data stored or delete from NGS projects.

Analysis type	Data type	Time	Remarks
Variant calling pipeline	Remove reads—raw and shuffled	When final/closed out	Difficult clusters being examined for research. Latest Complete Run Folder
	Keep cleaned reads, reference, log, and msa folders	Indefinite	
	Remove bam and vcf files	1 year	
Genome assembly (e.g., SPAdes)	Keep the scaffolds.fasta file delete the intermediate assembly directory	Immediately remove the directory	In-depth assembly projects
Taxonomic classification (e.g., Kraken)	Delete the Kraken.out file	Immediately remove the file	Difficult clusters or in-depth projects

7.1.2 Data collection and preparation

This step is critical to ensure the quality and quantity of the material obtained, which will directly impact the quality of the resulting NGS data. The DNA/RNA raw sequence reads/protein-encoding genes stored in FASTQ or unaligned BAM (uBAM) format have been submitted to Genome databases such as the Human Reference Genome database (https://www.ncbi.nlm.nih.gov/genome/guide/human/), HUGO (https://www.genenames.org/), and Human Metabolome Database (https://hmdb.ca/). Such formats provide short sequences in the plain text together with information about each raw sequence that includes the id number and base rankings. Users must download the raw file in any format and perform the alignment. The collected raw data in sequence format can be trimmed by TrimGalore (https://github.com/broadinstitute/picard) or Cutadapt (Martin, 2011).

7.1.3 Sequencing

DNA or RNA is sequenced using one of several NGS technologies, such as Illumina sequencing, Oxford Nanopore sequencing, or PacBio sequencing. This step generates raw sequencing data that must be processed further.

7.1.4 Quality control and preprocessing

Raw sequencing data must undergo quality control and preprocessing to remove contaminants, trim low-quality reads, and perform adapter trimming. This step is critical to ensure the quality of the resulting NGS data. The right protocols must be followed to filter out human-read data before keeping sequence files in any other format. This is crucial for guaranteeing regulatory compliance and cutting down on the amount of space needed to store sequencing data. SanitizeMe and CLC Microbial Genomics Module can be used for data QC and cleaning the host DNA.

7.1.5 Alignment and assembly

This step provides a comprehensive view of the genome or transcriptome, including information about gene expression levels, alternative splicing, and novel transcripts or genomic variants. The raw sequence reads or unaligned BAM files (Compressed files) were aligned against the human reference genome. This procedure gives the short reads in the reference genome positional context, which also creates various metadata fields with alignment properties, including matches, mismatches, and gaps in a compact, unique gapped alignment report format. Sequence alignment mapping (SAM), often called CRAM file format, is used to record the aligned sequences and the associated metadata. The downstream algorithms use the SAM/BAM file to identify various genomic abnormalities, such as tumor mutation load, single-nucleotide variants, insertions, and deletions. The sequence alignment can also be performed by HISAT2 (Kim et al., 2015), BWA (Li & Durbin, 2009), STAR (Dobin et al., 2013), TopHat2 (Kim et al., 2013), bowtie2 (Langdon, 2015), or SAMtools (Li et al., 2009). To mark the duplicates during alignment SAMblaster (Faust & Hall, 2014) program is used and SAMbamba (Tarasov et al., 2015) is used to sort the aligned reads. If you want to remove some duplicates found in alignment can be removed by Picard (https://github.com/broadinstitute/picard) and realignment by the GATK program (https://gatk.broadinstitute.org/hc/en-us). If you want to ignore reads and be marked as a duplicate automatically, two variant caller programs are used: FreeBayes (Garrison & Marth, 2012) and GATK Haplotype Caller (HC). RNA-seq data quantification can be done by HTSeq (Putri et al., 2022) or Cufflinks (http://cole-trapnell-lab.github.io/cufflinks/), and DNA copy number estimation is conducted by Sequenza program (Favero et al., 2015). To identify the variants from sequences data; the following variants calling programs can be used HISAT2, GATK, VarScan2 (Koboldt et al., 2012), MuSE (Fan et al., 2016), or SomaticSniper (Larson et al., 2012). Variant annotation can be done by VEP (McLaren et al., 2016) or ANNOVAR (Wang et al., 2010) program. In general, indels are underperformed by variant calling pipelines built on HISAT2 that BWA can manage to realign soft-clipped and unaligned reads.

7.1.6 Data analysis

The results of the alignment and assembly step are analyzed to gain insights into the biology of the system being studied. This may involve differential expression analysis, functional analysis, variant calling, or other types of data analysis, depending on the goals of the study. The mapped sequencing data can be used for various types of analysis. The single-nucleotide variant and small indel analysis (<10 bp) performed by FreeBayes and GATK HC extract the genome positions for which insertion or deletion is present on the cigar of at least read. This reduced the size of the genotype quality and speed up the processing. The large structural variants be identified by the two programs; Manta (Chen et al., 2016) and Expansion Hunter from Illumina (Dolzhenko et al., 2019). These two techniques pick up the more significant variants >50 bp, including insertions, deletions, translocations, duplications, and known repeat expansions.

7.1.7 Annotation by cross-omics data integration

Several genomics data standards such as the General Feature Format (GFF) and the Variant Call Format (VCF) help exchange and integrate genomics data between different platforms and software. This step can be used for data integration like metagenomics. This can be used by ClinVar (https://www.ncbi.nlm.nih.gov/clinvar/), GnomAD (https://gnomad.broadinstitute.org/), dbSNP (https://www.ncbi.nlm.nih.gov/snp/) and dbNSFP (http://database.liulab.science/dbNSFP). The output of NGS known as variants call is annotated using the Annover program (https://annovar.openbioinformatics.org/en/latest/). Generally, sequence reads in FASTQ format are used as an input that is correct and trims using Quake (Kelley et al., 2010), Sickle (https://github.com/najoshi/sickle), or Musket (Liu et al., 2013) applications then reassembling the reads using ABySS (https://github.com/bcgsc/abyss) or Velvelt (Zerbino & Birney, 2008) application, after that analysis is done by Quast (https://quast.sourceforge.net/) or KAT (Mapleson et al., 2017) software.

7.1.8 Assessment with cross-omics data

Several parameters are available to assess the NGS data using cross-omics data and methods, that is, diagnostic value. The gene prioritization was considered accurate when the disease-causing genes raked in the top hits of the NGS prioritized list of genes. This needs to repeat the NGS data analysis more than 100 times and then build a reference and validation set to check the parameter combination with higher values of diagnostic data. The other favorable parameters also help to analyze the differences, such as missed fraction, maximum distance, and extension stringency, Mendelian Inheritance in Man, EST, and biochemical stringency. After applying these parameters, gene-specific metabolites can be found that can help to treat specific diseases. The primary reaction of each encoded protein coupled using the R package PaxtoolsR (https://bioconductor.org/packages/release/bioc/html/paxtoolsr.html), PathwayCommons (https://www.pathwaycommons.org/); biological pathway and interaction database, mixOmics (http://mixomics.org/), provide a wide range of multivariate statistical methods. The reactions can be added using Recon3D (http://bigg.ucsd.edu/models/Recon3D), a metabolite-reaction database.

7.1.9 Data compression

Data compression is the process of reducing the amount of data needed to represent a given set of information. This is typically done by identifying and eliminating redundancy in the data, such as repeating patterns or common sequences of data. Data compression should be used wherever it makes sense, especially for large datasets that may not need to be accessed regularly. Compression of both local and archived data is a more efficient way to save space given limited resources. Specific software can be used to compress NGS datasets. CRAM, a compressed columnar file format for storing biological sequences aligned to a reference sequence, is one of the most popular options due to EMBL-EBI support. It reduces the storage requirements for a BAM file by 30%−60%.

7.1.10 Data management and sharing

NGS data must be properly managed and stored to ensure its long-term accessibility and reproducibility. There are many publicly accessible repositories, such as the National Center for Biotechnology Information (NCBI) or the European Nucleotide Archive (ENA). Open-access microarray data repositories include the European Bioinformatics Institute's (EBI: https://www.ebi.ac.uk/), ArrayExpressDB (https://www.ebi.ac.uk/biostudies/arrayexpress), and the NCBI's Gene Expression Omnibus (GEO: https://www.ncbi.nlm.nih.gov/geo/), Functional Genomics database (FGED: https://www.fged.org/), MIME (https://www.ncbi.nlm.nih.gov/geo/info/MIAME.html). A minimum of two backups, for a total of three copies of the data, should be maintained for critical datasets. If storage costs outweigh the advantages of additional backups, noncritical datasets may have a minimum of one backup and a maximum of two copies. Cloud storage can be utilized (AWS: https://aws.amazon.com/; Microsoft Azure: https://azure.microsoft.com/en-us).

7.2 Single-cell sequencing (transcriptomics)

The study of the entire set of RNA transcripts (molecules) produced by a cell or an organism, or the transcriptome, is known as transcriptomics. As part of the investigation of gene expression, all RNA molecules, including messenger RNA (mRNA), noncoding RNA (ncRNA), and microRNA, are identified and quantified (miRNA). Transcriptomics can help uncover prospective targets for drug development or biomarkers for illness diagnosis and treatment, as well as insights into the functional roles of genes and the regulation of gene expression. Microarray analysis, Single-cell RNA sequencing (scRNA-seq), and quantitative polymerase chain reaction are examples of transcriptomic approaches (qPCR). To provide a more thorough understanding of biological processes, the data produced by transcriptome analysis can be combined with data from other "omics" disciplines, including genomics, proteomics, and metabolomics. Transcriptomics data management is a multistep process that starts with sample preparation and collecting and ends with data processing and interpretation. Below is a high-level breakdown of the key phases in managing transcriptomics data:

In preparation for RNA extraction and sequencing, samples are collected and processed. This step is critical for ensuring the quality and quantity of RNA obtained, as the quality of the resulting transcriptomics data is directly affected by it. To sequence RNA, one of several RNA sequencing technologies, such as Illumina sequencing, Oxford Nanopore sequencing, or PacBio sequencing, is used. This step generates raw sequencing data that must be processed further.

7.2.1 Quality control and preprocessing

Before analyzing your scRNA-seq data, it is important to ensure that the quality is high. This can be done by assessing various quality metrics such as mapping rates, gene expression counts, and cell viability.

- The RNA-SeQC program in Java computes a set of quality control metrics for RNA-seq data. One or more BAM files can be used as input. The output includes HTML reports and tab-delimited files containing metrics data (https://software.broadinstitute.org/cancer/cga/rna-seqc).
- RNA-seq Quality Control: This is the R package-based workflow to control the quality of RNA-seq (https://cran.r-project.org/web/packages/RNAseqQC/vignettes/introduction.html).

It works based on the following steps.

7.2.1.1 Initiation

First load all related packages/libraries such as library("RNAseqQC"), library("DESeq2"), library("ensembldb"), library("dplyr"), library("ggplot2"), library("purrr"), library("tidyr"), library("tibble") and library("magrittr").

7.2.1.2 Input

The workflow's input is a genes samples count matrix and a *data.frame* of metadata annotating the samples, with the number of metadata rows equaling the number of matrix columns. Furthermore, the count matrix must have column names and row names that are ENSEMBL gene IDs.

7.2.1.3 Dataset

A *DESeqDataSet* builds from the input count matrix and metadata.

7.2.1.4 Plotting

This can be done by following R libraries.

- Total sample counts: *plot_total_counts(dds)*
- Library complexity: *plot_library_complexity(dds)*
- Gene detection: *plot_gene_detection(dds)*
- Gene biotypes: *plot_biotypes(dds)*
- Chromosome expression plots are used to find chromatin-related expression levels.

7.2.1.5 Alignment and assembly

Reads from the scRNA-seq process are aligned to a reference genome and assembled into transcripts. This step provides a comprehensive view of the transcriptome, including information about gene expression levels, alternative splicing, and novel transcripts. scRNA-seq data is high-dimensional at this step, making it difficult to analyze. Removing genes with low counts is a good intermediate step before moving on to quality control or the next step. As a result, the total number of genes, as well as the overall size of the dataset and computation time, are frequently reduced significantly. Determine the size of a biological group, that is, all samples under the same biological condition, and filter (keep) genes with a certain count at least as frequently as the biological group's size. This can be done by gene filtering and dimensionality reduction techniques such as principal component analysis (PCA) or t-SNE used to determine if biological or batch effects are the primary cause of the data's fluctuation.

7.2.1.6 Differential expression analysis

Differential expression analysis is performed to determine which genes are differentially expressed between different sample groups. This can be applied to the evaluation of quality. This begins by actually mapping the expression profile of intriguing genes before performing differential testing.

1. Plot a gene data
2. Differential testing: The MA-plot, which charts the mean normalized count versus the log2-fold change, is a useful diagnostic tool. Below R-based code helps to plot the differential testing.

```
dds$mutation <- as.factor(dds$mutation) dds$treatment <- as.factor(dds$treatment) design(dds) <- ~ mutation + treatmentdds <- DESeq(dds, parallel = T) plotDispEsts(dds)
```

7.2.1.7 Functional analysis

The results of differential expression analysis are integrated with other data, such as gene ontology, pathway analysis, and network analysis, to gain a deeper understanding of the biological processes and functions affected by the changes in gene expression. Many analyses may be performed to get a further biological understanding of the differentially expressed genes, including:

- Check to see whether any known biological relationships, activities, or pathways have been enriched.
- By putting genes in groups based on comparable tendencies, you can determine which genes are involved in new pathways or networks.
- Use external interaction data to visualize all genes that are significantly up- or downregulated globally to change gene expression.

Many packages are available in Bioconductor to perform the functional analysis such as

1. The edgeR (https://bioinf.wehi.edu.au/edgeR/) performed the differential expression analysis of digital gene expression data and Gene ontology analysis. This also used the Benjamini–Hochberg approach for multiple testing corrections to manage the false discovery rate (FDR).
2. The limma (https://bioinf.wehi.edu.au/limma/) evaluated studies on gene expression, particularly the use of linear models for the evaluation of planned experiments and the identification of differential expression.
3. The clusterProfiler (https://github.com/YuLab-SMU/clusterProfiler) is used to do overrepresentation analysis on GO keywords connected to our list of important genes. It uses hypergeometric testing to perform statistical enrichment analysis on a significant gene list and a background gene list.
4. Gene Set Enrichment Analysis (GSEA: https://www.gsea-msigdb.org/gsea/index.jsp) analyzes a group of genes to see if there are statistically significant changes between two biological states (e.g., phenotypes). It uses molecular profile data and a gene set database to analyze the results. It has MsigDB with annotated information on gene sets.
5. Signaling Pathway Impact Analysis (SPIA: http://bioconductor.org/packages/release/bioc/html/SPIA.html): It determines the pathways most pertinent to the situation under study using data from a list of differentially expressed genes and their log fold changes along with signaling network structure.
6. Coexpression clustering (https://horvath.genetics.ucla.edu/html/CoexpressionNetwork/Rpackages/WGCNA/): It is frequently used to identify genes of novel pathways or networks by grouping genes based on similar expression trends.
7. GeneMANIA (http://www.genemania.org/): It helps to predict the gene function of your gene of interest.

7.2.1.8 Data repository

Transcriptomics data must be properly managed and stored to ensure its long-term accessibility and reproducibility. This step also involves sharing the data with the scientific community, making it publicly available through repositories such as

- NCBI's Sequence Read Archive (SRA: https://www.ncbi.nlm.nih.gov/sra): It has the largest publicly accessible library of high-throughput sequencing, and it is made available through cloud service.
- Gene Expression Omnibus (GEO: https://www.ncbi.nlm.nih.gov/geo/): GEO is a free, open-access repository for functional genomics data that accepts submissions of MIAME-compliant data. Data based on arrays and sequences are acceptable. Users can download curated gene expression profiles and experiment results using tools that are made available to them.

- Human Cell Atlas Data Coordination Platform (HCA: https://data.humancellatlas.org/): Single-cell data contributed by labs from all around the world is stored and made available through the HCA Data Portal. Anyone can use the community's tools and applications, add data, or find data.

7.3 Metagenomics

Metagenomics is the study of collective genetic material from a complex assemblage of microorganisms in a specific environment, such as soil, water, or the gut microbiome. Without the need to cultivate particular microbes, metagenomic analysis entails sequencing, assembly, and annotation of DNA or RNA that has been directly isolated from ambient materials. Metagenomics can help us better understand the diversity and operation of microbial communities, as well as the interactions between different species and their environment. Metagenomic data analysis can include taxonomic profiling, gene function prediction, and a comparison of different samples or habitats. Metagenomics is a rapidly developing field that is heavily reliant on advances in DNA sequencing and bioinformatics tools. Metagenomics is expected to play an increasingly important role in understanding the diversity and function of microbial communities in a variety of environments as sequencing technologies improve and become more affordable. Metagenomics sequencing data is managed in the following steps.

7.3.1 Sample collection and preparation

Samples are collected from the environment and prepared for DNA extraction and sequencing (Venter et al., 2004). This step is critical to ensure the quality and quantity of DNA obtained, which will directly impact the quality of the resulting metagenomics data. This should be taken care of in these two steps for collecting the samples.

- If the target community is linked to a host (such as an invertebrate or plant), selective lysis or fractionation may be appropriate to ensure that only a small amount of host DNA is recovered (Burke et al., 2009).
- To increase DNA yield or prevent the coextraction of enzyme inhibitors (such as humic acids) that can interfere with further processing, physical separation and isolation of cells from the samples may also be crucial (Delmont et al., 2011).

7.3.2 DNA sequencing

DNA is sequenced using one of several DNA sequencing technologies, such as Illumina sequencing, Oxford Nanopore sequencing, or PacBio sequencing. This step generates raw sequencing data that must be processed further. It has been demonstrated that the ePCR generates synthetic duplicate sequences, which will affect any estimates of gene abundance (Niu et al., 2010). For the data quality of sequencing runs, it is essential to understand how many replicate sequences are present. Replicates can be found and removed using bioinformatics methods. The CloVR-metagenomics (CloVR: Cloud Virtual Resource), a desktop program that automates sequence analysis, requires two different inputs. For comparative analysis, the raw sequencing data (in fasta format) and the metadata file (tab-delimited) with sample-specific data are needed. A Virtual Machine (VM) player needs to run this program. To find an available Amazon Machine Image, users of Amazon Cloud can create a cloud-based instance and use the Request Instances Wizard (AMI). First, duplicate sequence reads are clustered using UCLUST (http://www.drive5.com/uclust) before functional and taxonomic identification is determined using BLAST (https://blast.ncbi.nlm.nih.gov/Blast.cgi) homology searches against the COG (http://www.ncbi.nlm.nih.gov/COG) and RefSeq (https://www.ncbi.nlm.nih.gov/refseq/) databases, respectively.

7.3.3 Quality control and preprocessing

Raw sequencing data must undergo quality control and preprocessing to remove contaminants, trim low-quality reads, and perform adapter trimming. This step is critical to ensure the quality of the resulting metagenomics data.

- FastQC (http://www.bioinformatics.babraham.ac.uk/projects/fastqc/): It assesses raw sequencing data per-base quality, per-base GC content, and sequence length distribution.
- Fastx-Toolkit (http://hannonlab.cshl.edu/fastx_toolkit/): It is a set of command-line tools for preprocessing short-read FASTA/FASTQ files, including read length trimming, identical reads collapsing, adapter removal, and format conversion.
- PRINSEQ (https://prinseq.sourceforge.net/): This tool can filter, reformat, and trim genomic and metagenomic sequence data.
- Trimmomatic (http://www.usadellab.org/cms/?page=trimmomatic): It is a tool for trimming and filtering Illumina reads. It can remove adapter sequences, low-quality bases, and reads with a length below a specified threshold.

- KneadData (https://huttenhower.sph.harvard.edu/kneaddata): It is used for quality control and preprocessing of metagenomic sequencing data. It uses several tools, including Trimmomatic and Bowtie2, to remove contaminants, host DNA, and low-quality reads.
- CheckM (https://ecogenomics.github.io/CheckM/): This tool is used for assessing the quality and completeness of metagenome-assembled genomes (MAGs). It can identify and remove low-quality MAGs and estimate their completeness and contamination.

7.3.4 Sequence assembly and binning

Reads from the DNA sequencing process are assembled into contigs and binned into genomic bins based on sequence similarity and completeness. This step provides a comprehensive view of the microbial community, including information about the abundance and diversity of different species.

7.3.5 Taxonomic and functional annotation

The genomic bins from the sequence assembly and binning step are annotated with taxonomic and functional information to identify the species present in the sample and their metabolic functions. This step is critical for understanding the underlying biology of a system and identifying potential biomarkers.

- BlobTools (https://blobtools.readme.io/): This tool is used for visualizing and analyzing metagenomic sequencing data. It can provide an overview of the taxonomic composition of the dataset, as well as identify and remove contaminants and other nontarget sequences.
- Mg-RAST (https://www.mg-rast.org/): This is a web-accessible tool having a set of tools for the analysis and display of metagenomic data. The system is an adaption of the RAST server system, which was initially put in place to allow for comprehensive microbial genome annotation in high quality utilizing SEED data. The mg-RAST analysis pipeline continues to use the microbial SEED data, but many other tools have been added to the system to improve microbial sequence taxonomic and functional classification. To facilitate 16s rRNA classification of metagenomic datasets, the green genes, RDP-II, and European ribosomal RNA databases have recently been added. All job-relevant resultant data is saved during pipeline installation in a flat file and SQLite (SQLite 7) formats.
- CAMAMED (https://github.com/mhnb/camamed): This provides a pipeline for mapping-based analysis of metagenomic data that takes composition into account. Metagenomic samples can be examined using this approach at the taxonomic and functional profiling levels.
- ezTree (https://github.com/yuwwu/ezTree): This is a Perl pipeline used to find marker genes and create phylogenetic trees for a collection of genomes. This script is used to align single copy marker genes for phylogenetic tree reconstruction from a set of genomes, including genome bins retrieved from metagenomes.
- metagenomeSeq (http://www.cbcb.umd.edu/software/metagenomeSeq): It identifies the characteristics that are differentially abundant between two or more groups of numerous samples, such as species and Operational Taxonomic Units (OTUs).

7.3.6 Data management

A few platforms are available to manage the metagenomic sequences from the first step to the last step such as Galaxy platform (https://usegalaxy.org/), and SqueezeMeta (https://github.com/jtamames/SqueezeMeta) and MetaHCR (https://github.com/metahcr). The Galaxy platform is a universal open-source framework being developed for large-scale data-intensive medical research that combines computational tools and databases into a unified workspace. It is combined with raw sequencing results (raw reads) and follows these steps.

- Examining the readings for filtration and quality (dada2)
- Text editing and data format conversion (Diamond makeDB)
- Looking up homology in the NCBI-NT database
- Taxonomic analysis using unique tools and specialized tools to visualize the results.

7.4 Proteomics

The SqueezeMeta is a very adaptable pipeline for investigating massive amounts of metagenomes. Everything is provided, starting with assembly, taxonomic/functional assignment of the resulting genes, and abundance estimation. SqueezeMeta reduces the strain of coassembling tens of metagenomes by using a sequential metagenomic assembly and later contig merging on computational infrastructures of a reasonable size. MetaHCR connects with the database back-end, and incorporates a user-friendly and contemporary front-end web interface, making the program accessible to nonexperts. It could be useful for several activities involving the management and analysis of complex metagenomics data and metadata for hydrocarbon-rich environments.

The study of the proteome, or proteomics, focuses on how various proteins interact and the functions they serve inside an organism. Mainly protein identification, quantification, localization, posttranslational modifications, functional and structural aspects, and protein–protein interactions study in this field. Various techniques are available for low-level proteomic studies, such as ELISA, 2D-PAGE, and chromatography, while high-throughput data-generated techniques are mass spectrometry and reverse-phase microarrays. Mass spectrometry is a rapidly leading method for protein identification. It is also a rapidly evolving field focusing on the large-scale quantification of specific proteins in specific cell types under defined conditions (Fig. 13.3). The rise of gel-free protein separation methods, combined with advances in MS instrument sensitivity and technology, has laid the groundwork for high-throughput approaches to protein research. The identification of parent proteins from derived peptides is now almost entirely dependent on search engine software, which can perform in silico digests of protein sequences to generate peptides. Their molecular mass is then compared to the protein fragments obtained experimentally. Proteomics data can be managed in the following steps, starting from the initial characterization of proteins to interactions of protein–proteins.

7.4.1 Characterization of proteins from sequences

Proteins are analyzed using one of several proteomics technologies, such as liquid chromatography–mass spectrometry (LC–MS) or tandem mass spectrometry (MS/MS). Raw proteomics data must undergo quality control and preprocessing to remove contaminants, perform signal-to-noise filtering, and perform mass calibration. Mass spectrometry (MS) is a

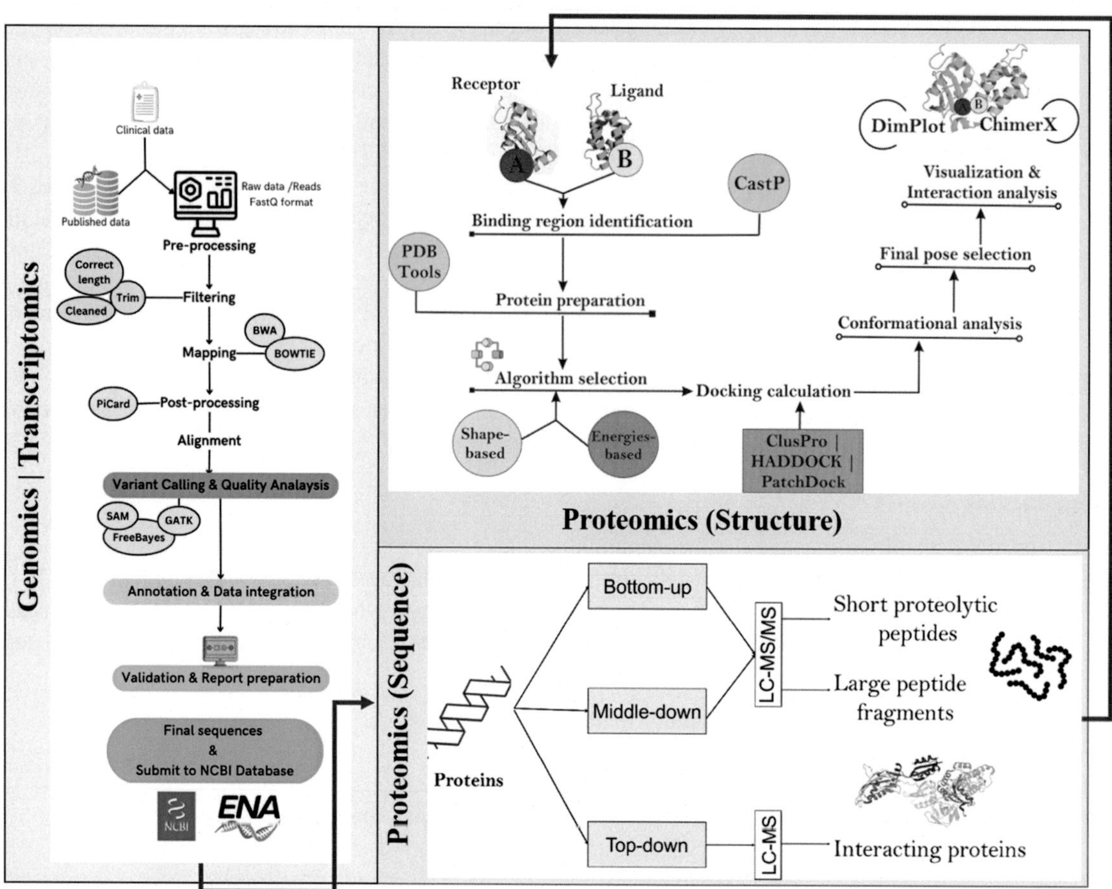

FIGURE 13.3 Cross-omics data integration. Genomics and transcriptomics connect with sequences-based and structure-based proteomics. The NGS (Next-Generation Sequencing) pipeline involves processing and analyzing raw data to extract the final gene that encodes proteins. The encoded proteins are utilized to construct a protein model, and these protein models can be used for protein-protein interaction analysis. The interaction analysis can be performed using docking calculation methods. In this, different omics fields are interconnected through cross-omics management.

rapidly developing international technique for classifying proteins. It is also a rapidly evolving discipline, with the current emphasis on extensively quantifying specific proteins in specific cell types under specified conditions. The advancement of gel-free protein separation methods and advances in the automation and sensitivity of MS instruments have paved the way for high-throughput approaches to protein research. Identifying parent proteins from generated peptides is now almost entirely the responsibility of search engine software, which can perform in silico digestions of protein sequences to generate peptides. The molecular mass of the protein fragments is then compared to the experimentally determined mass of the protein fragments (Fig. 13.3). The two majorly protein sequence databases are Uniprot and NCBI-Proteins. They have a lot of protein sequences with metadata. The identified proteins can be submitted to this database for further study. Several steps can be taken to manage proteomics data effectively. Using consistent and standardized data formats can make integrating and analyzing data from different experiments or sources easier.

- The SWISS 2-D PAGE (https://world-2dpage.expasy.org/swiss-2dpage) hosted by the Swiss Institute of Bioinformatics, provides one of the most comprehensive collections of proteomics data.
- Mass-spectrometry-based proteomics data can be retrieved and submitted on Open Proteomics Database (OPD: http://data.marcottelab.org/MSdata/OPD/) and Yeast genome-scale protein-localization data can be submitted in Yeast GFP Fusion Localization Database (http://yeastgfp.ucsf.edu).

There are numerous software tools for managing, analyzing, and visualizing proteomics data. The following tools and software can assist in automating data processing, identifying patterns, and performing statistical analyses.

- GeneSpring (www.agilent.com): This software allows users to perform statistical analyses of protein expression data and visualize the results in various plots and graphs.
- Progenesis Q.I (https://www.nonlinear.com/progenesis/qi/): This software is specifically designed for analyzing and visualizing proteomics data and includes tools for data normalization, statistical analysis, and visualization.
- R (https://www.r-project.org/about.html): This is a programming language and software environment for statistical computing and graphics. It has a wide range of packages available for analyzing and visualizing protein expression data.
- Partek Proteomics (https://www.partek.com/): This software is designed for analyzing and visualizing proteomics data and includes tools for data normalization, statistical analysis, and visualization.
- Perseus (https://maxquant.net/perseus/): This software is designed for analyzing and visualizing proteomics and other types of high-throughput data and includes tools for data normalization, statistical analysis, and visualization.

Several databases are specifically designed for storing and managing proteomics data, such as PRIDE (https://www.ebi.ac.uk/pride/) and PeptideAtlas (http://www.peptideatlas.org/). The following databases that can help to organize and annotate data and make it easier to share with other researchers are listed here.

- BioGRID (http://thebiogrid.org/): BioGRID is an interaction repository that archives and disseminates genetic and protein interaction data from model organisms and humans, which currently holds over 1,400,000 interactions curated from both high-throughput datasets and individual-focused studies.

7.4.2 Protein–protein interactions

By accurately defining a protein's function within a particular cell type, many methods frequently used to experimentally determine protein interactions lend themselves to high-throughput procedures, offering useful insights into numerous fields. Complementation assays, such as the 2-hybrid, evaluate the oligomerization-assisted complementation of two fragments of a single protein (Fig. 13.3). When joined, the two protein fragments are fused to possible bait/prey interaction partners, producing a straightforward biological readout.

- Database of interacting proteins (DIPs: http://dip.doe-mbi.ucla.edu/dip/Main.cgi): It lists protein pairs that are known to interact with each other, which combines information from a variety of sources to create a single, consistent set of protein–protein interactions.
- HitPredict (http://hintdb.hgc.jp/htp/): It is a resource of experimentally determined protein–protein interactions, which currently includes 12 species (like *Arabidopsis thaliana*, *Escherichia coli*, *Homo sapiens*, and *Saccharomyces cerevisiae*).
- IntAct (http://www.ebi.ac.uk/intact/): It is a freely available and open-source database system and analysis tool for molecular interaction data, and provides PPI information derived from literature curation or direct user submissions.
- Interactome3D (http://interactome3d.irbbarcelona.org/): It provides the structural annotation of protein–protein interaction networks.

- Pathway Commons (http://www.pathwaycommons.org/): It is a web resource for collecting and disseminating biological pathway and interaction data.
- STRING (http://string-db.org/): It is a database of known and predicted protein—protein interactions, including direct (physical) and indirect (functional) associations.

7.5 System biology/network modeling

It is a method for analyzing and representing complicated systems as entangled networks of nodes and edges. The nodes in a network model represent the components of the system (genes, proteins, or persons), while the edges show the connections or interactions between these components (physical interactions, regulatory relationships). Network models can be used to pinpoint potential sites for intervention or control or to forecast how a system will behave under various circumstances. Cross-omics data can be utilized to build statistical, machine learning, and computational models using a variety of network modeling techniques. These methods apply to the analysis of cross-omics data from diverse sources, such as simulations, observational studies, and experiments. In the field of systems biology, there are many methods and instruments for managing cross-omics data. Pathway modeling, which entails creating and analyzing mathematical models of biological pathways and processes, is possible thanks to the availability of numerous software tools. A standardized language for displaying and exchanging mathematical models of biological systems is called SBML (Systems Biology Markup Language). There are many software, tools, and databases that support SBML, including COPASI, JWS Online, and Cell Designer.

- COPASI (https://copasi.org/): It is a software tool for modeling, simulating, and analyzing biochemical and cellular networks. It is particularly well-suited for the analysis of large and complex systems and supports a wide range of modeling approaches.
- JWS Online (https://jjj.mib.ac.uk/): JWS Online is a web-based platform for building, simulating, and analyzing biochemical and cellular networks as well as for storing models. The builder supports the Minimum Information Needed In the Annotation of Models (MIRIAM) and Systems Biology Ontology (SBO) standards for annotation and closely complies with the SBML model definition. It also offers a helpful online annotation tool.
- Cell Designer (https://www.celldesigner.org/): It is a software tool for constructing and simulating biochemical and cellular networks. It has a graphical user interface that allows users to easily build and edit pathways and includes a range of tools for simulating and analyzing the behavior of the model.
- KEGG (https://www.genome.jp/kegg/): Kyoto Encyclopedia of Genes and Genomes is a comprehensive database of pathways and metabolic pathways in cells. It contains information about the genes, proteins, and small molecules involved in these pathways, as well as the enzymes that catalyze the reactions and the regulatory mechanisms that control the pathways.
- Reactome (https://reactome.org/): It is a database of pathways and processes that occur in human cells. It contains information about the genes, proteins, and small molecules involved in these pathways, as well as the enzymes that catalyze the reactions and the regulatory mechanisms that control the pathways.
- BioCyc (https://biocyc.org/): It is a database of pathways and metabolic pathways in cells. It contains information about the genes, proteins, and small molecules involved in these pathways, as well as the enzymes that catalyze the reactions and the regulatory mechanisms that control the pathways.

8. Cross-omics data integration

Cross-omics data integration is the process of combining and analyzing several omics data types, such as genomics, transcriptomics, proteomics, and metabolomics, to fully comprehend biological systems (Fig. 13.3). To gain a deeper knowledge of the underlying biological processes and discover new biomarkers and therapeutic targets, several omics datasets can be merged. Integration of cross-omics data faces a number of difficulties, including heterogeneity and sparsity of the data as well as the necessity for efficient computing techniques for data analysis. Among the frequently employed methods for integrating cross-omics data include the following:

8.1 Network-based integration

To discover important nodes and pathways that are shared by various datasets, this method includes building biological networks using the various omics datasets and merging them. It is a method for combining different kinds of omics data by building biological networks and combining them to find critical nodes and pathways that are shared by various datasets.

Building biological networks entails figuring out how various biological components are related based on experimental evidence such as gene expression, protein–protein interactions, and metabolic fluxes. Depending on the type of data being integrated, several networks can be built, including gene regulatory networks, protein interaction networks, and metabolic networks. Once the networks are constructed, they can be integrated using a variety of methods, such as clustering, network alignment, and module detection, to identify common nodes and pathways that are shared across different omics datasets. These common nodes and pathways can provide insights into the underlying biological processes and can help in the identification of novel biomarkers and therapeutic targets.

Many biological systems, including cancer, neurodegenerative illnesses, and infectious diseases, have benefited from network-based integration. This method can provide a more thorough understanding of biological systems than individual datasets alone since it can incorporate several forms of omics data. Due to the complexity of biological networks and the requirement for efficient computing tools for data processing, network-based integration can be difficult. Several omics datasets can be integrated using a variety of network-based integration methods to build biological networks and pinpoint important nodes and routes. A few of these tools include.

- Cytoscape (https://cytoscape.org/): A popular open-source platform for visualizing and analyzing biological networks, including protein–protein interaction networks, gene regulatory networks, and metabolic networks. Cytoscape also provides a range of plugins for network-based data integration, including plugins for gene set enrichment analysis, pathway analysis, and network clustering.
- Omics Integrator (http://fraenkel-nsf.csbi.mit.edu/omicsintegrator/): An integrative analysis tool that combines multiple omics datasets to generate a consensus network of genes and proteins. Omics Integrator uses a network propagation algorithm to integrate different data sources and identify key nodes and pathways.
- NetworkAnalyst (https://www.networkanalyst.ca/): An online tool for visualizing and analyzing biological networks, including protein–protein interaction networks, gene regulatory networks, and metabolic networks. NetworkAnalyst provides a range of tools for network-based data integration, including tools for pathway enrichment analysis, gene set enrichment analysis, and network clustering.
- String (https://string-db.org/): A database of known and predicted protein–protein interactions, which can be used to construct protein–protein interaction networks. String also provides tools for functional enrichment analysis and network visualization.
- Pathway Studio (https://mammalcedfx.pathwaystudio.com/app/search): A commercial software platform for pathway analysis and network visualization. Pathway Studio integrates data from multiple omics datasets, including gene expression, protein expression, and metabolomics data, to generate a comprehensive view of biological pathways and networks.
- iOmicsPASS (https://github.com/cssblab/iOmicsPASS): It is a network-based data integration and feature selection method for predictive data. The tool combines several -omics datasets over biological networks, discovers significant subnetworks as signatures for differentiating sample groups, and then combines these signatures (i.e., phenotypes). It is a supervised method in which the samples' classification is known from the training data.
- NetICS (https://github.com/cbg-ethz/netics): To prioritize genes based on their proximity to genetically abnormal and differentially expressed genes (cancer genes), it uses a per-sample bidirectional network diffusion technique.
- netOmics (https://github.com/abodein/netOmics): It is a tool for creating and exploring multiomics networks. It constructs multilayered networks by combining network inference algorithms and knowledge-based graphs.
- timeOmics (https://github.com/abodein/timeOmics): It is a data-driven framework for integrating multiomics longitudinal data from the same biological samples and identifying key temporal features with strong associations within the same sample group.

8.2 Machine learning-based integration

Using machine learning algorithms, this approach integrates multiple omics datasets to identify patterns and associations that are not readily apparent from individual datasets. Machine learning-based integration can help identify complex relationships between biological entities and provide insights into disease mechanisms. Some of the most commonly used machine learning-based integration approaches are as follows.

- Multiomics factor analysis (MOFA: https://biofam.github.io/MOFA2/): MOFA is a probabilistic factor analysis framework that can integrate multiple omics datasets, including transcriptomics, proteomics, and metabolomics data. MOFA can identify shared and dataset-specific variation across different omics datasets and can help identify key features that are important in biological systems.

- Joint and individual variation explained (JIVE: https://github.com/idc9/py_jive): JIVE is a dimensionality reduction technique that can be used to decompose multiomics datasets into joint and individual components. JIVE can help identify patterns and associations that are common to multiple omics datasets and can help identify novel features that are unique to individual datasets.
- MoGCN (https://github.com/Lifoof/MoGCN): A Graph Convolutional Network-Based Approach for Multi-Omics Integration for Cancer Subtype Analysis.
- BRANet (https://github.com/Surabhivj/BRANet): It is an embedding multilayer network for integrating omics data.

8.3 Deep learning-based integration

Deep learning-based integration is the process of integrating and analyzing various omics datasets using deep neural networks. Due to its capacity to recognize intricate patterns and relationships in sizable, high-dimensional datasets, deep learning-based techniques have grown in popularity in recent years. For combining various omics datasets, there are a number of deep learning-based integration tools available, including.

- DeepGx: A deep learning-based method for integrating gene expression and genetic data. DeepGx uses a convolutional neural network (CNN) to learn the relationships between genetic variants and gene expression patterns (42).
- DeepGestalt (https://www.face2gene.com/): A deep learning-based method for integrating genomic and clinical data to diagnose rare genetic disorders. DeepGestalt uses a CNN to learn the features that distinguish between different genetic disorders.
- DeepInsight (https://alok-ai-lab.github.io/DeepInsight/): A deep learning-based method for integrating epigenomic and transcriptomic data. DeepInsight uses a deep autoencoder to learn a low-dimensional representation of the data that captures the underlying relationships between different omics datasets.
- DeepCpG (https://github.com/cangermueller/deepcpg): A deep learning-based method for integrating DNA methylation and gene expression data. DeepCpG uses a CNN to learn the relationships between DNA methylation patterns and gene expression patterns.
- DeepCC (https://github.com/CityUHK-CompBio/DeepCC): A deep learning-based method for integrating multimodal neuroimaging data. DeepCC uses a CNN to learn the features that distinguish healthy and diseased brains.
- DeepMulti (https://github.com/SharifBioinf/DeePathology): The PreProcess folder contains all of the code for obtaining and preprocessing data. These are custom R scripts for downloading The Cancer Genome Atlas (TCGA) mRNA and miRNA expression profiles from Genomic Data Commons.
- CustOmics (https://github.com/HakimBenkirane/CustOmics): It is used for multiomics integration. TCGA datasets were used to evaluate classification and survival using multiple test cases.

These integration tools, which are based on deep learning, can support the development of precision medicine and personalized medicine by assisting in the identification of new therapeutic targets and biomarkers. Nevertheless, deep learning-based techniques demand a lot of data and computing power, and their output can be challenging to understand. Hence, thorough results validation and interpretation are required.

8.4 Correlation-based integration

Identifying correlations and coexpression patterns across different omics datasets and using these to infer functional relationships between genes, proteins, and metabolites is the goal of this approach. For integrating multiple omics datasets, several correlation-based integration methods are available, including.

- CNAmet (http://csbi.ltdk.helsinki.fi/CNAmet): It is an algorithm and R package that integrates copy number, DNA methylation, and expression data to prioritize putative cancer driver genes. Our algorithm generates scores for methylation alteration-induced expression, copy number aberration-induced expression, a combined effect score, and adjacent *P*-values that are multiple hypotheses corrected.
- Weighted Gene Coexpression Network Analysis (WGCNA: https://horvath.genetics.ucla.edu/html/CoexpressionNetwork/Rpackages/WGCNA/): A method for identifying modules of coregulated genes based on their patterns of expression across different conditions. WGCNA uses a correlation-based approach to identify groups of genes that are highly correlated and that share similar functions.
- Gene Set Enrichment Analysis (GSEA: https://www.gsea-msigdb.org/gsea/index.jsp): A method for identifying gene sets that are differentially expressed across different conditions. GSEA uses correlation analysis to identify groups of genes that are highly correlated and that are enriched in specific biological pathways or functions.

8.5 Bayesian-based integration

This approach involves using probabilistic models to integrate multiple omics datasets and identify causal relationships between different biological entities. The integration of cross-omics data is an important area of research in systems biology and can help in the development of personalized medicine and precision healthcare.

- BNCreator and BNetBuilder: These two plugins are available in Cytoscape for Bayesian network analysis. BNCreator constructs Bayesian networks from gene expression data, and BNetBuilder provides a graphical interface for Bayesian network modeling.
- GeneMANIA (https://genemania.org/): It is a web-based tool for gene function prediction and network analysis that integrates several data sources, including protein–protein interactions, gene coexpression, and functional annotations. GeneMANIA uses a Bayesian approach to predict gene function based on functional associations with other genes in the network.
- NetworkInference (https://www.networkinference.org/): It is a web-based tool for inferring gene regulatory networks from gene expression data using Bayesian networks. NetworkInference allows users to upload gene expression data and generate a network model that can be visualized and analyzed.
- Bayesian Factor Regression Models (https://bfarm.mssm.edu/): It is a web-based tool for integrating multiple genomic data sources to identify genetic associations with complex traits. Bayesian Factor Regression Models use a Bayesian framework to integrate genetic variants, gene expression, and other genomic data sources to identify causal variants and prioritize candidate genes for further study.
- BINOVA (https://bioconductor.org/packages/release/bioc/html/BINOVA.html): It is a web-based tool for integrating multiple omics data sources to identify functional gene modules and pathways. BINOVA uses a Bayesian framework to identify significant associations between omics data sources and generate functional networks that can be visualized and analyzed.

8.6 Similarity-based integration

- PINSPlus (https://cran.r-project.org/web/packages/PINSPlus/index.html): It is an algorithm that aids in determining how frequently patients are clustered into a single cluster. The technique is used to combine data and group patients who are closely associated with a disease subtype.
- Neighborhood-based multiomics clustering (NEMO: https://github.com/Shamir-Lab/NEMO): Creates a similarity matrix by first determining how similar the patients are across all of the omics datasets. This is after that combined into a single matrix and clustered.
- OmicsSIMLA (https://omicssimla.sourceforge.io/): It is a tool for simulating the production of multiomics data with disease status. It has four modules; SeqSIMLA, pWGBSSimla, RNA-seq, and RPPA.

9. Quality control for cross-omics data management

It is clear that computer power is necessary to make sense of the enormous amount of data generated by omics research and for that, Bioinformatics must not only supply the structures or store the data but also store it in a way that makes it retrievable and comparable to both other types of information and data of a similar nature. The parameters for data sorting, storage, and integration with other types of data should all be configurable by the user. The following is a list of data analysis parameters.

- Data quality: Making sure the data are of high quality is crucial for accurate and reliable analysis. This may comprise producing data in accordance with predetermined processes and subjecting the data to quality control inspections.
- Data integration: Due to variances in data kinds, scale, and formatting, integrating data from several omics technologies can be difficult. While integrating the data, it is critical to employ the proper tools and methods and to pay close attention to the constraints and biases of the various data sources.
- Data storage and management: It might be difficult to store and manage vast amounts of data from various omics technologies. To handle and store the data and make sure that it is logically and consistently arranged, it is crucial to employ the right software tools and database systems.
- Data analysis: Analyzing cross-omics data requires specialized tools and expertize. It is important to choose appropriate analysis methods suited to the study's specific goals and carefully consider the data's limitations and biases.
- Data sharing and dissemination: Sharing cross-omics data with others can facilitate collaboration and ensure the data is used to its full potential. It is important to use appropriate data-sharing platforms and to consider any ethical and legal issues related to data sharing.

- A precise, predictive transcription initiation and termination model can be set before running the experiment. This can be done by theoretical study; it can predict where and when transcription will occur in a genome.
- Exact, quantitative models of signal transduction pathways that can forecast cellular responses to environmental cues are needed in proteomics and system biology.
- Ascertain the basic protein model and folding using a bioinformatics program, and predict the protein structure and early details.
- Use text or data mining techniques to predict the chemical structure or search the literature to assist in the development of small molecule inhibitors.
- Look up homologs or templates in databases of protein sequences that can explain mutation and evolutionary studies.

The structure and functions of biological data can be understood through the prediction of primary data using computational methods. It reduces both the cost and time of the trials. When data come from various datasets, data integration also aids in matching cross-studies.

10. Challenges in cross-omics data management

We will not be able to gain perspectives into fundamental biological processes until the enormous volumes of data generated by omics experiments are systematically documented and preserved in databases that can be searched, compared, and evaluated. Data must be kept in a structured, defined manner to facilitate data exchange between various resources, the development of standardized tools, and the fusion of data generated by multiple technologies. Omics heavily relies on technology, and all instrument and software manufacturers only supply data in their proprietary formats, usually limiting users to a small number of supplemental instrumental techniques. Attempts have been made for many years to develop and promote standardized data transfer formats as well as standard methods for labeling such data to allow dataset comparability. Several challenges can arise when managing cross-omics data, including:

- Data integration: Integrating data from different omics technologies can be challenging due to differences in the data types, the data's scale, and the data's formatting. It can be difficult to align and merge data from different sources and to ensure that the data is consistent and accurate.
- Data storage and management: Storing and managing large volumes of data from multiple omics technologies can be challenging. It can be difficult to choose appropriate software tools and database systems to store and manage the data and to ensure that the data is organized logically and consistently.
- Data analysis: Analyzing cross-omics data requires specialized tools and expertize. It can be challenging to choose appropriate analysis methods suited to the study's specific goals and to accurately interpret the results in the context of the different data sources.
- Data sharing and dissemination: Sharing cross-omics data with others can facilitate collaboration and ensure the data is used to its full potential. However, there may be ethical and legal issues to consider when sharing data, such as issues related to confidentiality and privacy.
- Funding and resources: Managing cross-omics data can be resource-intensive, and it may be difficult to secure sufficient funding to support the necessary infrastructure and expertize.

11. Data sources (Table 13.2)

TABLE 13.2 Data repositories for cross-omics data.

#	Repository name	Data type	Website link
1	The Cancer Genome Atlas (TCGA)	RNA-Seq, DNA-Seq, miRNA-Seq, SNV, CNV	https://cancergenome.nih.gov/
2	ColPortal	Methylation 450 k arrays, 16 S sequencing, expression arrays, microRNA arrays	https://colportal.imib.es/
3	Clinical Proteomic Tumor Analysis Consortium (CPTAC)	Proteomics data	https://pdc.cancer.gov/pdc/browse
4	Personal Genome Project-UK	RNA-Seq, WGBS, 450 K methylomics, WGS	https://www.personalgenomes.org.uk/

TABLE 13.2 Data repositories for cross-omics data.—cont'd

#	Repository name	Data type	Website link
5	International Cancer Genomics Consortium (ICGC)	Genomic variations data	https://icgc.org/
6	Cancer Cell Line Encyclopedia (CCLE)	Gene expression, copy number, and sequencing data	https://sites.broadinstitute.org/ccle
7	Molecular Taxonomy of Breast Cancer International Consortium (METABRIC)	Clinical traits, gene expression, SNP, and CNV	https://www.bccrc.ca/dept/mo/
8	TARGET	Gene expression, miRNA expression, copy number, and sequencing data	https://ocg.cancer.gov/programs/target
9	Omics Discovery Index	Genomics, transcriptomics, proteomics, and metabolomics	https://www.omicsdi.org
10	Lifebit	Biological data	https://www.lifebit.ai/
11	Seven Bridges Genomics	Biomedical and informatics data	https://www.sevenbridges.com/
12	MaxQuant	Analyzing large mass-spectrometric datasets	https://www.maxquant.org/
13	Expasy	SWISS bioinformatics resource portal	https://www.expasy.org/
14	Cytoscape	Visualizing and integrating complex networks	http://www.cytoscape.org
15	Metagenomics		http://www.ncbi.nlm.nih.gov/Genbank/metagenome.html
16	Metagenomics		http://www.ebi.ac.uk/genomes/wgs.html
17	The European Genotype Archive		https://ega-archive.org/
18	ENCODE	A public research consortium	http://www.genome.gov/10005107
19	Ensembl genome browser		https://useast.ensembl.org/index.html
20	Uniprot		https://www.uniprot.org/

Abbreviations: *CNV*, copy number variation; *miRNA*, microRNA; *RPPA*, reverse-phase protein array; *SNP*, single-nucleotide polymorphism; *SNV*, single-nucleotide variant.

12. Conclusion

Cross-omics data management involves integrating and analyzing data from different levels of biological organization, such as genomics, transcriptomics, proteomics, and metabolomics. Effective management of cross-omics data requires robust and scalable data storage and management systems, as well as bioinformatics tools for data integration, analysis, and visualization. Additionally, it is important to have a clear understanding of the experimental design and the biological context of the data and appropriate statistical and computational methods for data analysis. Overall, cross-omics data management aims to gain a holistic understanding of the biological systems being studied and to facilitate the discovery of new insights and knowledge.

References

Bayley, H. (2015). Nanopore sequencing: From imagination to reality. *Clinical Chemistry, 61*(1), 25–31. https://doi.org/10.1373/clinchem.2014.223016

Burke, C., Kjelleberg, S., & Thomas, T. (2009). Selective extraction of bacterial DNA from the surfaces of macroalgaeδ. *Applied and Environmental Microbiology, 75*(1), 252–256. https://doi.org/10.1128/AEM.01630-08. http://aem.asm.org/cgi/reprint/75/1/252

Chen, X., Schulz-Trieglaff, O., Shaw, R., Barnes, B., Schlesinger, F., Källberg, M., Cox, A. J., Kruglyak, S., & Saunders, C. T. (2016). Manta: Rapid detection of structural variants and indels for germline and cancer sequencing applications. *Bioinformatics, 32*(8), 1220–1222. https://doi.org/10.1093/bioinformatics/btv710. http://bioinformatics.oxfordjournals.org/

Delmont, T. O., Robe, P., Clark, I., Simonet, P., & Vogel, T. M. (2011). Metagenomic comparison of direct and indirect soil DNA extraction approaches. *Journal of Microbiological Methods, 86*(3), 397–400. https://doi.org/10.1016/j.mimet.2011.06.013

Dobin, A., Davis, C. A., Schlesinger, F., Drenkow, J., Zaleski, C., Jha, S., Batut, P., Chaisson, M., & Gingeras, T. R. (2013). Star: Ultrafast universal RNA-seq aligner. *Bioinformatics, 29*(1), 15–21. https://doi.org/10.1093/bioinformatics/bts635

Dolzhenko, E., Deshpande, V., Schlesinger, F., Krusche, P., Petrovski, R., Chen, S., Emig-Agius, D., Gross, A., Narzisi, G., Bowman, B., Scheffler, K., van Vugt, J., French, C., Sanchis-Juan, A., Ibáñez, K., Tucci, A., Lajoie, B. R., Veldink, J. H., Raymond, F., … Birol, I. (2019). ExpansionHunter: A sequence-graph-based tool to analyze variation in short tandem repeat regions. *Bioinformatics, 35*(22), 4754–4756. https://doi.org/10.1093/bioinformatics/btz431

Fan, Y., Xi, L., Hughes, D. S. T., Zhang, J., Zhang, J., Futreal, P. A., Wheeler, D. A., & Wang, W. (2016). MuSE: Accounting for tumor heterogeneity using a sample-specific error model improves sensitivity and specificity in mutation calling from sequencing data. *Genome Biology, 17*(1), 178. https://doi.org/10.1186/s13059-016-1029-6

Faust, G. G., & Hall, I. M. (2014). Samblaster: Fast duplicate marking and structural variant read extraction. *Bioinformatics, 30*(17), 2503–2505. https://doi.org/10.1093/bioinformatics/btu314

Favero, F., Joshi, T., Marquard, A. M., Birkbak, N. J., Krzystanek, M., Li, Q., Szallasi, Z., & Eklund, A. C. (2015). Sequenza: Allele-specific copy number and mutation profiles from tumor sequencing data. *Annals of Oncology, 26*(1), 64–70. https://doi.org/10.1093/annonc/mdu479. https://www.journals.elsevier.com/annals-of-oncology

Garrison, & Marth. (2012). *Haplotype-based variant detection from short-read sequencing.*

Iacoangeli, A., Al Khleifat, A., Sproviero, W., Shatunov, A., Jones, A. R., Morgan, S. L., Pittman, A., Dobson, R. J., Newhouse, S. J., & Al-Chalabi, A. (2019). DNAscan: Personal computer compatible NGS analysis, annotation and visualisation. *BMC Bioinformatics, 20*(1). https://doi.org/10.1186/s12859-019-2791-8. http://www.biomedcentral.com/bmcbioinformatics/

Joo, T., Choi, J., Lee, J., Park, S. E., Jeon, Y., Jung, S., & Woo, H. (2019). SEQprocess: A modularized and customizable pipeline framework for NGS processing in R package. *BMC Bioinformatics, 20*(1). https://doi.org/10.1186/s12859-019-2676-x

Kelley, D. R., Schatz, M. C., & Salzberg, S. L. (2010). Quake: Quality-aware detection and correction of sequencing errors. *Genome Biology, 11*(11). https://doi.org/10.1186/gb-2010-11-11-r116. http://genomebiology.com/2010/11/11/R116

Kim, D., Langmead, B., & Salzberg, S. L. (2015). HISAT: A fast spliced aligner with low memory requirements. *Nature Methods, 12*(4), 357–360. https://doi.org/10.1038/nmeth.3317

Kim, D., Pertea, G., Trapnell, C., Pimentel, H., Kelley, R., & Salzberg, S. L. (2013). TopHat2: Accurate alignment of transcriptomes in the presence of insertions, deletions and gene fusions. *Genome Biology, 14*(4). https://doi.org/10.1186/gb-2013-14-4-r36. http://genomebiology.com/2013/14/4/R36

Koboldt, D. C., Zhang, Q., Larson, D. E., Shen, D., McLellan, M. D., Lin, L., Miller, C. A., Mardis, E. R., Ding, L., & Wilson, R. K. (2012). VarScan 2: Somatic mutation and copy number alteration discovery in cancer by exome sequencing. *Genome Research, 22*(3), 568–576. https://doi.org/10.1101/gr.129684.111. http://genome.cshlp.org/content/22/3/568.full.pdf+html

Langdon, W. B. (2015). Performance of genetic programming optimised Bowtie2 on genome comparison and analytic testing (GCAT) benchmarks. *BioData Mining, 8*(1). https://doi.org/10.1186/s13040-014-0034-0. http://www.biodatamining.org/

Larson, D. E., Harris, C. C., Chen, K., Koboldt, D. C., Abbott, T. E., Dooling, D. J., Ley, T. J., Mardis, E. R., Wilson, R. K., & Ding, L. (2012). Somaticsniper: Identification of somatic point mutations in whole genome sequencing data. *Bioinformatics, 28*(3), 311–317. https://doi.org/10.1093/bioinformatics/btr665

Li, H., & Durbin, R. (2009). Fast and accurate short read alignment with Burrows-Wheeler transform. *Bioinformatics, 25*(14), 1754–1760. https://doi.org/10.1093/bioinformatics/btp324

Li, H., Handsaker, B., Wysoker, A., Fennell, T., Ruan, J., Homer, N., Marth, G., Abecasis, G., & Durbin, R. (2009). The sequence alignment/map format and SAMtools. *Bioinformatics, 25*(16), 2078–2079. https://doi.org/10.1093/bioinformatics/btp352

Liu, Y., Schröder, J., & Schmidt, B. (2013). Musket: A multistage k-mer spectrum-based error corrector for Illumina sequence data. *Bioinformatics, 29*(3), 308–315. https://doi.org/10.1093/bioinformatics/bts690

Mapleson, D., Accinelli, G. G., Kettleborough, G., Wright, J., & Clavijo, B. J. (2017). KAT: A K-mer analysis toolkit to quality control NGS datasets and genome assemblies. *Bioinformatics, 33*(4), 574–576. https://doi.org/10.1093/bioinformatics/btw663. http://bioinformatics.oxfordjournals.org/

Martin, M. (2011). Cutadapt removes adapter sequences from high-throughput sequencing reads. *EMBnet.journal, 17*(1), 10. https://doi.org/10.14806/ej.17.1.200

McLaren, W., Gil, L., Hunt, S. E., Riat, H. S., Ritchie, G. R. S., Thormann, A., Flicek, P., & Cunningham, F. (2016). The ensembl variant effect predictor. *Genome Biology, 17*(1). https://doi.org/10.1186/s13059-016-0974-4. http://genomebiology.com/

Niu, B., Fu, L., Sun, S., & Li, W. (2010). Artificial and natural duplicates in pyrosequencing reads of metagenomic data. *BMC Bioinformatics, 11*. https://doi.org/10.1186/1471-2105-11-187. http://www.biomedcentral.com/1471-2105/11/187

Park, P. J. (2009). ChIP–seq: Advantages and challenges of a maturing technology. *Nature Reviews Genetics, 10*(10), 669–680. https://doi.org/10.1038/nrg2641

Putri, G. H., Anders, S., Pyl, P. T., Pimanda, J. E., Zanini, F., & Boeva, V. (2022). Analysing high-throughput sequencing data in Python with HTSeq 2.0. *Bioinformatics, 38*(10), 2943–2945. https://doi.org/10.1093/bioinformatics/btac166

Shendure, J., & Ji, H. (2008). Next-generation DNA sequencing. *Nature Biotechnology, 26*(10), 1135–1145. https://doi.org/10.1038/nbt1486

Tarasov, A., Vilella, A. J., Cuppen, E., Nijman, I. J., & Prins, P. (2015). Sambamba: Fast processing of NGS alignment formats. *Bioinformatics, 31*(12), 2032–2034. https://doi.org/10.1093/bioinformatics/btv098

Venter, J. C., Remington, K., Heidelberg, J. F., Halpern, A. L., Rusch, D., Eisen, J. A., Wu, D., Paulsen, I., Nelson, K. E., Nelson, W., Fouts, D. E., Levy, S., Knap, A. H., Lomas, M. W., Nealson, K., White, O., Peterson, J., Hoffman, J., Parsons, R., … Smith, H. O. (2004). Environmental genome shotgun sequencing of the Sargasso Sea. *Science, 304*(5667), 66–74. https://doi.org/10.1126/science.1093857

Wang, K., Li, M., & Hakonarson, H. (2010). Annovar: Functional annotation of genetic variants from high-throughput sequencing data. *Nucleic Acids Research, 38*(16), e164. https://doi.org/10.1093/nar/gkq603

Wang, Z., Gerstein, M., & Snyder, M. (2009). RNA-seq: A revolutionary tool for transcriptomics. *Nature Reviews Genetics, 10*(1), 57–63. https://doi.org/10.1038/nrg2484

Zerbino, D. R., & Birney, E. (2008). Velvet: Algorithms for de novo short read assembly using de Bruijn graphs. *Genome Research, 18*(5), 821–829. https://doi.org/10.1101/gr.074492.107. http://www.genome.org/cgi/reprint/18/5/821

Chapter 14

Omics and clinical data integration and data warehousing

Sanjay Kumar Singh, Ajay Singh Dhama, Jasmine Kaur, Naveen Sharma, Pulkit Verma and Harpreet Singh
Division of Biomedical Informatics, Indian Council of Medical Research, New Delhi, India

1. Clinical data

Clinical data are a collection of data related to patient diagnosis, demographics, exposures, laboratory tests, and family relationships. Health data from patients are collected by the treating physicians and stored within the treating hospitals. The medical record of patients acts as a crucial and comprehensive source of clinical data as it contains even the minutest details of diagnosis and treatment followed. Conventionally, clinical or health data referred to any information present in a patient's medical record which included the diagnosis of diseases and subsequent services provided by the healthcare providers along with the impact of those services as analyzed by clinical outcomes of those services. This information may be acquired from notes derived from a hospital admission or a doctor's visit. These data comes in various forms such as text or numbers (patient identification, demographics, history, laboratory data, etc), analog or digital signals (ECG, EEG, EMG, ENG, etc), images (histological, radiological, ultrasound, etc), and videos.

Furthermore, clinical studies involve specimen collection from multiple patients. Ensuring the privacy of patient identifying information has become even more crucial as it cannot be utilized for public purposes. Health data are very complex in nature. Fig. 14.1 represents the various components of health data (Lyman et al., 2008).

A massive amount of health records, related documents, and medical images created by clinical diagnostic equipment are generated daily.

The medical information is gathered during routine daily activities and is stored in a number of systems, such as the Hospital Information System (HIS), Picture Archiving and Communications System (PACS), Radiology Information System (RIS), and others (Osada & Nishihara, 1999). However, various medical systems store clinical data during the time the patient is visiting. These types of data include (Lober et al., 2002) the following:

- Demographic Information: Information is collected once to provide a rich data analysis environment.
- Clinical information: Information about the patient's life habits, which is used to enhance data analysis capabilities.
- Diagnostics information: Describes the diagnostic process.
- Treatment information: Information about the treatment process including treatment type, treatment process, and treatment risk information.
- Laboratory Information: The result of a laboratory test.

Primary care data are the single richest source of routinely collected healthcare data. However, its use, both in research and clinical work, often requires data from multiple clinical sites with different health record systems and integration with clinical trial data and other types of medical data (Thiru et al., 2003).

Therefore, the need of the hour is to have an effective integrated health management system that can integrate patient data from different sources to provide a comprehensive analysis of patient healthcare data resulting in better treatment approaches for patients (Azadi & García-Peñalvo, 2023; Heart et al., 2017; Dhayne et al., 2019). To achieve integration of patient data from different sources such as Laboratory Information System (LIS), RIS/PACS, electronic case report form (eCRF), and electronic health record (EHR) data interoperability is a crucial factor that can be implemented by the integration of standard data models, terminologies, and messaging standards (Azadi & García-Peñalvo, 2023; Heart et al.,

Integrative Omics. https://doi.org/10.1016/B978-0-443-16092-9.00014-X

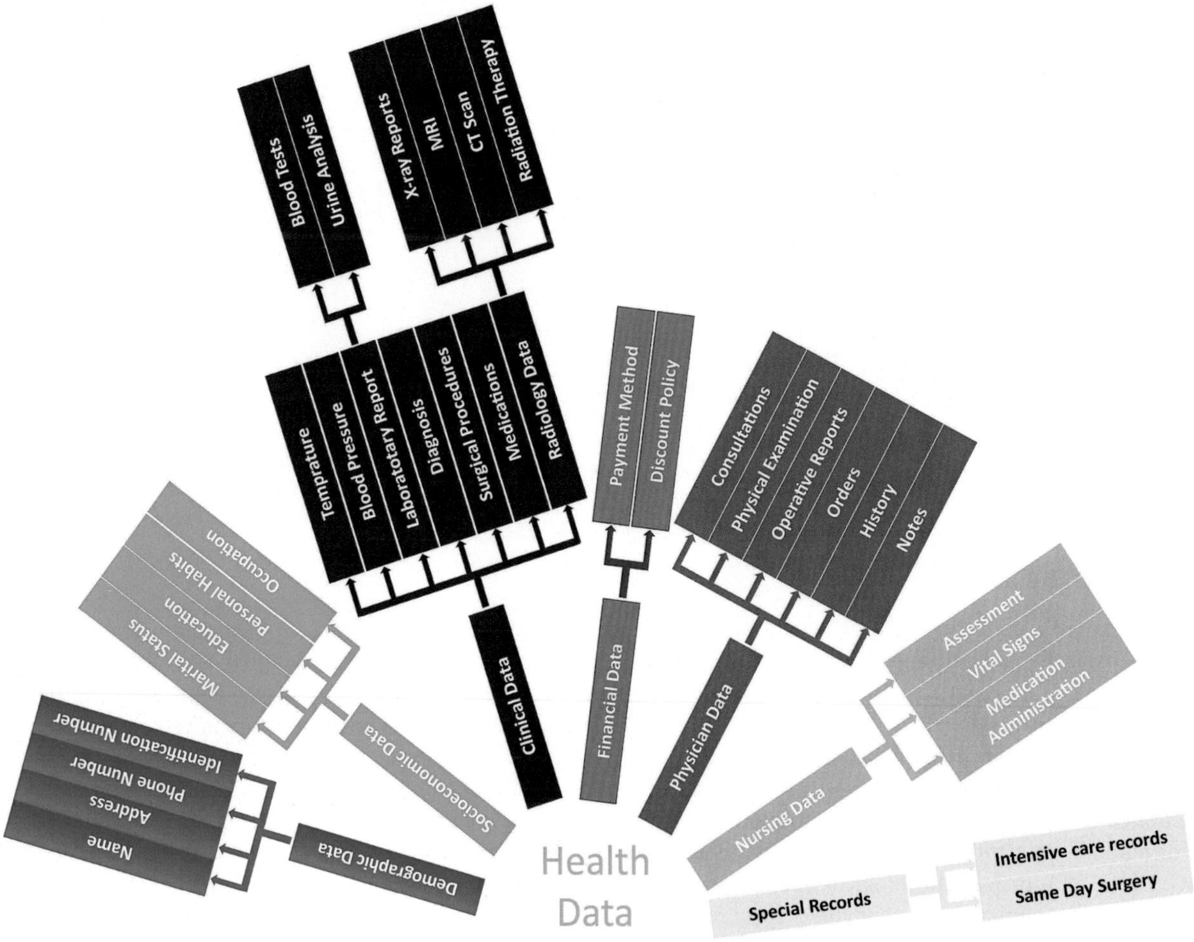

FIGURE 14.1 Components of health data.

2017; Dhayne et al., 2019). Another key aspect that needs to be kept in mind while ensuring data interoperability between different patient care data systems is ensuring that data sharing between different systems does not compromise the privacy/safety aspect of patient data; this can be achieved by utilizing innovative combinations of privacy-enhancing technologies (safe data, safe settings, safe outputs) and methods of distributed computing (Azadi & García-Peñalvo, 2023; Heart et al., 2017; Dhayne et al., 2019; Gagalova et al., 2020; Prasser et al., 2018).

2. Sources of clinical data

2.1 Hospital information system

An HIS management system acts as the backbone of the healthcare system in the hospital by ensuring seamless access to different aspects of the hospital's operation such as medical, financial, administration, etc. It acts as a bridge and integrates data between all departments to ensure improvement in information integrity, reducing the chances of duplication of information entries and ensuring a smooth workflow throughout the hospital. With the help of HIS comprehensive information about the patient's current health status be it diagnosis, the treatment offered, or even financial aspects such as billing information can be easily retrieved with one click.

2.2 Laboratory information system

For efficient management of different operational processes in a clinical laboratory, the utility of an LIS has gained prominence. The LIS is a highly customizable collection of software, operating systems, and hardware designed to cater to the specific operational need of an individual laboratory. The commercially available LIS are modular with the option of integrating separate software options, for example, in the case of general laboratory options include immunology, hematology, and microbiology, while other specialized LIS information includes blood bank information systems and anatomical pathology.

2.3 Radiology information system/picture archiving and communication system

To display, store, and transfer images generated from radiology techniques such as CT scan, MRI scan, and X-ray, we utilize a PACS, a medical imaging technology that assists in the electronic transmission of images and reports thereby eliminating the need for the physical transport and retrieval of traditional film.

A multilevel Digital Imaging and Communications in Medicine (DICOM) data model is utilized in the case of radiology datasets involving narrative radiology reports and associated imaging data. A detailed search for combinations of clinical and imaging data could also benefit from data integration at multiple grouping levels, i.e., to search for specific DICOM series instead of just discovering whether or not a patient or case has assigned images.

The DICOM data model is usually preferred for storing images in PACS. The DICOM data model is hierarchically structured and based on a real radiological examination: a single image (an instance), a series of images (e.g., images of a single MRI sequence), and a study containing all series required for the examination of a specific diagnosis. The radiology report in an RIS is generally linked to a DICOM study via the accession number (e.g., an order identifier) (Kaspar, Liman, et al., 2022).

3. Case report form

An eCRF reports health data that are crucial in the comprehensive analysis of patient health conditions such as patient medical history, any adverse event to medication, lab exam, and vital signs of the patient etc.

A powerful tool for collecting information is a case report form (CRF). The basic components of CRF are represented in Fig. 14.2. A CRF is a paper-based or digital questionnaire created to collect important data from study participants in accordance with the protocol's specifications. CRFs are usually required for the collection of primary data, but they can also be helpful for secondary data analysis to anticipate, define, and, if necessary, extract essential variables for the research question (Dhudasia et al., 2023).

CRFs offer some advanced features like setting validation rules to minimize errors during data collection. Designing an effective CRF upfront during the study planning phase helps to streamline the data collection process, and make it more efficient. Many types of inaccuracy may be present in the data collected, including typographical errors in data manually entered into a CRF. Checking for missing data is crucial. Values that are missing but would be useful for analysis if they were observed are referred to as missing data. By performing a number of data validation checks, errors in the data can be found. Data errors can either be flagged for deletion after identification or can be repaired, if that is possible (Daymont et al., 2017).

3.1 Electronic health record

An EHR system is utilized for the input, processing, storage, and retrieval of digital health data of patients. Fig. 14.3 illustrates the standard EHR workflow.

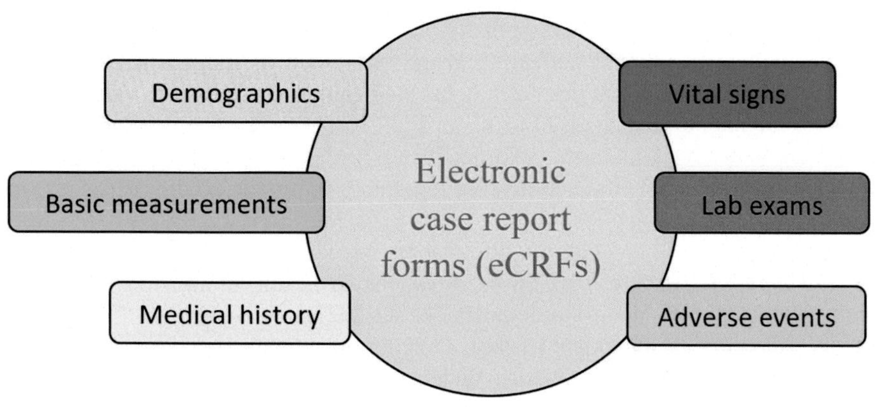

FIGURE 14.2 Electronic case report form.

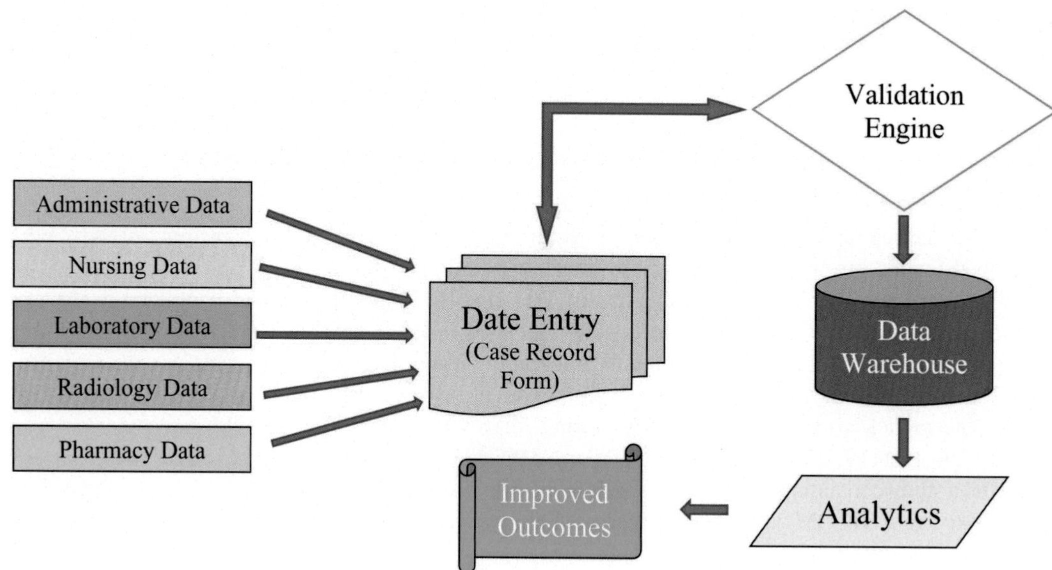

FIGURE 14.3 Standard workflow of EHR.

Since the last decade acceptance and adoption rate of EHR systems have significantly increased in the United States (Adler-Milstein et al., 2017) while in the case of several other countries, their use is gradually gaining acceptance in hospital and outpatient care settings (Lau et al., 2012; Schoen et al., 2009). The typical organization of an EHR is patient-centric and is a powerful utility tool that can store data in a time-dependent and longitudinal structure. Integration of EHR data into an enterprise data warehouse (DW) or integrated data repository (IDR) is also possible. The collection of heterogeneous data from multiple sources can be presented to the user in a comprehensive view by utilizing IDRs (MacKenzie et al., 2012). IDRs provide the advantage of specialized analytical tools for researchers or analysts to perform data analyses unlike EHRs that do not have any analytical tool.

The EHR is one such solution to support the healthcare facility, irrespective of levels and sizes to improve patient care by enabling functions that other types of records cannot deliver. The major requirements in healthcare facility are to use interoperability and standardization technique to enable easy sharing and exchange of healthcare data between the various levels. The main foundation for the interoperability is the standard terminology, which improves the effective communication between the two healthcare users (Warren et al., 2015).

The first component of the EHR system is the administrative system module, where data such as patient registration details (patient demographic data), admission, discharge, and transfer of record from one hospital to other are maintained. The second component of the system is nursing module, where mainly apparent information of the patient is recorded such as height, weight, blood pressure, and BMI values. The third component of EHR includes laboratory data such as microbiology, biochemistry etc., which plays an essential part in clinical process by furnishing relevant data to healthcare service provider. Approximately 60%–70% of the quality decisions are decided based on the lab results. The fourth module includes radiology-related data and image such as X-r ay, CT scan, and MRI. Medical imaging system called PACS is an imaging technology for managing digital images. The important and key component of EHR system is clinical documentation module to capture patient's clinical data such as diagnosis, procedure, complication, and medication. A clinical document in the electronic form helps the healthcare provider to validate and to provide quality care to the patient. One of the major advantages of using an EHR is it provides detailed information about the patient being treated thereby giving better insights to the next healthcare professional to make an informed decision about future course of treatment in a clinical/hospital setting that ultimately improves quality of treatment provided (Pai et al., 2021).

Content Exchange Standards (CES) are helpful to share clinical information including clinical summaries, prescriptions, etc. DICOM (2016) and Health Level-7 (2016) are the CES used for sharing health information with other healthcare levels when patient moves from one healthcare facility to other (Pai et al., 2021).

The EHRs are not only regulated in view of security requirement but also according to country-specific privacy laws. There are few guidelines suggested by the United States of America under the Health Insurance Portability and Accountability Act (HIPAA) (Act 1996) to safeguard the patient health data that are available and maintained using EHR.

HIPAA defines privacy rules on who has authority to access to health information system and specifies the security measures including administrative, physical, and technical safeguard (Pai et al., 2021).

4. Data integration

The process of merging data from various different data sources within one or more organizations into a single physical repository is known as data integration (Lenzerini, 2002). This massive amount of data are combined, reorganized, and aggregated to offer a unified view for data analysis (Lane et al., 2005).

A significant problem is the integration of data from disparate source data systems, which is not exclusive to the biomedical industry. Each system has unique metadata attributes that must first be taken into consideration. The data structures employed were created especially for the corresponding source data system. Due to proprietary concerns, this can make it challenging to reuse data that were gathered within particular source systems. The process of combining metadata from various data storages, archives, or repositories and storing it in a centralized database schema is known as metadata harvesting. In a data warehouse, it combines medical data with relevant metadata from clinical research databases, HIS, and LIS. It includes longitudinal data gathering, data integration, and data from datasets with various (meta) data types (Bönisch et al., 2022).

Data integration often relies on a combination of two types of models: information models (also called structural models) and terminological models (also referred to as semantic models). These two types of models, structural and terminological, are not independent as there are mutual constraints between the information models and coding systems (Qamar et al., 2007) requiring these two models to be bound in order to fully assert their content.

A subject's orientation and integration are strongly related. DWs are required to transform data from various sources into a uniform format. They must fix issues including nomenclature disputes and measurement unit inconsistencies. When they achieve this, they are said to be integrated (Pandey, 2014).

These integrated data are not yet turned into useful knowledge due to the lack of efficient analysis tools (Kerdprasop & Kerdprasop, 2011), also the lack of standardization between institutions which makes gathering data difficult (Baudot et al., 2009). Therefore, the data integration is an important issue in developing clinical data ware house (CDWH) (Lane et al., 2005).

Initially, the specific requirement for primary care data was finalized via discussions with field experts and through a sampling of various research criteria to get a detailed insight into the domain (Köpcke et al., 2013). The continuum of primary care aims at following the patients from birth to death, including disease treatment and preventive care. The healthcare domain has evolved in the last decades becoming a highly dynamic sector, especially regarding the Information Systems that support the care process, where a lot of information about patients is stored and used during the various workflows of health professionals. Over the years, the quality of data in health information systems has become increasingly important. The need of the hour is the development of new data models conforming to the latest based on patient privacy policies and data security so that it can promote the quality and safety of care provided to patients (Nogueira et al., 2019).

On the other hand, there is a lack of research relating to the introduction and impact of EHR in data quality and provision of care. Adler-Milstein et al., 2017 affirmed that the term "data quality" has issues in the area of health care and they identified an absence of a standardized definition (Thiru et al., 2003; Adler-Milstein et al., 2017).

Interoperability among the healthcare domain is of enormous importance, but at the same time, it is in itself very complex, where one of its greatest challenges is to find a solution for sharing information between different systems without loss of meaning or even propagation of error (Street & Arcement, 2014).

A relational database, which forms the foundation of the DW, stores the data and any associated metadata. The data tables and the metadata are linked by a primary/foreign key in a separate table that contains only the metadata. Thus, it is possible to store metadata in the same format in an n-dimensional repository. The medical source data is pseudonymized and transformed into the internal harmonized data format using the hospital and department information systems. Metadata is extracted and loaded using a specific data protocol during the Extract-Transform-Load (ETL) process's metadata harvesting phase to avoid duplication. All accessible data sources will eventually be connected by the DW as part of a continuous process (Bönisch et al., 2022).

5. Data warehouse

A relational database called a "data warehouse" is made for query and analysis rather than transaction processing. It often includes historical information generated from transaction data, but information from other sources is also possible. It allows an organization to compile data from several sources and isolates analysis workload from transaction workload (Shivtare & Shelar, 2015).

A DW environment includes a relational database in addition to an extraction, transportation, transformation, and loading (ETL) solution, an OLAP engine, client analysis tools, and other programs that control the process of gathering data and providing it to business users (Shivtare & Shelar, 2015).

DWs are characterized as "a subject-oriented, integrated, nonvolatile, and time-variant collection of data in support of management's decisions" (Watson et al., 2001) by Bill Inmon, who is regarded as the inventor of data warehousing. Based on this definition, the data in clinical data warehouse (CDW) have key features.

An effortless access and sharing of data from HIS for medical research has become a premium focus of the digitized century. Conceivable approaches to pursue a routine data access are a single-source data collection for research and routine and the direct extraction from an HIS after improved data search and standard support (Demski et al., 2016). However, CDWs currently appear to be the most promising, as they provide a technical basis for the comprehensive combination and use of intra- and interhospital secondary data. The potential of CDWs has been demonstrated to improve study feasibility analyses and enhance patient recruitment (Kaspar, Fette, et al., 2022).

CDWs enable quick queries on homogenized data of a large number of patients and data of a multitude of clinical subsystems as shown by many examples. CDWs can be used for a variety of reasons, including rapid feasibility testing or long-term data processing support for individual studies (Kaspar, Fette, et al., 2022). DWs serve as a central repository for storing and analyzing information to make better informed decisions. An organization's DW receives data from a variety of sources, typically on a regular basis, including transactional systems, relational databases, and other sources.

A DW architecture is usually of three types: single tier, two tier, and three tier. The three-level (tier) architecture of data warehouse is illustrated in Fig. 14.4. The results are typically presented to clients in the top tier using tools for reporting, analysis, and data mining. The analytics engine used to access and evaluate the data is included in the middle tier. Data are loaded and stored on the database server, which is the bottom tier of the system.

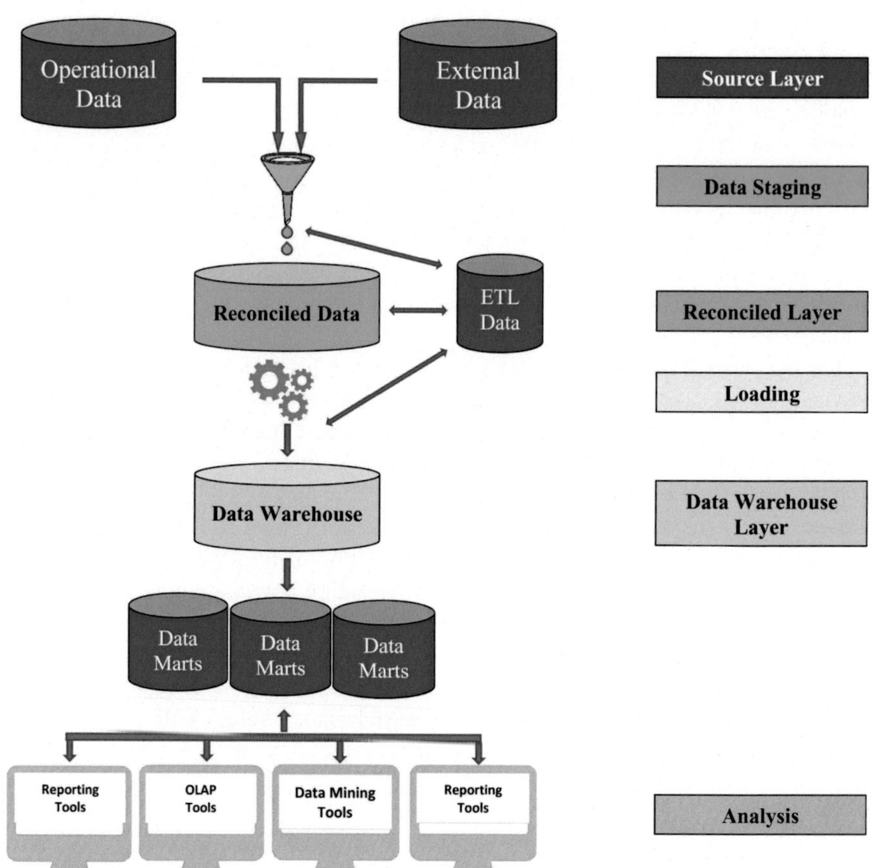

FIGURE 14.4 Three-tier architecture for a data warehouse.

The data warehouse (DWH) integrate data from two or more operational systems, which exist on one or several organizations. The integration process includes three steps:

1. Developing a unified model that can accommodate all information from single databases,
2. Transferring the data into developed model before loading them into the DW, and
3. Extract the data from the source databases and integrate it in one environment and provide access to obtain the required knowledge which the individual sources cannot provide it.

However, the integration steps require a set of hardware and software components that can be used to get better analysis of massive data as well as making better decisions making and research. Furthermore, the integration process requires architecture and tools to collect, analyze, clean, and present information. The DWHs technology provides many benefits that include enhancing the business intelligence and query performance, improving data quantity and quality, time-saving for users, and support the decision-making.

5.1 Key characteristics of data warehouse

The main characteristics of a DW are as follows:

- **Subject-Oriented**. A DW is subject-oriented since it provides topic-wise information rather than the overall processes of a business. Such subjects may be sales, promotion, inventory, etc. For example, if you want to analyze your company's sales data, you need to build a DW that concentrates on sales. Such a warehouse would provide valuable information like "who was your best customer last year?" or "who is likely to be your best customer in the coming year?"
- **Integrated**. A DW is developed by integrating data from varied sources into a consistent format. The data must be stored in the warehouse in a consistent and universally acceptable manner in terms of naming, format, and coding. This facilitates effective data analysis.
- **Non-Volatile**. Data once entered into a DW must remain unchanged. All data are read-only. Previous data are not erased when current data are entered. This helps you to analyze what has happened and when.
- **Time-Variant**. The data stored in a DW are documented with an element of time, either explicitly or implicitly. An example of time variance in data warehouse is exhibited in the primary key, which must have an element of time like the day, week, or month.

5.2 Benefits of data warehouse

- **Improved data consistency**

DWs are designed to apply a consistent format to all gathered data, making it simpler for decision-makers to assess and share data insights with their colleagues throughout the world. It also reduces the likelihood of a misinterpretation and increases overall accuracy to standardize data from various sources.

- **Accurate Reporting**

A consolidated data repository that provides real-time analytics will enable your healthcare facility to provide reports that are accurate and timely. For instance, you could effectively monitor patient conditions, medicine sales, and personnel efficiency. With sophisticated analytics and data visualization technologies, you can demonstrate data to key stakeholders, identify problem areas, or enhance clinical studies. Data collected from multiple sources and swiftly processed will ensure that you always have the most full and up-to-date information to act on.

- **Easier access to data for end-users**

The accessibility of a variety of data to end users is improved through data warehousing. Decision-makers frequently have to manually aggregate data by logging into each departmental system or seek reports from IT staff in order to obtain the information they require. Users can produce reports and queries independently by using a DW. Instead of requiring users to log into many systems, they may access all of the organization's data from a single interface. Less time is spent on data retrieval and more time is spent on data analysis when data are more easily accessible.

- **Improved patient experience and outcomes**

The entire patient journey can be seen when diagnostic data, follow-ups, and long-term results are combined with EHR/EMR information. Based on this information, it is possible to fill up service gaps, increase patient loyalty and happiness, and eventually raise the standard of treatment.

- **Better clinical decisions**

Using data effectively is a challenging task, but it pays off, especially when the appropriate insights are gained at the right moment. Analytics and storage technologies will be integrated with DWH. Get access to high-quality data, process them quickly and easily, and integrate them with clinical decision support systems that are correctly constructed, which offer an incredibly effective and flexible framework that can produce diagnostic judgments just when necessary.

- **Personalized value-based care**

Healthcare providers can advance toward providing value-based treatment by utilizing a DW in analytics. Hospitals can obtain deeper insights into the effectiveness of particular treatment regimens with the aid of machine learning (ML) and other advanced analytics tools. Patients can obtain treatment that is specifically tailored to their needs in this way, which helps prevent unforeseen costs.

5.3 ETL process in data warehouse

ETL is a process in data warehousing and it stands for extract, transform, and load. It is a process in which an ETL tool extracts the data from various data source systems, transforms them in the staging area, and then finally, loads them into the data warehouse system (Fig. 14.5).

Let us understand each step of the ETL process in depth:

1. **Extraction**: The first step of the ETL process is extraction. In this step, data from various source systems are extracted which can be in various formats like relational databases, NoSQL, XML, and flat files into the staging area. It is important to extract the data from various source systems and store them into the staging area first and not directly into the DW because the extracted data are in various formats and can be corrupted also. Hence loading it directly into the DW may damage it and rollback will be much more difficult. Therefore, this is one of the most important steps of ETL process.
2. **Transformation**: The second step of the ETL process is transformation. In this step, a set of rules or functions are applied on the extracted data to convert it into a single standard format. It may involve following processes/tasks:
   - Filtering—loading only certain attributes into the DW.
   - Cleaning—replacing the NULL entries with some default values and transforming America, the United States, and U.S.A. into USA, etc.
   - Joining—joining multiple attributes into one.
   - Splitting—splitting a single attribute into multiple attributes.
   - Sorting—sorting tuples on the basis of some attribute (generally key-attribute).
3. **Loading**: The third and final step of the ETL process is loading. In this step, the transformed data are finally loaded into the DW. Sometimes the data are updated by loading into the DW very frequently and sometimes it is done after longer but regular intervals. The rate and period of loading solely depends on the requirements and varies from system to system.

ETL process can also use the pipelining concept i.e., as soon as some data are extracted, they can transformed and during that period some new data can be extracted. And while the transformed data are being loaded into the DW, the already extracted data can be transformed.

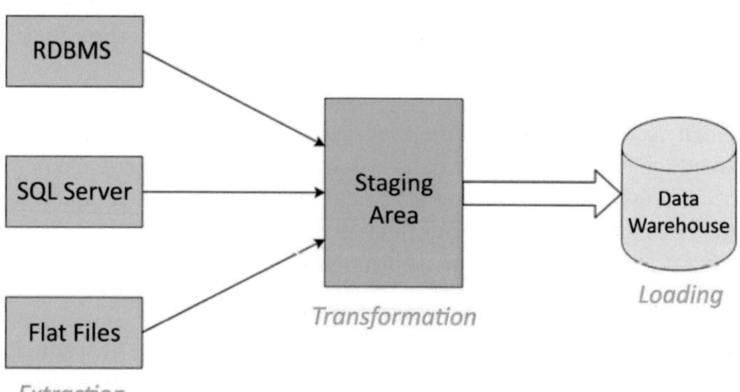

FIGURE 14.5 ETL process in data warehouse.

6. Clinical data integration in health care

Data integration of various healthcare datasets may enable improved access to more comprehensive patient data by doctors and reduce hospital and healthcare costs associated with increased handling of several EHRs and healthcare datasets.

- **Speed**. Especially in emergency cases, time is of the essence. Good integration enables speedy transmission of useful medical data. You can better identify risk factors and speed up diagnosis. Increasingly, these days, you can also consult a person's EHR, which consolidates much health and medical data about a person in one convenient digital form.
- **Accuracy**. Integrating data enables better evidence-based decision-making, leading to better health outcomes. That is crucial when it comes to your health. You need an accurate diagnosis, based on accurate information and good quality data, so that you can start the correct course of treatment.
- **Costs**. Integration can enable automation of business processes and more efficient use of resources for reduced costs.
- **Equity**. Getting access to fast, accurate medical data in publicly funded hospitals can mean fairer, more equitable health services for all, regardless of age, gender, ethnicity, neighborhood, or income level. Inequity in health services is real and involves many factors; good data connectivity is certainly one of them.

7. Data security and confidentiality

Data reliability for analysis, decision-making, and planning is a crucial attribute that is determined by data quality (Wang & Strong, 2015). It's possible that the data gathered are not in the format needed for analysis. The collection of data must be recategorized and recoded, along with the creation of new variables, as part of the data transformation process in order for it to be consistent with the study's analytical strategy. Since the data may contain protected health information about the study subjects, securely storing data is particularly crucial in clinical research (Health Insurance Portability and Accountability Act (HIPAA) Privacy Rule and the National Instant Criminal Background Check System (NICS). Final Rule, 2016).

Electronic data must never be transmitted using unencrypted portable media devices and must always be saved on a secure institutional server that has been approved. If some members of the study team do not need access to the study data, then the study team members should be given selective access based on their roles (Abouelmehdi et al., 2018).

The deidentification of data is another crucial component of data storage. In order to reduce privacy threats to individuals, data deidentification is the process of removing identifying attributes of the study participants from the data. Names, medical record numbers, dates of birth and deaths, and other details are used to identify study subjects.

Certain features should either be eliminated from the data or changed (for example, changing the medical record number to study IDs, changing dates to age/duration, etc.) in order to deidentify the data. Wherever possible, study data should be deidentified before storage (Dhudasia et al., 2023).

Security and privacy in clinical data are important issues. Privacy is often defined as having the ability to protect sensitive information about personally identifiable health care information.

According to the perspective of Abouelmehdi et al. (2018), security in clinical data has three major components i.e., information security, access control, and data security. Therefore, healthcare organizations must ensure security measures for protecting their clinical data associated hardware and software and also administrative information from internal and external risks.

Security, as is stated in the ISO/IEC 9126 International Standard, is one of the components of software quality. Information security can be defined as the preservation of confidentiality, integrity, and availability of information, in which confidentiality ensures that information is accessible only to those users with authorization privileges.

Integrity safeguards are implemented to ensure the accuracy and completeness of information and process methods. It also ensures that only authorized users have access to information and associated assets when required. Other modern definitions of information security also consider factors like accountability, authenticity, reliability, and nonrepudiation.

The requirement of standards generally arises when excessive diversity in parameters creates inefficiencies or impedes effectiveness.

A standard can be defined as a set of rules and definitions that specify how to transfer a process or produce a product. A standard is useful to provide a solution to a problem that people can use and they do not need to start from the beginning. With standards, people can work in a collaborative way. Some standards evolve over time; others are developed deliberately.

The main objective of the HL7 standard is to provide guidelines for the exchange of data between health information systems, helping to reduce communication errors and data quality problems thus increasing interoperability in a healthcare institution. HL7 v2.x is still today the most widely disseminated and implemented standard (Nogueira et al., 2019).

An HL7 v2.x message contains demographic information or traits that are used to help specifically identify a particular patient, such as name, address, social security number, date of birth, etc. This approach is done to facilitate flexibility and interoperability in implementations across multiple health information systems in intrainstitution or interinstitution scenarios (Nogueira et al., 2019).

DW security is referred to as the mechanism ensuring the integrity, confidentiality, and availability of the DW and its components. Confidentiality of DW upon deployment is one of the most crucial aspects to be considered. As SQL and OLAP queries by users are frequent in operations, the risk of security attacks is maximum therefore proper security measures are needed to ensure the confidentiality of data stored in the DW (Heart et al., 2017).

Protecting data and information from hackers and breaches becomes very specific and important when it comes to the healthcare industry. Health information security is an iterative process driven by enhancements in technology as well as changes to the health care environment.

HIPAA has given some standards which are aimed at protecting healthcare data and HL7 standards have smoothened clinical and administrative data communication between applications used by various healthcare providers.

8. Benefits of healthcare data integration and interoperability

- New standards of data integration by healthcare systems can facilitate the smooth exchange of electronic information.
- It also helps in the reduction of costs and problems of building interfaces between different systems.
- The important element of data sharing in the healthcare industry is interoperability.
- According to Healthcare Information and Management Systems Society, healthcare data interoperability is "the ability of different information technology systems and software applications to communicate, exchange data, and use the information that has been exchanged."
- The ease of sharing patient data and information available in the EHR with other medical providers in case of emergencies.
- Utilizing IoT to provide quality care to patients living in rural areas with the assistance of telemedicine and data analytics.
- The minutest details in the data can be analyzed which assists professionals to detect threats earlier and take preventive measures.
- Maintaining interoperability (IoT) and big data will provide for real opportunities to access quality, minimize expenses, and ensure that every decision is made correctly.

9. Applications and future prospects

Clinical registries are databases that routinely compile health-related data on people who have undergone a specific surgical procedure, used a specific medical device or drug, been given a specific diagnosis of illness, or have been managed by a particular healthcare resource. These databases are part of an overall governance and management structure. Images, signals, and textual data are just a few of the data kinds that can be found in patient medical records. Since all of these data are integrated into one system, it is possible to select the clinical data that are of relevance before deciding how to extract information from them. The wise use of medical data is becoming increasingly important to achieving this goal. For making current, patient-centered decisions and appropriately developing care processes, the continuously expanding, multidimensional health data from electronic medical records, imaging databases, and other multilayered, frequently fragmented IT systems are becoming increasingly crucial. Following is the list of some clinical registries/databases (Table 14.1).

Electronic Health Records (EHRs) have made it possible for medical professionals to access patient medical records more easily. Faster data retrieval is made possible by EHRs, which also make it easier to report important healthcare quality indicators to organizations. By immediately notifying authorities of disease outbreaks, EHRs significantly enhance public health surveillance. A vast amount of health-related medical information that is essential for patients is accessible because to EHRs and the internet combined.

By enhancing communication between various healthcare providers and patients, there would be a higher continuity of treatment and quicker interventions. It would be crucial to share data with other healthcare organizations. If the data are not interoperable, data sharing between different organizations may be severely limited. This might prevent clinicians from having access to crucial data needed to make decisions about patient follow-up and treatment plans.

Clinical data management is increasingly evolving toward a digital era of real-time data, collection, and management thanks to technological developments like artificial intelligence and ML. This implies that more data are being gathered than ever before, necessitating the need for a central site to collect, validate, and analyze the data.

TABLE 14.1 List of some clinical registries/databases.

#	Name	Link
1.	Antimicrobial Resistance Surveillance and Research Network (AMRSN)	https://iamrsn.icmr.org.in/
2.	Autoimmune Registry	https://www.autoimmuneregistry.org/
3.	Cerebral Palsy Research Network MyCP	https://cprn.org/mycp/
4.	Clinical Trials Registry—India	http://ctri.nic.in/Clinicaltrials/login.php
5.	India Dialysis Registry	https://www.indiadialysisregistry.org/
6.	Integrated Disease Surveillance Programme (IDSP)	https://idsp.mohfw.gov.in/
7.	National Cancer Registry Programmme	http://ncdirindia.org/
8.	National Center for Disease Informatics and Research	http://www.ncdirindia.org/
9.	National Injury Surveillance Trauma Registry and Capacity Building Center	http://www.nisc.gov.in/Default.aspx
10.	Neonatal and Maturity onset of youth registry India	http://monogenicdiabetes.in/mody.html
11.	Rare Diseases Registry Program (RaDaR)	https://registries.ncats.nih.gov/

References

Abouelmehdi, K., Beni-Hessane, A., & Khaloufi, H. (2018). Big healthcare data: Preserving security and privacy. *Journal of Big Data, 5*(1). https://doi.org/10.1186/s40537-017-0110-7. http://journalofbigdata.springeropen.com/

Adler-Milstein, J., Holmgren, A. J., Kralovec, P., Worzala, C., Searcy, T., & Patel, V. (2017). Electronic health record adoption in US hospitals: The emergence of a digital \advanced use\ divide. *Journal of the American Medical Informatics Association, 24*(6), 1142–1148. https://doi.org/10.1093/jamia/ocx080. http://jamia.oxfordjournals.org/content/22/e1

Azadi, A., & García-Peñalvo, F. J. (2023). Synergistic effect of medical information systems integration: To what extent will it affect the accuracy level in the reports and decision-making systems? *Informatics, 10*(1). https://doi.org/10.3390/informatics10010012. http://www.mdpi.com/journal/informatics

Bönisch, C., Kesztyüs, D., & Kesztyüs, T. (2022). Harvesting metadata in clinical care: A crosswalk between FHIR, OMOP, CDISC and openEHR metadata. *Scientific Data, 9*(1), 659. https://doi.org/10.1038/s41597-022-01792-7

Baudot, A., Gómez-López, G., & Valencia, A. (2009). Translational disease interpretation with molecular networks. *Genome Biology, 10*(6), 221. https://doi.org/10.1186/gb-2009-10-6-221

Daymont, C., Ross, M. E., Russell Localio, A., Fiks, A. G., Wasserman, R. C., & Grundmeier, R. W. (2017). Automated identification of implausible values in growth data from pediatric electronic health records. *Journal of the American Medical Informatics Association: JAMIA, 24*(6), 1080–1087. https://doi.org/10.1093/jamia/ocx037

Demski, H., Garde, S., & Hildebrand, C. (2016). Open data models for smart health interconnected applications: The example of openEHR. *BMC Medical Informatics and Decision Making, 16*(1). https://doi.org/10.1186/s12911-016-0376-2

Dhayne, H., Haque, R., Kilany, R., & Taher, Y. (2019). In search of big medical data integration solutions—A comprehensive survey. *IEEE Access, 7*, 91265–91290. https://doi.org/10.1109/ACCESS.2019.2927491. http://ieeexplore.ieee.org/xpl/RecentIssue.jsp?punumber=6287639

Dhudasia, M. B., Grundmeier, R. W., & Mukhopadhyay, S. (2023). Essentials of data management: An overview. *Pediatric Research, 93*(1), 2–3. https://doi.org/10.1038/s41390-021-01389-7

Gagalova, K. K., Leon Elizalde, M. A., Portales-Casamar, E., & Görges, M. (2020). What you need to know before implementing a clinical research data warehouse: Comparative review of integrated data repositories in health care institutions. *JMIR Formative Research, 4*(8), e17687. https://doi.org/10.2196/17687

Heart, T., Ben-Assuli, O., & Shabtai, I. (2017). A review of PHR, EMR and EHR integration: A more personalized healthcare and public health policy. *Health Policy and Technology, 6*(1), 20–25. https://doi.org/10.1016/j.hlpt.2016.08.002. http://www.journals.elsevier.com/health-policy-and-technology

Health insurance portability and accountability act (HIPAA) privacy rule and the national instant criminal background check system (NICS). Final rule. *Federal Register, 81*(3), (2016), 382–396.

Köpcke, F., Trinczek, B., Majeed, R. W., Schreiweis, B., Wenk, J., Leusch, T., Ganslandt, T., Ohmann, C., Bergh, B., Röhrig, R., Dugas, M., & Prokosch, H. U. (2013). Evaluation of data completeness in the electronic health record for the purpose of patient recruitment into clinical trials: A retrospective analysis of element presence. *BMC Medical Informatics and Decision Making, 13*(1). https://doi.org/10.1186/1472-6947-13-37. http://www.biomedcentral.com/bmcmedinformdecismak/

Kaspar, M., Fette, G., Hanke, M., Ertl, M., Puppe, F., & Störk, S. (2022). Automated provision of clinical routine data for a complex clinical follow-up study: A data warehouse solution. *Health Informatics Journal, 28*(1). https://doi.org/10.1177/14604582211058081

Kaspar, M., Liman, L., Morbach, C., Dietrich, G., Seidlmayer, L. K., Puppe, F., & Störk, S. (2022). Querying a clinical data warehouse for combinations of clinical and imaging data. *Journal of Digital Imaging*. https://doi.org/10.1007/s10278-022-00727-3. https://www.springer.com/journal/10278

Kerdprasop, N., & Kerdprasop, K. (2011). *Higher order programming to mine knowledge for a modern medical expert system.*

Lane, P., Schupmann, V., & Stuart, I. (2005). Oracle database data warehousing guide, 10g release 2 (10.2). *Oracle All Right Reserved.*

Lau, F., Price, M., Boyd, J., Partridge, C., Bell, H., & Raworth, R. (2012). Impact of electronic medical record on physician practice in office settings: A systematic review. *BMC Medical Informatics and Decision Making, 12*(1). https://doi.org/10.1186/1472-6947-12-10. http://www.biomedcentral.com/bmcmedinformdecismak/

Lenzerini, M. (2002). Data integration: A theoretical perspective. *Proceedings of the ACM SIGACT-SIGMOD-SIGART Symposium on Principles of Database Systems*, 233–246.

Lober, W. B., Karras, B. T., Wagner, M. M., Overhage, J. M., Davidson, A. J., Fraser, H., Trigg, L. J., Mandl, K. D., Espino, J. U., & Tsui, F. C. (2002). Roundtable on bioterrorism detection: Information system-based surveillance. *Journal of the American Medical Informatics Association, 9*(2), 105–115. https://doi.org/10.1197/jamia.M1052

Lyman, J. A., Scully, K., & Harrison, J. H. (2008). The development of health care data warehouses to support data mining. *Clinics in Laboratory Medicine, 28*(1), 55–71. https://doi.org/10.1016/j.cll.2007.10.003

MacKenzie, S. L., Wyatt, M. C., Schuff, R., Tenenbaum, J. D., & Anderson, N. (2012). Practices and perspectives on building integrated data repositories: Results from a 2010 CTSA survey. *Journal of the American Medical Informatics Association, 19*(1), e119–e124. https://doi.org/10.1136/amiajnl-2011-000508. http://jamia.bmj.com/content/19/e1/e119.full.pdf

Nogueira, A. C., Oliveira, R., Cruz-Correia, R., & Vieira-Marquesa, P. (2019). Validation of patient identification in an HL7 messages integrator for health data monitoring and portability. *Procedia Computer Science, 164*, 670–677. https://doi.org/10.1016/j.procs.2019.12.234. http://www.sciencedirect.com/science/journal/18770509

Osada, M., & Nishihara, E. (1999). Implementation and evaluation of workflow based on hospital information system/radiology information system/picture archiving and communications system. *Journal of Digital Imaging, 12*(2), 103–105. https://doi.org/10.1007/bf03168770

Pai, M. M. M., Ganiga, R., Pai, R. M., & Sinha, R. K. (2021). Standard electronic health record (EHR) framework for Indian healthcare system. *Health Services & Outcomes Research Methodology, 21*(3), 339–362. https://doi.org/10.1007/s10742-020-00238-0. http://www.wkap.nl/journalhome.htm/1387-3741

Pandey, R. K. (2014). Data quality in data warehouse: Problems and solution. *IOSR Journal of Computer Engineering, 16*(1), 18–24. https://doi.org/10.9790/0661-16141824

Prasser, F., Kohlbacher, O., Mansmann, U., Bauer, B., & Kuhn, K. A. (2018). Data integration for future medicine (DIFUTURE). *Methods of Information in Medicine, 57*(01), e57–e65. https://doi.org/10.3414/ME17-02-0022

Qamar, R., Kola, J., & Rector, A. L. (2007). Unambiguous data modeling to ensure higher accuracy term binding to clinical terminologies. *AMIA Annual Symposium Proceedings/AMIA Symposium*, 608–613.

Schoen, C., Osborn, R., Doty, M. M., Squires, D., Peugh, J., & Applebaum, S. (2009). A survey of primary care physicians in eleven countries, 2009: Perspectives on care, costs, and experiences. *Health Affairs, 28*(Suppl. 1), w1171–w1183. https://doi.org/10.1377/hlthaff.28.6.w1171

Shivtare, S., & Shelar, P. (2015). Data warehouse with data integration: Problems and solution. *IOSR Journal of Computer Engineering*, 67–71.

Street, J., & Arcement, R. (2014). *Health level seven (HL7) version 2.5.1 guidelines.*

Thiru, K., Hassey, A., & Sullivan, F. (2003). Systematic review of scope and quality of electronic patient record data in primary care. *British Medical Journal, 326*(7398), 1070–1072.

Wang, R. Y., & Strong, D. M. (2015). Beyond accuracy: What data quality means to data consumers. *Journal of Management Information Systems, 12*(4), 5–33. https://doi.org/10.1080/07421222.1996.11518099

Warren, J. J., Matney, S. A., Foster, E. D., Auld, V. A., & Roy, S. L. (2015). Toward interoperability: A new resource to support nursing terminology standards. *CIN: Computers, Informatics, Nursing, 33*(12), 515–519. https://doi.org/10.1097/CIN.0000000000000210. http://www.cinjournal.com

Watson, H., Ariyachandra, T., & Matyska, R. J. (2001). Data warehousing stages of growth. *Information Systems Management, 18*(3), 42–50. https://doi.org/10.1201/1078/43196.18.3.20010601/31289.6

Chapter 15

Integrative omics data mining: Challenges and opportunities

Swarna Kanchan[1,2], Minu Kesheri[1,2], Upasna Srivastava[3,7], Hiren Karathia[4], Ratnaprabha Ratna-Raj[2,5], Bhaskar Chittoori[2], Lydia Bogomolnaya[1], Rajeshwar P. Sinha[6] and James Denvir[1]

[1]*Marshall University, Huntington, WV, United States;* [2]*Boise State University, Boise, ID, United States;* [3]*University of California, San Diego, CA, United States;* [4]*Greenwood Genetic Centre, Greenwood, SC, United States;* [5]*Texas A&M University, College Station, TX, United States;* [6]*Banaras Hindu University, Varanasi, Uttar Pradesh, India;* [7]*School of Medicine, Yale University, New Haven, CT, United States*

1. Introduction

Rapid advances in high-throughput technologies have led to the fast accumulation of patient data. The surge in advancements of next-generation sequencing (NGS) technologies leads to high-throughput data generation for genomics (DNA or Genome of the organism), epigenomes (DNA methylation, histone modifications chromatin accessibility, and transcription factor binding), and transcriptomes (gene expression, alternative splicing, long noncoding RNAs, and small RNAs such as microRNAs), proteome (all the proteins produced by the organisms), metabolome (all the secondary metabolites produced by an organisms in our body), and microbiome (all the microbes present in our gut contributes to immunity and health) (Ritchie et al., 2015). Computational bioinformatics plays a crucial role and helps in knowledge discovery by dealing with storage, retrieval, and optimal use of omics data. Integrated omics refers to multiomics approaches integrating three or more omics datasets and data types, i.e., genomics, transcriptomics, proteomics, and metabolomics which has revolutionized biology and led to a better understanding of biological processes (Kumari et al., 2018; Quinn et al., 2016). With decreasing time and cost to generate these omics datasets, integration of these datasets creates exciting opportunities to reveal the mysteries of biology and on the other hand, causes immense challenges for researchers. The integrative omics data analysis uses different data sources to understand the system in a better way. Single omics data is used in many studies, but the causes of complex traits could not be explained in such studies. The growing significance of integrated omics can be perceived by its vast research panorama encompassing systems biology (Fukushima & Kusano, 2013; Mochida & Shinozaki, 2011), natural product discovery (Richa, Kannaujiya, et al., 2011; Richa, Rastogi, et al., 2011; Yang et al., 2011), disease biology (Kesheri et al., 2017, pp. 166−195; Pathak & Davé, 2014), environment (Priya et al., 2017, pp. 1044−1071), food and nutrition science (Kato et al., 2011), systems microbiology (Fondi & Liò, 2015; Kesheri et al., 2011; Kesheri, Kanchan, Richa, & Sinha, 2014), microbiomes (Kanchan et al., 2020; Muller et al., 2014), genotype−phenotype interactions (Ritchie et al., 2015), etc. In cancer research, the integration of multiomics is highly required because of the large-scale production of omics data production in various projects such as the International Cancer Genome Consortium, The Cancer Genome Atlas (TCGA), and Therapeutically Applicable Research to Generate Effective Treatments (Zhang et al., 2018). In bioinformatics, developing models utilizing large heterogenous multiomics datasets is very challenging for making predictions for improved diagnostics, prognostics, and therapeutics. In recent years, with the increase in datasets of multiomics data, many computational models, programs and applications are developed to integrate multiple levels of data through statistical and machine learning (ML) approaches (Kristensen et al., 2014). Data integration of multiomics data are mainly categorized into three types which are called concatenation-based integration, transformation based integration and model based integration. In concatenation-based integration, multiple types of omics data are combined, and then the combined matrix is analyzed by single omics tools to retrieve the results. Transformation-based integration is considered better compared to concatenation based integration, in this method; the data type is transformed into a matrix of graph or kernel which is merged to get integrative datasets. In model-based integration, the omics datasets are analyzed separately, and then the results are combined to get integrative results.

Integrative Omics. https://doi.org/10.1016/B978-0-443-16092-9.00015-1

In several review articles tools, resources, databases, and software for analysis and visualization of various omics datasets and data types such as genomics, transcriptomics, proteomics (Oveland et al., 2015), and metabolomics data (Misra, 2018; Misra et al., 2017; Misra & van der Hooft, 2016) are discussed, but this book chapter with a comprehensive review on integrative multiomics data mining is still the need of the hour. This chapter aims to review the status of the various available databases, tools, and methods being utilized in multiomics data analysis, their integration, data mining, and various challenges associated.

2. High-throughput multiomics in human health and diseases

2.1 Application of genomic resources in human health

Genomics in biomedical research aims to identify the phenotype associated with the genetic variants leading to disease, the response to treatment, diagnostics, therapeutics, gene therapy applications, pharmacogenomics, and disease prevention (Misra, 2018). Genomics includes sample preparation and collection, high-quality nucleic acid extraction, library preparation, clonal amplification, and finally sequencing using high-throughput technologies such as pyrosequencing, sequencing-by ligation, or sequencing by synthesis, etc. Following sequencing, genomics pipelines include data quality checkup, adapter trimming (cleaning), assembly, alignment (de novo or reference-based), variant calling, annotation, and functional predictions. Several genomic databases are very helpful in the annotation of human disease (Kumari et al., 2016). The Encyclopedia of DNA Elements (ENCODE) consortium is an important database funded by the National Human Genome Research Institute (NHGRI) is considered one of the comprehensive and largest resources to investigate the effects of somatic mutations in various types of human cancer and aims to build a comprehensive list of a functional element comprising of elements acting at protein, RNA, and regulatory elements that control cells and the human genome. Catalog of somatic mutations in cancer (COSMIC) is considered another large and comprehensive resource to explore the impacts of somatic mutations in various types of human cancer (Tate et al., 2019). COSMIC in November 2022 reported around 23 million coding mutations. These coding mutations were reported from 1.5 million tumor samples. The Human Gene Mutation Database is another central genomic database that records all disease-causing mutations reported in the literature and provides access to these data to academic, clinical, as well as commercial researchers (Stenson et al., 2020).

2.2 Genome-wide association studies resources

Genome-Wide Association Studies (GWAS) is an important and popular genomic approach that aims to identify numerous genetic variants associated with complex disease phenotypes in various human populations including thousands of people genotyped for millions of genetic markers. GWAS has also identified thousands of noncoding loci that are associated with human diseases and complex traits revealing the mechanisms of disease. GWAS aims to identify the statistically significant differences in allele frequencies between cases and controls. This helps us further to understand complex disease phenotypes in a better and more meaningful way. The variants investigated by GWAS help in exploring complex diseases leading to the development of new therapies. GWAS studies generate a Manhattan plot that illustrates the association between genetic variants and a trait (e.g., a disease) at a genome-wide level. It is interesting to integrate quantitative trait loci (QTL) maps with GWAS which helps in identifying potential molecular mechanisms associated with the disease. Despite the success of GWAS, the clinical insights derived from their results have been limited due to the difficulty of interpreting GWAS associations. GWAS catalog is a popular genomic database that includes various GWAS studies which were founded by the NHGRI, in response to the rapid increase in the number of published GWAS, available at https://www.ebi.ac.uk/gwas/home

2.3 Application of epigenomics in human diseases and health

Reversible alterations of nucleoproteins associated with DNA or DNA itself comprising acetylation of histone or methylation of DNA form the epicenter of epigenomics. Covalent modifications of DNA and histones are determined by genetic as well as environmental factors are major regulators of gene transcription (Gut & Verdin, 2013; Liu et al., 2008; Piunti & Shilatifard, 2016; Taudt et al., 2016). Epigenomics studies are used for investigating the disease status for various metabolic syndromes (Horvath, 2013; Multhaup et al., 2015), cardiovascular diseases (Kim et al., 2010), cancer (Baylin et al., 2001), and many other pathophysiologic states (Raghuraman et al., 2016). Since GWAS did not fulfill the initial promise of explaining complex diseases, epigenetics played a major role in resolving the missing heritability. Since epigenetic signatures are often tissue-specific (Zhu et al., 2013) aiming at platforms like roadmap epigenomics available at http://www.roadmapepigenomics.org/ program delivering comprehensive epigenomic maps from several human tissues and cells, developing tools, and methods for analyzing the human epigenome may seem beneficial. The roadmap epigenomics program was initiated in 2008 as a public resource of human epigenomic data to discover novel epigenetic marks, exploring methods to manipulate the epigenome, and illustrating epigenetic contributions to diverse human diseases and health.

2.4 Transcriptomics (RNA-seq) in exploring human diseases

Transcriptomics examines RNA levels both qualitatively and quantitatively. In humans, only ~3% of the genome codes for proteins while the rest of the genome serves as junk DNA. Transcriptome analysis allows the quantification of gene expression levels and allele-specific expression in a single experiment, as well as to identify novel genes, fusion transcripts, etc. In recent years, transcriptomics has been one of the most utilized omics approaches to investigate human diseases at the molecular level. RNA-seq studies identified thousands of novel isoforms which tells us about a larger protein-coding transcriptome. For instance, a recent RNA-seq study revealed abnormal alternative splicing in Huntington's disease by identifying 593 differential alternative splicing events between pathological and control brains (Lin et al., 2016). The quantitative reverse transcription PCR (qRT-PCR) method was often applied to validate the results obtained from RNA-seq based on high-throughput platforms. However, on the other hand, the noncoding RNA discipline has also become popular. Among noncoding RNA, pathogenic miRNA was first reported in chronic lymphocytic leukemia where the miRNA cluster, miR-15a/16−1, was observed to act as a tumor suppressor. Concerning mammalian cells, there are pieces of evidence for transcription of thousands of long noncoding RNAs that are shown to participate in playing essential roles in several processes associated with physiology, for instance, functions pertaining to regulating endocrine, differentiation of brown adipose tissue, and development of the neuronal system. Dysregulation of long noncoding RNA plays a major role in various diseases, such as myocardial infarction and diabetes. For example, well-known long noncoding RNA called *MALAT1* was observed to be responsible for metastasis associated in lung adenocarcinoma transcript. Further studies also demonstrated that *MALAT1* was also involved in other disorders such as diabetes (Gao et al., 2017). Transcriptomics studies also help in the identification of the distinct cellular composition among lesions and metastatic or recurrent patient-derived organoids (PDOs) of colon cancer organoids that contain fewer stem-like cell and differentiated-like cell clusters and more proliferating cell clusters (Okamoto et al., 2021).

2.5 Role of proteomics resources in human health and biomarker discovery

Proteomics is used to quantify proteins in multiple sample types using both shotgun and targeted approaches and aims to investigate peptide abundance, and posttranslational modification such as proteolysis, glycosylation, phosphorylation, nitrosylation, and ubiquitination (Beck et al., 2006; Mann & Jensen, 2003), etc. Posttranslational modifications also play important roles in intracellular signaling as well as enzymatic activity (Kesheri et al., 2021; Kesheri et al., 2022), protein turnover, transport, and maintaining overall cell structure (Wu et al., 2011). Recent developments in mass spectroscopy have dramatically increased sensitivity while decreasing the amount of sample requirement for proteomics analyses. Mass spectroscopy−based proteomics analysis of proteins is now well adapted for thousands of proteins in cells or body fluids at once using high-throughput technologies (Hein et al., 2013; Selevsek et al., 2015) and antioxidative enzymes produced in response to stress (Ghai et al., 2016; Kesheri & Kanchan, 2015a, 2015b; Kesheri, Kanchan, & Chowdhury, 2014; Kesheri et al., 2015). Extending to the proteomics studies, the 3D structures of various proteins are still unknown due to lagging in the experimental determination of 3D structures. Bioinformatics plays a major role in predicting the 3D structure of various proteins which are helping in structural drug design, evolution, drug resistance, etc., (Gahoi et al., 2013; Garg et al., 2009; Kanchan et al., 2014, 2015, 2019; Sahu et al., 2023). Homology modeling for HPV type 16 E seven proteins by modeler (Srivastava & Singh, 2013) program and in silico vaccine designing based on epitope prediction of HPV Type 16 E7 (Srivastava et al., 2017) contributed to enhance the knowledge of proteomics. Protein−protein interactions are traditionally investigated by yeast two-hybrid assays whereas interactions between proteins and nucleic acids are mainly investigated by ChIP-Seq data analysis. The proteomic approach includes sample collection, protein extraction, enzymatic digestion of proteins into peptides, and separation using liquid chromatography (LC) approaches, followed by MS, peptide, and protein identification. Moreover, proteomics is also being used in diagnosis, protein-based biomarker development, and therapeutics involving synthesis, characterization, and pharmacological evaluation of various natural compound derivatives (Mishra et al., 2015a, 2015b). Several proteomics data analysis programs and tools have been developed and used. Skyline is a freely available, open-source program to analyze proteomics data produced by cutting-edge technologies for targeted proteomics methods especially designed for large-scale quantitative mass spectrometry studies. Skyline automatically quantifies several hundred thousand peptides in analyzing data resulting from mass spectrometry, thus allowing statistically robust identification and quantification of >4000 proteins in mammalian cell lysates. PEAKS studio is another popular software program used for analyzing proteomics data. PEAKS studio analyzes datasets for protein/peptide identification, quantification, mutations, posttranslational modifications, as well as de novo sequencing. MSTracer is a new software program for analyzing proteomics data which is based on ML for detecting peptide features from MS data that incorporates two scoring functions for detecting the peptide features and providing them with a valid quality score.

2.6 Metabolomics resources in human health

Metabolomics is a rapidly growing omics discipline that aims to measure/quantify various small molecules such as amino acids, fatty acids, etc., (Ghai et al., 2015; Mishra et al., 2015a, 2015b; Saxena et al., 2015), carbohydrates, or other products of cellular metabolic functions at a system scale (Shin et al., 2014). Bioinformatics has immensely aided in the better interpretation of omics data (Kumari et al., 2016). Exploring the metabolites is crucial for understanding most cellular phenomena to gain a comprehensive insight into all the biological processes involved (Cambiaghi et al., 2017). Metabolites are often end products of complex biochemical cascades connecting genome, transcriptome, and proteome to phenotype and help in the discovery of the genetic basis of metabolic variation (Srivastava et al., 2023). Changes in metabolite levels and their relative ratios reflect metabolic function which could serve as disease indicators. Initial steps of metabolomics comprises of experimental design, sample collection, standardization of metabolite extraction, and their reconstruction from samples. This may further be followed by the reconstitution of chemical derivatives, analysis results of mass spectrometry, nuclear magnetic resonance spectroscopy, alignment of data, filtering of data of interest, and imputation. The final steps comprise the refinement of the data statistically, annotation as well as assorting specific pathways or related networks. Thus, an integrative approach to metabolomics data incorporates a spectrum of scientific analysis spanning statistics (Singla et al., 2019a, 2019b), mathematical calculations, computational programming, and bioinformatics. MetaCoreTM is a commercial tool available as standalone as well as web-based server for functional analysis of different kinds of high-throughput molecular data ranging from genomics data from NGS, transcriptomics results comprising of siRNA, microRNA as well as microarray-based gene expression. Moreover, data from the gene (Kesheri et al., 2016) expression, DNA arrays, and array-comparative genomic hybridization along with data comprising of proteomics as well as metabolomics profiling and screening also find suitable analysis platforms at MetaCoreTM. Another program is called MetaboAnalyst (a freely accessible web-based platform) is an online tool for metabolomic data analysis providing analysis results with functional and biological interpretation and visualization. Recently updated MetaboAnalyst 5.0 was optimized for LC–HRMS spectra processing and integrates metabolomics data with transcriptomics data combining multiple metabolomics datasets to conduct exploratory statistical analysis (Pang et al., 2022).

2.7 Application of metagenomics in exploring gut microbiome

Metagenomics is a rapidly growing field that explores the abundance and functions of the microorganisms of a given community together (Kanchan et al., 2020). Complex microbial communities that interact with the host to influence disease progression are well studied in metagenomics. Human skin, mucosal surfaces, and the human gut are colonized by microorganisms such as bacteria, viruses, and fungi, collectively known as the microbiota, that are rapidly investigated via metagenomics for their abundance and functionality. The human microbiome is very complex; for example, the human gut contains roughly 100 trillion bacteria from 1000 different species. Diet, environmental factors, lifestyle, drugs (Galande et al., 2014), age, etc., are responsible for variations in microbiome composition among individuals (Org, Mehrabian, & Lusis, 2015; Org, Parks, et al., 2015). Several studies investigated the changes in gut microbiome composition in a variety of disorders such as cancer, diabetes, obesity, autism, colitis, heart disease, etc. In targeted metagenomics studies, hypervariable regions of the bacterial 16S rRNA genes are amplified and then sequenced followed by clustering the sequences into operational taxonomic units. Several programs such as QIIME2, Dada2, etc., are developed for analyzing NGS data from targeted 16S or shotgun metagenomics experiments (Caporaso et al., 2010). In shotgun metagenomics sequencing, the total DNA is sequenced to identify genetically close microbial species. In shotgun metagenomics sequencing, the total DNA is sequenced which can provide additional resolution for distinguishing genetically close microbial species. Metagenomics investigates all the microorganisms in the community which could be correlated with disease or other phenotypes of interest (Org, Mehrabian, & Lusis, 2015; Org, Parks, et al., 2015). The Human Microbiome Project was initiated by the National Institutes of Health and aimed to develop comprehensive resources for the characterization of the human microbiota. It also aimed at deciphering the probable involvement and role of the human microbiome in various aspects of health and disease. Microbiome database (MDB) is another microbiome database (https://db.cngb.org/microbiome/) that includes the sequencing resource and metadata of ecological community samples of microorganisms such as host-associated or environmental microbes. Several multiomics resources helpful in annotating human health and diseases are listed in (Table 15.1).

European MetaHIT is a collaborative project of 15 institutes and 8 countries. MetaHIT project aims to create an exhaustive reference catalog enumerating genes and genomes of microbes inhabiting the human gut. MetaHIT project aims to develop tools to investigate genes and reference genomes of different individuals and their frequencies.

Moreover, single-cell multiomics technologies enable simultaneous profiling of the genome, epigenome, transcriptomics, proteome, and metabolome from a single cell has become possible (Srivastava & Singh, 2022, pp. 271–294).

The various applications of the integration of multiomics technologies are illustrated in Fig. 15.1.

TABLE 15.1 List of omics resources helpful in annotating human health and disease.

Database	Availability	Remark	References
ENCODE	https://www.encodeproject.org/	A list of functional elements in the human genome	Davis et al. (2018)
FANTOM6	https://fantom.gsc.riken.jp/zenbu/reports/#FANTOM6	List of functions of long noncoding RNAs (lncRNAs) in the human genome	Ramilowski et al. (2020)
COSMIC (Catalog of Somatic Mutations in Cancer)	https://cancer.sanger.ac.uk/cosmic	A comprehensive resource for exploring the impact of somatic mutations in human cancer	Tate et al. (2019)
Human Gene Mutation Database(HGMD)	http://www.hgmd.org.	A comprehensive collection of germ-line mutations associated with human inherited disease	Stenson et al. (2020)
Skyline	https://skyline.ms/project/home/software/Skyline/begin.view	Program for analyzing the resulting mass spectrometer data	MacLean et al. (2010)
Maxquant	https://www.maxquant.org/	A suite of programs developed for high-resolution, quantitative MS data analysis	Cox and Mann (2008)
PEAKS Studio Xpro	https://www.bioinfor.com/peaks-studio/	Deep learning-based program for peptide identification and quantification.	Tran et al. (2019)
MSTracer	https://github.com/waterlooms/ms-tracer	Tool for detecting peptide features by machine learning and scoring function	Zeng and Ma (2021)
MetaCore	https://clarivate.com/products/biopharma/discovery-clinical-regulatory/early-research-intelligence-solutions/	Program for functional analysis of different kinds of high-throughput molecular data	
MetaboAnalyst	http://www.metaboanalyst.ca	Program for univariate and multivariate statistical methods, extensive data visualization, and functional analysis	Xia et al. (2009)
InCroMA	http://www.ra.cs.uni-tuebingen.de/software/InCroMAP/index.htm	Tool for generic or pathway-based analysis and visualization of heterogeneous, cross-platform datasets.	Wrzodek et al. (2012)
3Omics	https://3omics.cmdm.tw/	Tool for integration and visualization of transcriptomic, proteomic and metabolomic human data	Kuo et al. (2013)
Human Microbiome Project (HMP)	https://hmpdacc.org/hmp/overview/	Comprehensive characterization of the human microbiome and its role in human health and disease	Turnbaugh et al. (2007)

Continued

TABLE 15.1 List of omics resources helpful in annotating human health and disease.—cont'd

Database	Availability	Remark	References
European MetaHIT	https://www.gutmicrobiotaforhealth.com/metahit/	Catalog enumerating genes and genomes of microbes inhabiting the human gut	Qin et al. (2010)
Integrative Human Microbiome Project (iHMP)	https://www.hmpdacc.org/ihmp/	Repository for diverse human microbiome datasets	"The Integrative Human Microbiome Project: Dynamic Analysis of Microbiome-Host Omics Profiles during Periods of Human Health and Disease," (2014)

Swarna Kanchan.

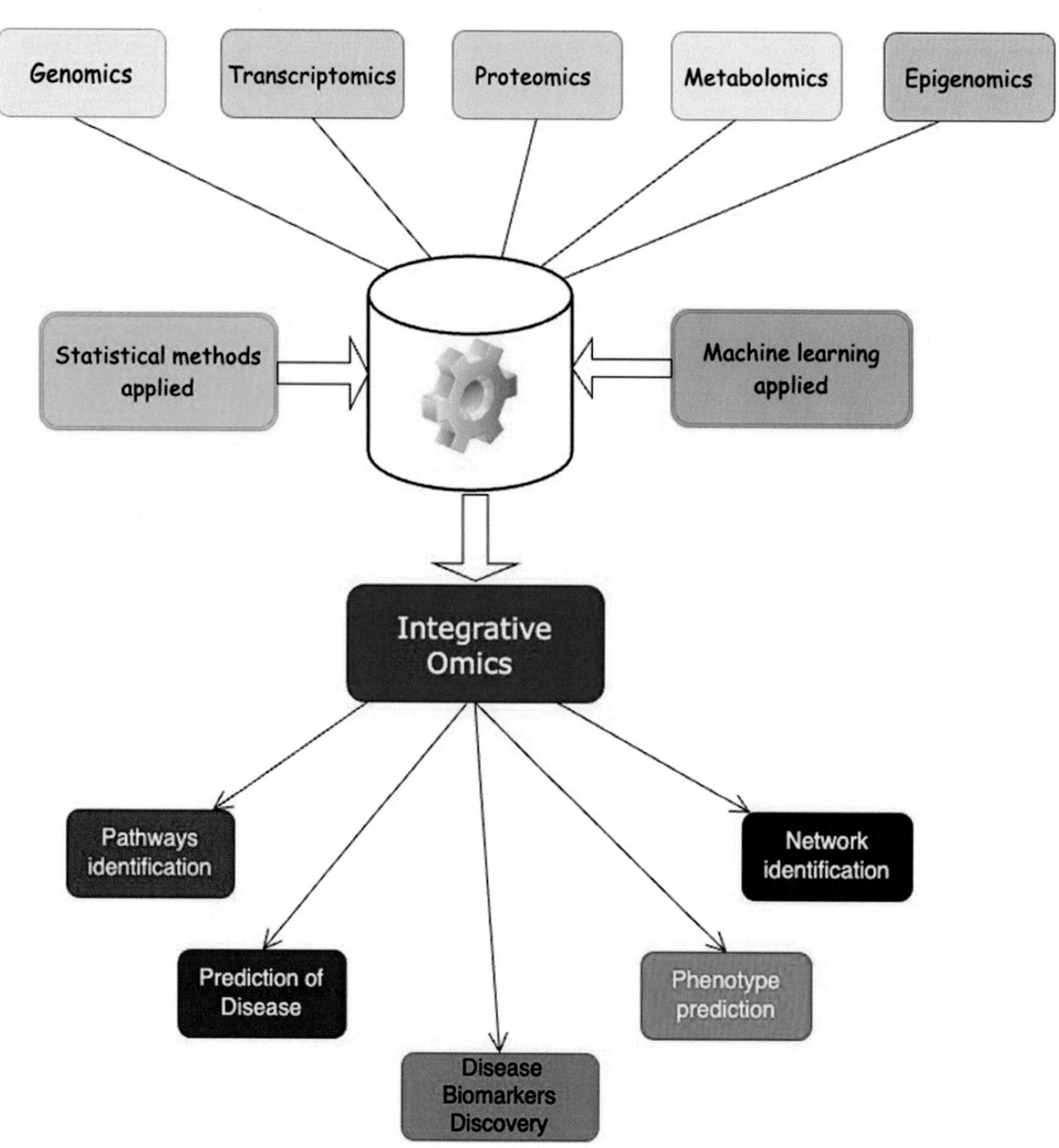

FIGURE 15.1 Various applications of integrative omics technologies in human health and diseases.

3. Learning and use of various programming languages

Learning and use of various programming languages are major challenges in multiomics data analysis, their integration, and data mining.

3.1 Advantages of using different programming languages

Programming languages such as Python and bash scripting are used mostly for designing the workflow. Bash scripting is most useful for handling large datasets, arranging, and transforming them working directly where they are found on your machine or server. Python is used especially for efficient processing and ML applications designed for ease of use and efficiency. C/C++/Java is being used for the development of methods and algorithms. R programming is considered the standard language for statistical programming and data visualization of datasets with specialized libraries designed for various high-throughput omics datasets. Perl is not only the most popular because of being good for working with manipulating strings such as large DNA molecules or genomes but also the most valuable programming language for biologists to know for generating reports. Using various Bioperl modules makes it easier for biologists to perform omics data analysis. Java is also useful and capable of handling large omics datasets that need both high performance and scalability. Ruby programming can also be used to process large omics datasets to extract relevant information. Shiny applications that aid in visualizing outlines are suitable for crafting web-based interactive multiomics functions. Shiny is useful in launching applications either on a local computer or a server.

4. Database resources in multiomics analysis

Numerous online databases and tools are utilized for integrative omics data analysis and data mining (Karczewski & Snyder, 2018). cBioPortal represents vast and comprehensive database that is significant for exploring cancer genomics through visualization, download and analysis of large cancer datasets (Gao et al., 2013). Another integrative, multiomics database UCSC Xena Browser (Goldman et al., 2018), provides a large resource for multiomics data analysis as well as clinical and phenotypes datasets. UK Biobank is a large-scale biomedical database accessible to global researchers undertaking vital research into life-threatening diseases which includes in-depth genetic and health information from around half a million UK participants (Sudlow et al., 2015). TCGA, another landmark cancer genomics database initiated by NCI and the National Human Genome Research Institute in 2006 includes molecularly characterized over 20,000 primary cancer and matched normal samples spanning 33 cancer types. The ENCODE projects pertains to the identification of functional elements encompassing the human genome (Davis et al., 2018); The Human Protein Atlas is another integrative database that was initiated in 2003 in Sweden to map all the human proteins in cells, tissues, and organs including antibody-based imaging, mass spectrometry—based proteomics, transcriptomics, and systems biology using an integration of various omics technologies (Thul et al., 2017).

5. Multiomics data integration methods

Recently, in order to facilitate the management and analysis of multiomics data and to identify novel biomarkers in human diseases, multiomics data integration approaches have been developed. Most of the multiomics data integration methods are based on PCA, correlation or Bayesian, or nonBayesian network-based methods. In data integration, it is very difficult to accurately estimate many parameters and correct them because multiple hypotheses are tested simultaneously. Susceptibility to elevated rates of false positives often observed in integrated omics data may be solved by subjecting multiple testing in the analytical pipeline to curb for type I error rate through Bonferroni corrections, Westfall, and Young permutations as well as false-positive rate by Benjamini—Hochberg.

5.1 Tools for multiomics data integration and analysis

Various multiomics tools have been introduced in recent decades (Misra et al., 2019), encompassing 3Omics (Kuo et al., 2013), SNF or Similarity network fusion (Wang et al., 2014), mixOmics (Rohart et al., 2017), Paintomics (Hernández-De-Diego et al., 2018), Multiomics Factor Analysis or MOFA (Argelaguet et al., 2018), and miodin (Ulfenborg, 2019). Based on the R package, MixOmics facilitates multivariate analysis prioritizing exploration, dimension reduction, and visualization of biological datasets. MixOmics statistically integrates several heterogeneous omics datasets to explore relationships among omics datasets using a systems biology approach. Similarity, network fusion works to combine multiomics data to create a comprehensive view of a given disease or a biological process constructing networks of samples (e.g., patients) for each available data type and then efficiently fusing these into one network that represents the full spectrum of patient or disease data. It helped in a combination of mRNA expression, DNA methylation, and microRNA (miRNA) expression in cancer datasets. Miodin is an excellent resource that is available as an R package that allows vertical and horizontal analysis of multiomics datasets across experiments using the same samples. PaintOmics four is another wonderful resource available as a web tool for the integrative visualization of multiple omics datasets that cover a comprehensive pathway analysis workflow, including automatic feature name conversion and multilayered feature matching. PaintOmics four also aims to explore pathway enrichment, interactive heatmaps, network analysis, trend charts, etc., onto Reactome, KEGG, and MapMan pathways. 3Omics is another web-server-based program for the integration, analysis, and visualization of transcriptomic, proteomic, and metabolomic omics datasets (Kuo et al., 2013). 3Omics offers four types of 'omics' data analyses:

1. Transcriptomics−Proteomics−Metabolomics (T−P−M),
2. Transcriptomics−Proteomics (T−P),
3. Proteomics−Metabolomics (P−M), and
4. Transcriptomics−Metabolomics (T−M).

Multiomics factor analysis (MOFA), a statistical framework for the integrative analysis of multiomics data from a common set of samples infers a low-dimensional representation of the data for capturing the global sources of variability. MOFA+ is an advanced version of MOFA that addresses various challenges by developing a stochastic variational inference framework amenable to graphical processing unit (GPU) computations, enabling the analysis of datasets with potentially millions of cells. MOFA + incorporated further flexibility, and structure regularization, thus facilitating the joint modeling of multiple groups and data modalities (Argelaguet et al., 2020).

Canonical correlation analysis is another program based on multiview feature selection algorithm which is applied to predict the survival of kidney renal clear cell carcinoma (KIRC) using gene expression, copy number alteration, RNA-seq data, etc. (El-Manzalawy, 2018).

6. Computational infrastructure, data sharing, and benchmarking

We also need to learn about standard or enterprise-level Linux distribution and open-source cloud-based Google Colab to run workflows for mutiomics data integration and data mining. Stand-alone workstations empowered with highly powered Central Processing Units, servers bestowed with GPU, or high-end computing infrastructures enabled with Tensor Processing Unit form a prerequisite for integration as well as mining of mutiomics data.

Storage as well as sharing of a wide range of omics data has been facilitated by public database repositories. Some of them are NCBI-SRA (Leinonen et al., 2011), GEO and EBI-ENA, and PRIDE and ProteomeXchange, MetaboLights, and Metabolomics Workbench as well as GNPS-MASSIVE provide an interface for metabolomics data. These public repositories must be integrated in a systematic manner, and therefore OmicsDI was designed to integrate these datasets in a discoverable manner.

Due to the increase in several tools for the integration of multiomics data mining, these computational methods must go through a systematic evaluation (benchmarking) (Mangul et al., 2019). The key challenges in benchmarking include the acquisition of "gold standard" datasets and ensuring reproducibility in the context of the increasing complexity of the software involved (Mangul et al., 2019; Marx, 2020; Weber et al., 2019).

7. Data mining methods

7.1 Clustering analysis

Clustering analysis is an important technique for explorative data mining for bioinformatics-based discovery. Clustering is defined as the statistical method to group numerous objects into a limited number of groups known as clusters. These clusters are combinations of objects having similar characteristics, which are separated from objects having different characteristics. In the biomedical field, clustering can be used to discover groups of patients suitable for treatment protocols, each group comprising all the patients who respond to the treatment in the same way. In biomedical studies, omics profiling is now a routine job that includes clustering as an essential step. Clustering analysis results can suggest the unknown functionalities of omics studies, in which the same clusters used to have related biological functions (Levine, 2013). Clustering also helps in dimension reduction in outcome model building (Shen et al., 2009). Several clustering methods have been developed for low-dimensional settings and high-dimensional settings for omics data (Langfelder et al., 2008).

Clustering analysis with multilayer omics data is challenging due to interconnections within layers and those across layers are different both biologically and statistically. For multiomics datasets, researchers can choose from a variety of clustering techniques, including distance-based clustering, hierarchical clustering, k-means clustering, etc.

Distance-based clustering is used to explore similarity/dissimilarity in terms of distance between data points of the same cluster or data points of other clusters measured by Pearson correlation, etc.

Hierarchical clustering was invented by Robert Sibson (from the United Kingdom) and Daniel Defays (from Belgium) (Defays, 1977; Sibson, 1973). A cluster is estimated by the distances calculated as the squared difference between two cases/values. The smallest distances are used to form the first clusters and this procedure continues and stops when all patients are in a cluster. In hierarchical clustering, a series of steps are performed to organize the data into a cluster. Based on the similarity/dissimilarity of the objects, the hierarchical cluster may begin with one cluster that contains *n* objects, or with *n* clusters that each contains one object. Hierarchical clustering is either agglomerative (bottom−up) or divisive (top−down). Researchers use hierarchical clustering techniques in bioinformatics to find the appropriate number of clusters for microarray data.

K-means clustering was invented by Stuart Lloyd, a physicist from New Jersey, and was first published in 1982 (Lloyd, 1982). *K*-means clustering partitions, *n* instances into *k* clusters by assigning each data point to the partition with the nearest centroid which is implemented in open-source data mining software WEKA. K-means clustering does not start with all patients/data points being a cluster of their own, but instead, randomly selects cluster centers. After selecting a cluster center, iteration tries and finds the best-fit centers for the data given, i.e., those with the shortest distances to the centers. In the k-means clustering method, clusters are equally sized, while in the hierarchical clustering method, clusters are of different sizes.

7.2 Feature selection and extraction methods (filter, wrapper, embedded, and hybrid) and genome mining

A feature (referred to as an attribute) is always measured by its ability to distinguish instances of the dataset with respect to the target class to which the instance belongs. Feature selection is an important data mining method involving selecting a subset of relevant features from a larger set of available features. The primary focus of the feature selection method is to identify the most informative and relevant features and retain them. Feature selection method usually discards or ignores the irrelevant or redundant ones. The purpose of the feature selection data mining method is to find a subset of factors that maximizes the performance of the prediction model. The performance of the prediction depends on how these methods incorporate the feature selection search with the classification algorithms. Feature selection reduces the dimensionality of the data, making it easier for the model to learn and it also reduces the risk of overfitting. By reducing the number of features using the feature selection method, feature selection can also help to reduce training time and computational costs. Different feature selection techniques, including filter, wrapper, embedded, and hybrid methods, can be used depending on the type of data and the modeling approach. Feature extraction uses feature construction, generating feature subsets, defining evaluation criteria, and estimating evaluation criteria. Based on these four steps, feature selection is further characterized by filter, wrapper, embedded, and hybrid methods.

The filter method ranks features based on statistical measures such as the correlation of individual features with respect to the target label of the dataset. Feature subset generation includes algorithms such as heuristic/stochastic search, exhaustive searches of features, a nested subset strategy for feature selection, forward selection, backward elimination, and single feature ranking. The filter method for feature selection uses the predictive power of many features collectively rather than independently.

The wrapper method is simple compared to the filter method. In this method, the performance of a learning algorithm is focused to evaluate the goodness of selected feature subsets by their information content. This method does not rely on optimizing the performance of a learning algorithm directly. Filter methods are computationally more efficient, but wrapper methods provide better results (Saeys et al., 2007). There are three basic mechanisms for the wrapper method which are called Forward Selection, Backward Elimination, and Recursive Feature Elimination.

Furthermore, a hybrid approach that is designed to reduce the bias introduced by each method is generated by the combination of information-gain and the chi-squared methods.

Embedded methods incorporate feature selection methods into the model training process, selecting the most relevant features during the training of the model.

Genome mining approaches have gained popularity in recent years by adding more fundamental insights into the genomic signature of an organism (Bhattarai et al., 2020). Genome mining has become a promising tool for the detection of natural compounds. Genome mining focuses on using genomic information for investigating the biosynthetic pathways of natural products and various interactions.

7.3 Association

Based on the concept of conditional probabilities, association rules were introduced by Rakesh Agrawal, from IBM Almaden Research Center, San Jose, California, introduced in 1993 (Agrawal et al., 1993). Currently, data mining association rule methodology has been observed as an important task of ML which replaced mental arithmetic with artificial intelligence and uses specialized statistical software. Traditionally, regression analyses (with the assumed predictor as an independent variable), paired chi-square tests, and other methods are applied for significance testing of the assumed effects. However, these methods are only sensitive to strong associations and weak associations. Association rule analysis based on conditional probabilities may be more sensitive because the weak associations use subsets of the data rather than the entire dataset.

7.4 Data cleaning and visualization

Many large biological omics databases are designed to store and retrieve biological information, but they contain several errors, incompleteness, redundancies, and ambiguities reducing the quality of biological information. These data cleaning methods are considered an important data mining tool that detects and removes data inconsistencies and errors to improve the data quality. The first step of these data cleaning methods is to identify the erroneous data which are due to missing values, inconsistent inputs, misspellings, etc. The second step of data cleaning is to detect duplicate records which is challenging, especially in large databases due to having complex schemas. Duplicates are more prevalent in large biological databases where people from the entire world submit large datasets obtained through various high-throughput experiments. Large biological databases such as Genebank (It is part of the International Nucleotide Sequence Database Collaboration, which comprises the DNA DataBank of Japan (DDBJ), the European Nucleotide Archive (ENA), and GenBank at NCBI which exchange data on a daily basis.) allow the resubmission of DNA sequences from throughout the world which includes a lot of redundancies (Benson et al., 2013). But another biological database such as SwissProt is nonredundant because it is manually curated by experts from the Swiss Institute of Bioinformatics (Bairoch & Apweiler, 2000). Data cleaning is very challenging for evolving large biological databases, but this could be done by data transformation. Data transformation is an important method that helps in the migration of legacy systems to new information systems. Evolving databases also face data quality issues which are associated with single-source and multisource databases. Single-source databases face integration problems that arise during the schema design. Single-source databases also face problems that are not prevented at the schemas level, such as missing values, and duplicate records are called instance-specific problems. In multiple sources databases, the main challenge of data cleaning is to identify the overlapping data information.

Visual representation of omics datasets is also considered one of the most important aspects in multiomics data analysis and data mining for other interpretations. Resources such as BioTools and WilsON are utilized for the visual representation of multiomics datasets. Nowadays, dashboards are very popular for the visual presentation of multiomics datasets due to the presence of multiple interactive panels. R, Python, Jupyter, and Tableau are frequently used for designing such dashboards. These dashboards can be installed locally or accessed online through a web browser based on servers or cloud platforms. Python-oriented Plotly Dash is utilized for the creation of complex dashboards. Cytoscape, as a stand-alone application, is a popular tool used mainly by systems biologists. Servers like Bokeh facilitate the creation of interactive web-based applications that can be used by Python users.

7.5 Classification

Classification is the process of classifying a record or identifying the category/class of a new observation. Classification is, therefore, an operation that places everyone from the population under study in one of several specified classes depending on the characteristics of the individual which are identified as independent variables. An individual is generally assigned to a class based on the explanatory characteristics by using a formula, an algorithm, or a set of rules, which forms a model, and which must be discovered.

The data mining methods can be divided into two large families: descriptive methods and predictive methods. In descriptive methods, we do not have any dependent variable (or privileged variable) for reducing, summarizing, and grouping data. In predictive methods, which explain data, there is a dependent variable (or a privileged variable). We can be more precise in predictive methods by distinguishing the differences in the type of variable, namely independent and dependent. There is a wide variety of classification techniques to choose from, and they perform differently under different data types and disciplines. It is very important for us to understand how these algorithms work to understand how the performances and results differ.

7.6 Machine learning

ML has become part of our everyday practice and has been widely employed in the fields of biomedical and bioinformatics computing. ML is an important field in computer science frequently used in biomedical research to make predictions, to investigate the efficacy of novel treatments, and to identify patterns, for making health decisions, etc. ML is much like traditional statistical methods such as time series, ANOVA, variance, correlation, and regression methods for small datasets. ML methods are better than traditional methods to handle big and complex data.

ML methods nowadays are being used for the statistical analysis of multiomics data and data with many variables. ML methods focus on integrating and analyzing large omics datasets for the discovery of new biomarkers. These biomarkers have the potential to help in accurate disease prediction, cancer subtype identification, patient stratification, and delivery of precision medicine (Reel et al., 2021). ML offers promising approaches to integrating and recognizing fine-grained patterns and relationships by rigorous optimizations to process large multiomics-derived datasets (Hesami et al., 2022). ML approaches also have the potential to reveal regular patterns in multilayered omics datasets. ML algorithms might improve the interpretation of large complex data and transform them into clinical decision-making. In a recent study role of ML was illustrated in the prognosis prediction of cardiovascular disease (Kresoja et al., 2023). Recently, various artificial intelligence techniques like ML based on multiomics analysis have been presented for cancer diagnosis and treatment in the era of precision medicine. Another ML-based program called meta-analytic support vector machine (Meta-SVM) was developed for detecting genes associated with disease by accommodating multiomics data which identifies potential biomarkers for multiomics data effectively (Kim et al., 2017).

7.7 Neural networks

Neural networks (NNs) have become widely used, due to their modeling power with excellent results across a broad range of disciplines. NNs are powerful ML models made of many neurons which are organized in many layers. A NN has an architecture based on that of the brain, organized in interconnected units called neurons and synapses. NNs contain an input layer corresponding to a unit at the first level, several hidden layers, and an output layer to make predictions. NNs are either supervised or unsupervised. Supervised NNs use the classification of the data and enable the NN to adjust its weights and thresholds to correctly classify new input data. Learning in supervised NNs is driven by a predefined training set that includes both normal and outlier data points. Unsupervised NNs are used in case of unavailability of the training set for NNs. ML is also introduced to develop a cross-modal NN architecture used for the integration of multiomics datasets for the dataset having a small training set. The proposed model investigates the most relevant genes and the influence of different types of omics datasets on each other providing an accuracy of 99% (Bica et al., 2018).

In a particular type of NN called artificial neural network (ANN), the model trained in a supervised learning manner can output high-accuracy prediction by giving enough training datasets. Such a multilayer ANN model was developed using a backpropagation algorithm which is considered the most used supervised learning NN (Kim & Kim, 2020; Kumari et al., 2018; Kanchan et al., 2024). ML-based ANN models are also developed for risk prediction of lung cancer survival in various laboratories (Chen et al., 2014). Deep learning, which is another type of NN, is composed of several hidden layers of NNs performing complex operations on large amounts of biological data. A deep learning algorithm was also implemented in autoencoder for integration of multiomics data and identification of two subtypes having significant survival differences, it generates a better result for classification and prognostic subtypes' identification in neuroblastoma when results are compared with PCA, iCluster, and DGscore (Poirion et al., 2018). Few NN architectures are specialized in learning, which are called Restricted Boltzmann Machines. MOLI method (**mu**ltiomics **l**ate **i**ntegration) was developed based on deep NNs which consider somatic mutation, copy number aberration, and gene expression data as input, and integrate them for drug response prediction (Sharifi-Noghabi et al., 2019). Malik and coworkers proposed a late multiomics integrative framework that robustly quantifies survival and drug response for breast cancer patients with a focus on the relative predictive ability of available omics data types (Malik et al., 2021). Recently, moBRCA-net, another deep learning-based breast cancer subtype classification framework was developed that uses multiomics datasets comprising gene expression, DNA methylation, and microRNA expression data that were integrated while considering the biological relationships among them (Choi & Chae, 2023).

The Correlation coefficient (R), Coefficient of determination (R^2), Mean Square Error (MSE), Root Mean Square Error (RMSE), and mean absolute error (MAE) are frequently used to evaluate NN/ANN model performance (Kanchan et al., 2024; Jierula et al., 2021). R refers to the Pearson correlation coefficient and corresponds to the degree of correlation between the actual and predicted variables. R ranges from -1 to $+1$, and an absolute value of one indicates the perfect correlation between actual and predicted values. The correlation coefficient $+1$ indicates a positive correlation while -1

suggests an inverse correlation between actual and predicted values (Jierula et al., 2021). The Coefficient of determination (R^2) is simply the square of Pearson's correlation coefficient ranging from 0 to 1. 0 indicates that the regression model explains none of the predicted variables (no correlation), while one suggests that the regression model explains all of the predicted variables (perfect correlation). MSE measures the mean squared error between the predicted value and the actual value. RMSE is the square root of MSE (Jierula et al., 2021). RMSE measures the average magnitude of error between the predicted value and the actual value. MAE measures the mean average error as the average magnitude of the absolute errors between the predicted value and actual value. MSE, RMSE, and MAE range from 0 to ∞, the smaller the value is, the higher the accuracy of the prediction model. The perfect value is 0, indicating that the prediction model is perfect.

7.8 Outlier detection

Outliers are defined as patterns (records) that do not conform to an expected behavior that must be detected to improve the data quality of large omics datasets. Thus, outlier detection is considered an important method that must be able to define a region representing normal behavior and declare an observation as an outlier. However, there are three fundamental approaches to outlier detection, determining the outliers with no prior knowledge of the data, modeling either normal/abnormal data, or both data. It is difficult to define and capturing the boundary between a normal region and an outlier due to the availability of diverse omics datasets is a further challenge. Defining an outlier that needs prior definition requires domain knowledge is another challenge in outlier detection and removal. To determine outliers with no prior knowledge of the data, unsupervised clustering should be applied because, in all clustering algorithms, this kind of outlier detection algorithm considers each record as a point in an n-dimensional space. These points are considered in clusters on the basis of their proximity flaging the remote points as outliers. Once the outliers are identified, the algorithm iteratively prunes the outliers until no more outliers are identified, and the system model is fitted to the remaining data that represent the normal data distribution.

7.9 Prediction in data mining

Prediction is an important aspect of data mining which corresponds to the analysis done to predict a future event (Shruti et al., 2016) or trends or novel biomarkers when multiomics datasets are used. Prediction refers to the use of omics or large biological datasets to predict outcomes. For example, in ANN, the training dataset contains the inputs, and an algorithm derives the model/predictor which generates corresponding numerical output values. The trained model should find a numerical output for the test dataset provided, and the performance of the model should be measured by some error or correlation coefficient-based metrics. Prediction from large biological omics datasets is challenging if datasets are inconsistent, incomplete, or redundant therefore before using such omics datasets, these datasets must be subjected to data cleaning. Prediction from large biological omics datasets is also challenging if the data transformation (such as normalization) or data dimensional reduction methods are not applied.

7.10 Methods for data warehousing

A data warehouse is a nonvolatile collection of data that is subject-oriented. As the information stored in a data warehouse is subject-oriented because it is focused on one subject related to an organization. Data warehouses are primarily focused on organizational decision-making. A data warehouse is also time-varying because each data is associated with a time stamp. Data warehouse accepts data for storage in a universally accepted format for storage. Data warehouses do not store updated information because their primary function is to enable high-level decision-making. Data warehousing methodologies share a common set of tasks, and they are focused on requirements-based analysis. Data warehousing methodologies are also focused on data and architecture design, implementation, and deployment. The performance metrics in data warehousing are focused on query throughput and response times. Data warehouses are of two types based on data processing types.

Since data warehouses emphasize high-level data integration and decision support, therefore, such data warehouses are focused on providing consolidating information rather than on providing specific details of individual transactions. Online analytical processing (OLAP) performs data consolidation and complex analysis of information which is well integrated into data warehouse creation.

Online transaction processing (OLTP) is a data warehouse system that focuses on fast query processing times. Such data warehouse systems are characterized by a large volume of short online transactions, such as data entry, deletion, update, and retrieval in typical databases. In contrast, OLAP systems are designed for a low volume of transactions while OLTP is designed for a high volume of transactions.

8. Challenges in multiomics data integration and data mining

The data integration and data mining of each individual omics analysis present multiple challenges. Analyzing multiple datasets deriving from different platforms and experimental procedures is challenging due to existing biases due to high-throughput sequencing platforms, laboratories, and analysis methods. The major limitation of multiomics data integration is the lack of data across multiple layers for the same experimental conditions. Vertical and horizontal integration is also important which will become more effective and effective as more datasets are added within and across multiple layers. Multiple-layer integration could further lead to lower false discovery rates. To obtain trustworthy results and conclusions, several samples for omics data are required. Reliability depends on the false discovery rate, which in turn is influenced by the number of measured entities in the form of transcripts, proteins, or metabolites. Statistical analysis of these omics data serves to pose yet another challenging task. Various challenges posed at different levels of data integration of various omics datasets and their types are illustrated in Fig. 15.2.

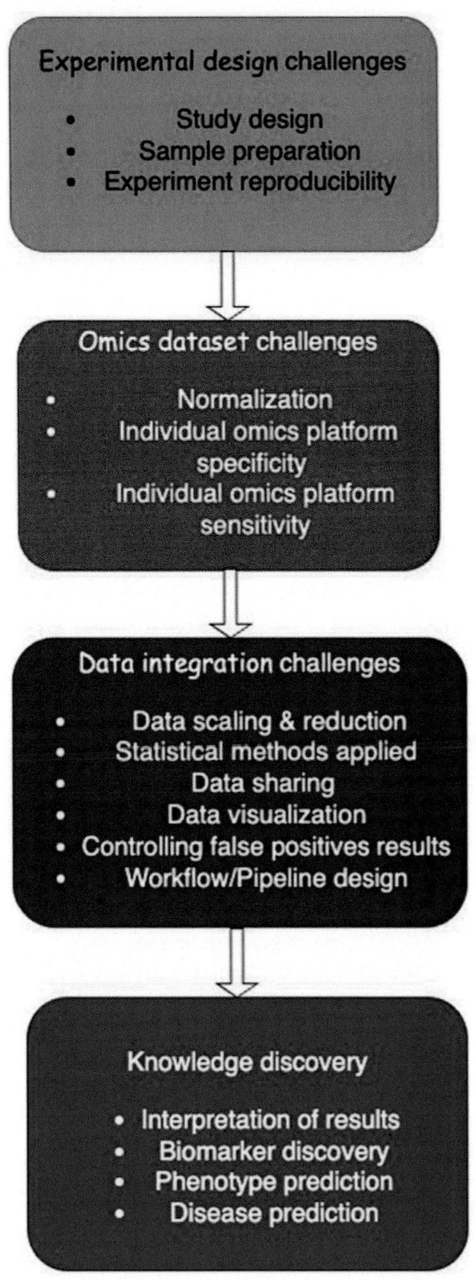

FIGURE 15.2 Challenges in multiomics data integration and analysis. Various challenges at different levels arise when we integrate multiomics datasets to analyze which are due to high-throughput sequencing platforms, laboratories, and methodologies used, etc.

Dimension reduction of features such as single nucleotide polymorphisms (SNPs), transcripts, proteins, and metabolites across samples, conditions, and different omics layers is the first and important step in dealing with multiomics datasets. Postprocessing of multiomics data after data normalization is applied to identify outliers and technical sources of variation—such as batch effects. This helps in the identification of biological patterns at each level of multiomics data analysis—such as feature identification, extraction, and selection. Challenges in multiomics analysis start from the single omics datasets and add new challenges of data integration using clustering, visualization, and functional characterization (Jamil et al., 2020). Data harmonization, data normalization, and data transformation further pose a big challenge for individual omics data integration. Integration of multiomics data further becomes difficult due to computational burden and huge storage space requirements. Data heterogeneity is another challenge that acts as another major bottleneck while dealing with multiomics datasets. Data heterogeneity in multiomics datasets is due to the use of varied technologies and high-throughput technologies platforms, etc. Therefore, there is a need for future work that incorporates and addresses the challenges of this field for a better platform facilitating integrated omics studies.

References

Agrawal, R., Imieliński, T., & Swami, A. (1993). Mining association rules between sets of items in large databases. *ACM SIGMOD Record, 22*(2), 207–216. https://doi.org/10.1145/170036.170072

Argelaguet, R., Arnol, D., Bredikhin, D., Deloro, Y., Velten, B., Marioni, J. C., & Stegle, O. (2020). MOFA+: A statistical framework for comprehensive integration of multi-modal single-cell data. *Genome Biology, 21*(1). https://doi.org/10.1186/s13059-020-02015-1

Argelaguet, R., Velten, B., Arnol, D., Dietrich, S., Zenz, T., Marioni, J. C., Buettner, F., Huber, W., & Stegle, O. (2018). Multi-Omics Factor Analysis—A framework for unsupervised integration of multi-omics data sets. *Molecular Systems Biology, 14*(6). https://doi.org/10.15252/msb.20178124

Bairoch, A., & Apweiler, R. (2000). The SWISS-PROT protein sequence database and its supplement TrEMBL in 2000. *Nucleic Acids Research, 28*(1), 45–48. https://doi.org/10.1093/nar/28.1.45

Baylin, S. B., Esteller, M., Rountree, M. R., Bachman, K. E., Schuebel, K., & Herman, J. G. (2001). Abberant patterns of DNA methylation, chromatin formation and gene expression in cancer. *Human Molecular Genetics, 10*(7), 687–692.

Beck, H. C., Nielsen, E. C., Matthiesen, R., Jensen, L. H., Sehested, M., Finn, P., Grauslund, M., Hansen, A. M., & Jensen, O. N. (2006). Quantitative proteomic analysis of post-translational modifications of human histones. *Molecular and Cellular Proteomics, 5*(7), 1314–1325. https://doi.org/10.1074/mcp.M600007-MCP200

Benson, D. A., Cavanaugh, M., Clark, K., Karsch-Mizrachi, I., Lipman, D. J., Ostell, J., & Sayers, E. W. (2013). GenBank. *Nucleic Acids Research, 41*(1), D36–D42. https://doi.org/10.1093/nar/gks1195

Bhattarai, K., Bastola, R., & Baral, B. (2020). Antibiotic drug discovery: Challenges and perspectives in the light of emerging antibiotic resistance. *Advances in Genetics, 105*, 229–292. https://doi.org/10.1016/bs.adgen.2019.12.002

Bica, I., Veličković, P., Xiao, H., & Liò, P. (2018). Multi-omics data integration using cross-modal neural networks. *ESANN 2018 - Proceedings, European Symposium on Artificial Neural Networks, Computational Intelligence and Machine Learning*, 385–390.

Cambiaghi, A., Ferrario, M., & Masseroli, M. (2017). Analysis of metabolomic data: Tools, current strategies and future challenges for omics data integration. *Briefings in Bioinformatics, 18*(3), 498–510. https://doi.org/10.1093/bib/bbw031

Caporaso, J. G., Kuczynski, J., Stombaugh, J., Bittinger, K., Bushman, F. D., Costello, E. K., Fierer, N., Pẽa, A. G., Goodrich, J. K., Gordon, J. I., Huttley, G. A., Kelley, S. T., Knights, D., Koenig, J. E., Ley, R. E., Lozupone, C. A., McDonald, D., Muegge, B. D., Pirrung, M., … Knight, R. (2010). QIIME allows analysis of high-throughput community sequencing data. *Nature Methods, 7*(5), 335–336. https://doi.org/10.1038/nmeth.f.303

Chen, Y. C., Ke, W. C., & Chiu, H. W. (2014). Risk classification of cancer survival using ANN with gene expression data from multiple laboratories. *Computers in Biology and Medicine, 48*(1), 1–7. https://doi.org/10.1016/j.compbiomed.2014.02.006

Choi, J. M., & Chae, H. (2023). moBRCA-net: A breast cancer subtype classification framework based on multi-omics attention neural networks. *BMC Bioinformatics, 24*(1). https://doi.org/10.1186/s12859-023-05273-5

Cox, J., & Mann, M. (2008). MaxQuant enables high peptide identification rates, individualized p.p.b.-range mass accuracies and proteome-wide protein quantification. *Nature Biotechnology, 26*(12), 1367–1372. https://doi.org/10.1038/nbt.1511

Davis, C. A., Hitz, B. C., Sloan, C. A., Chan, E. T., Davidson, J. M., Gabdank, I., Hilton, J. A., Jain, K., Baymuradov, U. K., Narayanan, A. K., Onate, K. C., Graham, K., Miyasato, S. R., Dreszer, T. R., Strattan, J. S., Jolanki, O., Tanaka, F. Y., & Cherry, J. M. (2018). The Encyclopedia of DNA elements (ENCODE): Data portal update. *Nucleic Acids Research, 46*(1), D794–D801. https://doi.org/10.1093/nar/gkx1081

Defays, D. (1977). An efficient algorithm for a complete link method. *The Computer Journal, 20*(4), 364–366. https://doi.org/10.1093/comjnl/20.4.364

El-Manzalawy, Y. (2018). CCA based multi-view feature selection for multiomics data integration. In *2018 IEEE Conference on Computational Intelligence in Bioinformatics and Computational Biology, CIBCB 2018* (pp. 1–8). United States: Institute of Electrical and Electronics Engineers Inc.. https://doi.org/10.1109/CIBCB.2018.8404968

Fondi, M., & Liò, P. (2015). Multi -omics and metabolic modelling pipelines: Challenges and tools for systems microbiology. *Microbiological Research, 171*, 52–64. https://doi.org/10.1016/j.micres.2015.01.003

Fukushima, A., & Kusano, M. (2013). Recent progress in the development of metabolome databases for plant systems biology. *Frontiers in Plant Science, 4*. https://doi.org/10.3389/fpls.2013.00073

Gahoi, Shachi, Mandal, Rahul Shubhra, Ivanisenko, Nikita, Shrivastava, Priyanka, Jain, Sriyans, Singh, Ashish Kumar, Raghunandanan, Muthukurrusi Varieth, Kanchan, Swarna, Taneja, Bhupesh, Mandal, Chhabinath, Ivanisenko, Vladimir A., Kumar, Anil, Kumar, Rita, Open Source Drug Discovery Consorti, & Ramachandran, Srinivasan (2013). Computational screening for new inhibitors of M. tuberculosis mycolyltransferases antigen 85 group of proteins as potential drug targets. *Journal of Biomolecular Structure and Dynamics, 31*(1), 30–43. https://doi.org/10.1080/07391102.2012.691343

Galande, S. H., Kanchan, S., & Kesheri, M. (2014). *Drug discovery and design: A bioinformatics approach.* LAP Lambert Academy Publishing.

Gao, J., Aksoy, B. A., Dogrusoz, U., Dresdner, G., Gross, B., Sumer, S. O., Sun, Y., Jacobsen, A., Sinha, R., Larsson, E., Cerami, E., Sander, C., & Schultz, N. (2013). Integrative analysis of complex cancer genomics and clinical profiles using the cBioPortal. *Science Signaling, 6*(269), pl1. https://doi.org/10.1126/scisignal.2004088

Gao, Jinning, Xu, Wenhua, Wang, Jianxun, Wang, Kun, & Li, Peifeng (2017). The role and molecular mechanism of non-coding RNAs in pathological cardiac remodeling. *International Journal of Molecular Sciences, 18*(3), 608. https://doi.org/10.3390/ijms18030608

Garg, S., Saxena, V., Kanchan, S., Sharma, P., Mahajan, S., Kochar, D., & Das, A. (2009). Novel point mutations in sulfadoxine resistance genes of Plasmodium falciparum from India. *Acta Tropica, 110*(1), 75–79. https://doi.org/10.1016/j.actatropica.2009.01.009

Ghai, R., Nagarajan, K., Kumar, V., Kesheri, M., & Kanchan, S. (2015). Amelioration of lipids by Eugenia caryophyllus extract in atherogenic diet induced hyperlipidemia. *International Bulletin of Drug Research, 5*(8), 90–101.

Ghai, R., Nagarajan, K., Singh, J., Swarup, S., & Kesheri, M. (2016). Evaluation of antioxidant status in-vitro and in-vivo in hydro-alcoholic extract of Eugenia caryophyllus. *International Journal of Pharmacology and Toxicology, 4*(1), 19–24.

Goldman, M., Craft, B., Hastie, M., Repečka, K., McDade, F., Kamath, A., Banerjee, A., Luo, Y., Rogers, D., Brooks, A. N., Zhu, J., & Haussler, D. (2018). The UCSC Xena platform for public and private cancer genomics data visualization and interpretation. *bioRxiv.* https://doi.org/10.1101/326470

Gut, P., & Verdin, E. (2013). The nexus of chromatin regulation and intermediary metabolism. *Nature, 502*(7472), 489–498. https://doi.org/10.1038/nature12752

Hein, M. Y., Sharma, K., Cox, J., & Mann, M. (2013). Proteomic analysis of cellular systems. In *Handbook of systems biology* (pp. 3–25). Elsevier Inc. https://doi.org/10.1016/B978-0-12-385944-0.00001-0

Hernández-De-Diego, R., Tarazona, S., Martínez-Mira, C., Balzano-Nogueira, L., Furió-Tarí, P., Pappas, G. J., & Conesa, A. (2018). PaintOmics 3: A web resource for the pathway analysis and visualization of multi-omics data. *Nucleic Acids Research, 46*(1), W503–W509. https://doi.org/10.1093/nar/gky466

Hesami, M., Alizadeh, M., Jones, A. M. P., & Torkamaneh, D. (2022). Machine learning: Its challenges and opportunities in plant system biology. *Applied Microbiology and Biotechnology, 106*(9–10), 3507–3530. https://doi.org/10.1007/s00253-022-11963-6

Horvath, Steve (2013). DNA methylation age of human tissues and cell types. *Genome Biology, 14*(10), R115. https://doi.org/10.1186/gb-2013-14-10-r115

Jamil, I. N., Remali, J., Azizan, K. A., Nor Muhammad, N. A., Arita, M., Goh, H. H., & Aizat, W. M. (2020). Systematic multi-omics integration (MOI) approach in plant systems biology. *Frontiers in Plant Science, 11.* https://doi.org/10.3389/fpls.2020.00944

Jierula, A., Wang, S., Oh, T. M., & Wang, P. (2021). Study on accuracy metrics for evaluating the predictions of damage locations in deep piles using artificial neural networks with acoustic emission data. *Applied Sciences, 11*(5), 1–21. https://doi.org/10.3390/app11052314

Kanchan, S., Mehrotra, R., & Chowdhury, S. (2014). Evolutionary pattern of four representative DNA repair proteins across six model organisms: An in silico analysis. *Network Modeling Analysis in Health Informatics and Bioinformatics, 3*(1). https://doi.org/10.1007/s13721-014-0070-1

Kanchan, S., Mehrotra, R., & Chowdhury, S. (2015). In silico analysis of the endonuclease III protein family identifies key residues and processes during evolution. *Journal of Molecular Evolution, 81*(1–2), 54–67. https://doi.org/10.1007/s00239-015-9689-5

Kanchan, S., Ogden, E., Kesheri, M., Skinner, A., Miliken, E., Lyman, D., Armstrong, J., Sciglitano, L., & Hampikian, G. (2024). COVID-19 hospitalizations and deaths predicted by SARS-CoV-2 levels in Boise, Idaho wastewater. *The Science of the Total Environment, 907*, Article 167742. https://doi.org/10.1016/j.scitotenv.2023.167742

Kanchan, S., Sharma, P., & Chowdhury, S. (2019). Evolution of endonuclease IV protein family: An in silico analysis. *3 Biotech, 9*(5). https://doi.org/10.1007/s13205-019-1696-6

Kanchan, S., Sinha, R. P., Chaudière, J., & Kesheri, M. (2020). Computational metagenomics: Current status and challenges. In *Recent trends in 'computational omics': Concepts and methodology* (pp. 371–395). Nova Science Publishers, Inc. https://novapublishers.com/shop/recent-trends-in-computational-omics-concepts-and-methodology/

Karczewski, K. J., & Snyder, M. P. (2018). Integrative omics for health and disease. *Nature Reviews Genetics, 19*(5), 299–310. https://doi.org/10.1038/nrg.2018.4

Kato, H., Takahashi, S., & Saito, K. (2011). Omics and integrated omics for the promotion of food and nutrition science. *Journal of Traditional and Complementary Medicine, 1*(1), 25–30. https://doi.org/10.1016/S2225-4110(16)30053-0

Kesheri, M., & Kanchan. (2015a). Computational methods and strategies for protein structure prediction. *Biological Sciences: Innovations and Dynamics,* 277–291.

Kesheri, M., & Kanchan. (2015b). Oxidative stress: Challenges and its mitigation mechanisms in cyanobacteria in. In *Biological sciences: Innovations and dynamics* (pp. 309–324). New India Publishing Agency.

Kesheri, Minu, Kanchan, Swarna, & Sinha, Rajeshwar P. (2017). *Exploring the potentials of antioxidants in retarding ageing* (pp. 166–195). IGI Global. https://doi.org/10.4018/978-1-5225-0607-2.ch008

Kesheri, Minu, Kanchan, Swarna, & Sinha, Rajeshwar P. (2021). Isolation and in silico analysis of antioxidants in response to temporal variations in the cyanobacterium Oscillatoria sp. *Gene Reports, 23*, 101023. https://doi.org/10.1016/j.genrep.2021.101023

Kesheri, M., Kanchan, S., & Chowdhury, S. (2014). *Cyanobacterial stresses: An ecophysiological, biotechnological, and bioinformatic approach.* LAP Lambert Academy Publishing.

Kesheri, M., Kanchan, S., Chowdhury, S., & Sinha, R. P. (2015). Secondary and tertiary structure prediction of proteins: A bioinformatic approach. *Studies in Fuzziness and Soft Computing, 319*, 541–569. https://doi.org/10.1007/978-3-319-12883-2_19

Kesheri, M., Kanchan, S., Richa, & Sinha, R. P. (2014). Isolation and in silico analysis of Fe-superoxide dismutase in the cyanobacterium Nostoc commune. *Gene, 553*(2), 117–125. https://doi.org/10.1016/j.gene.2014.10.010

Kesheri, M., Kanchan, S., & Sinha, R. P. (2022). Responses of antioxidants for resilience to temporal variations in the cyanobacterium Microcystis aeruginosa. *South African Journal of Botany, 148*, 190–199. https://doi.org/10.1016/j.sajb.2022.04.017

Kesheri, M., Richa, & Sinha, R. P. (2011). Antioxidants as natural arsenal against multiple stresses in Cyanobacteria. *International Journal of Pharma and Bio Sciences, 2*(2), 168–187.

Kesheri, M., Sinha, R. P., & Kanchan, S. (2016). Advances in soft computing approaches for gene prediction: A bioinformatics approach. *Studies in Computational Intelligence, 651*, 383–405. https://doi.org/10.1007/978-3-319-33793-7_17

Kim, Cho Hwe, & Kim, Young Chul (2020). Application of artificial neural network over Nickel-based catalyst for combined steam-carbon dioxide of methane reforming (CSDRM). *Journal of Nanoscience and Nanotechnology, 20*(9), 5716–5719. https://doi.org/10.1166/jnn.2020.17627

Kim, M., Long, T. I., Arakawa, K., Wang, R., Yu, M. C., & Laird, P. W. (2010). DNA methylation as a biomarker for cardiovascular disease risk. *PLoS One, 5*(3). https://doi.org/10.1371/journal.pone.0009692

Kim, S., Jhong, J. H., Lee, J., & Koo, J. Y. (2017). Meta-analytic support vector machine for integrating multiple omics data. *BioData Mining, 10*(1). https://doi.org/10.1186/s13040-017-0126-8

Kresoja, K. P., Unterhuber, M., Wachter, R., Thiele, H., & Lurz, P. (2023). A cardiologist's guide to machine learning in cardiovascular disease prognosis prediction. *Basic Research in Cardiology, 118*(1). https://doi.org/10.1007/s00395-023-00982-7

Kristensen, V. N., Lingjærde, O. C., Russnes, H. G., Vollan, H. K. M., Frigessi, A., & Børresen-Dale, A. L. (2014). Principles and methods of integrative genomic analyses in cancer. *Nature Reviews Cancer, 14*(5), 299–313. https://doi.org/10.1038/nrc3721

Kumari, A., Kanchan, S., Sinha, R. P., & Kesheri, M. (2016). Applications of bio-molecular databases in bioinformatics. *Studies in Computational Intelligence, 651*, 329–351. https://doi.org/10.1007/978-3-319-33793-7_15

Kumari, A., Kesheri, M., Sinha, R. P., & Kanchan, S. (2018). Integration of soft computing approach in plant biology and its applications in agriculture. In *Soft Computing for Biological Systems* (pp. 265–281). Springer Singapore. https://doi.org/10.1007/978-981-10-7455-4_16

Kuo, T. C., Tian, T. F., & Tseng, Y. J. (2013). 3Omics: A web-based systems biology tool for analysis, integration and visualization of human transcriptomic, proteomic and metabolomic data. *BMC Systems Biology, 7*. https://doi.org/10.1186/1752-0509-7-64

Langfelder, P., Zhang, B., & Horvath, S. (2008). Defining clusters from a hierarchical cluster tree: The Dynamic Tree Cut package for R. *Bioinformatics, 24*(5), 719–720. https://doi.org/10.1093/bioinformatics/btm563

Levine, Douglas A. (2013). Integrated genomic characterization of endometrial carcinoma. *Nature, 497*(7447), 67–73. https://doi.org/10.1038/nature12113

Lin, L., Park, J. W., Ramachandran, S., Zhang, Y., Tseng, Y. T., Shen, S., Waldvogel, H. J., Curtis, M. A., Richard, R. L., Troncoso, J. C., Pletnikova, O., Ross, C. A., Davidson, B. L., & Xing, Y. (2016). Transcriptome sequencing reveals aberrant alternative splicing in Huntington's disease. *Human Molecular Genetics, 25*(16), 3454–3466. https://doi.org/10.1093/hmg/ddw187

Liu, L., Li, Y., & Tollefsbol, T. O. (2008). Gene-environment interactions and epigenetic basis of human diseases. *Current Issues in Molecular Biology, 10*(1), 25–36.

Lloyd, S. P. (1982). Least squares quantization in PCM. *IEEE Transactions on Information Theory, 28*(2), 129–137. https://doi.org/10.1109/TIT.1982.1056489

MacLean, B., Tomazela, D. M., Shulman, N., Chambers, M., Finney, G. L., Frewen, B., Kern, R., Tabb, D. L., Liebler, D. C., & MacCoss, M. J. (2010). Skyline: An open source document editor for creating and analyzing targeted proteomics experiments. *Bioinformatics, 26*(7), 966–968. https://doi.org/10.1093/bioinformatics/btq054

Malik, V., Kalakoti, Y., & Sundar, D. (2021). Deep learning assisted multi-omics integration for survival and drug-response prediction in breast cancer. *BMC Genomics, 22*(1). https://doi.org/10.1186/s12864-021-07524-2

Mann, M., & Jensen, O. N. (2003). Proteomic analysis of post-translational modifications. *Nature Biotechnology, 21*(3), 255–261. https://doi.org/10.1038/nbt0303-255

Mishra, P., Saxena, V., Kesheri, M., & Saxena, A. (2015a). Synthesis, characterization and antiinflammatory activity of cinnolines (pyrazole) derivatives. *IOSR Journal of Pharmacy and Biological Sciences, 10*(6), 77–82.

Mishra, P., Saxena, V., Kesheri, M., & Saxena, A. (2015b). Synthesis, characterization and pharmacological evaluation of cinnoline (thiophene) derivatives. *The Pharma Innovation Journal, 4*(10), 68–73.

Misra, B. B., Fahrmann, J. F., & Grapov, D. (2017). Review of emerging metabolomic tools and resources: 2015–2016. *Electrophoresis, 38*(18), 2257–2274. https://doi.org/10.1002/elps.201700110

Misra, B. B., Langefeld, C., Olivier, M., & Cox, L. A. (2019). Integrated omics: Tools, advances and future approaches. *Journal of Molecular Endocrinology, 62*(1), R21–R45. https://doi.org/10.1530/JME-18-0055

Misra, B. B. (2018). New tools and resources in metabolomics: 2016–2017. *Electrophoresis, 39*(7), 909–923. https://doi.org/10.1002/elps.201700441

Misra, B. B., & van der Hooft, J. J. J. (2016). Updates in metabolomics tools and resources: 2014-2015. *Electrophoresis, 37*(1), 86–110. https://doi.org/10.1002/elps.201500417

Mochida, K., & Shinozaki, K. (2011). Advances in omics and bioinformatics tools for systems analyses of plant functions. *Plant and Cell Physiology, 52*(12), 2017–2038. https://doi.org/10.1093/pcp/pcr153

Muller, E. E. L., Pinel, N., Laczny, C. C., Hoopmann, M. R., Narayanasamy, S., Lebrun, L. A., Roume, H., Lin, J., May, P., Hicks, N. D., Heintz-Buschart, A., Wampach, L., Liu, C. M., Price, L. B., Gillece, J. D., Guignard, C., Schupp, J. M., Vlassis, N., Baliga, N. S., … Wilmes, P. (2014). Community-integrated omics links dominance of a microbial generalist to fine-tuned resource usage. *Nature Communications, 5*. https://doi.org/10.1038/ncomms6603

Multhaup, M. L., Seldin, M. M., Jaffe, A. E., Lei, X., Kirchner, H., Mondal, P., Li, Y., Rodriguez, V., Drong, A., Hussain, M., Lindgren, C., McCarthy, M., Näslund, E., Zierath, J. R., Wong, G. W., & Feinberg, A. P. (2015). Mouse-human experimental epigenetic analysis unmasks dietary targets and genetic liability for diabetic phenotypes. *Cell Metabolism, 21*(1), 138–149. https://doi.org/10.1016/j.cmet.2014.12.014

Okamoto, T., duVerle, D., Yaginuma, K., Natsume, Y., Yamanaka, H., Kusama, D., Fukuda, M., Yamamoto, M., Perraudeau, F., Srivastava, U., Kashima, Y., Suzuki, A., Kuze, Y., Takahashi, Y., Ueno, M., Sakai, Y., Noda, T., Tsuda, K., Suzuki, Y., Nagayama, S., & Yao, R. (2021). Comparative analysis of patient-matched PDOs revealed a reduction in OLFM4-associated clusters in metastatic lesions in colorectal cancer. *Stem Cell Reports, 16*(4), 954–967. https://doi.org/10.1016/j.stemcr.2021.02.012

Org, E., Mehrabian, M., & Lusis, A. J. (2015). Unraveling the environmental and genetic interactions in atherosclerosis: Central role of the gut microbiota. *Atherosclerosis, 241*(2), 387–399. https://doi.org/10.1016/j.atherosclerosis.2015.05.035

Org, E., Parks, B. W., Joo, J. W. J., Emert, B., Schwartzman, W., Kang, E. Y., Mehrabian, M., Pan, C., Knight, R., Gunsalus, R., Drake, T. A., Eskin, E., & Lusis, A. J. (2015). Genetic and environmental control of host-gut microbiota interactions. *Genome Research, 25*(10), 1558–1569. https://doi.org/10.1101/gr.194118.115

Oveland, E., Muth, T., Rapp, E., Martens, L., Berven, F. S., & Barsnes, H. (2015). Viewing the proteome: How to visualize proteomics data? *Proteomics, 15*(8), 1341–1355. https://doi.org/10.1002/pmic.201400412

Pang, Z., Zhou, G., Ewald, J., Chang, L., Hacariz, O., Basu, N., & Xia, J. (2022). Using MetaboAnalyst 5.0 for LC–HRMS spectra processing, multi-omics integration and covariate adjustment of global metabolomics data. *Nature Protocols, 17*(8), 1735–1761. https://doi.org/10.1038/s41596-022-00710-w

Pathak, R. R., & Davé, V. (2014). Integrating omics technologies to study pulmonary physiology and pathology at the systems level. *Cellular Physiology and Biochemistry, 33*(5), 1239–1260. https://doi.org/10.1159/000358693

Piunti, A., & Shilatifard, A. (2016). Epigenetic balance of gene expression by polycomb and compass families. *Science, 352*(6290). https://doi.org/10.1126/science.aad9780

Poirion, O. B., Chaudhary, K., & Garmire, L. X. (2018). Deep Learning\ndata integration for better risk stratifcation models of bladder\ncancer. In *AMIA summits on translational science proceedings* (pp. 197–206).

Priya, Prerna, Kesheri, Minu, Sinha, Rajeshwar P., & Kanchan, Swarna (2017). *Molecular dynamics simulations for biological systems*. IGI Global. https://doi.org/10.4018/978-1-5225-1762-7.ch040

Qin, J., Li, R., Raes, J., Arumugam, M., Burgdorf, K. S., Manichanh, C., Nielsen, T., Pons, N., Levenez, F., Yamada, T., Mende, D. R., Li, J., Xu, J., Li, S., Li, D., Cao, J., Wang, B., Liang, H., Zheng, H., … Zoetendal, E. (2010). A human gut microbial gene catalogue established by metagenomic sequencing. *Nature, 464*(7285), 59–65. https://doi.org/10.1038/nature08821

Quinn, R. A., Navas-Molina, J. A., Hyde, E. R., Song, S. J., Vázquez-Baeza, Y., Humphrey, G., Gaffney, J., Minich, J. J., Melnik, A. V., Herschend, J., Dereus, J., Durant, A., Dutton, R. J., Khosroheidari, M., Green, C., Da Silva, R., Dorrestein, P. C., & Knight, R. (2016). From sample to multi-omics conclusions in under 48 hours. *mSystems, 1*(2). https://doi.org/10.1128/mSystems.00038-16

Raghuraman, S., Donkin, I., Versteyhe, S., Barrès, R., & Simar, D. (2016). The emerging role of epigenetics in inflammation and immunometabolism. *Trends in Endocrinology and Metabolism, 27*(11), 782–795. https://doi.org/10.1016/j.tem.2016.06.008

Ramilowski, J. A., Yip, C. W., Agrawal, S., Chang, J. C., Ciani, Y., Kulakovskiy, I. V., Mendez, M., Ooi, J. L. C., Ouyang, J. F., Parkinson, N., Petri, A., Roos, L., Severin, J., Yasuzawa, K., Abugessaisa, I., Akalin, A., Antonov, I. V., Arner, E., Bonetti, A., … Carninci, P. (2020). Functional annotation of human long noncoding RNAs via molecular phenotyping. *Genome Research, 30*(7), 1060–1072. https://doi.org/10.1101/gr.254219.119

Reel, Parminder S., Reel, Smarti, Pearson, Ewan, Trucco, Emanuele, & Jefferson, Emily (2021). Using machine learning approaches for multi-omics data analysis: A review. *Biotechnology Advances, 49*, 107739. https://doi.org/10.1016/j.biotechadv.2021.107739

Richa, Kannaujiya, V. K., Kesheri, M., Singh, G., & Sinha, R. P. (2011). Biotechnological potentials of phycobiliproteins. *International Journal of Pharma and Bio Sciences, 2*(4), 446–454.

Richa, Rastogi, R. P., Kumari, S., Singh, K. L., Kannaujiya, V. K., Singh, G., Kesheri, M., & Sinha, R. P. (2011). Biotechnological potential of mycosporine-like amino acids and phycobiliproteins of cyanobacterial origin. *Biotechnology, Bioinformatics and Bioengineering, 1*(2), 159–171.

Ritchie, M. D., Holzinger, E. R., Li, R., Pendergrass, S. A., & Kim, D. (2015). Methods of integrating data to uncover genotype-phenotype interactions. *Nature Reviews Genetics, 16*(2), 85–97. https://doi.org/10.1038/nrg3868

Rohart, Florian, Gautier, Benoît, Singh, Amrit, Lê Cao, Kim-Anh, & Schneidman, Dina (2017). mixOmics: An R package for 'omics feature selection and multiple data integration. *PLoS Computational Biology, 13*(11), e1005752. https://doi.org/10.1371/journal.pcbi.1005752

Saeys, Y., Inza, I., & Larrañaga, P. (2007). A review of feature selection techniques in bioinformatics. *Bioinformatics, 23*(19), 2507–2517. https://doi.org/10.1093/bioinformatics/btm344

Sahu, N., Mishra, S., Kesheri, M., Kanchan, S., & Sinha, R. P. (2023). Identification of cyanobacteria-based natural inhibitors against SARS-CoV-2 druggable target ACE2 using molecular docking study, ADME and toxicity analysis. *Indian Journal of Clinical Biochemistry, 38*, 361–373. https://doi.org/10.1007/s12291-022-01056-6

Saxena, A., Saxena, V., Kesheri, M., & Mishra. (2015). Comparative hypoglycemic effects of different extract of clitoriaternatea leaves on rats. *IOSR Journal of Pharmacy and Biological Sciences, 10*(2), 60–65.

Selevsek, N., Chang, C. Y., Gillet, L. C., Navarro, P., Bernhardt, O. M., Reiter, L., Cheng, L. Y., Vitek, O., & Aebersold, R. (2015). Reproducible and consistent quantification of the saccharomyces cerevisiae proteome by SWATH-mass spectrometry. *Molecular and Cellular Proteomics, 14*(3), 739–749. https://doi.org/10.1074/mcp.M113.035550

Sharifi-Noghabi, H., Zolotareva, O., Collins, C. C., & Ester, M. (2019). Moli: Multi-omics late integration with deep neural networks for drug response prediction. *Bioinformatics, 35*(14), i501–i509. https://doi.org/10.1093/bioinformatics/btz318

Shen, R., Olshen, A. B., & Ladanyi, M. (2009). Integrative clustering of multiple genomic data types using a joint latent variable model with application to breast and lung cancer subtype analysis. *Bioinformatics, 25*(22), 2906–2912. https://doi.org/10.1093/bioinformatics/btp543

Shin, S. Y., Fauman, E. B., Petersen, A. K., Krumsiek, J., Santos, R., Huang, J., Arnold, M., Erte, I., Forgetta, V., Yang, T. P., Walter, K., Menni, C., Chen, L., Vasquez, L., Valdes, A. M., Hyde, C. L., Wang, V., Ziemek, D., Roberts, P., … Soranzo, N. (2014). An atlas of genetic influences on human blood metabolites. *Nature Genetics, 46*(6), 543–550. https://doi.org/10.1038/ng.2982

Shruti, Millerjyothi, N. K., & Kesheri, M. (2016). Forecast analysis of the potential and availability of renewable energy in India: A review. *International Journal of Industrial Electronics and Electrical Engineering, 4*(10), 17–22.

Sibson, R. (1973). Slink: An optimally efficient algorithm for the single-link cluster method. *The Computer Journal, 16*(1), 30–34. https://doi.org/10.1093/comjnl/16.1.30

Singla, S., Kesheri, M., Kanchan, S., & Aswath, S. (2019). Current status and data analysis of diabetes in India. *International Journal of Innovative Technology and Exploring Engineering, 8*(9), 1920–1934. https://doi.org/10.35940/ijitee.i8403.078919

Singla, S., Kesheri, M., Kanchan, S., & Mishra, A. (2019). Impact of diwali firecrackers on air quality in India and its effect on the health. *International Journal of Pharma and Bio Sciences, 10*(2). https://doi.org/10.22376/ijpbs.2019.10.2.b155-169

Srivastava, U., Kanchan, S., Kesheri, M., & Singh, S. (2023). Nutrimetabolomics: Metabolomics in nutrition research. *Metabolomics* (pp. 241–268). Springer Nature.

Srivastava, Upasna, & Singh, Gurmit (2013). Comparative homology modelling for HPV type 16 E 7 proteins by using MODELLER and its validations with SAVS and ProSA web server. *Journal of Computational Intelligence in Bioinformatics, 6*(1), 27. https://doi.org/10.37622/jcib/6.1.2013.27-33

Srivastava, Upasna, & Singh, Satendra (2022). *Approaches of single-cell analysis in crop improvement.* Springer Science and Business Media LLC. https://doi.org/10.1007/978-1-0716-2533-0_14

Srivastava, Upasna, Singh, Satendra, Gautam, Budhyash, Yadav, Pramod, Yadav, Madhu, Thomas, George, & Singh, Gurmit (2017). Linear epitope prediction in HPV type 16 E7 antigen and their docked interaction with human TMEM 50A structural model. *Bioinformation, 13*(05), 122–130. https://doi.org/10.6026/97320630013122

Stenson, P. D., Mort, M., Ball, E. V., Chapman, M., Evans, K., Azevedo, L., Hayden, M., Heywood, S., Millar, D. S., Phillips, A. D., & Cooper, D. N. (2020). The human gene mutation database (HGMD®): Optimizing its use in a clinical diagnostic or research setting. *Human Genetics, 139*(10), 1197–1207. https://doi.org/10.1007/s00439-020-02199-3

Sudlow, Cathie, Gallacher, John, Allen, Naomi, Beral, Valerie, Burton, Paul, Danesh, John, Downey, Paul, Elliott, Paul, Green, Jane, Landray, Martin, Liu, Bette, Matthews, Paul, Ong, Giok, Pell, Jill, Silman, Alan, Young, Alan, Sprosen, Tim, Peakman, Tim, & Collins, Rory (2015). UK Biobank: An open access resource for identifying the causes of a wide range of complex diseases of middle and old age. *PLoS Medicine, 12*(3), e1001779. https://doi.org/10.1371/journal.pmed.1001779

Tate, J. G., Bamford, S., Jubb, H. C., Sondka, Z., Beare, D. M., Bindal, N., Boutselakis, H., Cole, C. G., Creatore, C., Dawson, E., Fish, P., Harsha, B., Hathaway, C., Jupe, S. C., Kok, C. Y., Noble, K., Ponting, L., Ramshaw, C. C., Rye, C. E., … Forbes, S. A. (2019). Cosmic: The catalogue of somatic mutations in cancer. *Nucleic Acids Research, 47*(1), D941–D947. https://doi.org/10.1093/nar/gky1015

Taudt, A., Colomé-Tatché, M., & Johannes, F. (2016). Genetic sources of population epigenomic variation. *Nature Reviews Genetics, 17*(6), 319–332. https://doi.org/10.1038/nrg.2016.45

The integrative human microbiome project: Dynamic analysis of microbiome-host omics profiles during periods of human health and disease. *Cell Host & Microbe, 16*(3), (2014), 276–289. https://doi.org/10.1016/j.chom.2014.08.014

Thul, P. J., Akesson, L., Wiking, M., Mahdessian, D., Geladaki, A., Ait Blal, H., Alm, T., Asplund, A., Björk, L., Breckels, L. M., Bäckström, A., Danielsson, F., Fagerberg, L., Fall, J., Gatto, L., Gnann, C., Hober, S., Hjelmare, M., Johansson, F., … Lundberg, E. (2017). A subcellular map of the human proteome. *Science, 356*(6340). https://doi.org/10.1126/science.aal3321

Tran, N. H., Qiao, R., Xin, L., Chen, X., Liu, C., Zhang, X., Shan, B., Ghodsi, A., & Li, M. (2019). Deep learning enables de novo peptide sequencing from data-independent-acquisition mass spectrometry. *Nature Methods, 16*(1), 63–66. https://doi.org/10.1038/s41592-018-0260-3

Turnbaugh, P. J., Ley, R. E., Hamady, M., Fraser-Liggett, C. M., Knight, R., & Gordon, J. I. (2007). The human microbiome project. *Nature, 449*(7164), 804–810. https://doi.org/10.1038/nature06244

Ulfenborg, B. (2019). Vertical and horizontal integration of multi-omics data with miodin. *BMC Bioinformatics, 20*(1). https://doi.org/10.1186/s12859-019-3224-4

Wang, B., Mezlini, A. M., Demir, F., Fiume, M., Tu, Z., Brudno, M., Haibe-Kains, B., & Goldenberg, A. (2014). Similarity network fusion for aggregating data types on a genomic scale. *Nature Methods, 11*(3), 333–337. https://doi.org/10.1038/nmeth.2810

Wrzodek, C., Eichner, J., & Zell, A. (2012). Pathway-based visualization of cross-platform microarray datasets. *Bioinformatics, 28*(23), 3021–3026. https://doi.org/10.1093/bioinformatics/bts583

Wu, R., Haas, W., Dephoure, N., Huttlin, E. L., Zhai, B., Sowa, M. E., & Gygi, S. P. (2011). A large-scale method to measure absolute protein phosphorylation stoichiometries. *Nature Methods, 8*(8), 677–683. https://doi.org/10.1038/nmeth.1636

Xia, J., Psychogios, N., Young, N., & Wishart, D. S. (2009). MetaboAnalyst: A web server for metabolomic data analysis and interpretation. *Nucleic Acids Research, 37*(2), W652–W660. https://doi.org/10.1093/nar/gkp356

Yang, J. Y., Karr, J. R., Watrous, J. D., & Dorrestein, P. C. (2011). Integrating "-omics" and natural product discovery platforms to investigate metabolic exchange in microbiomes. *Current Opinion in Chemical Biology, 15*(1), 79–87. https://doi.org/10.1016/j.cbpa.2010.10.025

Zeng, X., & Ma, B. (2021). MSTracer: A machine learning software tool for peptide feature detection from liquid chromatography-mass spectrometry data. *Journal of Proteome Research, 20*(7), 3455–3462. https://doi.org/10.1021/acs.jproteome.0c01029

Zhang, L., Lv, C., Jin, Y., Cheng, G., Fu, Y., Yuan, D., Tao, Y., Guo, Y., Ni, X., & Shi, T. (2018). Deep learning-based multi-omics data integration reveals two prognostic subtypes in high-risk neuroblastoma. *Frontiers in Genetics, 9*. https://doi.org/10.3389/fgene.2018.00477

Zhu, J., Adli, M., Zou, J. Y., Verstappen, G., Coyne, M., Zhang, X., Durham, T., Miri, M., Deshpande, V., De Jager, P. L., Bennett, D. A., Houmard, J. A., Muoio, D. M., Onder, T. T., Camahort, R., Cowan, C. A., Meissner, A., Epstein, C. B., Shoresh, N., & Bernstein, B. E. (2013). Genome-wide chromatin state transitions associated with developmental and environmental cues. *Cell, 152*(3), 642–654. https://doi.org/10.1016/j.cell.2012.12.033

Chapter 16

Data science and analytics, modeling, simulation, and issues of omics dataset

Sanjay Kumar[1] and Manish Kumar Gupta[2]
[1]Bioinformatics Center, Biotech Park, Lucknow, Uttar Pradesh, India; [2]Department of Biotechnology, Faculty of Science, Veer Bahadur Singh Purvanchal University, Jaunpur, Uttar Pradesh, India

1. What is data science?

Since the creation of computers, the term "data" has been used to describe information sent or stored by computers. Data can be written words or numbers on paper, bytes or bits stored in the memory of a computer or other technological equipment, or facts that are stored in a person's memory. To make working with data more accessible, data science is applied. Data science is a field that uses scientific methods, processes, algorithms, and systems to extract knowledge and insights from structured and unstructured data. It involves using statistics, machine learning, and computer science techniques to analyze and interpret complex datasets. Data science can be applied in a variety of fields, including business, finance, healthcare, and science, to help organizations make informed decisions, predict outcomes, and optimize processes. Data science can be used in various tools and techniques to collect, process, and analyze data. They often work closely with data engineers and data analysts to design and implement data systems.

1.1 How it works?

Data science plays a crucial role in understanding complex omics data by processing and analyzing large and diverse datasets. It works in various stages, which are given in the following.

- *Data Acquisition:* The first step is to acquire Big Data from different omics technologies such as microarrays, next-generation sequencing, Single-cell sequencing technologies, and mass spectrometry.
- *Data Preprocessing*: The raw data obtained from different omics technologies may contain errors and inconsistencies, and data preprocessing is required to remove these errors and normalize the data.
- *Data Integration*: Data integration is a critical step in omics, as it involves combining data from multiple sources to gain a comprehensive view of the complex networks under study. Integration may involve combining data from different omics technologies or integrating data from multiple studies.
- *Statistical Analysis*: Once the data is preprocessed and integrated, statistical methods are used to identify differentially expressed genes/proteins/metabolites, cluster analysis, and pathway analysis to gain insights into the biological data under study. Statistical techniques such as *t*-tests, ANOVA, and linear regression are commonly used in omics data study.
- *Machine Learning:* Machine learning techniques are used in omics to build predictive models for different omics data. For example, machine learning can be used to predict disease outcomes or identify drug targets. Common machine learning algorithms used in omics include decision trees, random forests, support vector machines (SVMs), and neural networks.
- *Clustering:* Clustering techniques are used to group similar genes, proteins, or metabolites together based on their expression profiles. Clustering algorithms such as hierarchical clustering, k-means clustering, and self-organizing maps (SOMs) are commonly used in omics study.
- *Data visualization*: The last step involves visualizing the results of the data analysis, which can be done using various tools such as heatmaps, network graphs, and interactive visualizations.

Integrative Omics. https://doi.org/10.1016/B978-0-443-16092-9.00016-3

2. What is data analytics?

The term "data analytics" is general and encompasses a variety of data analysis methods. Data analytics techniques can be applied to any type of information to gain insight that can be utilized to make things better. Data analytics techniques can help reveal trends and indicators that could otherwise be lost in a flood of data.

3. Difference between data science and data analytics

Data analytics and data science are two related but distinct fields that both deal with data. The main variations between the two are as follows.

- Data analytics is focused on analyzing data to answer specific questions or solve specific problems. It involves using statistical methods, data mining, and machine learning techniques to identify patterns and insights in data. Data science, on the other hand, is a more broad and interdisciplinary field that encompasses a wide range of activities, including data collection, data cleaning, data modeling, and visualization.
- Data analytics typically requires strong skills in statistics, programming, and data visualization. Data scientists, on the other hand, require a broader set of skills, including programming, statistics, machine learning, domain expertize, and communication skills.
- Data analytics is primarily concerned with structured data, which is data that is organized in a specific format, such as a spreadsheet or database. Data science, on the other hand, deals with a broader range of data types, including unstructured data, such as text and images.
- Data science plays a central role in data analytics, which uses data to gain insights and make informed decisions. Analytics can be applied in healthcare and science.

4. What is omics?

Omics is an area of study that focuses on the thorough examination of biological molecules present in a biological system, including genes, proteins, metabolites, and other biomolecules. The suffix "-omics," which denotes a thorough examination of a certain kind of biological molecule, is where the word "omics" originates. The information/data generated or collected from high-throughput technologies and biological methods with the suffix -omics, such as genomics, transcriptomics, proteomics, metabolomics, and others is known as omics data. These technologies generate vast amounts of data that provide detailed information on the various molecules present in a biological sample, such as DNA, RNA, proteins, and metabolites (Subedi et al., 2022). Omics data can be used to inform the development and validation of models of molecular systems. For example, genomic data can be used to build models of genetic regulatory networks, and proteomic data can be used to build models of protein—protein interactions. Additionally, omics data can be used to test and refine existing models of molecular systems and make predictions about these systems' behavior (Katara et al., 2023).

4.1 Types of omics data

Data can be categorized in many ways it could be text, video, and audio. Omics data could be from DNA, RNA, protein, chemical compounds, images, density maps, and so on. Based on these data sizes and quality, the name has been given to large size, called Big Data. It can also be described as 5Vs: variety, volume, value, veracity, and velocity. Nowadays, web-based servers designed for specific purposes generate lots of data and evolve and treat data as an asset. Big Data have many benefits, such as reducing the experiment's cost and enhancing the efficiency of server or software to predict precise information (Yadav et al., 2020).

Any information that characterizes the biological, genetic, biochemical, and/or physical traits, makes-up, or functions of the substances to be examined utilizing the Goods is referred to as biological data. Refined nucleotide sequence data will be included in biological data. However, instrumentation data will not be included. The biological data includes primary sequence data obtained from automated sequencing technology stored in text strings. Sequencing technology generates vast data related to different species and has dramatically increased. The secondary biological data are graphs such as biochemical pathways, gene regulatory networks, and structured taxonomies. The additional biological data type belongs to high-dimensional data, such as gene expression profiles that store thousands of genes information in data points.

There are many different types of biological omics data (Fig. 16.1), and they can be used for various purposes. Here are a few examples of the kinds of biological data and some of the ways they can be used (Katara et al., 2023).

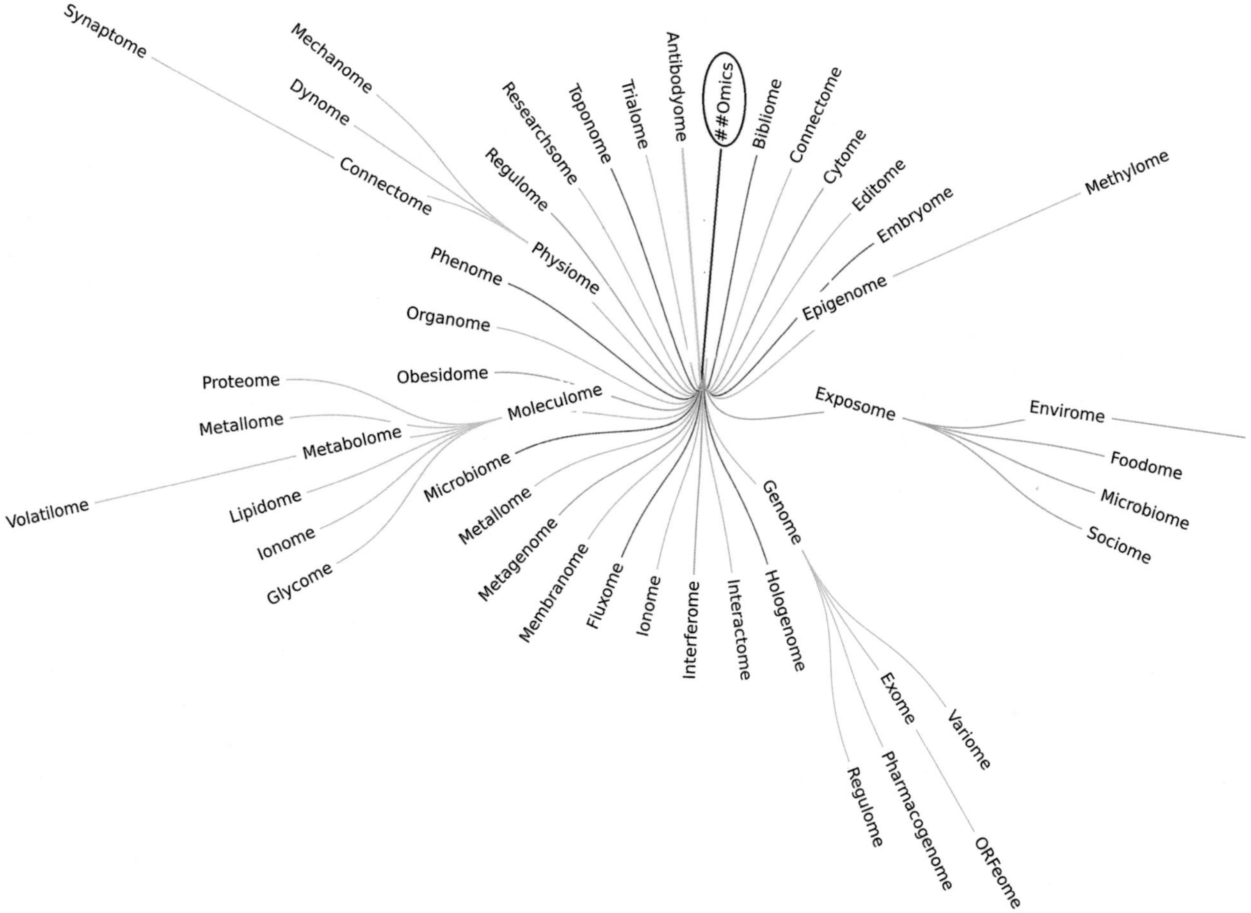

FIGURE 16.1 A hierarchy-based mind map of omics data. Omics data provides a comprehensive understanding of the underlying structure of a biological system to expand scientific research.

- **Genomic data:** This refers to data related to the DNA sequences of organisms. It can be used to study the genetic basis of traits and diseases, develop personalized medicine approaches, and understand species' evolutionary history.
- **Transcriptomic data:** This refers to data related to the expression levels of mRNA molecules involved in regulating gene expression. It can be used to study the regulation of gene expression, identify gene expression changes associated with particular diseases or conditions, and understand the mechanisms underlying cellular processes.
- **Proteomic data:** This refers to data related to protein expression levels and functions. It can be used to study the functions of proteins and their roles in biological processes, identify protein expression changes associated with particular diseases or conditions, and understand the mechanisms underlying cellular processes (Kumar et al., 2023).
- **Metabolomic data:** This refers to data related to the concentrations of small molecules, such as carbohydrates, lipids, and amino acids, in cells or tissues. It can be used to study the metabolism of cells and organisms, identify changes in metabolite concentrations associated with particular diseases or conditions, and understand the mechanisms underlying cellular processes.
- **Lipidomic data:** This refers to data related to the concentrations and structures of lipids in cells or tissues. It can be used to study the roles of lipids in biological processes, to identify changes in lipid concentrations that are associated with particular diseases or conditions, and to understand the mechanisms underlying cellular processes.
- **Epigenomic data:** This refers to data related to the modifications of DNA and histones that regulate gene expression. It can be used to study the regulation of gene expression, identify changes in epigenetic modifications associated with particular diseases or conditions, and understand the mechanisms underlying cellular processes.
- **Phenomics:** This refers to the study of the phenotypes of organisms, which are the observable characteristics and traits of an organism that result from the interaction of its genotype (its genetic makeup) with the environment. Phenomics can involve the study of a wide range of phenotypes, such as physical, biochemical, and behavioral traits. It can be used to understand the genetic basis of traits, identify the environmental factors that influence traits' expression, and predict how traits will change over time.

5. Application of data science in omics data analysis

Data analytics is the process of assessing data collections to make decisions based on the information they contain. It is increasingly carried out with specialized software and systems. Commercial sectors use tools for data analytics to assist businesses in making decisions. Similar tools or platforms can be used for biological data analysis such as Juypter Notebook (https://jupyter.org/), Power BI (https://powerbi.microsoft.com/en-us/), and Python (https://www.python.org/) help in data analysis and visualization. This helps to understand the significant changes and improvements in the performance of omics data.

The major application of data science in omics data analysis is precision medicine, personalized nutrition, agricultural biotechnology, and treating age-related diseases and disabilities. When diagnosing and treating age-related diseases and disabilities, data science uses many techniques, including reviewing the patient's medical history, performing a physical exam, and more. By also analyzing the condition at the molecular level, the precision and efficacy of treatment can be considerably increased. Phenotyping or genome sequencing can diagnose patients with complex cases and unusual disorders much more quickly and efficiently than doing numerous regular procedures. Researchers can also use patient data and samples for additional research, guiding new understanding and treatment.

Data science can be used in the field of systems biology to analyze and interpret biological data, such as protein—protein interaction (PPI) data. This can involve using machine learning algorithms to identify patterns and relationships in the data or statistical methods to understand the functional significance of specific genes or pathways. Data science can also be used to build predictive models of biological systems, which can help researchers understand how these systems might respond to different perturbations or interventions. Additionally, data science can be used to visualize and communicate the results of systems biology analyses, such as by creating interactive graphs or visualizations.

There are several ways that can use system biology-generated data utilizing data science techniques.

- System biology data such as PPI data can be analyzed using machine learning algorithms to identify patterns and relationships between different proteins. This can help researchers understand how different proteins function within a biological system and how they interact with each other.
- Building predictive models.
- Visualizing the data also can be used to create network diagrams that show the connections between different proteins.
- Analyzing the functional significance of specific proteins.
- Generated data can also be used to identify potential therapeutic targets for the treatment of diseases.

6. How to apply data science in omics data analysis?

Data science can be applied in omics data analysis to help users to extract meaningful insights from large and complex biological datasets. Here, four ways that can be used in omics data analysis are categorized.

1. Data preprocessing
2. Statistical analysis
3. Data visualization
4. Data interpretation or analysis

Data preprocessing: This step involves cleaning and formatting the raw data, such as removing any noise or outliers, and transforming it into a format that can be easily analyzed. it is a crucial stage in the study of omics data. The data can be prepared for analysis using data science techniques such as cleansing, standardization, and transformation. The steps of NGS data preprocessing are as follows:

- *Quality control:* The first step is to perform quality control checks to identify any technical issues or artifacts that could affect the accuracy of the analysis. This includes assessing the quality of sequencing reads, checking for batch effects, and identifying outlier samples.
- *Data normalization:* Omics data can be influenced by technical and biological variations, which can lead to biased results. This step can be used to correct for these variations and ensure that the data is comparable between samples. Common normalization methods include mean centering, scaling, and quantile normalization.
- *Feature selection:* Omics data often contains a large number of features, such as genes, proteins, or metabolites. The most informative features that are most relevant to the research question can be identified. This can help to reduce the dimensionality of the data and improve the accuracy of the analysis.

- *Missing value imputation:* Data may have some missing values due to technical issues or biological variability. This step can help to add and correct the missing values.
- *Data transformation:* Omics data can be transformed to improve the accuracy of the analysis or to meet specific assumptions of statistical or machine learning models. Common transformations include logarithmic or square root transformations for count data and *z*-score normalization for continuous data.
- *Batch effect correction:* Omics data can be influenced by batch effects, which are nonbiological variations that occur due to differences in experimental conditions or protocols. Tools can be used to remove these variations and ensure that the data is comparable between samples.

Statistical analysis: Genes, proteins, or metabolites that are differentially expressed between groups can be found using statistical analytic techniques, as can relationships between multiple variables. To conduct statistical analysis on Omics data, data science techniques including regression analysis, hypothesis testing, and clustering can be employed.

- *Differential expression analysis:* This method is used to determine which genes, proteins, or metabolites are expressed differentially among multiple groups of samples. *t*-tests, ANOVA, and regression analysis and more advanced methods include edgeR, DESeq2, and limma, which take into account sample variability, normalization, and multiple testing correction.
- *Correlation analysis:* Correlation analysis is used to identify associations between variables in Omics data, such as genes or metabolites that are coexpressed or coregulated. The widely used methods are Pearson correlation, Spearman correlation, and partial correlation analysis.
- *Machine learning:* Machine learning techniques can be used to build predictive models from omics data. Logistic regression, SVMs, random forests, and neural networks are the common methods to predict disease outcomes, identify potential drug targets, and classify patients into different disease subtypes.
- *Network analysis:* Network analysis techniques can be used to identify interactions between genes, proteins, and metabolites, and to identify key pathways and regulators involved in biological processes. Data analysis based on correlation-based network analysis, gene set enrichment analysis (GSEA), and pathway analysis using tools such as Cytoscape, STRING, and Reactome.
- *Survival analysis:* Survival analysis is used to analyze time-to-event data, such as time-to-death or time-to-recurrence. Omics data can be used to identify biomarkers that are associated with survival outcomes. Important methods are Kaplan–Meier analysis, Cox regression analysis, and machine learning–based survival analysis.
- *Clustering analysis:* Clustering analysis is used to group similar samples or features based on their expression patterns. The important methods include hierarchical clustering, k-means clustering, and fuzzy clustering.

Data visualization: This step involves creating visual representations of the data, such as heat maps, scatter plots, dimensionality reduction, clustering, or network diagrams, to make it easier to understand and interpret the results. Examples: ggplot and Ploty.

Data interpretation or analysis: This step involves using statistical methods (SPSS and NCSS) and machine learning algorithms such as K-means clustering to identify patterns and relationships in the data. This can include identifying differentially expressed genes or proteins, clustering samples based on similarity, or building predictive models. This step involves drawing conclusions from the data and interpreting the results in the context of the biological question being studied.

7. How to collect data to use in omics data processing?

Raw omics data can be obtained from different platforms such as pathway databases, biological sequence databases, chemical databases, and experimental sections such as microarrays, RNA-seq, mass spectrometry, and metabolomics. A pathway database is an online repository of information related to biochemical and signaling pathways. These databases store information about the interactions between biological molecules, such as proteins, metabolites, and genes, that participate in a particular pathway. They provide a valuable resource for researchers studying cellular and molecular biology, as they allow the identification of the components of a pathway and their interactions, and help to understand how these pathways are regulated and how they contribute to disease.

A big source of Genomics data is the National Center for Biotechnology Information (NCBI: https://www.ncbi.nlm.nih.gov/), which is a part of the United States National Library of Medicine (NLM), a branch of the National Institutes of Health (NIH). It has biological information and research, providing access to a vast collection of biomedical and genomic data, including DNA and protein sequences, gene expression data, and genetic variation data. It has various databases containing gene expression data, including microarray and next-generation sequencing data, genome, and single nucleotide variation data (Table 16.1). The omics data related to proteomics can be downloaded from KEGG, which has currently 562 pathway maps with references available in this database (Kanehisa & Goto, 2000). It stores chemical information with their biochemical reactions, enzyme nomenclature information, and disease-related network elements with associated genes and drug information. More than 8000 organisms have genome information with their orthology annotations. Downloadable resources from EcoCyc include the transcriptional regulation of the whole genome, transporters, and metabolic pathways (Karp et al., 2018). This database has *Escherichia coli* K-12 MG1655 experimental data including metabolic maps and genomes. Access can be gained by setting up an academic login for temporary usage while they offer premium access. The Comprehensive Enzyme Information information is provided by the BRENDA database (Scheer et al., 2011). It has 8423 enzymes based on organisms including metabolic pathways and structural classification of ligands and proteins. This database shows the enzyme-ligand interactions with associated disease and molecular properties. To get biochemical kinetics information, the SABIO-RK database can be used, which is a web-based application based on the SABIO relational database that contains information about biochemical reactions, their kinetic equations with their parameters, and the experimental conditions under which these parameters were measured. Similar information also can be fetched from the

TABLE 16.1 List of omics databases with different kinds of data that can be stored and examined.

#	Name of database	Type of data	Website
	GenBank	DNA/RNA	https://www.ncbi.nlm.nih.gov/genbank/
	NCBI Gene Expression Omnibus (GEO)	High-throughput gene expression and other functional genomics	https://www.ncbi.nlm.nih.gov/gds/
	dbSNP	Human single nucleotide variations	https://www.ncbi.nlm.nih.gov/snp/
	Genome	The data on genomes, including sequences, maps, chromosomes, assemblies, and annotations, is organized in this site	https://www.ncbi.nlm.nih.gov/genome/
	gnomAD	Exome and genome sequencing data from large-scale projects	https://gnomad.broadinstitute.org/
	RefSeq	Reference sequences including genomic, transcript, and protein	https://www.ncbi.nlm.nih.gov/refseq/
	UniProt	Protein sequence	https://www.uniprot.org/
	The Cancer Genome Atlas (TCGA)	Cancerous genomic, transcriptomic, and proteomic data	https://portal.gdc.cancer.gov/
	1000 Genomes Project	Catalog of common human genetic variation	https://www.internationalgenome.org/
	Human Protein Atlas	Expression and localization of human proteins in various tissues and cell types	https://www.proteinatlas.org/
	KEGG	Pathway maps	https://www.genome.jp/kegg/
	EcoCyc	*Escherichia coli*	https://ecocyc.org/
	BRENDA	Enzyme information	https://www.brenda-enzymes.org/
	Reactome	A curated knowledge base of biological pathways	https://reactome.org/
	BiGG Database	Metabolic networks	http://bigg.ucsd.edu/
	Lynx	Medicine database	https://lynx.ci.uchicago.edu/
	STRING	Protein–protein interactions	https://string-db.org/
	MetaboLights	A public repository for metabolomics	https://www.cbi.ac.uk/metabolights/
	PRIDE	Public repository for proteomics data	https://www.ebi.ac.uk/pride/
	ENCODE	Functional elements in the human genome	https://www.encodeproject.org/

Reactome database, which has a curated knowledgebase of biological pathways including classical intermediary metabolism, signaling, transcriptional regulation, apoptosis, and disease information (Wittig et al., 2018). They provide access to interaction, reaction, and route data such as flat, Neo4j GraphDB, MySQL, BioPAX, SBML, and PSI-MITAB downloadable files. The metabolomics data can be downloaded from the BiGG database, which has a metabolic reconstruction of human metabolism designed for systems biology simulation and metabolic flux balance modeling. It combines more than 70 documented genome-scale metabolic networks into a single database. The NCBI genome annotations are used to map the genes in the BiGG models, and numerous additional databases are connected to the metabolites (KEGG, PubChem, and many more). The protein–protein interactions data in sequences and interactions can be downloaded from the STRING database (Szklarczyk et al., 2023). It can be used to integrate data from a variety of sources such as high-throughput lab experiment data, coexpression data, automated text-mining, and previous knowledge in databases. The interactions come from computational prediction, knowledge transfer between species, and interactions gathered from other (primary) databases; they comprise direct (physical) and indirect (functional) correlations. Currently, ~6.8 bn proteins from 14,094 species are included in the STRING database. Similar data types can also be downloaded from Pathway Commons that disseminate biological pathway and interaction data. Currently, it has information on 2,424,055 interactions between 5772 pathways and 22 databases.

8. Pathway modeling, visualization, and simulation

After collecting the omics data, the next step is to understand the complex interactions between biological molecules in cellular processes. Network modeling is an important method also called pathway modeling used to visualize and analyze the protein–protein interaction networks, gene regulatory networks, and other types of biological networks (Kwoh & Ng, 2007).

The following steps can be used to design networks:

- *Identify the pathway of interest:* The first step in pathway modeling is to identify the pathway that the user wants to model, which is called data mining. This could be a signaling pathway, a metabolic pathway, or a gene regulation pathway.
- *Gather data:* Once the pathway of interest has been identified, the next step is to gather data about the pathway's components and their interactions. This data can come from a variety of sources, such as literature, databases, or experimental data.
- *Create a representation of the pathway:* Once the data has been gathered, the next step is to create a representation of the pathway. This can be done using a variety of software programs, such as CellDesigner, Pathway Studio, or KEGG Pathway.
- *Add components:* The next step is to add the different components of the pathway, such as proteins, genes, and small molecules. This can be done by dragging and dropping the components from the palette onto the canvas.
- *Create interactions:* Once the components are added to the canvas, users can create interactions between them by drawing arrows between the components to represent the different types of interactions.
- *Add annotations:* To make the pathway model more informative, users can add annotations to the components and interactions. This includes adding information such as gene names, protein names, and molecular details.
- *Simulate the model:* Once the pathway model is complete, the user can simulate the model to see how it behaves under different conditions. This can be done by adjusting the parameters of the model, such as the rate constants of the interactions, and running the simulation.
- *Evaluate the results:* After simulating the model, the user can evaluate the results to see how the pathway behaves. This includes analyzing the time course of the different components and checking for any steady-state solutions.
- *Refine the model:* If the results are not as expected, it may be necessary to refine the model by adjusting the parameters or adding new interactions. Once the model is refined, it can be simulated again to check for any changes in behavior.
- *Share and collaborate:* CellDesigner provides the option to share the model with others by exporting it to a variety of file formats and sharing it with other researchers through collaboration tools.

8.1 Tools for designing pathway

To design a pathway, required software or tools need to be installed in the system. Then, define or design the pathway the user wants to model. This involves identifying the biochemical reactions and components that are involved in the pathway the user wants to design. The commonly used software is Cell Designer, which is a structured diagram editor for creating biochemical and gene regulatory networks. It supports SBML and XML format and also integrates the pathway with ODE

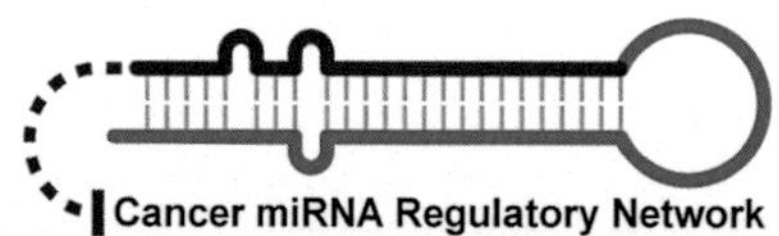

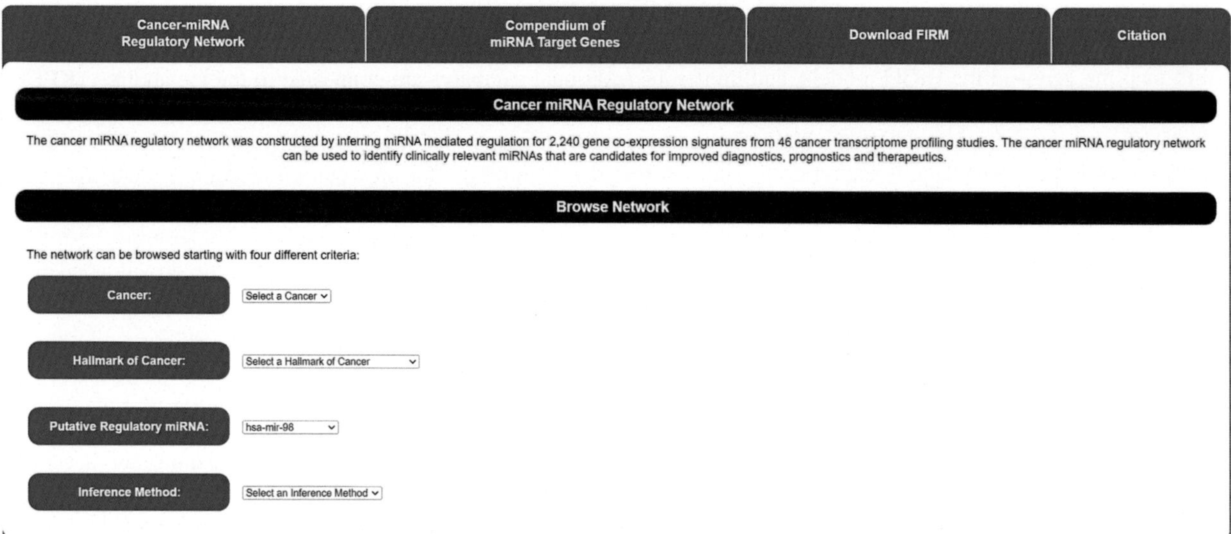

FIGURE 16.2 Homepage of the cancer miRNA regulatory network. Finding clinically important miRNAs that are potential candidates for better diagnostics, prognostics, and therapies can be done using the cancer miRNA regulatory network. *From https://cmrn.systemsbiology.net/.*

solver and COPASI for simulation (Funahashi et al., 2003). It has a wide variety of SBGN graphical notations to represent nodes, proteins, genes, reactions, and others. After designing the pathway, add the reactions that are involved in the pathway. This involves adding reaction arrows between the components in the graphical representation, specifying the reactants and products, and defining the reaction rate and kinetic parameters. After that add constraints to the pathway, such as regulatory interactions, kinetic constants, and steady-state constraints. After that validate the pathway by running simulations and analyzing the results. This involves using the software to simulate the behavior of the pathway under different conditions, such as changes in concentration, temperature, and pH. This can be done by the COPASI plugin available in Cell Designer, which is used to simulate and analyze the dynamics of biochemical networks (Hoops et al., 2006). It supports models that adhere to the SBML standard and can mimic the behavior of those models using ODEs or Gillespie's stochastic simulation technique; such simulations can contain any number of discrete events. The designed validated pathway can be exported in other formats for further analysis such as integration and visualization with other biochemical pathways using Cytoscape, which is a popular open-source software tool used for visualizing and analyzing complex networks, such as molecular and genetic networks (Shannon et al., 2003). It provides a range of tools for visualizing networks, including various layout algorithms, node and edge styling options, and the ability to view large networks using zooming and panning. It also allows users to integrate multiple types of data, such as gene expression data, protein–protein interaction data, and genetic variation data, to create integrated networks. The network analysis can also be done through SBEToolbox, which is an open-source Matlab toolbox for biological network analysis (Konganti et al., 2013). It takes a network file as input and calculates several centralities and topological metrics, organizes nodes into modules, and then utilizes several graph layout methods to display the network. The homepage of the cancer miRNA regulatory network has been shown in Fig. 16.2.

Some more powerful tools and software used for pathway simulation, protein–protein interactions, simulation, and analysis are listed in Table 16.2.

9. Statistical method for genomics data analysis

Genomics data analysis involves analyzing large amounts of genomic data to extract meaningful insights. Techniques such as *t*-tests, ANOVA, and regression analysis can be used to identify correlations between variables and to test hypotheses about the relationships between different types of omics data.

TABLE 16.2 List of software, tools, and server used for pathway modeling and analysis.

#	Softwares/tools	Aim	Web address
	Cell Designer	Design and edit biochemical pathway	https://www.celldesigner.org/
	Cytoscape	It is a software platform for visualizing and analyzing complex networks, and pathways with many plugins and resources specifically designed for analyzing omics data.	https://cytoscape.org/
	ContextNet	Identifying signaling paths linking proteins to their downstream target genes	http://netbio.bgu.ac.il/ContextNet
	GRN2SBML	Intends to make it easier to encode gene regulatory networks (GRNs) in the systems biology markup language (SBML).	https://www.leibniz-hki.de/en/grn2sbml.html
	MeltDB	It is used to examine and annotate metabolomics experiment datasets.	https://meltdb.cebitec.uni-bielefeld.de/cgi-bin/login.cgi
	Metingear	It is an open-source desktop program for building and maintaining metabolic networks with chemical structures at the genome size.	http://johnmay.github.io/metingear/
	ODEion	A tool that helps identify the structural components of ordinary differential equations using any user-defined function implicitly define the model space. The program looks for potential interactions between the variables and offers an appropriate ODEs form, along with parameter estimations.	http://www.odeidentification.org/identification/ODEion/
	SBEToolbox	It takes a network file as input, computes various centralities and topological metrics, classifies nodes into modules, and then utilizes several graph layout methods to display the network.	https://github.com/biocoder/SBEToolbox/releases
	SPNConverter	It is a tool for GraphML conversion of graphs. Users can automatically change networks using a basic systems biology markup language with this Cytoscape plugin for study with the SPN simulator.	https://apps.cytoscape.org/apps/spnconverter
	Gephi	It is used as a graph and network exploration and visualization software.	https://gephi.org/
	BioLayout Express3D	It is an open-source visual analytics tool created to facilitate the analysis of enormous and intricate biological datasets represented as networks.	http://biolayout.org/
	COPASI	Through the use of ODEs, SDEs, or Gillespie's stochastic simulation method, it imitates the behavior of models in the SBML standard.	http://copasi.org/
	JSim	It is a Java-based simulation system for creating numerical models that are quantitative in nature and analyzing those models in relation to experimental reference data.	https://www.imagwiki.nibib.nih.gov/physiome
	LibSBMLSim	It is a library for modeling ordinary differential Equations (ODEs) in SBML models.	https://github.com/libsbmlsim
	Morpheus	A modeling and simulation platform for the combination of cell-based models including reaction-diffusion systems, ordinary differential equations, and multiscale systems.	https://imc.zih.tu-dresden.de/wiki/morpheus/?animal=morpheus
	BioTapestry	It is a program that allows users to construct, visualize, and simulate genetic regulatory networks interactively.	http://www.biotapestry.org/

Continued

TABLE 16.2 List of software, tools, and server used for pathway modeling and analysis.—cont'd

#	Softwares/tools	Aim	Web address
	Cell phone simulation	The structure and characteristics of a cell phone network are modeled and examined with this application utilizing Cytoscape.	https://see.isbscience.org/resources/cell-phone-simulation/
	Bio-SPICE	The goal of Bio-SPICE is to simulate and model spatiotemporal processes in living cells.	https://biospice.sourceforge.net/
	BioUML	This comprehensive Java platform is open-source and could be used to create virtual physiological humans and cells. It offers access to databases with experimental data, tools for formalized descriptions of the structure and operation of biological systems, as well as tools for their visualization, simulation, parameter fitting, and analysis.	http://wiki.biouml.org/index.php/Landing
	SIMScells	It is an online tool for biomedical researchers in academic and medical facilities. Simulating the entire behavior of the cell, SIMScells provide researchers with a systems biology analysis of their findings, placing their genes, proteins, medications, or pathologies of interest in the context of the overall biological system of a human being or another distinct biological system, like a cell or tissue.	http://www.simscells.com/
	SBRT	It is a freely available, integrated software framework that supports the computational aspects of systems biology. The SBRT currently employs a number of techniques for studying stoichiometric networks as well as techniques from the domains of graph theory, geometry, algebra, and combinatorics.	https://www.ieu.uzh.ch/wagner/software/SBRT/
	SBSI	It is a collection of modular applications that makes parameter fitting easier. SBSI is made up of three main parts: SBSIDispatcher, a middleware application to track experiments and submit jobs to back-end servers; SBSINumerics, a high-performance library with parallelized algorithms for parameter fitting; and SBSIVisual, an extensible client application to set up optimization experiments and view results.	http://www.sbsi.ed.ac.uk/
	SBW	A software platform that enables heterogeneous application components to communicate and utilize one another's capabilities via a quick binary encoded-message system. These components are written in various programming languages and are running on various platforms.	https://sbw.sourceforge.net/
	The cancer miRNA regulatory network	The 46 cancer transcriptome profiling studies identified 2240 gene coexpression signatures that were used to infer miRNA-mediated regulation, which was then used to build the cancer miRNA regulatory network.	https://cmrn.systemsbiology.net/
	cMonkey	It is a machine learning algorithm and data integration framework that aims to discover coregulated modules, or bicluster in gene expression profiles.	https://baliga.systemsbiology.net/projects/cmonkey/
	BioSmalltalk	A pure object system and bioinformatics library. With the help of the software, users can create bioinformatics programs and scripts in the Smalltalk programming language.	https://biosmalltalk.github.io/web/

9.1 Differential gene expression

The process of analyzing normalized read count data statistically to discover quantitative differences in expression levels between experimental groups is known as differential expression analysis (Conesa et al., 2016). This method is used to identify genes that are differentially expressed between two or more groups of samples. It involves statistical tests such as *t*-tests, ANOVA, and regression analysis.

T-tests: These tests can be used to compare the means of two groups and determine whether there is a significant difference between them. They are often used to identify differences in the expression levels of genes or proteins between two groups of samples, such as healthy and diseased samples.

The most widely software used for *t*-tests are NCCS (https://www.ncss.com), WINKS (https://www.texasoft.com/winksst.html), GraphPad (https://www.graphpad.com), and DataTab (https://datatab.net/).

ANOVA: This test can be used to compare the means of more than two groups and determine whether significant differences exist between them. It is often used to identify differences in the expression levels of genes or proteins between multiple groups of samples, such as samples from different treatments or conditions. OriginLab (https://www.originlab.com/), and XLSTAT (https://www.xlstat.com/) are highly recommended software for ANOVA analysis.

Regression analysis: This technique can be used to identify the strength and direction of the relationship between two variables. It is often used to identify correlations between the expression levels of genes or proteins and other variables, such as clinical outcomes or environmental factors. Alteryx (https://www.alteryx.com/), Origin Lab (https://www.originlab.com/), RegressIt (https://regressit.com/), and XLSTAT (https://www.xlstat.com/) are the standard software for regression analysis.

Principal component analysis (PCA): This technique can be used to reduce the dimensionality of omics datasets and identify the underlying patterns and trends in the data. It is often used to identify clusters of samples that are similar to each other based on their omics profiles. The common tools used in the analysis are Matlab (https://www.mathworks.com/), XLSTAT (https://www.xlstat.com/), Minitab (https://www.minitab.com), and Q (https://www.qresearchsoftware.com/) software, which are majorly used in PCA analysis.

9.2 One-by-one assessment

Big Data including, numerous genes, genetic variations, chemicals, biomarkers, cells, and people are all considered in omics data analysis. Analysis of each aspect separately and obtaining a variety of outputs is one way to manage these components. This strategy makes use of both transcriptomic RNA-seq analysis, in which all coding genes are measured, in which a large number of single nucleotide variations (SNVs) are examined one by one (Wu et al., 2019).

9.3 Dimension reduction

To evaluate individual items to find patterns is another method for evaluating many items. This method uses PCA to find informative components made up of a weighted sum of individual items and clustering samples with expression patterns (Jolliffe & Cadima, 2016). When the observed dataset has numerous variables or is spread in a high-dimensional space, dimension reduction is helpful because the variables are localized in a subspace rather than being dispersed randomly throughout the high-dimensional space. In reality, omics research is carried out with the assumption that biological phenomena are complicated and should be explained by fewer key principles. The dataset can be expressed in lower-dimensional space because the subspace is smaller than the entire space. Finding fewer components to describe the entire dataset is achieved by collapsing the entire set of data with a high dimension (Huang et al., 2019).

Bioconductor is a popular open-source software project for the analysis and comprehension of high-throughput genomic data. It provides a large collection of packages and tools for the analysis and visualization of genomic and transcriptomic data. Table 16.3 summarizes some Bioconductor packages that are commonly used in genomics statistical data analysis.

9.4 Error calculation methods

- **P-values:** In hypothesis testing, a *P* value is used to assist the acceptance or rejection of a null hypothesis. The *P* value acts as evidence that refutes the null hypothesis. The stronger the argument is against accepting the null hypothesis, the smaller the *P* value. The *P*-values can be obtained from probability methods such as *t*-test distributions.

TABLE 16.3 List of important packages available in bioconductor used for genomics statistical data analysis.

#	Package	Role	Web address
	EdgeR	Bioconductor packages based on negative binomial (NB) distributions, or baySeq and EBSeq, which are Bayesian approaches based on a negative binomial model, are two methods for differential expression analysis.	https://bioconductor.org/packages/release/bioc/html/edgeR.html
	HTSeq	This is used to summarize and aggregate the mapped reads across genes.	https://htseq.readthedocs.io/en/master/
	iRAP	It calculates the fragments per kilobase of exon per million mapped fragments (FPKMs) from the raw counts for baseline expression	https://github.com/nunofonseca/irap
	DESeq	This is used to identify genes that are expressed differently between the test and reference groups in each pairwise contrast.	https://bioconductor.org/packages/release/bioc/html/DESeq2.html
	maSigPro		
	limma	Differential expression analysis of microarray and other types of data	http://www.bioconductor.org/packages/release/bioc/html/limma.html
	GenomicRanges	Handling and manipulating genomic intervals and ranges	https://bioconductor.org/packages/release/bioc/html/GenomicRanges.html
	Biostrings	Manipulation of biological sequences such as DNA and RNA	https://bioconductor.org/packages/release/bioc/html/Biostrings.html
	Biobase	Reading and processing microarray data and other genomic data formats	https://www.bioconductor.org/packages/release/bioc/html/Biobase.html
	IRanges	Handling and manipulating interval ranges for large datasets	https://www.bioconductor.org/packages/release/bioc/html/IRanges.html
	ChIPseeker	Analysis and visualization of ChIP-seq data	https://www.bioconductor.org/packages/release/bioc/html/ChIPseeker.html
	GOstats	Gene ontology analysis and enrichment analysis	https://www.bioconductor.org/packages/release/bioc/html/GOstats.html

- **False Discovery rates:** The expected percentage of type I mistakes is known as the false discovery rate (FDR). Incorrectly rejecting the null hypothesis, or getting a false positive, is a type I error. The FDR formula is:

$$FDR = E(V/R \mid R > 0)\ P(R > 0)$$

Where:
V = Number of Type I errors (i.e., false positives)
R = Number of rejected hypotheses.

10. Machine learning–based genomics software and tools

Machine learning algorithms such as decision trees, random forests, and SVMs can be trained on omics datasets to make predictions about the behavior of molecular systems, such as the likelihood of a particular gene being expressed or the likelihood of a particular protein being involved in a specific process. Many machine learning software packages can be used for data analytics. The commonly used software is MiBiOmics which is a web-based platform used to analyze and interpret data from several omics, including genomes, transcriptomics, proteomics, and metabolomics. It offers tools for the discovery of biomarkers and the research of biological pathways, as well as an intuitive interface for integrating, analyzing, and displaying complicated omics data. It allows the study of a wide range of data types, including unprocessed data from several open databases and raw sequencing data. It enables researchers to carry out a variety of analyses, including route, network, and differential expression analysis, and to show their findings using interactive graphs and charts. To identify illness biomarkers and forecast disease outcomes, MiBiOmics also offers a number of machine learning–based tools. By using their own data or preexisting data from public databases, researchers can train machine learning models and then use these models to find or predict biomarkers in their own datasets. Tools that are commonly used for genomics omics data analysis are listed.

11. Omics data visualization

Omics datasets can be combined with other data types, such as clinical or environmental data, to understand complex data better. Many software tools can be used to integrate omics datasets with other types of data to gain a more complete understanding of molecular systems. There are various types of genomic data that can be visualized, including DNA sequences, gene expression data, protein interactions, and genomic variations. The most commonly used visualization techniques include heatmaps, scatter plots, bar graphs, line graphs, and network graphs. One of the most popular tools for genomics data visualization is the Integrative Genomics Viewer (IGV), which allows users to visualize and explore genomic data from a variety of sources, including next-generation sequencing data, microarrays, and annotations (Robinson et al., 2011). The ggplot2 is a popular data visualization package in the R programming (https://www.r-project.org/) language that allows users to create high-quality, customizable graphics using a declarative syntax (Wickham, 2016). The package is based on the grammar of graphics, which provides a framework for constructing complex graphics from simple building blocks. Omics data can be plotted in layers of visual elements, such as data points, lines, and labels. This allows users to easily customize and modify their graphics by adding or removing layers. It also allows users to map variables in their data to visual aesthetics, such as color, size, and shape. This makes it easy to create visualizations that highlight patterns and relationships in the data. Other tools are also available in public domains that can be used for omics data visualization (Table 16.4).

TABLE 16.4 List of machine learning–based software, tools, and servers for genomics data analysis.

#	Tools	Aims	Web address
	scikit-learn	This is an open-source machine learning library for Python that provides a range of algorithms for tasks such as classification, regression, clustering, and dimensionality reduction.	https://scikit-learn.org/stable/
	TensorFlow	This is an open-source machine learning library developed by Google that can be used to build and train machine learning models.	https://www.tensorflow.org/
	PyTorch	This is an open-source machine learning library developed by Facebook that can be used to build and train machine learning models.	https://pytorch.org/
	Weka	This is a group of tools for data mining, predictive modeling, and machine learning that were created at the University of Waikato in New Zealand. It has a graphical user interface and can be used for classification, regression, clustering, and association rule mining tasks.	https://www.weka.io/
	RapidMiner	This commercial machine learning platform provides a range of algorithms and tools for data preparation, modeling, evaluation, and deployment.	https://rapidminer.com/
	Bioconductor	It is an open-source software project for analyzing and comprehending genomic data developed for the R programming language.	https://www.bioconductor.org/
	GenePattern	It is a platform for analyzing and interpreting genomic data, developed by the Broad Institute of MIT and Harvard.	https://www.genepattern.org/
	CLC Genomics Workbench	It is a software platform for analyzing and visualizing genomic data, developed by Partek Inc.	https://digitalinsights.qiagen.com/
	Partek Genomics Suite	It is statistical analytic software that enables desktop computer-based examination of microarray, qPCR, and preprocessed NGS data.	https://www.partek.com/partek-genomics-suite/

Continued

TABLE 16.4 List of machine learning–based software, tools, and servers for genomics data analysis.—cont'd

#	Tools	Aims	Web address
	Galaxy	It is an open-source platform for data-intensive research, with many tools and resources specifically designed for omics data analysis.	https://usegalaxy.org/
	QIIME	It is an open-source software package for analyzing and visualizing microbial community data.	https://qiime2.org/
	ATAQS	Targeted mass spectrometry based on SRM is a key method for producing consistent, sensitive, and quantitatively precise data from biological material. And developed proteomic techniques for discovery.	http://tools.proteomecenter.org/ATAQS/ATAQS.html

12. Omics data integration

The omics data management and integration of biological data across different species and databases is offered by various open-source, free software, such as BioMart, MiBiOmics, 3Omics, PaintOmics, and so on, which is being used for omics data analysis (Table 16.5). Roles and features of important software used in data integration and analysis are discussed in Table 16.6.

The BioMart was initially created as a cooperative initiative between the European Bioinformatics Institute and the Ontario Institute for Cancer Research (OICR) (EBI). It provides a centralized repository for biological data, including genomic sequences, gene annotations, and functional annotations (Fig. 16.3). BioMart is widely used in the genomics community for data mining and integration, and it is supported by a large number of biological databases, including Ensembl, UniProt, and Reactome.

Some key features of BioMart include the following:

It allows users to integrate data from multiple sources and perform complex queries to extract relevant information. This is particularly useful for large-scale data mining and analysis projects.

It provides a wide range of output formats, including HTML, CSV, and XML, which can be customized to meet the needs of the user. This makes it easy to extract and manipulate data for downstream analysis.

BioMart has a user-friendly interface that allows users to easily browse and search for data, as well as create and save queries for future use.

It supports cross-species queries, which allow users to extract data from multiple species and compare them in a single analysis. This is particularly useful for evolutionary and comparative genomics studies.

Biographer (https://biographer.biologie.hu-berlin.de/): It comprises a JavaScript web component (biographer-display) for interactive in-browser visualization of those networks and a layout algorithm C/C++ library (biographer-layout) that

TABLE 16.5 List of machine learning–based omics data visualization tools and servers.

#	Tools	Aims	Web address
	ggplot2	This is an open-source data visualization package for the R programming language that provides a wide range of plots and charts for visualizing data. It is particularly well-suited for creating publication-quality figures.	https://ggplot2.tidyverse.org/
	Matplotlib	This is an open-source data visualization library for Python that provides a wide range of plots and charts for visualizing data.	https://matplotlib.org/
	Plotly	This commercial data visualization platform provides a wide range of plots and charts for visualizing data. It has a web-based interface and can be used to create interactive figures that can be shared online.	https://plotly.com/
	Tableau	This commercial data visualization platform provides a wide range of plots and charts for visualizing data. It has a graphical user interface and can be used to create interactive dashboards and reports.	https://www.tableau.com/
	D3.js	It is possible to create interactive data visualizations in web browsers using this free JavaScript package.	https://d3js.org/

TABLE 16.6 List of omics data integration and analysis software.

#	Name	Aims	Web address
	OMERO server	It manages image data from biomedical research, arranged, analyzed, and shared from any location with an internet connection. OpenBIS platform can be used for data integration.	https://github.com/qbicsoftware/omero-lib
	KNIME	This open-source data integration and analysis platform provides a graphical user interface for building workflows for data manipulation, visualization, and analysis. It can be used to integrate and analyze a wide range of data types, including omics data.	https://www.knime.com/
	Talend	This commercial data integration platform provides a graphical user interface for building data pipelines and integrating data from a wide range of sources. It can be used to integrate and analyze omics data and other types of data.	https://www.talend.com/
	Trifacta	This commercial data integration platform provides a graphical user interface for cleaning and preparing data for analysis. It can be used to integrate and prepare omics data and other types of data for further analysis.	https://www.trifacta.com/
	Omics Mock Generator (OMG) library	The Omics Mock Generator (OMG) library is used to generate synthetic multiomics data that can be used to test computational tools for bioengineering metabolic models.	https://www.osti.gov/biblio/1765948
	3Omics	A quick visualization and integration of various inter- or intratranscriptomic, proteomic, and metabolomic human datasets. It contains five widely used studies, including correlation network, coexpression, phenotype generation, KEGG/HumanCyc pathway enrichment, and GO enrichment. It encompasses and links cascades originating from transcripts, proteins, and metabolites.	https://3omics.cmdm.tw/
	The Model Composition Tool	Users can create their own systems biology models, which simplifies the difficult task of combining systems biology models. It can be accessed through a VPN as it is available in a restricted firewall.	http://nashua.case.edu/PathwaysSB/Web/
	MiBiOmics	It is a web-based platform used to analyze and interpret data from several omics, including genomes, transcriptomics, proteomics, and metabolomics.	https://shiny-bird.univ-nantes.fr/app/Mibiomics
	UCSC Xena	It is an online application that enables users to browse and view datasets from many UCSC databases, including TCGA, ICGC, TARGET, GTEx, and CCLE.	https://xena.ucsc.edu/

generates a graph layout suited for biological networks, particularly for bipartite response graphs. All subprojects are combined by a server component (biographer-server), which also supports network import from multiple sources and subsequent visualization.

Paintomics is used for the integrated analysis of multiomics datasets. It allows researchers to explore the functional relationships between different molecular layers, such as genes, proteins, metabolites, and pathways and to identify key biological processes that are dysregulated in disease. Transcriptomics, proteomics, metabolomics, and pathway databases are just a few of the sources of data that PaintOmics incorporates. It also offers a user-friendly interface for data display and analysis (Fig. 16.4). It makes use of a color-coding system to draw attention to the different ways that genes, proteins, or metabolites are expressed across several samples or situations, enabling researchers to spot recurring patterns and pathologies. PaintOmics also provides several analytical tools, such as enrichment analysis, pathway analysis, and network analysis, to help researchers identify key regulatory pathways and molecular networks that are involved in disease. It can be used for a wide range of applications, including biomarker discovery, drug target identification, and personalized medicine.

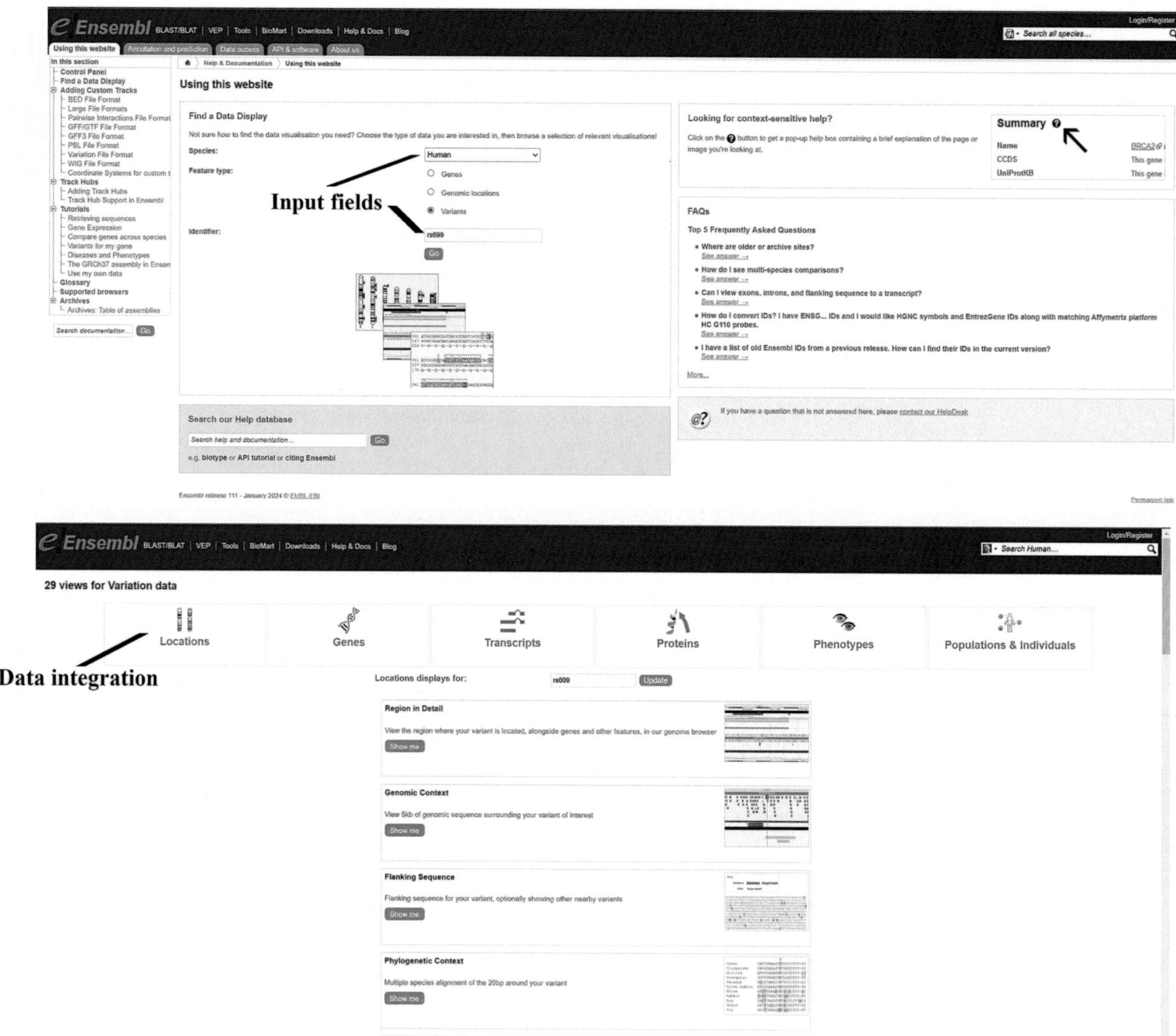

FIGURE 16.3 Omics data integration using BioMart. With the help of the user-friendly web application BioMart, data may be extracted without the need for programming expertize or a thorough understanding of the underlying database structure. Using the left panel, users may explore the BioMart web interface. In the right panel, filters and qualities can be chosen. On the left side, there is also a summary of your selections. *From https://useast.ensembl.org/info/data/biomart/how_to_use_biomart.html.*

PhenoTimer is used for the analysis of high-throughput phenotyping data in plants. It allows researchers to quantify the phenotypic differences between different plant genotypes, treatments, or environmental conditions and to identify key genes or pathways that are involved in the regulation of plant growth and development (Fig. 16.5). It uses time-lapse imaging and machine learning algorithms to extract quantitative phenotypic data from plants over time. It can analyze a wide range of plant phenotypes, such as leaf size, shape, color, and movement, as well as root growth, architecture, and response to environmental stimuli. It also provides several analytical tools, such as growth curve analysis, PCA, and clustering analysis, to help researchers identify significant differences in plant phenotypes and correlate these phenotypes with genetic or environmental factors. It also allows researchers to generate interactive visualizations of their data, such as heat maps, scatter plots, and trajectory plots, to help them better understand the complex relationships between different phenotypic traits and their underlying genetic or environmental factors.

POMO (https://pomo.cs.tut.fi/): POMO (Protein–Protein Interaction prediction with Multi-Omics integration) is used for predicting protein–protein interactions (PPIs) using a multiomics approach. It integrates data from various sources, such as gene expression data, protein structural data, and functional annotations, to predict novel PPIs and to improve the accuracy of existing PPI predictions. It first preprocesses the input data to extract relevant features, such as coexpression patterns, domain interactions, and gene ontology annotations. It then trains a classifier to predict PPIs based on these features, using a set of positive and negative training examples.

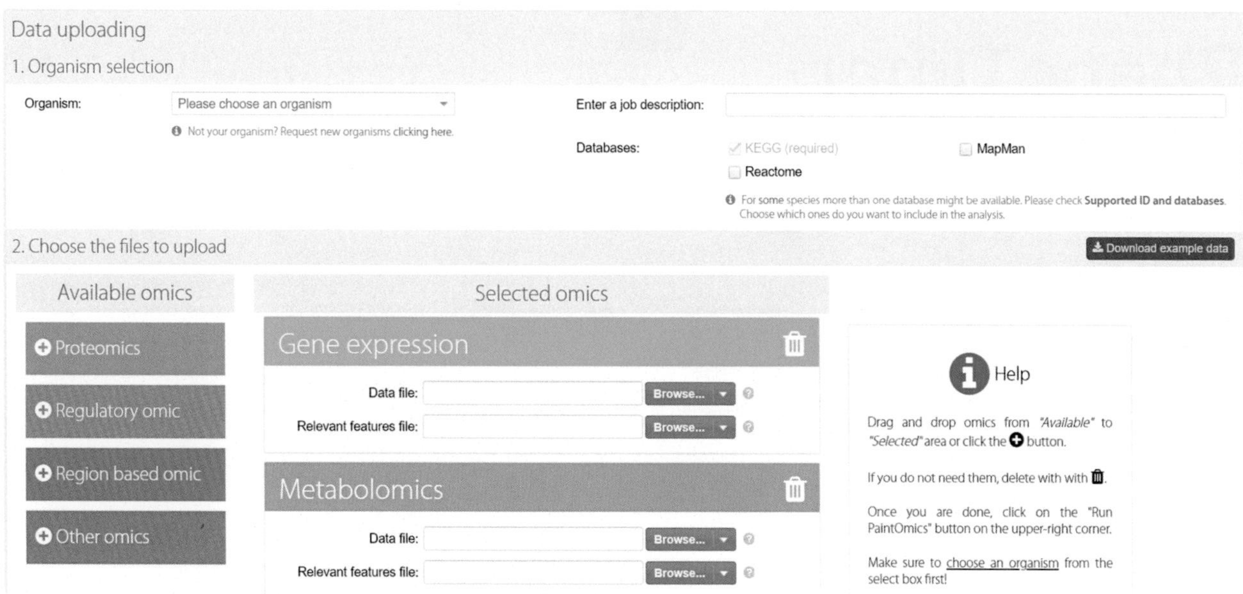

FIGURE 16.4 Homepage of PaintOmics. KEGG, Reactome, and MapMan biological pathway maps can be integrated with various omic datasets using the web program PaintOmics. *From https://www.paintomics.org/.*

Network Portal is used for the visualization and analysis of molecular interaction networks, such as protein–protein interactions, gene regulatory networks, and metabolic pathways. It provides a user-friendly interface for exploring complex networks and for identifying key nodes and pathways that are involved in biological processes (Fig. 16.6). It integrates data from various sources, including public databases and user-generated data, and allows researchers to upload and analyze their own network data. It provides several visualization options, such as force-directed layout, circular layout, and hierarchical layout, to help researchers explore and interpret their network data. It also allows researchers to filter and highlight nodes and edges based on various criteria, such as degree, centrality, and functional annotation. It also provides several analytical tools, such as pathway enrichment analysis, network clustering, and module identification, to help researchers identify key biological pathways and functional modules within their network data. It also allows researchers to perform network-based drug target prediction and to visualize drug–protein interaction networks.

BioFabric is a network visualization tool designed to simplify the representation of complex biological networks, such as protein–protein interactions, gene regulatory networks, and metabolic pathways (Fig. 16.7). It provides a unique visual style for network diagrams that allows for a clear and intuitive representation of relationships between nodes. It uses a unique visualization style that represents edges as ribbons rather than lines, with each ribbon representing the interaction between two nodes. The width and color of the ribbon reflect the strength and type of interaction, making it easy to distinguish between different types of interactions. The nodes are arranged in a simple grid layout, with more highly connected nodes placed closer together. It also has several analytical tools, such as clustering, filtering, and highlighting, to help researchers explore and interpret their network data. It also allows for interactive exploration of network data, with the ability to zoom and pan to focus on specific areas of the network.

13. Hardware acceleration in omics data analysis

The majority of hardware accelerators are coprocessors with CPUs rather than separate platforms. Graphics processing units (GPUs), field programmable gate arrays (FPGAs), and custom integrated circuits (ICs) are the three most common categories of hardware platforms. The required hardware acceleration along with current data analytics applications are listed below.

- Computers with high processing power and memory, such as servers or desktop workstations, for running analytical software and storing large amounts of data.
- Storage devices such as hard drives or solid-state drives for storing and accessing data.
- Networking equipment for connecting devices and transferring data.

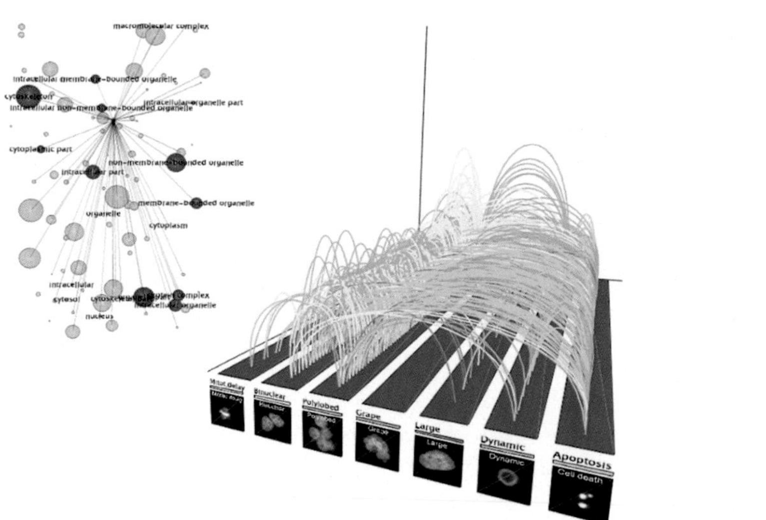

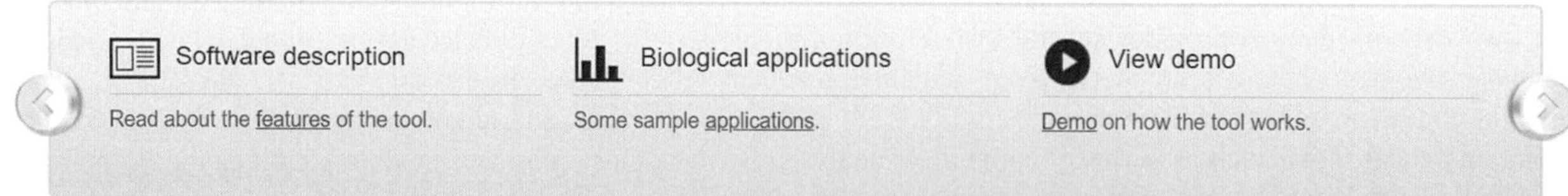

FIGURE 16.5 Homepage of PhenoTimer. A tool for mapping time-resolved phenotypic interactions in a genetic setting in 2D or 3D. To facilitate quick pattern identification and hypothesis formulation, it employs an innovative visualization method to illustrate relationships between morphological flaws, pathways, or diseases. *From http://phenotimer.org/.*

- GPUs for running complex calculations and machine learning and deep learning algorithms.
- FPGAs, are common tools for hardware prototyping. They have advancements in speed and density, high-performance systems are using FPGAs more and more frequently. FPGAs have advantages over CPUs (and GPUs) in terms of memory bandwidth, power consumption, and capability for streaming dataflow computations. They are most suited for simple, repetitive tasks that require high performance and low power.
- ICs have the fastest accelerators available right now and deliver speedups at several orders of magnitude over CPU single-threaded software performance.

In general, users can process a limited amount of omics data using a laptop, desktop PC, or tablet—not a Chromebook. Users should use the most recent processor generation and the Linux operating system for data processing; the P1 or M1 processors are not recommended. However, here are some general guidelines for a PC configuration suitable for data analytics.

- **Operating system:** Windows's latest version preferred (Windows 10 or 11) for data visualization and Linux for data processing.
- **Processor:** A high-end processor, such as an Intel Core i9 or AMD Ryzen 9, with multiple cores (>16) and high clock speeds (>2.3 GHz) is recommended.
- **Memory:** A minimum of 16 GB of RAM is recommended, but more is better for more complex tasks or larger datasets.

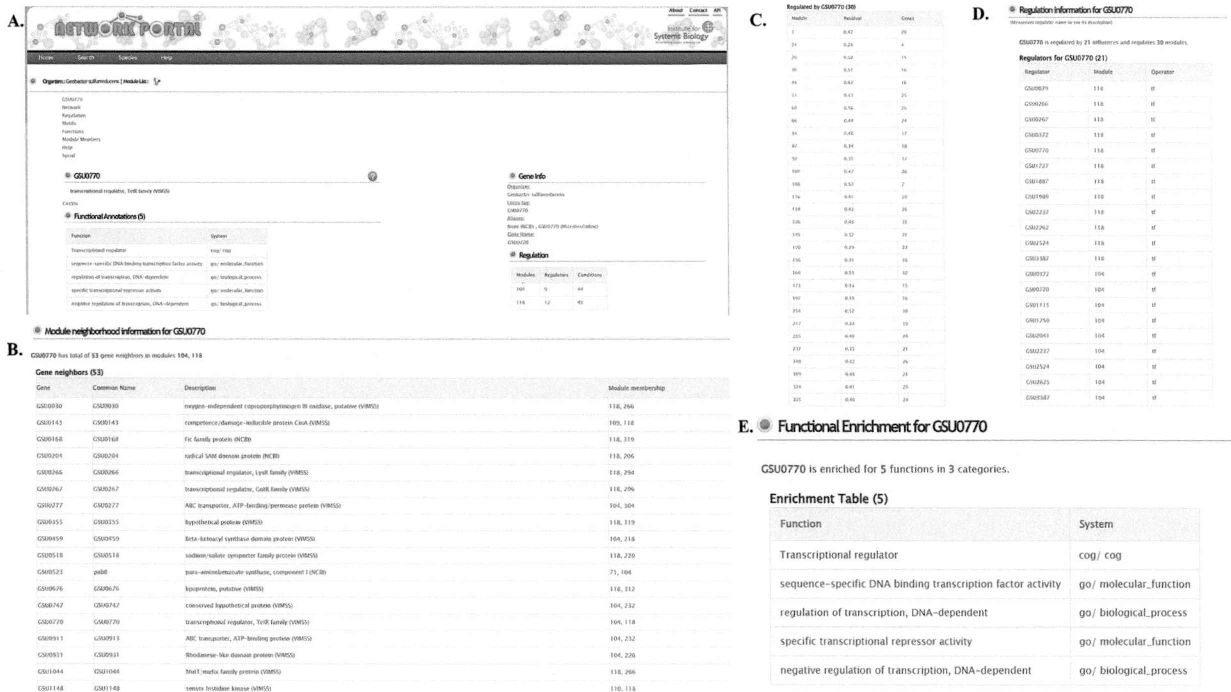

FIGURE 16.6 Data analysis pages from network portal: (A) gene expression profile, (B) list of expressed genes, (C) interaction map of expressed genes, (D) list of regulators, and (E) list of identified motifs. To assist researchers in the discovery of biology and hypothesis formulation, the Network portal offers analysis and visualization tools for a few chosen gene regulatory networks. *From http://networks.systemsbiology.net/.*

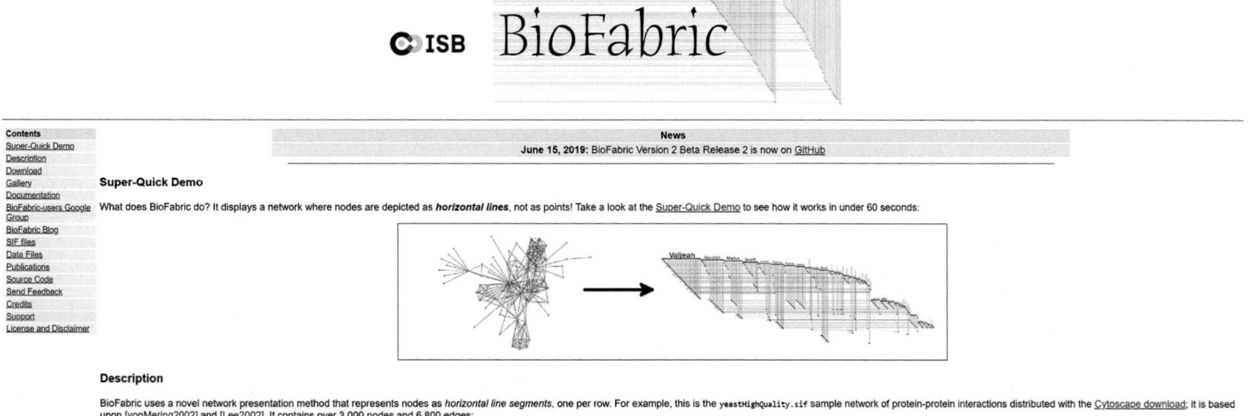

FIGURE 16.7 Homepage of BioFabric. It uses Cytoscape in network visualization. An open-source program for drawing graphs has been designated BioFabric. It displays graphs as a node-link picture, but despite other graph drawing tools, it shows nodes using horizontal lines rather than discrete symbols. *From http://www.biofabric.org/.*

- **Storage:** A minimum 500 GB NVMe solid-state drive (SSD) (>1 TB) is recommended for the operating system and software, and a minimum 1 TB hard disk drive (HDD) or additional SSDs for storing data.
- **Graphics:** A 2 GB or 4 GB dedicated graphics card, such as an NVIDIA GeForce RTX or AMD Radeon RX, is recommended for running machine learning algorithms.
- **Network:** A wired Ethernet connection (minimum speed: 10 mbps) or a wireless card with high speeds is recommended for transferring data.
- **Backup Drive:** It is strongly advised to connect a backup drive to a PC because there is always the possibility of data loss in computers. As a result, the PC should have at least double storage (4 TB or more).

- **Cloud services:** It will prefer cloud services provided by many companies such as Amazon Web Services (AWS: https://aws.amazon.com/), Microsoft Azure (https://azure.microsoft.com/en-us), and Google Cloud Platform (https://cloud.google.com/).

It is also important to note that, depending on the task, specific software requirements may need to be met and a powerful GPU (4 × 4 GB Graphic card memory) will be needed. It is important to keep in mind that as technology advances and datasets grow, these requirements may change and it is always a good idea to consult with experts and stay updated on the most recent scientific developments.

14. Issues in omics data visualization

Human sight and cognition are the main issues with data visualization. To be fair, deceptive visualizations are not usually created with malicious purposes, but even innocent errors may deceive viewers. Humans tend to skim information fast in quest of important details because their eyes are impressionable. Every data visualization must be designed with sight and cognition in mind. The visualization mistakes can be avoided by selecting good color contrast or picking the color schemes, using 3D graphics in 2D as well as in 3D format. When the foreground color or background is changed, some data also changes; therefore, before plotting the color code, check first the foreground and background color. First, select a sort range of data from the large dataset, make a reference result, and go for large omics data to analyze and get the final conclusion. Eliminate the biased text descriptions such as residues labels, titles, and captions of structure files. Be sure to pick the right information from the files. Choose the right software for the data visualization before analyzing the result. Several issues can arise when visualizing data.

- Lack of context: It is important to provide enough context in the visualization so the viewer can understand the presented data.
- Cluttered or confusing visualizations: A too cluttered or confusing visualization can be difficult to interpret.
- Misleading visualizations: It is important to be careful not to present the data in a way that is intentionally or unintentionally misleading.
- Lack of scalability: Some visualizations may not work well when the data being visualized is very large or has many dimensions.
- Limited customization: Some visualization tools may not allow for much customization, making it difficult to create the desired visualization.
- Incompatibility with different devices: Some visualizations may not display properly on certain devices, such as smartphones or tablets.
- Limited accessibility: Some visualizations may not be accessible to users with visual impairments or other disabilities.
- Lack of interactivity: Some visualizations may not be interactive, making it difficult for the viewer to explore the data more deeply.

15. Issues in omics dataset

Omics dataset can have a number of challenges and issues. A few issues that were faced in data analysis are listed below.

- Data volume: Omics datasets can be extremely large, with many millions or billions of data points. This can make them difficult to manage and analyze and may require specialized software and hardware.
- Data complexity: Omics data can be complex and multidimensional, with many different variables and factors that need to be considered. This can make interpreting and drawing meaningful conclusions from the data challenging.
- Data quality: Ensuring the quality of omics data is critical, as errors or biases in the data can lead to incorrect conclusions. However, it can be difficult to ensure the quality of such large and complex datasets.
- Data integration: Omics datasets are often generated from multiple sources and using different technologies, which can make it difficult to integrate and compare the data.
- Data sharing and reproducibility: There is a growing emphasis on the importance of sharing and reproducing omics data, but this can be challenging due to the size and complexity of the datasets, as well as data security and privacy issues.

16. Overcoming challenges in omics data analytics

Extrapolating useful insights from massive, high-dimensional datasets from various sources is one of the difficulties in evaluating and interpreting omics data. Every stage of data handling, from preprocessing to statistical analysis to interpreting and drawing conclusions, depends on the capacity to treat the datasets as a whole. However, omics analysis is significantly more difficult than traditional studies since the number of variables in omics studies can be very vast in comparison to the number of variables in normal multivariate analyses. Additionally, multiomics datasets frequently contain missing results because of technical restrictions and numerous experimental constraints. These missing variables may hamper the Omics data consolidative analysis. Omics data analysis involves sophisticated data reduction and visualization techniques, which can only be accomplished with strong data analytics tools, methodologies, and software. Several strategies can help overcome challenges in omics data analytics.

- *Choose the right tools:* It is important to use the right tools for the task at hand. There are many different software packages and platforms available for analyzing omics data, and choosing the one most suitable for the needs is important.
- *Preprocess the data:* Preprocessing the data can help to make it more manageable and easier to analyze. This may include tasks such as filtering, normalization, or imputation.
- *Use appropriate statistical methods:* Choosing the right statistical methods is important for accurately analyzing omics data. It is important to choose appropriate methods for the type of data being analyzed and the research question being addressed.
- *Visualize the data:* Visualizing the data can help to reveal patterns and trends that may not be apparent in the raw data. Many different tools and techniques are available for visualizing omics data, and it is important to choose the most appropriate for the task at hand.
- *Use multiple approaches:* It is often useful to use multiple approaches to analyze omics data. This can help to confirm findings and provide a more complete understanding of the data.
- *Collaborate with experts:* Collaborating with experts in omics data analysis can help overcome challenges and advance the research.
- *Keep up to date with new methods and tools:* The field of omics data analysis is rapidly evolving, and it is essential to keep up to date with new methods and tools to take advantage of the latest advances.
- *Hosting and accessing multiomics data*: For hosting multiomics data, there is currently no single public infrastructure. It is challenging to locate and access the many omics data repositories since there are no uniform standards or methods.

17. Conclusion

The analysis of omics data, which refers to the study of large datasets generated from different biological fields, such as genomics, proteomics, transcriptomics, and other "omics" fields, can provide valuable insights into the underlying mechanisms of biological systems and processes. However, the analysis of omics data can also present a number of challenges and issues, including the need to manage and analyze large and complex datasets, ensure the quality of the data, and integrate data from multiple sources. Despite these challenges, the use of advanced computational and statistical methods that were added based on data science methods step by step is enabling researchers to make significant progress in the analysis of omics data, leading to new discoveries and advances in fields such as medicine and agriculture.

References

Conesa, A., Madrigal, P., Tarazona, S., Gomez-Cabrero, D., Cervera, A., McPherson, A., Szcześniak, M. W., Gaffney, D. J., Elo, L. L., Zhang, X., & Mortazavi, A. (2016). A survey of best practices for RNA-seq data analysis. *Genome Biology, 17*(1). https://doi.org/10.1186/s13059-016-0881-8. http://genomebiology.com/

Funahashi, A., Morohashi, M., Kitano, H., & Tanimura, N. (2003). CellDesigner: A process diagram editor for gene-regulatory and biochemical networks. *Biosilico, 1*(5), 159–162. https://doi.org/10.1016/s1478-5382(03)02370-9

Hoops, S., Sahle, S., Gauges, R., Lee, C., Pahle, J., Simus, N., Singhal, M., Xu, L., Mendes, P., & Kummer, U. (2006). COPASI—a COmplex PAthway SImulator. *Bioinformatics, 22*(24), 3067–3074. https://doi.org/10.1093/bioinformatics/btl485

Huang, X., Wu, L., & Ye, Y. (2019). A review on dimensionality reduction techniques. *International Journal of Pattern Recognition and Artificial Intelligence, 33*(10). https://doi.org/10.1142/S0218001419500174. http://www.worldscinet.com/ijprai/

Jolliffe, I. T., & Cadima, J. (2016). Principal component analysis: A review and recent developments. *Philosophical Transactions of the Royal Society A: Mathematical, Physical and Engineering Sciences, 374*(2065), 20150202. https://doi.org/10.1098/rsta.2015.0202

Kanehisa, M., & Goto, S. (2000). KEGG: Kyoto encyclopedia of genes and genomes. *Nucleic Acids Research, 28*(1), 27–30. https://doi.org/10.1093/nar/28.1.27. https://academic.oup.com/nar/issue

Karp, Ong, W. K., Paley, S., Billington, R., Caspi, R., Fulcher, C., Kothari, A., Krummenacker, M., Latendresse, M., Midford, P. E., Subhraveti, P., Gama-Castro, S., Muñiz-Rascado, L., Bonavides-Martinez, C., Santos-Zavaleta, A., Mackie, A., Collado-Vides, J., Keseler, I. M., & Paulsen, I. (2018). The EcoCyc database. *EcoSal Plus, 8*. https://doi.org/10.1128/ecosalplus.ESP-0006-2018

Katara, P., Tyagi, S., & Gupta, M. K. (2023). Integrative omics: Trends and scope for agriculture. In A. Mani, & S. Kushwaha (Eds.), *Genomics of Plant Pathogen Interaction and the Stress Response* (first, pp. 01–16). CRC Press. https://dx.doi.og/10.1201/9781003153481-1.

Konganti, K., Wang, G., Yang, E., & Cai, J. J. (2013). SBEToolbox: A Matlab toolbox for biological network analysis. *Evolutionary Bioinformatics, 2013*(9), 355–362. https://doi.org/10.4137/EBO.S12012. http://www.la-press.com/redirect_file.php?fileId=5185&filename=3852-EBO-SBEToolbox:-A-Matlab-Toolbox-for-Biological-Network-Analysis.pdf&fileType=pdf

Kumar, S., Gupta, M. K., Gupta, S. K., & Katara, P. (2023). Investigation of molecular interaction and conformational stability of disease concomitant to HLA-DRβ3. *Journal of Biomolecular Structure and Dynamics, 41*(17), 8417–8431. https://doi.org/10.1080/07391102.2022.2134211

Kwoh, C. K., & Ng, P. Y. (2007). Genetic studies of diseases: Network analysis approach for biology. *Cellular and Molecular Life Sciences, 64*(14), 1739–1751. https://doi.org/10.1007/s00018-007-7053-7

Robinson, J. T., Thorvaldsdóttir, H., Winckler, W., Guttman, M., Lander, E. S., Getz, G., & Mesirov, J. P. (2011). Integrative genomics viewer. *Nature Biotechnology, 29*(1), 24–26. https://doi.org/10.1038/nbt.1754

Scheer, M., Grote, A., Chang, A., Schomburg, I., Munaretto, C., Rother, M., Söhngen, C., Stelzer, M., Thiele, J., & Schomburg, D. (2011). BRENDA, the enzyme information system in 2011. *Nucleic Acids Research, 39*(1), D670–D676. https://doi.org/10.1093/nar/gkq1089

Shannon, P., Markiel, A., Ozier, O., Baliga, N. S., Wang, J. T., Ramage, D., Amin, N., Schwikowski, B., & Ideker, T. (2003). Cytoscape: A software environment for integrated models of biomolecular interaction networks. *Genome Research, 13*(11), 2498–2504. https://doi.org/10.1101/gr.1239303

Subedi, P., Moertl, S., & Azimzadeh, O. (2022). Radiation biology: Surprised but not disappointed. *Radiation, 2*, 124–129.

Szklarczyk, D., Kirsch, R., Koutrouli, M., Nastou, K., Mehryary, F., Hachilif, R., Gable, A. L., Fang, T., Doncheva, N. T., Pyysalo, S., Bork, P., Jensen, L. J., & von Mering, C. (2023). The STRING database in 2023: Protein–protein association networks and functional enrichment analyses for any sequenced genome of interest. *Nucleic Acids Research, 51*(D1), D638–D646. https://doi.org/10.1093/nar/gkac1000

Wickham, H. (2016). *ggplot2: Elegant graphics for data analysis*. Springer-Verlag.

Wittig, U., Rey, M., Weidemann, A., Kania, R., & Müller, W. (2018). SABIO-RK: An updated resource for manually curated biochemical reaction kinetics. *Nucleic Acids Research, 46*(D1), D656–D660. https://doi.org/10.1093/nar/gkx1065

Wu, C., Zhou, F., Ren, J., Li, X., Jiang, Y., & Ma, S. (2019). A selective review of multi-level omics data integration using variable selection. *High-Throughput, 8*(1). https://doi.org/10.3390/ht8010004. https://www.mdpi.com/2571-5135/8/1/4/pdf

Yadav, A., Vishwakarma, S., Krishna, N., & Katara, P. (2020). Integrative omics: Current status and future directions. In P. Katara (Ed.), *Recent Trends in Computational Omics: Concepts and Methodology* (pp. 01–46). USA: Nova Science Publisher.

Chapter 17

Emerging trends in translational omics

Sapna Pandey[1], Sarika Sahu[2] and Dev Bukhsh Singh[3]

[1]Department of Computational Biology & Bioinformatics, Jacob Institute of Biotechnology & Bio-Engineering, Sam Higginbottom University of Agriculture, Technology and Science (SHUATS), Allahabad, Uttar Pradesh, India; [2]ICAR-Indian Agricultural Statistics Research Institute, New Delhi, India; [3]Department of Biotechnology, Siddharth University, Kapilvastu, Uttar Pradesh, India

1. Omics data sources and analysis

Ever since its beginnings, the field of sequencing has received a lot of attention and has grown in significance. This field has seen impressive improvements, particularly in the areas of DNA and RNA sequencing. RNA sequencing has made significant breakthroughs in recent decades and has become a vital method for transcriptome profiling. Single-cell transcriptome techniques have made it possible to resolve individual cells with ever-greater accuracy and also included spatial data. The discovery of new novel molecular diagnostics may be facilitated through these new approaches. Small noncoding RNAs, RNA editing events, and alternative splice variants that could not be studied with previous hybridization-based technologies like microarray might someday be applied to develop omics-based testing. For example, in cancer research and treatment, RNA sequencing has become an indispensable tool for biomarker discovery and validation, cancer subtypes, and progress assessment, understanding drug resistance mechanisms, cancer neoantigens, and cancer immune microenvironment analysis, and immunotherapy (Hong et al., 2020).

Similarly, DNA sequencing is important in many areas such as biology, molecular biology, genetics, forensic sciences, archeology, and others. DNA sequencing is categorized into three generations. The second and the third generation are often called "Next-generation sequencing" (NGS). DNA sequencing is enabling the discovery of unusual or undefined mutations that may have significant clinical ramifications. In addition to understanding entire DNA and RNA sequences, NGS techniques offer great potential for the high-throughput detection of epigenetic and posttranscriptional alterations to DNA or RNA. NGS brings to the challenge of the multidimensional dataset and the potential consequences of modeling error of computational models to the available data because these new technologies produce even more statistics per sample than conventional methods. Large-scale meta-analyses of sequencing datasets taken from many areas may help minimize these problems and enable the development of omics-based assessments that are helpful for medical applications (Wei Zhang et al., 2022).

Proteomics has evolved significantly during the past 2 decades, predominantly as a result of significant developments in mass spectrometry, high-throughput manufacturing of antibodies, and biostatistics algorithms and bioinformatics. To discover and describe at least one protein product for each of the 20,300 protein-coding genes and to describe mass spectra, the Human Proteome Project was initiated in September 2010 (Legrain et al., 2011). Selected Reaction Monitoring (SRM) proteomics is one example of such development (Shi et al., 2012). SRM assays are generated whether from experimental proteomics identification data as well as from readily available databases like the generalized proteomics data meta-analysis (GPM) proteome database (Craig et al., 2004) or Peptide Atlas (Deutsch et al., 2008). Commercially available software such as Pinpoint (Thermo Fisher Inc.) and MRMPilot (AB Sciex Inc.), as well as license-free Skyline (MacLean et al., 2010), MRMaid (Mead et al., 2009) mProphet (Reiter et al., 2011), and SRMCollider (Röst et al., 2012), help to develop SRM assay. Due to their sensitivity, high-throughput capabilities, and combinatorial possibilities, SRM assays remain to be the best option for proteomic analysis and biomarker identification among all MS techniques.

The fast-developing metabolomics field produces a significant amount of meaningful data. For these data to be properly understood, other omics data must be merged and studied simultaneously. The most popular strategy used today is to constantly monitor protein, transcript, and metabolite levels by using integration methods to analyze structural and dynamic changes in the underlying biological system of focus. To integrate, analyze, visualize, and map the vast amount of

Integrative Omics. https://doi.org/10.1016/B978-0-443-16092-9.00017-5

metabolites data, appropriate statistical and computational methods are therefore required. Multiomics analysis provides a better understanding of metabolites and their role in biological processes (Chen et al., 2022).

The main role of high-throughput screening and combinatorial protein expression monitoring in both health and disease is played via antibody-based proteomic techniques. Such antibody-based methods will offer (highly advanced) setups that can simultaneously profile many proteins in a quick, and efficient way, emphasizing both high- and low-abundant proteins, even in unstructured proteomes such as serum. To identify the molecular composition of the target proteome, the generated protein snapshots and protein expression patterns can then have converted into proteomics maps, or molecular fingerprints (Wingren, 2016). The Human Protein Atlas portal (www.proteinatlas.org) provides a free interactive resource that permits users to examine tissue-elevated proteomes in organs and tissues and study tissue profiles for particular protein classes (Uhlen et al., 2010). Of the 20,300 gene-coded proteins, the Human Protein Atlas contains antibody data for more than 12,000 proteins.

The absence of highly sensitive and accurate protein capture agents for specific proteins is a big challenge in the successful implementation of large-scale proteomic techniques (including variants due to alternative splicing, post-translational modifications (PTM), and single-nucleotide polymorphisms (SNPs) or fusions gene). The complexity of combinations like blood, whose protein concentrations can fluctuate by more than 10 orders of magnitude, makes this problem even more difficult. "Click Chemistry" is one of the most versatile reactions that can be used in this regard. The potential to profile cells at the individual level on a variety of omics layers (genomes, transcriptomes, epigenomes, and proteomes) has recently become possible due to improved technology.

2. Translational omics

It is essential to use the multiomics data integration approaches to emphasize the interactions of the various biomolecules and their functions to understand complex and difficult biological processes completely. Several potential tools and approaches have been created for data integration and interpretation as a result of the development of high-throughput techniques and the availability of multiomics data generated from a large number of samples. To understand the interaction of molecules, integrated techniques combine various omics data in both sequential and simultaneous ways. Major areas which define the applications of translational omics are shown in Fig. 17.1. They aid in interpreting the information flow from one omics level to the next and, in doing so, that help in filling the gap between genotype and phenotype. Some major points which cover the application or translation of various types of omics or multiomics analysis are as given here.

- Microarray data analysis can be used for biomarker discovery, cancer diversity or subtypes analysis, understanding drug resistance, and the impact or outcome of various factors or drugs on the outcome of diseases.
- Subtype identification helps in finding different heterogeneous patient groups in the same disease population, which differs in the mechanism of disease progression or therapeutic response. Subtype identification promotes targeted therapy through biological drugs, hormonal therapy, and immunotherapy.
- High-throughput detection of epigenetic and posttranscriptional alterations to DNA or RNA may be useful in understanding the mechanism of disease, which may help in finding a biomarker for prognostic, diagnostic, and therapeutic purposes.
- Approximately 1000 proteins from blood or other sources can be quantitatively measured using recently developed protein capture-agent aptamer chips. Nucleic acid aptamers capable of specific binding to their targets are being used for diagnostic and therapeutic applications (Shatunova et al., 2020). The discovery of a protein-based profile of fibrosis (liver scarring) in nonalcoholic fatty liver disease, one of the most common causes of liver disease worldwide, is the most recent research example of aptamer-based proteomics.
- The human transcriptome data are widely used to evaluate gene expression in host–pathogen interaction and immunological perturbation. Researchers also studied the pancreatic cancer patient's serum exosomes and identified some biomarkers like miRNA, lincRNA, tRNAs, and piRNAs specific to pancreatic cancer patients.
- Further, uncontrolled/deregulated expression of viral genes is mainly responsible for the highly deadly disease of cancer (Huang et al., 2021). The long-reads transcriptome data of gastric cancer reveal many cancer-associated isoforms and new variants of cancer-related genes.
- Through "click chemistry," it is now possible to functionally study crucial protein/small-molecule interactions and posttranslational changes in their biological systems contexts without being constrained by the limitation of biomedical applications. The mapping of biomolecular interactions and various PTMs are the common applications of click chemistry (Parker & Pratt, 2020).

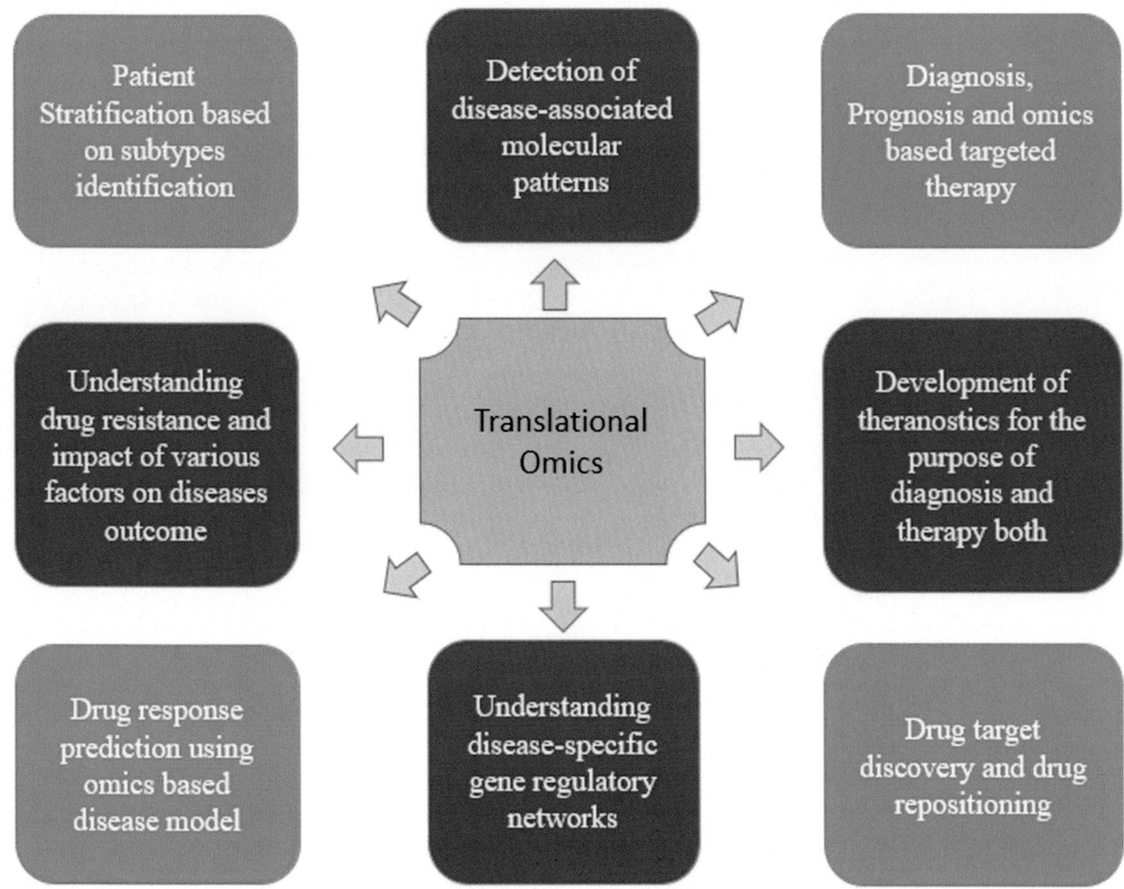

FIGURE 17.1 Major applications/areas of translational omics analysis. The information flows from one omics level to the next.

- Multiomics integration analysis can certainly help in the easy recognition of important metabolites and biological processes if they are involved in certain research instances (Chen et al., 2022).
- Patients may have a different response to a drug therapy due to differences in their gut microbiomes. Therefore, a better understanding of a patient's microbiome may help in recommending the drug and its dose to different patient subgroups (Xu Zhang et al., 2019).
- Recent research showed that emerging omics data sets lead to a greater understanding of the biological system that is studied. To identify genes in both rectal and colon cancers, it is crucial to integrate proteomics, genomic, and transcriptome data. The integration of both transcriptomics and metabolomics data led to the discovery of prostate cancer–related molecular variations. These integration studies clearly show that multiomics data analysis provides more valuable and clinically important information as compared to single omics analysis (Subramanian et al., 2020).
- Theranostics deals with the use of two identical or very closely related radiopharmaceuticals for therapy and diagnosis purposes. In cancer biology, tumor-specific substrates, ligands, or drugs are labeled with specific radionuclides for imaging or therapy. Here, the structure of diagnostic and therapeutic radiopharmaceuticals is very similar or identical; as a result, one serves the purpose of diagnosis, while another plays role in therapy.

3. Omics-based tests and precision medicine

Recent years have seen a significant increase in the new omics-based technologies and omics-based tests, which involve the use of computational techniques to omics-based measures to provide a clinically effective result. High-throughput omics technology is very useful in the identification and development of biomarkers. Only a small subset of discovered biomarkers has been translated into clinical applications. There is a need to validate and translate the proteomics, genomics, metabolomics, transcriptomics, and microbiomics biomarkers of cancer and other diseases to enhance the care of patients with neoplasms. The most commonly used omics-based test development strategies are NGS and mass spectroscopy

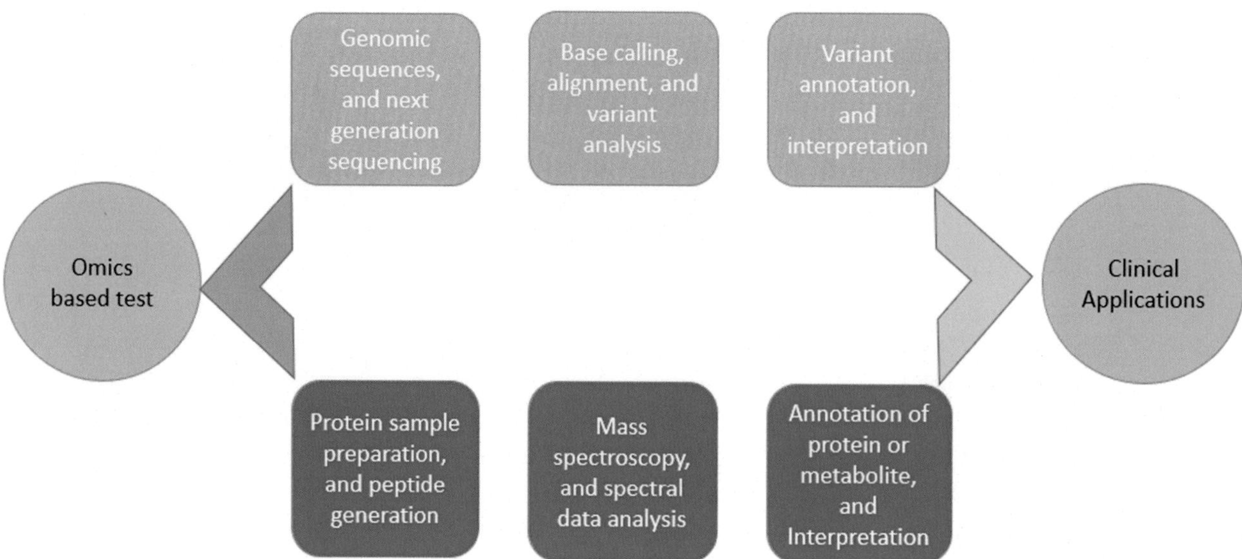

FIGURE 17.2 Omics-based test: NGS and mass spectroscopy analysis approaches for the annotation of genetic variants, proteins, and metabolites.

analysis, which are used for the annotation of genetic variants, proteins, and metabolites (Fig. 17.2). Development of potential biomarkers having the capability of targeted therapies with fewer side effects and beneficial clinical outcomes is hard due to the intrinsic complexity of neoplasms and the requirement to develop well-designed biomarker discovery strategies that are based on omics technology (Quezada et al., 2017).

Various challenges to the success of candidate omics-based tests are data complexity, the need for data integration and analyses tool, and the unknown biological mechanism behind the tests. These challenges must be overcome to utilize full applications of tests. A potential omics-based test's difficulty to advance clinical application has two main scientific causes:

1. A candidate omics-based test might not be sufficiently constructed to address a particular, concise, and essential clinical issue.
2. The likelihood or even impossibility that a potential omics-based test will be proved to be clinically valid or efficient depends on how rigorously statistical and bioinformatics-based discovery studies are carried out.

The first component of the committee's recommended test development and evaluation is confirmation of a candidate omics-based test. The committee assumes that a well-defined and clinically relevant question has been framed, and omics dataset from a set of patients and an associated clinical outcome is known. For instance, a researcher might look into whether gene expression data may significantly outperform traditional clinical prognostic factors like tumor size and grade in identifying reappearance in node-negative breast cancer samples. The researcher may have data on disease-free survival time for each patient postsurgery, as well as breast cancer gene expression analyses from individuals with node-negative breast cancer.

The main objective of precision medicine is to provide accurate and personalized therapy to patients. Precision medicine is based on diverse types of biomarkers such as gene expression products, proteins, and metabolites. High-throughput omics approaches are very useful in finding the whole picture of biological systems in a very accurate and precise manner. For a better understanding of the disease process, a complete set of data-driven techniques and statistical approaches are applied to multiple levels of biological information. High-throughput analytical techniques are used to find a large number of omics-based markers and their association with a phenotype and disease, which direct the way to implement a precise and accurate therapeutic intervention against a disease (Tebani et al., 2016). Metabolomics is more easily introduced into clinical applications as compared to other omics approaches because metabolite analysis is more frequently used for drug monitoring and screening of inborn errors of metabolism.

Supervised and unsupervised machine learning approaches are used for omics-based patient stratification, which is based on clinically validated biomarker signatures. Many clinically validated biomarker models have been developed for patient stratification, and are useful in precision medicine (Glaab et al., 2021). These biomarker models are derived from omics data and are very useful in guiding treatment decisions, monitoring treatment success and disease progression, and recommending new/alternate drugs. Omics-based signatures have the potential to provide specific and sensitive information for a disease. Some clinically approved omics-derived diagnostic or prognostic tests used for personalized therapies are MammaPrint (breast cancer risk), ColoPrint (colon cancer) Prosigna assay/PAM50 (breast cancer risk), Oncotype DX (breast cancer risk), Afirma Gene Expression Classifier (benign and cancerous thyroid nodules), Vectra DA (rheumatoid arthritis), and AlloMap Heart (heart transplant rejection) (Glaab et al., 2021). Omics-based tests are also supported by software and databases such as (Decipher Genomic Resource Information Database), which analyzes many expression markers per patient to help in the successful implementation of personalized therapy. Treatment-related manipulations may be made depending on the changes observed in the disease outcome.

Integrative omics approaches can be used for patient stratification and also for finding the potential prognosis markers of different types of cancers including pan-gastrointestinal cancers (Jiangzhou et al., 2021). Molecular features of different subtypes of cancer are used to differentiate cancer subtypes and may help to develop a model for prognosis. Significant prognostic markers may be identified using statistical and machine learning approaches, which can be effectively used to measure the risk of cancer or disease development. Omics-based test development strategies and steps have already been discussed in this chapter. Characteristics of successful omics-derived biomarker studies are study design and sample size used to build the model, study documentation with reproducibility, model interpretability to explain the underlying mechanism, interpretation of prior knowledge (pathways, omics data), and statistical evaluation using cross-validation and multiple testing corrections (Glaab et al., 2021).

4. Omics test development process

The objective is to create a method for data collection and a computational model to accurately predict whether a cancer patient will revert based on gene expression measurements on a new clinical sample. It is important to take into consideration if indeed the test must have clinical validity and value before commencing an omics-based project. Considerable consideration should be given to, for instance, the required sensitivity and specificity, especially given the prevalence of the disease in the population under study. The process of omics-based development involves data collection, data filtering for feature extraction, model development, cross-validation, and model refinement (Fig. 17.3). Various research components involved in the development, validation, and assay validation of omics tests are shown in Fig. 17.4.

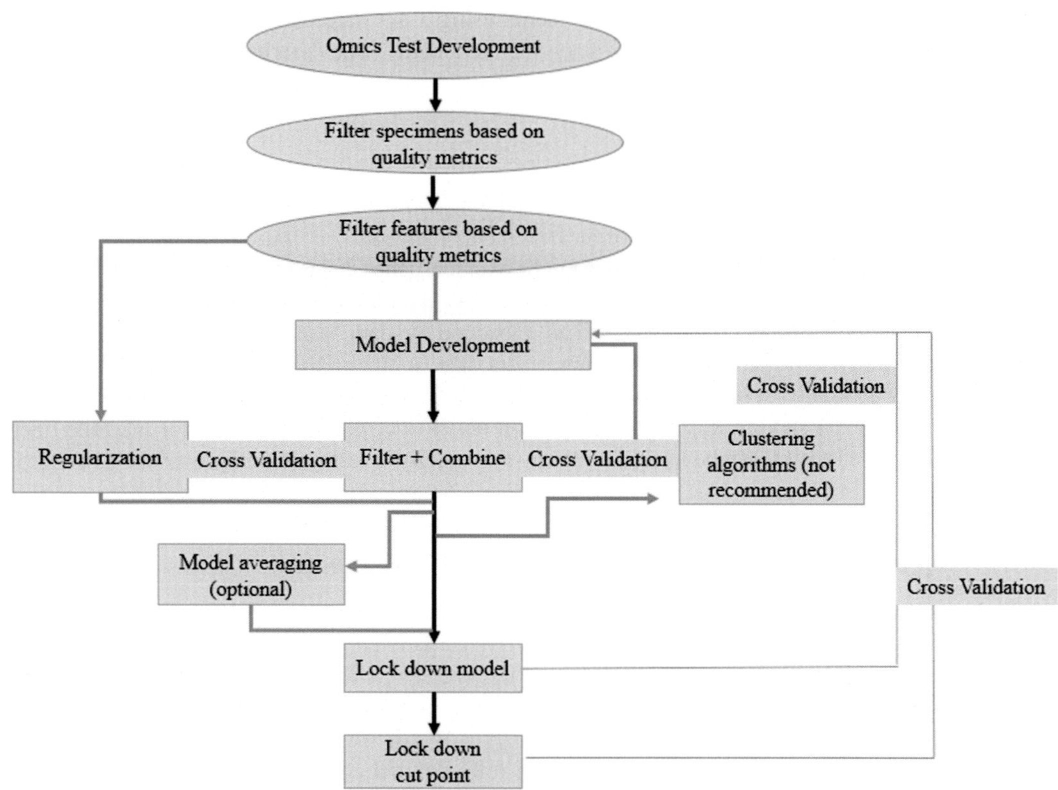

FIGURE 17.3 Schematic illustrating the omics-based test development process.

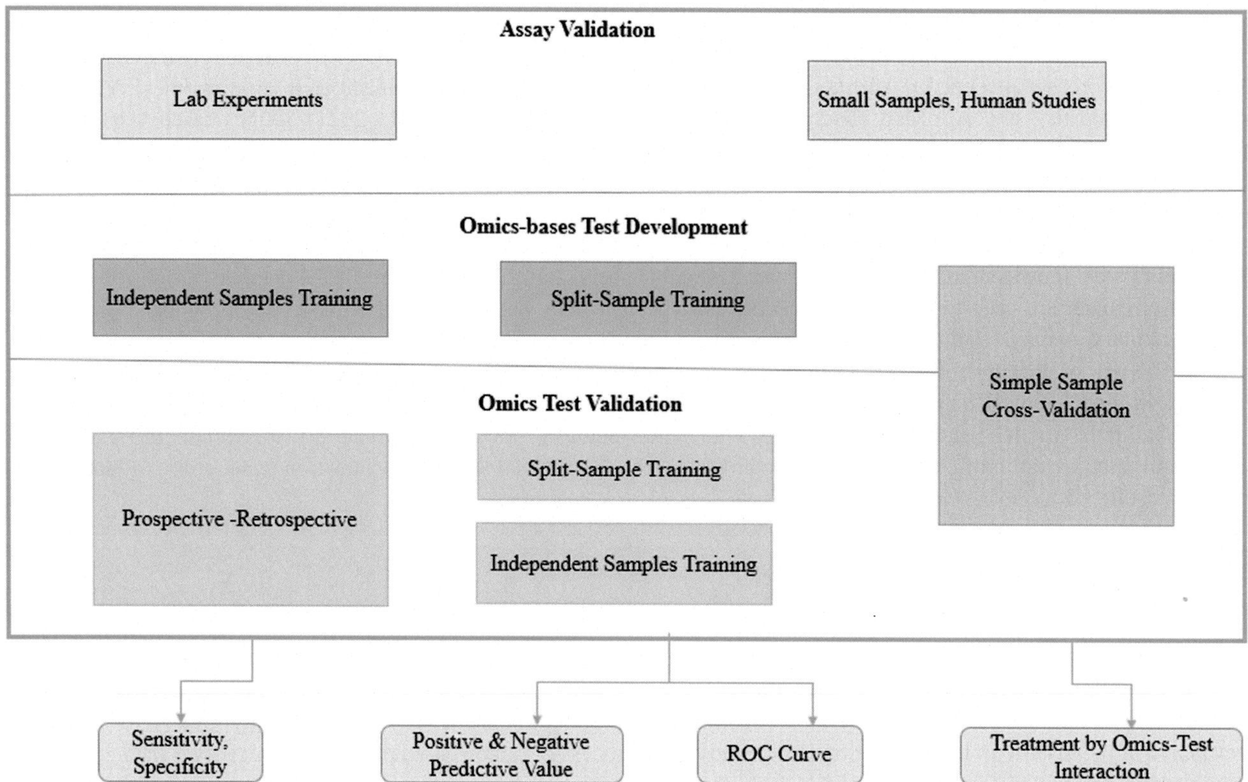

FIGURE 17.4 Diagram demonstrating the various research components involved in the development, validation, and assay validation of omics tests.

To achieve this, many steps must be taken: (1) data quality management; (2) development and cross-validation of the computational model; (3) validation of the computational model using a different dataset; and (4) dissemination of data, code, and fully described computational methods to the scientific community. The description of each of these is discussed below.

4.1 Step 1: Data quality management

Data quality management is an important step, as it is in many fields of science. Data quality management is frequently carried out computationally due to the large number of measurements that make up omics datasets, which can range from thousands to millions. For instance, a research scientist might exclude genes from a microarray that were expressed at conditions close to or below reference values. It is also possible to evaluate the consistency of the values from run to run (technical variance). To determine whether these have had an impact on the data, it may also be helpful to closely examine various aspects of the experimental design, such as the sample run date, source of the tissue, and potential heterogeneities related to tissues. This is important because parameters of biological importance like duration to disease recurrence or cancer type can frequently have a significantly higher impact on omics data than the run date or machine operator. The treatment outcomes of the patients must not be identified during this quality assessment analysis of the data that need to be done in a blinded way.

4.2 Step 2: Development and cross-validation of computational model

A possible omics-based test linked to an important phenotype can be developed based on the omics measurements once researchers have shown that the data are of high enough quality. Since there are almost numerous statistical tools that can be used to complete this work. But all of these approaches have a few essential features and difficulties, which are covered in more detail below. Feature selection involves choosing a subset of measurements that is related to the biologically relevant characteristic or result based on past knowledge.

To predict the clinical outcome based on the omics data, a fully defined computational model can be developed by using this measurement subset. The possibility that an omics-based test requiring a large number of measures will not be clinically viable in the future for financial or technical reasons can be avoided by reducing the number of required measurements. Cross-validation is the process of selecting features of a computer model to evaluate the performance of the computational model. Overfitting the data is a big concern because omics datasets include an outsized number of molecular measurements. In reality, the data may be overfitted when suitable statistical techniques are not followed. It is therefore always possible to build a computational model that properly fits the data for a typical omics dataset and a related medical outcome, even in the absence of a real tie between the omics measurements and the treatment outcomes.

The overall utility of a computational model is defined by how effectively it performs on patients as compared to how well it acts on the patients who produced the original dataset to develop the computational model. Therefore, an overfitting model will perform extremely poorly on the test dataset. Because of the enormous amount of feature measurements, overfitting can result in an inaccurate analysis of omics data, even while it is crucial to use the biggest sample size available to improve the "power" of the statistical analysis. The simplest approach to avoid overfitting issues is to adopt a training set/test set strategy for creating the model and to avoid creating models with several parameters that are not supported by the sample sizes. When sufficient samples are not available, developers will often choose to apply a less rigorous method called cross-validation.

The cross-validation approach is explained after the training set/test set approach. A training set and a test set are the ideal datasets for an investigator to use while building a computer model; both sets are made up of independent samples that were collected and processed by various researchers at various institutions. All selections for a tuning parameter, standardization technique, as well as other factors can be regarded as a different computational model because every computational model involves several potential tuning parameters and there are different ways to standardize the data. Researchers first fitted each computer model that is considered on the training set. The features must be chosen solely based on the training samples if feature selection is carried out before or as part of a model-fitting procedure. It's crucial to fit the model on the training set without using any knowledge of the test set to get an accurate estimate of the error. Researchers then assess the model's performance by using a test sample set after it has been trained.

The final model of the candidate omics-based test should make use of all available data. The completely described computational techniques are now locked down, and the investigation moves on to Step 3, where the selected model is assessed on a different dataset. A carefully planned set of computation operations are carried out to process the raw data, generate the mathematical model, and use data to predict relevant phenotypes. The availability of few samples, however, makes it impossible to create separate test and training sets in many investigations.

Cross-validation is used to test the model's performance using a single dataset and is an alternative to creating designated training and test sets. The data are divided into multiple segments, and the model is iteratively equipped to all but one segment before even being evaluated on its effectiveness on the remaining segment. Cross-validation can be expected to reduce overfitting if done correctly, although it does not always do so. In more detail, K sample sets are created by dividing the samples that made up a single dataset. After that, several computational models are fit to K-1 of these sample sets, and the models are assessed on the remaining, held-out sample subset. Finally, a single computation model has achieved that fit on the total set of samples and also performed well on sample subsets. The investigator then moves on to Step 3 when this computational model has been confirmed. Contrary to possessing overfit of the data used to build the model, cross-validation provides a measure of how well a given computational model predicted future observations. Consequently, a computational model which works well in cross-validation seems to be more likely than one that fails to perform well on patient samples. In addition, there is a minimal probability that an omics-based test will be successful using a computational model that performs poorly in cross-validation. Although cross-validation is a straightforward method, if it is not carried out carefully, errors might occur. For instance, based on all of the samples included in the dataset, the subset of omics measurements in certain published research shows a significant correlation with the clinical result of interest.

With that subset of omics measurements, cross-validation is then carried out. When using partial cross-validation, only a portion of all potential "leave-one-out" validation tests are examined to cut down computing costs in a lot of circumstances. For instance, Simon et al. (2003) explanation of the cross-validation process used to create the Mamma Print test demonstrates this. As a result, omics-based tests that show good outcomes in cross-validation also perform well on patient samples in the future and it can lead to significant overfitting of the data. For cross-validation to be done correctly, just one K-1 sample sets used to build the computational model in each cross-validation cycle must be utilized for feature selection and data processing.

The computational model's accuracy can be tested on random test samples. Since the data are randomly divided, the training and test sets may certainly belong to the same population distribution, which leads to unrealistically optimistic cross-validation error rates. This means that a plethora of pertinent sources of variance that have an impact on clinical performance are overlooked. Validation on an independent dataset cannot be substituted by cross-validation or the training set/test set approach. The computational model must be completely locked down so that there is no longer any element of randomization before continuing to Step 3; hence, a simple random sample seed should be selected. A significant error in the development of tests created at Duke University was the failure to select a random sample and lock down the model.

The computational techniques created in Step 2 and further explored in Step 3 should be properly specified and published by the researcher. Before moving on to Step 3, researchers should specify objectives for a computational model's good testing. A computational model may have poor cross-validation performance in particular circumstances. In this situation, the computational model might need to be further enhanced or refined. The refined computational model must therefore have evaluated once more using cross-validation or the training set/test set approach, or even both. That is, before moving on to Step 3, the computational model needs to be completely established and locked down and refinement and enhancement should not continue into Step 3.

4.3 Step 3: An independent dataset's validation

As was noted in the previous stage, to prevent overfitting of the data, an error estimation strategy should be used in designing a candidate omics-based test. However, using this method typically results in an underestimation of the error rate when used to analyze new patient samples. This error is because the samples used to create the model and the samples used to evaluate the model differ. Samples often represent a homogenous group of patients who were gathered at one or more institutions and ran on the same machine roughly at the same time. Variability in omics data due to changes in person, place, and time is of scientific importance and affects the clinical outcome (Leek et al., 2010).

In addition, the high variation of cross-validation estimates and the fact that only the best tuning parameters and models are usually chosen can lead to bias. Because of this, cross-validation error rates do not offer sufficient proof of how well a candidate test works. A computational model on an independent test set should be validated during the discovery to avoid wasting time, effort, money, and resources on a test that has little chance of succeeding later on in the development process. The independent set of samples should not have been used in the development of the computational model. Neither the independent dataset nor the clinical utility assessment of any samples must be used to build the computational model in Step 2. Two tests used overlapping training and datasets confirmation in the cases development processes: AlloMap (Deng et al., 2006) and MammaPrint (Buyse et al., 2006). AlloMap is a blood test that measures the RNA expression of 11 genes to obtain a single score on a scale of 0–40, with a lower score reflecting a lower probability of acute cellular rejection. Before AlloMap, the standard of care was a more invasive method for monitoring heart transplant patients for ACR. MammaPrint is a prognostic test designed to predict the risk of recurrence of distant breast cancer following surgery. In 2007, MammaPrint became the first Food and Drug Administration (FDA)–cleared molecular test profiling genetic activity. It was developed by investigators at the Netherlands Cancer Institute and commercialized by Agendia.

The clinical dataset and independent samples must be appropriate to the candidate omics-based test's implementation. To be more precise, the potential omics-based test must be independently confirmed with patients of the same kind of disease, stage, and clinical environment for whom it is planned to be utilized in the future. To show the applicability of the test and also it is not overfitting to any specific case, the samples (independent confirmation) were taken at different times, at different labs, from various patient groups, and samples were analyzed in different research labs. If the samples are taken in the same lab, prepared, and analyzed by the same technician, then characteristics or properties of that data, machine, and lab are used across the samples to generate and evaluate the model. Because of this, this model may not perform similarly when applied to samples from patients at different hospitals and handled by different technicians in separate labs, etc. The significance of independent datasets for confirmation is highlighted by the OvaCheck case study. The institution that contributed the specimens needed to build the computational model was the majority of the tissue specimens (Baggerly et al., 2004; Baggerly et al., 2004; Petricoin et al., 2002; Petricoin et al., 2002). The performance of an omics-based computational model on an independent specimen set is evaluated by two "levels of evidence" that are provided below.

1. Lower Level of Evidence: from the same patient group's independent clinical data and specimens are collected from a single institution under well-regulated conditions.
2. Higher Level of Evidence: from the same patient group's independent clinical data and specimens are collected from multiple institutions under well-regulated conditions.

Under these conditions, the locked-down computational method performed well, indicating that it is effective in the specific environment that was studied when used in connection with the institution's protocols and patient profile, etc. A slightly different patient set, different laboratory processing, or technique would make this proposed omics-based diagnostic less effective. The fact that the omics measurements and locked-down computational method worked well in this environment implies that future patients will benefit from them as well. It demonstrates that the test is resistant to the kinds of variables that could differ between sites, such as biological characteristics of the populations (hospitals, collection and handling procedures, and other techniques), and these differences frequently have higher effects on the acquired omics measurements than differences related to the relevant phenotypes.

Further modifications to the computational model are unacceptable once Step 3 has been started since they can result in overfitting and a very high chance that the model will not function effectively in future stages of the development process. If additional model refinement takes place after Step 3 has begun, this must be stated clearly when reporting the computational model's evolution and efficiency in Step 3.

4.4 Step 4: Data, code, and fully specified computational methods release to the scientific community

The potential omics-based test is ready to go on to the test validation phase (analytical and clinical or biological validation), once an omics-based measurement and the locked-down computational methods are tested correctly. At this phase, a database that is separately managed should contain the data and metadata required to develop the potential omics-based test (e.g., dbGaP [database of Genotypes and Phenotypes]). dbGaP is a National Institutes of Health–sponsored repository tasked with collecting, preserving, and disseminating data generated by studies of genotype and phenotype interactions. The data in dbGaP is a hierarchical structure that contains the accessioned objects, phenotypes (as variables and datasets), various molecular assay data (Sequence, Expression Array data and SNPs, and Epigenomic markers), analyses, and documentation. The dbGaP website offers free access to publicly available metadata about submitted research, summary-level data, and study-related papers. Scientists from all around the world get access to individual-level data through the Controlled Access application (Tryka et al., 2014).

Additionally, it is important to make the computer code and fully described computational processes used to develop the potential omics-based test sustainably available. This implies that computer code and fully described computational techniques of publicly funded research should be made public at the time of publication. If the creators of commercial tests are looking for FDA approval or clearance, code and fully specified computational techniques will be submitted to the FDA for assessment. A laboratory-developed publication is in this circumstance essentially necessary since laboratories must support the claims made for the test.

The distributed computer code should ideally include all phases of computational analysis, including the data preparation procedures. Transparent reporting of the entire analysis is required. If a portion of the code or data cannot be made available, then an explanation of the reason must be presented. For results to be independently verified, it's important to fully describe release data, source code, and computational processes. This suggestion strengthens the demand for reporting transparency voiced in previous National Research Council studies. Implementation of an electronic health record system improves the quality of health care by managing the health data and treatment records and also reduces medication errors and the risk of adverse drug reactions (Campanelle et al., 2016).

Comprehensive access to the complete sample data, handling procedures or analysis methods, and computer codes for published omics studies has recently been demanded by many others (Ince et al., 2012; Ioannidis & Khoury, 2011; Morin et al., 2012). Release of this data is especially important in the omics context because: It can be challenging to reproduce results due to the complexity of the data and analyses, but thorough scientific community verification of results is essential to verify that candidate omics-based tests are statistically and scientifically acceptable. Future researchers will be able to undertake additional analysis to gain new scientific insights if the data are made available. The qualities of potential omics-based test concepts that raise the potential of making a fake test are listed below. A few of these problems have already been brought up. Others are not specifically related to the development of omics-based testing, but they are equally vital to the process.

1. High dimensionality data
2. Biological plausibility
3. Data variability unrelated to the clinical outcome of interest
4. Need for multiple datasets
5. Study design and batch effects
6. Computational procedure lock-down
7. Role of biostatistics and bioinformatics experts

Measurements (such as genes, RNA, and proteins) that are much bigger than the number of clinical samples available are typical characteristics of genomics datasets. If appropriate steps are not taken, such as cross-validation combined with confirmation on an independent dataset, the data may end up getting overfit. The potential omics-based test can be perused even in absence of a well-known biological mechanism. For example, a breast cancer effect modifier based on a group of genes already known to play a role in the disease is more likely to stand up in further research than a breast cancer effect modifier based on a group of genes not yet known to play an important role in the disease. The process of developing omics-based tests can frequently start with a subset of data for which a biological mechanism is known, leading to better outcomes.

Computational models that are fitted to many datasets in Step 2 will typically perform better afterward. In other words, rather than fitting a model on just one dataset, researchers are argued to generate a model from datasets collected from samples and associated clinical outcomes collected at various laboratories. A good study design is essential in all fields of biomedical research. Batch effects (Leek et al., 2010) can occur if the dataset used in Step 2 to create the computational model was produced as a result of poor experimental design (for example, if the patient's samples were processed at a different time, by a different expert, or in different institutions). This may produce a model that generates better results for the data for which it was generated.

Before moving on to Step 3's confirmation on an independent test set, Step 2's fully stated computational techniques must be locked down. A transcriptomics-based test, for example, cannot simply provide the set of genes that are part of the computational model used to support it because this does not represent a fully described computational approach. It is impossible to foresee every potential hazard that researchers can experience during the discovery phase in a discipline that is still relatively young and in flux like omics. An additional precaution is the participation of suitably qualified biostatistician or bioinformatician colleagues who are completely integrated into all stages of the discovery and assessment process. Depending on what step or phase of test development is being used, a certain type of biostatistician may be needed. For instance, professionals in designing clinical trial designs might not have the necessary knowledge of those who construct computer models for omics-based assessments.

5. Completion of the discovery phase of omics-based test development

In the test validation phase, molecular measurements, computational methods, and the future usage of the developed test must be explained. The fast-changing spectrum of omics technologies offers tremendous potential for discovering measures with potential clinical utility. The transition from the first discovery of the potentially important differences in omics measures to reliable, validated clinical diagnostics, however, is full of challenges. These difficulties include the potential for overfitting the data during the construction of the computer model and the tremendous heterogeneity between various investigations of what are purportedly the same disease conditions (for both biological and technical reasons). Moving forward, it is crucial to present all the aspects of the designed omics-based test transparently i.e., the measurements are taken, the applied preprocessing methods, and the fully described computing method. To develop increasingly reliable omics-based tests to help health care, it is crucial to provide enough metadata with publications to identify potential omics-based tests that apply to various sites.

To create a clinically viable, affordable, and reliable assay for use in medical practice, the methodologies utilized to retrieve the omics measures from patient samples may be modified in the subsequent phase of test development. In all future test design processes, the fully described computational procedures that were created in the discovery stage must be locked down and unchangeable. Before clinical applications, test methods, including the omics measurements approaches, and computational procedures must be locked down at the conclusion stage of the validation phase.

6. Issues and limitations of translational omics analysis

At the beginning of omics data analysis, many issues and limitations have been faced by the experts, but with the development in computational technologies, data analysis, and integration approaches, many issues have already been resolved. The collection of clinically relevant and disease-associated data is the first and most requisite condition in developing an accurate and efficient prediction model. Some issues and limitations that arise in the way of omics analysis and its translational implementation are as follows:

1. Experimental and analytical noise: Noise in experimental and analytical approaches results in inaccuracy in decision-making (Church et al., 2015). For example, various versions of microarray, sequencing, proteome analysis tools, and packages are available for genomics, transcriptomics, epigenomics, and proteomics-related studies.
2. Analytical accuracy and clinical relevance: Inaccurate omics data may lead to the development of a poor prediction model, which may limit the clinical applicability of omics-based tests (Vinaixa et al., 2016).

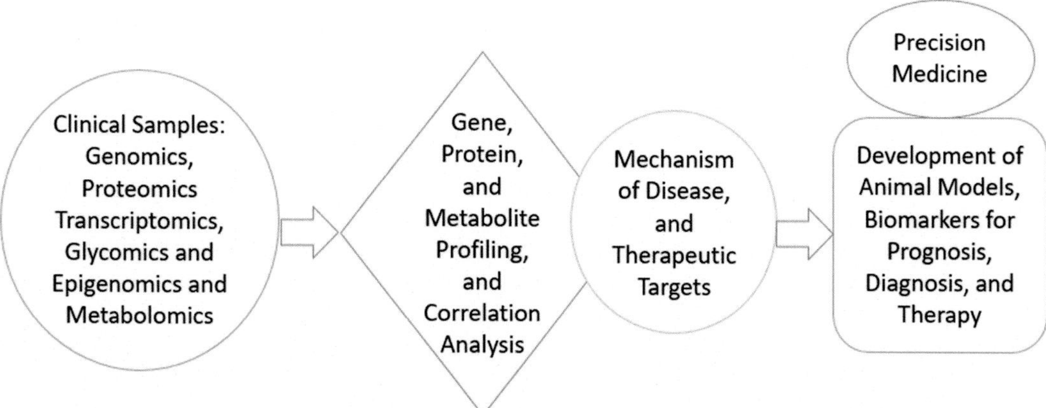

FIGURE 17.5 Use of omics data for applying the therapeutic guidelines related to precision medicine.

3. Data visualization in various formats limits their clinical implementations: For example, the use of various mass spectroscopy data formats limits the sharing and processing of data during analysis and interpretation.
4. Challenges and opportunities in translational and clinical contexts.

There are many challenges in omics data analysis and interpretation, which limits its applicability in translational and clinical contexts. Standard protocols and regulatory approaches are required for omics data collection, integration, and sharing to achieve the goal of personalized and precision medicine (Tenenbaum et al., 2014). Bioinformatics professionals with diverse skill sets are required to bridge the gap between omics research and their clinical utility. Overcoming these challenges and issues may result in the precise and efficient implementation of translational omics in personalized therapy of cancer and many other complex diseases.

1. Data integrity, standardization, and sharing: Large amounts of healthcare data create challenges of sample collection, handling, storage, transport, multiomics data analyses, collecting electronic medical record data, and integration of medical and biological data. Data sharing among stakeholders may also create legal, ethical, and privacy issues.
2. Turning data into knowledge: Biomarkers associated with pathophysiological mechanisms of disease should be clinically important, relevant, and feasible in terms of public health. Functional characterization of a clinical biomarker must be assessed through pathways and network analysis. There is still a need to address the issues in the omics-data integration, and simulation of disease modeling.
3. Informatics and new pathways to clinical actionability: Biological databases and related annotations help in the development of decision support systems that can provide accurate patient-specific reports (Sirintrapun et al., 2016). Multilevel evidence should be required to establish the fact that a detected molecular abnormality may serve as a clinically relevant biomarker (Fig. 17.5).
4. Interaction of multidisciplinary sciences and experts: To achieve the goal of precision medicine, experts from diverse fields such as medical science, computer, and data science, bioinformatics, biostatistics, sociology, and law are required for more effective interactions with healthcare partners (Henricks et al., 2016).

7. Summary

A significant improvement has been made in the techniques and approaches used for genomics, transcriptomics, proteomics, metabolomics, and phenomics studies. The individual omics analysis as well as the integration of multiomics data has made it possible to understand the omics basis of patient subtypes, genomic mutants/variants related to drug response, mechanism of drug resistance, the impact of varied responses to the drug and precision medicine, and biomarkers roles in prognosis, diagnosis, and therapy. There are still many limitations and challenges in analyzing the omics data and translating omics data for the development of omics-based tests and useful knowledge. Identification of biomarkers for various analyses and omics-based tests for prognosis, diagnosis, and therapy of cancer and many other complex diseases may play a potential role in the implementation of personalized therapy.

References

Baggerly, K. A., Morris, J. S., & Coombes, K. R. (2004). Reproducibility of SELDI-TOF protein patterns in serum: Comparing datasets from different experiments. *Bioinformatics, 20*(5), 777–785. https://doi.org/10.1093/bioinformatics/btg484. http://bioinformatics.oxfordjournals.org/

Buyse, M., Loi, S., van't Veer, L., Viale, G., Delorenzi, M., Glas, A. M., d'Assignies, M. S., Bergh, J., Lidereau, R., Ellis, P., Harris, A., Bogaerts, J., Therasse, P., Floore, A., Amakrane, M., Piette, F., Rutgers, E., Sotiriou, C., Cardoso, F., … Straehle, C. (2006). Validation and clinical utility of a 70-gene prognostic signature for women with node-negative breast cancer. *Journal of the National Cancer Institute, 98*(17), 1183–1192. https://doi.org/10.1093/jnci/djj329

Campanella, P., Lovato, E., Marone, C., Fallacara, L., Mancuso, A., Ricciardi, W., & Specchia, M. L. (2016). The impact of electronic health records on healthcare quality: A systematic review and meta-analysis. *European Journal of Public Health, 26*(1), 60–64. https://doi.org/10.1093/eurpub/ckv122

Chen, Y., Li, E. M., & Xu, L. Y. (2022). Guide to metabolomics analysis: A bioinformatics workflow. *Metabolites, 12*(4). https://doi.org/10.3390/metabo12040357. https://www.mdpi.com/2218-1989/12/4/357/pdf

Church, D. M., Schneider, V. A., Steinberg, K. M., Schatz, M. C., Quinlan, A. R., Chin, C. S., Kitts, P. A., Aken, B., Marth, G. T., Hoffman, M. M., Herrero, J., Mendoza, M. L. Z., Durbin, R., & Flicek, P. (2015). Extending reference assembly models. *Genome Biology, 16*(1). https://doi.org/10.1186/s13059-015-0587-3. http://genomebiology.com/

Craig, R., Cortens, J. P., & Beavis, R. C. (2004). Open source system for analyzing, validating, and storing protein identification data. *Journal of Proteome Research, 3*(6), 1234–1242. https://doi.org/10.1021/pr049882h

Deng, M. C., Eisen, H. J., Mehra, M. R., Billingham, M., Marboe, C. C., Berry, G., Kobashigawa, J., Johnson, F. L., Starling, R. C., Murali, S., Pauly, D. F., Baron, H., Wohlgemuth, J. G., Woodward, R. N., Klingler, T. M., Walther, D., Lal, P. G., Rosenberg, S., & Hunt, S. (2006). Noninvasive discrimination of rejection in cardiac allograft recipients using gene expression profiling. *American Journal of Transplantation, 6*(1), 150–160. https://doi.org/10.1111/j.1600-6143.2005.01175.x

Deutsch, E. W., Lam, H., & Aebersold, R. (2008). PeptideAtlas: A resource for target selection for emerging targeted proteomics workflows. *EMBO Reports, 9*(5), 429–434. https://doi.org/10.1038/embor.2008.56

Glaab, E., Rauschenberger, A., Banzi, R., Gerardi, C., Garcia, P., & Demotes, J. (2021). Biomarker discovery studies for patient stratification using machine learning analysis of omics data: A scoping review. *BMJ Open, 11*(12). https://doi.org/10.1136/bmjopen-2021-053674

Henricks, W. H., Karcher, D. S., Harrison, J. H., Sinard, J. H., Riben, M. W., Boyer, P. J., Plath, S., Thompson, A., & Pantanowitz, L. (2016). Pathology informatics essentials for residents: A flexible informatics curriculum linked to accreditation council for graduate medical education milestones. *Journal of Pathology Informatics, 7*(1). https://doi.org/10.4103/2153-3539.185673. www.jpathinformatics.org/browse.asp?sabs=n

Hong, M., Tao, S., Zhang, L., Diao, L. T., Huang, X., Huang, S., Xie, S. J., Xiao, Z. D., & Zhang, H. (2020). RNA sequencing: New technologies and applications in cancer research. *Journal of Hematology and Oncology, 13*(1). https://doi.org/10.1186/s13045-020-01005-x. http://www.jhoonline.org/

Huang, K. K., Huang, J., Wu, J. K. L., Lee, M., Tay, S. T., Kumar, V., Ramnarayanan, K., Padmanabhan, N., Xu, C., Tan, A. L. K., Chan, C., Kappei, D., Göke, J., & Tan, P. (2021). Long-read transcriptome sequencing reveals abundant promoter diversity in distinct molecular subtypes of gastric cancer. *Genome Biology, 22*(1). https://doi.org/10.1186/s13059-021-02261-x. http://genomebiology.com/

Ince, D. C., Hatton, L., & Graham-Cumming, J. (2012). The case for open computer programs. *Nature, 482*(7386), 485–488. https://doi.org/10.1038/nature10836. http://www.nature.com/nature/index.html

Ioannidis, J. P. A., & Khoury, M. J. (2011). Improving validation practices in "omics" research. *Science, 334*(6060), 1230–1232. https://doi.org/10.1126/science.1211811. http://www.sciencemag.org/content/334/6060/1230.full.pdf

Jiangzhou, H., Zhang, H., Sun, R., Fahira, A., Wang, K., Li, Z., Shi, Y., & Wang, Z. (2021). Integrative omics analysis reveals effective stratification and potential prognosis markers of pan-gastrointestinal cancers. *iScience, 24*(8). https://doi.org/10.1016/j.isci.2021.102824

Leek, J. T., Scharpf, R. B., Bravo, H. C., Simcha, D., Langmead, B., Johnson, W. E., Geman, D., Baggerly, K., & Irizarry, R. A. (2010). Tackling the widespread and critical impact of batch effects in high-throughput data. *Nature Reviews Genetics, 11*(10), 733–739. https://doi.org/10.1038/nrg2825

Legrain, P., Aebersold, R., Archakov, A., Bairoch, A., Bala, K., Beretta, L., Bergeron, J., Borchers, C. H., Corthals, G. L., Costello, C. E., Deutsch, E. W., Domon, B., Hancock, W., He, F., Hochstrasser, D., Marko-Varga, G., Salekdeh, G. H., Sechi, S., Snyder, M., … Omenn, G. S. (2011). The human proteome project: Current state and future direction. *Molecular and Cellular Proteomics, 10*(7). https://doi.org/10.1074/mcp.M111.009993. http://www.mcponline.org/content/10/7/M111.009993.full.pdf+html

MacLean, B., Tomazela, D. M., Shulman, N., Chambers, M., Finney, G. L., Frewen, B., Kern, R., Tabb, D. L., Liebler, D. C., & MacCoss, M. J. (2010). Skyline: An open source document editor for creating and analyzing targeted proteomics experiments. *Bioinformatics, 26*(7), 966–968. https://doi.org/10.1093/bioinformatics/btq054

Mead, J. A., Bianco, L., Ottone, V., Barton, C., Kay, R. G., Lilley, K. S., Bond, N. J., & Bessant, C. (2009). MRMaid, the web-based tool for designing multiple reaction monitoring (MRM) transitions. *Molecular and Cellular Proteomics, 8*(4), 696–705. https://doi.org/10.1074/mcp.M800192-MCP200. http://www.mcponline.org/content/8/4/585.full.pdf+html

Morin, A., Urban, J., Adams, P. D., Foster, I., Sali, A., Baker, D., & Sliz, P. (2012). Shining light into black boxes. *Science, 336*(6078), 159–160. https://doi.org/10.1126/science.1218263

Parker, C. G., & Pratt, M. R. (2020). Click chemistry in proteomic investigations. *Cell, 180*(4), 605–632. https://doi.org/10.1016/j.cell.2020.01.025. https://www.sciencedirect.com/journal/cell

Petricoin, E. F., Ardekani, A. M., Hitt, B. A., Levine, P. J., Fusaro, V. A., Steinberg, S. M., Mills, G. B., Simone, C., Fishman, D. A., Kohn, E. C., & Liotta, L. A. (2002). Use of proteomic patterns in serum to identify ovarian cancer. *Lancet, 359*(9306), 572–577. https://doi.org/10.1016/S0140-6736(02)07746-2. http://www.journals.elsevier.com/the-lancet/

Quezada, H., Guzmán-Ortiz, A. L., Díaz-Sánchez, H., Valle-Rios, R., & Aguirre-Hernández, J. (2017). Omics-based biomarkers: Current status and potential use in the clinic. *Boletin Medico del Hospital Infantil de Mexico, 74*(3), 219–226. https://doi.org/10.1016/j.bmhimx.2017.03.003. http://www.bmhim.com/

Röst, H., Malmström, L., & Aebersold, R. (2012). A computational tool to detect and avoid redundancy in selected reaction monitoring. *Molecular and Cellular Proteomics, 11*(8), 540–549. https://doi.org/10.1074/mcp.m111.013045

Reiter, L., Rinner, O., Picotti, P., Hüttenhain, R., Beck, M., Brusniak, M. Y., Hengartner, M. O., & Aebersold, R. (2011). MProphet: Automated data processing and statistical validation for large-scale SRM experiments. *Nature Methods, 8*(5), 430–435. https://doi.org/10.1038/nmeth.1584

Shatunova, E. A., Korolev, M. A., Omelchenko, V. O., Kurochkina, Y. D., Davydova, A. S., Venyaminova, A. G., & Vorobyeva, M. A. (2020). Aptamers for proteins associated with rheumatic diseases: Progress, challenges, and prospects of diagnostic and therapeutic applications. *Biomedicines, 8*(11), 1–44. https://doi.org/10.3390/biomedicines8110527. https://www.mdpi.com/2227-9059/8/11/527/pdf

Shi, T., Su, D., Liu, T., Tang, K., Camp, D. G., Qian, W. J., & Smith, R. D. (2012). Advancing the sensitivity of selected reaction monitoring-based targeted quantitative proteomics. *Proteomics, 12*(8), 1074–1092. https://doi.org/10.1002/pmic.201100436

Simon, R., Radmacher, M. D., Dobbin, K., & McShane, L. M. (2003). Pitfalls in the use of DNA microarray data for diagnostic and prognostic classification. *Journal of the National Cancer Institute, 95*(1), 14–18. https://doi.org/10.1093/jnci/95.1.14

Sirintrapun, S. J., Zehir, A., Syed, A., Gao, J. J., Schultz, N., & Cheng, D. T. (2016). Translational bioinformatics and clinical research (biomedical) informatics. *Clinics in Laboratory Medicine, 36*(1), 153–181. https://doi.org/10.1016/j.cll.2015.09.013. http://www.elsevier.com/inca/publications/store/6/2/3/3/1/6/index.htt

Subramanian, I., Verma, S., Kumar, S., Jere, A., & Anamika, K. (2020). Multi-omics data integration, interpretation, and its application. *Bioinformatics and Biology Insights, 14*. https://doi.org/10.1177/1177932219899051

Tebani, A., Afonso, C., Marret, S., & Bekri, S. (2016). Omics-based strategies in precision medicine: Toward a paradigm shift in inborn errors of metabolism investigations. *International Journal of Molecular Sciences, 17*(9). https://doi.org/10.3390/ijms17091555

Tenenbaum, J. D., Sansone, S. A., & Haendel, M. (2014). A sea of standards for omics data: Sink or swim? *Journal of the American Medical Informatics Association, 21*(2), 200–203. https://doi.org/10.1136/amiajnl-2013-002066. http://jamia.bmj.com/content/21/2/200.full.pdf

Tryka, K. A., Hao, L., Sturcke, A., Jin, Y., Wang, Z. Y., Ziyabari, L., Lee, M., Popova, N., Sharopova, N., Kimura, M., & Feolo, M. (2014). NCBI's database of genotypes and phenotypes: DbGaP. *Nucleic Acids Research, 42*(1), D975–D979. https://doi.org/10.1093/nar/gkt1211

Uhlen, M., Oksvold, P., Fagerberg, L., Lundberg, E., Jonasson, K., Forsberg, M., Zwahlen, M., Kampf, C., Wester, K., Hober, S., Wernerus, H., Björling, L., & Ponten, F. (2010). Towards a knowledge-based human protein Atlas. *Nature Biotechnology, 28*(12), 1248–1250. https://doi.org/10.1038/nbt1210-1248

Vinaixa, M., Schymanski, E. L., Neumann, S., Navarro, M., Salek, R. M., & Yanes, O. (2016). Mass spectral databases for LC/MS- and GC/MS-based metabolomics: State of the field and future prospects. *TrAC, Trends in Analytical Chemistry, 78*, 23–35. https://doi.org/10.1016/j.trac.2015.09.005. www.elsevier.com/locate/trac

Wingren, C. (2016). Antibody-based proteomics. *Advances in Experimental Medicine and Biology, 926*, 163–179. https://doi.org/10.1007/978-3-319-42316-6_11. http://www.springer.com/series/5584

Zhang, W., Wan, Z., Li, X., Li, R., Luo, L., Song, Z., Miao, Y., Li, Z., Wang, S., Shan, Y., Li, Y., Chen, B., Zhen, H., Sun, Y., Fang, M., Ding, J., Yan, Y., Zong, Y., Wang, Z., … Nie, C. (2022). A population-based study of precision health assessments using multi-omics network-derived biological functional modules. *Cell Reports Medicine, 3*(12). https://doi.org/10.1016/j.xcrm.2022.100847

Zhang, X., Li, L., Butcher, J., Stintzi, A., & Figeys, D. (2019). Advancing functional and translational microbiome research using meta-omics approaches. *Microbiome, 7*(1). https://doi.org/10.1186/s40168-019-0767-6

Chapter 18

Omics technologies for crop improvement

Arvind Kumar Yadav[1,2], Bharti Shree[2], Deepika Lakhwani[2] and Amit Kumar Singh[2]
[1]*Institute of Life Sciences (ILS), Bhubaneswar, Odisha, India;* [2]*ICAR-National Bureau of Plant Genetic Resources (ICAR-NBPGR), New Delhi, India*

1. Introduction

Plants are an important source of food production and there is a need to increase food production due to forever growing population. The world population is projected to exceed nine billion individuals by 2050. Therefore, the future food security is a major concern nowadays which is placing pressure on breeding programs to enhance the yield and nutritive qualities of major crops in order to meet the food requirements (Grote et al., 2021; Kopittke et al., 2019). High-throughput omics technologies have emerged as novel approaches for studying and manipulating the economic traits of crop plants for enhancing yield and nutritional qualities. Omics technologies, including genomics, transcriptomics, proteomics, and metabolomics, involve the study of the genome sequences, gene expression profiles, protein content, and metabolites of an organism, respectively (Shukla et al., 2020; Yang et al., 2021). Several studies have been done that integrated the multiple omics approaches to elucidate gene functions and networks under environmental stress that contribute to the improved productivity and quality of modern crop varieties (Ali et al., 2022).

The advancement of high-throughput technologies such as next-generation sequencing (NGS) opened the doors for researchers to generate huge amounts of data and accessibility to genomic information of model crop plants (Singh et al., 2018). "Omics" is a common term for disciplines that aim to gather and evaluate such huge amounts of biological data (Van Emon, 2016). Various omics methods may be employed individually or combined to each other to elucidate the biological functions linked with the genetic information. This has helped discover new genes linked with desired phenotypic characters and understanding of gene functions and their complex networks (Yang et al., 2021). It is now possible to investigate intricate biological processes like plant growth, development, photosynthesis, and biotic and abiotic stress tolerance because of the identification of target loci linked to specific phenotypic traits (Villalobos-Lópcz ct al., 2022). The most crop traits are complicated quantitative traits regulated by numerous genes (Zhang, Liu, et al., 2020). Multiomics techniques with network and system-based approaches have been used to identify and decipher the significant traits in various important crops (Abdullah-Zawawi et al., 2022; Muthamilarasan et al., 2019; Scossa et al., 2021). In addition, the artificial intelligence (AI) approach has been engaged in diverse crop genomics and phenomics research (Khan et al., 2022). Moreover, the application of such advanced approaches in plant breeding programs has enhanced the productivity and sustainability of crop plants so that they provide elite cultivars that can adapt to a range of environments without compromising agronomic performance, grain quality, or disease resistance (Gao, 2021; Singh et al., 2021).

In this chapter, we have covered a variety of omics techniques, their function, and their major use in crop science. The goal is to increase crop yields with increased biotic and abiotic stress tolerance. We have also discussed about the network-based and system biology approaches along with AI in crop breeding science that help to improve genetic development, crop yield, and crop resistance.

2. OMICS-based technologies

With the advancement in high-throughput technologies, which include research in DNA, RNA, proteins, ions, and metabolites level, crop research has advanced significantly over these years. Several high-throughput approaches such as genomics, transcriptomics, proteomics, metabolomics, and phenomics constitute the OMICS-based approach for crop

Integrative Omics. https://doi.org/10.1016/B978-0-443-16092-9.00018-7

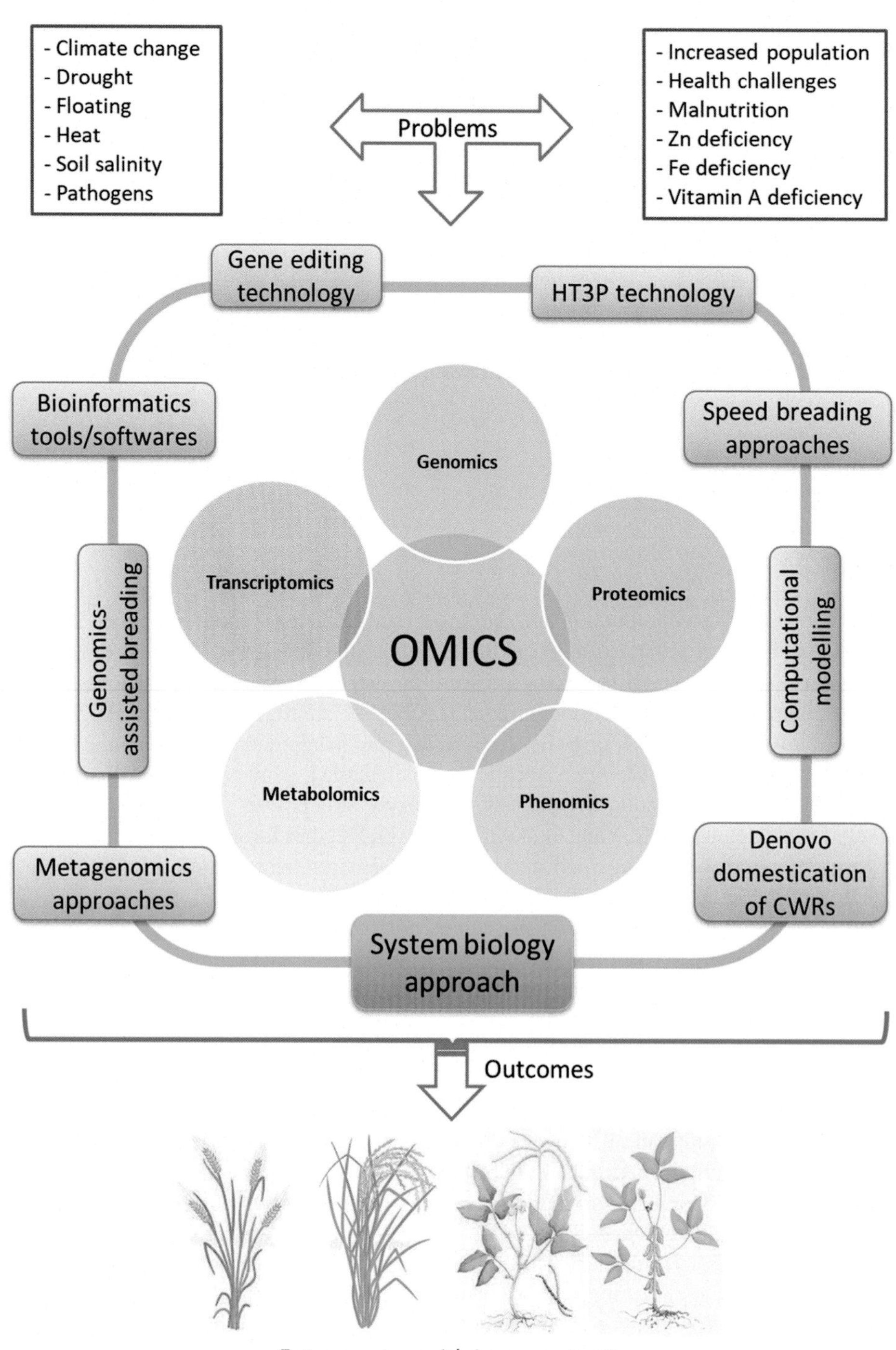

FIGURE 18.1 Representation of the role of the omics approach in crop improvement.

improvement (Fig. 18.1). In this section, we have focused on these omics-based approaches and how it is helping to improve crop quality and productivity.

2.1 Genomics

It is possible to understand how genes have been modified or adapted by selection pressure as a result of environmental restrictions by understanding the structure and function dynamics of genomes in plants. Genomics approach includes functional genomics, structural genomics, and comparative genomics. Functional genomics primarily uses either sequence-based methods (expressed sequence tag (EST) sequencing, Serial Analysis of Gene Expression (SAGE), massive parallel signature sequencing) or hybridization-based methods (microarray). Genome sequencing, mapping, and molecular marker identification are some examples of structural genomics. So far, few plant/crop species have been sequenced and are called model plants (*Arabidopsis thaliana* L., *Oryza sativa* L., *Zea mays* L.). In the past decade, we have advanced in sequencing technologies which, as a result, led to sequence several other crops such as black cottonwood, grapevine, soybean, pigeon pea, etc (Schmutz et al., 2010; Singh et al., 2012; Varshney et al., 2012; Velasco et al., 2007). NGS, as well as third-generation sequencing technology, has evolved substantially, leading to a continuous higher throughput and lower sequencing costs. Gene identification has now become comparatively easier, and this would further help in understanding the genetic properties of crops and plants. Identification and selection of molecular markers (simple-sequence repeat (SSR), single nucleotide polymorphism (SNP), restriction fragment length polymorphism (RFPL), random amplified polymorphic DNA (RAPD), etc.) associated with particular trait(s) have enabled the selection of genotypes or variants and thereby assisted in marker-assisted breeding of crops (McCouch et al., 1997). Markers help in mapping the trait(s) such as grain length, width, yield, and resistance to bacteria, fungi etc. and this information can be exploited for future breeding (Ramakrishnan et al., 2015). Genomics-based approaches assist in finding genetic diversity that will be helpful in breeding various agronomically important crops, ultimately leading to sustainable agriculture. Crops (sorghum, rice, wheat) (Habash et al., 2009; Ismail et al., 2007; Mace et al., 2019) and vegetables such as *Brassica* vegetables (Saha et al., 2022, pp. 153–185; Xiao et al., 2014) have been designed to combat abiotic stresses using genomic information such as position of quantitative trait loci (QTLs) responsible for resistance genes, number of molecular markers, and SNPs.

2.2 Transcriptomics

Transcriptomics technologies are used to examine the transcriptome, which is the totality of all of its RNA transcripts. The entire molecular information of a crop/plant species is encoded in its genome and expressed through its RNA, where both coding and noncoding RNAs have specified functions in plants (Chekanova, 2015). Transcriptomics-based approach involves not only the genome of crop/plant species but also the process by which the information in the genome is utilized by the cell (Rao et al., 2021). Transcriptome profiling of several major crops, such as rice (Matsumura et al., 1999), *Arabidopsis* (Lee & Lee, 2003), maize (Poroyko et al., 2005), etc., has been performed to study gene expression by the SAGE method. Later on, in the late 90s, a high-throughput sequencing technique also known as digital gene expression analysis was used, which is a higher variant of SAGE (Wan et al., 2019). High-throughput sequencing of whole transcripts offered more details on transcript structure, such as splice variants, mainly supplanted previous techniques. Analysis of plant responses to various stimuli and its growth and development, makes extensive use of transcriptome sequencing. Additionally, transcriptome sequencing has been used for a number of functional genomics goals, including the identification of noncoding RNA, genome annotation, and gene expression monitoring (Guo et al., 2021).

2.3 Proteomics

Proteomics-based approach for crop improvement includes the investigation of proteins and their interactions within a cell. The study of various facets of plant activities is becoming more and more essential due to the explosion of genetic data and advancements in analytical technologies. Studies at the protein level are crucial as the proteomes of plants are extremely dynamic and complicated to elucidate the molecular mechanisms driving plant growth, development, and interactions with the environment because proteins play a significant role in various signaling and metabolic pathways. Understanding protein folding and protein–protein interaction is essential for understanding biological processes under changing environmental and development conditions. Protein identification, separation, or purification can be performed by techniques such as 2D-gel electrophoresis, SDS PAGE, chromatic methods (ion exchange chromatography, HPLC, thin layer chromatography), yeast hybrid system, immunoblotting, protein microarray, western blot, nuclear magnetic resonance (NMR), matrix-assisted laser desorption ionization–time of flight (MALDI-TOF) mass spectrometry, and circular dichroism spectrometry (Pang et al., 2010; Timperio et al., 2008). For example, overexpression of rice protein OsDi19 shows involvement in drought resistance by using yeast hybrid system for protein–protein interaction (Wang et al., 2014). Liu et al. identified a novel heat shock protein (*Hsa32*) in tomato using a heterologous expression system and northern analysis, suggesting its role in abiotic (heat) stress conditions (Liu et al., 2006). Proteomics together with physiological and biochemical study of legumes, such as soybean, allowed for the identification of tolerance pathways in response to abiotic stresses (Katam et al., 2020). In order to expedite crop development initiatives globally, a combinatorial strategy of rapid gene identification using genomics, proteomics, and other related -omic branches of biotechnology is proven to be a successful method.

2.4 Metabolomics

The systematic identification and measurement of all metabolites and their dynamic changes during development and stress conditions in a specific species or biological sample is known as metabolomics. Metabolomics is the most universal of the -omics technologies and may be used on a variety of species with little to no adjustments. To identify specific patterns linked to stress tolerance, it has been effectively used to study plants' molecular phenotypes in response to abiotic stress (Arbona et al., 2013). Metabolites are more closely associated with the plant phenotype than mRNA transcripts or proteins alone because they integrate gene expression, protein interaction, and other regulatory mechanisms (Schauer & Fernie, 2006). It includes analytical techniques such as high-performance liquid chromatography (HPLC), gas chromatography (GC), and ultra-performance liquid chromatography (UPLC) coupled to NMR and MS (Zhang et al., 2012). Plant metabolites such as polyamines, alkaloids, secondary metabolites, etc, play an essential roles in growth, development, defense, stress response, nutrition, and yield (Chen, Shao, et al., 2019). Metabolomics analysis in crops like maize revealed metabolites (smiglaside and smilaside) associated with biotic (*Fusarium graminearum*) stress (Zhou et al., 2019). Metabolite profiling to demonstrate developmental process has been seen in crops like rice (Turhadi et al., 2019), wheat (Byeon et al., 2022; Lahuta et al., 2022), maize (Zhao et al., 2019). Metabolomics is therefore significantly important for crop improvement by combining with other omics-based approaches that can provide further insight into developing better crops in the future.

2.5 Phenomics

Crop phenomics has significantly advanced in the past decade, providing multidimensional phenotypic data at several levels, from cell to population levels (Jin et al., 2021). The phenomics-based approach includes collecting, measuring, and analyzing the phenotypic information of crops such as root architecture, shoot development, plant growth and root−shoot composition, leaf traits, leaf area index, weed detection, pathogen infection, yield estimation, etc under normal as well as stressed condition (Petrozza et al., 2014). In this omics-based approach, sensors such as proximal hyperspectral and optical sensors are used simultaneously for accurate data generation and interpretation (Sun et al., 2022; Tao et al., 2022). Along with sensory techniques, imaging techniques such as X-ray microcomputed tomography, laser microscopy, and time-lapse 3D imaging are used to study the root architecture, soil water and root relation, and tissue expansion (Bao et al., 2014). Recently, there has been a significant increase in the number of plants and crops that can be analyzed with the help of new imaging technologies such as robotic and conveyer-belt systems in greenhouses, as well as ground-based and aerial imaging platforms in fields (Fahlgren et al., 2015). These sensory automated technologies are now being used to record the dynamic response of crops that could further be correlated with sequence (Yang et al., 2020). Crop responses to several biotic and abiotic stresses can now be monitored and analyzed with various sensory and imaging techniques.

3. How omics technologies help in crop improvement?

Omics technologies such as genomics, transcriptomics, proteomics, and metabolomics can help identify genes and pathways involved in plant growth and development, as well as responses to biotic and abiotic stresses. This information can be used to develop crops with improved yield, quality, and resistance to environmental stresses. These technologies can also be used in the development of molecular markers for crop breeding, improve crop nutrition and development of stress-tolerant crops. In recent years, genomics and molecular phenotyping approaches have advanced significantly, leading to new and improved strategies for crop improvement (Varshney et al., 2021). These technologies are being used to enhance crop yield and quality. For example, genome editing technologies such as CRISPR-Cas9 have revolutionized the field of crop improvement (Wang et al., 2019). These technologies allow for precise modifications to crop genomes, such as knocking out genes or introducing desirable traits. Such technologies allow a deeper understanding of the biological processes involved in crop growth and development. Therefore, with the help of these technologies, it is possible to develop crops with improved yield, quality, and resistance to environmental stresses and meet the growing demand for food in a sustainable and efficient way. A number of crops have been improved through omics techniques using diverse agronomical traits in response to abiotic and biotic stresses (Tables 18.1 and 18.2).

Additionally, in the context of marine plants, omics technology has been used to study a variety of organisms, including seaweeds, seagrasses, and microalgae. These studies have provided insights into the biology and ecology of these organisms, as well as their potential applications in fields such as aquaculture, biotechnology, and pharmaceuticals (Abid et al., 2018). For example, genomics and transcriptomics have been used to identify genes associated with desirable traits in seaweeds, such as fast growth and high yield. Proteomics has been used to study the protein composition of seaweeds and identify potential food sources for humans and animals. Metabolomics has been used to identify bioactive compounds

TABLE 18.1 List of some important crops that are improved by omics technologies for abiotic stresses.

Abiotic stress	Crop	Tissue	Technique	Significant outcomes	References
Transcriptomics					
Drought	*Oryza sativa*	Leaf	Illumina	APXs, GSTs, LEA proteins, ROS scavengers, and genes that are resistant to drought were all substantially expressed.	Liang et al. (2021)
Drought	*Zea mays*	Leaf	Illumina	Genes that respond to drought in the tolerant genotype YE8112 were mostly associated with stress signal transduction.	Zenda et al. (2019)
Drought	*Glycine max*	Leaf	Illumina	A large number of DEGs and different pathways suggested that soybean uses intricate mechanisms to deal with drought.	Xu et al. (2018)
Drought	*Triticum aestivum*	Root	Illumina	Pyrroline-5-carboxylate reductase, as well as late-embryogenesis-abundant (LEA) proteins, was increased.	Derakhshani et al. (2020)
Heat	*Glycine max*	Leaf	Illumina	The expression of several genes involved in metabolism, photosynthesis, and the defense response varied in response to heat.	Wang et al. (2018)
Heat	*Oryza sativa*	Seedling	Microarray-based	As part of rice's response to heat stress, auxin, and ABA were linked with a group of uniquely regulated genes and associated pathways.	Sharma et al. (2021)
Heat	*Zea mays*	Seedling leaf	Illumina	MYB, AP2-EREBP, b-ZIP, bHLH, NAC, and WRKY were among the TF families linked to maize heat stress response.	Qian et al. (2019)
Salinity stress	*Triticum aestivum*	Leaf and root	Illumina	Most significant genes were related with polyunsaturated fatty acid (PUFA) metabolism in leaf tissues.	Luo et al. (2019)
Salinity stress	*Zea mays*	Seedling roots	Illumina	Aux/IAA, SAUR, and CBL-interacting kinases genes involved in salt tolerance.	Zhang et al. (2021)
Cold stress	*Oryza sativa*	Seed	Illumina	The upregulated DEGs were associated with cold stress.	Pan et al. (2020)
Cold stress	*Zea mays*	Root	Illumina	Different genotypes responded to cold stress in their transcriptomes in wildly different ways.	Frey et al. (2020)
Cold stress	*Triticum aestivum*	Leaf	Illumina	Identified genes that play a role in the response to cold stress.	Konstantinov et al. (2021)
Proteomics					
Drought	*Phaseolus vulgaris*	Leaf	iTRAQ	Drought stress was reacted to by changes in energy metabolism, photosynthesis, ATP interconversions, protein synthesis and proteolysis, stress, and defense-related DAPs.	Zadražnik et al. (2013)
Drought	*Zea mays*	Seedling root	iTRAQ	The capacity for greater water absorption was attributed to the root system.	Zeng et al. (2019)
Water deficit	*Vigna unguiculata*	Leaf	2D-PAGE	108 DAPs associated with drought response were identified.	Habash et al. (2009)
Drought	*Sorghum bicolor*	Seedling root	iTRAQ	Unique proteins that accumulated in the reaction to water deficiency were discovered by studying the root proteome.	Goche et al. (2020)
Heat	*Oryza sativa*	Anthers	iTRAQ	Higher expression of sHSP, β-expansins, and lipid transfer proteins was responsible for heat stress.	Mu et al. (2017)

Continued

TABLE 18.1 List of some important crops that are improved by omics technologies for abiotic stresses.—cont'd

Abiotic stress	Crop	Tissue	Technique	Significant outcomes	References
Heat	*Capsicum annuum*	Seedling leaf	iTRAQ	Proteins that are sensitive to heat include 1591 DAPs.	Wang et al. (2021)
Heat	*Triticum aestivum*	Leaf	iTRAQ	258 heat-responsive proteins (HRPs) that are active in many biological processes.	Lu et al. (2017)
Metabolomics					
Heat and drought	*Glycine max*	Leaf	GC-MS, LC-MS	Under conditions of heat stress and drought, the metabolism of sugar, nitrogen, and phytochemicals is extremely important.	Das et al. (2017)
Drought	*Cicer arietinum*	Leaf	UPLC-HRMS	Key drought-responsive metabolites were found to be 20 recognized compounds.	Khan et al. (2019)
Drought	*Sorghum bicolor*	Leaf	FY-IRS, nontargeted GC-MS	There were found to be 188 chemicals associated with drought, including 142 recognized metabolites and 46 unidentified small molecules.	Ogbaga et al. (2016)
Salinity	*Hordeum vulgare*	Root	GC-MS	Key salt-responsive metabolites were found to be 76 recognized metabolites, including 29 amino acids and amines, 20 organic acids and fatty acids, and 19 sugars and sugar phosphates.	Shelden et al. (2016)
Cold	*Oryza sativa*	Germinating seeds	LC-MS/MS, LC-ESI-MS/MS	35 diverse metabolites responsible for cold stress were identified.	Yang et al. (2019)

TABLE 18.2 List of some important crops that are improved by omics technologies for biotic stresses.

Crop	Tissue	Technique	Stress condition	Key outcomes	References
Transcriptomics					
Triticum aestivum	Seedling leaf	Illumine HiSeq 4000	*S. graminum* aphids	Within hours following the start of aphid feeding, defense-related metabolic pathways and oxidative stress were quickly triggered in the tolerant genotype.	Zhang, Fu, et al. (2020)
Glycine max	Seedling roots	Illumine	*Bacillus* simplex (strain Sneb545) soybean cyst nematode (SCN)	It has been proposed that important metabolic processes, including phenylpropanoid production and cysteine and methionine metabolism, contribute to the soybean response to SCN.	Kang et al. (2018)
Zea mays	Seedling leaf	Illumine	*Fusarium verticillioides*	TPS1 and cytochrome P450 genes were upregulated, suggesting that kauralexins were involved in *Fusarium* ear rot defense response.	Baldwin et al. (2014)
Proteomics					
Triticum aestivum	Leaf	iTRAQ	Pst infection	The most important DAPs in controlling wheat's immunological response to Pst infection were peptidyl-prolyl cis-trans isomerases (PPIases), RNA-binding proteins (RBPs), and chaperonins.	Yang et al. (2016)

TABLE 18.2 List of some important crops that are improved by omics technologies for biotic stresses.—cont'd

Crop	Tissue	Technique	Stress condition	Key outcomes	References
Sorghum bicolor	Leaf	LC-MS/MS	*Chilo partellus*	There were many DAPs responding to the *C. partellus* infestation.	Tamhane et al. (2021)
Metabolomics					
Oryza sativa	Leaf	CE/TOF-MS in negative ion mode	*Rhizoctonia solani* infection	Effects of R. *solani* infection were linked to the activation of metabolic pathways in plants.	Suharti et al. (2016)
Solanum lycopersicum	Leaf	NMRS	CEVd and *Pseudomonas syringae* infection	In response to viral and bacterial infection, a number of primary and secondary metabolites were found.	López-Gresa et al. (2010)

in seaweeds that have potential applications in the food and pharmaceutical industries. Thus, omics technology has the potential to contribute significantly to food security and high yield in marine plants by enabling the development of more sustainable and efficient aquaculture practices, as well as the discovery of new sources of food and nutritional supplements from marine plants.

4. Systems biology approaches for crop improvement

Understanding the cellular components and complex behaviors of biological systems requires an integration of the numerous omics approaches to predict the responses of a given organism under a given set of circumstances. It has become common to refer to this new field of study as systems biology or integrated biology (Shukla et al., 2021). A systems biology approach that incorporates modeling and cellular process prediction, as well as integration of various omics data, is required to understand the flow of biological information that underlies complex features. It generates predictions for how all components, including interactions between genes, proteins, and metabolites, will respond to outside stimuli (Kumar et al., 2015). Systems biology's solid basis and multiomics have been coupled to provide a holistic understanding of an organism's growth and adaptation (Pinu et al., 2019). Systems biology—related plant stress research has made use of multiomics methods (Mosa et al., 2017, pp. 21—34). Comprehensive studies utilizing the three omics technologies of transcriptomics, metabolomics, and proteomics have, however, also advanced our knowledge of the systems biology underpinning plant responses to abiotic stress (Cramer et al., 2011). Phenotypic responses and metabolic pathways can be predicted using systems biology—based techniques that combine multi omics with genetics and/or metabolomics as a baseline (Pinu et al., 2019).

Integrative systems biology has been used to find agriculturally significant features utilizing high-throughput multiomics data analysis using systems biology and bioinformatics techniques (Kumar et al., 2015). The finding of molecular regulatory networks for salt stress tolerance in grapevine crops is the outcome of the merging of multiomics and systems biology techniques (Daldoul et al., 2016). Additionally, network and testing models for abiotic stress responses in crop plants have been proposed using systems biology coupled with omics methods (Gupta, 2013). Systems biology thus integrates networks of biochemical events using experimental and computational methods to provide a thorough understanding of complex agricultural features related to agricultural production in the context of crop protection and agricultural productivity. Computational approaches are needed for curation, pathway modeling, analysis, and visualization of agricultural traits because they improve understanding of the dynamics of systems under various physiological and environmental situations. For modeling and analyzing high throughput data produced by NGS technologies for the identification of novel features in crop plants, in silico systems biology approaches perform best (Pazhamala et al., 2017). To combine multidimensional omics data, a number of platforms are available, including mixOmics, Integromics, OnPLS modeling, sparse multiblock partial least squares, and COVAIN (Misra et al., 2019; Sun & Weckwerth, 2012). Numerous software programs and databases are made possible by developments in the field of bioinformatics and the growing data sets produced by high-throughput sequencing technology, as summarized by Kumar et al. (2015). For plant systems biology, several approaches, software tools, web applications, and databases, as well as a systematic multi-data integration approach, have recently been proposed (Jamil et al., 2020; Pinu et al., 2019). The visualization and study of complex features in crop plants can be greatly aided by these software programs and databases.

5. Databases and software tools for crop omics analysis

Data from the genome, proteome, transcriptome, metabolome, and epigenome are all included in the category of "multiomics data" in general. In addition to making it feasible to store, catalog, and analyze the data that is already available, computational resources have also made it simple to access user-friendly databases. Multiomics analysis is crucial to understand the link between various omics (called panomics) data types. Panomics offers a platform for integrating many complicated omics, including metabolomics, transcriptomics, proteomics, and phenomics (Weckwerth et al., 2020). Here, different omics-specific databases are compiled and tabulated under the umbrella of panomics databases (Table 18.3). These data can help in the unraveling of the mechanisms behind the biological condition of interest by offering important insights into the flow of biological information at various levels. Moreover, there are a number of publicly available multiomics databases that offer integrated multiomics data sets for crops (Table 18.4). Integrating complex "omics" datasets could reduce the number of false positives that single data sources for genotype–phenotype prediction produce (Ritchie et al., 2015).

The use of computational technology in bioinformatics is more valuable for managing and analyzing biological data. The use of computational software to interpret biological questions is beneficial (Raza et al., 2021). So, for data structuring and mining in support of diverse omics technologies, bioinformatics is crucial (Ambrosino et al., 2020). Omics data are big data, and omics data analysis is huge data as well. In order to manage data and traverse analysis results, a user-friendly interface is crucial. As a result, numerous programs and tools have been created and are freely accessible for multiomics analysis in order to produce insightful results (Table 18.5).

TABLE 18.3 List of publicly available plant panomics databases.

Database	Target organism	Brief description	Links
Genomics databases			
MaizeGDB	Maize	It provides access to genomic data from over 27 diverse maize lines.	https://www.maizegdb.org/
SoyBase	Soybean	It provides access to genomic data from over 100 diverse soybean lines.	https://www.soybase.org/
SoyGD	Soybean		http://soybeangenome.siu.edu/
CottonGen	Cotton	It provides access to genomic data from over 350 diverse cotton lines.	https://www.cottongen.org/
WheatIS	Wheat	It provides access to genomic data from multiple wheat varieties, including wild relatives of wheat.	http://www.wheatis.org/
WGI	Wheat	Genomics	http://wheatgenome.info/
Barley Pan-genome	Barley	It provides access to genomic data from multiple barley varieties, including wild relatives of barley.	
BARLEX	Barley	BARLEX provides access to the draft genome sequence of barley.	http://barlex.barleysequence.org
Ricebase	Rice	Ricebase provides a comprehensive collection of genomic, genetic, and phenotypic data for rice.	https://ricebase.org/
The 1001 Genomes Project	*Arabidopsis thaliana*	It provides access to genomic data from over 1000 diverse *Arabidopsis* lines.	https://1001genomes.org/
MTGD	*Medicago truncatula*	It provides a comprehensive collection of genomic, genetic, and biological information on the model legume species, *Medicago truncatula*.	http://www.MedicagoGenome.org
PLAZA	Multiplant species	A pan-genomics database for multiple plant species, including *Arabidopsis*, tomato, maize, and rice, which provides access to genomic data from multiple cultivars and wild relatives of these species.	https://bioinformatics.psb.ugent.be/plaza/
ECPD	Potato	It provides the data on genetic resources of potato.	https://www.europotato.org/

TABLE 18.3 List of publicly available plant panomics databases.—cont'd

Database	Target organism	Brief description	Links
Sol Genomics Network	Tomato	A pan-genomics database for tomato, which provides access to genomic data from over 5000 diverse tomato lines.	https://solgenomics.net/
SGH	Sugarcane	It provides access to the reference genome sequences and genomic resources of sugarcane.	https://sugarcane-genome.cirad.fr/
PeanutBase	Peanut	It provides access to genomic data from multiple peanut cultivars and wild relatives.	https://www.peanutbase.org/home
Banana Genome Hub	Banana	It provides access to genomic data from multiple banana cultivars and wild relatives.	https://banana-genome-hub.southgreen.fr/
Transcriptomics databases			
GreenPhylDB	Multiplant		https://www.greenphyl.org/cgi-bin/index.cgi
PlantExpress	46 plant species	This database contains expression data from more than 3000 samples across 46 plant species.	https://plantsexpress.com/
PlantGenIE	Multiplant species	It contains data from both RNA-seq and microarray experiments from a range of plant species.	https://plantgenie.org/
Phytozome	Multiplant species	It contains data from over 50 plant species and allows for comparisons of gene expression between species.	https://phytozome-next.jgi.doe.gov/
EMBL-EBI Expression Atlas	Multiplant species	It allows for the exploration of gene expression patterns across different conditions and tissues from a wide range of organisms.	https://www.ebi.ac.uk/gxa/home
CANTATAdb	Multiplant species	This database contains data from a variety of experimental conditions, including abiotic stress, pathogen infection, and developmental stages for several important crop species, including wheat, maize, and rice.	http://cantata.amu.edu.pl/
PlantTFDB	Multiplant species	It is a database of transcription factors (TFs) and their target genes in several plant species, including *Arabidopsis thaliana, Oryza sativa,* and *Solanum lycopersicum.*	http://planttfdb.gao-lab.org/
Metabolomics databases			
Golm Metabolome Database	Multiplant species	It provides a comprehensive collection of metabolomic data on plants and microorganisms.	http://gmd.mpimp-golm.mpg.de/
PMN	Multiplant species	It is a comprehensive database that provides access to a range of plant metabolic data, including metabolite structures, biochemical pathways, and enzyme information.	https://plantcyc.org/
MassBank	Multiplant species	It is a public database of mass spectral data that includes information on metabolites from a variety of plant species.	https://massbank.eu/MassBank/
MetaboLights	Multiplant species	It is a database that contains metabolomics data from a range of organisms, including plants. It includes information on metabolite identification, quantification, and annotation.	https://www.ebi.ac.uk/metabolights/
PhytoMetaSyn	Multiplant species	This database contains information on the metabolic pathways of over 400 plant species.	https://pubmed.ncbi.nlm.nih.gov/23602801/
CropPAL	Multicrops species	This database contains information on the phytochemicals found in various crops, including fruits, vegetables, grains, and herbs.	https://crop-pal.org/
SoyMetDB	Soybean	This database contains metabolomic data for soybean and related species.	http://soymetdb.org/

Continued

TABLE 18.3 List of publicly available plant panomics databases.—cont'd

Database	Target organism	Brief description	Links
CottonGen Metabolomics	Cotton	This database contains metabolomic data for cotton and related species.	https://www.cottongen.org/
Proteome databases			
PlantPan 2.0	Multiplant species	It is a comprehensive pan-genomic database that integrates genomic, transcriptomic, and proteomic data from 13 plant species.	http://plantpan2.itps.ncku.edu.tw/
Uniprot Plant Proteome Portal	Multiplant species	Even though there are still many experimental and technical design-related problems to be solved, attention needs to be directed toward the development of cutting-edge software tools and databases, and more importantly, the sharing of these resources. It provides access to proteomic data for a wide range of plant species.	https://www.uniprot.org/help/Plants
Phytozome	Multiplant species	It is a comprehensive database of plant genomes, transcriptomes, and proteomes. It includes data from over 30 plant species.	https://phytozome-next.jgi.doe.gov/
PRIDE	Multiplant species	The PRIDE database is a public repository of proteomic data from a variety of organisms, including plants.	https://www.ebi.ac.uk/pride/
CropPAL	Multiplant species	It is a pan-crop proteome database that contains proteomic data from over 20 crop species, including maize, rice, wheat, and soybean.	https://crop-pal.org/
CropEKB	Multiplant species	The Crop EST Knowledge Base is a comprehensive database of EST data from over 70 crop species, including rice, wheat, maize, and sorghum.	
CropQuant	Multiplant species	It provides quantitative information on the abundance of proteins in different crop species. It includes data from over 10 crops, including wheat, maize, and soybean.	https://www.agri-tech-e.co.uk/tag/cropquant/#
Soybean Proteome Database	Soybean	It is a comprehensive resource for studying the soybean proteome. It includes information on protein function, localization, and expression.	http://proteome.dc.affrc.go.jp/Soybean/
MaizeGDB	Maize	It is a resource for studying the maize genome and proteome. It includes data on gene expression, functional annotations, and genetic variation.	https://www.maizegdb.org/

TABLE 18.4 Databases available for crop multiomics analysis.

Database	Crop species	Features and functionality	Availability/URL
BBDG	Blueberry	For Genomics and transcriptomics data	http://bioinformatics.towson.edu/BBGD/
BRAD	*Brassica* species	For Genomics and transcriptomics data	http://brassicadb.org/brad/
CerealsDB	Wheat	Functional genomics related data	http://www.cerealsdb.uk.net/cerealgeno/mics/
CottonFGD	Cotton	Functional genomics data	https://cottonfgd.org/
CottonQTLdb	Cotton	For QTL data	http://www.cottonqtldb.org
CropSNPdb	*Brassica* species and wheat	Collection of SNPs data	http://snpdb.appliedbioinformatics.com.au/

TABLE 18.4 Databases available for crop multiomics analysis.—cont'd

Database	Crop species	Features and functionality	Availability/URL
CSRDB	Maize and rice	Small RNAs database	http://sundarlab.ucdavis.edu/smrnas/
CTDB	Chickpea	Functional genomics and transcriptomics data	http://www.nipgr.ac.in/ctdb.html
GabiPD	Multispecies	Collection of Multiomics data	http://www.gabipd.org/
GrainGenes	Multispecies	Genomics dataset	http://www.graingenes.org
Gramene	Multispecies	Data for comparative functional genomics, transcriptomics, and metabolic pathways	http://www.gramene.org/
KaPPA-View4 KEGG	Multispecies	Collection of transcriptomic data for metabolome and genomic pathway	http://kpv.kazusa.or.jp/kpv4/kegg
KNApSAcK	Multispecies	Metabolomics data	http://kanaya.naist.jp/KNApSAcK/Family/
KOMICS	Multispecies	For metabolomics data	http://www.kazusa.or.jp/komics/en/
MaizeDIG	Maize	Phenomics and genomics dataset	https://maizedig.maizegdb.org/
MaizeGDB	Maize	Collection of multiomics data	https://www.maizegdb.org/
MaizeSNPDB	Maize	SNPs data	https://github.com/venyao/MaizeSNPDB
MMAD	Maize	Microarray dataset	http://maizearrayannot.bi.up.ac.za/
Oryzabase	Rice	Database for genome and integrated biological information	https://shigen.nig.ac.jp/rice/oryzabase/
PCD	Multispecies	Data for Genomics-assisted breeding	https://www.pulsedb.org/
Plant Reactome	Multispecies	Data for genomics, proteomics, transcriptomics, and functional integrated metabolic pathways	http://plants.reactome.org/
PlantGDB	Multispecies	Comparative genomics data	http://www.plantgdb.org/
PlantTFDB	Multispecies	Plant transcriptomic factor data	http://planttfdb.cbi.pku.edu.cn/
PMND	Multispecies	Genomic pathway and metabolome integrated transcriptome data	https://www.plantcyc.org/
PPDB	Multispecies	Proteomics data	http://ppdb.tc.cornell.edu/
RAP-DB	Rice	Genomics integrated multiomics data	http://rapdb.dna.affrc.go.jp/
RGPDB	Maize, soybean, and sorghum	For multiomics data	http://sysbio.unl.edu/RGPDB/
RiceVarMap	Rice	Functional annotation for Genomic variation	http://ricevarmap.ncpgr.cn
RiceXPro	Rice	Transcriptomics and functional genomics data	http://ricexpro.dna.affrc.go.jp/
RicyerDB	Rice	Proteomics and integrative genomics data	http://server.malab.cn/Ricyer/index.html
SFGD	Soybean	Functional genomics, transcriptomics, and metabolic pathways	http://bioinformatics.cau.edu.cn/SFGD/
SFGD	Sunflower	Genomics and transcriptomics data	https://www.sunflowergenome.org
SiFGD	Foxtail millet	Collection of functional genomics integrated transcriptomics, and metabolomics	http://structuralbiology.cau.edu.cn/SIFGD/
SNP-Seek II	Rice	SNP-seek database	http://snp/seek.irri.org
SorghumFDB	Sorghum	Functional genomics data	http://structuralbiology.cau.edu.cn/sorghum/index.html
SorGSD	Sorghum	SNPs data	http://sorgsd.big.ac.cn/
SoyBase	Soybean	Multiomics data for soybean	https://soybase.org/

Continued

TABLE 18.4 Databases available for crop multiomics analysis.—cont'd

Database	Crop species	Features and functionality	Availability/URL
SoyKB	Soybean	Multiomics data for soybean	http://soykb.org/
SoyNet	Soybean	Functional genomics and transcriptomics data for soybean	https://www.inetbio.org/soynet/
TFGD	Tomato	Functional genomics, integrated transcriptomics, and metabolomics	http://ted.bti.cornell.edu/
TOMATOMICS	Tomato	Data for tomato multiomics	http://plantomics.mind.meiji.ac.jp/tomatomics/index.html
wDBTF	Wheat	Transcription factors data	http://wwwappli.nantes.inra.fr:8180/wDBFT/
WheatGPE	Wheat	Phenotype–genotype and environment data	http://www.wheatdb.org/

TABLE 18.5 List of software available for multispecies crop omics analysis.

Software	Features and functionality	Availability/URL
AMDIS	GC-MS data interpretation related to metabolomics	http://www.amdis.net/
BioLeaf	For leaf surface and disease analysis related to phenomics data	http://www.plant/image/analysis.org/software/bioleaf
EasyPCC	For field crop canopy measurement of phenomics data	http://www.plant/image/analysis.org/software/easypcc
GAP4	Analysis for structural genomic sequence assembly	http://stadensourceforge.net/overview.html
Gromacs	Genomics, proteomics and metabolomics	https://omictools.com/gromacs/tool
GSDS	To visualizes gene structure (exons, introns, and UTRs).	http://gsds.cbi.pku.edu.cn/
LemnaLauncher	For the image-based measurements of length, width, surface of seeds and color	http://www.plant/image/analysis.org/software/lemnalauncher
ProteinProspector	Used for sequence mining with MS	http://prospector.ucsf.edu/
PTools	Multiomics analysis	https://omictools.com/ptools/tool
SIMCA-P 14.0	For principal metabolic component analysis of integrated ionomics data	https://umetrics.com/kb/simca/online/140
SPPS	For the prediction of protein–protein interaction partners	http://mdl.shsmu.edu.cn/SPPS/
STRING	For the prediction of protein interactions with their functional associations	http://string.embl.de
VISTA	Comparative genomics analysis	http://genome.lbl.gov/vista/index.shtml

6. Network-based approaches in crop improvement

In biological systems, molecules like DNA, RNA, and proteins interact with one another to build a complex network that regulates a variety of an organism's functions. To properly exploit massive omics datasets for a greater understanding of the plethora of molecular pathways underpinning complex biological features is the main issue of the postgenomics era (Yadav et al., 2021). Such knowledge will ultimately aid in the engineering of plant improvement or the breeding of germplasm (Chen, Wang, et al., 2019). To integrate global measurements at various molecular levels and develop models representing biological systems, a network-based analysis is a useful strategy (Ko & Brandizzi, 2020). In order to comprehend the mechanisms underpinning biological systems, this method stresses understanding the interactions between molecular components (for example, genes and TFs) rather than the function of components alone. Transcriptional regulation in plants is one of the scientific fields where network-based strategies are used a lot. Transcriptional networks can be utilized to make in silico predictions, develop hypotheses, and give instructions for carrying out in vivo studies.

In order to achieve this, RiceNet, a genome-scale gene network for rice, was built and experimentally validated. RiceNet accurately predicted gene functions in monocotyledonous species (Lee et al., 2011). Additionally, a coexpressed gene network for barley (*Hordeum vulgare* L.) created using transcriptome data showed gene clusters linked to the response to drought stress and cellulose biosynthesis (Mochida et al., 2011). A flowering gene network in soybeans could reveal the controlling functions of GmCOL1a and GmCOL1b in flowering (Wu et al., 2019). Gene coexpression networks were created using weighted-gene coexpression network analysis in order to identify regulatory networks and key genes controlling seed set and size (Du et al., 2017) and nodulation and nitrogen fixation (Wu et al., 2019) in soybean, as well as seed set and pollen fertility in pigeonpea (Pazhamala et al., 2017) and acquisition of desiccation tolerance in *Boea hygrometrica* (Lin et al., 2019) among many other plants. A metabolic network was used in a study to clearly show the impact of various concentrations of plant growth regulators and the agro-ecosystem environment on the tomato metabolome (Fatima et al., 2016). Using integrated omics techniques, complex network interactions in nitrogen metabolism and signaling in crop plants have been discovered (Fukushima & Kusano, 2014). To determine the link between a gene and a metabolite in tobacco (*Nicotiana tabacum* L), coexpression gene modules and metabolite modules were combined. Furthermore, gene regulatory networks (GRNs) help us comprehend how a plant's genotype and environment interact to determine the physiological responses that follow. GRNs are a reliable method for categorizing genes with or associated with a specific biological process or pathway (Bulbul Ahmed & Humayan Kabir, 2022). For the purpose of creating crops resistant to abiotic stress, transcription factors (TFs) can be crucial targets. Therefore, it is crucial to understand how various biological processes, such as growth, development, or stress responses, are transcriptionally controlled in plants. This can be done by effectively identifying regulatory links between TF and target gene in order to delineate GRNs. High-throughput assays have been employed in numerous research to systematically investigate the GRNs underlying particular circumstances in the model plant *Arabidopsis thaliana* (Kulkarni & Vandepoele, 2020). Understanding GRN can be useful in modifying breeding strategies. A comprehensive review was performed by Bulbul Ahmed and Humayan Kabir (2022), that represents the various aspects of GRNs in crop improvement (Bulbul Ahmed & Humayan Kabir, 2022).

7. Artificial intelligence in crop improvement

The "omics" approaches have given plant scientists the precise instruments to quickly assess the critical agronomic features for larger-sized germplasm in the early growth phases. AI has emerged as a cutting-edge field that has the potential to significantly improve crop breeding (Khan et al., 2022). Breeding crops are capable of coping with numerous biotic and abiotic challenges while maintaining or improving agricultural yields and quality under a variety of climatic and environmental variations.

The rising interest in AI in the breeding industry is a result of the technological maturity gained, or the capacity to quickly and efficiently analyze massive amounts of data to uncover unexpected relationships. AI reduces processing and identification times for data, which is a significant advantage. Marker-assisted selection, genomic selection, and genomic prediction have all seen an increase in the usage of AI techniques (Budhlakoti et al., 2022; Sandhu et al., 2022). Utilizing modern tools and information systems boost agriculture's total production (El Bilali & Allahyari, 2018). Research on AI technologies, including machine learning, deep learning, and predictive analysis, aims to create a machine that can replicate human thought processes and actions, such as planning, learning, reasoning, thinking, and action-taking abilities (Shaw et al., 2019). Systems are being created by plant breeders to help with a better knowledge of how plants behave in various climatic conditions. AI has the ability to revolutionize agriculture in various ways (Fig. 18.2) and ensure global food security in the near future (Streich et al., 2020).

Recent research has demonstrated that crop phenotyping using AI improves crop phenotyping and predictions (Nabwire et al., 2021). Additionally, AI-assisted high-throughput phenotyping systems have been successfully used in the following fields: oilseed crops for semantic segmentation of the crops and weeds (Abdalla et al., 2019), the phenotyping of disease resistance of crops (Mochida et al., 2019), wheat and maize to identify the plant growth stage (Sadeghi-Tehran et al., 2017), plant image segmentation (Brichet et al., 2017), and the phenotyping of disease resistance of crops (Mahlein et al., 2019) to improvement of plant productivity. AI algorithms could be applied to comparative genomic studies or the transfer of knowledge from a model plant to a crop of interest (Caudai et al., 2021). To predict and assess the genetic traits, two deep learning models have recently been developed such as DeepSEA (Zhou & Troyanskaya, 2015) and DeepBind (Alipanahi et al., 2015). Additional deep learning techniques have been used in genomic selection to speed up the breeding cycle and enable quick selection of superior genotypes. Deep learning algorithms have the potential to capture complicated higher-order interactions and achieve higher predictability, which makes them advantageous for studying genomic

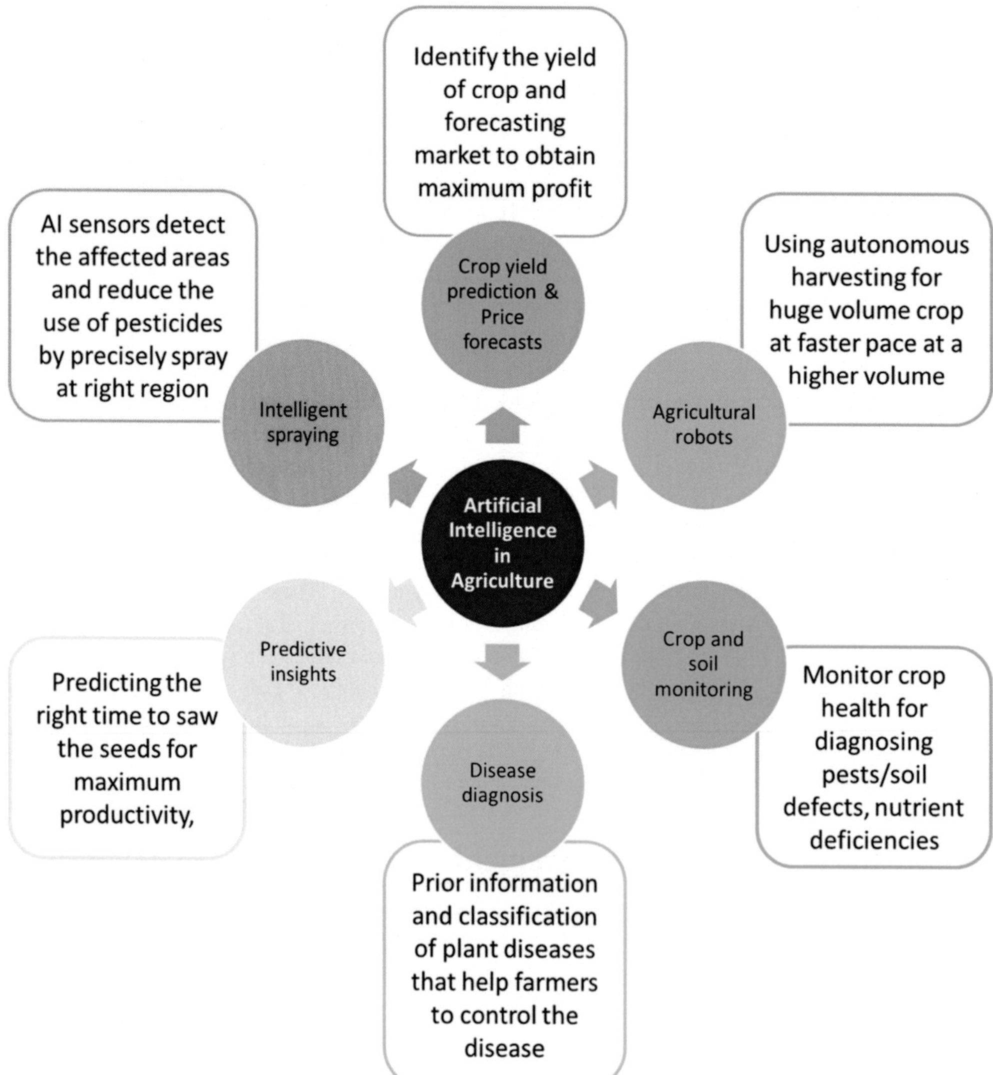

FIGURE 18.2 Application of artificial intelligence agriculture to improve the crop production.

selection (González-Camacho et al., 2012). Convolutional neural networks (CNNs) were found to outperform other deep learning techniques when it comes to predicting crop phenotypes from genotypic data (Pérez-Enciso & Zingaretti, 2019). When used to forecast features linked to grains in a population of wheat, the CNN-based model DeepGS performed best when compared to other statistical models, such as rrBLUP (Ma et al., 2018). The performance of a dual-stream CNN strategy, which was proposed in a different study, demonstrated improved results when compared to a single-stream CNN approach for predicting five traits in soybean (Yang Liu et al., 2019). In classification situations, deep learning techniques are also used to predict phenotypes (Montesinos-López et al., 2019). In maize and wheat populations, the probabilistic neural network (PNN), which forecasts the likelihood that genotypes will belong to good or bad groups, performed better than multilayer perceptron models, leading researchers to draw the conclusion that the PNN is a promising method of genomic selection in crop breeding (González-Camacho et al., 2012). A study suggested a deep learning calibration method for genomic selection that outperformed more established deep learning techniques (Montesinos-López et al., 2021). At present, modern breeding strategies are concentrated on correctly and precisely matching the genotype with the crop phenotype. A lot of data are produced by crop diversity, SNP detection and selection, QTL analysis, genome-wide association study analysis, genomic selection, and sequences; AI can assess and link the phenomics and genomics data from these big data to improve breeding methods. As a result, the use of AI with phenomics and genomics techniques can facilitate the quick discovery of genes linked to agricultural phenotypes, which will ultimately speed up crop development programs.

8. Challenges and complexities associated with OMICS-based approach for crop improvement

OMICS technologies (such as genomes, transcriptomics, proteomics, metabolomics, and epigenomics) are becoming more common in the field of plant sciences due to lower sequencing costs and improved knowledge. Therefore, independent of the availability of the genome sequence, the discovery of novel alleles from various sources was made possible by the sequencing and resequencing information that is generated for numerous crops. Furthermore, the quick development of omics technology offers the chance to provide fresh, useful datasets on many plant species. The identification of genes and pathways responsible for significant agronomic features would result from an effective integration of these genomes and functional omics data with genetic and phenotypic information. But due to the inherent data discrepancies among them, integrating several omics platforms continues to be difficult. The most important multiomics data collection for systems modeling and prediction, for which there are no accepted worldwide standards, is data on various molecular and functional components (Macklin, 2019). Additionally, the use of model-based integration is frequently limited to specific model organisms (Pinu et al., 2019). The study of multiomics data is especially difficult since it necessitates not only the analysis of many omics data sets but also the discovery of intricate relationships between various genetic components. In addition, quite a bit of computing infrastructure is required for the analysis of multiomics data. Additionally, it frequently necessitated repeating the study from several angles in order to evaluate the results of the analysis, which required coordination among numerous scientists.

Different studies provide a thorough description of a variety of technological difficulties encountered during testing, data creation, integrations, management, sharing, and analysis (Macklin, 2019; Misra et al., 2019). Though modern technologies like AI, system and network biology approaches have enormous potential to change agricultural research by helping us better understand how plants respond to environmental constraints and development (Muthuramalingam et al., 2019; Talaviya et al., 2020). Although there are promising possibilities, implementing a really AI-assisted agricultural enhancement has many difficulties. For major grain crops, high-throughput phenotyping is currently regarded as a crucial part of crop improvement (Singh et al., 2019). Nevertheless, the genotype–phenotype association analysis via AI cannot be completed without the phenotype. Researchers face difficulties because of the high-throughput phenotyping's extensive applicability in vegetable and specialty crop development. Despite the advantages of AI for sustainable farming, there are still some serious challenges, such as a lack of familiarity with AI machines, lack of experience with emerging technologies and privacy and security issues for using AI.

9. Conclusion and future prospects

The growing human population and climate change are imposing unprecedented challenges for the global food supply. To cope with these pressures, crop improvement demands enhancing agronomical traits such as yield, resistance, and nutritional value by applying advanced approaches. Various omics approaches have employed high-throughput techniques to dissect significant developmental clues and gene characterization that help in crop improvement. These omics approaches along with high-throughput phenotyping are providing information to researchers and farmers that are required for advanced breeding methods. In the coming years, strategies need to be developed to maximize the limited resources and their utilization for crop improvement. Hence, there is an urgent need to direct future research toward the discovery of key players, molecular networks, and AI-based models that could unravel the complex cellular and molecular functions to enhance agronomic traits in crops. Systems biology and multiomics approach provide a more inclusive molecular perspective of plant biology than individual approaches. However, the constant evolution of databases and data analysis tools would help meaningful biological interpretation of multiomics data. Moreover, it is high time to form a highly connected community for crop improvement research, instead of isolated efforts. Even though there are still many experimental and technical design-related problems to be solved, attention needs to be directed toward the development of cutting-edge software tools and databases, and more importantly, the sharing of these resources. This is even more crucial because of the urgency to double crop yields by 2050 under uncertain climate scenarios. From this perspective, we are confident this chapter will provide an overview to readers about the role of omics technologies along with system biology and AI-based approaches that significantly contribute to crop improvement.

References

Abdalla, A., Cen, H., Wan, L., Rashid, R., Weng, H., Zhou, W., & He, Y. (2019). Fine-tuning convolutional neural network with transfer learning for semantic segmentation of ground-level oilseed rape images in a field with high weed pressure. *Computers and Electronics in Agriculture, 167*. https://doi.org/10.1016/j.compag.2019.105091

Abdullah-Zawawi, M. R., Govender, N., Harun, S., Muhammad, N. A. N., Zainal, Z., & Mohamed-Hussein, Z. A. (2022). Multi-omics approaches and resources for systems-level gene function prediction in the plant kingdom. *Plants, 11*(19). https://doi.org/10.3390/plants11192614. www.mdpi.com/journal/plants

Abid, F., Zahid, M. A., Abedin, Z. U., Nizami, S. B., Abid, M. J., Kazmi, S. Z. H., Khan, S. U., Hasan, H., Ali, M., & Gul, A. (2018). Omics approaches in marine biotechnology: The treasure of ocean for human betterments. *Omics Technologies and Bio-Engineering: Towards Improving Quality of Life, 1*, 47–61. https://doi.org/10.1016/B978-0-12-804659-3.00003-8. http://www.sciencedirect.com/science/book/9780128046593

Ali, A., Altaf, M. T., Nadeem, M. A., Karaköy, T., Shah, A. N., Azeem, H., Baloch, F. S., Baran, N., Hussain, T., Duangpan, S., Aasim, M., Boo, K. H., Abdelsalam, N. R., Hasan, M. E., & Chung, Y. S. (2022). Recent advancement in OMICS approaches to enhance abiotic stress tolerance in legumes. *Frontiers in Plant Science, 13*. https://doi.org/10.3389/fpls.2022.952759. https://www.frontiersin.org/journals/plant-science

Alipanahi, B., Delong, A., Weirauch, M. T., & Frey, B. J. (2015). Predicting the sequence specificities of DNA- and RNA-binding proteins by deep learning. *Nature Biotechnology, 33*(8), 831–838. https://doi.org/10.1038/nbt.3300. http://www.nature.com/nbt/index.html

Ambrosino, L., Colantuono, C., Diretto, G., Fiore, A., & Chiusano, M. L. (2020). Bioinformatics resources for plant abiotic stress responses: State of the art and opportunities in the fast evolving-omics era. *Plants, 9*(5). https://doi.org/10.3390/plants9050591

Arbona, V., Manzi, M., de Ollas, C., & Gómez-Cadenas, A. (2013). Metabolomics as a tool to investigate abiotic stress tolerance in plants. *International Journal of Molecular Sciences, 14*(3), 4885–4911. https://doi.org/10.3390/ijms14034885. http://www.mdpi.com/1422-0067/14/3/4885/pdf

Baldwin, T. T., Zitomer, N. C., Mitchell, T. R., Zimeri, A. M., Bacon, C. W., Riley, R. T., & Glenn, A. E. (2014). Maize seedling blight induced by fusarium verticillioides: Accumulation of fumonisin B1 in leaves without colonization of the leaves. *Journal of Agricultural and Food Chemistry, 62*(9), 2118–2125. https://doi.org/10.1021/jf5001106

Bao, Y., Aggarwal, P., Robbins, N. E., Sturrock, C. J., Thompson, M. C., Tan, H. Q., Tham, C., Duan, L., Rodriguez, P. L., Vernoux, T., Mooney, S. J., Bennett, M. J., & Dinneny, J. R. (2014). Plant roots use a patterning mechanism to position lateral root branches toward available water. *Proceedings of the National Academy of Sciences of the United States of America, 111*(25), 9319–9324. https://doi.org/10.1073/pnas.1400966111. http://www.pnas.org/content/111/25/9319.full.pdf+html

Brichet, N., Fournier, C., Turc, O., Strauss, O., Artzet, S., Pradal, C., Welcker, C., Tardieu, F., & Cabrera-Bosquet, L. (2017). A robot-assisted imaging pipeline for tracking the growths of maize ear and silks in a high-throughput phenotyping platform. *Plant Methods, 13*(1). https://doi.org/10.1186/s13007-017-0246-7

Budhlakoti, N., Kushwaha, A. K., Rai, A., Chaturvedi, K. K., Kumar, A., Pradhan, A. K., Kumar, U., Kumar, R. R., Juliana, P., Mishra, D. C., & Kumar, S. (2022). Genomic selection: A tool for accelerating the efficiency of molecular breeding for development of climate-resilient crops. *Frontiers in Genetics, 13*. https://doi.org/10.3389/fgene.2022.832153. https://www.frontiersin.org/journals/genetics#

Bulbul Ahmed, M., & Humayan Kabir, A. (2022). Understanding of the various aspects of gene regulatory networks related to crop improvement. *Gene, 833*. https://doi.org/10.1016/j.gene.2022.146556

Byeon, Y. S., Hong, Y. S., Kwak, H. S., Lim, S. T., & Kim, S. S. (2022). Metabolite profile and antioxidant potential of wheat (*Triticum aestivum* L.) during malting. *Food Chemistry, 384*. https://doi.org/10.1016/j.foodchem.2022.132443. www.elsevier.com/locate/foodchem

Caudai, C., Galizia, A., Geraci, F., Le Pera, L., Morea, V., Salerno, E., Via, A., & Colombo, T. (2021). AI applications in functional genomics. *Computational and Structural Biotechnology Journal, 19*, 5762–5790. https://doi.org/10.1016/j.csbj.2021.10.009

Chekanova, J. A. (2015). Long non-coding RNAs and their functions in plants. *Current Opinion in Plant Biology, 27*, 207–216. https://doi.org/10.1016/j.pbi.2015.08.003. www.elsevier.com/inca/publications/store/6/0/1/3/1/4/index.htt

Chen, D., Shao, Q., Yin, L., Younis, A., & Zheng, B. (2019). Polyamine function in plants: Metabolism, regulation on development, and roles in abiotic stress responses. *Frontiers in Plant Science, 9*. https://doi.org/10.3389/fpls.2018.01945

Chen, K., Wang, Y., Zhang, R., Zhang, H., & Gao, C. (2019). CRISPR/Cas genome editing and precision plant breeding in agriculture. *Annual Review of Plant Biology, 70*(1), 667–697. https://doi.org/10.1146/annurev-arplant-050718-100049

Cramer, G. R., Urano, K., Delrot, S., Pezzotti, M., & Shinozaki, K. (2011). Effects of abiotic stress on plants: A systems biology perspective. *BMC Plant Biology, 11*(1). https://doi.org/10.1186/1471-2229-11-163

Daldoul, S., Ben Amar, A., Guillaumie, S., & Mliki, A. (2016). Integration of omics and system biology approaches to study grapevine (*Vitis vinifera* L.) response to salt stress: A perspective for functional genomics—A review. *OENO One, 48*(3). https://doi.org/10.20870/oeno-one.2014.48.3.1573

Das, A., Rushton, P. J., & Rohila, J. S. (2017). Metabolomic profiling of soybeans (Glycine max L.) reveals the importance of sugar and nitrogen metabolism under drought and heat stress. *Plants, 6*(2), 199–208. https://doi.org/10.3390/plants6020021. http://www.mdpi.com/2223-7747/6/2/21/pdf

Derakhshani, B., Ayalew, H., Mishina, K., Tanaka, T., Kawahara, Y., Jafary, H., & Oono, Y. (2020). Comparative analysis of root transcriptome reveals candidate genes and expression divergence of homoeologous genes in response to water stress in wheat. *Plants, 9*(5). https://doi.org/10.3390/plants9050596

Du, J., Wang, S., He, C., Zhou, B., Ruan, Y. L., & Shou, H. (2017). Identifcation of regulatory networks and hub genes controlling soybean seed set and size using RNA sequencing analysis. *Journal of Experimental Botany, 68*(8), 1955–1972. https://doi.org/10.1093/jxb/erw460. http://jxb.oxfordjournals.org/

El Bilali, H., & Allahyari, M. S. (2018). Transition towards sustainability in agriculture and food systems: Role of information and communication technologies. *Information Processing in Agriculture, 5*(4), 456–464. https://doi.org/10.1016/j.inpa.2018.06.006. http://www.elsevier.com/journals/information-processing-in-agriculture/2214-3173#

Fahlgren, N., Gehan, M. A., & Baxter, I. (2015). Lights, camera, action: High-throughput plant phenotyping is ready for a close-up. *Current Opinion in Plant Biology, 24*, 93–99. https://doi.org/10.1016/j.pbi.2015.02.006. www.elsevier.com/inca/publications/store/6/0/1/3/1/4/index.htt

Fatima, T., Sobolev, A. P., Teasdale, J. R., Kramer, M., Bunce, J., Handa, A. K., & Mattoo, A. K. (2016). Fruit metabolite networks in engineered and non-engineered tomato genotypes reveal fluidity in a hormone and agroecosystem specific manner. *Metabolomics, 12*(6). https://doi.org/10.1007/s11306-016-1037-2. http://www.kluweronline.com/issn/1573-3890/

Frey, F. P., Pitz, M., Schön, C. C., & Hochholdinger, F. (2020). Transcriptomic diversity in seedling roots of European flint maize in response to cold. *BMC Genomics, 21*(1). https://doi.org/10.1186/s12864-020-6682-1. http://www.biomedcentral.com/bmcgenomics

Fukushima, A., & Kusano, M. (2014). A network perspective on nitrogen metabolism from model to crop plants using integrated 'omics' approaches. *Journal of Experimental Botany, 65*(19), 5619–5630. https://doi.org/10.1093/jxb/eru322

Gao, C. (2021). Genome engineering for crop improvement and future agriculture. *Cell, 184*(6), 1621–1635. https://doi.org/10.1016/j.cell.2021.01.005

Goche, T., Shargie, N. G., Cummins, I., Brown, A. P., Chivasa, S., & Ngara, R. (2020). Comparative physiological and root proteome analyses of two sorghum varieties responding to water limitation. *Scientific Reports, 10*(1). https://doi.org/10.1038/s41598-020-68735-3. www.nature.com/srep/index.html

González-Camacho, J. M., de los Campos, G., Pérez, P., Gianola, D., Cairns, J. E., Mahuku, G., Babu, R., & Crossa, J. (2012). Genome-enabled prediction of genetic values using radial basis function neural networks. *Theoretical and Applied Genetics, 125*(4), 759–771. https://doi.org/10.1007/s00122-012-1868-9

Grote, U., Fasse, A., Nguyen, T. T., & Erenstein, O. (2021). Food security and the dynamics of wheat and maize value chains in Africa and Asia. *Frontiers in Sustainable Food Systems, 4*. https://doi.org/10.3389/fsufs.2020.617009. www.frontiersin.org/journals/sustainable-food-systems#

Guo, J., Huang, Z., Sun, J., Cui, X., & Liu, Y. (2021). Research progress and future development trends in medicinal plant transcriptomics. *Frontiers in Plant Science, 12*. https://doi.org/10.3389/fpls.2021.691838

Gupta, B. (2013). Plant abiotic stress: 'Omics' approach. *Journal of Plant Biochemistry a Physiology, 01*(03). https://doi.org/10.4172/2329-9029.1000e108

Habash, D. Z., Kehel, Z., & Nachit, M. (2009). Genomic approaches for designing durum wheat ready for climate change with a focus on drought. *Journal of Experimental Botany, 60*(10), 2805–2815. https://doi.org/10.1093/jxb/erp211

Ismail, A. M., Heuer, S., Thomson, M. J., & Wissuwa, M. (2007). Genetic and genomic approaches to develop rice germplasm for problem soils. *Plant Molecular Biology, 65*(4), 547–570. https://doi.org/10.1007/s11103-007-9215-2

Jamil, I. N., Remali, J., Azizan, K. A., Nor Muhammad, N. A., Arita, M., Goh, H. H., & Aizat, W. M. (2020). Systematic multi-omics integration (MOI) approach in plant systems biology. *Frontiers in Plant Science, 11*. https://doi.org/10.3389/fpls.2020.00944. https://www.frontiersin.org/journals/plant-science

Jin, S., Sun, X., Wu, F., Su, Y., Li, Y., Song, S., Xu, K., Ma, Q., Baret, F., Jiang, D., Ding, Y., & Guo, Q. (2021). Lidar sheds new light on plant phenomics for plant breeding and management: Recent advances and future prospects. *ISPRS Journal of Photogrammetry and Remote Sensing, 171*, 202–223. https://doi.org/10.1016/j.isprsjprs.2020.11.006

Kang, W., Zhu, X., Wang, Y., Chen, L., & Duan, Y. (2018). Transcriptomic and metabolomic analyses reveal that bacteria promote plant defense during infection of soybean cyst nematode in soybean. *BMC Plant Biology, 18*(1). https://doi.org/10.1186/s12870-018-1302-9

Katam, R., Shokri, S., Murthy, N., Singh, S. K., Suravajhala, P., Khan, M. N., Bahmani, M., Sakata, K., & Reddy, K. R. (2020). Proteomics, physiological, and biochemical analysis of cross tolerance mechanisms in response to heat and water stresses in soybean. *PLoS One, 15*(6). https://doi.org/10.1371/journal.pone.0233905. https://journals.plos.org/plosone/article/file?id=10.1371/journal.pone.0233905&type=printable

Khan, M. H. U., Wang, S., Wang, J., Ahmar, S., Saeed, S., Khan, S. U., Xu, X., Chen, H., Bhat, J. A., & Feng, X. (2022). Applications of artificial intelligence in climate-resilient smart-crop breeding. *International Journal of Molecular Sciences, 23*(19). https://doi.org/10.3390/ijms231911156. http://www.mdpi.com/journal/ijms

Khan, N., Bano, A., Rahman, M. A., Rathinasabapathi, B., & Babar, M. A. (2019). UPLC-HRMS-based untargeted metabolic profiling reveals changes in chickpea (Cicer arietinum) metabolome following long-term drought stress. *Plant, Cell and Environment, 42*(1), 115–132. https://doi.org/10.1111/pce.13195. http://onlinelibrary.wiley.com/journal/10.1111/(ISSN)1365-3040

Ko, D. K., & Brandizzi, F. (2020). Network-based approaches for understanding gene regulation and function in plants. *The Plant Journal, 104*(2), 302–317. https://doi.org/10.1111/tpj.14940. http://onlinelibrary.wiley.com/journal/10.1111/(ISSN)1365-313X

Konstantinov, D. K., Zubairova, U. S., Ermakov, A. A., & Doroshkov, A. V. (2021). Comparative transcriptome profiling of a resistant vs susceptible bread wheat (*Triticum aestivum* L.) cultivar in response to water deficit and cold stress. *PeerJ, 9*. https://doi.org/10.7717/peerj.11428. https://peerj.com/articles/11428/

Kopittke, P. M., Menzies, N. W., Wang, P., McKenna, B. A., & Lombi, E. (2019). Soil and the intensification of agriculture for global food security. *Environment International, 132*. https://doi.org/10.1016/j.envint.2019.105078. www.elsevier.com/locate/envint

Kulkarni, S. R., & Vandepoele, K. (2020). Inference of plant gene regulatory networks using data-driven methods: A practical overview. *Biochimica et Biophysica Acta (BBA)—Gene Regulatory Mechanisms, 1863*(6). https://doi.org/10.1016/j.bbagrm.2019.194447

Kumar, A., Pathak, R. K., Gupta, S. M., Gaur, V. S., & Pandey, D. (2015). Systems biology for smart crops and agricultural innovation: Filling the gaps between genotype and phenotype for complex traits linked with robust agricultural productivity and sustainability. *OMICS: A Journal of Integrative Biology, 19*(10), 581–601. https://doi.org/10.1089/omi.2015.0106. www.liebertonline.com/omi

López-Gresa, M. P., Maltese, F., Bellés, J. M., Conejero, V., Kim, H. K., Choi, Y. H., & Verpoorte, R. (2010). Metabolic response of tomato leaves upon different plant-pathogen interactions. *Phytochemical Analysis, 21*(1), 89–94. https://doi.org/10.1002/pca.1179. http://www3.interscience.wiley.com/cgi-bin/fulltext/122662804/PDFSTART

Lahuta, L. B., Szablińska-Piernik, J., Stałanowska, K., Głowacka, K., & Horbowicz, M. (2022). The size-dependent effects of silver nanoparticles on germination, early seedling development and polar metabolite profile of wheat (*Triticum aestivum* L.). *International Journal of Molecular Sciences, 23*(21). https://doi.org/10.3390/ijms232113255

Lee, I., Seo, Y. S., Coltrane, D., Hwang, S., Oh, T., Marcotte, E. M., & Ronald, P. C. (2011). Genetic dissection of the biotic stress response using a genome-scale gene network for rice. *Proceedings of the National Academy of Sciences of the United States of America, 108*(45), 18548–18553. https://doi.org/10.1073/pnas.1110384108. http://www.pnas.org/content/108/45/18548.full.pdf+html

Lee, J. Y., & Lee, D. H. (2003). Use of serial analysis of gene expression technology to reveal changes in gene expression in Arabidopsis pollen undergoing cold stress. *Plant Physiology, 132*(2), 517–529. https://doi.org/10.1104/pp.103.020511. http://www.plantphysiol.org/

Liang, Y., Tabien, R. E., Tarpley, L., Mohammed, A. R., & Septiningsih, E. M. (2021). Transcriptome profiling of two rice genotypes under mild field drought stress during grain-filling stage. *AoB Plants, 13*(4). https://doi.org/10.1093/aobpla/plab043. http://aobpla.oxfordjournals.org/

Lin, C. T., Xu, T., Xing, S. L., Zhao, L., Sun, R. Z., Liu, Y., Moore, J. P., & Deng, X. (2019). Weighted Gene Co-expression Network Analysis (WGCNA) reveals the hub role of protein ubiquitination in the acquisition of desiccation tolerance in boea hygrometrica. *Plant and Cell Physiology, 60*(12), 2707–2719. https://doi.org/10.1093/pcp/pcz160. http://pcp.oxfordjournals.org/

Liu, N. Y., Ko, S. S., Yeh, K. C., & Charng, Y. Y. (2006). Isolation and characterization of tomato Hsa32 encoding a novel heat-shock protein. *Plant Science, 170*(5), 976–985. https://doi.org/10.1016/j.plantsci.2006.01.008

Lu, Y., Li, R., Wang, R., Wang, X., Zheng, W., Sun, Q., Tong, S., Dai, S., & Xu, S. (2017). Comparative proteomic analysis of flag leaves reveals new insight into wheat heat adaptation. *Frontiers in Plant Science, 8*. https://doi.org/10.3389/fpls.2017.01086

Luo, Q., Teng, W., Fang, S., Li, H., Li, B., Chu, J., Li, Z., & Zheng, Q. (2019). Transcriptome analysis of salt-stress response in three seedling tissues of common wheat. *The Crop Journal, 7*(3), 378–392. https://doi.org/10.1016/j.cj.2018.11.009

Ma, W., Qiu, Z., Song, J., Li, J., Cheng, Q., Zhai, J., & Ma, C. (2018). A deep convolutional neural network approach for predicting phenotypes from genotypes. *Planta, 248*(5), 1307–1318. https://doi.org/10.1007/s00425-018-2976-9

Mace, E., Innes, D., Hunt, C., Wang, X., Tao, Y., Baxter, J., Hassall, M., Hathorn, A., & Jordan, D. (2019). The sorghum QTL atlas: A powerful tool for trait dissection, comparative genomics and crop improvement. *Theoretical and Applied Genetics, 132*(3), 751–766. https://doi.org/10.1007/s00122-018-3212-5

Macklin, P. (2019). Key challenges facing data-driven multicellular systems biology. *GigaScience, 8*(10). https://doi.org/10.1093/gigascience/giz127

Mahlein, A. K., Kuska, M. T., Thomas, S., Wahabzada, M., Behmann, J., Rascher, U., & Kersting, K. (2019). Quantitative and qualitative phenotyping of disease resistance of crops by hyperspectral sensors: Seamless interlocking of phytopathology, sensors, and machine learning is needed! *Current Opinion in Plant Biology, 50*, 156–162. https://doi.org/10.1016/j.pbi.2019.06.007. www.elsevier.com/inca/publications/store/6/0/1/3/1/4/index.htt

Matsumura, H., Nirasawa, S., & Terauchi, R. (1999). Transcript profiling in rice (Oryza sativa L.) seedlings using serial analysis of gene expression (SAGE). *The Plant Journal, 20*(6), 719–726. https://doi.org/10.1046/j.1365-313x.1999.00640.x

McCouch, S. R., Chen, X., Panaud, O., Temnykh, S., Xu, Y., Cho, Y. G., Huang, N., Ishii, T., & Blair, M. (1997). Microsatellite marker development, mapping and applications in rice genetics and breeding. *Plant Molecular Biology, 35*(1–2), 89–99. https://doi.org/10.1007/978-94-011-5794-0_9. https://link.springer.com/journal/11103

Misra, B. B., Langefeld, C., Olivier, M., & Cox, L. A. (2019). Integrated omics: Tools, advances and future approaches. *Journal of Molecular Endocrinology, 62*(1), R21–R45. https://doi.org/10.1530/JME-18-0055. https://jme.bioscientifica.com/downloadpdf/journals/jme/62/1/JME-18-0055.xml

Mochida, K., Koda, S., Inoue, K., Hirayama, T., Tanaka, S., Nishii, R., & Melgani, F. (2019). Computer vision-based phenotyping for improvement of plant productivity: A machine learning perspective. *GigaScience, 8*(1). https://doi.org/10.1093/gigascience/giy153

Mochida, K., Uehara-Yamaguchi, Y., Yoshida, T., Sakurai, T., & Shinozaki, K. (2011). Global landscape of a co-expressed gene network in barley and its application to gene discovery in Triticeae crops. *Plant and Cell Physiology, 52*(5), 785–803. https://doi.org/10.1093/pcp/pcr035

Montesinos-López, O. A., Martín-Vallejo, J., Crossa, J., Gianola, D., Hernández-Suárez, C. M., Montesinos-López, A., Juliana, P., & Singh, R. (2019). New deep learning genomic-based prediction model for multiple traits with binary, ordinal, and continuous phenotypes. *G3: Genes, Genomes, Genetics, 9*(5), 1545–1556. https://doi.org/10.1534/g3.119.300585. http://www.g3journal.org/

Montesinos-López, O. A., Montesinos-López, A., Mosqueda-González, B. A., Bentley, A. R., Lillemo, M., Varshney, R. K., & Crossa, J. (2021). A new deep learning calibration method enhances genome-based prediction of continuous crop traits. *Frontiers in Genetics, 12*. https://doi.org/10.3389/fgene.2021.798840. https://www.frontiersin.org/journals/genetics#

Mosa, K. A., Ismail, A., & Helmy, M. (2017). *Omics and system biology approaches in plant stress research* (pp. 21–34). Springer Science and Business Media LLC. https://doi.org/10.1007/978-3-319-59379-1_2

Mu, Q., Zhang, W., Zhang, Y., Yan, H., Liu, K., Matsui, T., Tian, X., & Yang, P. (2017). iTRAQ-based quantitative proteomics analysis on rice anther responding to high temperature. *International Journal of Molecular Sciences, 18*(9). https://doi.org/10.3390/ijms18091811

Muthamilarasan, M., Singh, N. K., & Prasad, M. (2019). Multi-omics approaches for strategic improvement of stress tolerance in underutilized crop species: A climate change perspective. *Advances in Genetics, 103*, 1–38. https://doi.org/10.1016/bs.adgen.2019.01.001. http://www.elsevier.com/wps/find/bookdescription.cws_home/703716/description#description

Muthuramalingam, P., Jeyasri, R., Krishnan, S. R., Pandian, S. T. K., Sathishkumar, R., & Ramesh, M. (2019). Integrating the bioinformatics and omics tools for systems analysis of abiotic stress tolerance in Oryza sativa (L.). *Advances in Plant Transgenics: Methods and Applications*, 59–77. https://doi.org/10.1007/978-981-13-9624-3

Nabwire, S., Suh, H. K., Kim, M. S., Baek, I., & Cho, B. K. (2021). Review: Application of artificial intelligence in phenomics. *Sensors, 21*(13). https://doi.org/10.3390/s21134363. https://www.mdpi.com/1424-8220/21/13/4363/pdf

Ogbaga, C. C., Stepien, P., Dyson, B. C., Rattray, N. J. W., Ellis, D. I., Goodacre, R., & Johnson, G. N. (2016). Biochemical analyses of sorghum varieties reveal differential responses to drought. *PLoS One, 11*(5). https://doi.org/10.1371/journal.pone.0154423. http://journals.plos.org/plosone/article/asset?id=10.1371%2Fjournal.pone.0154423.pdf

Pérez-Enciso, M., & Zingaretti, L. M. (2019). A guide for using deep learning for complex trait genomic prediction. *Genes, 10*(7). https://doi.org/10.3390/genes10070553

Pan, Y., Liang, H., Gao, L., Dai, G., Chen, W., Yang, X., Qing, D., Gao, J., Wu, H., Huang, J., Zhou, W., Huang, C., Liang, Y., & Deng, G. (2020). Transcriptomic profiling of germinating seeds under cold stress and characterization of the cold-tolerant gene LTG5 in rice. *BMC Plant Biology, 20*(1). https://doi.org/10.1186/s12870-020-02569-z

Pang, Q., Chen, S., Dai, S., Chen, Y., Wang, Y., & Yan, X. (2010). Comparative proteomics of salt tolerance in *Arabidopsis thaliana* and Thellungiella halophila. *Journal of Proteome Research, 9*(5), 2584–2599. https://doi.org/10.1021/pr100034f

Pazhamala, L. T., Purohit, S., Saxena, R. K., Garg, V., Krishnamurthy, L., Verdier, J., & Varshney, R. K. (2017). Gene expression atlas of pigeonpea and its application to gain insights into genes associated with pollen fertility implicated in seed formation. *Journal of Experimental Botany, 68*(8), 2037–2054. https://doi.org/10.1093/jxb/erx010. http://jxb.oxfordjournals.org/

Petrozza, A., Santaniello, A., Summerer, S., Di Tommaso, G., Di Tommaso, D., Paparelli, E., Piaggesi, A., Perata, P., & Cellini, F. (2014). Physiological responses to Megafol® treatments in tomato plants under drought stress: A phenomic and molecular approach. *Scientia Horticulturae, 174*(1), 185–192. https://doi.org/10.1016/j.scienta.2014.05.023

Pinu, F. R., Beale, D. J., Paten, A. M., Kouremenos, K., Swarup, S., Schirra, H. J., & Wishart, D. (2019). Systems biology and multi-omics integration: Viewpoints from the metabolomics research community. *Metabolites, 9*(4). https://doi.org/10.3390/metabo9040076. https://www.mdpi.com/2218-1989/9/4/76/pdf

Poroyko, V., Hejlek, L. G., Spollen, W. G., Springer, G. K., Nguyen, H. T., Sharp, R. E., & Bohnert, H. J. (2005). The maize root transcriptome by serial analysis of gene expression. *Plant Physiology, 138*(3), 1700–1710. https://doi.org/10.1104/pp.104.057638

Qian, Y., Ren, Q., Zhang, J., & Chen, L. (2019). Transcriptomic analysis of the maize (*Zea mays* L.) inbred line B73 response to heat stress at the seedling stage. *Gene, 692*, 68–78. https://doi.org/10.1016/j.gene.2018.12.062

Ramakrishnan, A. P., Ritland, C. E., Blas Sevillano, R. H., & Riseman, A. (2015). Review of potato molecular markers to enhance trait selection. *American Journal of Potato Research, 92*(4), 455–472. https://doi.org/10.1007/s12230-015-9455-7. http://springerlink.com/content/1099-209X

Rao, A., Barkley, D., França, G. S., & Yanai, I. (2021). Exploring tissue architecture using spatial transcriptomics. *Nature, 596*(7871), 211–220. https://doi.org/10.1038/s41586-021-03634-9. http://www.nature.com/nature/index.html

Raza, A., Tabassum, J., Kudapa, H., & Varshney, R. K. (2021). Can omics deliver temperature resilient ready-to-grow crops? *Critical Reviews in Biotechnology, 41*(8), 1209–1232. https://doi.org/10.1080/07388551.2021.1898332. http://www.tandfonline.com/loi/ibty20

Ritchie, M. D., Holzinger, E. R., Li, R., Pendergrass, S. A., & Kim, D. (2015). Methods of integrating data to uncover genotype-phenotype interactions. *Nature Reviews Genetics, 16*(2), 85–97. https://doi.org/10.1038/nrg3868. http://www.nature.com/reviews/genetics

Sadeghi-Tehran, P., Sabermanesh, K., Virlet, N., & Hawkesford, M. J. (2017). Automated method to determine two critical growth stages of wheat: Heading and flowering. *Frontiers in Plant Science, 8*. https://doi.org/10.3389/fpls.2017.00252. http://journal.frontiersin.org/article/10.3389/fpls.2017.00252/full

Saha, P., Singh, S., Aditika, Bhatia, R., Dey, S. S., Das Saha, N., Ghoshal, C., Sharma, S., Shree, B., Kumar, P., & Kalia, P. (2022). *Genomic designing for abiotic stress resistant Brassica vegetable crops* (pp. 153–185). Springer Science and Business Media LLC. https://doi.org/10.1007/978-3-031-03964-5_5

Sandhu, K. S., Shiv, A., Kaur, G., Meena, M. R., Raja, A. K., Vengavasi, K., Mall, A. K., Kumar, S., Singh, P. K., Singh, J., Hemaprabha, G., Pathak, A. D., Krishnappa, G., & Kumar, S. (2022). Integrated approach in genomic selection to accelerate genetic gain in sugarcane. *Plants, 11*(16). https://doi.org/10.3390/plants11162139. www.mdpi.com/journal/plants

Schauer, N., & Fernie, A. R. (2006). Plant metabolomics: Towards biological function and mechanism. *Trends in Plant Science, 11*(10), 508–516. https://doi.org/10.1016/j.tplants.2006.08.007

Schmutz, J., Cannon, S. B., Schlueter, J., Ma, J., Mitros, T., Nelson, W., Hyten, D. L., Song, Q., Thelen, J. J., Cheng, J., Xu, D., Hellsten, U., May, G. D., Yu, Y., Sakurai, T., Umezawa, T., Bhattacharyya, M. K., Sandhu, D., Valliyodan, B., … Jackson, S. A. (2010). Genome sequence of the palaeopolyploid soybean. *Nature, 463*(7278), 178–183. https://doi.org/10.1038/nature08670

Scossa, F., Alseekh, S., & Fernie, A. R. (2021). Integrating multi-omics data for crop improvement. *Journal of Plant Physiology, 257*. https://doi.org/10.1016/j.jplph.2020.153352

Sharma, E., Borah, P., Kaur, A., Bhatnagar, A., Mohapatra, T., Kapoor, S., & Khurana, J. P. (2021). A comprehensive transcriptome analysis of contrasting rice cultivars highlights the role of auxin and ABA responsive genes in heat stress response. *Genomics, 113*(3), 1247–1261. https://doi.org/10.1016/j.ygeno.2021.03.007. http://www.elsevier.com/inca/publications/store/6/2/2/8/3/8/index.htt

Shaw, J., Rudzicz, F., Jamieson, T., & Goldfarb, A. (2019). Artificial intelligence and the implementation challenge. *Journal of Medical Internet Research, 21*(7). https://doi.org/10.2196/13659

Shelden, M. C., Dias, D. A., Jayasinghe, N. S., Bacic, A., & Roessner, U. (2016). Root spatial metabolite profiling of two genotypes of barley (Hordeum vulgare L.) reveals differences in response to short-term salt stress. *Journal of Experimental Botany, 67*(12), 3731–3745. https://doi.org/10.1093/jxb/erw059. http://jxb.oxfordjournals.org/

Shukla, R., Yadav, A. K., & Singh, T. R. (2020). Application of deep learning in biological big data analysis. *Large-Scale Data Streaming, Processing, and Blockchain Security*, 117–148. https://doi.org/10.4018/978-1-7998-3444-1.ch006. https://www.igi-global.com/book/large-scale-data-streaming-processing/242357

Shukla, R., Yadav, A. K., Sote, W. O., Junior, M. C., & Singh, T. R. (2021). Systems biology and big data analytics. *Bioinformatics: Methods and Applications*, 425–442. https://doi.org/10.1016/B978-0-323-89775-4.00005-5. https://www.sciencedirect.com/book/9780323897754

Singh, D., Wang, X., Kumar, U., Gao, L., Noor, M., Imtiaz, M., Singh, R. P., & Poland, J. (2019). High-throughput phenotyping enabled genetic dissection of crop lodging in wheat. *Frontiers in Plant Science, 10*. https://doi.org/10.3389/fpls.2019.00394. https://www.frontiersin.org/articles/10.3389/fpls.2019.00394/pdf

Singh, N. K., Gupta, D. K., Jayaswal, P. K., Mahato, A. K., Dutta, S., Singh, S., Bhutani, S., Dogra, V., Singh, B. P., Kumawat, G., Pal, J. K., Pandit, A., Singh, A., Rawal, H., Kumar, A., Prashat, G. R., Khare, A., Yadav, R., Raje, R. S., … Sharma, T. R. (2012). The first draft of the pigeonpea genome sequence. *Journal of Plant Biochemistry and Biotechnology, 21*(1), 98–112. https://doi.org/10.1007/s13562-011-0088-8

Singh, R. K., Muthamilarasan, M., & Prasad, M. (2021). Biotechnological approaches to dissect climate-resilient traits in millets and their application in crop improvement. *Journal of Biotechnology, 327*, 64–73. https://doi.org/10.1016/j.jbiotec.2021.01.002. www.elsevier.com/locate/jbiotec

Singh, S., Rao, A., Mishra, P., Yadav, A. K., Maurya, R., Kaur, S., & Tandon, G. (2018). Bioinformatics in next-generation genome sequencing. *Current Trends in Bioinformatics: An Insight*, 27–38. https://doi.org/10.1007/978-981-10-7483-7_2. https://www.springer.com/in/book/9789811074813

Streich, J., Romero, J., Gazolla, J. G. F. M., Kainer, D., Cliff, A., Prates, E. T., Brown, J. B., Khoury, S., Tuskan, G. A., Garvin, M., Jacobson, D., & Harfouche, A. L. (2020). Can exascale computing and explainable artificial intelligence applied to plant biology deliver on the United Nations sustainable development goals? *Current Opinion in Biotechnology, 61*, 217–225. https://doi.org/10.1016/j.copbio.2020.01.010. http://www.elsevier.com/locate/copbio

Suharti, W. S., Nose, A., & Zheng, S. H. (2016). Metabolomic study of two rice lines infected by Rhizoctonia solani in negative ion mode by CE/TOF-MS. *Journal of Plant Physiology, 206*, 13–24. https://doi.org/10.1016/j.jplph.2016.09.004. www.urbanfischer.de/journals/jpp/p_physio.htm

Sun, D., Xu, Y., & Cen, H. (2022). Optical sensors: Deciphering plant phenomics in breeding factories. *Trends in Plant Science, 27*(2), 209–210. https://doi.org/10.1016/j.tplants.2021.06.012

Sun, X., & Weckwerth, W. (2012). COVAIN: A toolbox for uni- and multivariate statistics, time-series and correlation network analysis and inverse estimation of the differential Jacobian from metabolomics covariance data. *Metabolomics, 8*(S1), 81–93. https://doi.org/10.1007/s11306-012-0399-3

Talaviya, T., Shah, D., Patel, N., Yagnik, H., & Shah, M. (2020). Implementation of artificial intelligence in agriculture for optimisation of irrigation and application of pesticides and herbicides. *Artificial Intelligence in Agriculture, 4*, 58–73. https://doi.org/10.1016/j.aiia.2020.04.002

Tamhane, V. A., Sant, S. S., Jadhav, A. R., War, A. R., Sharma, H. C., Jaleel, A., & Kashikar, A. S. (2021). Label-free quantitative proteomics of Sorghum bicolor reveals the proteins strengthening plant defense against insect pest Chilo partellus. *Proteome Science, 19*(1). https://doi.org/10.1186/s12953-021-00173-z. http://www.proteomesci.com/start.asp

Tao, H., Xu, S., Tian, Y., Li, Z., Ge, Y., Zhang, J., Wang, Y., Zhou, G., Deng, X., Zhang, Z., Ding, Y., Jiang, D., Guo, Q., & Jin, S. (2022). Proximal and remote sensing in plant phenomics: 20 years of progress, challenges, and perspectives. *Plant Communications, 3*(6). https://doi.org/10.1016/j.xplc.2022.100344

Timperio, A. M., Egidi, M. G., & Zolla, L. (2008). Proteomics applied on plant abiotic stresses: Role of heat shock proteins (HSP). *Journal of Proteomics, 71*(4), 391–411. https://doi.org/10.1016/j.jprot.2008.07.005

Turhadi, T., Hamim, H., Ghulamahdi, M., & Miftahudin, M. (2019). Iron toxicity-induced physiological and metabolite profile variations among tolerant and sensitive rice varieties. *Plant Signaling and Behavior, 14*(12). https://doi.org/10.1080/15592324.2019.1682829. http://www.tandfonline.com/loi/kpsb20

Van Emon, J. M. (2016). The omics revolution in agricultural research. *Journal of Agricultural and Food Chemistry, 64*(1), 36–44. https://doi.org/10.1021/acs.jafc.5b04515. http://pubs.acs.org/journal/jafcau

Varshney, R. K., Bohra, A., Yu, J., Graner, A., Zhang, Q., & Sorrells, M. E. (2021). Designing future crops: Genomics-assisted breeding comes of age. *Trends in Plant Science, 26*(6), 631–649. https://doi.org/10.1016/j.tplants.2021.03.010. www.elsevier.com/inca/publications/store/3/0/9/6/0/index.htt

Varshney, R. K., Chen, W., Li, Y., Bharti, A. K., Saxena, R. K., Schlueter, J. A., Donoghue, M. T. A., Azam, S., Fan, G., Whaley, A. M., Farmer, A. D., Sheridan, J., Iwata, A., Tuteja, R., Penmetsa, R. V., Wu, W., Upadhyaya, H. D., Yang, S. P., Shah, T., … Jackson, S. A. (2012). Draft genome sequence of pigeonpea (Cajanus cajan), an orphan legume crop of resource-poor farmers. *Nature Biotechnology, 30*(1), 83–89. https://doi.org/10.1038/nbt.2022

Velasco, R., Zharkikh, A., Troggio, M., Cartwright, D. A., Cestaro, A., Pruss, D., Pindo, M., FitzGerald, L. M., Vezzulli, S., Reid, J., Malacarne, G., Iliev, D., Coppola, G., Wardell, B., Micheletti, D., Macalma, T., Facci, M., Mitchell, J. T., Perazzolli, M., … Viola, R. (2007). A high quality draft consensus sequence of the genome of a heterozygous grapevine variety. *PLoS One, 2*(12). https://doi.org/10.1371/journal.pone.0001326. http://www.plosone.org/article/fetchObjectAttachment.action?uri=info%3Adoi%2F10.1371%2Fjournal.pone.0001326&representation=PDF

Villalobos-López, M. A., Arroyo-Becerra, A., Quintero-Jiménez, A., & Iturriaga, G. (2022). Biotechnological advances to improve abiotic stress tolerance in crops. *International Journal of Molecular Sciences, 23*(19). https://doi.org/10.3390/ijms231912053

Wan, J., Wang, R., Wang, R., Ju, Q., Wang, Y., & Xu, J. (2019). Comparative physiological and transcriptomic analyses reveal the toxic effects of ZnO nanoparticles on plant growth. *Environmental Science and Technology, 53*(8), 4235–4244. https://doi.org/10.1021/acs.est.8b06641

Wang, L., Liu, L., Ma, Y., Li, S., Dong, S., & Zu, W. (2018). Transcriptome profilling analysis characterized the gene expression patterns responded to combined drought and heat stresses in soybean. *Computational Biology and Chemistry, 77*, 413–429. https://doi.org/10.1016/j.compbiolchem.2018.09.012

Wang, L., Yu, C., Chen, C., He, C., Zhu, Y., & Huang, W. (2014). Identification of rice Di19 family reveals OsDi19-4 involved in drought resistance. *Plant Cell Reports, 33*(12), 2047–2062. https://doi.org/10.1007/s00299-014-1679-3

Wang, T., Zhang, H., & Zhu, H. (2019). CRISPR technology is revolutionizing the improvement of tomato and other fruit crops. *Horticulture Research, 6*(1). https://doi.org/10.1038/s41438-019-0159-x

Weckwerth, W., Ghatak, A., Bellaire, A., Chaturvedi, P., & Varshney, R. K. (2020). PANOMICS meets germplasm. *Plant Biotechnology Journal, 18*(7), 1507–1525. https://doi.org/10.1111/pbi.13372. http://onlinelibrary.wiley.com/journal/10.1111/(ISSN)1467-7652

Wu, F., Kang, X., Wang, M., Haider, W., Price, W. B., Hajek, B., & Hanzawa, Y. (2019). Transcriptome-enabled network inference revealed the GmCOL1 feed-forward loop and its roles in photoperiodic flowering of soybean. *Frontiers in Plant Science, 10*. https://doi.org/10.3389/fpls.2019.01221. https://www.frontiersin.org/journals/plant-science

Xiao, D., Wang, H., Basnet, R. K., Zhao, J., Lin, K., Hou, X., & Bonnema, G. (2014). Genetic dissection of leaf development in Brassica rapa using a genetical genomics approach. *Plant Physiology, 164*(3), 1309–1325. https://doi.org/10.1104/pp.113.227348. http://www.plantphysiol.org/content/164/3/1309.full.pdf

Xu, C., Xia, C., Xia, Z., Zhou, X., Huang, J., Huang, Z., Liu, Y., Jiang, Y., Casteel, S., & Zhang, C. (2018). Physiological and transcriptomic responses of reproductive stage soybean to drought stress. *Plant Cell Reports, 37*(12), 1611–1624. https://doi.org/10.1007/s00299-018-2332-3

Yadav, A. K., Shukla, R., & Singh, T. R. (2021). Topological parameters, patterns, and motifs in biological networks. *Bioinformatics: Methods and Applications*, 367–380. https://doi.org/10.1016/B978-0-323-89775-4.00012-2. https://www.sciencedirect.com/book/9780323897754

Yang, M., Yang, J., Su, L., Sun, K., Li, D., Liu, Y., Wang, H., Chen, Z., & Guo, T. (2019). Metabolic profile analysis and identification of key metabolites during rice seed germination under low-temperature stress. *Plant Science, 289.* https://doi.org/10.1016/j.plantsci.2019.110282

Yang, W., Feng, H., Zhang, X., Zhang, J., Doonan, J. H., Batchelor, W. D., Xiong, L., & Yan, J. (2020). Crop phenomics and high-throughput phenotyping: Past decades, current challenges, and future perspectives. *Molecular Plant, 13*(2), 187–214. https://doi.org/10.1016/j.molp.2020.01.008. https://www.cell.com/molecular-plant/home

Yang, Y., Saand, M. A., Huang, L., Abdelaal, W. B., Zhang, J., Wu, Y., Li, J., Sirohi, M. H., & Wang, F. (2021). Applications of multi-omics technologies for crop improvement. *Frontiers in Plant Science, 12.* https://doi.org/10.3389/fpls.2021.563953. https://www.frontiersin.org/journals/plant-science

Yang, Y., Yu, Y., Bi, C., & Kang, Z. (2016). Quantitative proteomics reveals the defense response of wheat against Puccinia striiformis f. sp. tritici. *Scientific Reports, 6*(1). https://doi.org/10.1038/srep34261

Zadražnik, T., Hollung, K., Egge-Jacobsen, W., Meglič, V., & Šuštar-Vozlič, J. (2013). Differential proteomic analysis of drought stress response in leaves of common bean (Phaseolus vulgaris L.). *Journal of Proteomics, 78*, 254–272. https://doi.org/10.1016/j.jprot.2012.09.021

Zenda, T., Liu, S., Wang, X., Liu, G., Jin, H., Dong, A., Yang, Y., & Duan, H. (2019). Key maize drought-responsive genes and pathways revealed by comparative transcriptome and physiological analyses of contrasting inbred lines. *International Journal of Molecular Sciences, 20*(6). https://doi.org/10.3390/ijms20061268

Zeng, W., Peng, Y., Zhao, X., Wu, B., Chen, F., Ren, B., Zhuang, Z., Gao, Q., & Ding, Y. (2019). Comparative proteomics analysis of the seedling root response of drought-sensitive and drought-tolerant maize varieties to drought stress. *International Journal of Molecular Sciences, 20*(11). https://doi.org/10.3390/ijms20112793

Zhang, A., Sun, H., Wang, P., Han, Y., & Wang, X. (2012). Modern analytical techniques in metabolomics analysis. *The Analyst, 137*(2), 293–300. https://doi.org/10.1039/c1an15605e

Zhang, M., Liu, Y. H., Xu, W., Smith, C. W., Murray, S. C., & Zhang, H. B. (2020). Analysis of the genes controlling three quantitative traits in three diverse plant species reveals the molecular basis of quantitative traits. *Scientific Reports, 10*(1). https://doi.org/10.1038/s41598-020-66271-8. www.nature.com/srep/index.html

Zhang, X., Liu, P., Qing, C., Yang, C., Shen, Y., & Ma, L. (2021). Comparative transcriptome analyses of maize seedling root responses to salt stress. *PeerJ, 9.* https://doi.org/10.7717/peerj.10765

Zhang, Y., Fu, Y., Wang, Q., Liu, X., Li, Q., & Chen, J. (2020). Transcriptome analysis reveals rapid defence responses in wheat induced by phytotoxic aphid *Schizaphis graminum* feeding. *BMC Genomics, 21*(1). https://doi.org/10.1186/s12864-020-6743-5

Zhao, L., Zhang, H., White, J. C., Chen, X., Li, H., Qu, X., & Ji, R. (2019). Metabolomics reveals that engineered nanomaterial exposure in soil alters both soil rhizosphere metabolite profiles and maize metabolic pathways. *Environmental Science: Nano, 6*(6), 1716–1727. https://doi.org/10.1039/c9en00137a. http://pubs.rsc.org/en/journals/journal/en

Zhou, J., & Troyanskaya, O. G. (2015). Predicting effects of noncoding variants with deep learning-based sequence model. *Nature Methods, 12*(10), 931–934. https://doi.org/10.1038/nmeth.3547. http://www.nature.com/nmeth/

Zhou, S., Zhang, Y. K., Kremling, K. A., Ding, Y., Bennett, J. S., Bae, J. S., Kim, D. K., Ackerman, H. H., Kolomiets, M. V., Schmelz, E. A., Schroeder, F. C., Buckler, E. S., & Jander, G. (2019). Ethylene signaling regulates natural variation in the abundance of antifungal acetylated diferuloylsucroses and Fusarium graminearum resistance in maize seedling roots. *New Phytologist, 221*(4), 2096–2111. https://doi.org/10.1111/nph.15520. http://onlinelibrary.wiley.com/journal/10.1111/(ISSN)1469-8137

Chapter 19

Ecology and environmental omics

Minu Kesheri[1,2], Swarna Kanchan[1,2], Upasna Srivastava[3,4], Bhaskar Chittoori[2], Ratnaprabha Ratna-Raj[5], Rajeshwar P. Sinha[6], Akhouri Vaishampayan[6], Rajesh P. Rastogi[7] and Donald A. Primerano[1]

[1]*Marshall University, Huntington, WV, United States;* [2]*Boise State University, Boise, ID, United States;* [3]*University of California, San Diego, CA, United States;* [4]*School of Medicine, Yale University, New Haven, CT, United States;* [5]*Texas A&M University, College Station, TX, United States;* [6]*Banaras Hindu University, Varanasi, Uttar Pradesh, India;* [7]*Ministry of Environment, Forest, and Climate Change, IPB, New Delhi, India*

1. Introduction

Omics alludes to an interdisciplinary science encompassing genomics, transcriptomics, proteomics, metabolomics, metagenomics, etc., which was introduced in the early 2000s and became popular rapidly in recent years due to advancements in high-throughput analysis technologies for DNA, RNA, proteins, and metabolites. Although "omics" poses to be an *au courant* boisterous buzzword in the lexicon of biological world, yet there are tremendous speculations related to its origin. Dr. Thomas H. Roderick, a geneticist, wins the credit for the contrivances of the word "genomics" in the year 1986 (Kuska et al., 1998) and his proposal being accepted by Victor McKusick and Frank Ruddle, as the catchword for their new journal, "Genomics" (Lederberg & McCray, 2001). In a similar instance, Marc Wilkins, a Ph.D. student at Macquarie University in Sydney, Australia, first coined the term "proteome" to represent the phrase "the protein complements of the genome" (Swinbanks, 1995) that might have led to the innovation of the term "proteomics" (Yadav, 2007). However, the dogma of origin of "omics" still spans from being rooted to the Sanskrit word "OM" that implies to holistic approach, inclusiveness, divine wholeness etc., incorporating the whole universe in its unlimitedness. In total, omics analysis can provide a comprehensive picture of gene expression, protein expression, and metabolite pattern, characterize the microbiomes in specific habitats and provide a deeper insight into how the organisms cope with external environmental stressors. Application of omics technologies to ecology, environmental toxicology, and health research have resulted in a better understanding of the environmental and genetic factors, chemical toxicity mechanisms, and modes of action in response to exposure to environmental chemicals; these studies ultimately improved our understanding of the development of diseases (Singla et al., 2019b) with long-term effects. Environmental omics aims to investigate environmental monitoring enabling risk assessment, diverse human health outcomes, environmental impacts, ecological functions, and environmental adaptation. Environmental and ecological omics assess acceptable levels of various chemical toxicants and the potential impacts on environmental target species and ecosystems. For example, various omics technologies such as genomics, proteomics, and metabolomics have been used extensively to study the molecular mechanisms of arsenic toxicity (Pershagen, 1981; Srivastava et al., 2023).

Environmental genomics explores the interspecific as well as intraspecific alterations in an organism or group of organisms under the influence of their abiotic as well as biotic microenvironment, niche, or ecosystem at the genetic level. In transcriptomics-based studies, RNA from individual organisms or microbial communities are sequenced, to determine changes in gene expression due to chemical exposure in the environment. Proteomics was also applied frequently in environmental research to investigate differential proteomes of environmental bacteria to understand antibiotic resistance mechanisms (Ralston−Hooper et al., 2011). In environmental metagenomics, DNA sequences extracted from samples collected from environmental habitats such as ponds, lakes, rivers, wastewater, etc. of microbial communities and analyzed to understand the bacterial and other major organism's composition and functional characterization of a habitat (Aguiar-Pulido et al., 2016; White et al., 2016).

Integrative Omics. https://doi.org/10.1016/B978-0-443-16092-9.00019-9

Metagenomics studies in environmental research received more attention compared with studies focusing on either toxicity mechanisms or ecosystem toxicity. We made substantial progress in certain areas of environmental omics; further improvements will facilitate the incorporation of multiomics rather than using a single omics into environmental toxicology and health research. In environmental omics research, many biological samples can come from different individuals, populations, and communities, which are diverse and heterogenous in nature and often involve multiple target organisms and multifactorial experiments. These features can lead to an extremely large number of measurable parameters and make the omics analysis very challenging.

In this chapter, we discuss high-throughput molecular methods being used in environmental toxicity and health research, the role of omics in assessing the ecosystems, the effect of dietary and environmental exposures on organism's genomes, the characterization of single or chemical mixtures, and their effects on target organisms.

2. High-throughput molecular technologies in environmental research

Various high-throughput experimental molecular methods are used for the analysis of the environmental data which are categorized into traditional, cultural, and noncultural methods (i.e., random amplified polymorphic DNA, RAPD; real-time polymerase chain reaction, RT-PCR; restriction fragment length polymorphism, RFLP; denaturing gradient gel electrophoresis, DGGE) providing preliminary knowledge of environmental organisms or communities (Botero et al., 2005; Feinstein et al., 2009).

The use of molecular phylogeny in the late 1960s was frequent in microbial ecology research. Advances in conventional Sanger DNA-sequencing technology led to a large-scale, wide range of applications (Hajibabaei et al., 2007). Genomic analysis of complex environmental samples has emerged as an important tool for understanding the functional and ecological biodiversity of the environment. Specific gene markers such as species-specific DNA barcodes were used for the analysis of the environment by utilizing the next-generation sequencing technologies in ecological and environmental research. DNA barcodes which are species-specific genomic regions that are used for the identification of unknown environmental specimens, for example, 16S ribosomal RNA (16S) can be used for bacterial identification (Flanagan et al., 2007; Sogin et al., 2006), whereas the internal transcribed spacer (ITS) region of the nuclear ribosomal DNA genes can be used for identification of fungal species.

Since Sanger sequencing (Sanger et al., 1977) can only sequence genes/regions from individual organisms, it is inadequate for analyzing complex environmental samples, especially for large-scale studies. Since 2005, advances in next-generation sequencing technologies have revolutionized biological, ecological, and environmental sciences (Liu et al., 2012; Slatko et al., 2018). Whole genomes or specific regions from thousands of species present in an environmental bulk sample can be sequenced accurately, rapidly, and affordably with NGS technology. Low-cost next-generation sequencing technologies such as Illumina, Pacific Biosciences (aka PacBio), and Oxford Nanopore Technologies are capable of parallel sequencing millions of DNA molecules with different yields and sequence lengths. Presently, the Illumina platform can sequence a maximum length of 300 bp in either single-end read or as paired-end read formats. Long-read sequencing platforms such as PacBio and Oxford Nanopore can sequence hundreds of kilobases per template and are also gaining popularity due to their lower cost compared with Illumina (Hu et al., 2021; Jain et al., 2018; Nellimarla & Kesanakurti, 2023). Long reads also have other advantages such as the improved taxonomic assignment of identified organisms, epigenetic marker detection, and better characterization of repetitive sequences in the genome (MacKenzie & Argyropoulos, 2023; Nellimarla & Kesanakurti, 2023; van Dijk et al., 2018). These long-read technologies, however, suffer drawbacks such as greater error rates, and lower output compared with short-read technologies, making the former less popular than their counterparts such as Illumina (Besser et al., 2018; MacKenzie & Argyropoulos, 2023; Nellimarla & Kesanakurti, 2023).

2.1 Genomics in ecology and environmental research

Genomics provides us with a better understanding of the implications of the genetic makeup of various environmental organisms, particularly genomic databases that play a major role in achieving the same (Kumari et al., 2016). Genomics also includes genome assembly, annotation, and prediction of genomic regions or genes using various soft computing tools such as machine learning (Kanchan et al., 2024; Kesheri et al., 2016; Kumari et al., 2018). Genomics was probably the first omics approach to be applied to toxicity studies due to chemicals or pesticides and investigating toxicity pathways. These studies involved environmental chemicals and their mixtures focused on the characterization of gene expression that is shown by the presence of specific mRNA transcripts in a biological sample (Iwahashi et al., 2006). Gene expression profiling such as transcriptomics was used to compare the expression profiles of several genes with and without exposure to environmental chemicals or a mixture of chemicals. This could be because there are important changes in proteins and metabolites within cells that are not detectable by just studying various genes. Therefore, omics technologies such as transcriptomics, proteomics, metabolomics, and metagenomics are mostly used to complement genomics for ecological and environmental studies (Kishi et al., 2006).

2.2 Transcriptomics, proteomics, and metabolomics in ecological and environmental studies

Transcriptomics helps us have a better understanding of the relationships between the transcriptome with the phenotype across a wide range of environmental target organisms. Various transcriptome studies aim to identify differentially expressed genes in response to different chemicals or a mixture of chemicals exposed to the environment in living organisms in the ecosystem. While transcriptomics studies dominated until 2016, a shift toward proteomics is apparent. Proteomics studies mainly focus on one organism or cell type, and the effects on the environment are investigated by comparing different conditions. Many of the proteins produced by environmental microorganisms have antioxidative, anticancer, antiinflammatory capacities, and antiaging effects (Richa et al., 2011a, b; Kesheri & Kanchan, 2015b; Kesheri et al., 2017, pp. 166−195). Proteins are ultimately functional molecules, but still 3D structures of many of these proteins are not known; therefore, bioinformatics tools for structure prediction, molecular docking, molecular dynamics simulations, etc. play a major role in exploring the mysteries of protein function (Kesheri et al., 2015; Kesheri & Kanchan, 2015a; Priya et al., 2017, pp. 286−313; Sahu et al., 2022). Protein modeling not only helps in function prediction but also helps in exploring protein evolution in all three domains of life, drug resistance, etc. (Garg et al., 2009; Kanchan et al., 2014, 2015, 2019). Comparative homology modeling (Srivastava & Singh, 2013) and in silico vaccine designing based on epitope prediction (Srivastava et al., 2017) also contribute toward enhancing knowledge of proteomics. The omics approach is also involved in the proteomics study of HPV type 16 E7 protein expressed in cervical cancer using proteomics and sequence analysis. Toxic/unfavorable responses triggered by chemicals or UV radiations are driven by interactions between chemicals and biomolecular targets in the form of DNA/proteins.

Environmental metabolomics is the latest addition to the omics family, which is the application of metabolomic techniques to analyze the interactions of organisms with their environment. Few metabolomic studies have investigated environmental stressors such as xenobiotic exposure and temperature from multiple model organisms (Bundy et al., 2008).

2.3 Environmental metagenomics for studying large communities

Studies on the microbial community structure based on 16S rRNA gene sequences that started in the late 1980s have revealed enormous diversity of uncultured microorganisms (bacterial/other small eukaryotic organisms) in nature. Environmental metagenomics has been responsible for substantial advances in microbial ecology, as well as diversity over the past 10 years. Metagenomics uses high-throughput genomic technologies, followed by statistical, and bioinformatics analysis to directly access the genetic content of entire communities of organisms present in the environment. Metagenomics provides access to the functional gene composition of microbial communities providing a broader overview than phylogenetic relationships, which are often based only on the diversity of one gene, such as the 16S rRNA gene (Kanchan et al., 2020; Thomas et al., 2012). A typical metagenomic pipeline includes sample processing, sequencing technology, assembly, binning, annotation, statistical analysis, and data storage. Environmental metagenomics is further grouped into two categories called shotgun and targeted metagenomics. Shotgun metagenomics refers to the random sequencing of all genomic content of all the microorganisms in a community commonly known as the microbiome (Cai & Zhang, 2013). Shotgun metagenomics has an advantage over targeted metagenomics due to advancements in sequencing technology to profile genes as well as all the microbial species in the microbiome. Shotgun metagenomics doesn't involve the amplification of marker genes such as 16S rRNA, provides a more precise reflection of the composition of the community, has greater taxonomic resolution than 16S rDNA/rRNA sequencing, and identifies multiple markers in the community of microorganisms by sequencing. However, most of the metagenomic studies have focused on DNA sequencing of entire communities using either a targeted approach such as PCR-amplicon sequencing of 16S rRNA genes or other marker genes due to lower cost. Therefore, targeted metagenomics has become a standard tool for many laboratories and scientists working in the field of microbial ecology.

2.4 Single-cell omics in ecological and environmental studies

Single-cell omics have revolutionized our ability to identify and characterize various subpopulations of cells within the target environmental organism. Single-cell sequencing is an important methodology to study cell heterogeneity and has been widely applied in many areas of research (AlJanahi et al., 2018; Srivastava & Singh, 2022, pp. 271–294). However, the use of single-cell omics is still limited in environmental science and ecology, which is obvious by the publication of only 85 articles (related to single-cell sequencing) in the past 5 years, compared with 2917 articles published related to high-throughput sequencing (Liu et al., 2022). Single-cell sequencing is also important in toxicogenomic studies because it can label each cell through a barcode and record specific changes in different cell types during cellular development. Single-cell toxicogenomics allows in-depth analysis of toxicological mechanisms at the level of cell subtype (Liu et al., 2022). A recent single-cell RNA-Seq study on the effect of exposure to perfluorooctanoic acid illustrated the significantly altered functions of zebrafish heart cells, including cardiac muscle contraction, oxygen binding, and actin cytoskeleton (Yu et al., 2022).

A single RNA-Seq study based on the exposure of diesel exhaust particles using single-cell RNA analysis illustrated the dysregulated genes enriched in immune system pathways in murine macrophages. This study revealed the mechanisms of particulate matter inhalation and systemic vascular effects, leading to worsened atherosclerosis (Bhetraratana et al., 2019). Single-cell sequencing pipeline includes isolation of the individual cells followed by DNA extraction, whole-genome amplification, DNA sequencing, and quality check of the sequence reads, alignment and assembly, etc. The major advantage of single-cell sequencing technology is that it provides a good-quality genome for species with low abundance, which could be missed by metagenomic sequencing (Cheng et al., 2019). Another advantage of single-cell genomics is its power to discriminate and validate the functions of individuals by linking these functions to specific species within the community (Cheng et al., 2019). Single-cell omics have also proved their ability to quantify the relative abundances of the gut microbiome (Props et al., 2017). A few limitations of single-cell omics are the removal of DNA contamination, cell sorting, and isolating cells from solid mediums such as swabs, and tissues that are truly challenging and time-consuming (Tolonen & Xavier, 2017). Identification and removal of chimeras are also considered major limitations of single-cell omics technology, which is due to the absence of reference genomes. In such cases, metagenomics-based contigs could be used as a reference that can help in chimera identification and removal (Cheng et al., 2019). Uneven read coverage in single-cell sequencing is rectified by normalizing the reads using trimming based on their k-mer depth, which is available to several assembly algorithms such as SPAdes (Bankevich et al., 2012).

2.5 Spatial omics in environmental research

Spatial omics is a rapidly evolving high-throughput omics approach for exploring tissue microenvironments, which could be used for exploring toxicity mechanisms in environmental research. Spatial omics are used for exploring cellular networks by integrating spatial knowledge with protein expression. Commercially available platforms for spatial transcriptomics now facilitate spatially resolved, high-dimensional assessment of gene transcription. In recent years, traditional experimental methods, such as fluorescent in situ hybridization (FISH), barcoding with reporters (Cho et al., 2022), and immunohistochemistry (IHC), provided ways for spatial omics technologies to cover several transcripts. Moreover, spatial omics technologies vary in their spatial resolution, coverage, scale, throughput, and multiplexing capacity (Park et al., 2022). Depending on the objective of the research, the spatial omics methods can be divided into (1) targeted or multiplexed probe- or antibody-based and (2) transcriptome-wide or next-generation sequencing (NGS)–based approaches (Ståhl et al., 2016). Targeted spatial omics methods are used to identify cellular states and functions for specific molecular entities (Park et al., 2022). However, spatial omics have a few limitations, which are challenging, such as large volumes of data, heterogenous data types, and the lack of unified spatially aware data structures.

Among the many branches of omics methodologies, genomics, transcriptomics, proteomics, metabolomics, and metagenomics, there is a need for multiomics and integromics approaches for the progression of ecological and environmental sciences research to investigate the ecological drivers/biomarkers of our environment (Fig. 19.1).

Multiomics need the cooperation of researchers from interdisciplinary sciences such as bioinformatics, mathematics, statistics (Shruti et al., 2016; Singla et al., 2019a), machine learning, etc. for analyzing and interpreting results obtained from high-throughput sequencing techniques. Moreover, with the discovery of single-cell multiomics technologies, simultaneous profiling of the genome, epigenome, transcriptomics, proteome, and metabolome from a single cell has become possible (Srivastava & Singh, 2022, pp. 271–294).

3. Environmental omics in the context of toxicology

Omics technologies applied to environmental toxicology and health research growth reached to next level due to powerful genomics, transcriptomics, proteomics, and metabolomics tools to explore novel toxicity pathways/signatures/biomarkers to better understand toxicity mechanisms and modes of action. Environmental omics aims to identify environmental

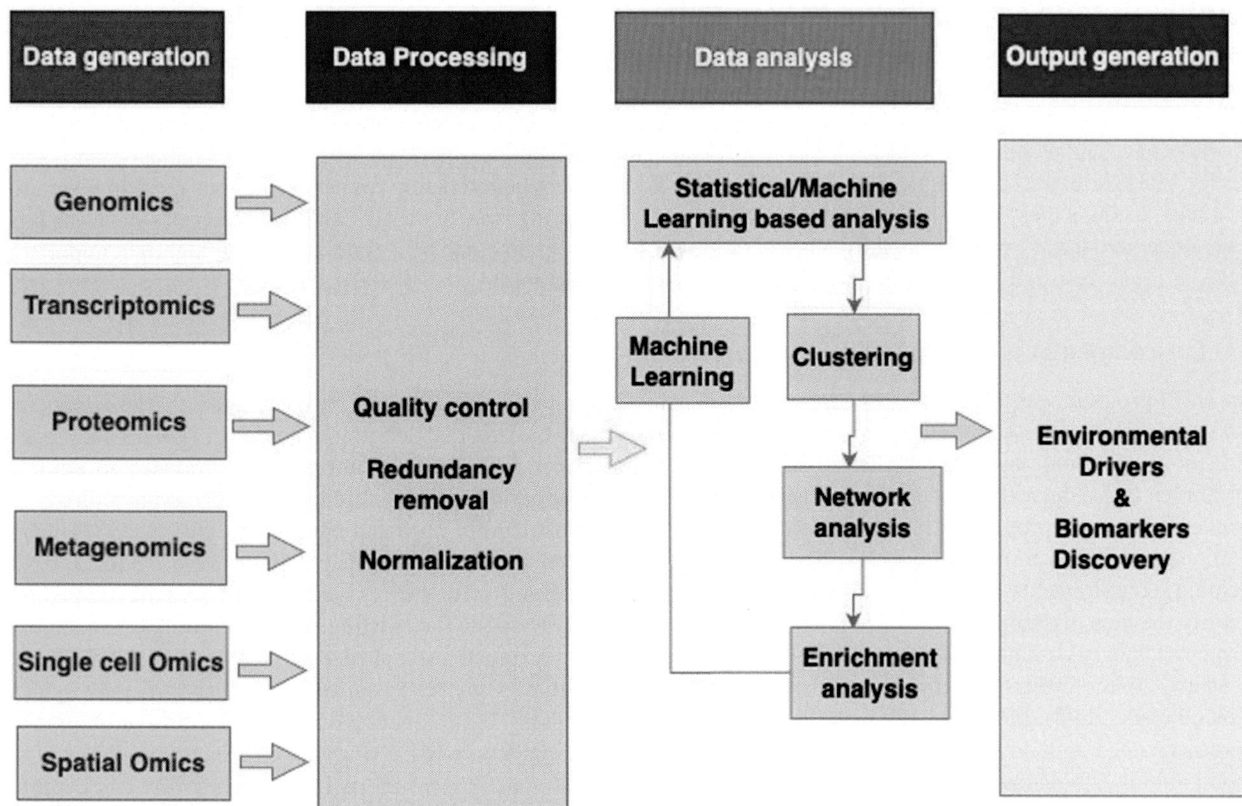

FIGURE 19.1 Integrative omics approaches applied to environmental research. Workflow of multiomics and integrative omics to discover environmental drivers and biomarkers.

chemicals and monitor as well as predict the risks associated with exposure to environmental chemicals on human health and the environment. Environmental omics give us a clearer picture of ecosystem health for more comprehensive water safety management, and environmental monitoring enabling risk assessment. Environmental omics also investigates adverse human health outcomes and environmental impacts. It also explores ecological functions and environmental adaptation (Kesheri, Kanchan, & Chowdhury, 2014). One of the important purposes of environmental omics is to predict the effects of pollution and diagnose the stressors of polluted ecosystems. Environmental omics not only investigates toxicity mechanisms but also explores short- and long-term effects of various environmental chemicals on human and ecosystem health. To fulfill this goal, environmental omics studies based on molecular, cellular, tissue, organs, and individuals, all the way to the population, community, ecosystem and, ultimately, the biosphere were considered (Kramer et al., 2011). Environmental omics investigates the acceptable levels and potential impacts of environmentally toxic chemicals on various species and ecosystems. Ecological omics is broadly used to investigate chemical toxicity mechanisms and modes of action (Mei et al., 2007; Ortiz et al., 2010; Wang et al., 2012). It focuses on identifying biomarkers of exposure and toxicity (Matheis et al., 2011) and studying toxic effects and environmental diseases (Vineis et al., 2009; Yasokawa & Iwahashi, 2010). Environmental omics not only helps in expression profiling studies but also helps in the identification of modifications in gene, protein, and metabolite levels to address environmental toxicology studies and human health research (Mezhoud et al., 2008; Yi & Pan, 2011). However, concentration-dependent transcriptomics studies have been performed earlier using inconsistent bioinformatic methods for data filtering, concentration–response modeling, and quantitative characterization of genes, and pathways. These inconsistent bioinformatics methods made the comparisons across studies problematic; therefore, future studies should focus first on the development of standardized protocols for concentration-dependent transcriptomic characterization of chemicals. Among the various high-throughput omics technologies, proteomics also plays a major role in ecotoxicological research because toxic responses are due to various proteins produced, which are responsible for various interactions between chemicals and biomolecular targets. Therefore, proteomic-based toxicity studies and biomarkers are highly relevant to biological functions, adverse health outcomes, and human health risk assessment.

4. Role of omics in assessing ecosystems

Omics technologies also aim to study ecosystems to develop monitoring tools for regulatory purposes. Several studies (Fent & Sumpter, 2011; Hook, 2010; Van Aggelen et al., 2010) have outlined various necessary steps to move omics technologies to be more useful and to be applied to regulatory or environmental monitoring approaches. Several researchers have addressed various basic factors in environmental science (i.e., the impact of variables such as season, sex, species, and nutritional status on normal biological variation) that are required before further discussions regarding the use of omics technologies in environmental regulation and monitoring can proceed. The development of Canada's government-mandated Environmental Effects Monitoring (EEM) program could be a good example to highlight important issues in omics research for inclusion in regulatory monitoring programs.

4.1 Environmental effects monitoring program

The EEM program was a cyclical, effects-based monitoring program designed to identify effluent discharge effects on fish or benthic macroinvertebrates. In the late 1980s, Environment Canada decided to update the 1971 Pulp and Paper Effluent Regulations under the Fisheries Act. The Pulp and Paper Effluent Regulations were proposed to control discharges and reduce effects on fish and fish habitats. During that period, several studies investigated physiological, biochemical, and reproductive responses in fish populations downstream of pulp and paper mills in Canada (Hodson et al., 1992; Munkittrick et al., 1991). Revisions of the Pulp and Paper Effluent Regulations were made in 1992 including requirements for conducting cyclical monitoring. Looking into the revisions, EEM was developed to identify the new discharge limits for pulp and paper industries. It was due to the existing discharge limits being unable to protect the environment adequately. The EEM program was designed in such a way as to enforce mandatory measures on the effects of effluents on fish, fish habitat, and use of fisheries resources in 3-year to 6-year cycles (Lowell et al., 2005). Moreover, the EEM program was aimed to assess, the frequency with which mills in compliance were associated with downstream impacts on a national level. The first cycle of EEM (completed in 1996) was used to collect initial data on environments. Initial data was focused on estimates of variability in fish. Suitability and capture of sentinel species were the focus of EEM during the first cycle of EEM. Optimal sampling methods and reference sites were also focused during the first cycle of the EEM program (Munkittrick et al., 2002). The next two cycles (completed in 2000 and 2004) were used to assess and finalize the effects of pulp and paper effluents on fish populations and habitats. The EEM program was further amended in 2004 to incorporate cycle four to include requirements for pulp and paper industries to confirm effects of a magnitude above the critical effect size. It was intended to investigate of case study to assist in understanding the cause of confirmed effect(s). In 2008, the EEM program was further amended to incorporate cycle 5 (completed in 2010), which required mills to investigate solutions to eliminate environmental effects if the cause of the effect was investigated in any recent studies. Under EEM programs, observations were made at >100 pulp and paper mills throughout Canada over more than 20 years. More than 60 species have been used in EEM programs in Canada, due to varied receiving environments (Barrett & Munkittrick, 2010). In cycle 2, the critical effect size for fish was defined as a 25% change in gonad size, because that was the level of change seen at Jackfish Bay (ON, Canada), and gonad size was the most variable endpoint (Munkittrick et al., 1991). By cycle 3, sufficient data were available to look at the distribution of changes, and a 25% change in most fish endpoints was ascertained as unusual (Munkittrick et al., 2002). Finally, these effect sizes were standardized after an international review of monitoring to set critical effect sizes (Munkittrick et al., 2009).

In various studies, assessing transcriptome responses in laboratory-reared animals exposed to water or sediment was observed as the common method used in environmental assessments (Hasenbein et al., 2014; He et al., 2012; Osachoff et al., 2013). Using transcriptomics data (Martyniuk & Houlahan, 2013) showed that some molecular processes contain genes that exhibit low (stable) or high (unstable) variability in expression within the pathway. This could be used to monitor a defined molecular pathway or network showing high stability over a range of reproductive stages and environmental conditions responsive to chemical stressors.

Moreover, studies show that changes in the transcriptome do not always correlate with changes in the proteome, making proteomics studies important. Few studies illustrated the positive relationship between transcriptional and proteomic endpoints in Atlantic salmon (Kanerva et al., 2014). There are good examples of studies demonstrating the use of metabolomics in the laboratory to study environmental chemical stressors (Van Scoy et al., 2010; Li et al., 2014); therefore, metabolomics is considered a powerful tool in understanding the effects of environmental stressors in aquatic species.

5. Omics approaches in ecotoxicology and stress biology

With the growth of industrialization, human activities have increased resulting in the release of novel chemicals into the environment. Freshwater and marine organisms are affected by exposure to chemicals such as pollutants, which might result in changes in physiology, reproduction, morphology, nutrition, migration, extinction, and death. By integrating the concepts of toxicology with ecology, a new discipline called ecotoxicology arose, which studies the effects of toxic substances or other harmful agents such as algal toxins on the health of ecosystems and their constituent species as an individual or a population (Truhaut, 1977; Kesheri et al., 2011; Kesheri et al., 2021; Kesheri et al., 2011; Kesheri et al., 2022). Ecotoxicological studies use a wide range of biological concepts, starting from the molecular, cellular, tissue, organs, individuals, population, community, ecosystem, and, ultimately, the biosphere (Kramer et al., 2011). Ecotoxicology and stress ecology (Kesheri et al., 2017, pp. 166−195; Kesheri et al., 2021; Kesheri et al., 2022) aim to explore and predict the effects of environmental stressors on the ecosystem (Straalen, 2003; Walker et al., 2012, pp. 978−1). Toxic substances in the environment generate adverse effects at all levels starting from the molecular level to the biosphere. Exposure to environmental stressors is responsible for changes at multiple levels resulting in the detection and measurement of negative impacts, regulatory molecular pathways involved in these responses, interactions, etc. (Bludau & Aebersold, 2020).

In the past years, environmental research was mainly focused on the use of single omics, and there is a need for environmental studies based on the integration of these omics technologies. For example, transcriptomics studies help in the identification of the distinct cellular composition among lesions and metastatic or recurrent PDOs of colon cancer organoids that contain fewer stem-like cell and differentiated-like cell clusters and more proliferating cell clusters (Okamoto et al., 2021).

Each year, various environmental studies investigated many environmental stressors compared with that of species. In various environmental omics-based studies, many nonmodel organisms were investigated compared with model organisms, but cross-species comparisons are still rare. In recent ecotoxicology studies using the same sample, multiomics were implemented to investigate the complex processes (Faugere et al., 2023; Nam et al., 2023). Environmental multiomics studies also investigated molecular changes at all levels of biological organization, enabling us to understand the impacts of environmental stressors in detail. Omics technologies also help in integrating with adverse outcome pathways (AOPs) that identify the sequence of molecular and cellular events required to produce a toxic effect when an organism is exposed to a substance making it easy to understand the potential effects at higher levels of biological organization (Ankley et al., 2010; Brockmeier et al., 2017; Sauer et al., 2017). The use of omics technologies doesn't require prior assumptions to provide an unbiased and clear picture of the ecotoxicological effect at an early stage of the investigation. Omics technologies also have a few limitations in environmental studies because each environmental stressor in the environment cannot be studied for each individual target organism present on earth. Since these individual target organisms within classes and families often contain similar genomic architectures and conserved cellular pathways, knowledge about available data is greatly beneficial for extrapolating results (Luo et al., 2015; Siepel et al., 2005).

An extensive study was performed, which included articles from 2000−20 (648 studies were selected in total), which included various environmental stressors ($n = 259$) and species ($n = 184$) (Ebner, 2021). Among the various omics technologies, transcriptomics was observed as the most frequently used (43%), followed by proteomics (30%), metabolomics (13%), and finally, multiomics (13%). A combination of transcriptomics and proteomics (38%) ranked top followed by transcriptomics and metabolomics (33%), while proteomics and metabolomics (21%) ranked last in considering all environmental omics studies. In transcriptomics analysis, samples are prepared from the chemically exposed tissues and control/normal tissues to be processed further for library preparation and sequencing, which results in the generation of fastq files that are subjected to be processed through the pipeline to explore chemical toxicity mechanisms, biomarker genes, affected pathways, etc. (Fig. 19.2).

Out of 184 investigated species, *Danio rerio* (11%), *Daphnia magna* (7%), *Mytilus edulis* (4%), *Oryzias latipes* (3%), and *Pimephales* were observed as the top five organisms belonging to Chordata (44%) (Ablain & Zon, 2013). Studies based on ecotoxicology are nowadays publicly available, facilitating the analysis and integration of data. Another amphipod organism called *Gammarus fossarum* has also become more common in ecotoxicological studies. Within the Chlorophyta, *Chlamydomonas reinhardtii* is studied most often because of the availability of high-quality genome information. The availability of sex-specific transcriptomes for *C. reinhardtii* made it popular in environmental omics studies. Well-established laboratory-based cultivation protocol for *C. reinhardtii* also makes it popular for studies based on integration of multiomics data at various levels (May et al., 2008; Shrager et al., 2003). Among mollusks, *Mytilus edulis* and *Mytilus galloprovincialis* were observed as top investigated target organisms. Among various fungal target organisms,

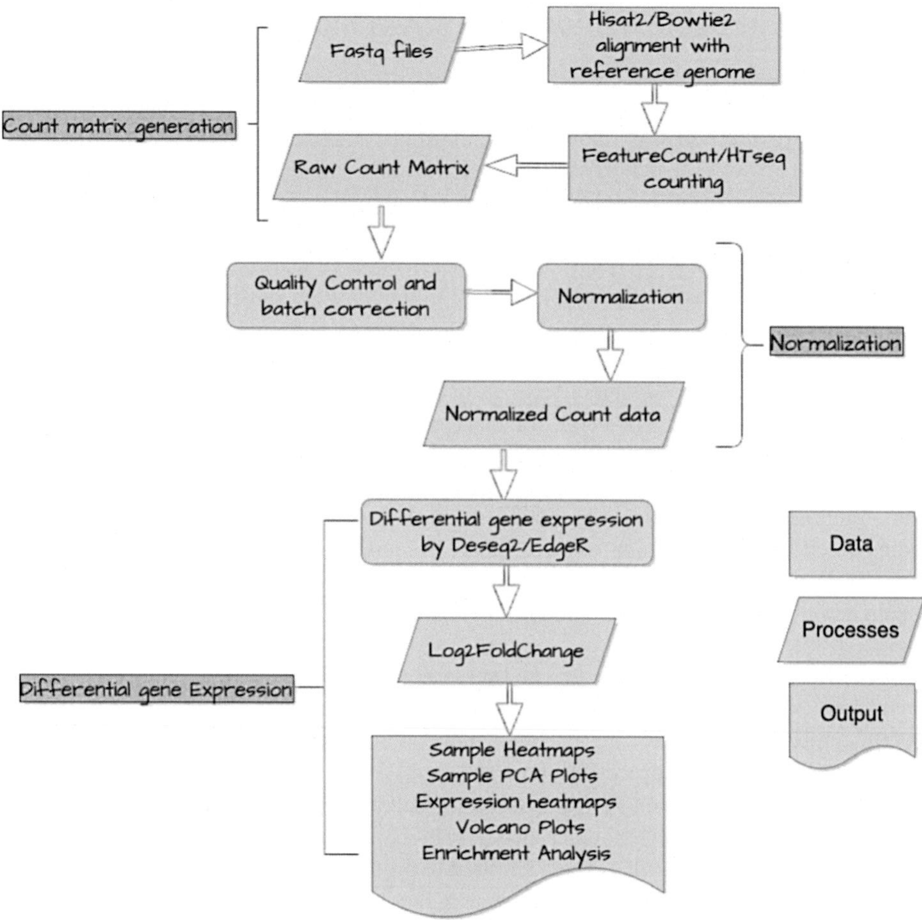

FIGURE 19.2 Transcriptomics (RNA-Seq) pipeline to investigate the chemical toxicity mechanisms, biomarker genes, affected pathways, etc. Workflow of transcriptomics (RNA-Seq) data analysis.

Aspergillus niger (ascomycota), and *Pleurotus ostreatus* were frequently studied (Boontawon et al., 2021; Brandl & Andersen, 2017). The most frequently studied environmental stressors were temperature (8%), copper (5%), cadmium (2%), oil (2%), etc.

6. Organism's genome related to dietary and environmental exposures

Genome, genes, mRNA, proteins, and metabolites are frequently modified in response to dietary, lifestyle, and exposure to environmental stresses.

6.1 Effects of diet and lifestyle on human health

Dietary, lifestyle, and metabolic risk factors concluded that smoking and high blood pressure are responsible for the largest number of deaths with each accounting for about one in five or six deaths in adults in the United States (Danaei et al., 2009). Obesity, physical inactivity, alcohol consumption, and poor diet were also major risk factors for premature death, cardiovascular diseases, cancers (Ali et al., 2024), respiratory diseases, etc. (Danaei et al., 2009). A study of 20,244 middle-aged and older people in the United Kingdom found that the combined effects of just four health behaviors (current nonsmoking, not physically inactive, moderate alcohol intake, and plasma vitamin C concentration >50 μM predicted a fourfold difference in total mortality (Khaw et al., 2008). Use of tobacco is associated with a high incidence of carcinomas of the oral cavity and esophagus. Tobacco also causes tumors of the larynx, pancreas, kidney, and bladder and, in conjunction with alcohol consumption. Moreover, diets that contain a high proportion of plant foods (fruits, vegetables, relatively unprocessed cereals, and pulses) are associated with a lower risk of several common cancers. On the other hand, higher intakes of red, and especially processed meats cause colorectal cancer. Although there is strong evidence from various studies that higher intakes of specific micronutrients, for example, vitamin A, folate, selenium, and vitamin D, are associated with lower cancer risk.

Heavy metals notably lead, cadmium, mercury, and arsenic pose threats to human health (Järup, 2003). The presence of arsenic in drinking water in the Ganges delta region of Bangladesh has kept around 40 million people at risk of arsenic poisoning-related diseases such as skin cancer. Multiomics studies could help us to understand how these exposures are responsible for the accumulation of genetic damage and aberrant gene expression, which are fundamental hallmarks of tumor development.

6.2 Effect of environmental exposure on our genome

Environmental exposure to nutritional, chemical, and physical factors can alter gene expression and affect adult phenotype by not only mutating promoter and coding regions of genes but also by modifying CpG methylation and other epigenetic modifications at critical epigenetically labile genomic regions (Tiffon, 2018). Multiple studies illustrate that the epigenetic marks (including DNA methylation, histone modifications, and microRNA (miRNA) expression) are influenced by environmental exposures, including diet, tobacco, alcohol, physical activity, stress, environmental carcinogens, genetic factors, and infectious agents, which play important roles in the etiology of cancer (Mathers et al., 2010). DNA modifications and a few epigenetic modifications change the expression of tumor suppressor genes and oncogenes and are thus responsible for the cause of different varieties of cancer. To date, there have been few studies that have examined the effects of dietary (and other environmental) factors on epigenetic markings in intervention studies in humans. Epigenetic marks are attractive candidates, which could be used in dietary or lifestyle intervention studies for cancer prevention. Transposable elements, the promoter regions of housekeeping genes, and cis-acting regulatory elements of imprinted genes are considered as epigenetic susceptibility targets that contain CpG islands that are methylated, unmethylated, or differentially methylated (Dolinoy & Jirtle, 2008). Diets rich in the methyl-donating group rapidly affect gene expression, especially during early development. Few studies have reported that diets poor in methyl-donating folate before or just after birth cause permanent hypomethylation of parts of the genome especially in animals. Folic acid and vitamin B12 are considered epigenetically active ingredients that play important roles in DNA metabolism and the maintenance of DNA methylation patterns (Tiffon, 2018). Intake of diets that are having low folate is also associated with hypomethylation and an increased risk of colorectal and pancreatic cancers. Diets rich in fruits and vegetables containing natural antioxidants can protect against cancer. The potential epigenetic effects of several nutritional components in addition to folate, mostly derived from vegetables, have been examined in several studies, which have shown reductions in DNA hypermethylation of critical genes resulting in tumor suppression.

The main consequences of environmental exposure to chemicals such as endocrine-disrupting chemicals are adverse effects on the growth and development of reproductive organs, effects on neuronal differentiation, and an increase in the combined incidence of all cancers. Endocrine-disrupting chemicals are exogenous agents that alter one or more endocrine system axes and, consequently, impose adverse health effects at the level of the organism, its progeny, and/or subpopulations (Diamanti-Kandarakis et al., 2009).

7. Environmental omics of single chemicals versus chemical mixtures

Environmental chemicals damage human health directly or indirectly through food chains. Chemicals (Ghai et al., 2015; Roma Ghai et al., 2016; Saxena et al., 2015) such as hormones or drugs (Gahoi et al., 2013; Galande et al., 2014), tobacco smokes, pesticides, etc. can act at very low doses to disrupt ecological balance and threaten our biodiversity and ecosystem (Mishra et al., 2015a, 2015b; Campos et al., 2012). Metabolomics was successfully applied to the marine mussel *Perna viridis* to investigate changes in the metabolic profiles of soft tissue as a response to copper (Cu) and cadmium (Cd), both as single metal and as a mixture (Wu & Wang, 2010).

7.1 Environmental omics associated with exposure to a single chemical

Most of the environmental omics research is focused on the toxicity of single chemicals. Due to the focus on environmental omics, experimental methodologies and technologies to study the toxicity of single chemicals and their risks to human health are well established. Arsenic is considered one of the most common and well-studied single environmental chemicals (Marshall et al., 2007). Arsenic is commonly found in our environment; it could be present in drinking water, foods, soil, and airborne particles. In a study, arsenic ingestion is observed as associated with cancer incidence

(Baastrup et al., 2008). Arsenic was also investigated to activate environmental stress gene expression (Li et al., 2002; Martin-Chouly et al., 2011) and to induce cell proliferation and apoptosis pathways (Cheung et al., 2007). Various proteins such as enolase, HSP90, pyruvate kinase, aldolase reductase, GAPDH, phosphoglycerate mutase B, Cu–Zn SOD (Kesheri et al., 2011; Kesheri, Kanchan, Richa, & Sinha, 2014; Kesheri, Kanchan, & Chowdhury, 2014), and thioredoxin were investigated in arsenic-induced cancer cells and as part of antioxidative systems. Various genes responsible to produce proteins such as peripherin, cytokeratin 14, and cytokeratin 8 were observed as downregulated in arsenic-induced transformed cells (Li et al., 2011; Sun et al., 2009). Single chemical omics mainly focused on the identification of toxic pathways, mechanisms, and biomarkers of chemicals such as arsenic. Previous studies on arsenic investigated toxicity pathways and toxicity mechanisms in various tissues, organs, or whole organisms. Such studies helped us in incorporating omics data into the regulatory frameworks for environmental chemicals. Omics studies are important for the evaluation of the persistent behavior of single chemicals. However, genomics, proteomic, and metabolomic responses are dynamic and only capture the expression of mRNAs, proteins, and metabolites at a certain time point; therefore, sampling at multiple time points will make environmental omics studies on a single chemical more meaningful for society. The omics profiles for single chemicals collected at different time points also help in the identification of true toxicity-specific changes for chemicals. It will be more meaningful and informative if samples treated with different dosages of toxicants are analyzed to identify toxicity-specific changes. Moreover, computational biology, bioinformatics, and system biology tools would be helpful to integrate and model the complex omics data sets for such chemicals to understand the toxic processes of other chemicals that lead to environmental toxicity and risks to human health.

7.2 Environmental omics associated with exposure to chemical mixtures

Humans are exposed to mixtures of chemicals where exposure of multiple chemicals occurs in each environmental medium or is encountered from multiple environmental media via multiple exposure routes (Celander, 2011). For example, chemical disinfectants react with organic and inorganic matter in water to produce many disinfection by-products (DBPs). To date, more than 500 DBPs have been identified and reported in several studies. Another example is tobacco smoke, which is considered as one of the most important drivers responsible for damage to human health. Tobacco smoke contains more than 4000 chemicals, and at least 200 of them are toxic to humans. Tobacco smoke contains chemicals of which more than 70 are recognized as known or probable human carcinogens (Husgafvel-Pursiainen, 2004). Diesel exhausts, welding fumes, oven emissions, metal, working gas, and fluids are also considered drivers for damage to human health. Exposure to chemical mixtures is therefore a major issue in the environmental and health risk assessment of chemicals due to the higher risks of chemical mixtures compared with individual chemicals to the ecosystem. Toxicity mechanisms by which a single chemical induces carcinogenicity have been well established, which cannot be extended to chemical mixtures. High-throughput omics technologies and methods applied to explore toxicology mechanisms address these challenges. High-throughput omics technologies have the potential to improve our understanding and predictability of combined effects on human health due to the mixture of chemicals. The complex cross-talk between environmental stresses and the dynamic responses of organisms at the levels of genes, RNAs, proteins, and metabolites can be efficiently determined using high-throughput omics technologies. High-throughput omics analysis can identify multiple molecular targets, pathways, and environmental responses to exposed targeted organisms. Omics technologies also enable us to understand the modes of action of chemicals and determine various components that cause toxicity in a mixture. High-throughput omics technologies also provide a great opportunity to identify novel toxic substances for categorization and prioritization (Hook et al., 2008). Proteomics and metabolomics have also been used to characterize chemical interactions and chemical mixture toxicity (Merhi et al., 2010; Pelletier et al., 2009). These studies focused on the identification of protein or metabolite signatures associated with the toxicity of both individual chemicals and mixtures. Moreover, the integration of genomic, proteomic, and metabolomic approaches is very important for improving our understanding and predictive capability of the combined effects of chemical mixtures. Well-accepted conceptual frameworks and standards should be formulated and implemented for experimental design, data interpretation, and computational modeling for the use of OMICS tools to study chemical mixture toxicity and associated risks to our environment and human health.

8. Traditional versus omics-based environmental toxicology

Traditional environmental toxicology and environmental health research mainly focus on the effects of environmental chemicals on the functions of various organ systems, phenotypes, and modes of action of the chemicals at cellular levels in a particular animal model. Traditional environmental toxicology and environmental health research are time-, labor-, and resource-intensive. Traditional methods are unable to characterize the full spectrum of targets and toxicity mechanisms for chemicals that affect multiple systems. To understand the toxicity mechanisms and the modes of action, there is a need to understand the toxic processes at the cellular and molecular levels, which needs to integrate genomic, proteomic, and metabolomic technologies for ecological toxicology research (Gaytán & Vulpe, 2014).

Genomics was probably the first omics approach for studies of toxicity and toxicity pathways in ecotoxicological research. Genomic studies involved environmental chemicals and their mixtures to be characterized for gene expression and reflected by the abundance of specific mRNA transcripts in a biological sample (Iwahashi et al., 2006). Gene expression profiling refers to the comparison of groups of genes with and without exposure to environmental chemicals in terms of gene expression. However, the microarray technology revolutionized biological research and enabled thousands of gene transcripts to be monitored simultaneously, but this is being replaced by RNA-Seq, an omics-based high-throughput technology. Most of the time, transcriptional responses often do not accurately reflect important toxicologically relevant biological responses because of important changes in proteins and metabolites within cells that are not detectable by just studying the levels of mRNAs. Therefore, proteomics and metabolomics are often used to complement genomics for toxicological and human health studies. Omics-based high-throughput technologies, when used in totality, can perform qualitative and quantitative measurement of changes from the molecular, cell, tissue, individual to population, to the community level, and thus forecast a story of toxic substances released to the environment and its effect on the ecosystem and human health. Omics technologies aim to provide a global view of the cellular processes of an individual and the changes occurring in organisms in response to environmental change. Therefore, using high-throughput omics technologies can increase efficacy in less time while conducting an environmental assessment, in addition to exploring novel mechanisms contributing to environmental degradation. Genomics, transcriptomics, proteomics, and metabolomics indirectly correlate chemical exposure with gene products (e.g., mRNA, protein, metabolites), whereas using functional genomics, which can directly link genes and their functions, which could explain various critical genes and/or pathways modulated by chemicals (North & Vulpe, 2010; Shen et al., 2015). The development of reproducible high-throughput omics protocols for chemical testing is urgently needed to support the incorporation of omics technologies into chemical risk assessment.

9. Environmental monitoring and health risk assessment

Environmental monitoring and health risk assessment of ecosystems are important in the development of effective strategies for not only protecting the environment but human health as well. In aquatic ecosystems from oceans to river basins, water quality assessment has focused on providing specific information regarding the pollutants released into water, and dynamics of pollutants, and their effects on the health of different aquatic species and food chains including humans. Traditional biomonitoring programs directly quantify the bioaccumulation of pollutants in the tissues of exposed organisms or analyze the effects on those exposed organisms. Bivalves have been extensively used organisms to study the bioaccumulation of pollutants in tissues due to their sessile nature, filter feeder habits, and high capacity to accumulate contaminants (Kolpin et al., 2002). In a study in 2006, proteomics in bivalves investigated some specific and sensitive markers of pollution, thereby providing an accurate estimation of ecosystem health (Apraiz et al., 2006).

10. Future directions of ecology and environmental omics

There has been substantial progress in studying environmental stressors and environmental single or mixture of chemicals as toxicants, due to the advances and increased usage of high-throughput molecular technologies, which helped in identifying, characterizing these chemicals, and exploring the short-term and long-term effects of these chemicals on human health. Most of the high-throughput genomics, transcriptomics, proteomics, epigenomics, and metagenomics studies today are highly automated for efficiently processing many samples; they have become more efficient and less expensive tools for research. Despite the progress, the use of multiomics technologies and integration of high-throughput multiomics technologies—based studies are still few due to the broad impact of toxicants at molecular, cellular, tissue, organ, individual, population, ecosystem, and biosphere. Multiomics technologies should also be used to explore a greater number of single or mixture of chemicals for their toxicity mechanisms, gene expression, pathways affected, and adverse outcome pathways, as well as short- and long-term effects due to exposure to chemicals on environmental target organisms. Several model organisms are still unexplored for the effects of various environmental stressors/chemicals using multiomics technologies, which should be investigated. Several model organism's genomes should also be explored for changes due to dietary and environmental factors. There is also a need for investigating several model organisms to be used for environmental monitoring owing to the heath of information in this field.

References

Ablain, J., & Zon, L. I. (2013). Of fish and men: Using zebrafish to fight human diseases. *Trends in Cell Biology, 23*(12), 584–586. https://doi.org/10.1016/j.tcb.2013.09.009

Aguiar-Pulido, V., Huang, W., Suarez-Ulloa, V., Cickovski, T., Mathee, K., & Narasimhan, G. (2016). Metagenomics, metatranscriptomics, and metabolomics approaches for microbiome analysis. *Evolutionary Bioinformatics, 12*(Suppl. 1), 5–16. https://doi.org/10.4137/EBO.S36436. PMID: 27199545; PMCID: PMC4869604.

Ali, N., Wolf, C., Kanchan, S., Veerabhadraiah, S. R., Bond, L., Turner, M. W., Jorcyk, C. L., & Hampikian, G. (2024). 9S1R nullomer peptide induces mitochondrial pathology, metabolic supression, and enhanced cell infilteration in triple negative breast cancer mouse model. *Biomedicine & Pharmacotherapy, 170*, Article 115997. https://doi.org/10.1016/j.biopha.2023.115997

AlJanahi, A. A., Danielsen, M., & Dunbar, C. E. (2018). An introduction to the analysis of single-cell RNA-sequencing data. *Molecular Therapy—Methods and Clinical Development, 10*, 189–196. https://doi.org/10.1016/j.omtm.2018.07.003. https://www.journals.elsevier.com/molecular-therapy-methods-and-clinical-development/

Ankley, G. T., Bennett, R. S., Erickson, R. J., Hoff, D. J., Hornung, M. W., Johnson, R. D., Mount, D. R., Nichols, J. W., Russom, C. L., Schmieder, P. K., Serrrano, J. A., Tietge, J. E., & Villeneuve, D. L. (2010). Adverse outcome pathways: A conceptual framework to support ecotoxicology research and risk assessment. *Environmental Toxicology and Chemistry, 29*(3), 730–741. https://doi.org/10.1002/etc.34. http://www.interscience.wiley.com/jpages/0730-7268

Apraiz, I., Mi, J., & Cristobal, S. (2006). Identification of proteomic signatures of exposure to marine pollutants in mussels (*Mytilus edulis*). *Molecular and Cellular Proteomics, 5*(7), 1274–1285. https://doi.org/10.1074/mcp.M500333-MCP200

Baastrup, R., Sørensen, M., Balstrøm, T., Frederiksen, K., Larsen, C. L., Tjønneland, A., Overvad, K., & Raaschou-Nielsen, O. (2008). Arsenic in drinking-water and risk for cancer in Denmark. *Environmental Health Perspectives, 116*(2), 231–237. https://doi.org/10.1289/ehp.10623. http://www.ehponline.org/members/2007/10623/10623.pdf

Bankevich, A., Nurk, S., Antipov, D., Gurevich, A. A., Dvorkin, M., Kulikov, A. S., Lesin, V. M., Nikolenko, S. I., Pham, S., Prjibelski, A. D., Pyshkin, A. V., Sirotkin, A. V., Vyahhi, N., Tesler, G., Alekseyev, M. A., & Pevzner, P. A. (2012). SPAdes: A new genome assembly algorithm and its applications to single-cell sequencing. *Journal of Computational Biology, 19*(5), 455–477. https://doi.org/10.1089/cmb.2012.0021

Barrett, T. J., & Munkittrick, K. R. (2010). Seasonal reproductive patterns and recommended sampling times for sentinel fish species used in environmental effects monitoring programs in Canada. *Environmental Reviews, 18*(1), 115–135. https://doi.org/10.1139/A10-004. http://article.pubs.nrc-cnrc.gc.ca/RPAS/rpv?hm=HInit&calyLang=eng&journal=er&volume=18&afpf=a10-004.pdf

Besser, J., Carleton, H. A., Gerner-Smidt, P., Lindsey, R. L., & Trees, E. (2018). Next-generation sequencing technologies and their application to the study and control of bacterial infections. *Clinical Microbiology and Infection, 24*(4), 335–341. https://doi.org/10.1016/j.cmi.2017.10.013. http://onlinelibrary.wiley.com/journal/10.1111/(ISSN)1469-0691

Bhetraratana, M., Orozco, L. D., Hong, J., Diamante, G., Majid, S., Bennett, B. J., Ahn, I. S., Yang, X., Lusis, A. J., & Araujo, J. A. (2019). Diesel exhaust particles dysregulate multiple immunological pathways in murine macrophages: Lessons from microarray and scRNA-seq technologies. *Archives of Biochemistry and Biophysics, 678*. https://doi.org/10.1016/j.abb.2019.108116. http://www.elsevier.com/inca/publications/store/6/2/2/7/8/7/index.htt

Bludau, I., & Aebersold, R. (2020). Proteomic and interactomic insights into the molecular basis of cell functional diversity. *Nature Reviews Molecular Cell Biology, 21*(6), 327–340. https://doi.org/10.1038/s41580-020-0231-2. http://www.nature.com/molcellbio

Boontawon, T., Nakazawa, T., Inoue, C., Osakabe, K., Kawauchi, M., Sakamoto, M., & Honda, Y. (2021). Efficient genome editing with CRISPR/Cas9 in Pleurotus ostreatus. *AMB Express, 11*(1). https://doi.org/10.1186/s13568-021-01193-w. http://www.amb-express.com/

Botero, L. M., D'Imperio, S., Burr, M., McDermott, T. R., Young, M., & Hassett, D. J. (2005). Poly(A) polymerase modification and reverse transcriptase PCR amplification of environmental RNA. *Applied and Environmental Microbiology, 71*(3), 1267–1275. https://doi.org/10.1128/AEM.71.3.1267-1275.2005

Brandl, J., & Andersen, M. R. (2017). Aspergilli: Models for systems biology in filamentous fungi. *Current Opinion in Structural Biology, 6*, 67–73. https://doi.org/10.1016/j.coisb.2017.09.005. https://www.journals.elsevier.com/current-opinion-in-systems-biology/

Brockmeier, E. K., Hodges, G., Hutchinson, T. H., Butler, E., Hecker, M., Tollefsen, K. E., Garcia-Reyero, N., Kille, P., Becker, D., Chipman, K., Colbourne, J., Collette, T. W., Cossins, A., Cronin, M., Graystock, P., Gutsell, S., Knapen, D., Katsiadaki, I., Lange, A., … Falciani, F. (2017). The role of omics in the application of adverse outcome pathways for chemical risk assessment. *Toxicological Sciences, 158*(2), 252–262. https://doi.org/10.1093/toxsci/kfx097. http://toxsci.oxfordjournals.org/

Bundy, J. G., Sidhu, J. K., Rana, F., Spurgeon, D. J., Svendsen, C., Wren, J. F., Stürzenbaum, S. R., Morgan, A. J., & Kille, P. (2008). 'Systems toxicology' approach identifies coordinated metabolic responses to copper in a terrestrial non-model invertebrate, the earthworm *Lumbricus rubellus*. *BMC Biology, 6*. https://doi.org/10.1186/1741-7007-6-25

Cai, L., & Zhang, T. (2013). Detecting human bacterial pathogens in wastewater treatment plants by a high-throughput shotgun sequencing technique. *Environmental Science and Technology, 47*(10), 5433–5441. https://doi.org/10.1021/es400275r

Campos, A., Tedesco, S., Vasconcelos, V., & Cristobal, S. (2012). Proteomic research in bivalves. Towards the identification of molecular markers of aquatic pollution. *Journal of Proteomics, 75*(14), 4346–4359. https://doi.org/10.1016/j.jprot.2012.04.027

Celander, M. C. (2011). Cocktail effects on biomarker responses in fish. *Aquatic Toxicology, 105*(3–4), 72–77. https://doi.org/10.1016/j.aquatox.2011.06.002

Cheng, M., Cao, L., & Ning, K. (2019). Microbiome big-data mining and applications using single-cell technologies and metagenomics approaches toward precision medicine. *Frontiers in Genetics, 10*. https://doi.org/10.3389/fgene.2019.00972. https://www.frontiersin.org/journals/genetics#

Cheung, W. M. W., Chu, P. W. K., & Kwong, Y. L. (2007). Effects of arsenic trioxide on the cellular proliferation, apoptosis and differentiation of human neuroblastoma cells. *Cancer Letters, 246*(1–2), 122–128. https://doi.org/10.1016/j.canlet.2006.02.009

Cho, N. H., Cheveralls, K. C., Brunner, A. D., Kim, K., Michaelis, A. C., Raghavan, P., Kobayashi, H., Savy, L., Li, J. Y., Canaj, H., Kim, J. Y. S., Stewart, E. M., Gnann, C., McCarthy, F., Cabrera, J. P., Brunetti, R. M., Chhun, B. B., Dingle, G., Hein, M. Y., ... Leonetti, M. D. (2022). OpenCell: Endogenous tagging for the cartography of human cellular organization. *Science, 375*(6585). https://doi.org/10.1126/science.abi6983. https://www.science.org/doi/10.1126/science.abi6983

Danaei, G., Ding, E. L., Mozaffarian, D., Taylor, B., Rehm, J., Murray, C. J. L., & Ezzati, M. (2009). The preventable causes of death in the United States: Comparative risk assessment of dietary, lifestyle, and metabolic risk factors. *PLoS Medicine, 6*(4). https://doi.org/10.1371/journal.pmed.1000058. http://www.plosmedicine.org/article/fetchObjectAttachment.action?uri=info%3Adoi%2F10.1371%2Fjournal.pmed.1000058&representation=PDF

Diamanti-Kandarakis, E., Bourguignon, J. P., Giudice, L. C., Hauser, R., Prins, G. S., Soto, A. M., Zoeller, R. T., & Gore, A. C. (2009). Endocrine-disrupting chemicals: An Endocrine Society scientific statement. *Endocrine Reviews, 30*(4), 293–342. https://doi.org/10.1210/er.2009-0002. http://edrv.endojournals.org/cgi/reprint/30/4/293

Dolinoy, D. C., & Jirtle, R. L. (2008). Environmental epigenomics in human health and disease. *Environmental and Molecular Mutagenesis, 49*(1), 4–8. https://doi.org/10.1002/em.20366

Ebner, J. N. (2021). Trends in the application of "omics" to ecotoxicology and stress ecology. *Genes, 12*(10), 1481. https://doi.org/10.3390/genes12101481

Faugere, J., Brunet, T. A., Clément, Y., Espeyte, A., Geffard, O., Lemoine, J., Chaumot, A., Degli-Esposti, D., Ayciriex, S., & Salvador, A. (2023). Development of a multi-omics extraction method for ecotoxicology: Investigation of the reproductive cycle of Gammarus fossarum. *Talanta, 253*, 123806. https://doi.org/10.1016/j.talanta.2022.123806

Feinstein, L. M., Woo, J. S., & Blackwood, C. B. (2009). Assessment of bias associated with incomplete extraction of microbial DNA from soil. *Applied and Environmental Microbiology, 75*(16), 5428–5433. https://doi.org/10.1128/AEM.00120-09. http://aem.asm.org/cgi/reprint/75/16/5428

Fent, K., & Sumpter, J. P. (2011). Progress and promises in toxicogenomics in aquatic toxicology: Is technical innovation driving scientific innovation? *Aquatic Toxicology, 105*(3–4), 25–39. https://doi.org/10.1016/j.aquatox.2011.06.008

Flanagan, J. L., Brodie, E. L., Weng, L., Lynch, S. V., Garcia, O., Brown, R., Hugenholtz, P., DeSantis, T. Z., Andersen, G. L., Wiener-Kronish, J. P., & Bristow, J. (2007). Loss of bacterial diversity during antibiotic treatment of intubated patients colonized with *Pseudomonas aeruginosa*. *Journal of Clinical Microbiology, 45*(6), 1954–1962. https://doi.org/10.1128/JCM.02187-06

Gahoi, S., Mandal, R. S., Ivanisenko, N., Shrivastava, P., Jain, S., Singh, A. K., Raghunandanan, M. V., Kanchan, S., Taneja, B., Mandal, C., Ivanisenko, V. A., Kumar, A., Kumar, R., Consorti, Open Source Drug Discovery, Ramachandran, S. (2013). Computational screening for new inhibitors of *M. tuberculosis* mycolyltransferases antigen 85 group of proteins as potential drug targets. *Journal of Biomolecular Structure and Dynamics, 31*(1), 30–43. https://doi.org/10.1080/07391102.2012.691343

Galande, S. H., Kanchan, S., & Kesheri, M. (2014). *Drug discovery and design: A bioinformatics approach.* LAP Lambert Academy Publishing.

Garg, S., Saxena, V., Kanchan, S., Sharma, P., Mahajan, S., Kochar, D., & Das, A. (2009). Novel point mutations in sulfadoxine resistance genes of Plasmodium falciparum from India. *Acta Tropica, 110*(1), 75–79. https://doi.org/10.1016/j.actatropica.2009.01.009

Gaytán, B. D., & Vulpe, C. D. (2014). Functional toxicology: Tools to advance the future of toxicity testing. *Frontiers in Genetics, 5*(MAY). https://doi.org/10.3389/fgene.2014.00110. http://journal.frontiersin.org/Journal/10.3389/fgene.2014.00110/full

Ghai, R., Nagarajan, K., Kumar, V., Kesheri, M., & Kanchan, S. (2015). Amelioration of lipids by Eugenia caryophyllus extract in atherogenic diet induced hyperlipidemia. *International Bulletin of Drug Research, 5*(8), 90–101.

Ghai, R., Nagarajan, K., Singh, J., Swarup, S., & Keshari, M. (2016). Evaluation of anti-oxidant status in-vitro and in-vivo in hydro-alcoholic extract of Eugenia caryophyllus. *International Journal of Pharmacology and Toxicology, 4*(1), 19. https://doi.org/10.14419/ijpt.v4i1.5880

Hajibabaei, M., Singer, G. A. C., Clare, E. L., & Hebert, P. D. N. (2007). Design and applicability of DNA arrays and DNA barcodes in biodiversity monitoring. *BMC Biology, 5*. https://doi.org/10.1186/1741-7007-5-24

Hasenbein, M., Werner, I., Deanovic, L. A., Geist, J., Fritsch, E. B., Javidmehr, A., Foe, C., Fangue, N. A., & Connon, R. E. (2014). Transcriptomic profiling permits the identification of pollutant sources and effects in ambient water samples. *Science of the Total Environment, 468–469*, 688–698. https://doi.org/10.1016/j.scitotenv.2013.08.081. http://www.elsevier.com/locate/scitotenv

He, Y., Patterson, S., Wang, N., Hecker, M., Martin, J. W., El-Din, M. G., Giesy, J. P., & Wiseman, S. B. (2012). Toxicity of untreated and ozone-treated oil sands process-affected water (OSPW) to early life stages of the fathead minnow (*Pimephales promelas*). *Water Research, 46*(19), 6359–6368. https://doi.org/10.1016/j.watres.2012.09.004. http://www.elsevier.com/locate/watres

Hodson, P. V., Thivierge, D., Levesque, M.-., McWhirter, M., Ralph, K., Gray, B., Whittle, D. M., Carey, J. H., & Van Der Kraak, G. (1992). Effects of bleached kraft mill effluent on fish in the St. Maurice River, Quebec. *Environmental Toxicology and Chemistry, 11*(11), 1635–1651. https://doi.org/10.1002/etc.5620111113

Hook, S. E. (2010). Promise and progress in environmental genomics: A status report on the applications of gene expression-based microarray studies in ecologically relevant fish species. *Journal of Fish Biology, 77*(9), 1999–2022. https://doi.org/10.1111/j.1095-8649.2010.02814.x

Hook, S. E., Skillman, A. D., Gopalan, B., Small, J. A., & Schultz, I. R. (2008). Gene expression profiles in rainbow trout, onchorynchus mykiss, exposed to a simple chemical mixture. *Toxicological Sciences, 102*(1), 42–60. https://doi.org/10.1093/toxsci/kfm293

Hu, T., Chitnis, N., Monos, D., & Dinh, A. (2021). Next-generation sequencing technologies: An overview. *Human Immunology, 82*(11), 801–811. https://doi.org/10.1016/j.humimm.2021.02.012. http://www.elsevier.com/locate/humimm

Husgafvel-Pursiainen, K. (2004). Genotoxicity of environmental tobacco smoke: A review. *Mutation Research: Reviews in Mutation Research, 567*(2–3), 427–445. https://doi.org/10.1016/j.mrrev.2004.06.004

Iwahashi, Y., Hosoda, H., Park, J.-H., Lee, J.-H., Suzuki, Y., Kitagawa, E., Murata, S. M., Jwa, N.-S., Gu, M.-B., & Iwahashi, H. (2006). Mechanisms of patulin toxicity under conditions that inhibit yeast growth. *Journal of Agricultural and Food Chemistry, 54*(5), 1936–1942. https://doi.org/10.1021/jf052264g

Järup, L. (2003). Hazards of heavy metal contamination. *British Medical Bulletin, 68*, 167–182. https://doi.org/10.1093/bmb/ldg032

Jain, M., Koren, S., Miga, K. H., Quick, J., Rand, A. C., Sasani, T. A., Tyson, J. R., Beggs, A. D., Dilthey, A. T., Fiddes, I. T., Malla, S., Marriott, H., Nieto, T., O'Grady, J., Olsen, H. E., Pedersen, B. S., Rhie, A., Richardson, H., Quinlan, A. R., ... Loose, M. (2018). Nanopore sequencing and assembly of a human genome with ultra-long reads. *Nature Biotechnology, 36*(4), 338–345. https://doi.org/10.1038/nbt.4060. http://www.nature.com/nbt/index.html

Kanchan, S., Mehrotra, R., & Chowdhury, S. (2014). Evolutionary pattern of four representative DNA repair proteins across six model organisms: An in silico analysis. *Network Modeling Analysis in Health Informatics and Bioinformatics, 3*(1). https://doi.org/10.1007/s13721-014-0070-1. http://www.springer.com/new+%26+forthcoming+titles+%28default%29/journal/13721

Kanchan, S., Mehrotra, R., & Chowdhury, S. (2015). In silico analysis of the endonuclease III protein family identifies key residues and processes during evolution. *Journal of Molecular Evolution, 81*(1–2), 54–67. https://doi.org/10.1007/s00239-015-9689-5. http://link.springer.de/link/service/journals/00239/index.htm

Kanchan, S., Ogden, E., Kesheri, M., Skinner, A., Miliken, E., Lyman, D., Armstrong, J., Sciglitano, L., & Hampikian, G. (2024). Covid-19 hospitalizations and deaths predicted by SARS-CoV-2 levels in Boise, Idaho wastewater. *The Science of the Total Environment, 907*, Article 167742. https://doi.org/10.1016/j.scitotenv.2023.167742

Kanchan, S., Sharma, P., & Chowdhury, S. (2019). Evolution of endonuclease IV protein family: An in silico analysis. *3 Biotech, 9*(5). https://doi.org/10.1007/s13205-019-1696-6. http://www.springerlink.com/content/2190-572x/

Kanchan, S., Sinha, R. P., Chaudière, J., & Kesheri, M. (2020). Computational metagenomics: Current status and challenges. *Recent Trends in Computational Omics: Concepts and Methodology*, 371–395. https://novapublishers.com/shop/recent-trends-in-computational-omics-concepts-and-methodology/.

Kanerva, M., Vehmas, A., Nikinmaa, M., & Vuori, K. A. (2014). Spatial variation in transcript and protein abundance of Atlantic salmon during feeding migration in the Baltic sea. *Environmental Science and Technology, 48*(23), 13969–13977. https://doi.org/10.1021/es502956g. http://pubs.acs.org/journal/esthag

Kesheri, M., Kanchan, S., & Sinha, R. P. (2017). *Exploring the potentials of antioxidants in retarding ageing* (pp. 166–195). https://doi.org/10.4018/978-1-5225-0607-2.ch008

Kesheri, M., Kanchan, S., & Sinha, R. P. (2021). Isolation and in silico analysis of antioxidants in response to temporal variations in the cyanobacterium Oscillatoria sp. *Gene Reports, 23*, 101023. https://doi.org/10.1016/j.genrep.2021.101023

Kesheri, M., Kanchan, S., & Chowdhury, S. (2014). *Cyanobacterial stresses: An ecophysiological, biotechnological and bioinformatic approach.* LAP Lambert Academy Publishing.

Kesheri, M., Kanchan, S., Chowdhury, S., & Sinha, R. P. (2015). Secondary and tertiary structure prediction of proteins: A bioinformatic approach. *Studies in Fuzziness and Soft Computing, 319*, 541–569. https://doi.org/10.1007/978-3-319-12883-2_19

Kesheri, M., Kanchan, S., Richa, & Sinha, R. P. (2014). Isolation and in silico analysis of Fe-superoxide dismutase in the cyanobacterium Nostoc commune. *Gene, 553*(2), 117–125. https://doi.org/10.1016/j.gene.2014.10.010. http://www.elsevier.com/locate/gene

Kesheri, M., Kanchan, S., & Sinha, R. P. (2022). Responses of antioxidants for resilience to temporal variations in the cyanobacterium Microcystis aeruginosa. *South African Journal of Botany, 148*, 190–199. https://doi.org/10.1016/j.sajb.2022.04.017. http://www.elsevier.com

Kesheri, M., Richa, & Sinha, R. P. (2011). Antioxidants as natural arsenal against multiple stresses in Cyanobacteria. *International Journal of Pharma and Bio Sciences, 2*(2), 168–187. http://ijpbs.net/volume2/issue2/bio/17.pdf.

Kesheri, M., & Kanchan. (2015a). Computational methods and strategies for protein structure prediction. *Biological Science: Innovations and Dynamics. Chapter, 12*, 277–291.

Kesheri, M., & Kanchan. (2015b). Oxidative stress: Challenges and its mitigation mechanisms in cyanobacteria in. *Biological Science: Innovations and Dynamics*, 309–324.

Kesheri, M., Sinha, R. P., & Kanchan, S. (2016). Advances in soft computing approaches for gene prediction: A bioinformatics approach. *Studies in Computational Intelligence, 651*, 383–405. https://doi.org/10.1007/978-3-319-33793-7_17. http://www.springer.com/series/7092

Khaw, K.-T., Wareham, N., Bingham, S., Welch, A., Luben, R., Day, N., & Lopez, A. (2008). Combined impact of health behaviours and mortality in men and women: The EPIC-Norfolk prospective population study. *PLoS Medicine, 5*(1), e12. https://doi.org/10.1371/journal.pmed.0050012

Kishi, K., Kitagawa, E., Onikura, N., Nakamura, A., & Iwahashi, H. (2006). Expression analysis of sex-specific and 17β-estradiol-responsive genes in the Japanese medaka, *Oryzias latipes*, using oligonucleotide microarrays. *Genomics, 88*(2), 241–251. https://doi.org/10.1016/j.ygeno.2006.03.023

Kolpin, D. W., Furlong, E. T., Meyer, M. T., Thurman, E. M., Zaugg, S. D., Barber, L. B., & Buxton, H. T. (2002). Pharmaceuticals, hormones, and other organic wastewater contaminants in U.S. Streams, 1999–2000: A national reconnaissance. *Environmental Science and Technology, 36*(6), 1202–1211. https://doi.org/10.1021/es011055j

Kramer, V. J., Etterson, M. A., Hecker, M., Murphy, C. A., Rocsijadi, G., Spade, D. J., Spromberg, J. A., Wang, M., & Ankley, G. T. (2011). Adverse outcome pathways and ecological risk assessment: Bridging to population-level effects. *Environmental Toxicology and Chemistry, 30*(1), 64–76. https://doi.org/10.1002/etc.375

Kumari, A., Kanchan, S., Sinha, R. P., & Kesheri, M. (2016). Applications of bio-molecular databases in bioinformatics. *Studies in Computational Intelligence, 651*, 329–351. https://doi.org/10.1007/978-3-319-33793-7_15. http://www.springer.com/series/7092

Kumari, A., Kesheri, M., Sinha, R. P., & Kanchan, S. (2018). Integration of soft computing approach in plant biology and its applications in agriculture. *Soft Computing for Biological Systems*, 265–281. https://doi.org/10.1007/978-981-10-7455-4_16. http://www.springer.com/in/book/9789811074547

Kuska, B., Beer, Bethesda, & biology. (1998). How "genomics" came into being. *Journal of the National Cancer Institute, 90*(2).

Lederberg, J., & McCray, A. T. (2001). 'Ome sweet' omics—A geneological treasure of words. *The Scientist, 15*(7).

Li, G., Lee, L. S., Li, M., Tsao, S. W., & Chiu, J. F. (2011). Molecular changes during arsenic-induced cell transformation. *Journal of Cellular Physiology, 226*(12), 3225–3232. https://doi.org/10.1002/jcp.22683

Li, M., Cai, J. F., & Chiu, J. F. (2002). Arsenic induces oxidative stress and activates stress gene expressions in cultured lung epithelial cells. *Journal of Cellular Biochemistry, 87*(1), 29–38. https://doi.org/10.1002/jcb.10269

Li, M., Wang, J., Lu, Z., Wei, D., Yang, M., & Kong, L. (2014). NMR-based metabolomics approach to study the toxicity of lambda-cyhalothrin to goldfish (*Carassius auratus*). *Aquatic Toxicology, 146*, 82–92. https://doi.org/10.1016/j.aquatox.2013.10.024. http://www.elsevier.com/wps/find/journaldescription.cws_home/505509/description#description

Liu, L., Li, Y., Li, S., Hu, N., He, Y., Pong, R., Lin, D., Lu, L., & Law, M. (2012). Comparison of next-generation sequencing systems. *Journal of Biomedicine and Biotechnology, 2012*. https://doi.org/10.1155/2012/251364

Liu, Y., Chen, L., Yu, J., Ye, L., Hu, H., Wang, J., & Wu, B. (2022). Advances in single-cell toxicogenomics in environmental toxicology. *Environmental Science and Technology, 56*(16), 11132–11145. https://doi.org/10.1021/acs.est.2c01098. http://pubs.acs.org/journal/esthag

Lowell, R., Ring, B., Pastershank, G., Walker, S., Trudel, L., & Hedley, K. (2005). *National assessment of pulp and paper environmental effects monitoring data: Findings from cycles 1 through 3.*

Luo, H., Gao, F., & Lin, Y. (2015). Evolutionary conservation analysis between the essential and nonessential genes in bacterial genomes. *Scientific Reports, 5*. https://doi.org/10.1038/srep13210. http://www.nature.com/srep/index.html

MacKenzie, M., & Argyropoulos, C. (2023). An introduction to nanopore sequencing: Past, present, and future considerations. *Micromachines, 14*(2), 459. https://doi.org/10.3390/mi14020459

Marshall, G., Ferreccio, C., Yuan, Y., Bates, M. N., Steinmaus, C., Selvin, S., Liaw, J., & Smith, A. H. (2007). Fifty-Year study of lung and bladder cancer mortality in Chile related to arsenic in drinking water. *Journal of the National Cancer Institute, 99*(12), 920–928. https://doi.org/10.1093/jnci/djm004

Martin-Chouly, C., Morzadec, C., Bonvalet, M., Galibert, M. D., Fardel, O., & Vernhet, L. (2011). Inorganic arsenic alters expression of immune and stress response genes in activated primary human T lymphocytes. *Molecular Immunology, 48*(6–7), 956–965. https://doi.org/10.1016/j.molimm.2011.01.005

Martyniuk, C. J., & Houlahan, J. (2013). Assessing gene network stability and individual variability in the fathead minnow (*Pimephales promelas*) transcriptome. *Comparative Biochemistry and Physiology Part D: Genomics and Proteomics, 8*(4), 283–291. https://doi.org/10.1016/j.cbd.2013.08.002. https://www.journals.elsevier.com/comparative-biochemistry-and-physiology-part-d-genomics-and-proteomics

Matheis, K., Laurie, D., Andriamandroso, C., Arber, N., Badimon, L., Benain, X., Bendjama, K., Clavier, I., Colman, P., Firat, H., Goepfert, J., Hall, S., Joos, T., Kraus, S., Kretschmer, A., Merz, M., Padro, T., Planatscher, H., Rossi, A., ... Molac, B. (2011). A generic operational strategy to qualify translational safety biomarkers. *Drug Discovery Today, 16*(13–14), 600–608. https://doi.org/10.1016/j.drudis.2011.04.011

Mathers, J. C., Strathdee, G., & Relton, C. L. (2010). Induction of epigenetic alterations by dietary and other environmental factors. *Advances in Genetics, 71*(C), 3–39. https://doi.org/10.1016/B978-0-12-380864-6.00001-8. http://www.elsevier.com/wps/find/bookdescription.cws_home/703716/description#description

May, P., Wienkoop, S., Kempa, S., Usadel, B., Christian, N., Rupprecht, J., Weiss, J., Recuenco-Munoz, L., Ebenhöh, O., Weckwerth, W., & Walther, D. (2008). Metabolomics- and proteomics-assisted genome annotation and analysis of the draft metabolic network of *Chlamydomonas reinhardtii*. *Genetics, 179*(1), 157–166. https://doi.org/10.1534/genetics.108.088336. http://www.genetics.org/cgi/reprint/179/1/157

Mei, N., Guo, L., Liu, R., Fuscoe, J. C., & Chen, T. (2007). Gene expression changes induced by the tumorigenic pyrrolizidine alkaloid riddelliine in liver of Big Blue rats. *BMC Bioinformatics, 8*(7). https://doi.org/10.1186/1471-2105-8-S7-S4

Merhi, M., Demur, C., Racaud-Sultan, C., Bertrand, J., Canlet, C., Estrada, F. B. Y., & Gamet-Payrastre, L. (2010). Gender-linked haematopoietic and metabolic disturbances induced by a pesticide mixture administered at low dose to mice. *Toxicology, 267*(1–3), 80–90. https://doi.org/10.1016/j.tox.2009.10.024

Mezhoud, K., Praseuth, D., Francois, J. C., Bernard, C., & Edery, M. (2008). Global quantitative analysis of protein phosphorylation status in fish exposed to microcystin. *Advances in Experimental Medicine and Biology, 617*, 419–426. https://doi.org/10.1007/978-0-387-69080-3_40

Mishra, P., Saxena, V., Kesheri, M., & Saxena, A. (2015a). Synthesis, characterization and pharmacological evaluation of cinnoline (thiophene) derivatives. *The Pharma Innovation Journal, 4*(10), 68–73.

Mishra, Saxena, V., Kesheri, M., & Saxena, A. (2015b). Synthesis, characterization and antiinflammatory activity of cinnolines (pyrazole) derivatives. *IOSR Journal of Pharmacy and Biological Sciences, 10*(6), 77–82.

Munkittrick, K. R., Arens, C. J., Lowell, R. B., & Kaminski, G. P. (2009). A review of potential methods of determining critical effect size for designing environmental monitoring programs. *Environmental Toxicology and Chemistry, 28*(7), 1361–1371. https://doi.org/10.1897/08-376.1

Munkittrick, K. R., McGeachy, S. A., McMaster, M. E., & Courtenay, S. C. (2002). Overview of freshwater fish studies from the pulp and paper environmental effects monitoring program. *Water Quality Research Journal of Canada, 37*(1), 49–77. https://doi.org/10.2166/wqrj.2002.005. http://www.iwaponline.com/wqrjc/toc.htm

Munkittrick, K. R., Portt, C. B., Van Der Kraak, G. J., Smith, I. R., & Rokosh, D. A. (1991). Impact of bleached kraft mill effluent on population characteristics, liver MFO activity, and serum steroid levels of a Lake Superior white sucker (*Catostomus commersoni*) population. *Canadian Journal of Fisheries and Aquatic Sciences, 48*(8), 1371–1380. https://doi.org/10.1139/f91-164

Nam, S. E., Bae, D. Y., Ki, J. S., Ahn, C. Y., & Rhee, J. S. (2023). The importance of multi-omics approaches for the health assessment of freshwater ecosystems. *Molecular and Cellular Toxicology, 19*(1), 3–11. https://doi.org/10.1007/s13273-022-00286-2. https://www.springer.com/journal/13273

Nellimarla, S., & Kesanakurti, P. (2023). Next-generation sequencing: A promising tool for vaccines and other biological products. *Vaccines, 11*(3), 527. https://doi.org/10.3390/vaccines11030527

North, M., & Vulpe, C. D. (2010). Functional toxicogenomics: Mechanism-centered toxicology. *International Journal of Molecular Sciences, 11*(12), 4796–4813. https://doi.org/10.3390/ijms11124796. http://www.mdpi.com/1422-0067/11/12/4796/pdf

Okamoto, T., duVerle, D., Yaginuma, K., Natsume, Y., Yamanaka, H., Kusama, D., Fukuda, M., Yamamoto, M., Perraudeau, F., Srivastava, U., Kashima, Y., Suzuki, A., Kuze, Y., Takahashi, Y., Ueno, M., Sakai, Y., Noda, T., Tsuda, K., Suzuki, Y., Nagayama, S., & Yao, R. (2021). Comparative analysis of patient-matched PDOs revealed a reduction in OLFM4-associated clusters in metastatic lesions in colorectal cancer. *Stem Cell Reports, 16*(4), 954–967. https://doi.org/10.1016/j.stemcr.2021.02.012. http://www.elsevier.com/journals/stem-cell-reports/2213-6711

Ortiz, P. A., Bruno, M. E., Moore, T., Nesnow, S., Winnik, W., & Ge, Y. (2010). Proteomic analysis of propiconazole responses in mouse liver: Comparison of genomic and proteomic profiles. *Journal of Proteome Research, 9*(3), 1268–1278. https://doi.org/10.1021/pr900755q

Osachoff, H. L., Van Aggelen, G. C., Mommsen, T. P., & Kennedy, C. J. (2013). Concentration-response relationships and temporal patterns in hepatic gene expression of Chinook salmon (*Oncorhynchus tshawytscha*) exposed to sewage. *Comparative Biochemistry and Physiology Part D: Genomics and Proteomics, 8*(1), 32–44. https://doi.org/10.1016/j.cbd.2012.10.002. https://www.journals.elsevier.com/comparative-biochemistry-and-physiology-part-d-genomics-and-proteomics

Park, J., Kim, J., Lewy, T., Rice, C. M., Elemento, O., Rendeiro, A. F., & Mason, C. E. (2022). Spatial omics technologies at multimodal and single cell/subcellular level. *Genome Biology, 23*(1). https://doi.org/10.1186/s13059-022-02824-6. https://genomebiology.biomedcentral.com

Pelletier, G., Masson, S., Wade, M. J., Nakai, J., Alwis, R., Mohottalage, S., Kumarathasan, P., Black, P., Bowers, W. J., Chu, I., & Vincent, R. (2009). Contribution of methylmercury, polychlorinated biphenyls and organochlorine pesticides to the toxicity of a contaminant mixture based on Canadian Arctic population blood profiles. *Toxicology Letters, 184*(3), 176–185. https://doi.org/10.1016/j.toxlet.2008.11.004

Pershagen, G. (1981). The carcinogenicity of arsenic. *Environmental Health Perspectives, 40*, 93–100. https://doi.org/10.1289/ehp.814093

Priya, P., Kesheri, M., Sinha, R. P., & Kanchan, S. (2017). Molecular dynamics simulations for biological systems. *Pharmaceutical sciences: Breakthroughs in research and practice* (pp. 286–313). IGI Global.

Props, R., Kerckhof, F.-M., Rubbens, P., De Vrieze, J., Hernandez Sanabria, E., Waegeman, W., Monsieurs, P., Hammes, F., & Boon, N. (2017). Absolute quantification of microbial taxon abundances. *The ISME Journal, 11*(2), 584–587. https://doi.org/10.1038/ismej.2016.117

Ralston-Hooper, K. J., Sanchez, B. C., Adamec, J., & Sepúlveda, M. S. (2011). Proteomics in aquatic amphipods: Can it be used to determine mechanisms of toxicity and interspecies responses after exposure to atrazine? *Environmental Toxicology and Chemistry, 30*(5), 1197–1203. https://doi.org/10.1002/etc.475. Epub 2011.

Richa, Kannaujiya, V. K., Kesheri, M., Singh, G., & Sinha, R. P. (2011). Biotechnological potentials of phycobiliproteins. *International Journal of Pharma and Bio Sciences, 2*(4), 446–454. http://www.ijpbs.net/vol-2_issue-4/bio_science/50.pdf.

Richa, R. R. P., Kumari, S., Singh, K. L., Kannaujiya, V. K., Singh, G., Kesheri, M., & Sinha, R. P. (2011b). Biotechnological potential of mycosporine-like amino acids and phycobiliproteins of cyanobacterial origin. *Biotechnology Bioinformatics Bioengineering, 1*(2), 159–171.

Sahu, N., Mishra, S., Kesheri, M., Kanchan, S., & Sinha, R. P. (2022). Identification of cyanobacteria-based natural inhibitors against SARS-CoV-2 druggable target ACE2 using molecular docking study, ADME and toxicity analysis. *Indian Journal of Clinical Biochemistry*. https://doi.org/10.1007/s12291-022-01056-6. http://www.springer.com/life+sci/biochemistry+and+biophysics/journal/12291

Sanger, F., Nicklen, S., & Coulson, A. R. (1977). DNA sequencing with chain-terminating inhibitors. *Proceedings of the National Academy of Sciences of the United States of America, 74*(12), 5463–5467. https://doi.org/10.1073/pnas.74.12.5463

Sauer, U. G., Deferme, L., Gribaldo, L., Hackermüller, J., Tralau, T., van Ravenzwaay, B., Yauk, C., Poole, A., Tong, W., & Gant, T. W. (2017). The challenge of the application of 'omics technologies in chemicals risk assessment: Background and outlook. *Regulatory Toxicology and Pharmacology, 91*, S14–S26. https://doi.org/10.1016/j.yrtph.2017.09.020. http://www.elsevier.com/inca/publications/store/6/2/2/9/3/9/index.htt

Saxena, A., Saxena, V., Kesheri, M., & Mishra. (2015). Comparative hypoglycemic effects of different extract of clitoriaternatea leaves on rats. *IOSR Journal of Pharmacy and Biological Sciences, 10*(2), 60–65.

Shen, H., McHale, C. M., Smith, M. T., & Zhang, L. (2015). Functional genomic screening approaches in mechanistic toxicology and potential future applications of CRISPR-Cas9. *Mutation Research: Reviews in Mutation Research, 764*, 31–42. https://doi.org/10.1016/j.mrrev.2015.01.002. http://www.sciencedirect.com/science/journal/13835742

Shrager, J., Hauser, C., Chang, C. W., Harris, E. H., Davies, J., McDermott, J., Tamse, R., Zhang, Z., & Grossman, A. R. (2003). Chlamydomonas reinhardtii genome project. A guide to the generation and use of the cDNA information. *Plant Physiology, 131*(2), 401–408. https://doi.org/10.1104/pp.016899. http://www.plantphysiol.org/

Shruti, Millerjoth, N. K., & Kesheri, M. (2016). Forecast analysis of the potential and availability of renewable energy in India: A review. *International Journal of Industrial Electronics and Electrical Engineering, 4*(10), 21–26.

Siepel, A., Bejerano, G., Pedersen, J. S., Hinrichs, A. S., Hou, M., Rosenbloom, K., Clawson, H., Spieth, J., Hillier, L. D. W., Richards, S., Weinstock, G. M., Wilson, R. K., Gibbs, R. A., Kent, W. J., Miller, W., & Haussler, D. (2005). Evolutionarily conserved elements in vertebrate, insect, worm, and yeast genomes. *Genome Research, 15*(8), 1034–1050. https://doi.org/10.1101/gr.3715005. http://www.genome.org/cgi/reprint/15/8/1034

Singla, S., Kesheri, M., Kanchan, S., & Aswath, S. (2019b). Current status and data analysis of diabetes in India. *International Journal of Innovative Technology and Exploring Engineering, 8*(9), 1920–1934. https://doi.org/10.35940/ijitee.I8403.078919

Singla, S., Kesheri, M., Kanchan, S., & Mishra, A. (2019). Data analysis of air pollution in India and its effects on health. *Intern. J. Pharma Biosci, 10*(2), 155–169.

Slatko, B. E., Gardner, A. F., & Ausubel, F. M. (2018). Overview of next-generation sequencing technologies. *Current Protocols in Molecular Biology, 122*(1). https://doi.org/10.1002/cpmb.59. http://onlinelibrary.wiley.com/book/10.1002/0471142727

Sogin, M. L., Morrison, H. G., Huber, J. A., Welch, D. M., Huse, S. M., Neal, P. R., Arrieta, J. M., & Herndl, G. J. (2006). Microbial diversity in the deep sea and the underexplored \rare biosphere\. *Proceedings of the National Academy of Sciences of the United States of America, 103*(32), 12115–12120. https://doi.org/10.1073/pnas.0605127103

Srivastava, U., Kanchan, S., Kesheri, S., & Singh, S. (2023). Nutrimetabolomics: Metabolomics in nutrition research. In V. Soni, & T. E. Hartman (Eds.), *Metabolomics* (pp. 241–268). Cham: Springer.

Srivastava, U., & Singh, G. (2013). Comparative homology modelling for HPV type 16 E 7 proteins by using MODELLER and its validations with SAVS and ProSA web server. *Journal of Computational Intelligence in Bioinformatics, 6*(1), 27. https://doi.org/10.37622/jcib/6.1.2013.27-33

Srivastava, U., & Singh, S. (2022). *Approaches of single-cell analysis in crop improvement* (pp. 271–294). Springer Science and Business Media LLC. https://doi.org/10.1007/978-1-0716-2533-0_14

Srivastava, U., Singh, S., Gautam, B., Yadav, P., Yadav, M., Thomas, G., & Singh, G. (2017). Linear epitope prediction in HPV type 16 E7 antigen and their docked interaction with human TMEM 50A structural model. *Bioinformation, 13*(05), 122–130. https://doi.org/10.6026/97320630013122

Ståhl, P. L., Salmén, F., Vickovic, S., Lundmark, A., Navarro, J. F., Magnusson, J., Giacomello, S., Asp, M., Westholm, J. O., Huss, M., Mollbrink, A., Linnarsson, S., Codeluppi, S., Borg, Å., Pontén, F., Costea, P. I., Sahlén, P., Mulder, J., Bergmann, O., Lundeberg, J., & Frisén, J. (2016). Visualization and analysis of gene expression in tissue sections by spatial transcriptomics. *Science, 353*(6294), 78–82. https://doi.org/10.1126/science.aaf2403. http://science.sciencemag.org/content/sci/353/6294/78.full.pdf

Straalen, N. M. V. (2003). Peer reviewed: Ecotoxicology becomes stress ecology. *Environmental Science and Technology, 37*(17), 324A–330A. https://doi.org/10.1021/es0325720

Sun, Y., Pi, J., Wang, X., Tokar, E. J., Liu, J., & Waalkes, M. P. (2009). Aberrant cytokeratin expression during arsenic-induced acquired malignant phenotype in human HaCaT keratinocytes consistent with epidermal carcinogenesis. *Toxicology, 262*(2), 162–170. https://doi.org/10.1016/j.tox.2009.06.003

Swinbanks, D. (1995). Government backs proteome proposal. *Nature, 378*(6558), 653. https://doi.org/10.1038/378653b0

Thomas, T., Gilbert, J., & Meyer, F. (2012). Metagenomics—A guide from sampling to data analysis. *Microbial Informatics and Experimentation, 2*(1). https://doi.org/10.1186/2042-5783-2-3

Tiffon, C. (2018). The impact of nutrition and environmental epigenetics on human health and disease. *International Journal of Molecular Sciences, 19*(11), 3425. https://doi.org/10.3390/ijms19113425

Tolonen, A. C., & Xavier, R. J. (2017). Dissecting the human microbiome with single-cell genomics. *Genome Medicine, 9*(1). https://doi.org/10.1186/s13073-017-0448-7. http://www.genomemedicine.com/

Truhaut, R. (1977). Ecotoxicology: Objectives, principles and perspectives. *Ecotoxicology and Environmental Safety, 1*(2), 151–173. https://doi.org/10.1016/0147-6513(77)90033-1

Van Aggelen, G., Ankley, G. T., Baldwin, W. S., Bearden, D. W., Benson, W. H., Chipman, J. K., Collette, T. W., Craft, J. A., Denslow, N. D., Embry, M. R., Falciani, F., George, S. G., Helbing, C. C., Hoekstra, P. F., Iguchi, T., Kagami, Y., Katsiadaki, I., Kille, P., Liu, L., … Yu, L. (2010). Integrating omic technologies into aquatic ecological risk assessment and environmental monitoring: Hurdles, achievements, and future outlook. *Environmental Health Perspectives, 118*(1), 1–5. https://doi.org/10.1289/ehp.0900985. http://ehp03.niehs.nih.gov/article/fetchObjectAttachment.action?uri=info%3Adoi%2F10.1289%2Fehp.0900985&representation=PDF

van Dijk, E. L., Jaszczyszyn, Y., Naquin, D., & Thermes, C. (2018). The third revolution in sequencing technology. *Trends in Genetics, 34*(9), 666–681. https://doi.org/10.1016/j.tig.2018.05.008. http://www.elsevier.com/locate/tig

Van Scoy, A. R., Yu Lin, C., Anderson, B. S., Philips, B. M., Martin, M. J., McCall, J., Todd, C. R., Crane, D., Sowby, M. L., Viant, M. R., & Tjeerdema, R. S. (2010). Metabolic responses produced by crude versus dispersed oil in Chinook salmon pre-smolts via NMR-based metabolomics. *Ecotoxicology and Environmental Safety, 73*(5), 710–717. https://doi.org/10.1016/j.ecoenv.2010.03.001

Vineis, P., Khan, A. E., Vlaanderen, J., & Vermeulen, R. (2009). The impact of new research technologies on our understanding of environmental causes of disease: The concept of clinical vulnerability. *Environmental Health, 8*(1). https://doi.org/10.1186/1476-069X-8-54

Walker, C. H., Sibly, R. M., Hopkin, S. P., & Peakall, D. B. (2012). *Principles of ecotoxicology* (pp. 978–981). CRC Press.

Wang, J., Wang, Y. Y., Lin, L., Gao, Y., Hong, H. S., & Wang, D. Z. (2012). Quantitative proteomic analysis of okadaic acid treated mouse small intestines reveals differentially expressed proteins involved in diarrhetic shellfish poisoning. *Journal of Proteomics, 75*(7), 2038–2052. https://doi.org/10.1016/j.jprot.2012.01.010

White, R. A., 3rd, Chan, A. M., Gavelis, G. S., Leander, B. S., Brady, A. L., Slater, G. F., Lim, D. S., & Suttle, C. A. (2016). Metagenomic analysis suggests modern freshwater microbialites harbor a distinct core microbial community. *Frontiers in Microbiology, 6*, 1531. https://doi.org/10.3389/fmicb.2015.01531

Wu, H., & Wang, W. X. (2010). NMR-based metabolomic studies on the toxicological effects of cadmium and copper on green mussels Perna viridis. *Aquatic Toxicology, 100*(4), 339–345. https://doi.org/10.1016/j.aquatox.2010.08.005. http://www.elsevier.com/wps/find/journaldescription.cws_home/505509/description#description

Yadav, S. P. (2007). The wholeness in suffix -omics, -omes, and the word om. *Journal of Biomolecular Techniques, 18*(5), 277. http://www.ncbi.nlm.nih.gov/pmc/articles/PMC2392988/pdf/jbt-18-277.pdf.

Yasokawa, D., & Iwahashi, H. (2010). Toxicogenomics using yeast DNA microarrays. *Journal of Bioscience and Bioengineering, 110*(5), 511–522. https://doi.org/10.1016/j.jbiosc.2010.06.003

Yi, C., & Pan, T. (2011). Cellular dynamics of RNA modification. *Accounts of Chemical Research, 44*(12), 1380–1388. https://doi.org/10.1021/ar200057m

Yu, J., Cheng, W., Jia, M., Chen, L., Gu, C., Ren, H. Q., & Wu, B. (2022). Toxicity of perfluorooctanoic acid on zebrafish early embryonic development determined by single-cell RNA sequencing. *Journal of Hazardous Materials, 427*. https://doi.org/10.1016/j.jhazmat.2021.127888. http://www.elsevier.com/locate/jhazmat

Chapter 20

Current trends and approaches in clinical metagenomics

Shivani Tyagi and Pramod Katara

Computational Omics Lab, Centre of Bioinformatics, IIDS, University of Allahabad, Prayagraj, Uttar Pradesh, India

1. Introduction

It is estimated that infectious diseases are the leading cause of approximately 19% of global deaths (Lozano et al., 2010). According to the World Health Organization, the most common communicable disease is the "lower respiratory tract infection" causing 3.2 million deaths in 2015; enteric, tuberculosis, and HIV/AIDS were responsible for 1.4 and 1.1 million deaths. Pathogen identification and characterization is a cornerstone for the diagnosis of infectious diseases and guides treatment duration and antibiotic choice (Kolb et al., 2019). The identification and characterization of infectious pathogens (such as bacteria, viruses, parasites, and fungi) are critical for the clinical regulation of patients and the prevention of transmission (Forbes et al., 2018). It aims to develop customized diagnostic, treatment, and prevention approaches according to the individual characteristics of patients. It allows deeper understanding and better-targeted strategies that are differing from person to person. The metagenomic approach has been used to learn more about the hidden diversity of microbes living with us. Initially, this discipline had the major challenges of sequencing and understanding over 100 trillion microorganisms living in and on the human body (Hadrich, 2018). In recent years, metagenomic analysis by using high-throughput culture-independent techniques based on next-generation sequencing (NGS) has emerged as a new field to identify and characterize the pathogenic microorganisms of infectious diseases and opened new perspectives for microbiological diagnosis in the clinical setting (Wu et al., 2018).

2. Clinical microbiome

Humans are 99.9% identical in their genetic materials, yet the tiny difference in our genetic material gives rise to enormous phenotypic diversity across the human population. Contrariwise, the human metagenome is quite more variable, with only a third of its constituent genes found in a majority of healthy individuals. Understanding this variability of the human microbiome has become a major challenge in clinical microbiome research (Lloyd-Price et al., 2016).

It has been considered that humans have 100 trillion cells in their body but hardly one-tenth are human cells. The human body is a shelter of trillions of microbes such as bacteria, viruses, fungi, and other microorganisms that create complex, body—habitat-specific; adaptive ecosystems that are constantly changing host physiology. It has been known for a long time that the human microbiome is an essential component of immunity that influences metabolism and modulates drug interactions and plays an important role in maintaining human health (Katara et al., 2011). These microbes inhabit various sites of the human body such as the skin, nose, mouth, vagina, and digestive gut.

1. Skin: The most diverse population of microbes lives on the human skin, which is the point of contact with the world. There are at least 1000 different species of fungi, bacteria, viruses, and other microbes present on the skin. One example of bacteria that protect the skin is Bacillus subtilis which produces bacitracin on the skin, a toxin that helps it in fighting with other microbes, and other skin colonizers are species of Corynebacterium, Propionibacterium, and Brevibacterium. The most commonly isolated fungal species is *Malassezia* sp. which is present in the areas of the skin enriched in sebaceous glands. The Demodex mites viz. *Demodex folliculorum* and *Demodex brevis* are microscopic arthropods that are also a part of normal skin (Kumar & Chordia, 2017).

Integrative Omics. https://doi.org/10.1016/B978-0-443-16092-9.00020-5

2. Nasal cavity: Little-known microbes are present in the nasal cavity. Microorganisms of the nasal cavity play a crucial role in determining the reaction patterns of the mucosal and systemic immune systems. Different parts of the nasal cavity have different microorganisms such as Corynebacterium, Aureobacterium, Rhodococcus, Staphylococcus, and Streptococci (Kumar & Chordia, 2017).
3. Oral cavity: The human oral cavity or mouth have several distinct habitats for the microbial community such as gingival sulcus, teeth, tongue, gingival, lip, cheek, hard plate, and soft plate of gingival sulcus. There are over 600 microbial species including Actinobacteria, Streptococcus, Chloroflexi, Chlamydiae, Euryarchaeota, Firmicutes, Fusobacteria, Proteobacteria, Spirochetes, Bacteroidetes, Synergistetes, and Tenericutes (Kumar & Chordia, 2017).
4. Gut: The GI tract has two main functions: nutrition and defense. It absorbs nutrients, digests food, and assists with waste excretion. At the same time, the intestine shelters a huge population of microorganisms that help in digestion and guard against pathogenic microbes. The population of gut microflora is influenced by various factors such as age, diet, and socioeconomic conditions (Kumar & Chordia., 2017).
5. Vaginal microbiome: A healthy vagina carries 250 species of bacteria, and the sharp cooperation between the host and the microorganisms serves as the first line of defense against the invasion of opportunistic infections. This harmonious balance is known as eubiosis (Kalia et al., 2020; Mendling, 2016). Infant, puberty, pregnancy, and menopause are just a few of the life phases that can cause changes in the microbiota makeup of a human vagina. Menstruation, uncontrolled antibiotic use, hormonal shifts, and vaginal douching are the main causes of the temporal alterations in the human vaginal microbiota (Chee et al., 2020).

Dysbiosis in the metagenome of the human microbiome in above mentioned sites has been associated with numerous diseases, including allergies, autism, diabetes, inflammatory bowel disease, multiple sclerosis, and cancer (Lloyd-Price et al., 2016). It has been confirmed that the human gut microbiome is highly correlated with human health and diseases. The human gut microbiome consists of approximately 15,000 to 36,000 species of bacteria and also contains more than 100 times more genes, compared with 25,000 genes in humans (Cheng et al., 2019). The study of over 14,000 autism individuals reveals the higher prevalence of inflammatory bowel disease (IBD) and other GI disorders in autism patients compared to controls. The causes of autism-associated GI problems remain unclear, but may be linked to gut bacteria as a number of studies report that autism individuals exhibit altered composition of the intestinal microbiota (Hsiao et al., 2013). Although cancer is generally a disease caused by host genetics and environmental factors, microbes present at mucosal sites can be a part of tumors of intratumoral microbes and aerodigestive tract malignancies. Contrariwise gut microflora also functions as a detoxifying of dietary components, maintaining balance and reducing inflammation and proliferation. A large amount of a microbe at a tumor site doesn't mean that a microbe is directly associated with disease (Garrett, 2015).

3. Clinical metagenomics

Metagenomics is defined as the more powerful and reliable technique for direct analysis of the genome for the entire community of microbes within an environmental sample, overruling the need to isolate and culture individual microbial species. The field began with the cloning of environmental DNA, adhered by functional expression screening, and then quickly complemented by shotgun sequencing of environmental DNA (Thomas et al., 2012). Bacteria are the most abundant microbes in the human body and have been shown to play a crucial role in human health, affecting the development of growth, nutrition, weight, the immune system, and other physiological aspects. Changes in the global balance of the microbiome might play a more crucial function instead of distinct microbial species, being either beneficial or harmful to human health. Therefore, it is significant to understand how microbial communities may vary under different conditions (Masha et al., 2019). Using the metagenomic approach (Fig. 20.1), the Human Microbiome Project (HMP) illustrated that millions of microbes cohabitate with their healthy host. Several instances reflect that many microbes maintain symbiotic relationships with their hosts. The different role played by these microbes includes assisting in food digestion, and immunity modifications and might cause disease conditions in a healthy host at a later time. Many opportunistic microbes such as *Staphylococcus aureus* and *Candida albicans* often live as commensalisms but sometimes cause severe modifications to the functioning of immune systems (Zhou et al., 2016). Meta-analytic techniques are useful to identify statistically significant changes, but it is limited to understand the biological significance of microbial variation (Perz et al., 2019).

3.1 Sequencing for clinical metagenomics

Over the past decade, sequencing technology has revolutionized infectious diseases research. A comprehensive spectrum of potential causes (such as viral, bacterial, fungal, and parasitic) for the diagnosis of infectious diseases can be identified by a promising approach known as NGS (Wilson et al., 2019). Further, it involves methods such as

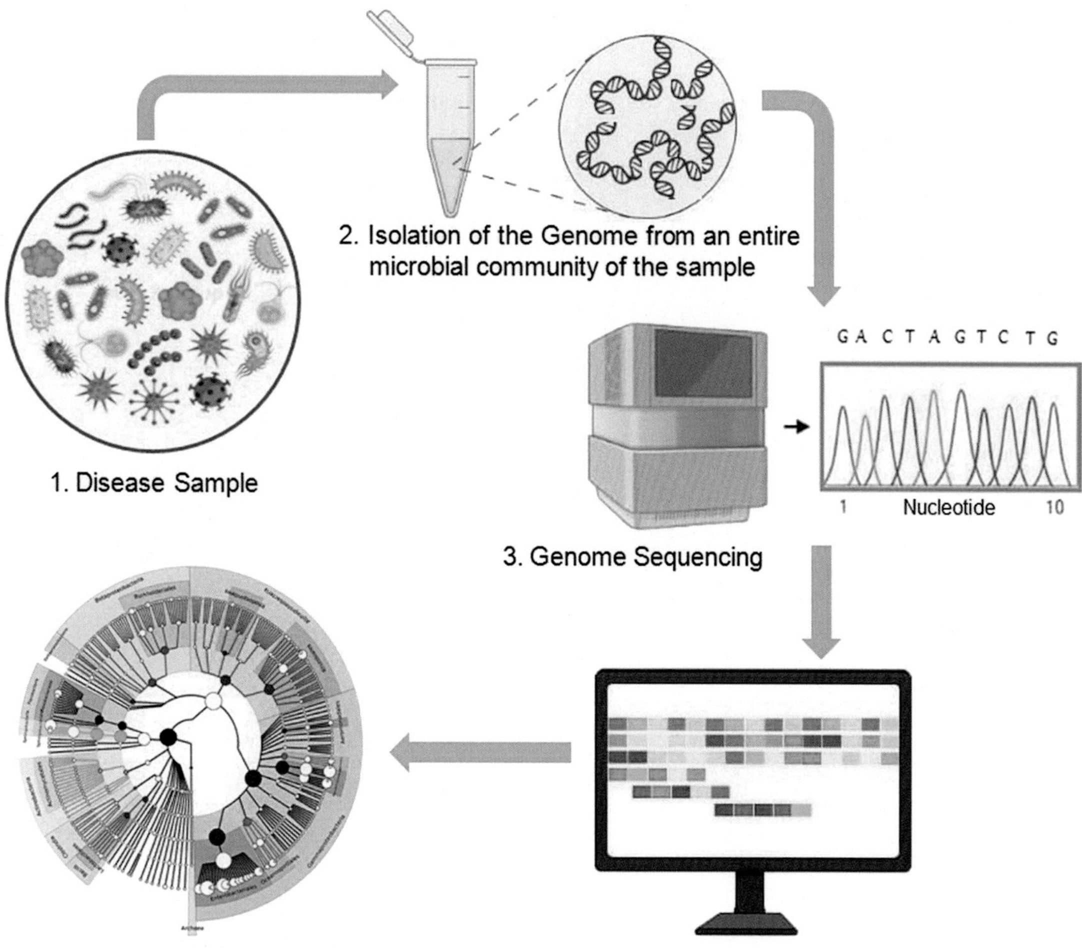

FIGURE 20.1 Overview of computational metagenomics.

whole-genome sequencing (WGS) of pure cultures that have been used for evolutionary analysis, outbreak detection, transmission investigations, and pathogen characterization. This approach for cultured isolation is now being implemented at a vast level. For instance, molecular epidemiology and antibiotic resistance gene prediction are two major applications of whole-genome sequencing concerning cultured strains in clinical diagnostic microbiology (Köser et al., 2012).

Contrary to this, the metagenomic approach analyzes the community of microorganisms but eliminates the isolation step of a pure cultured technique of WGS which is done by involving specific conserved genes, such as the 16S rRNA gene, or by the metagenomic shotgun sequencing (MSS) of total microbial nucleic acids within samples. Metagenomics can refer to both the approaches of NGS to produce large quantities of data in a relatively short time, either 16S rRNA gene sequencing or MSS. Apart from MSS and 16S rRNA sequencing, a few more NGS-based targeted sequencing methods (e.g., amplicon and ITS sequencing) are utilized by the scientist in a few purpose-specific metagenomic studies (Table 20.1).

Despite being less expensive than MSS, 16S rRNA sequencing suffers from potential PCR-related bias, whereas MSS, without the biasing nature inherent to PCR, is capable of classifying bacteria to the species level. The classification based on taxonomic level by partial 16S rRNA gene sequencing is limited from phylum to genera level specificity but certain species that are highly heterogeneous within certain genera can be differentiated (Ravel et al., 2011). In addition to the relatively high cost, certain computing resources on a significant level for data handling, processing, and storage are required for a large amount of sequenced data, generated by MSS (Zhou et al., 2016). MSS has also been used to detect and identify known and novel microbes some of which cannot be cultured yet and to identify (Wylie et al., 2012).

TABLE 20.1 Common sequencing techniques for clinical metagenomics.

S. no.	Technique name	Description
1.	Targeted sequencing (amplicon sequencing)	• The targeted sequencing approach is extensively used to characterize microbial populations (Navgire et al., 2022). • It's PCR-dependent technique utilizes primer specific to the defined regions.
2.	16S rRNA gene amplicon sequencing	• Metagenomic sequencing technique specific to 16S rRNA (a most conserved taxonomic marker for bacterial). • Due to the small sample size (16S rRNA), it's fast. • Suitable for metagenomic studies to identify the sample's taxonomic profile of microbial communities. • It's PCR-dependent technique utilizes a primer specific to the 16S rRNA region.
3.	ITS sequencing	• It utilizes the internal transcribed spacer (ITS) of the nuclear ribosomal DNA to identify the fungal community in the metagenomic samples (Rossmann et al., 2021). • It's PCR-based technique utilizes primer specific to the region flanked by ITS2 region.
4.	Whole genome sequencing (metagenomics shotgun sequencing)	• Characterizing the abundance of microorganisms, identifies the microbial species and generates information about the genes (including 16S rRNA) present in the metagenomic sample. • It's PCR independent, thus no chance of biasing due to primer binding. • Helpful for finding unknown microbes and identifying and discovering novel viruses in the given disease sample.

3.2 Data analysis

Once the metagenome sequence data through any of NGS techniques are produced it needs analysis to get a meaningful conclusion. Various steps are used for the analysis of metagenomic shotgun sequencing data.

3.2.1 Evaluation of quality check and quality control

The first step involves quality checking and quality control, which identify and remove low-quality sequences and contaminants. The quality check provides information about the sequencing output (number of reads, length, GC content, overrepresented sequences, etc.), which is usually done by a program such as FastQC Table 20.2. Based on the quality score of the data, adapter removal, quality filtering or trimming can be done through Cutadapt. Trimmomatic to modify the sequence reads according to the researcher's desire. After quality control, the reads can either be assembled into longer contiguous sequences called contigs that are passed directly to taxonomic classifiers. Direct taxonomic classification is useful in the identification of organisms with close relatives in the database.

3.2.2 Metagenomic assembly

Genome assembly is the reconstruction of genomes of interest from the smaller DNA segments called sequence reads which are generated by an NGS sequencing experiment. Genome assembly is of two types de novo genome assembly and comparative genome assembly. De novo assembly involves reconstructing genomes directly from sequence read without any prior knowledge, and comparative assembly uses the sequences of previously sequenced closely related organisms to guide the construction of a new genome. De novo assembly is an "NP-hard" problem, which mostly relies on a number of heuristic methods such as overlap-layout-consensus (OLC) and De Bruijn graph. Metagenomic assembly is used to reconstruct multiple genomes from sequencing reads. It involves the reconstruction of genomes from sequencing reads originating from multiple species which are present at variable abundances. Metagenomic assembly is a complicated and multi-step process, it needs to handle multiple overlapping reads and visualization as well. Ranges of assembly tools, called assemblers, are available to assemble contigs Table 20.3.

TABLE 20.2 List of tools utilized for preprocessing of metagenomic data.

S. no.	Software/tool	Description/comments
1.	AmpliconNoise	• AmpliconNoise is a collection of programs for the removal of noise from 454 sequenced PCR amplicons (Quince et al., 2011)
2.	FastQC	• A quality control tool for high throughput sequence data • Excellent visualization; exhaustive functions
3.	FASTX Toolkit	• The FASTX-Toolkit is a collection of command line tools for short-reads FASTA/FASTQ files preprocessing
4.	PRINSEQ	• PRINSEQ can be used to filter, reformat, or trim metagenomic sequence data (Schmieder & Edwards, 2011)
5.	Trimmomatic	• Trimmomatic performs trimming tasks for Illumina paired-end and single-ended data (Bolger et al., 2014) • Its flexible and exhaustive functions

TABLE 20.3 List of tools for metagenome assembly.

S. no.	Softwares/tool	Description/comments
1.	MEGAHIT	• MEGAHIT enables an efficient assembly of large and complex metagenomics data on a single server • MEGAHIT is available in both CPU-only and GPU-accelerated versions (Li et al., 2015) • It is based on de Bruijn graph
2.	metaFlye	• It facilitates long-read metagenomic assembly • It designs to handle uneven bacterial composition and intra-species heterogeneity (Kolmogorov et al., 2020) • It is based on overlap-layout consensus
3.	Meta-IDBA	• Meta-IDBA is a de-novo assembler for metagenomic data • It can assemble high-complexity datasets that contain more branches due to the presence of more genomes in the dataset (Peng et al., 2011) • It is based on de Bruijn graph
4.	MetaVelvet	• It performs de novo metagenome assembly from short sequence reads (Namiki et al., 2012) • It is based on de Bruijn graph
5.	MetaVelvet-DL	• It provides more accurate de novo assemblies of whole metagenome data using deep learning (Liang et al., 2021) • It is based on de Bruijn graph
6.	Omega	• Assembling and scaffolding Illumina sequencing data of microbial communities (Haider et al., 2014) • It is based on overlap-layout consensus

Assembly concept: All assemblers use the assumption that highly similar sequencing reads originate from the same genome region and attempt to merge overlapping sequencing reads into a longer contiguous consensus sequence, most commonly called a contig. These contigs can subsequently be further assembled into scaffolds. Metagenomic experiments might generate hundreds/millions/billions of reads from a single sample. Depending on the number of reads and the complexity of the microbial species in the sample, some genomes are assembled as an entire original genome sequence, or parts of it, from the short reads. Genome assembly is a challenging problem, even for a single microbial genome. Genome assembly of a mixed sample with many species in different abundances is more challenging and needs special assembly algorithms.

Challenges: Metagenomics assembly is really difficult when similar sequences don't originate from the same region of a genome, for example, repetitive sequences and sequencing errors. Assemblers find it difficult to distinguish where repetitive genomic regions originate, especially when repeats are longer than reads. Sequencing errors are also problematic for assembly and lead to erroneous sequence alignments. The highly uneven sequencing depth of different organisms in a metagenomics sample is another challenge in metagenomic assembly. Depth of coverage of a particular species is rarely high, unless that species is present in high quantities in the sample. Metagenomics samples mostly contain deep coverage of more than one or two organisms. Therefore, it can be concluded that the metagenomic assembly can never be as good as those from the assembly of a single, clonal organism. Standard assemblers assume that a depth of coverage is approximately uniform across a genome which helps the algorithm in resolving repeats as well as it also removes erroneous reads.

Assembly QC: Poor assembly-based studies can provide a false analysis and produce a wrong clinical hypothesis. It is very important to check the quality of produced assemblies before their further use. Tools are available to perform the quality check of assemblies. Few of these tools need reference genomes and others work on an ab initio basis Table 20.4.

3.2.3 Taxonomic binning and classification

Taxonomic binning: Various methods have been developed to address the complexity of analyzing metagenome datasets. Taxonomic binning is a method to group sequence reads/contigs into different clusters/bins, which is done by grouping the reads into bins corresponding to their taxon ID, where each cluster/bin represents different taxonomic units. Taxonomic structure and diversity are evaluated based on the abundance of each taxonomic bin. The majority of binning methods are mainly applied to assembled contigs because these methods generally perform better on longer sequences (Sangwan et al., 2016). Binning algorithms are designed to group contigs from the same or closely related/neighbor organisms. After binning, downstream analysis (taxonomic assignment and functional analysis) is then performed on the bins instead of individual contigs. Binning also clusters contigs for rare bacterial species and can recover draft genomes from previously uncultivated bacteria.

Binning strategies: Binning algorithms use different strategies to address the taxonomic assignment: (a) sequence composition; (b) sequence alignment against references. One of strategy is based on k-mer frequencies methods to retrieve the similarity among all words in the query which may be called "genomic signature" and was widely used to explore evolutionary conservation among species. A range of current methods uses a combination of these features to address the taxonomic-binning (Table 20.5). When classifying contigs instead of reads, the search space is much smaller where phylogenetic methods can be used for taxonomic binning.

Metagenomic classification: Metagenomic classification is used to match reads/assembled contigs against a microbial genomes database to identify the taxon of each sequence reads/contigs. BLAST is being used since early days of metagenomics to compare each read with all sequences in GenBank. Day by day the size of reference databases has grown due to which alignment using BLAST has become computationally difficult. To reduce the time and data handling issues, in the

TABLE 20.4 List of tools for assembly QC.

S. no.	Tools	Description
1.	CheckM	• CheckM tool assessing the quality of a metagenome assembly using a broader set of marker genes. • CheckM utilizes a reference genome, but it can also act without using reference genomic information (Parks et al., 2015).
2.	DeepMAsED	• DeepMAsED detects misassembled contigs through a deep learning model. • It does not require reference genomes for this purpose (Mineeva et al., 2020).
3.	MetaQUAST	• MetaQUAST is a reference-based tool for assembly QC, it rapidly calculates the basic statistics (reference-based statistics, such as genome coverage, NA50, and NGA50 values) for the contigs in the "meta" mode (Mikheenko et al., 2016).
4.	REAPR	• REAPR is designed to identify errors in metagenome assemblies in a precise manner, without relying on a reference sequence. • It also offers quantitative comparison across multiple assemblies using sequencing reads information (Hunt et al., 2013).
5.	VALET	• VALET relooks at metagenome binning before QC. • It mainly checks the effect of uneven read depth on assemblies and results in false positives and false negatives (Olson et al., 2019).

TABLE 20.5 Taxonomic binning and classification.

S. no.	Softwares/tool	Description/comments
1.	Kraken	Kraken is an ultrafast and highly accurate program for assigning taxonomic labels to metagenomic DNA sequences. It utilizes K-mer alignment for fast and accurate classification (Wood et al., 2014)
2.	LCAClassifier	A program for taxonomical classification, using the Lowest Common Ancestor (LCA) algorithm (Lanzén et al., 2012)
3.	MetaPhlAn	MetaPhlAn performs taxonomic profiling, which uses rely on detecting the presence and estimating the coverage of a collection of species-specific marker genes to estimate the relative abundance of known and unknown microbial taxa in shotgun metagenomic samples (Segata et al., 2012)
4.	PhyloSift	PhyloSift is a suite of software tools to conduct phylogenetic analysis of genomes and metagenomes. If required, it uses a protein sequence for phylogenetic profiling of the new sequence (Darling et al., 2014)
5.	Taxator-tk	Taxator-tk is a taxonomic assignment software package that generates very precise taxonomic assignments with few errors for metagenome shotgun (Dröge et al., 2015)

recent past a variety of alternative strategies have been proposed for the matching of sequence reads/contigs: aligning marker genes only, using complete genomes, mapping k-mers, translating the DNA, and aligning to protein sequences.

Direct versus assembly-based taxonomic classification: The choice of direct taxonomic classification of reads versus assembly-based analyses depends on the research question. For quantitative community profiling and identification of organisms with close relatives in the database, direct taxonomic classification is more useful than MSS, whereas MSS enables the detection of organisms across all domains of life, assuming that DNA can be extracted from the target environment and alleviates biases from primer choice (Breitwieser et al., 2019). By using biogeographical and ecological measures such as species diversity, richness, and uniformity of the communities, researchers can quantify the structure of microbial communities (Chiarucci et al., 2011). Clinical metagenomics often focuses on the absence and presence of pathogens involved in infectious diseases, which can be identified by matching reads against a reference database (Langelier et al., 2018). Further, the functional potential of the microbiome can be identified by matching the reads against pathway and gene databases (Truong et al., 2017).

Metagenomic bins versus metagenome-assembled genomes (MAGs): Metagenomic Bins with a complete collection of the contings are designated as metagenome-assembled genomes (MAGs). The designation of metagenome bins as MAGs relies on their quality and accuracy-related parameters. The most crucial parameters are the completeness of marker genes and the contamination of single-copy genes. In general, on the basis of completeness, level of contamination, and rRNA/tRNA prediction the bins are classified as finished, high-quality (completeness >90%; contamination <5%), medium-quality (completeness >50%; contamination <10%), or low-quality drafts based. Generally, only those metagenomic bins with relatively high quality are considered as the MAGs and included for subsequent assembly and annotation (Yang et al., 2021).

3.3 Metagenome annotation

3.3.1 Gene prediction/identification

After metagenome assembly, gene prediction and functional annotation can be done. Single genome/ gene prediction tools are not well suited for metagenomic datasets due to the diversity of sequence composition, sequence errors, and sequence length. Preferential bias in codon usage, di-codons frequency, patterns in the use of start and stop codons, Open Reading Frame (ORF) length, ribosome-binding sites patterns, and GC content of coding-sequences are utilized in algorithms to predict genes from metagenomics datasets.

Microbial gene assignment/annotation uses the similarity of sequences to reference genes or proteins for identifying potential gene functions. This process relies on the existence and quality of current reference datasets or databases. Generalized databases are available which contain gene or protein sequences that represent functions that have been validated in the laboratory or whose function is unknown (e.g., hypothetical proteins). Universal Protein Resource (UniProt: http://www.uniprot.org/) is the most widely used database for gene assignment/annotation.

A range of bioinformatics software and tools are available for gene prediction purposes which are based on the gene's inherent signals, i.e., Open Reading Frame (ORF) length, ribosome-binding sites patterns, and GC content of coding sequences (Katara, 2014). On the basis of algorithms, these tools are mainly categorized as (i) Model-based gene prediction tools, and (ii) Deep learning–based gene prediction tools (Table 20.6).

3.3.2 Gene function annotation

Function assignment of predicted ORFs is one of the very crucial steps of metagenomic studies, especially when the predicted genome is novel. The function annotation process is mainly performed on either nucleotide or translated sequences against known gene or protein databases. Various tools and software are available and already in practice for function annotation purposes (Table 20.7), which mainly rely on (i) Homology, (ii) Motif, and (iii) Gene context.

1. Homology-based approach: In the case of the presence of gene homologs, homology-based function prediction is the easiest and most frequent annotation method. Traditionally, homology-based tools rely on different variants of the blast to screen query reads (genes) against known/reference gene databases. Blast-based tools are reliable, but they have limitations in handling large data sets, thus are time-consuming. Modern methods with optimized alignment strategies enable 100- to 1000-fold faster alignment of query gene sequences to databases. They are generally utilized precomputed sequence clusters and phylogenic information to compare query genes, e.g., eggNOG. Few of these modern tools provide complete automatic annotation pipelines and connect query genes with metabolic pathway annotations as well, e.g., GhostKOALA. Tools like MG-RAST provide GUI interface for online metagenomic analysis that includes data uploading, QC, and alignment with reference databases.
2. Motif-based approach: Due to noise or other factors, sometimes protein sequence is partially decoded from metagenomic reads. These partial protein sequences do not have enough information to perform homology-based annotations. Despite poor alignment homologies, these partial protein sequences can be utilized for functional annotation against the database of protein sequence patterns or specific motifs (e.g., Interpro, ProSite, PRINTS, etc.), which mainly focus on sequence patterns/motifs, not on the whole protein sequence. Motif-based annotation tools generally performed systematic searches against these databases and are relying on motif-based statistical inference (Table 20.7).
3. Gene context–based approach: Metagenomic sequencing produces a huge amount of sequence data, and analysis of such data resulted in a large number of known genes along with a high number of novel genes. There are chances that these novel genes may not share homology with any of the known genes, in such cases their functional annotation through homology-based or motif-based approaches are not possible. In such a scenario, context-based tools play a vital role, they are based on models, e.g., HMM, and perform functional annotation of novel genes to some extent.

Use of protein structure information: Some proteins share a common function but at the sequence level they are diverged. Many homology-based functional predictions rely on a threshold fixed for sequence identity/similarity. Overall

TABLE 20.6 List of tools for gene prediction and identification from metagenomic data.

S. no.	Tools	Description
1.	Balrog	It is a machine learning (Convolutional Neural Network) based tool which utilized as a universal protein model for prokaryotic gene prediction (Sommer et al., 2021).
2.	FragGeneScan	FragGeneScan (model-based, HMM) tool provides the prediction of protein-coding regions in short reads. It combines sequencing error models and codon usages in a hidden Markov model for better results (Rho et al., 2010).
3.	Glimmer-MG	Glimmer-MG (model-based, Interpolated Markov Model) tool provides genes finding facilities in environmental shotgun DNA sequences. It uses interpolated Markov models (IMMs) to identify the coding regions and distinguish them from noncoding DNA (Kelley et al., 2012).
4.	MetaGene Annotator	MetaGene Annotator is a model-based (dynamic programming) tool that detects Species-Specific Patterns of Ribosomal Binding Site for Precise Gene Prediction in metagenome data (Noguchi et al., 2008).
5.	Meta-MFDL	Meta-MFDL is a machine learning (Deep Neural Network)-based tool that relies on multiple features, such as monocodon usage, mono-amino acid usage, etc., for training, to distinguish between coding and noncoding ORFs (Zhang et al., 2017).

TABLE 20.7 List of tools for Gene annotation.

S. no.	Tools	Descriptions
1.	eggNOG-mapper	Homology-based tool for functional annotation of large sets of gene sequences using pre-computed clusters and phylogenies from the eggNOG database (Huerta-Cepas et al., 2017).
2.	MetaPath	MetaPath uses a combination of metagenomic sequence data and prior metabolic pathway knowledge to identify differentially abundant pathways in metagenomic data sets (Liu et al., 2011).
3.	GhostKOALA	It is a Homology-based KEGG Tool for Functional Characterization of Genome and Metagenome Sequences (Kanehisa et al., 2017).
4.	FunGeCo	Gene context–based web tool for estimation of the functional potential of microbiomes (Anand et al., 2020).
5.	GeConT	GeConT (Gene Context Tool) is a web interface that provides visualization of the genome context of a group of genes. The graphical information of GeConT can be used to perform functional annotation of a set of genes from a metagenome (Ciria et al., 2004).

sequence similarity of such proteins is usually lower than the usual threshold value. However, they contain one/more common sequence/structural patterns/motifs which are important to maintain their structure and function.

Challenges: Length of sequence reads plays a critical role; it is observed that reads with longer sequences are more informative. The use of short reads for homology-based strategies for gene prediction and annotation can result in a number of false negatives hits. It has been observed that functional annotation is easier and more reliable for longer sequence reads. Therefore, searching databases using short reads has low sensitivity and specificity for gene prediction from metagenomic datasets. To retrieve correct annotation results or to minimize the false positive results with short reads need to adjust the E-value threshold, which needs experience.

3.3.3 Metabolic pathway reconstruction

After gene function annotation, pathway reconstruction is one of the major annotation goals in the present scenario. Information flows through different species and their physiological aspects can be better understood by metabolic pathways. Therefore, the term "inter-organismic meta-pathways" has been proposed for this kind of analysis. Metagenomic sequencing provides a tool to study diverse microbiota within different ecological niches, including different organs and disease conditions in the human body that differ in microbial composition and are responsible for health. As it has been well established that the human gut microbiome plays an important role in host health, and largely contributes to metabolism and immune response. Good functional annotation must be achieved to perform a reliable metabolic reconstruction (Nicholson et al., 2012). Metabolic pathways are used to find each gene in an appropriate metabolic context, filling missing enzymes in pathways to perform the best pathway reconstructions.

Challenges: Since all the enzymes or pathways are conserved among all species or environments, therefore it is difficult to reconstruct metabolic pathways for metagenomic datasets. It has also been observed that most of the metabolic information comes from model organisms. Current approaches are using model organism information as reference/background information to reconstruct pathways, therefore failing in the metabolic reconstruction of variant pathways.

3.4 Pipelines for metagenomic data analysis

As discussed, metagenomic data analysis includes various steps and for each step, various computational tools are available which help to solve the particular objective. In the last decade, there has been a flood of metagenomic data at the same time observations conclude that to get reliable conclusions, metagenomics studies need to consider large data sets. In such cases, the use of different tools for each step is not a straightforward task, the user has to face various data handling and format-related issues. At the same time, due to the large size user needs to have programming skills at different stages. Considering the situation, a number of metagenomic pipelines have been developed in the last few years, which provide a straightforward way to deal with all steps of metagenomic data analysis in a user-friendly manner (Fig. 20.2). These pipelines took shotgun metagenomics data through the needed steps to perform data analysis and annotation tasks and provide valuable results. Though a list of pipelines is available, the frequently used pipelines are (i) CloVR-metagenomics, (ii) Galaxy

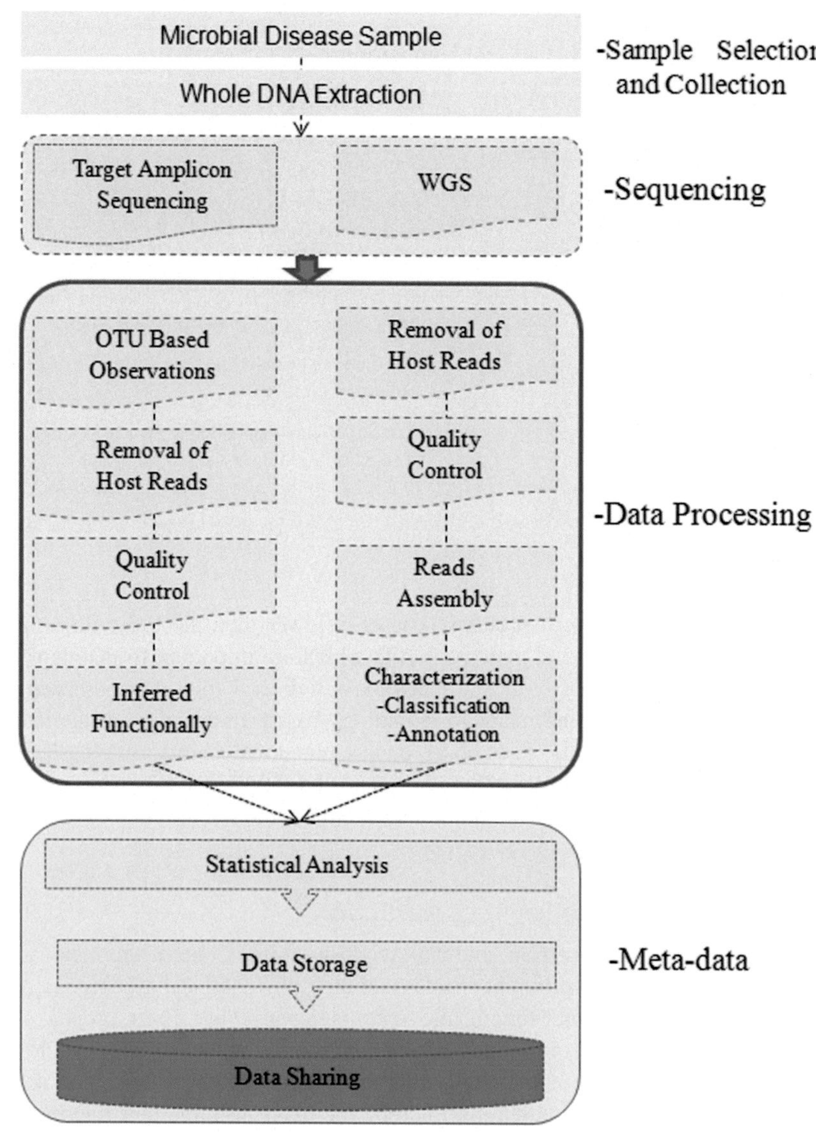

FIGURE 20.2 Workflow of metagenomics with the whole process from specimen handling to data sharing.

platform, (iii) IMG/M, (iv) MetAMOS, and (v) MG-RAST. The basic task which is facilitated by most of the metagenomic pipelines is Quality control, Assembly, Gene detection, Functional annotation, Taxonomic analysis, Comparative analysis, and Data management (Navgire et al., 2002).

3.5 Metagenomic database of clinical importance

With the emergence NGS and its frequent use for metagenome sequencing, a huge amount of metagenomic data has been generated in the last decades. Scientist uses the advantage of metagenomic concepts and clinical data for clinical purpose. As microbiomes vary from organ to organ and condition to condition, the produced metagenomic data for clinical purposes are disease and organ-specific, e.g., intestine, vagina, etc. Bioinformaticians compiled these data in database form, and in most cases, linked them with a web-based interface that provides user-friendly handling and accessing of these data. A few databases are also connected with data analysis pipelines where users directly can analyze these data for the purpose. Following are a few of the databases which are frequently utilized in clinical studies.

1. HMPC database: The HMPC database is a user-friendly, web-based interface that provides a species-level, standardized phylogenetic classification of over 1800 human gastrointestinal metagenomic samples with the ability to search for health or disease states and community structure associated with a microbial group and the enrichment of a microbial gene or

sequence and their functional annotations. It allows comprehensive analysis of the human microbiome and supports clinical research from basic microbiology and immunology to therapeutic development in health and disease (Forster et al., 2016).

2. MAHMI database: The resources provided by MAHMI such as the antiproliferative bioactivity of new amino acidic sequences with their bioactive peptides have a huge biotechnological potential, and will be useful to those clinical researching of gastrointestinal disorders of autoimmune and inflammatory nature, such as inflammatory bowel diseases (Blanco-Míguez et al., 2017).
3. IBDM database: Inflammatory Bowel Disease Multi'omics Database provides the most comprehensive description of host and microbial activities in inflammatory bowel diseases and a detailed view of functional dysbiosis in the gut microbiome during inflammatory bowel disease activity (Lloyd-Price et al., 2019).
4. eHOMD: The eHOMD is a vital resource for enhancing the clinical relevance of 16S rRNA gene-based microbiome studies and the individual microbes in body sites in the human aerodigestive tract, which includes throat, mouth, esophagus, sinuses, the nasal passage, and the lower respiratory tract, in human health and disease (Escapa et al., 2018).
5. VMH database: The VMH database encloses current knowledge of human metabolism with 5180 unique metabolites, 255 Mendelian diseases, 632,685 microbial genes, and 8790 food items, 17,730 unique reactions, and 818 microbes (Noronha et al., 2019).
6. HMP: Human Microbiome Project examines the role of microbiome related to human conditions. The study focuses on (1) gut disease onset, using inflammatory bowel disease as a model; (2) respiratory viral infection and onset of type 2 diabetes; and (3) pregnancy, including those resulting in preterm birth (Integrative HMP (iHMP) Research Network Consortium, 2014).
7. Vaginal 16s: This database and method provide accurate species-level classifications of metagenomic 16S rDNA sequence reads that will be useful for the analysis and comparison of microbiome profiles from vaginal samples. STIRRUPS can be used to classify 16S rDNA sequence reads from other ecological niches if an appropriate reference database of 16S rDNA sequences is available (Fettweis et al., 2012).
8. SRA: The NIH Sequence Read Archive (SRA), a database archived for the storage, retrieval, and analysis of next-generation nucleotide sequencing data, is maintained by NCBI. SRA has global coverage because of daily data exchanges with the DNA Data Bank of Japan (DDBJ) and the European Nucleotide Archive (ENA) in Europe (Sayers et al., 2022). This archive currently contains 4.6 Petabytes of dbGaP data with controlled access in addition to 8.8 Petabytes of data that is openly accessible. Although the collection has a lot of potential for scientific research, its bulk makes it challenging to keep, retrieve, and analyze (Sayers et al., 2021).
9. ENA: The European Nucleotide Archive (ENA) is a part of the International Nucleotide Sequence Database Collaboration (INSDC), which works with global organizations like the DNA Data Bank of Japan (DDBJ) of the Japanese National Institute of Genetics and the National Center for Biotechnology Information (NCBI) of the United States to ensure that data are recorded and mirrored globally. In order to give the most comprehensive and accurate data network possible, this active collaboration focuses on developing shared data standards and exchange systems. In the field of nucleotide sequencing data, the European Nucleotide Archive has played a significant role in the storage, contextualization, and discovery of these ever-expanding datasets (Cummins et al., 2022).

4. Clinical metagenomic projects

Before the 20th century, clinical studies were limited to bacterial cultures enabling the detection of pathogenic microorganisms. Information about the relationship between human hosts and microorganisms has increased during the 20th century. With the introduction of culture-free analysis methods, extensive enlisting of the human microbiome was possible for the first time. Since then, many projects have been introduced to study healthy and diseased individuals. The era of metagenomics research of microorganisms in the human gastrointestinal tract had been coming with two international research projects such as the Human Microbiome Project (HMP) and Metagenomics of the Human Intestinal Tract (MetaHIT), which were officially launched in 2007 (Lin et al., 2016).

The Human Microbiome Project Consortium (HMP) was funded by The National Institutes of Health (NIH), which brought together a broad collection of scientific experts to explore the communities of microbes and their relationship with their human hosts. To provide a critical framework for subsequent metagenomic annotation and analysis, the HMP has focused on producing reference genomes (viral, bacterial, and eukaryotic) with generating a baseline of the microbial community's structure and function from an adult cohort. As a consequence of this, the Human Microbiome Project Consortium (HMP) complements other projects such as the Metagenomics of the Human Intestinal Tract (MetaHIT) project, which focused on the examination of the gut microbiome using WGS data including samples from cohorts (Human Microbiome Project Consortium. A Framework for Human Microbiome Research, 2012).

Since then, various projects have been introduced day by day such as MOMS-PI consortium and IBDMDB projects. The Multi-Omic Microbiome Study: Pregnancy Initiative (MOMS-PI) was conducted by The Vaginal Microbiome Consortium team at Virginia Commonwealth University with the collaboration of Global Alliance to Prevent Prematurity and Stillbirth (GAPPS) to better understand the role of microbiome and host profiles change throughout pregnancy and its components in the etiology of preterm birth, which occurs in over 10% of pregnancies and which is the leading cause of death (Fettweis et al., 2019).

5. Role of clinical metagenomics in human health

High-throughput sequencing technologies allow the generation of massive amounts of data and have pushed forward metagenomic studies. As a result of this, metagenomic data has become a more complex and computationally intensive task. Functional analyses offer an opportunity to improve our understanding to study the microorganism's ecology (Pereira-Flores et al., 2019). In almost every clinical specialty, accurate and replicable profiling of the microbiota is now becoming a growing interest for use in diagnostic and therapeutic applications (Schlaberg, 2020).

The human body performs as a host to the microorganism's community that outnumbers the body's own cells. Over the past few years, research on the human microbiome has become an area of interest due to the intimate linkage of the microbiome with human health. The major focus of host–microbiome research is the identification and characterization of the microbes inhabiting the human host and their distinct host phenotypes and the biochemical pathways by which microbes impact their hosts (Malla et al., 2019).

Analyses of unknown or novel microbial genes of interest, validation of metabolic hypothesis, and potential associations between microbial changes and human disease can be fulfilled by the computational metagenomics approach (Perz et al., 2019). The emerging field of computational metagenomics can also be advantageous to revolutionize pathogen detection by simultaneous detection of all microorganisms in a clinical sample through the use of next-generation DNA sequencing. Metagenomics analysis has the potential to uncover the role of dysbiotic microbiomes in infectious and chronic human diseases. Advances in sequencing platforms and bioinformatics tools, can even determine the whole-genome sequences of pathogens, allowing inferences about evolution, transmission, antibiotic resistance, and virulence (Miller et al., 2013). NGS approaches are invaluable for exploring the genetic, functional, and metabolic properties of the microbial community and the exploration of the composition of the microbiome. NGS on the human microbiome has revealed possible links between the gut microbiome and human diseases such as arthritis, rheumatoid, depression, and diabetes (Malla et al., 2019). Cancer is a chronic disease caused by changes between host genetic and environmental factors. The development of sequencing and bioinformatics technologies has changed the faith of cancer. Studies in bioinformatics have identified oncology pathways and genetic alterations that affect the progression of cancers. Apart from genetic alteration, various studies have revealed the effect of microbes on cancer biology. Some microbes such as hepatitis B and C viruses, *Helicobacter pylori,* and human *papillomaviruses* have been identified as carcinogenic agents, which have been estimated to cause $\sim$20% of all cancers. With the increasing number of researchers, the analysis of large, annotated cancer datasets helps in identifying the microbial signatures in tissue and blood in different types of cancer. However, it still remains unclear that, how microbiota is associated with cancer due to its dual role of promoting or inhibiting cancer progression. Undoubtedly, microbes at certain levels can influence tumor growth and also show antiinflammatory activity. This complexity can be summarized by defining three ways of microbes contributing to carcinogenesis: (i) modulating the balance between cell proliferation and death, (ii) steering the immune system, and (iii) influencing the metabolism of the host (Rossi et al., 2020).

6. Concern's and issues of clinical data handling

Advancement in NGS platforms has empowered rapid analysis of metagenomic and meta-transcriptomic data for clinical applications, from risk assessment, and diagnosis, to biomarker discovery and drug therapy choices. In clinical settings, these sequencing platforms pose different bioinformatics challenges; disrupting the translation of personal metadata into useful information. In spite of the potential of metagenomic Next-Generation Sequencing (mNGS), there are many hurdles to clear out before it can be a part of the downstream analysis. During specimen collection, sample contamination is the largest concern of mNGS and it needs to be validated by quality control processes step by step. Furthermore, some Illumina platforms generated wrong barcodes which leads to false-positive sequencing data. A common perceptiveness is that mNGS is highly sensitive that it can reveal a diagnosis when other testing is negative. But mNGS is more analytically sensitive than standard culturing methods, removal of a vast amount of human genome during sequencing preparation and during postanalysis is very crucial that can decrease the sensitivity for many organisms.

7. Applications of clinical metagenomics

7.1 Role in the detection of unusual or fastidious pathogenic microorganisms

Pathogens have already been identified in biological samples using clinical metagenomics (CMgs). The majority of studies report estimation of CMgs compared to traditional cultural methods. However, in many cases, CMgs analysis has provided a diagnosis that was not previously possible using culture or traditional molecular techniques. According to clinical metagenomic studies, the intestinal microbiota is associated with a wide range of illnesses, including cancers (Andrews et al., 2021), autoimmune diseases (Svoboda, 2021), and nonalcoholic fatty liver (Aron-Wisnewsky et al., 2020). Drug-resistant pathogenic genes can be found, and infectious disease outbreaks in hospitals and communities can also be tracked by using clinical metagenomics. An expert consensus was published in 2021 by the Laboratory Medicine Branch of the Chinese Medical Association, recommending the use of NGS on samples from suspected infection sites, in cases, where no etiological information is observed by routine biochemical analysis or microbial culture, and empirical anti-infection therapy fails (Zhang et al., 2021). Now metagenomic sequencing has been widely used in the detection of medical pathogens. Pathogenic microbes, such as bacteria, viruses, fungi, and parasites, can be handled using CMgs. CMgs can be used to diagnose respiratory RNA viruses including Picornaviridae, Coronaviridae, Paramyxoviridae, and Orthomyxoviridae (Thorburn et al., 2015). By using CMgs testing, the cerebrospinal fluid has been found to contain five different organisms: *Neisseria meningitidis, Streptococcus agalactiae, Candida albicans, Mycobacterium fortuitum*, and *Mycobacterium abscessus* (Miller et al., 2019).

7.2 CMgs may be useful in a forensic realm

Forensic medicine is another application of metagenomics in clinical microbiology. Microbiota can be used as a form of personal identification. Salivary microbiota, for example, can be used to distinguish between two people (Leake et al., 2016). Different microbiota live in different parts of the human body or in different types of fluid. Much more bacteria are deposed on a touched object than human DNA, and metagenomic analysis can help to identify or exclude suspects in criminal cases, or to determine the nature of a sample (Schmedes et al., 2016).

7.3 CMgs reduces the time period between pathogenicity and the start of effective antibiotic therapy

The time between the onset of an infection (e.g., bacterial) and the initiation of effective antibiotic therapy has frequently been linked to the outcome, particularly in intensive care (Kollef, 2000). Culture-based methods typically produce results within 24—72 h of sampling, whereas sequencing using nanopore technologies (Oxford Nanopore Technologies, Oxford, UK) can produce results in less than 10 h (Charalampous et al., 2019). Two pneumonia cases caused by *Pseudomonas aeruginosa* and *Staphylococcus aureus* were the first cases of respiratory infection diagnosed by CMgs faster than the culture method (Pendleton et al., 2017).

7.4 CMgs helps in finding novel or unexpected viruses of clinical importance

Traditional methods, such as PCR (including multiplex PCR) or serology, cannot be extended to find new evolving viruses, unforeseen viruses attributable to a given pathology, or viruses uncommon to a specific geographic area (d'Humières et al., 2021). Unlike bacteria, which have a 16S rRNA conserved gene that can be targeted, viruses have no common gene, and only shotgun sequencing provides unbiased detection of all viral genomes. In theory, CMg can be used to detect unidentified or unexpected viruses in clinical samples. In practice, multiple clinical cases or series have proven the effectiveness of this technique in diagnosing viruses in various types of infections such as encephalitis, fever/sepsis, and pneumonia (Rodriguez-Stanley et al., 2020; Doan et al., 2016).

7.5 CMgs may change the course of therapy toward nonantimicrobial drugs

In cases of infection, CMgs diagnosis of an etiological agent can save patients from unnecessary investigations, invasive procedures, and empirical antibiotic therapy. CMgs can also guide the use of immunomodulatory therapies by the therapist. In other words, the absence of infection as demonstrated by CMg, in addition to conventional tests, provides a supplementary argument for the clinician to initiate immunosuppressive therapy to treat a possible autoimmune disease (Wilson et al., 2019). However, it is important to note that CMgs may be less sensitive than targeted techniques for detecting a specific pathogen, expressing that CMg should not be used in place of conventional methods (Miller et al., 2019).

7.6 CMg may help in describing genetic variation

Another interesting application of CMg is the ability to characterize viral genetic diversity. Viruses, particularly RNA viruses and, to a lesser extent, DNA viruses, adapt quickly to their surroundings (immune pressure, antiviral treatment) (d'Humières et al., 2021). CMg has been extremely useful in characterizing the full-length genome of circulating variants in the SARS-CoV-2 pandemic, allowing researchers to investigate the selection of specific mutations toward immune response. It has been indicated that it is useful in characterizing divergent genotypes within a virus genus, as Sanger sequencing may miss them. The ability of CMg to sequence the entire genome provides an excellent opportunity to detect fitness-associated substitutions in HCV-infected patients who are failing treatment (Fourati et al., 2020).

7.7 CMg may play a beneficial role in outbreaks research

Taxonomic and phylogenetic analyses can also be performed for epidemiological studies. CMg has been fruitfully used to classify the epidemiology of Zika and Ebola viruses in several regions around the world, which is critical for monitoring emerging viruses and implementing large-scale public health measures quickly (Faria et al., 2017; Grubaugh et al., 2017). The current SARS-CoV-2 pandemic, discovered and characterized by CMg, demonstrates the importance of nontargeted screening and rapid characterization of potential pathogens (Wu et al., 2018).

8. Future aspects

Due to emerging technologies, clinical metagenomics laboratories are in the middle of a diagnostic revolution across the globe. The execution of computational diagnostic tools such as DNA sequence–based analyses has radically altered the approach to pathogen detection. However, most molecular methods by using specific primers or probes target only a selected number of pathogens. Computational approaches have been a great success in explaining the evidence to understand human pathogen interactions at a level never possible before. Increasing implementation of computational metagenomics in present-day clinical microbiology laboratories can be used to understand the diagnostic of clinical metagenomics (Jacob et al., 2019).

Clinical Metagenomics can also be used in the fields of bioremediation, personalized medicine, xenobiotic metabolism, and so forth. It can transform diagnostics and can detect emergent pathogens or novel genetic variants. Reduced costs, turnaround time, and increased sensitivity can promote the incorporation of computational metagenomics in clinical practice in the future for various syndromes (Forbes et al., 2018). An advantage of clinical metagenomics is that new tools are created, and discoveries made the sequencing data more utilized in such ways, which are previously unknown. As research and clinical labs have become more and more dependent on sequencing data, a skilled bioinformatician will become more important (Mulcahy et al., 2016).

Acknowledgment

The contribution of Shivani Tyagi is a part of her Ph.D. literature survey.

References

Anand, S., Kuntal, B. K., Mohapatra, A., Bhatt, V., & Mande, S. S. (April 15, 2020). FunGeCo: A web-based tool for estimation of functional potential of bacterial genomes and microbiomes using gene context information. *Bioinformatics, 36*(8), 2575–2577. https://doi.org/10.1093/bioinformatics/btz957

Andrews, M. C., Duong, C. P. M., Gopalakrishnan, V., Iebba, V., Chen, W. S., Derosa, L., Khan, M. A. W., Cogdill, A. P., White, M. G., Wong, M. C., Ferrere, G., Fluckiger, A., Roberti, M. P., Opolon, P., Alou, M. T., Yonekura, S., Roh, W., Spencer, C. N., Curbelo, I. F., … Wargo, J. A. (2021). Gut microbiota signatures are associated with toxicity to combined CTLA-4 and PD-1 blockade. *Nature Medicine, 27*(8), 1432–1441. https://doi.org/10.1038/s41591-021-01406-6. http://www.nature.com/nm/index.html

Aron-Wisnewsky, J., Vigliotti, C., Witjes, J., Le, P., Holleboom, A. G., Verheij, J., Nieuwdorp, M., & Clément, K. (2020). Gut microbiota and human NAFLD: Disentangling microbial signatures from metabolic disorders. *Nature Reviews Gastroenterology and Hepatology, 17*(5), 279–297. https://doi.org/10.1038/s41575-020-0269-9. http://www.nature.com/nrgastro/index.html

Blanco-Míguez, A., Gutiérrez-Jácome, A., Fdez-Riverola, F., Lourenço, A., & Sánchez, B. (2017). MAHMI database: A comprehensive MetaHit-based resource for the study of the mechanism of action of the human microbiota. *Database, 2017*(1). https://doi.org/10.1093/database/baw157

Bolger, A. M., Lohse, M., & Usadel, B. (2014). Trimmomatic: A flexible trimmer for Illumina sequence data. *Bioinformatics, 30*(15), 2114–2120. https://doi.org/10.1093/bioinformatics/btu170

Breitwieser, Florian P., Lu, Jennifer, & Salzberg, Steven L. (2019). A review of methods and databases for metagenomic classification and assembly. *Briefings in Bioinformatics, 20*(4), 1125–1136. https://doi.org/10.1093/bib/bbx120

Charalampous, T., Kay, G. L., Richardson, H., Aydin, A., Baldan, R., Jeanes, C., Rae, D., Grundy, S., Turner, D. J., Wain, J., Leggett, R. M., Livermore, D. M., & O'Grady, J. (2019). Nanopore metagenomics enables rapid clinical diagnosis of bacterial lower respiratory infection. *Nature Biotechnology, 37*(7), 783–792. https://doi.org/10.1038/s41587-019-0156-5. http://www.nature.com/nbt/index.html

Chee, W. J. Y., Chew, S. Y., & Than, L. T. L. (2020). Vaginal microbiota and the potential of Lactobacillus derivatives in maintaining vaginal health. *Microbial Cell Factories, 19*(1). https://doi.org/10.1186/s12934-020-01464-4

Cheng, M., Cao, L., & Ning, K. (2019). Microbiome big-data mining and applications using single-cell technologies and metagenomics approaches toward precision medicine. *Frontiers in Genetics, 10*. https://doi.org/10.3389/fgene.2019.00972. PMID: 31649735; PMCID: PMC6794611.

Chiarucci, A., Bacaro, G., & Scheiner, S. M. (2011). Old and new challenges in using species diversity for assessing biodiversity. *Philosophical Transactions of the Royal Society B: Biological Sciences, 366*(1576), 2426–2437. https://doi.org/10.1098/rstb.2011.0065. http://rstb.royalsocietypublishing.org/content/366/1576/2426.full.pdf+html

Ciria, R., Abreu-Goodger, C., Morett, E., & Merino, E. (2004). GeConT: Gene context analysis. *Bioinformatics, 20*(14), 2307–2308. https://doi.org/10.1093/bioinformatics/bth216

Cummins, C., Ahamed, A., Aslam, R., Burgin, J., Devraj, R., Edbali, O., Gupta, D., Harrison, P. W., Haseeb, M., Holt, S., Ibrahim, T., Ivanov, E., Jayathilaka, S., Kadhirvelu, V., Kay, S., Kumar, M., Lathi, A., Leinonen, R., Madeira, F., ... Cochrane, G. (2022). The European nucleotide archive in 2021. *Nucleic Acids Research, 50*(1), D106–D110. https://doi.org/10.1093/nar/gkab1051. https://academic.oup.com/nar/issue

d'Humières, C., Salmona, M., Dellière, S., Leo, S., Rodriguez, C., Angebault, C., Alanio, A., Fourati, S., Lazarevic, V., Woerther, P. L., Schrenzel, J., & Ruppé, E. (2021). The potential role of clinical metagenomics in infectious diseases: Therapeutic perspectives. *Drugs, 81*(13), 1453–1466. https://doi.org/10.1007/s40265-021-01572-4. http://rd.springer.com/journal/40265

Darling, A. E., Jospin, G., Lowe, Matsen, F. A., Bik, H. M., & Eisen, J. A. (2014). PhyloSift: phylogenetic analysis of genomes and metagenomes. *PeerJ, 2*. https://doi.org/10.7717/peerj.243

Doan, T., Akileswaran, L., Andersen, D., Johnson, B., Ko, N., Shrestha, A., Shestopalov, V., Lee, C. S., Lee, A. Y., & Van Gelder, R. N. (October 1, 2016). Paucibacterial microbiome and resident DNA virome of the healthy conjunctiva. *Investigative Ophthalmology and Visual Science, 57*(13), 5116–5126. https://doi.org/10.1167/iovs.16-19803

Dröge, J., Gregor, I., & McHardy, A. C. (2015). Taxator-tk: Precise taxonomic assignment of metagenomes by fast approximation of evolutionary neighborhoods. *Bioinformatics, 31*(6), 817–824. https://doi.org/10.1093/bioinformatics/btu745. http://bioinformatics.oxfordjournals.org/

Escapa, I. F., Chen, T., Huang, Y., Gajare, P., Dewhirst, F. E., & Lemona, K. P. (2018). New insights into human nostril microbiome from the expanded human oral microbiome database (eHOMD): A resource for the microbiome of the human aerodigestive tract. *mSystems, 3*(6). https://doi.org/10.1128/MSYSTEMS.00187-18. https://msystems.asm.org/content/3/6/e00187-18

Faria, N. R., Quick, J., Claro, I. M., Thézé, J., de Jesus, J. G., Giovanetti, M., Kraemer, M. U. G., Hill, S. C., Black, A., da Costa, A. C., Franco, L. C., Silva, S. P., Wu, C. H., Raghwani, J., Cauchemez, S., du Plessis, L., Verotti, M. P., de Oliveira, W. K., Carmo, E. H., ... Pybus, O. G. (June 15, 2017). Establishment and cryptic transmission of Zika virus in Brazil and the Americas. *Nature, 546*(7658), 406–410. https://doi.org/10.1038/nature22401

Fettweis, J. M., Serrano, M. G., Sheth, N. U., Mayer, C. M., Glascock, A. L., Brooks, & Jefferson, K. K. (2012). Vaginal Microbiome Consortium (additional members), Buck GA. Species-level classification of the vaginal microbiome. *BMC Genomics, 13*. https://doi.org/10.1186/1471-2164-13-S8-S17

Fettweis, J. M., Serrano, M. G., Brooks, J. P., Edwards, D. J., Girerd, P. H., Parikh, H. I., Huang, B., Arodz, T. J., Edupuganti, L., Glascock, A. L., Xu, J., Jimenez, N. R., Vivadelli, S. C., Fong, S. S., Sheth, N. U., Jean, S., Lee, V., Bokhari, Y. A., Lara, A. M., ... Buck, G. A. (2019). The vaginal microbiome and preterm birth. *Nature Medicine, 25*(6), 1012–1021. https://doi.org/10.1038/s41591-019-0450-2. http://www.nature.com/nm/index.html

Forbes, J. D., Knox, N. C., Peterson, C. L., & Reimer, A. R. (2018). Highlighting clinical metagenomics for enhanced diagnostic decision-making: A step towards wider implementation. *Computational and Structural Biotechnology Journal, 16*, 108–120. https://doi.org/10.1016/j.csbj.2018.02.006. www.csbj.org

Forster, S. C., Browne, H. P., Kumar, N., Hunt, M., Denise, H., Mitchell, A., Finn, R. D., & Lawley, T. D. (2016). HPMCD: The database of human microbial communities from metagenomic datasets and microbial reference genomes. *Nucleic Acids Research, 44*(1), D604–D609. https://doi.org/10.1093/nar/gkv1216. http://nar.oxfordjournals.org/

Fourati, M., Smaoui, S., Hlima, H. B., Elhadef, K., Braïek, O. B., Ennouri, K., Mtibaa, A. C., & Mellouli, L. (December, 2020). Bioactive compounds and pharmacological potential of pomegranate (*Punica granatum*) seeds—A review. *Plant Foods for Human Nutrition, 75*(4), 477–486. https://doi.org/10.1007/s11130-020-00863-7

Garrett, W. S. (2015). Cancer and the microbiota. *Science, 348*(6230), 80–86. https://doi.org/10.1126/science.aaa4972

Grubaugh, N. D., Ladner, J. T., Kraemer, M. U. G., Dudas, G., Tan, A. L., Gangavarapu, K., Wiley, M. R., White, S., Thézé, J., Magnani, D. M., Prieto, K., Reyes, D., Bingham, A. M., Paul, L. M., Robles-Sikisaka, R., Oliveira, G., Pronty, D., Barcellona, C. M., Metsky, H. C., ... ersen, K. G. (June 15, 2017). Genomic epidemiology reveals multiple introductions of Zika virus into the United States. *Nature, 546*(7658), 401–405. https://doi.org/10.1038/nature22400

Hadrich, D. (2018). Microbiome research is becoming the key to better understanding health and nutrition. *Frontiers in Genetics, 9*, 212. https://doi.org/10.3389/fgene.2018.00212

Haider, B., Ahn, T. H., Bushnell, B., Chai, J., Copeland, A., & Pan, C. (2014). Omega: An Overlap-graph de novo Assembler for Metagenomics. *Bioinformatics, 30*(19), 2717–2722. https://doi.org/10.1093/bioinformatics/btu395. http://bioinformatics.oxfordjournals.org/

Hsiao, E. Y., McBride, S. W., Hsien, S., Sharon, G., Hyde, E. R., McCue, T., Codelli, J. A., Chow, J., Reisman, S. E., Petrosino, J. F., Patterson, P. H., & Mazmanian, S. K. (2013). Microbiota modulate behavioral and physiological abnormalities associated with neurodevelopmental disorders. *Cell, 155*(7), 1451–1463. https://doi.org/10.1016/j.cell.2013.11.024. https://www.sciencedirect.com/journal/cell

Human Microbiome Project Consortium. (2012). A framework for human microbiome research. *Nature, 486*(7402), 215–221. https://doi.org/10.1038/nature11209

Huerta-Cepas, J., Forslund, K., Coelho, L. P., Szklarczyk, D., Jensen, L. J., Von Mering, C., & Bork, P. (2017). Fast genome-wide functional annotation through orthology assignment by eggNOG-mapper. *Molecular Biology and Evolution, 34*(8), 2115–2122. https://doi.org/10.1093/molbev/msx148

Hunt, M., Kikuchi, T., Sanders, M., Newbold, C., Berriman, M., & Otto, T. D. (2013). REAPR: A universal tool for genome assembly evaluation. *Genome Biology, 14*(5). https://doi.org/10.1186/gb-2013-14-5-r47. http://www.genomebiology.com/content/pdf/gb-2013-14-5-r47.pdf

Jacob, J. J., Veeraraghavan, B., & Vasudevan, K. (2019). Metagenomic next-generation sequencing in clinical microbiology. *Indian Journal of Medical Microbiology, 37*(2), 133–140. https://doi.org/10.4103/ijmm.IJMM_19_401. https://www.sciencedirect.com/journal/indian-journal-of-medical-microbiology/issues

Köser, C. U., Ellington, M. J., Cartwright, E. J. P., Gillespie, S. H., Brown, N. M., Farrington, M., Holden, M. T. G., Dougan, G., Bentley, S. D., Parkhill, J., & Peacock, S. J. (2012). Routine use of microbial whole genome sequencing in diagnostic and public health microbiology. *PLoS Pathogens, 8*(8). https://doi.org/10.1371/journal.ppat.1002824. http://www.plospathogens.org/article/fetchObjectAttachment.action?uri=info%3Adoi%2F10.1371%2Fjournal.ppat.1002824&representation=PDF

Kalia, N., Singh, J., & Kaur, M. (2020). Microbiota in vaginal health and pathogenesis of recurrent vulvovaginal infections: A critical review. *Annals of Clinical Microbiology and Antimicrobials, 19*(1). https://doi.org/10.1186/s12941-020-0347-4

Kanehisa, M., Furumichi, M., Tanabe, M., Sato, Y., & Morishima, K. (2017). KEGG: New perspectives on genomes, pathways, diseases and drugs. *Nucleic Acids Research, 45*(1), D353–D361. https://doi.org/10.1093/nar/gkw1092

Katara, P. (2014). Potential of Bioinformatics as functional genomics tool: An overview. *Network Modeling and Analysis in Health Informatics and Bioinformatics, 3*(1), 1–7. https://doi.org/10.1007/s13721-014-0052-3. http://www.springer.com/new+%26+forthcoming+titles+%28default%29/journal/13721

Katara, P., Grover, A., Kuntal, H., & Sharma, V. (2011). In silico prediction of drug targets in Vibrio cholerae. *Protoplasma, 248*(4), 799–804. https://doi.org/10.1007/s00709-010-0255-0

Kelley, D. R., Liu, B., Delcher, A. L., Pop, M., & Salzberg, S. L. (2012). Gene prediction with Glimmer for metagenomic sequences augmented by classification and clustering. *Nucleic Acids Research, 40*(1). https://doi.org/10.1093/nar/gkr1067

Kolb, M., Lazarevic, V., Emonet, S., Calmy, A., Girard, M., Gaïa, N., Charretier, Y., Cherkaoui, A., Keller, P., Huber, C., & Schrenzel, J. (2019). Next-generation sequencing for the diagnosis of challenging culture-negative endocarditis. *Frontiers of Medicine, 6*. https://doi.org/10.3389/fmed.2019.00203

Kollef, M. H. (2000). Inadequate antimicrobial treatment: An important determinant of outcome for hospitalized patients. *Clinical Infectious Diseases, 31*(4), S131–S138. https://doi.org/10.1086/314079

Kolmogorov, M., Bickhart, D. M., Behsaz, B., Gurevich, A., Rayko, M., Shin, S. B., Kuhn, K., Yuan, J., Polevikov, E., Smith, T. P. L., & Pevzner, P. A. (2020). metaFlye: scalable long-read metagenome assembly using repeat graphs. *Nature Methods, 17*(11), 1103–1110. https://doi.org/10.1038/s41592-020-00971-x. http://www.nature.com/nmeth/

Kumar, A., & Chordia, N. (2017). Role of microbes in human health. *Applied Microbiology: Open Access, 03*(02). https://doi.org/10.4172/2471-9315.1000131

Langelier, C., Zinter, M. S., Kalantar, K., Yanik, G. A., Christenson, S., O'Donovan, B., White, C., Wilson, M., Sapru, A., Dvorak, C. C., Miller, S., Chiu, C. Y., & DeRisi, J. L. (2018). Metagenomic sequencing detects respiratory pathogens in hematopoietic cellular transplant patients. *American Journal of Respiratory and Critical Care Medicine, 197*(4), 524–528. https://doi.org/10.1164/rccm.201706-1097LE. https://www.atsjournals.org/doi/pdf/10.1164/rccm.201706-1094LE

Lanzén, A., Jørgensen, S. L., Huson, D. H., Gorfer, M., Grindhaug, S. H., Jonassen, I., Øvreås, L., & Urich, T. (2012). CREST - classification resources for environmental sequence tags. *PLoS One, 7*(11). https://doi.org/10.1371/journal.pone.0049334. http://www.plosone.org/article/fetchObjectAttachment.action?uri=info%3Adoi%2F10.1371%2Fjournal.pone.0049334&representation=PDF

Leake, S. L., Pagni, M., Falquet, L., Taroni, F., & Greub, G. (2016). The salivary microbiome for differentiating individuals: Proof of principle. *Microbes and Infection, 18*(6), 399–405. https://doi.org/10.1016/j.micinf.2016.03.011. http://www.journals.elsevier.com/microbes-and-infection/

Li, D., Liu, C. M., Luo, R., Sadakane, K., & Lam, T. W. (2015). MEGAHIT: An ultra-fast single-node solution for large and complex metagenomics assembly via succinct de Bruijn graph. *Bioinformatics, 31*(10), 1674–1676. https://doi.org/10.1093/bioinformatics/btv033. http://bioinformatics.oxfordjournals.org/

Liang, K.c., & Sakakibara, Y. (2021). MetaVelvet-DL: A MetaVelvet deep learning extension for de novo metagenome assembly. *BMC Bioinformatics, 22*. https://doi.org/10.1186/s12859-020-03737-6

Lin, Z., Zu, Xie, H. P., Jin, H. Z., Yang, Liu, X. R., & Zhang, W. D. (2016). *Yao Xue Xue Bao*.

Liu, B., & Pop, M. (2011). MetaPath: Identifying differentially abundant metabolic pathways in metagenomic datasets. *BMC Proceedings, 5*(S2). https://doi.org/10.1186/1753-6561-5-s2-s9

Lloyd-Price, J., Abu-Ali, G., & Huttenhower, C. (2016). The healthy human microbiome. *Genome Medicine, 8*(1). https://doi.org/10.1186/s13073-016-0307-y

Lloyd-Price, J., Arze, C., Ananthakrishnan, A. N., Schirmer, M., Avila-Pacheco, J., Poon, T. W., Andrews, E., Ajami, N. J., Bonham, K. S., Brislawn, C. J., Casero, D., Courtney, H., Gonzalez, A., Graeber, T. G., Hall, A. B., Lake, K., Landers, C. J., Mallick, H., Plichta, D. R., … Huttenhower, C. (2019). Multi-omics of the gut microbial ecosystem in inflammatory bowel diseases. *Nature, 569*(7758), 655–662. https://doi.org/10.1038/s41586-019-1237-9. http://www.nature.com/nature/index.html

Lozano, R., Naghavi, M., & Foreman, K. (2010). Global and regional mortality from 235 causes of death for 20 age groups in 1990 and 2010: A systematic analysis for the global burden of disease study. *Lancet, 380*(9859). https://doi.org/10.1016/S0140-6736

Malla, M. A., Dubey, A., Kumar, A., Yadav, S., Hashem, A., & Allah, E. (2019). Exploring the human microbiome: The potential future role of next-generation sequencing in disease diagnosis and treatment. *Frontiers in Immunology, 10*. https://doi.org/10.3389/fimmu.2018.02868. www.frontiersin.org/Immunology

Masha, S. C., Owuor, C., Ngoi, J. M., Cools, P., Sanders, E. J., Vaneechoutte, M., Crucitti, T., & de Villiers, E. P. (2019). Comparative analysis of the vaginal microbiome of pregnant women with either *Trichomonas vaginalis* or *Chlamydia trachomatis*. *PLoS One, 14*(12). https://doi.org/10.1371/journal.pone.0225545. https://journals.plos.org/plosone/article/file?id=10.1371/journal.pone.0225545&type=printable

Mendling, W. (2016). Vaginal microbiota. *Advances in Experimental Medicine and Biology, 902*, 83–93. https://doi.org/10.1007/978-3-319-31248-4_6. http://www.springer.com/series/5584

Mikheenko, A., Saveliev, V., & Gurevich, A. (2016). MetaQUAST: Evaluation of metagenome assemblies. *Bioinformatics, 32*(7), 1088–1090. https://doi.org/10.1093/bioinformatics/btv697

Miller, R. R., Montoya, V., Gardy, J. L., Patrick, D. M., & Tang, P. (2013). Metagenomics for pathogen detection in public health. *Genome Medicine, 5*(9). https://doi.org/10.1186/gm485. http://www.advancesindifferenceequations.com/content/5/9/81

Miller, S., Naccache, S. N., Samayoa, E., Messacar, K., Arevalo, S., Federman, S., Stryke, D., Pham, E., Fung, B., Bolosky, W. J., Ingebrigtsen, D., Lorizio, W., Paff, S. M., Leake, J. A., Pesano, R., DeBiasi, R., Dominguez, S., & Chiu, C. Y. (2019). Laboratory validation of a clinical metagenomic sequencing assay for pathogen detection in cerebrospinal fluid. *Genome Research, 29*(5), 831–842. https://doi.org/10.1101/gr.238170.118. https://genome.cshlp.org/content/29/5/831.full.pdf+html

Mineeva, O., Rojas-Carulla, M., Ley, R. E., Schölkopf, B., Youngblut, N. D., & Luigi Martelli, P. (2020). DeepMAsED: Evaluating the quality of metagenomic assemblies. *Bioinformatics, 36*(10), 3011–3017. https://doi.org/10.1093/bioinformatics/btaa124

Mulcahy, Grady, H., & Workentine, M. L. (2016). The challenge and potential of metagenomics in the clinic. *Frontiers in Immunology, 7*. https://doi.org/10.3389/fimmu.2016.00029

Namiki, T., Hachiya, T., Tanaka, H., & Sakakibara, Y. (2012). MetaVelvet: An extension of Velvet assembler to de novo metagenome assembly from short sequence reads. *Nucleic Acids Research, 40*(20). https://doi.org/10.1093/nar/gks678

Navgire, G. S., Goel, N., Sawhney, G., Sharma, M., Kaushik, P., Mohanta, Y. K., Mohanta, T. K., & Al-Harrasi, A. (2022). Analysis and interpretation of metagenomics data: An approach. *Biological Procedures Online, 24*(1). https://doi.org/10.1186/s12575-022-00179-7. https://biologicalproceduresonline.biomedcentral.com/

Nicholson, J. K., Holmes, E., Kinross, J., Burcelin, R., Gibson, G., Jia, W., & Pettersson, S. (2012). Host-gut microbiota metabolic interactions. *Science, 336*(6086), 1262–1267. https://doi.org/10.1126/science.1223813. http://www.sciencemag.org/content/336/6086/1262.full.pdf

Noguchi, H., Taniguchi, T., & Itoh, T. (2008). MetaGeneAnnotator: Detecting species-specific patterns of ribosomal binding site for precise gene prediction in anonymous prokaryotic and phage genomes. *DNA Research, 15*(6), 387–396. https://doi.org/10.1093/dnares/dsn027

Noronha, A., Modamio, J., Jarosz, Y., Guerard, E., Sompairac, N., Preciat, G., Daníelsdóttir, A. D., Krecke, M., Merten, D., Haraldsdóttir, H. S., Heinken, A., Heirendt, L., Magnúsdóttir, S., Ravcheev, D. A., Sahoo, S., Gawron, P., Friscioni, L., Garcia, B., Prendergast, M., … Thiele, I. (2019). The virtual metabolic human database: Integrating human and gut microbiome metabolism with nutrition and disease. *Nucleic Acids Research, 47*(D1), D614–D624. https://doi.org/10.1093/nar/gky992

Olson, N. D., Treangen, T. J., Hill, C. M., Cepeda-Espinoza, V., Ghurye, J., Koren, S., & Pop, M. (2019). Metagenomic assembly through the lens of validation: Recent advances in assessing and improving the quality of genomes assembled from metagenomes. *Briefings in Bioinformatics, 20*(4), 1140–1150. https://doi.org/10.1093/bib/bbx098

Parks, D. H., Imelfort, M., Skennerton, C. T., Hugenholtz, P., & Tyson, G. W. (2015). CheckM: Assessing the quality of microbial genomes recovered from isolates, single cells, and metagenomes. *Genome Research, 25*(7), 1043–1055. https://doi.org/10.1101/gr.186072.114. http://genome.cshlp.org/content/25/7/1043.full.pdf+html

Pendleton, K. M., Erb-Downward, J. R., Bao, Y., Branton, W. R., Falkowski, N. R., Newton, D. W., Huffnagle, G. B., & Dickson, R. P. (2017). Rapid pathogen identification in bacterial pneumonia using real-time metagenomics. *American Journal of Respiratory and Critical Care Medicine, 196*(12), 1610–1612. https://doi.org/10.1164/rccm.201703-0537LE. http://www.atsjournals.org/doi/pdf/10.1164/rccm.201703-0537LE

Peng, Y., Leung, H. C. M., Yiu, S. M., & Chin, F. Y. L. (2011). Meta-IDBA: A de Novo assembler for metagenomic data. *Bioinformatics, 27*(13), i94–i101. https://doi.org/10.1093/bioinformatics/btr216

Pereira-Flores, E., Glöckner, F. O., & Fernandez-Guerra, A. (2019). Fast and accurate average genome size and 16S rRNA gene average copy number computation in metagenomic data. *BMC Bioinformatics, 20*(1). https://doi.org/10.1186/s12859-019-3031-y

Perz, A. I., Giles, C. B., Brown, C. A., Porter, H., Roopnarinesingh, X., & Wren, J. D. (2019). MNEMONIC: MetageNomic experiment mining to create an OTU network of inhabitant correlations. *BMC Bioinformatics, 20*. https://doi.org/10.1186/s12859-019-2623-x. http://www.biomedcentral.com/bmcbioinformatics/

Quince, C., Lanzen, A., Davenport, R. J., & Turnbaugh, P. J. (2011). Removing noise from pyrosequenced amplicons. *BMC Bioinformatics, 12*. https://doi.org/10.1186/1471-2105-12-38

Ravel, J., Gajer, P., Abdo, Z., Schneider, G. M., Koenig, S. S. K., McCulle, S. L., Karlebach, S., Gorle, R., Russell, J., Tacket, C. O., Brotman, R. M., Davis, C. C., Ault, K., Peralta, L., & Forney, L. J. (2011). Vaginal microbiome of reproductive-age women. *Proceedings of the National Academy of Sciences of the United States of America, 108*(1), 4680–4687. https://doi.org/10.1073/pnas.1002611107. http://www.pnas.org/content/108/suppl.1/4680.full.pdf+html

Rho, M., Tang, H., & Ye, Y. (2010). FragGeneScan: Predicting genes in short and error-prone reads. *Nucleic Acids Research, 38*(20). https://doi.org/10.1093/nar/gkq747

Rodriguez-Stanley, J., Alonso-Ferres, M., Zilioli, S., & Slatcher, R. B. (October, 2020). Correction to Rodriguez-Stanley et al. (2020) *Journal of Family Psychology, 34*(7), 845. https://doi.org/10.1037/fam0000800

Rossi, T., Vergara, D., Fanini, F., Maffia, M., Bravaccini, S., & Pirini, F. (2020). Microbiota-derived metabolites in tumor progression and metastasis. *International Journal of Molecular Sciences, 21*(16). https://doi.org/10.3390/ijms21165786

Rossmann, M., Thomas, A. M., Guima, S. S., Martins, L. F., Inderbitzin, P., Knight-Connoni, V., da Silva, A. M., & Setubal, J. C. (2021). Microbiomes of field-grown maize and Soybean in southeastern and central Brazil inferred by high-throughput 16s and internal transcribed spacer amplicon sequencing. *Microbiology Resource Announcements, 10*(31). https://doi.org/10.1128/MRA.00528-21. https://journals.asm.org/doi/10.1128/MRA.00528-21

Sangwan, N., Xia, F., & Gilbert, J. A. (2016). Recovering complete and draft population genomes from metagenome datasets. *Microbiome, 4.* https://doi.org/10.1186/s40168-016-0154-5

Sayers, E. W., Beck, J., Bolton, E. E., Bourexis, D., Brister, J. R., Canese, K., Comeau, D. C., Funk, K., Kim, S., Klimke, W., Marchler-Bauer, A., Landrum, M., Lathrop, S., Lu, Z., Madden, T. L., O'Leary, N., Phan, L., Rangwala, S. H., Schneider, V. A., ... Sherry, S. T. (2021). Database resources of the national center for Biotechnology information. *Nucleic Acids Research, 49*(1), D10–D17. https://doi.org/10.1093/nar/gkaa892. https://academic.oup.com/nar/issue

Sayers, E. W., O'Sullivan, C., & Karsch-Mizrachi, I. (2022). Using GenBank and SRA. *Methods in Molecular Biology, 2443*, 1–25. https://doi.org/10.1007/978-1-0716-2067-0_1. http://www.springer.com/series/7651

Schlaberg, R. (2020). Microbiome diagnostics. *Clinical Chemistry, 66*(1), 68–76. https://doi.org/10.1373/clinchem.2019.303248

Schmedes, S. E., Sajantila, A., & Budowle, B. (2016). Expansion of microbial forensics. *Journal of Clinical Microbiology, 54*(8), 1964–1974. https://doi.org/10.1128/JCM.00046-16. http://jcm.asm.org/content/54/8/1964.full.pdf

Schmieder, R., & Edwards, R. (2011). Quality control and preprocessing of metagenomic datasets. *Bioinformatics, 27*(6), 863–864. https://doi.org/10.1093/bioinformatics/btr026

Segata, N., Waldron, L., Ballarini, A., Narasimhan, V., Jousson, O., & Huttenhower, C. (2012). Metagenomic microbial community profiling using unique clade-specific marker genes. *Nature Methods, 9*(8), 811–814. https://doi.org/10.1038/nmeth.2066

Sommer, M. J., Salzberg, S. L., & Ouzounis, C. A. (2021). Balrog: A universal protein model for prokaryotic gene prediction. *PLoS Computational Biology, 17*(2). https://doi.org/10.1371/journal.pcbi.1008727

Svoboda, E. (2021). Gut feeling yields evidence of microbial involvement in autoimmunity. *Nature, 595*(7867), S54–S55. https://doi.org/10.1038/d41586-021-01837-8

Thomas, T., Gilbert, J., & Meyer, F. (2012). Metagenomics - a guide from sampling to data analysis. *Microbial Informatics and Experimentation, 2.* https://doi.org/10.1186/2042-5783-2-3

Thorburn, Bennett, Modha, S., Murdoch, Gunson, & Murcia, P. R. (2015). Erratum to \The use of next generation sequencing in the diagnosis and typing of respiratory infections. *Journal of Clinical Virology, 69.* https://doi.org/10.1016/j.jcv.2015.06.101

Truong, D. T., Tett, A., Pasolli, E., Huttenhower, C., & Segata, N. (2017). Microbial strain-level population structure and genetic diversity from metagenomes. *Genome Research, 27*(4), 626–638. https://doi.org/10.1101/gr.216242.116. http://genome.cshlp.org/content/27/4/626.full.pdf+html

Wilson, M. R., Sample, H. A., Zorn, K. C., Arevalo, S., Yu, G., Neuhaus, J., Federman, S., Stryke, D., Briggs, B., Langelier, C., Berger, A., Douglas, V., Josephson, S. A., Chow, F. C., Fulton, B. D., DeRisi, J. L., Gelfand, J. M., Naccache, S. N., Bender, J., ... Chiu, C. Y. (2019). Clinical metagenomic sequencing for diagnosis of meningitis and encephalitis. *New England Journal of Medicine, 380*(24), 2327–2340. https://doi.org/10.1056/NEJMoa1803396. http://www.nejm.org/medical-index

Wood, D. E., & Salzberg, S. L. (2014). Kraken: Ultrafast metagenomic sequence classification using exact alignments. *Genome Biology, 15*(3). https://doi.org/10.1186/gb-2014-15-3-r46

Wu, S. C., Rau, C. S., Liu, H. T., Kuo, P. J., Chien, P. C., Hsieh, T. M., Tsai, C. H., Chuang, J. F., Huang, C. Y., Hsieh, H. Y., & Hsieh, C. H. (2018). Metagenome analysis as a tool to study bacterial infection associated with acute surgical abdomen. *Journal of Clinical Medicine, 7*(10). https://doi.org/10.3390/JCM7100346. http://www.mdpi.com/2077-0383/7/10/346/pdf

Wylie, K. M., Truty, R. M., Sharpton, T. J., Mihindukulasuriya, K. A., Zhou, Y., Gao, H., Sodergren, E., Weinstock, G. M., & Pollard, K. S. (2012). Novel bacterial Taxa in the human microbiome. *PLoS One, 7*(6). https://doi.org/10.1371/journal.pone.0035294. http://www.plosone.org/article/fetchObjectAttachment.action?uri=info%3Adoi%2F10.1371%2Fjournal.pone.0035294&representation=PDF

Yang, C., Chowdhury, D., Zhang, Z., Cheung, W. K., Lu, A., Bian, Z., & Zhang, L. (2021). A review of computational tools for generating metagenome-assembled genomes from metagenomic sequencing data. *Computational and Structural Biotechnology Journal, 19*, 6301–6314. https://doi.org/10.1016/j.csbj.2021.11.028. www.csbj.org

Zhang, L., Chen, F. X., Zeng, Z., Xu, M., Sun, F., Yang, L., Bi, X., Lin, Y., Gao, Y. J., Hao, H. X., Yi, W., Li, M., & Xie, Y. (2021). Advances in metagenomics and its application in environmental microorganisms. *Frontiers in Microbiology, 12.* https://doi.org/10.3389/fmicb.2021.766364. https://www.frontiersin.org/journals/microbiology#

Zhang, S. W., Jin, X. Y., & Zhang, T. (2017). Gene prediction in metagenomic fragments with deep learning. *BioMed Research International, 2017.* https://doi.org/10.1155/2017/4740354. http://www.hindawi.com/journals/biomed/

Zhou, Y., Wylie, K. M., Feghaly, R. E. E., Mihindukulasuriya, K. A., Elward, A., Haslam, D. B., Storch, G. A., & Weinstock, G. M. (2016). Metagenomic approach for identification of the pathogens associated with diarrhea in stool specimens. *Journal of Clinical Microbiology, 54*(2), 368–375. https://doi.org/10.1128/JCM.01965-15. http://jcm.asm.org/content/54/2/368.full.pdf+html

Chapter 21

Biomolecular networks

Shiv Kumar Yadav[1], Atifa Hafeez[2], Raj Kumar[1], Manish Kumar Gupta[2] and Ravi Kumar Gutti[3]
[1]Department of Mathematics, Faculty of Engineering & Technology, Veer Bahadur Singh Purvanchal University, Jaunpur, Uttar Pradesh, India; [2]Department of Biotechnology, Faculty of Science, Veer Bahadur Singh Purvanchal University, Jaunpur, Uttar Pradesh, India; [3]Department of Biochemistry, School of Life Sciences, University of Hyderabad, Hyderabad, Telangana, India

1. Introduction

Biological molecules, also known as biomolecules, are produced by cells and living organisms. The molecular mechanisms of the cellular processes that constitute life depend heavily on DNA, RNA, proteins, and metabolites. Living organisms require the intricate process of gene expression. The crucial process of translating information contained in a gene into a useful product is known as gene expression. Transcription factors (TFs) are the main regulator of gene expression. TFs are proteins with the ability to bind certain DNA sequences and control the expression of genes (Mitsis et al., 2020). It becomes vital to understand the structure of biomolecules and interaction between them for a number of reasons, including the creation of novel drugs and the identification of disease pathways. A graph, which consists a collection of nodes and a set of edges reflecting the connections between nodes, can be used to depict both the structure and interactions of these entities. Many biological processes can be represented mathematically using nodes for the entities and edges for their interactions or relationships. Networks offer a straightforward and understandable description of diverse and complex biological processes (Muzio et al., 2021). The interactions between biomolecules are essential for life-sustaining processes. Every molecule has a major relationship between its function and structure, which is modified by the surrounding biomolecules. Weak interactions between the subunits help to stabilize the overall structure of biomolecules. The functions of biomolecules are diverse and include building organisms from single cells to complex living beings such as humans, transporting nutrients and other molecules in and out of cells, and serving as enzymes and catalysts for the vast majority of chemical reactions that take place in living organisms. Proteins also form antibodies and hormones, and they influence gene activity. Several functions of these biological molecules are still a mystery, and current advanced techniques are being used to discover more molecules and understand their role in life-sustaining processes.

Complex collections of binary interactions or relationships between different biological entities are known as biomolecular networks (Koutrouli et al., 2020; Menon & Krishnan, 2021). Systems biology, synthetic biology, biological engineering, and systems chemistry all heavily rely on biomolecular networks. The potential of biomolecular networks to describe intricate biological systems based on the interactions between pairs of biomolecules has already proven to be very useful.

Biological networks include gene regulatory networks (GRNs) and protein—protein interaction networks (PPINs) to extract useable information from these networks; it's critical to comprehend the statistical and mathematical methods used to discover linkages among them (Lü & Wang, 2020). Protein—DNA, protein—ligand, and supramolecular assemblies of noncovalent protein—protein interactions (PPIs) are frequently seen in molecular networks (Giampà & Sgobba, 2020).

These networks are crucial for comprehending how these systems handle information and how cellular life is structured (Menon & Krishnan, 2021). The potential of biomolecular networks to illuminate crucial biological mechanisms from a systems biology perspective is the significance of biomolecular networks in systems biology (Chen et al., 2009). Researchers can comprehend the crucial mechanisms used by living creatures by comparing biomolecular networks between species or situations (Zhang et al., 2008).

Integrative Omics. https://doi.org/10.1016/B978-0-443-16092-9.00021-7

2. What are networks?

Networks function by allowing nodes that are connected to one another to communicate and share information. There are some general concepts that apply to various types of networks, yet the specifics of how networks operate might vary based on the type of network and the underlying technology utilized.

Network approaches are now regularly used in many academic fields. They enable a deeper understanding of complex relationships between entities as well as group-level dynamics and characteristics. Numerous networks change over time, whether it takes a few seconds or billions of years depending on the type of network. The collection of interactions between a group of entities is referred to as a network or graph (Gysi & Nowick, 2020; Newman, 2010). In network notation, each term is referred to as a node and a vertex, respectively (Barabási, 2013). Thus, the connections between two things are referred to as edges or links (Barabási, 2013). The total number of nodes and interactions in a network are usually denoted by the letters Ni and Li, respectively. Although links can be marked, nodes can also be given labels (Barabási, 2013).

In any field of study, network analysis is extremely helpful for understanding complex systems. Complex biological or medical systems include networks in the fields of neuropsychology, ecology, and gene regulation.

Predicting economic crises or opportunities is frequently at the heart of network analysis' attractiveness in finance. The study of disease transmission, the control of pandemics and epidemics, and the identification of the patient "zero" are all topics covered by network science in epidemiology (Liu et al., 2018; Vespignani, 2012; Viboud & Vespignani, 2019). Network dynamics have also been used to understand how chaos could spread within a system, such as during a storm (Gysi & Nowick, 2020).

3. What are biological networks?

By incorporating biological omics data, biological interactions and correlations, statistical measures, graph theory, and visualizations, network biology is capable of integrating, representing, interpreting, and modeling complex biological systems. Recent research has demonstrated the value of biological networks for studies that explain biological processes and disease etiologies as well as for studies that forecast therapeutic responses, both at the molecular and system platforms (Zhang & Itan, 2019).

To answer biological challenges, there are plenty of approaches to build and utilize biological networks.

The following are a few of the primary objectives of biological networks:

- Analyzing and sorting disease-causing candidate genes.
- Determination of subnetworks involved in diseases and systematic disease complications.
- Monitoring therapeutic effects helps to accelerate up target analysis as well as drugs discovery.

3.1 Network analysis

Research in the rapidly growing field of "network biology" acknowledges that molecular interactions form a complex system-level network that regulates biological processes rather than being primarily regulated by discrete, disconnected linear pathways or individual proteins. To comprehend complex phenotypes in well-being and disease, it is essential to figure out how these molecular interaction networks provide way to newly developed biological processes and to discover the significant nodes along with additional topological properties that are vital for controlling them. A very potent and helpful alternative to conventional enrichment analysis techniques is network analysis. A number of advantages of this approach consist of the fact that proteome-scale maps of the interactome are now accessible for multiple species, especially humans, and that network-based analyses are less bound by the limitations of present functional annotations. As a result, network analyses have a considerably greater application of known genes and proteins and are fewer unfair toward well understood pathways (Charitou et al., 2016).

3.2 Types of biological network

The common biological network includes GRN, gene coexpression network (GCN), PPIN, RNA network, metabolic network, cell signaling network, and neural network.

3.2.1 Gene regulatory networks

These networks depict the relationships between genes and the regulators that control gene expression, such as TFs. GRNs are essential for controlling cellular functions and growth (Davidson & Levine, 2008).

In the majority of GRN maps, the nodes are potential DNA regulatory elements or TFs, and directed edges demonstrate the physical interactions between TFs and such regulatory elements.

3.2.1.1 Gene regulatory networks methods

GRNs are currently mapped on a broad scale using two general methods. The first method, yeast one-hybrid (Y1H) method, a putative cis-regulatory DNA sequence typically a presumed promoter region, is utilized as traps to identify TFs that bound to that sequence (Vidal et al., 2011). A second method called chromatin immunoprecipitation (ChIP) makes it possible to remove a particular protein–DNA chromatin complex from the cell nucleus, including DNA-binding proteins such as TFs (Gade & Kalvakolanu, 2012). In chromatin immunoprecipitation (ChIP) methods, antibodies produced opposed to TFs of interest, as well as against a peptide tag utilized in fusion with probable TFs, can be utilized to immunoprecipitated possibly interacting cross-linked DNA fragments. ChIP methods are generally protein-centric in which their tendency to start with TFs and aim to capture related gene areas while Y1H is described as being gene-centric, since it moves forward from genes and catches related proteins (Vidal et al., 2011).

3.2.2 Gene coexpression networks

These networks reflect genes that frequently coexpress one another, presumably indicating functional links. They are built using patterns of gene expression across various situations or data sets. It is achievable to comprehend evolution and the emergence of novel phenotypes by using GCNs, which are built from high-throughput gene expression data. GCNs are a useful tool for exploring evolution in nonmodel species due to the expanding availability of gene expression data. Gene–gene interactions are represented by GCNs, which are undirected graphs with nodes for genes and edges for coexpression strength between nodes. Currently, a variety of methodologies are used to compare GCNs across species, such as functional annotation transfer, inter- and intramodular hub discovery, and differential coexpression network analysis (Ovens et al., 2021). Fig. 21.1 shows a general network of gene coexpression.

3.2.3 Protein–protein interaction

These networks show how proteins physically interact, which is essential for biological processes such as signaling, metabolism, and gene expression. PPINs shed light on how proteins work and how biological processes are structured (Barabási & Oltvai, 2004; Rolland et al., 2014) for instance a Bcl2 PPIN has been shown in Fig. 21.2.

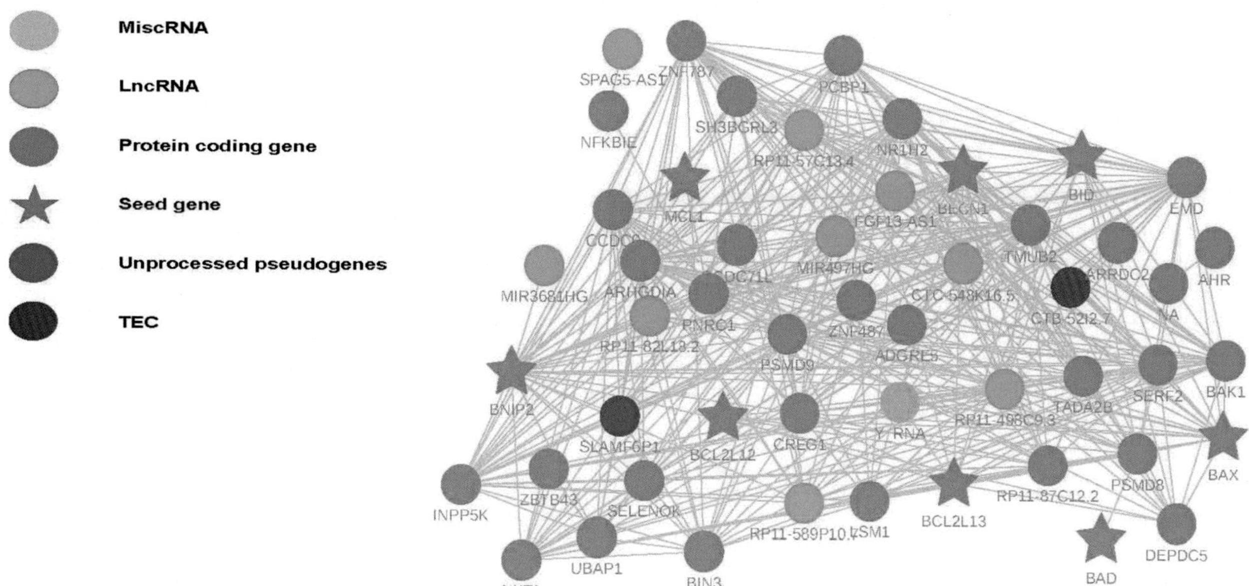

FIGURE 21.1 A general network gene coexpression. The analysis of this network may be used to find gene clusters that have comparable expression patterns under various situations (https://genefriends.org/). In this network yellow nodes represent miscellaneous RNA (MiscRNA), green color nodes represents long noncoding RNA (LncRNA), light blue represents protein coding gene, dark blue represents unprocessed pseudogenes, and brown represents to be experimentally verified genes (TEC).

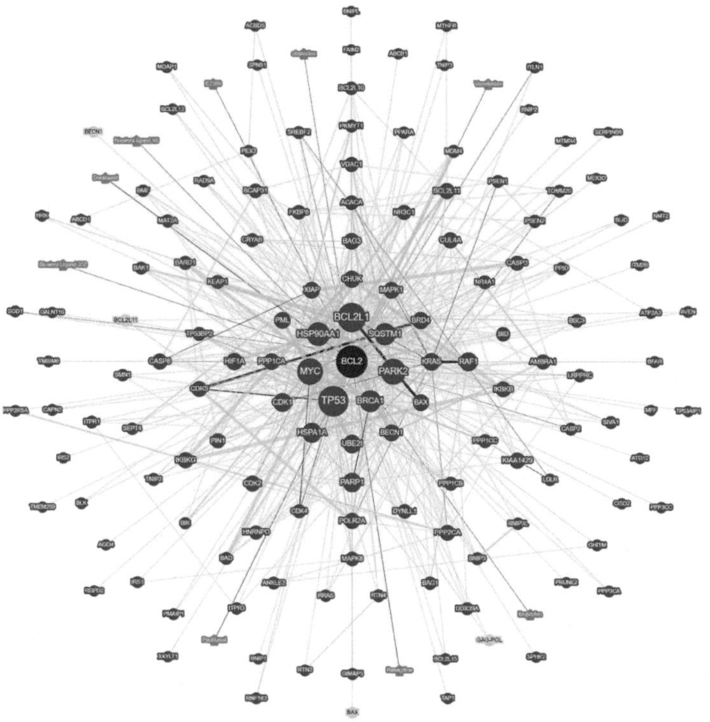

FIGURE 21.2 Protein–protein interaction network of Bcl2 protein. To facilitate the assembly of proteins into protein complexes, physical connections between proteins can be analyzed through the PPI network approach (http://thebiogrid.org/).

Proteins are represented as nodes in PPIN maps, while PPIs are shown as edges. The edges are nondirected because it is unknown which protein interacts the other, i.e., which partner affects the other functionally. Both physical and predicted interactions are possible in a PPIN (Vidal et al., 2011).

The BioGRID, MINT, BIND, DIP, IntAct, and HPRD database are well-known repositories, which contain PPIs for many organisms. The PPINs are scale-free networks that adhere to the small-world characteristic in terms of topology. While cliques (completely connected subgraphs) have been proven to have significant functional value, central hubs frequently indicate evolutionarily conserved proteins (Koutrouli et al., 2020).

3.2.3.1 PPI detection methods

Methods for detecting PPIs are categorically divided into three group: *in vitro, in vivo, and in silico approaches* as shown in Fig. 21.3.

Two of the numerous methods for mapping PPIs are used extensively for large-scale mapping. The yeast two-hybrid (Y2H) system, which is constantly being improved, is used to map binary interactions. Affinity purification is used to isolate protein complexes, and then mass spectrometry (AP/MS) is used to identify the proteins that make up these complexes. This process maps participation in protein complexes and offers indirect relationships between proteins. The majority of binary interactions in Y2H data sets are direct interactions, whereas the majority of indirect linkages are seen in AP/MS complex data sets. As a result, the graphs created using these two methods demonstrate many global properties, such as the connections between the importance of a gene and the quantity of associated proteins. A lot of progress has been made in the past 10 years toward creating detailed maps of PPINs (Vidal et al., 2011).

3.2.3.2 Comparative analysis of different PPI detection methods

Large-scale development of practical methods for the identification of PPIs among specified proteins, which might happen in various combinations, is being turned possible by the yeast two-hybrid (Y2H) system and other in vitro and in vivo approaches. The direct recognition of PPI between protein pairings is made possible by Y2H analysis. The approach may, however, result in a significant number of false-positive interactions. On the flip side of that point, the Y2H test may fail to detect many real interactions, producing misleading negative results. Due to the lack availability of PPIs, the data created using these methodologies could not be trustworthy. Several in silico approaches have been created to further support the interactions that have been found through experimental methods. Computational methods will reduce the collection of

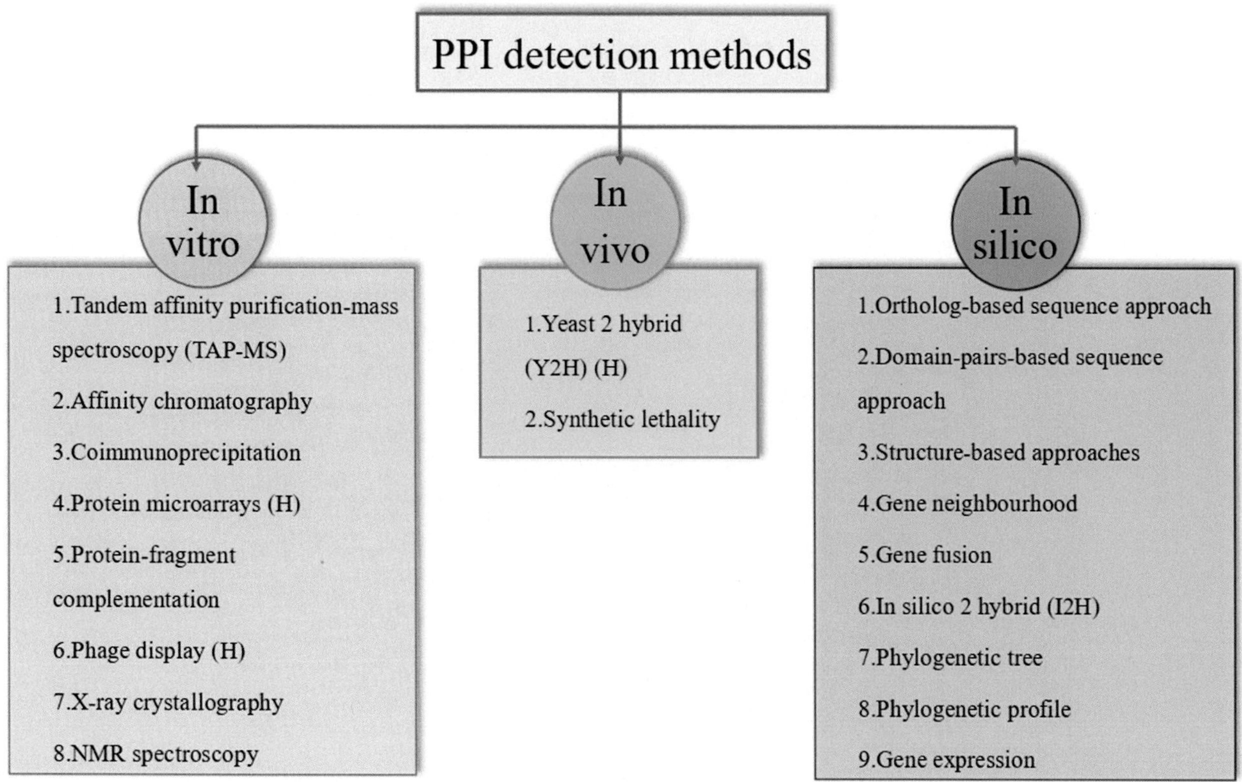

FIGURE 21.3 Different methods of protein–protein interaction detection analysis. Techniques for identifying PPIs in vivo and in vitro embrace: Protein arrays, protein fragment complementation, phage display, affinity chromatography, tandem affinity purification (TAP), coimmunoprecipitation, and yeast two-hybrid (Y2H) are a few examples of analytical techniques, while in silico also comprises different methods.

probable interactions to a subset of the most likely interactions, even though the methods now in use lack the ability to predict interactions with 100% accuracy (Rao et al., 2014).

3.2.4 RNA networks

The relationships between various RNA molecules, including coding RNAs, noncoding RNAs, and RNA–protein interactions, are represented by these networks, which are crucial for controlling gene activity and cellular functions (Mattick et al., 2023). Fig. 21.4 shows molecular network of different miRNA with breast cancer.

3.2.5 Metabolic networks

Graphs are used in metabolic networks to describe metabolism, which is the collection of all chemical reactions that take place inside a living thing to sustain life. Metabolites are metabolic actors that stand in for the intermediate and finished byproducts of metabolic events. Due to their complexity, metabolic networks are typically broken down into metabolic pathways, which are collections of chemical reactions that carry out a particular metabolic activity (Muzio et al., 2021). Comprehensive descriptions of all potential biochemical reactions to a specific cell or organism have been tried through metabolic network maps. The nodes of various metabolic network representations are biological metabolites, whereas the edges are either the enzymes that catalyze these events or the reactions that change one metabolite into another. For instance, glycolysis pathway (ID: BIOMD0000000211) has been shown in Fig. 21.5. Even while direct experimental studies will still need to fill in a significant number of gaps, metabolic network maps are perhaps the most complete of all biological networks (Vidal et al., 2011).

3.2.6 Cell signaling networks

These networks serve as a representation of the signaling routes and interactions between proteins, receptors, and other signaling molecules (Fig. 21.6) that move information throughout cells and control biological functions such as cell division, growth, and response to external stimuli (Armingol et al., 2021; Kholodenko, 2006).

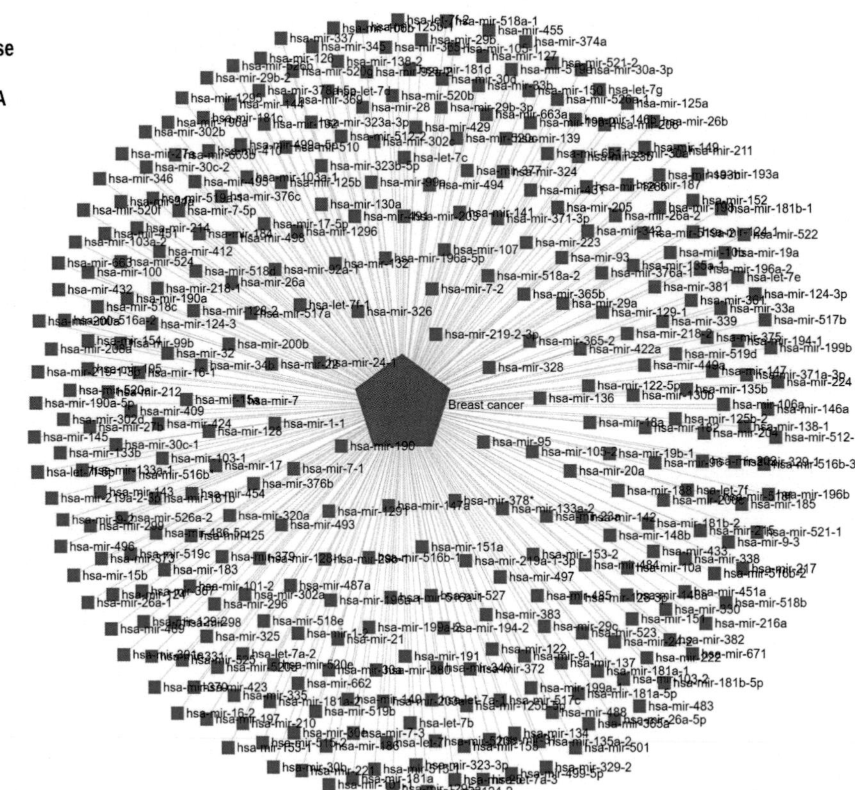

FIGURE 21.4 Molecular interaction network of RNA. RNA network involves miRNA (blue color nodes), and disease: Breast cancer (pink color nodes) (https://www.mirnet.ca/).

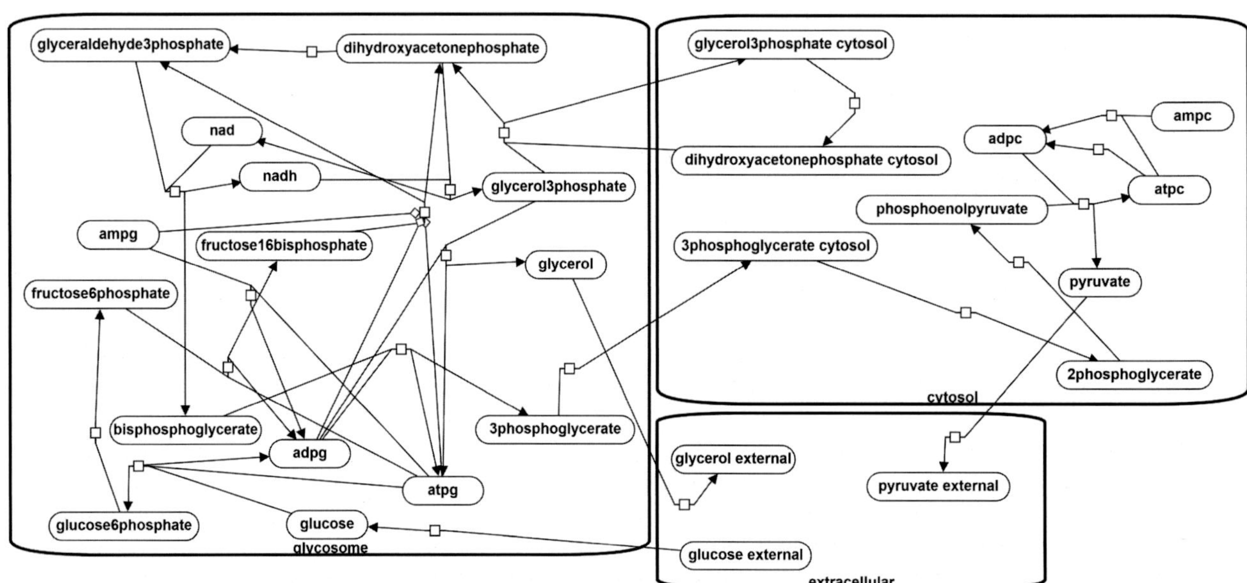

FIGURE 21.5 Metabolic network of glycolysis. This metabolic network shows that how glucose converts into pyruvate with the interaction of different molecules (https://www.ebi.ac.uk/biomodels/).

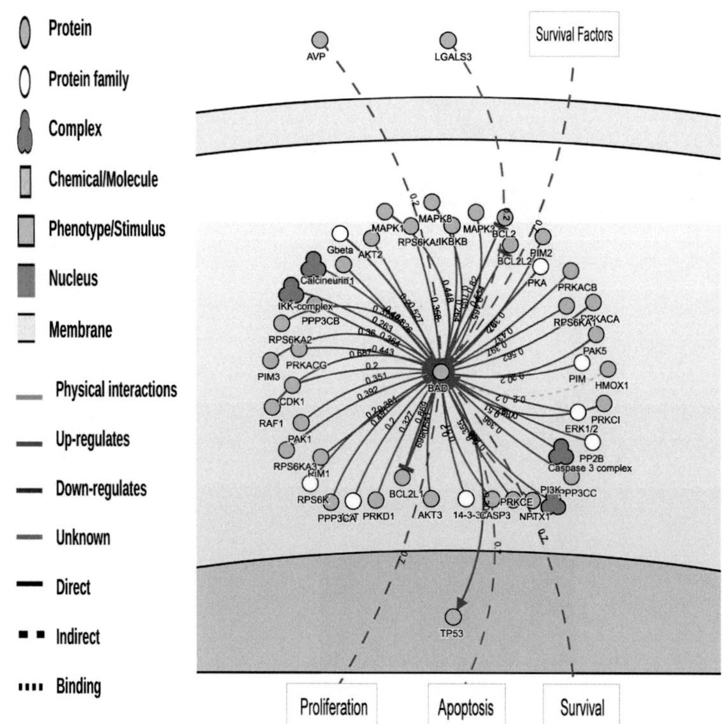

FIGURE 21.6 Signaling network of BCL2-associated agonist of cell death (BAD). This network shows an interaction between BCL2-associated agonist of cell death (BAD) and other molecules (https://signor.uniroma2.it/).

3.2.7 Neuronal networks

Neuronal networks, sometimes referred to as neural networks or brain networks, are intricate networks that depict how neurons in the nervous system are connected to one another. These networks participate in a variety of cognitive, sensory, motor, and behavioral processes as well as information processing and signaling in the brain (Sporns, 2011).

3.2.8 Human disease interaction network

The links between diseases based on shared genetic components, biochemical processes, and phenotypic similarity are represented by the human disease interaction network. It offers insights on disease comorbidity and aids in the discovery of disease linkages. One such study built a disease network to identify illness modules and probable disease correlations utilizing PPIs and disease–gene associations (Barabási et al., 2011).

For instance, a network of protein (Bcl2 family) associated with different diseases in human has been shown in Fig. 21.7.

3.2.9 Disease-drug network

On the basis of their biological targets, therapeutic indications, and drug efficacy, the disease–drug network depicts the connections between diseases and medications. It aids in the exploration of drug repurposing possibilities and the identification of possible therapeutic candidates for certain disorders. To build a disease–drug network, for instance, a study combined medication–target interactions and disease–gene connections. Fig. 21.8 highlighted possible drug candidates for various disorders and revealed chances for drug repositioning (Kakoti et al., 2022).

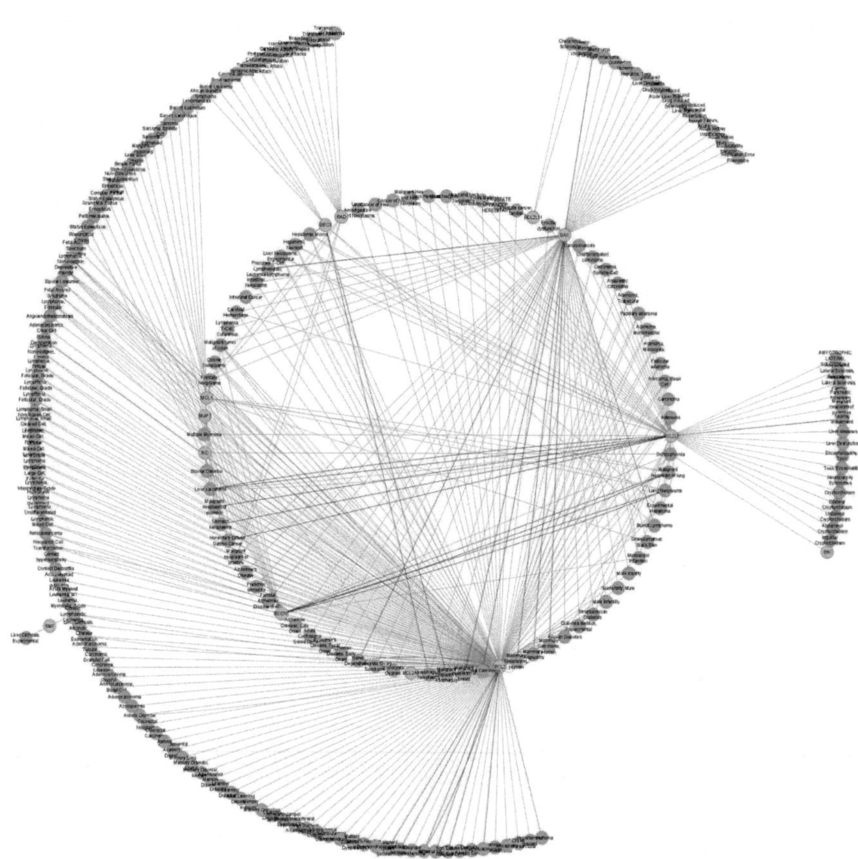

FIGURE 21.7 Human–disease interaction network. Network of Bcl2 family proteins associated with diseases in human.

4. Network modules and its importance in the network

4.1 Network modules

The modular nature and strong community structure of many molecular networks are crucial additional characteristics. Specific modules (subnetworks that are relatively densely connected) of genes or proteins tend to be enriched for particular biological functions. Therefore, identifying modules in networks might reveal coordinated biological roles or functions and lack adequate representation in existing canonical pathway annotations (Charitou et al., 2016).

Topological modules are frequently thought to have particular functions in cells, which gives rise to the notion of a functional module. A functional module is an assembly of nodes with comparable or related functions in the same network neighborhood. The study of disease modules, which are collections of network elements affected by a specific disease phenotype in humans, is becoming more popular (Vidal et al., 2011).

4.1.1 Topological network modules

According to their coordinates in a topological space, proteins known as topological network modules (TNMs) occupy around network places. By detecting proteins from several groups, such as proteins disrupted across networks, proteins inside a complex, proteins with shared biological roles, these modules offer novel perspectives into networks. TNMs that are different from the conventional concept of a module could represent a collection of proteins inside a single complex. TNMs can detect proteins in a variety of circumstances (Sardiu et al., 2017).

Numerous nodes are a part of topological modules, which are relatively compact neighborhoods, and nodes are more likely to be interconnected to other nodes within that area than outside of it. Network clustering techniques that are unaware of the function of individual nodes may discover a portion of the global network diagram, which relates to a probable topological module. These topological modules have certain cellular functions, providing way to the notion of a functional module, an assembly of nodes with related or identical functions in the common network area. Based on

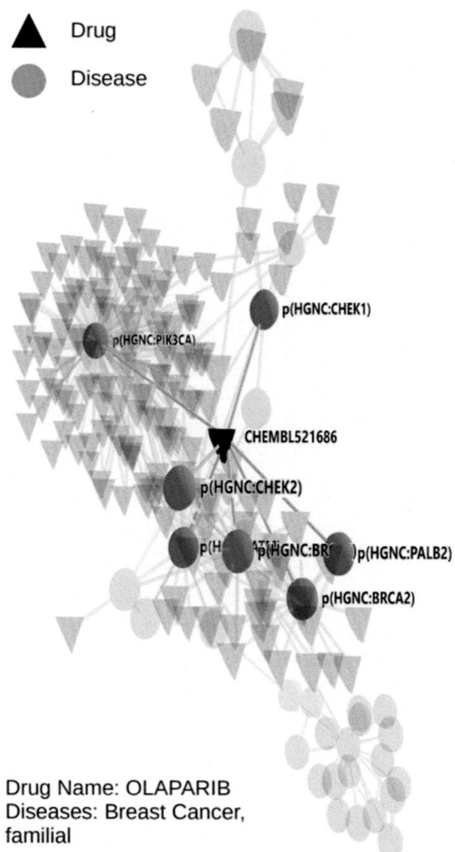

FIGURE 21.8 Drug–disease network. Network shows interaction between diseases (breast cancer, familial) and drug (OLAPARIB) (http://candis.insilab.org/).

information from the scientific literature, there is an implicit assumption that cellular elements comprising topological modules possess closely connected roles, which correspond to functional modules. The identification of topological and functional clustering's continues to be carried out using novel, possibly effective techniques. These modules are able to be applied as tools for developing hypotheses to find interactome regions that are probably relevant to specific cellular processes or diseases (Vidal et al., 2011).

4.1.2 Functional modules

A set of cellular components as well as their interactions that are associated with a particular biological process is referred to as a functional module.

The term "functional module" refers to a collection of genes or their by-products that are connected by a number of genetic or cellular interactions including coexpression, coregulation, metabolic or signaling pathway, membership in a protein complex, or a cellular aggregate (such as a ribosome, chaperone, protein transport facilitator).

4.1.3 Disease modules

It has also been demonstrated that proteins selectively connect to one another in "disease modules" when they are engaged in the same disease or in diseases exhibiting comparable characteristics. It is possible to recognize network modules that are abundant in genes or proteins that are linked to a particular disease of interest. These modules additionally include other proteins that have potential as disease-associated candidate genes but are not yet known to be connected to the disease. To find disease-associated modules for a variety of human disorders, this network module technique has now been extensively utilized (Charitou et al., 2016).

4.2 Biological network properties/network measures

4.2.1 Degree

The number of edges (or connections) is a node that has interactions with other nodes in the network is referred to as the node's degree as shown in Fig. 21.9. It is a fundamental indicator of a node's connectedness and is essential to comprehending the dynamics and structure of networks. Higher-degree nodes typically have more connections and are viewed as being more central nodes in the network. Degree is frequently used to research the significance or impact of nodes in diverse networks, including as social, biological, and technical networks (Barabási & Albert, 1999).

4.2.2 Degree distribution

Degree distribution is a statistical metric that expresses the frequency or distribution of degrees among all the network nodes. It offers information about a network's overall connectivity pattern and can highlight significant features such as the existence of hubs (nodes with significantly higher degrees than others), the presence of community structures, and the network's resilience to random failures or targeted attacks.

4.2.3 Scale-free networks

A class of complex network called scale-free networks, which has a power-law distribution for the distribution of node degrees. In other words, a power of the degree determines how likely a node is to possess that degree. A few nodes, referred to as "hubs," in a scale-free network that have disproportionately higher number of connections than the bulk of nodes, which have relatively few connections. These hubs are essential in determining the dynamics and structure of the network.

Barabasi and Albert showed that many real-world networks exhibit a power-law degree distribution and suggested a paradigm for building scale-free networks. Additionally, they demonstrated that while scale-free networks are resilient to random node deletions, they are weak to focused attacks on high-degree nodes.

4.2.4 Network shortest path and mean length

The fewest edges (or connections) that must be crossed to get from one node to another are referred to as the shortest path in a network. The average of the shortest pathways connecting all network node pairs is known as the mean length or average path length of a network. Understanding the effectiveness of information flow or communication in a network requires knowledge of these metrics. A network with efficient connectivity typically has shortest pathways and shorter mean lengths. The length of the shortest path between two nodes is measured by the number of links. The average path length is the sum of the shortest paths between all pairs of nodes (West, 2014).

The path length, which displays how many links we must travel through to reach one node from another in a network, is used to determine distance. Given the wide variety of possible routes that can link two nodes, the shortest path, or the route with the fewest links linking the selected nodes, plays a crucial role. Fig. 21.10 shows the graph of directed network (Barabási & Oltvai, 2004).

Clustering coefficient: In a network, a node's clustering coefficient calculates how interconnected its neighbors are. It assesses the likelihood that a node's neighbors are connected to one another and measure the local clustering or clustering tendency in a network. The presence of clusters or communities in networks, as well as the resilience and robustness of networks to failures or attacks, is frequently studied using the clustering coefficient (Watts & Strogatz, 1998).

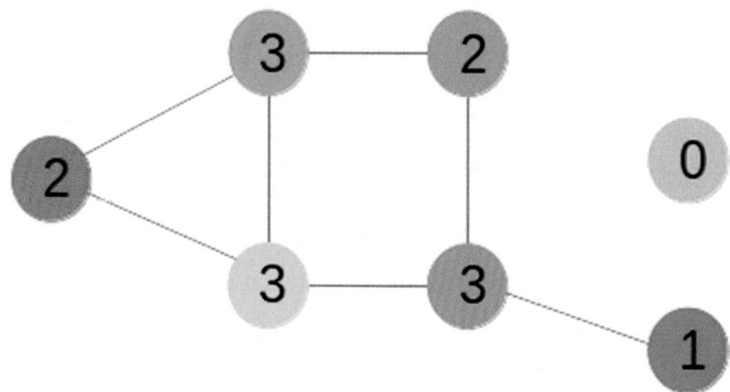

FIGURE 21.9 Degree of nodes. The number of nodes a node interacts with in its neighborhood determines its degree.

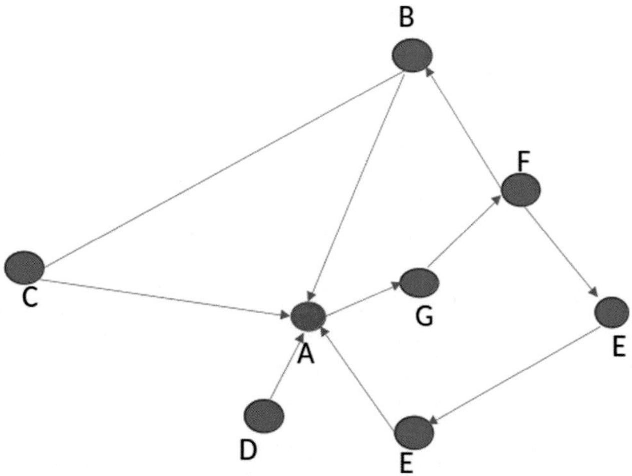

FIGURE 21.10 A graph of directed network. In directed networks, the distance AB between nodes A and B and the distance BA between nodes B and A commonly vary. As an example, $L_{BA} = 1$ and $L_{AB} = 3$. Often, there is no direct path between two nodes. There is a path from C to A, but not one from A to C. The average of the shortest paths between all pairs of nodes is represented by the mean path length, <L>, which provides a gauge of a network's overall navigability.

The clustering coefficient gives an indication of a node's connectedness to each of its neighbors. The global clustering coefficient measures the total number of triangles in the network. By averaging the clustering coefficients over all network nodes, the average clustering coefficient is determined.

5. Network models

Network models are mathematical or computational depictions of complex systems as networks, with nodes denoting entities and edges denoting interactions or relationships between things. To explore the structure, dynamics, and characteristics of complex systems, network models are frequently employed in a wide range of disciplines, including physics, computer science, social sciences, biology, and many more (Albert & Barabási, 2002).

5.1 Basic concepts of networks

5.1.1 Nodes and hub

A node in a network is an individual component or entity, frequently depicted as a point or a vertex. Nodes can stand in for a number of different entities, including people in social networks, genes in biological networks, and computers in technology networks. Some networks may have nodes with a disproportionately high number of connections to other nodes; these nodes are frequently referred to as "hub nodes" as shown in Fig. 21.11. Due to their ability to affect

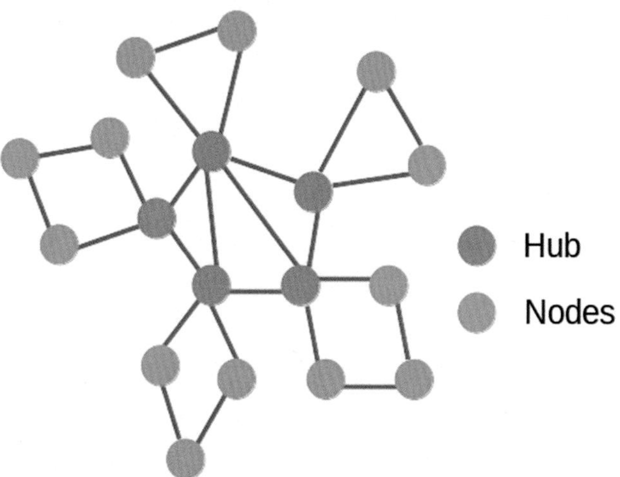

FIGURE 21.11 Nodes and hub nodes. Graphic model explains how nodes are interact with each other.

information flow, the diffusion of influence, and overall network resilience, hub nodes play a significant role in network structure and dynamics (Barabási & Albert, 1999).

5.1.2 Robustness of network and its role in evolution

The ability of a network to preserve its structural integrity and functionality in the face of disturbances or failures is referred to as robustness in the area of network research. Robust networks can endure targeted attacks and random failures without losing functionality, making them less vulnerable to interruptions. To comprehend the resilience and stability of complex systems, such as transportation networks, social networks, and biological networks, it is crucial to grasp a network's robustness.

"Robustness and evolvability in the functional anatomy of a transcriptional regulatory network" by Maslov and Sneppen is a study on the robustness of networks and its role in evolution. The authors used computational modeling to investigate the relationship between the robustness and evolvability of a transcriptional regulatory network. According to MacNeil and Walhout (2011) the networks that were more resistant to random mutations were also more adaptable to changing environmental conditions.

The authors focused on the factors that contribute to the robustness of these networks as they examined the importance of robustness in the evolution of GRNs (Kim et al., 2014). They also explored how robustness can limit the evolution of these networks and how it can be overcome by acquiring additional regulatory relationships.

5.1.3 Modularity

The degree to which a network can be separated into discrete, nonoverlapping modules or communities—where nodes within the same module are more densely connected to one another than nodes in different modules—are known as the modularity of the system. In community detection, which is the procedure of locating clusters of nodes with similar connectivity patterns in a network, modularity is used to measure the presence of structure or organization in a network. To locate functional modules or communities within social, biological, and technical networks, modularity is frequently utilized in network analysis (Newman, 2006).

5.1.4 Subgraph

A subgraph is a smaller network made up of a subset of the nodes and connections that make up a larger network, which are shown in Fig. 21.12. Subgraphs offer a way to examine more manageable, smaller sections of larger networks and can be used to research the local structure and connectivity patterns inside a network. According to Watts and Strogatz (1998), subgraphs can represent certain areas or clusters within a network, and their analysis can shed light on the local network's characteristics and behaviors (Watts & Strogatz, 1998).

5.1.5 Motifs

Networks often contain motifs, which are small subgraph structures or recurrent patterns. They serve as the structural building elements of networks and can provide crucial details regarding the dynamics and organizational functionality of networks. Simple or complicated themes are also acceptable, such as feed-forward loops or cliques. Simple themes include triangles and loops. Insights into the mechanisms of information flow, diffusion, and robustness in complex systems can be gained by studying network motifs (Milo et al., 2002).

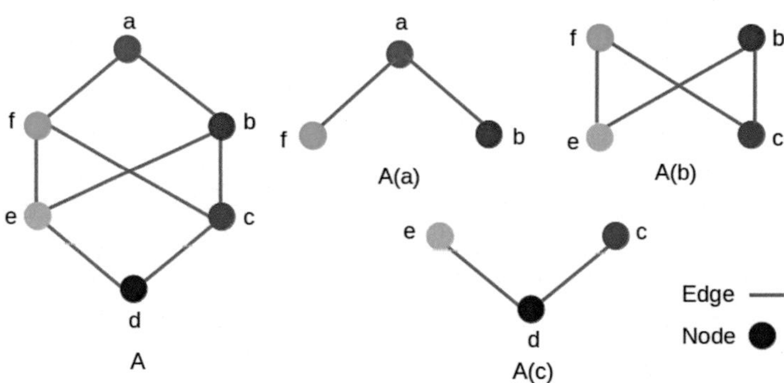

FIGURE 21.12 **(A) Graph (Aa, Ab, Ac) subgraph.** This graph shows a large network that comprises smaller networks.

5.1.6 Motif cluster

A group of motifs that are discovered to appear together more frequently than would be predicted by chance in a network is known as a motif cluster. When groupings of motifs are discovered to be connected to particular activities or behaviors, motif clusters are frequently utilized to identify structural or functional modules in networks. Motif clusters are frequently utilized in the research of biological, social, and technological networks because they can reveal information about the structure and functionality of networks (Shen-Orr et al., 2002).

5.1.7 Modules

In the context of network science, modules are groups of nodes that have sparser connections with nodes outside the module but are tightly coupled to one another within a broader network. Modules are groups of nodes that have comparable characteristics, capabilities, or behaviors. They are often referred to as communities, clusters, or subnetworks. Insights into the structure and dynamics of complex systems can be gained by identifying modules using a variety of community detection techniques (Ravasz et al., 2002).

5.1.8 Betweenness centrality

Based on a node's placement along the shortest paths connecting pairs of nodes, betweenness centrality quantifies a node's significance or centrality in a network. According to Freeman (1977), nodes with a high betweenness centrality serve as bridges.

Betweenness centrality is the total number of shortest paths between pairs of nodes that pass through a particular node (Freeman, 1977).

5.1.9 Network topology

The entire configuration or arrangement of nodes and edges in a network is referred to as network topology. It covers a number of characteristics that describe a network's structural characteristics, including degree distribution, clustering coefficient, and path lengths. Network topology is used to comprehend the overall structure and characteristics of networks and can provide light on either their dynamic or functional behavior (Newman, 2003).

Two nodes are deemed to be connected if they have a topological overlap (TO) of at least one, as determined by their shared neighbors. The section on gene (regulatory) networks contains more details.

6. Visualization of biomolecular network tools used for network analysis

In the subject of network medicine, there are numerous methods available for the visualization and analysis of biomolecular networks.

Cytoscape: A popular open-source tool for visualizing and analyzing complicated networks is called Cytoscape. It offers an user-friendly interface and covers a variety of network analysis activities, such as network integration, network topology analysis, pathway enrichment analysis, and network visualization. For sophisticated network research and visualization, Cytoscape also provides a wide range of plugins and extensions (Shannon et al., 2003).

Gephi: An open-source network exploration and visualization tool is Gephi. For visualizing and analyzing massive networks, it offers a versatile and dynamic platform. To efficiently explore and analyze biomolecular networks, Gephi provides a variety of layout techniques, filtering choices, and visualization features (Bastian et al., 2009).

Network analysis, visualization, and graphing toronto (navigators): A software program called NAViGaTOR (network analysis, visualization, and graphing toronto) was created specifically for the analysis and visualization of biological networks. It offers a wide range of network visualization choices, network analysis formulas, and integrated data analysis capabilities. Incorporating interactive tools for examining network attributes and locating crucial network components, NAViGaTOR also facilitates the inclusion of experimental data (Brown et al., 2009).

Visualization and analysis of networks (VisANT): VisANT (visualization and analysis of networks) is a web-based platform for visualizing and analyzing biological networks. It offers a range of network visualization options, including circular, force-directed, and hierarchical layouts. VisANT also provides several network analysis tools, such as network clustering, pathway enrichment analysis, and integration of gene expression data (Hu et al., 2009).

7. Network medicine

To comprehend the molecular pathways behind the pathophysiology of diseases, network medicine is a branch of science that primarily concentrates on the interactions between biological elements such proteins, microRNAs, or metabolites. For the purpose of elucidating the mechanisms underlying clinical diseases, network medicine has these days extended to incorporate molecular data with phenotypic aspects (Maron et al., 2020).

Network medicine is expanding into the world of networks of networks as well as utilizing ever-larger phenotypic and molecular data sets. Realizing the area's greatest possibility in biomedicine will depend on both dynamic and static characteristics of these more complex structures. Although diversity in network medicine analytics has been profitable, it has also created novel issues that must be overcome with the goal to fully utilize the vast amount of knowledge and data that are currently available on a global scale. The chance to fully progress the area has been constrained by the fact that multinational initiatives in network medicine have been rather dispersed (Barabási et al., 2011).

7.1 Network medicine in context of human disease network

An interdisciplinary subject known as network medicine combines network theory, systems biology, and medicine to better comprehend complex biological systems such as the human body and disease. In the context of the human illness network, network medicine seeks to pinpoint disease modules or communities, that is, collections of diseases that have similar genetic, molecular, and physiological characteristics. Network medicine can assist in identifying fresh disease linkages, predicting disease risk, and creating specialized therapeutic approaches by outlining the connections between various modules (Menche et al., 2015).

Network pharmacology, which uses network-based methods for drug development, is another area where network medicine is crucial. Network pharmacology can aid in the identification of new drug targets, the prediction of therapeutic efficacy and adverse effects, and the exploration of drug repurposing potential by integrating data on drug–target interactions, disease–gene connections, and biological pathways. To forecast drug–disease connections and prospective therapeutic targets for diverse diseases, one study, for instance, constructed a network pharmacology framework that combined a disease–gene network, a drug–target network, and a PPIN (Xuan et al., 2019).

7.2 Application of network medicine in network pharmacology

Network pharmacology uses network-based methods to investigate the connections between treatments, targets, and illnesses. Network pharmacology strives to predict medication efficacy, discover potential therapeutic targets, and promote drug repurposing by combining various data sources such as drug–target interactions, protein–protein interaction networks, and disease networks (Hopkins, 2008).

8. Biological network study by deep machine learning

Deep learning's capacity to discover intricate patterns and connections in huge datasets has made it a potent tool for studying biological networks.

Graph neural networks (GNNs) are one way to integrate deep learning with biological networks. A particular kind of deep learning model called GNNs can function with graph-structured data, including biological networks. To anticipate node labels or network features, GNNs can learn node and edge representations that encapsulate the structural and functional characteristics of the network.

For instance, using information about the network topology and protein sequences, GNNs have been utilized to forecast PPIs (Zitnik et al., 2018). To determine regulatory relationships and forecast gene expression levels, GNNs have also been applied to gene expression data (Ma et al., 2018).

Generative models, such as variational autoencoders (VAEs) and generative adversarial networks (GANs), can also be used to apply deep learning to biological networks. Researchers may run simulations and test ideas using these models because they can learn to create synthetic networks that mimic the statistical characteristics of actual biological networks.

For instance, synthetic GRNs that replicate the degree distribution and clustering coefficient of actual GRNs have been created using VAEs. Synthetic PPINs that mimic the degree distribution, clustering coefficient, and modularity of actual networks have been created using GANs (Yang et al., 2022).

To analyze biological networks, deep machine learning algorithms have thus far showed promise, allowing researchers to obtain insights into the structure, operation, and behavior of complex biological systems.

9. Scope and application of biomolecular network in biological sciences

Molecular networks, also referred to as biological networks, are potent tools used in a variety of biological sciences to represent and analyze complex interactions among molecules. These interactions include those between proteins, genes, nucleic acids, and metabolites. These networks give researchers a systems-level view of biological processes, allowing them to comprehend how molecules are organized, dynamic, and related to one another within a cell, organism, or ecosystem. Here are a few examples of the range and uses of molecular networks in several biological sciences fields:

Systems biology: Systems biology frequently uses molecular networks to investigate the complexity and dynamics of biological systems at the molecular level. It offers a framework for comprehending how molecules interact and how emergent characteristics of biological systems result from these interactions. To acquire insights into system-level behavior and function, molecular networks are used to simulate and analyze complicated biological processes such cell signaling, gene control, and metabolic pathways (Kitano, 2002).

Functional genomics: Large-scale genomic data, including information on gene expression, DNA–protein interactions, and epigenetic alterations, are analyzed using molecular networks to find functional modules and pathways linked to certain biological processes, illnesses, or phenotypes. These networks can offer insights into the prediction of gene function, drug development, and personalized medicine. They can also help in the discovery of gene functions, regulatory mechanisms, and disease-associated pathways (Barabási & Oltvai, 2004).

Drug discovery and development: To find prospective drug targets, forecast drug responses, and comprehend the processes of drug action, molecular networks are employed in drug discovery and development. To find new drug candidates and improve drug design, network pharmacology approaches, which combine molecular networks with drug information, use their use. The investigation of drug–target interactions, drug side effects, and drug resistance mechanisms is also made easier by molecular networks (Hopkins, 2008).

Personalized medicine: In personalized medicine, molecular networks are used to comprehend the molecular causes of diseases, forecast their outcomes, and create individualized treatment plans. In network-based approaches to patient stratification, biomarker discovery, and precision medicine, which take into account the unique molecular profiles of individuals for more individualized and efficient therapies, they are used (Ideker & Sharan, 2008).

Synthetic biology: In synthetic biology, molecular networks are used to create and construct unique biological systems with desired features or behaviors. They serve as a model for creating synthetic metabolic pathways, gene circuits, and other biological modules as well as for comprehending the dynamics and control of manmade biological systems (Khalil & Collins, 2010).

Proteomics and metabolomics: The studies of proteomics and metabolomics, metabolic pathways, and interactions between enzymes and substrates are examined using molecular networks. These networks can be utilized to find biomarkers, therapeutic targets, and metabolic pathways linked to diseases or environmental circumstances as well as to study PPIs, enzyme activities, and metabolic fluxes (Walther, 2023, pp. 179–197).

Ecology and environmental sciences: For the study of microbial communities, food webs, and ecosystem dynamics, molecular networks are employed in ecology and environmental studies. These networks can help with environmental monitoring, biodiversity preservation, and ecosystem management by helping us to better understand how different species interact, how nutrients flow through ecosystems, and what functions different types of microorganisms provide (Shade, 2023).

Neurobiology: To understand neural networks, synapse development, and brain connectivity, molecular networks are employed in neurobiology. These networks can aid in deciphering the molecular underpinnings of neurological illnesses, understanding the structure and operation of neural circuits, and locating important signaling molecules in the brain (Mandal et al., 2023).

10. Conclusion

Integrated biomolecules show synergistic effect on the cellular functions as a holistic way. The entity of biomolecules function like which perform same functions in one cluster act as functional modules (a set of genes or the products of those genes that are connected by a number of cellular or genetic interactions), topological modules (can assist to find significant network substructures), and disease modules (set of components of a network that operate a cellular function together whose interruption results in a specific disease phenotype). These modules interact with one another and frequently overlap in a network system. The comprehensive identification of all molecules and their interactions with others within a living cell is a major goal of post-genomic biomedical research. The PPI study helps to understand the uncharacterized protein, role of individuals molecules on cell signaling, effect of genes on the phenotype, and associationship of protein in multicomplex process such as cell death. The prioritizing nodes such as genes are a common use of biological networks, particularly when trying to find disease-causing genes or potential treatment targets in a network that is specific to a particular circumstance. Network interaction assists in the downstream analysis such as hypothesis testing, clustering, or prediction on the preprocessed data such as normal cell and cancer cell interaction and wild-type and mutant gene and protein interaction.

References

Albert, R., & Barabási, A.-L. (2002). Statistical mechanics of complex networks. *Reviews of Modern Physics, 74*(1), 47–97. https://doi.org/10.1103/RevModPhys.74.47

Armingol, E., Officer, A., Harismendy, O., & Lewis, N. E. (2021). Deciphering cell–cell interactions and communication from gene expression. *Nature Reviews Genetics, 22*(2), 71–88. https://doi.org/10.1038/s41576-020-00292-x

Barabási, A.-L., & Albert, R. (1999). Emergence of scaling in random networks. *Science, 286*(5439), 509–512. https://doi.org/10.1126/science.286.5439.509

Barabási, A.-L., Gulbahce, N., & Loscalzo, J. (2011). Network medicine: A network-based approach to human disease. *Nature Reviews Genetics, 12*(1), 56–68. https://doi.org/10.1038/nrg2918

Barabási, A.-L., & Oltvai, Z. N. (2004). Network biology: Understanding the cell's functional organization. *Nature Reviews Genetics, 5*(2), 101–113. https://doi.org/10.1038/nrg1272

Barabási, A. L. (2013). Network science. *Philosophical Transactions of the Royal Society of London. Series A, Mathematical and Physical Sciences.*

Bastian, M., Heymann, S., & Jacomy, M. (2009). Gephi: An open source software for exploring and manipulating networks. *Proceedings of the International AAAI Conference on Web and Social Media, 3*(1), 361–362. https://doi.org/10.1609/icwsm.v3i1.13937

Brown, K. R., Otasek, D., Ali, M., McGuffin, M. J., Xie, W., Devani, B., Toch, I. L. van, & Jurisica, I. (2009). NAViGaTOR: Network analysis, visualization and graphing Toronto. *Bioinformatics, 25*(24), 3327–3329. https://doi.org/10.1093/bioinformatics/btp595

Charitou, T., Bryan, K., & Lynn, D. J. (2016). Using biological networks to integrate, visualize and analyze genomics data. *Genetics Selection Evolution, 48*(1), 27. https://doi.org/10.1186/s12711-016-0205-1

Chen, L., Wang, R., & Zhang, X. (2009). *Biomolecular networks.* Wiley. https://doi.org/10.1002/9780470488065

Davidson, E. H., & Levine, M. S. (2008). Properties of developmental gene regulatory networks. *Proceedings of the National Academy of Sciences, 105*(51), 20063–20066. https://doi.org/10.1073/pnas.0806007105

Freeman, L. C. (1977). A set of measures of centrality based on betweenness. *Sociometry, 40*(1), 35. https://doi.org/10.2307/3033543

Gade, P., & Kalvakolanu, D. V. (2012). Chromatin immunoprecipitation assay as a tool for analyzing transcription factor activity. *Methods in Molecular Biology, 809*, 85–104. https://doi.org/10.1007/978-1-61779-376-9_6.

Giampà, M., & Sgobba, E. (2020). Insight to functional conformation and noncovalent interactions of protein-protein assembly using MALDI mass spectrometry. *Molecules, 25*(21), 4979. https://doi.org/10.3390/molecules25214979

Gysi, D. M., & Nowick, K. (2020). Construction, comparison and evolution of networks in life sciences and other disciplines. *Journal of The Royal Society Interface, 17*(166), 20190610. https://doi.org/10.1098/rsif.2019.0610

Hopkins, A. L. (2008). Network pharmacology: The next paradigm in drug discovery. *Nature Chemical Biology, 4*(11), 682–690. https://doi.org/10.1038/nchembio.118

Hu, Z., Hung, J.-H., Wang, Y., Chang, Y.-C., Huang, C.-L., Huyck, M., & DeLisi, C. (2009). VisANT 3.5: Multi-scale network visualization, analysis and inference based on the gene ontology. *Nucleic Acids Research, 37*(Suppl. l_2), W115–W121. https://doi.org/10.1093/nar/gkp406

Ideker, T., & Sharan, R. (2008). Protein networks in disease. *Genome Research, 18*(4), 644–652. https://doi.org/10.1101/gr.071852.107

Kakoti, B. B., Bezbaruah, R., & Ahmed, N. (2022). Therapeutic drug repositioning with special emphasis on neurodegenerative diseases: Threats and issues. *Frontiers in Pharmacology, 13*, 1007315. https://doi.org/10.3389/fphar.2022.1007315

Khalil, A. S., & Collins, J. J. (2010). Synthetic biology: Applications come of age. *Nature Reviews Genetics, 11*(5), 367–379. https://doi.org/10.1038/nrg2775

Kholodenko, B. N. (2006). Cell-signalling dynamics in time and space. *Nature Reviews Molecular Cell Biology, 7*(3), 165–176. https://doi.org/10.1038/nrm1838

Kim, J., Vandamme, D., Kim, J.-R., Munoz, A. G., Kolch, W., & Cho, K.-H. (2014). Robustness and evolvability of the human signaling network. *PLoS Computational Biology, 10*(7), e1003763. https://doi.org/10.1371/journal.pcbi.1003763

Kitano, H. (2002). Systems biology: A brief overview. *Science, 295*(5560), 1662–1664. https://doi.org/10.1126/science.1069492

Koutrouli, M., Karatzas, E., Paez-Espino, D., & Pavlopoulos, G. A. (2020). A guide to conquer the biological network era using graph theory. *Frontiers in Bioengineering and Biotechnology, 8*. https://doi.org/10.3389/fbioe.2020.00034

Lü, J., & Wang, P. (2020). Reconstruction of bio-molecular networks. In *Modeling and analysis of bio-molecular networks* (pp. 53–105). Springer Singapore. https://doi.org/10.1007/978-981-15-9144-0_2

Liu, Q.-H., Ajelli, M., Aleta, A., Merler, S., Moreno, Y., & Vespignani, A. (2018). Measurability of the epidemic reproduction number in data-driven contact networks. *Proceedings of the National Academy of Sciences, 115*(50), 12680–12685. https://doi.org/10.1073/pnas.1811115115

Ma, J., Yu, M. K., Fong, S., Ono, K., Sage, E., Demchak, B., Sharan, R., & Ideker, T. (2018). Using deep learning to model the hierarchical structure and function of a cell. *Nature Methods, 15*(4), 290–298. https://doi.org/10.1038/nmeth.4627

MacNeil, L. T., & Walhout, A. J. M. (2011). Gene regulatory networks and the role of robustness and stochasticity in the control of gene expression. *Genome Research, 21*(5), 645–657. https://doi.org/10.1101/gr.097378.109

Mandal, A. S., Brem, S., & Suckling, J. (2023). Brain network mapping and glioma pathophysiology. *Brain Communications, 5*(2). https://doi.org/10.1093/braincomms/fcad040

Maron, B. A., Altucci, L., Balligand, J.-L., Baumbach, J., Ferdinandy, P., Filetti, S., Parini, P., Petrillo, E., Silverman, E. K., Barabási, A.-L., Loscalzo, J., Maron, B. A., Altucci, L., Balligand, J.-L., Baumbach, J., Ferdinandy, P., Filetti, S., Parini, P., Petrillo, E., … Loscalzo, J. (2020). A global network for network medicine. *Npj Systems Biology and Applications, 6*(1), 29. https://doi.org/10.1038/s41540-020-00143-9

Mattick, J. S., Amaral, P. P., Carninci, P., Carpenter, S., Chang, H. Y., Chen, L.-L., Chen, R., Dean, C., Dinger, M. E., Fitzgerald, K. A., Gingeras, T. R., Guttman, M., Hirose, T., Huarte, M., Johnson, R., Kanduri, C., Kapranov, P., Lawrence, J. B., Lee, J. T., … Wu, M. (2023). Long non-coding RNAs: Definitions, functions, challenges and recommendations. *Nature Reviews Molecular Cell Biology, 24*(6), 430–447. https://doi.org/10.1038/s41580-022-00566-8

Menche, J., Sharma, A., Kitsak, M., Ghiassian, S. D., Vidal, M., Loscalzo, J., & Barabasi, A.-L. (2015). Uncovering disease-disease relationships through the incomplete interactome. *Science, 347*(6224). https://doi.org/10.1126/science.1257601, 1257601–1257601.

Menon, G., & Krishnan, J. (2021). Spatial localisation meets biomolecular networks. *Nature Communications, 12*(1), 5357. https://doi.org/10.1038/s41467-021-24760-y

Milo, R., Shen-Orr, S., Itzkovitz, S., Kashtan, N., Chklovskii, D., & Alon, U. (2002). Network motifs: Simple building blocks of complex networks. *Science, 298*(5594), 824–827. https://doi.org/10.1126/science.298.5594.824

Newman, M. E. J. (2006). Modularity and community structure in networks. *Proceedings of the National Academy of Sciences, 103*(23), 8577–8582. https://doi.org/10.1073/pnas.0601602103

Mitsis, T., Efthimiadou, A., Bacopoulou, F., Vlachakis, D., Chrousos, G. P., & Eliopoulos, E. (2020). Transcription factors and evolution: An integral part of gene expression (Review). *World Academy of Sciences Journal, 2*(1), 3–8. https://doi.org/10.3892/wasj.2020.32.

Muzio, G., O'Bray, L., & Borgwardt, K. (2021). Biological network analysis with deep learning. *Briefings in Bioinformatics, 22*(2), 1515–1530. https://doi.org/10.1093/bib/bbaa257.

Newman, M. E. J. (2003). The structure and function of complex networks. *SIAM Review, 45*(2), 167–256. https://doi.org/10.1137/S003614450342480

Newman, M. (2010). *Networks*. Oxford University Press. https://doi.org/10.1093/acprof:oso/9780199206650.001.0001

Ovens, K., Eames, B. F., & McQuillan, I. (2021). Comparative analyses of gene Co-expression networks: Implementations and applications in the study of evolution. *Frontiers in Genetics, 12*. https://doi.org/10.3389/fgene.2021.695399

Rao, V.S., Srinivas, K., Sujini, G.N., & Kumar, G.N. (2014). Protein-protein interaction detection: methods and analysis. *International Journal of Proteomics*, 2014, 147648. https://doi.org/10.1155/2014/147648.

Ravasz, E., Somera, A. L., Mongru, D. A., Oltvai, Z. N., & Barabási, A.-L. (2002). Hierarchical organization of modularity in metabolic networks. *Science, 297*(5586), 1551–1555. https://doi.org/10.1126/science.1073374

Rolland, T., Taşan, M., Charloteaux, B., Pevzner, S. J., Zhong, Q., Sahni, N., Yi, S., Lemmens, I., Fontanillo, C., Mosca, R., Kamburov, A., Ghiassian, S. D., Yang, X., Ghamsari, L., Balcha, D., Begg, B. E., Braun, P., Brehme, M., Broly, M. P., … Vidal, M. (2014). A proteome-scale map of the human interactome network. *Cell, 159*(5), 1212–1226. https://doi.org/10.1016/j.cell.2014.10.050

Sardiu, M. E., Gilmore, J. M., Groppe, B., Florens, L., & Washburn, M. P. (2017). Identification of topological network modules in perturbed protein interaction networks. *Scientific Reports, 7*(1), 43845. https://doi.org/10.1038/srep43845

Shade, A. (2023). Microbiome rescue: Directing resilience of environmental microbial communities. *Current Opinion in Microbiology, 72*, 102263. https://doi.org/10.1016/j.mib.2022.102263

Shannon, P., Markiel, A., Ozier, O., Baliga, N. S., Wang, J. T., Ramage, D., Amin, N., Schwikowski, B., & Ideker, T. (2003). Cytoscape: A software environment for integrated models of biomolecular interaction networks. *Genome Research, 13*(11), 2498–2504. https://doi.org/10.1101/gr.1239303

Shen-Orr, S. S., Milo, R., Mangan, S., & Alon, U. (2002). Network motifs in the transcriptional regulation network of *Escherichia coli*. *Nature Genetics, 31*(1), 64–68. https://doi.org/10.1038/ng881

Sporns, O. (2011). The human connectome: A complex network. *Annals of the New York Academy of Sciences, 1224*(1), 109–125. https://doi.org/10.1111/j.1749-6632.2010.05888.x

Vespignani, A. (2012). Modelling dynamical processes in complex socio-technical systems. *Nature Physics, 8*(1), 32–39. https://doi.org/10.1038/nphys2160

Viboud, C., & Vespignani, A. (2019). The future of influenza forecasts. *Proceedings of the National Academy of Sciences, 116*(8), 2802–2804. https://doi.org/10.1073/pnas.1822167116

Vidal, M., Cusick, M. E., & Barabási, A.-L. (2011). Interactome networks and human disease. *Cell, 144*(6), 986–998. https://doi.org/10.1016/j.cell.2011.02.016

Walther, D. (2023). *Specifics of metabolite-protein interactions and their computational analysis and prediction* (pp. 179–197). https://doi.org/10.1007/978-1-0716-2624-5_12

Watts, D. J., & Strogatz, S. H. (1998). Collective dynamics of 'small-world' networks. *Nature, 393*(6684), 440–442. https://doi.org/10.1038/30918

West, B. J. (2014). A mathematics for medicine: The Network Effect. *Frontiers in Physiology, 5*. https://doi.org/10.3389/fphys.2014.00456

Xuan, P., Song, Y., Zhang, T., & Jia, L. (2019). Prediction of potential drug–disease associations through deep integration of diversity and projections of various drug features. *International Journal of Molecular Sciences, 20*(17), 4102. https://doi.org/10.3390/ijms20174102

Yang, H., Gu, F., Zhang, L., & Hua, X.-S. (2022). Using generative adversarial networks for genome variant calling from low depth ONT sequencing data. *Scientific Reports, 12*(1), 8725. https://doi.org/10.1038/s41598-022-12346-7

Zhang, P., & Itan, Y. (2019). Biological network approaches and applications in rare disease studies. *Genes, 10*(10), 797. https://doi.org/10.3390/genes10100797

Zhang, S., Zhang, X.-S., & Chen, L. (2008). Biomolecular network querying: A promising approach in systems biology. *BMC Systems Biology, 2*(1), 5. https://doi.org/10.1186/1752-0509-2-5

Zitnik, M., Agrawal, M., & Leskovec, J. (2018). Modeling polypharmacy side effects with graph convolutional networks. *Bioinformatics, 34*(13), i457–i466. https://doi.org/10.1093/bioinformatics/bty294

Chapter 22

Machine learning fundamentals to explore complex *omics* data

Tapobrata Lahiri[1], Rajkrishna Mondal[2] and Asmita Tripathi[1]

[1]*Department of Applied Sciences, Indian Institute of Information Technology, Allahabad, Prayagraj, Uttar Pradesh, India;* [2]*Department of Biotechnology, Nagaland University, Dimapur, Nagaland, India*

1. Backdrop

The word "learning" is the most important component of intelligence. Simply saying, a natural human like learning leads to gain of natural intelligence, and an artificial machine−based learning leads to gain of artificial intelligence to drive machines behave and perform like an intelligent machine. In this regard, the simple question that comes in mind is: what is the utility of these buzzwords, machine learning (ML) and artificial intelligence (AI) in the fields of science and technology? The answer can be backtracked from the benefits of natural or human intelligence in everyday life that gives us the ability to compute, predict, recognize, identify, and compare real-life systems within the macro-, micro-, and nanouniverse surrounding us in our quest for gaining knowledge about them and in solving many problems associated to them. During the early age of automation, although the reference of these terms would yet to be obvious, the tasks being accomplished then were similar to the current applications of ML and AI. For example, in the routine production line of a factory, the decision on acceptance or rejection of items was based on application of rules on their simple measurable characteristics. However, the need of building applications based on artificial intelligence was felt more with the enhanced power of sensors or data acquisition systems to provide high-throughput data with more details of the system in question, for example, automatic surveillance using high-definition images, automated diagnostics, etc. In this regard, the whole spectrum of biomedical science is considered to be a fertile field to sow ML-based applications in yielding a wealth of knowledge to enrich the fields of diagnostics, especially those that are based on the analysis of high-throughput gene expression data, and for many other purposes. In this piece of discussion, the target will be primarily to help the unversed understand the fundamentals of ML on the foundation of various computational predictive methods. Finally, its utility will be demonstrated through two examples; first, through showcasing the use of clustering in high-throughput gene expression data and secondly through demonstrating the power of context in predicting secondary structure of proteins.

2. An easy explanation of machine learning from commonly used data standardization process

As already pointed out that learning is the most important part of intelligence, the foundation of it has its root from the simple *standardization* processes we have been using since time immemorial. In light of this, the fundamentals of ML can be derived from the perspective of predictive modeling of a system. For the first time visitor of the subject, ML, it is convenient to look at it from the angle of the already-known computational paradigm, although, in principle, ML and AI are considered to have wider scope in including the cognitive basis of the human. In the context of ML, the first pertinent question comes in mind is as follows.

Integrative Omics. https://doi.org/10.1016/B978-0-443-16092-9.00022-9

2.1 What is a machine in ML?

The simple answer from the computational paradigm is: machine in the context of ML is a mathematical model capable to be integrated within a machine for its performance, which can be presented in the following simple functional form:

$$y = f(x) \tag{22.1}$$

where f is the model or machine, which takes x as input and produces y as output, whereas y should approximately match with the desired output. The very question that follows next is: what is learning for this machine? In this regard, learning can be considered as nothing but a *standardization* process that gives a good estimation of f while all of us, engineers, scientists, and management folks, know what *standardization* is. To be precise, let us consider the following example of a regression model to optimize a simple function, i.e., a line from some experimentally obtained data (Mitchell, 1997).

2.2 Example of standard estimation of blood glucose

In this example, the very word *estimation* stands for prediction in the ML parlance. In a standard laboratory setup, the reagents used for the measurement of blood glucose produces reduced nicotinamide adenine dinucleotide (NADH), the increase of which is directly proportional to the glucose concentration and can be measured spectrophotometrically as optical density at 340 nm as shown in Fig. 22.1.

From this example it is clear that a line can be fitted through the points generated by the plot of blood glucose (say, Y) versus optical density (say, X) through a standard linear regression that can be mathematically represented as follows:

$$Y = mX + C \tag{22.2}$$

In Eq. (22.2), the constants of the line, m (slope), and C (intercept) can be estimated from the experimentally obtained set of X (i.e., optical density) values for known amount of glucose in samples. If we just move ahead a bit from this simple *standardization* process, we may present the same estimation of the previously referred known samples as a training set of points or data that help us to build our so-called ML model (or machine as such) as shown in Eq. (22.2). For a sample of unknown glucose level, we need to first obtain its optical density (i.e., X) and feed it to Eq. (22.2) (i.e., the ML model) to get the predicted amount of glucose as desired.

2.3 Machine learning beyond simple standardization

In the previously cited example if we want to demarcate the subjects for whom the blood glucose level is abnormally high (say, patients of class G) from the healthy control (say, class H) on the basis of a cut-off glucose level, say, Glu, we may move further to classify the subjects on the basis of the following simple rule that serves as our next stage machine to predict about the glucose based health status (i.e., class):

$$\text{if } X > \text{Glu; } Y \text{ belongs to class G else class H} \tag{22.3}$$

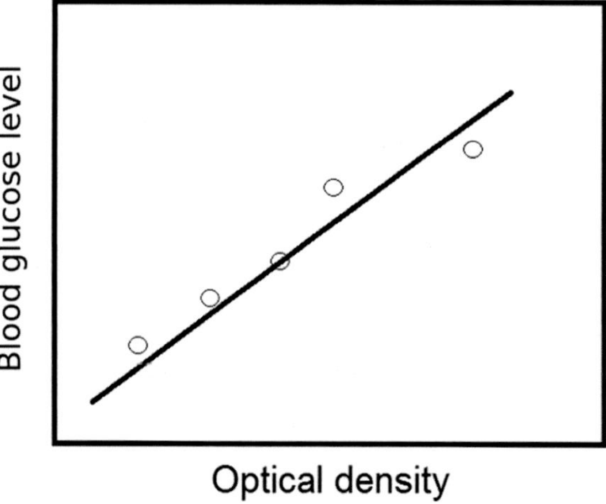

FIGURE 22.1 Blood glucose level in standard unit is plotted against optical density of NADH. *Standard glucose concentration* curves versus *optical density*, measured with the spectrophotometer.

The inequation Eq. (22.3) is an example of a rule-based classifier model, i.e., a little higher level of ML that helps to predict a typical output we refer as *class*. In this example to predict the class of X, one needs to feed it into Eq. (22.2) and subsequently to inequation Eq. (22.3) to finally get the class prediction for X. The given example reasonably convinces us to consider simple standardization protocol as ML. However, it is a highly simplified version of an ML protocol for a rather more simplified one dimensional data that are fittable into a line. However, if we extend the concept of classification further to incorporate

(i) multidimensional or multivariable data that cannot be visualized to understand its spread within a data space (since it is not possible beyond a dimension of 3).
(ii) higher degree of difficulty that cannot be fitted into a line or any other mathematical function (forget about any nonlinear types of regression),

we certainly need a protocol that should be much more powerful and accommodative than simple line- or curve-fitting based standardization to model these types of complex real-life problems for predicting its desired characteristics. The advent of the ML models happened from this requirement only.

2.4 Types of learning

The example so far discussed depends on a set of data samples with known outputs (i.e., training data) supplied beforehand by, as if, a supervisor. The model built out of this learning is thus referred to as the *supervised learning* model of ML. The ML models based on this paradigm are referred as *classifiers*. There are plenty of important applications in the field of *omics* to harness the utility of classifiers to extract knowledge from many biological events or physical constituents, e.g.,

(i) prediction of secondary structure of a protein from its sequence
(ii) prediction about the group of genes causing a specific type of disease from the gene expression data of healthy subjects and patients of such diseases

Examples of standard classifiers are nearest neighbor classifiers, probabilistic classifier (e.g., Bayes classifier, Hidden Markov Model), backpropagation network (BPN) as a subclass of artificial neural network (ANN), and its extension as deep learning model (Bishop, 2006).

Naturally, the other type of learning is referred to as unsupervised learning where no supervisor supported us with the known class information of the set of data samples. Basically it solves the question: who is the first learner to serve as a first supervisor? This means, in this initial phase of learning, our task is to create classes or groups or clusters from our interaction with a set of data. Normally the association between a pair of data is ascertained through their distance, and this step is repeated to group the data on the basis of this rule of association. The examples are K-means clustering, hierarchical clustering, Kohonen or self-organized map, fuzzy c-means clustering, etc. (Mitchell, 1997). For the problem of predicting a small group of genes responsible for a particular pathophysiological condition, normally samples of gene expression data are collected from both healthy control and patients of the specific disease. Subsequently, data indicating differential gene expression profile are created and utilized for clustering to find the groups of genes having common differential gene expression profile. From a few of these groups, the groups signifying those genes responsible for causing the disease are identified using various statistical methods based on null hypothesis.

3. Application of classifiers for predicting the secondary structure of a protein from its sequence

Before solving this problem with the help of a classifier model, first it is necessary to check how the problem fits into the paradigm of a classifier based model. Naturally, for this purpose, we need to check the following items:

(i) What do we need to predict? This means, what is going to be the nature of output Y specifically in a computational environment (please refer to Eq. (22.1) for more clarity).
(ii) What will be the nature of input, X?

The answers of the aforementioned questions lead to a beautiful fitting of the scientific part of the problem into a classifier model. Starting with simplified versions of secondary structures, e.g., helix (H), strand (S), and coil (C) For a protein, with a careful observation of the problem, it transpires that

(i) sequential arrangements of successive residues form a secondary structure, which further encourages us to ascertain the secondary structure class of an individual residue one after another.

(ii) it is highly difficult to decide on the secondary structure class of an individual residue since the same amino acid may serve as the member of different secondary structure classes.

This problem, however, can be solved by applying a typical supervised learning based on context-based learning following the similar learning capabilities of a human as explained in Fig. 22.2.

Similarly, a context window may be built by keeping the residue in question at the center of the context window containing its neighborhood residues. The basic framework is described in Fig. 22.3 where Fig. 22.3A and B signifies part of a protein sequence, i.e., the chain of successive residues and the corresponding secondary structures they belong to. For an individual residue shown by its one-letter representation, P, this correspondence is highlighted by a red circle to show its corresponding secondary structure membership, i.e., H. The context window data are, however, highlighted by a green circle, which, instead of P isolatedly, will serve as a data representing P.

Usually this string data is further converted into a numerical array by assigning a 5-digit binary number for each of the amino acids. Finally, this context window data serve as the input training data, X for a given set of proteins with known secondary structure classes (i.e., the outputs, Y) for computation of model parameters for a specific ML model. Once these parameters are computed, this model is further utilized for predicting the secondary structures of a protein from its sequence (Drozdetskiy et al., 2015).

4. Protein secondary structure prediction using artificial neural network–based classifier

This problem can also be solved using BPN model of ANN (Gose et al., 1999). The foundation of BPN is found in the linear discriminant function (LDF) model, which is a well-known statistical classifier where the set of M number of equations (called as LDFs) corresponding to the set of M number of classes for a data with N number of variables, x_1, x_2, …, x_N are as follows:

$$y1 = A_{10}x_0 + A_{11}x_1 + \ldots\ldots\ldots\ldots\ldots + A_{1,N}x_N$$

$$y2 = A_{20}x_0 + A_{21}x_1 + \ldots\ldots\ldots\ldots\ldots + A_{2,N}x_N$$

$$\ldots\ldots\ldots\ldots\ldots$$

$$yM = A_{M,0}x_0 + A_{M,1}x_1 + \ldots\ldots\ldots\ldots\ldots + A_{M,N}x_N \tag{22.4}$$

In this regard, a forced mapping of any ith class for a training data, x_1, x_2, …, x_N are obtained following the equation given below:

$$y1 = 0 = A_{10}x_0 + A_{11}x_1 + \ldots\ldots\ldots\ldots\ldots + A_{1,N}x_N$$

$$y2 = 0 = A_{20}x_0 + A_{21}x_1 + \ldots\ldots\ldots\ldots\ldots + A_{2,N}x_N$$

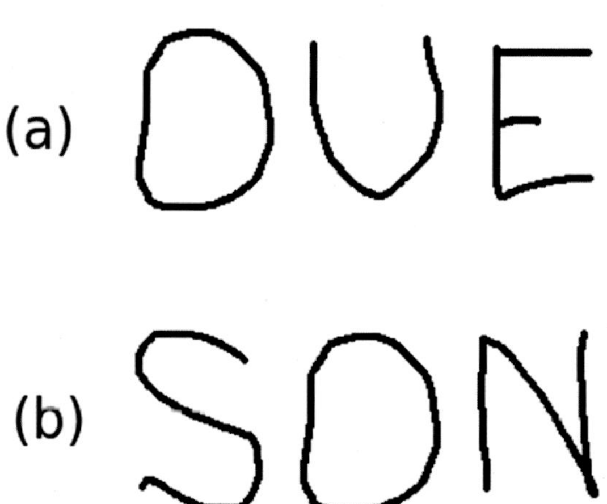

FIGURE 22.2 Example of context-based learning where (A) handwritten character in the first position is considered as D and (B) the same character in second position is considered as O. A learning approach that emphasizes understanding information and concepts within the context in which they are presented.

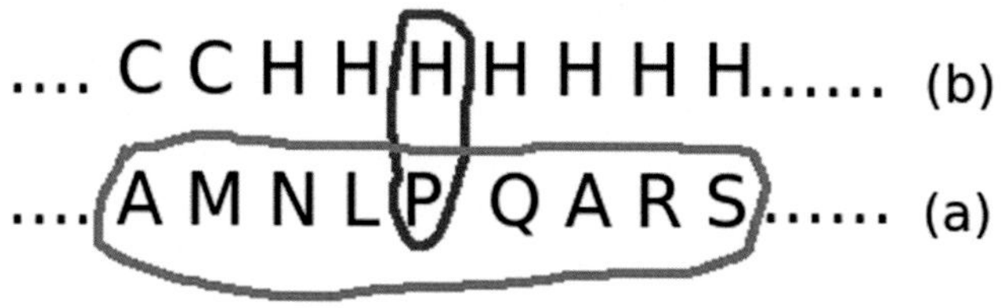

.... C C H H H H H H H...... (b)

....A M N L P Q A R S...... (a)

FIGURE 22.3 (A) and (B) signifies part of a protein sequence i.e., the chain of successive residues and the corresponding secondary structures they belong to. For an individual residue shown by its one-letter representation, P, this correspondence is highlighted by a red circle to show its corresponding secondary structure membership, i.e., H. Mapping of the amino acid sequence into the secondary structure.

$$\ldots\ .\ \ldots\ \ldots\ \ldots\ \ldots.$$

$$y2 = 1 = A_{20}x_0 + A_{21}x_1 + \ldots\ \ldots\ \ldots\ \ldots\ \ldots + A_{2,N}x_N$$

$$\ldots\ .\ \ldots\ .\ \ldots\ \ldots$$

$$yM = 0 = A_{M,0}x_0 + A_{M,1}x_1 + \ldots\ \ldots\ \ldots\ \ldots\ \ldots + A_{M,N}x_N \tag{22.5}$$

Usually a one-time statistical solution is obtained using a method similar to Cramer's rule in case of LDF. However, in case of BPN, a gradient descent method is applied for the purpose of getting the solution for the coefficients A (W or weights in case of ANN) through an iterative optimization method to minimize the error function (i.e., the sum of the square of the difference between LHS and RHS). Because of the similarity of this iterative optimization method with human reasoning or functioning, this learning is referred to as *soft computing*. Fig. 22.4 shows the BPN analog of Eq. (22.5).

From Fig. 22.4, it is evident that the multivariate system−based classifier presented in Eq. (22.5) can be presented as a two-layer system where the second one is the result of sum of the inputs through their corresponding weights. Normally the y's are further subjected to a nonlinear activation function with a limit of 0−1. It is also understandable that we can insert more layers in between these two layers that are referred to as the hidden layers. In case of deep neural networks (DNNs), it is expected that a good agreement between the expected (i.e., the 1,0 form of outputs as shown in Eq. 22.4) and the observed outputs can be obtained, which is not always true due to presence of a vanishing as well as an exploding gradient problem (discussed in the later section).

4.1 Classification rule of BPN

Naturally, the classification rule in this method expects that the data belong to the i-th class, the output y_i should be 1 and the outputs of remaining LDFs will be zero. However, in real-life scenario we cannot expect the observed solution as binary one and therefore answer for the class can be obtained from the following rule:

$$\text{For } Y = \{y_i\}_{i=1}^{M} \text{ if } \operatorname{argmax}(Y) \text{ is i, then the data belong to class i}$$

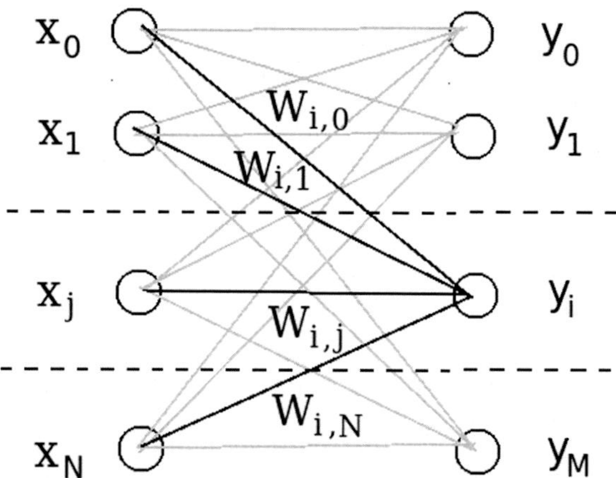

FIGURE 22.4 BPN analog of LDF. The use of a linear activation function in the output layer of the neural network for tasks that entail linear classification or regression is the backpropagation network equivalent of the linear discriminant function.

Following the previously explained rule, in Fig. 22.4, the output of i-th node of the second layer should be maximum and thus referred to as the winner rule. As already mentioned, in this regard, even after the application of DNN, we may face problem of getting good solution of W due to vanishing or exploding gradients, which lead to the further improvement of ANN and referred as the deep learning method.

5. A simple description of deep learning method

Let us start with the problem of vanishing or exploding gradients.

To understand it, we need to consider the iterative optimization process (usually referred as gradient descent optimization process only). As explained in the (a), for an unimodal differentiable function, the global minimum can be obtained from the following simple equation:

$$\frac{df}{dx} = 0; \text{ where f is short form of } f(x)$$

The minimum point (i.e., the value of x for which f is minimum) can also be obtained for a nondifferentiable function by adopting the following steps:

Step 1: Differentiate f(x) at any point x where x is designated as the starting point. If the value is very close to zero stop otherwise follow next steps.

Step 2: Move toward the negative gradient direction with a step-length equals to the gradient value.

If we consider the case 1 of (b), it appears that for a small step-length (i.e., almost vanishing gradient value), we will reach a wrong minimum and cannot come out of it. Similarly, as shown in case 2, for an exploding gradient, the step-length will be very high to jump over the global minimum again.

Therefore, to avoid this problem and also the huge computational complexity associated with DNN, the deep learning (DL) method is introduced to reasonably increase the accuracy of classification problems in comparison with DNN (Courville et al., 2016). It can be better understood from the popular example of convolutional neural network (CNN). The usual framework of a CNN are made of the following components:

1. A two-dimensional array of input data nodes of dimension say, $M \times N$. Please note the usual single-dimensional nodes of ANN can also be used as a special case of array with a dimension of $1 \times N$.
2. The data are divided into small overlapping windows (referred as receptive fields) of same size. However, the weights connected to the next layer (referred to as feature map) are kept same throughout the iterative steps, thus effecting a huge reduction of computational cost.
3. The feature maps are actually extracting hidden feature information for the data, thus lifting the burden of choosing appropriate features manually.
4. There is an opportunity for the creation of multiple features maps associated with different sets of weights to the receptive fields.
5. The hidden layers describing the feature-maps are connected to a pooling layer, which pools the values of selectively prioritized nodes of the feature maps through the rules of max-pooling, average pooling, etc., thus effecting a further reduction in the computational cost.
6. The pooling layers are normally connected to the output layers.

The way a CNN of a DL works can be understood from Fig. 22.5.

As depicted in Fig. 22.5 a CNN of DL system works in the following manner:

- The network begins with 28×28 input neurons, which are used to encode the pixel intensities for the images each of the size 28×28.
- This is then followed by a convolutional layer using a 5×5 local receptive field and three feature maps (i.e., set of weights corresponding two three parallel convolutional layers).
- The result is a layer of $3\times24\times24$ hidden feature neurons.
- The next step is a max-pooling layer, applied to 2×2 regions, across each of the three feature maps. The result is a layer of $3\times12\times12$ hidden feature neurons. The final layer of connections in the network is a fully connected layer. That is, this layer connects every neuron from the max-pooled layer to every one of the 10 output neurons.

6. Use of multiple ML models (multiple classifiers) to analyze multiomics data

Quite often the target of single mode *omics* data analysis focuses on identifying the molecular cause of a disease by discovering novel biomarkers for a specific pathological condition (Rajnish Kumar & Lahiri, 2018). In its broad definition,

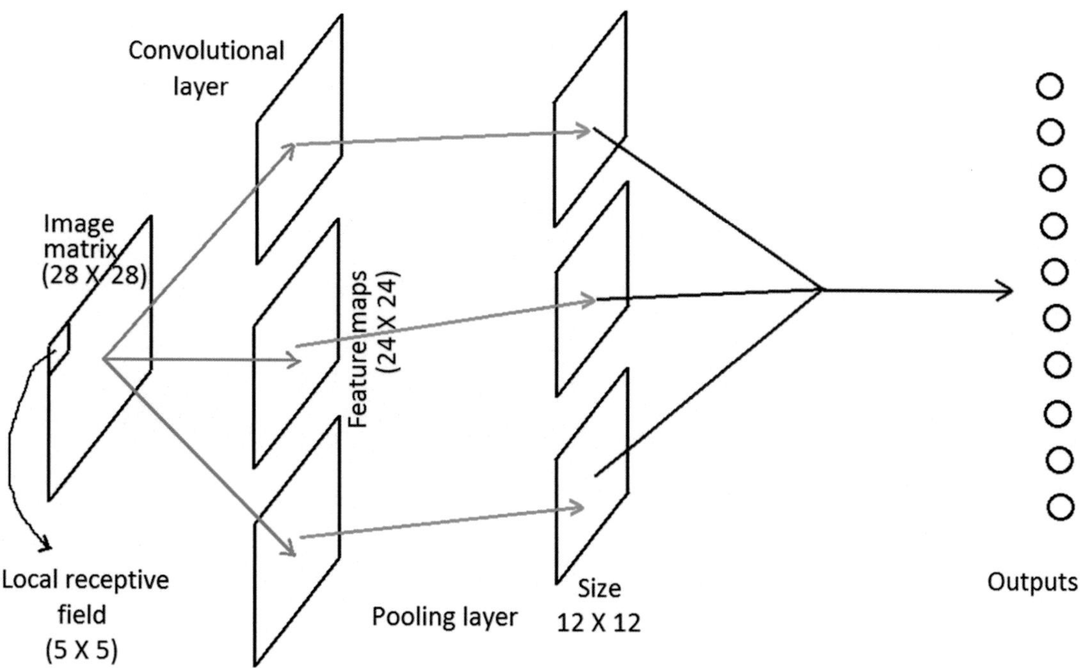

FIGURE 22.5 The way a CNN of a DL works.

a biomarker can be interpreted as a biomolecule or a biomolecular complex, molecular structure, or a full or partial metabolic process, leading to a pathophysiological manifestation. Genomics, epigenomics, transcriptomics, proteomics, and metabolomics have been applied in cancer detection to various extents through the use of different modalities of *omics* data (Momeni et al., 2020). The survey of the published work on ML techniques used in omics data analysis indicates that there exists a specific limitation in their application. As one can guess, most of these ML models (usually the unsupervised types, i.e., the clustering methods) substitute the earlier statistical methods used for hypothesis testing (e.g., t-test) for given two sets of *omics* data, such as gene expression data for healthy controls and patients of a particular disease. In this context, it is obvious that to zero-in the biomarkers from such a high-throughput *omics* data, the single mode data (i.e., the gene expression data) may not be sufficient. This is the limitation posed by such single-mode *omics* data that points out a need of cross-checking the findings with that resulted in through the analysis of other *omics* data. The other common example is the knowledge gap in genotype—phenotype correlation for certain types of diseases (e.g., cancer), which is not a simple one-to-one relationship and rather an outcome of a complex set of interactions among different types of molecular mechanisms. Therefore, for better understanding of disease processes, the current studies are shifting from a single mode *omics* data analysis to multimodal *omics* data-interpretable techniques (Fig. 22.6) through its convergence by multiple classifier-based ML. In this regard, data integration remains the fundamental challenge for multiomics data analysis to find relationships between various modes of omics data. Multiomics data integration can be classified into three popular groups: concatenation-based integration, transformation-based integration, and model-based integration (Lancaster et al., 2020; Momeni et al., 2020; Reel et al., 2021; Sidak et al., 2022). The matrixes of different omics data are combined to form a large data matrix before the model is constructed in case of concatenation-based integration that is also known as "early integration" method. However, in the case of the transform-based integration method, data integration is done in the intermediate step after initial transformation of data generated from individual *omics* experiments. Model-based integration (late integration) is performed to identify interactions among different levels of omics data related to a specific disease or phenotype through generation of a joint model followed by analysis. Secondly, a large number of features to samples ratio of *omics* data often cause overfitting, generating nonreproducible results and is affected by high variance (McCabe et al., 2020). Finding the most effective features is very crucial in the field of single-omics data analysis. Also, quite often the universal applicability of common computational techniques including deep learning ML models poses confusion in regard to offering a biologically meaningful feature and its usability for the intended purpose. It becomes more problematic for multiomics data analysis. The recent publication of Kumar et al. (2023) has shown a great promise to deal with multi*omics* data through a new systems biology approach, integrated data geography (IDG), a terminology borrowed and adopted from the terminology integrated geography to assimilate the findings from different forms of biological data under the paradigm of an inductive reasoning (Kumar et al., 2023).

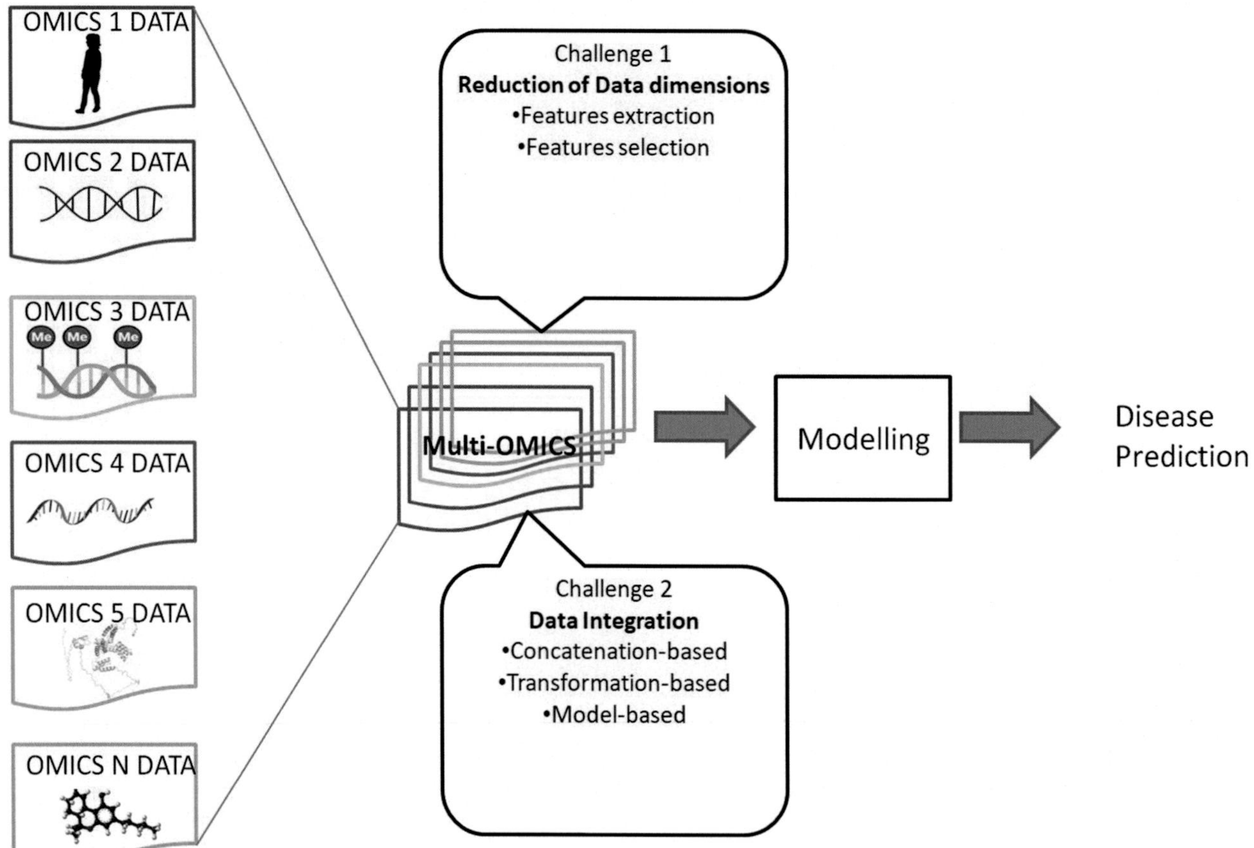

FIGURE 22.6 Schematic representation of multi*omics*-based disease prediction. The link between single- and multi*omics* data analysis issues is depicted in the overall diagram.

7. Concluding words

In today's computational world, application of ML approach is paramount for various forms of biological data analysis. The lack of primary knowledge in ML may lead to its spurious use resulting in interpretations that are not biologically meaningful. Therefore, analysis of the high-throughput *omics* data by ML models to extract biologically relevant information appears to be constrained by limitations quite often due to information gaps. To compensate for this knowledge, currently a multimodal data analysis approach is being adopted by various researchers using multi*omics* data.

References

Bishop, C. M. (2006). *Pattern recognition and machine learning*. New York: Springer New York.

Courville, A., Goodfellow, I., & Bengio, Y. (2016). *Deep learning*. MIT Press.

Drozdetskiy, A., Cole, C., Procter, J., & Barton, G. J. (2015). JPred4: A protein secondary structure prediction server. *Nucleic Acids Research, 43*(W1), W389–W394. https://doi.org/10.1093/nar/gkv332

Gose, E., Johnsonbaugh, R., & Jost, S. (1999). *Pattern recognition and image analysis*. New Delhi: Prentice Hall of India Pvt Ltd.

Kumar, R., & Lahiri, T. (2018). EcircPred: Sequence and secondary structural property based computational identification of exonic circular RNAs. *Computational Biology and Chemistry, 77*, 28–35. https://doi.org/10.1016/j.compbiolchem.2018.08.002

Kumar, R., Mondal, R., Lahiri, T., & Pal, M. K. (2023). Application of sequence semantic and integrated cellular geography approach to study alternative biogenesis of exonic circular RNA. *BMC Bioinformatics, 24*(1). https://doi.org/10.1186/s12859-023-05279-z. https://bmcbioinformatics.biomedcentral.com/

Lancaster, S. M., Sanghi, A., Wu, S., & Snyder, M. P. (2020). A customizable analysis flow in integrative multi-omics. *Biomolecules, 10*(12). https://doi.org/10.3390/biom10121606

McCabe, S. D., Lin, D.-Y., & Love, M. I. (2020). Consistency and overfitting of multi-omics methods on experimental data. *Briefings in Bioinformatics, 21*(4), 1277–1284.

Mitchell, T. (1997). *Machine learning*, 1997.

Momeni, Z., Hassanzadeh, E., Saniee Abadeh, M., & Bellazzi, R. (2020). A survey on single and multi omics data mining methods in cancer data classification. *Journal of Biomedical Informatics, 107.* https://doi.org/10.1016/j.jbi.2020.103466

Reel, P. S., Reel, S., Pearson, E., Trucco, E., & Jefferson, E. (2021). Using machine learning approaches for multi-omics data analysis: A review. *Biotechnology Advances, 49.* https://doi.org/10.1016/j.biotechadv.2021.107739

Sidak, D., Schwarzerová, J., Weckwerth, W., & Waldherr, S. (2022). Interpretable machine learning methods for predictions in systems biology from omics data. *Frontiers in Molecular Biosciences, 9.* https://doi.org/10.3389/fmolb.2022.926623

Chapter 23

Omics technology policy and society research

Manjusa Singh[1], Athaven Sukunathan[2], Swati Jain[2], Sunil Kumar Gupta[3], Ram Lakhan Singh[4] and Manish Kumar Gupta[1]

[1]*Department of Biotechnology, Faculty of Science, Veer Bahadur Singh Purvanchal University, Jaunpur, Uttar Pradesh, India;* [2]*North Middlesex University Hospital, London, United Kingdom;* [3]*Department of Pharmacoinformatics, National Institute of Pharmaceutical Education and Research, Hyderabad, Telangana, India;* [4]*Department of Biochemistry, Dr. Rammanohar Lohia Avadh University, Ayodhya, India*

1. Introduction

The last 10 years have seen a dramatic change in biological sciences due to the tremendous technological advancements in molecular biology. For the entire documentation of molecular biological data, several technologies have emerged, including genes, proteins, or small metabolites, and are named by appending the common suffix −"omics" (Greek: −ομική), as in the terms "genomics" for the collection of an organism's entire genetic code or "proteomics" for the study of proteins. These high-throughput technologies make it possible to produce huge amounts of data quickly and at significantly lower prices than was previously achievable. The generated data can be analyzed using advanced bioinformatics tools, which are stored in databases (Raupach et al., 2016).

The last 2 decade marked the exponential rise in omics technology upon the advent of discovering the human genome project and the evolution of DNA sequencing techniques. The cascade of omics that followed as proteomics, lipidomics, cytomics, interactomics, transcriptomics, and metabolomics can be grouped under taxonomics or Omics research. As time passes and the knowledge underlying these fields is strengthened, there becomes more cooperation and compatibility between the data sets of these fields, i.e., they can facilitate each other to strengthen a finding also known as "harmonization" or "interoperability." Crucially, improved information handling capabilities have allowed concepts such as artificial intelligence and machine learning to enter the research world (D'Adamo et al., 2021) and provide significant solutions such as machine learning is complex, computer algorithms may enhance themselves after receiving relevant information, and artificial intelligence is computer systems arc capable of handling activities that often need humans, such as decisions making. Four criteria for omics studies are taxonomically important such as type and number of genomic loci in research study, number of species and biological samples, and the type of omics technology, e.g., genomics, proteomics, etc., omics technology application type (e.g., pharmacogenomics and nutrigenomics phenotypes) (Pirih & Kunej, 2018). If used properly, omics technologies have the potential to enable qualitative and quantitative measurement of changes at the molecular, cellular, tissue, individual, population, and community levels, offering a historic opportunity to revolutionize our understanding of the effects of toxic substance release into the environment (Zhang et al., 2018).

Single cell multiomics profiling can answer issues that are too complex or challenging for other approaches. As an example, in addition to the traditional method based on fluorescence-activated cell sorting and lineage tracing, it is now possible to analyze complicated tissues and cellular lineage hierarchies. For single-cell tissue profiling, the sample of interest is divided into single-cell suspension, where each cell is profiled for the applicable omics dimensions before being bioinformatically ordered into a precise map of the system under study. Although technically and bioinformatically challenging, such data-driven dissection of cellular differentiation provides the possibility to solve some of the problems of the common approach, especially its dependence on surface markers and labor-intensive nature (Bock et al., 2016). Genome-wide study of DNA and RNA in single cells has become available due to the development of next-generation sequencing technology. Scientists have created a variety of techniques to measure other omics at the single cell level,

Integrative Omics. https://doi.org/10.1016/B978-0-443-16092-9.00023-0

including single cell DNA methylation, single cell chromatin sequencing, and single cell proteome analysis. These techniques were inspired by the very first reports of single cell DNA sequencing and single cell RNA sequencing (Hu, Scheben, & Edwards, 2018; Hu, An, et al., 2018).

"The health care system can adapt to new technologies at speed which would have been inconceivable a decade ago." A series of breakthroughs based on multiomics methods has helped us comprehend the complexity and heterogeneity of COVID-19 in order to provide reference for the diagnosis, surveillance, and clinical decision for COVID complications (Li et al., 2022). Technology advancements in data storage, bioinformatics skills, and laboratory-based techniques have allowed it to produce a significant quantity of "omics" data quickly and affordably. The enormous amount of COVID-19 study data generated in a short period of time serves as evidence for this. However, only a small number of clinical settings have adopted these technologies widely too far. Although they are frequently used, idealistic terms include "precision medicine" and "pharmacogenomics" remain ill-defined and mostly aspirational (D'Adamo et al., 2021). Precision medicines are personalized medical care for each patient based on their unique phenotypes or traits. The old words "personalized" medicine is frequently used interchangeably, and pharmacogenomics are the study of using a person's genetic profile to predict their reaction to drugs. However, the widespread novel increase in use of omics technology, while immensely beneficial to the scientific research field, also raises a few barriers such as ethical, legal, and social implications as well as inconsistencies between the omics practices of various companies thus posing as a cause for concern in the eyes of the media and society. In this chapter, policies and differing practices and their social implications is explored.

2. What are omics technology policies?

Standard omics practices when it comes to modern day biodata sharing and reporting include filtering source material through assays, i.e., genomics sequencing, transcriptomics analysis, protein—protein interaction assays, metabolomics assays, where all data must be presented as comprehensible, reusable, and sufficient to reproduce including communication of experimental information such as checklists, nomenclature of scientific terminology, and certain syntax formats. These strict standardizations are followed by journals and biocurators across the board to develop, preserve, and manage the datasets and increase any potential funding for omics technology, while enhancing the dataset of biodata and advancing the field. Even the Food and Drug Administration (FDA) and commercial sectors have noted that research and omics portfolios must adhere to a standard acquisition and must conform to the rules of multiple regulatory participants. The reporting of omics biodata in such a way can lead to increased efficiency, avoids risk of loss when data is turned over, enhanced security in data, and an overall agreement between multi-organization collaborators or journals when publishing (Wilkinson et al., 2016). The main benefit of OMICS technologies is the ability to obtain a significant amount of information at a relatively low cost and effort, which allows us to expand our understanding of the complex biological system (Egea et al., 2014; Lombardi et al., 2011). Omics science and technologies hold out the promise of improving the imprecise and reductionist experimental models that, up to this point, had only offered a temporal snapshot of the much more complex, longitudinal, and dynamic nature biological networks and the differences that result from exposure to social and environmental factors that latest version and fundamentally govern human health and disease. Far beyond immediate scope of the individual scientists, laboratories, scientific consortia, or governments that design, use, and regulate them, standards in omics science are of interest to (and shaped by) others. Indeed, the social, ethical, and legal environment in which novel technologies are governed is affected by scientific standards, which is why policy makers and citizens are so interested in them. With a focus on potential applications and ongoing knowledge such as (i) for standardization to be useful at the level of scientific practice in various locations, it must be adaptable; (ii) it also depends on who creates the standards as well as appear to be acceptable so that they can be used; (iii) and it is crucial to verify the legitimacy of a field of scientific study through the standardization process itself, and transfer in the particular setting of omics science (Holmes et al., 2010).

Furthermore, there are much more intricate levels of harmonization in omics, as certain systems aspire for the multiple domains of omics to provide data that can support each other and facilitate interoperability such as Functional Genomics Experiment (FuGE), Health Information Technology Standards Panel (HITSP), Minimal Information for Biological and Biomedical Investigations (MIBBI), etc. These user-friendly, interoperable systems can integrate a broad range of multiomic domains and accommodate a much more intricate research question which could potentially be used for clinical and diagnostic scenarios (Chervitz et al., 2011). One key implementation of omics technology and data standards policy is an organization called "BioSharing." BioSharing is a website that "centralize bioscience data policies and link portals" between various levels of omics data processing, i.e., intra-harmonization between funders and stakeholders, allowing for interoperable reporting standards without overlap or duplication or use of incompatible tools. The overall harmonization and sharing of standardized data via these strict policies are much more embraced by funding authorities, allowing for a much better quality of omics research to take place (McQuilton et al., 2016).

For governing individual of healthcare systems, the leaders, i.e., stakeholders and policy makers, are educated to the use of omics in terms of predicting incidence of a pathology, treatment, therapy measures, and its notorious use for diagnosing diseases. Hence, the inclusion of current healthcare professionals and leaders is vital if the field of omics is to reach a much more personalized level of medicine, which is highlighted by a statement released by the World Health Organization, wherein "policy makers can capture the potential of genomics to meet public health goals." A leading example of omics policy assimilated by healthcare legal frameworks is in Italy, national plan for public health genomics was published and approved, the impact it may have on the economy, to the use of big data in personalized medicine. The overall plan however has three main approaches: to increase awareness of all stakeholders to the ethical, social, and legal innovative aspects, to enforce a strategy to govern said innovation and to critically evaluate any omics opportunities that may present themselves all the more aiming to reshape the National Healthcare System (NHS). Furthermore, Italy has reinforced ideas about innovation from the standpoint of the Italian government and members of the chief medical office where data and computational medicine, and pharmacogenomic research can raise possibilities for sustainability by perhaps removing drugs from circulation and overall reducing disease incidence. These omics-related governance have already started taking effect in Italy, however, it is imperative to note that cooperation of the respective professional and social authorities is paramount (Boccia et al., 2017).

2.1 Enablers of policy

Science-based, predictable, and time-limited policies and laws will be simplified in response to stakeholder needs and the changing policy atmosphere under the guiding principle of "minimal government, maximum governance." To make sure that there were no barriers to scientific study, guidelines for exchanging biospecimens and data for COVID-19-related studies were also established. Future initiatives in this area will consist of the following:

- Create an "Ease of Doing Science Index" to assure efficient use of both the researcher's time and disbursed funding (including flexibility of budget utilization).
- Establish regulatory standards for the use of cutting-edge technology like gene editing.
- Modify the first indigenous biological data centrally and define policy on sharing biological data, particularly data collected through nucleic acid sequencing and microarrays, bio-molecular structures, and flow cytometry in today's high-throughput, high-volume environment.
- Building a nationwide "Network of infrastructure under Biosecurity and Biosafety" to protect the country and get it ready for any pandemics and epidemics.
- To tackle infections affecting humans, cattle, mammals, and crops by having a molecular surveillance system with cutting-edge diagnostic tools and conventional network facilities.
- Enhance the capacity of the Indian biotechnology sector to compete on a global scale through policy and funding reforms.
- Create frameworks and policies to support the use of biotechnology-based products in important areas that have value for society but not usually for the market.
- Transfer of technology and innovation to act as a go-to source for information on issues relating to innovation and technology transfer.
- Policies to support and build the ecosystem for bio manufacturing.
- Policies for the ethical use of emerging technologies and synthetic biology.
- The implementation of a frugal innovation policy through collaboration with state legislatures and science and technology councils (National Biotechnology Development Strategy & Department of Biotechnology, 2021), (India).

3. Omics technology policy debates

The advent of whole human genome sequencing was crucial for the development of omics and medicine as a whole. It was a revolutionary finding that generated omics research. Over the years, as technologies advance, it is increasingly cheaper and more cost effective to sequence the whole genome. However, a review article highlights the probable challenges: issue of human genome sequencing and especially the accessibility of it to the general public. They argue that the so-called "mainstreaming" of omics data is giving society access to powerful and extremely sensitive data, and a fear that people will become less aware of the fact that privacy is something that must be upheld. From a more scientific standpoint, policies must be driven from an ethical, legal, and social implicative view also referred to as Ethical, Legal, and Social Implications (ELSI) in order to undertake omics and ethics studies at the next level. While human omics studies have had an enormous positive effect on understanding the complexities of genetic diversity the population of African people are not accurately portrayed as part of these databases. Omics study grounded in understanding the genotype–phenotype correlations of various African ethnicities to improve omics classification, and as there was a historical false-positive rate and failures to read African genetic material. The pressing issues in modern omics research such as legal or socioeconomic standpoints. The data generated from this project was heavy on costs for African omics institutions specifically training of specialized personnel and experimental facilities. They added that there was a need for increased financial support but recognize the rising establishment of nonprofit and profit organizations that can collaborate and integrate networks of multiomics data (Hamdi et al., 2021). Omics was 80 times worldwide in comparison to African countries, who only spent a quarter of a billion in US currency (Sokolov-Mladenović et al., 2016). This socioeconomic gap is seen to be hindered by geopolitical factors such research and development (R&D) being the least important priority in comparison to education sectors, and private sector infrastructures. From a legal perspective, it is noted that an increasing number of countries are writing their own omics legislation such as (i) intellectual property rights, and (ii) storage of human data, etc. Despite this, it is alarming as to whether these governing bodies and rules are strict enough to avoid regular infringements and or take responsibility—for particular example, "African legislation [and IP] is absent, obsolete, restrictive, or difficult to navigate," due to their lack of funding, also consequentially leading to poor data management, and unstable internet thus slow data transfer and ultimately standardization.

Ethical concerns of omics data policy in the study stem from the privacy and confidentiality of data, the legal owner of the data, language barriers, and various methods of consent. The association of ethical omics standards should be mandated to promote collaboration and absolute transparency between scientists who work with each other around the world. From another ethical standpoint, consent and reconsenting or the continual checking of consent throughout a procedure is a topic of debate but can also be argues that is heavily inconveniencing members of family or next of kin of a participant who passed away. Revisiting socioeconomic, legal, and ethical arguments as well as omics policy—remarking that guidelines for continental harmonization of omics data and technology standards must be discussed at the highest level (Hamdi et al., 2021; Jao et al., 2015).

More on a consumer-based level, there is concern of omics policies in terms of the direct-to-consumer (DTC) microbiomics testing (MT), epigenetic testing (ET), and genetic testing (GT) (DTC-MT, DTC-ET, DTC-GT, respectively), as there are several ELSI concerns such as informed consent and discrimination issues. It is explained it can be a misleading exchange of omics technologies, paired with inadequate counseling measures from the appropriate healthcare professionals; all the more adding that the absence of counseling can consequent in consumers misinterpreting or on the other hand, have poor mindsets toward the relevance of certain genetic tests. Simultaneously, the lack of omics companies' policies are advise to the general public the risks of certain genetic conditions, the incomprehensible manner in which their policies are formatted for the general public, and scarce but still alarming instances of violations of privacy policy and unconsented research for the gain of "commercialization." However, they argue that there is no outstanding golden rule of a regulation or policy method of tackling the aforementioned concerns as 26 European countries (Romania, Germany, Hungary, Italy, Slovenia, Luxembourg, Poland, Lithuania, Slovakia, Sweden, the Netherlands, Belgium, Latvia, Cyprus, the Czech Republic, Greece, Austria, Denmark, Estonia, Portugal, Ireland, Finland, Spain, France, Norway, and the United Kingdom) have outright banned direct to consumer omics testing, and some have a variety of healthcare and privacy protection legislation which include vigorous medical and informed consent processes (Kalokairinou et al., 2018; Knoppers et al., 2021).

Moreover, the regulation and curation of new omics policies and new omics companies facilitates the scientists to be further informed on any unfortunate security situations where upload of any and all data is under the mandate that it be uploaded for any type of analytical studies. The ELSI issues, so far as to infer that there is a nonexhaustive list of compliance related restrictions when concerned with ELSI, for example, data security as when scientist uses and develops a multi omics tool and provides it open source for other people to see, there is often a level of restrictions that prevent the upload of said data. This has almost a "bottleneck" effect, confining the scientific community to "non-open-source enterprise level analytics platforms," endorsed by their compliance with guidelines enforced by The General Data Protection Regulation (GDPR) and the Health Insurance Portability and Accountability Act (HIPAA), while also being Good Manufacturing Practices (GMP) certified. With the condition that the cloud or network-based server has the aforementioned compliances any patient data can be interpolated under "strict vigilance"—which are the current practices that new omics companies are bringing into habit, which include Amazon and Microsoft (MS) omics (Krassowski et al., 2020).

Services for direct-to-consumer genetic testing (DTC-GT) have been made available online for the past 20 years by private businesses. The convenience of ordering, receiving, preparing, and reviewing these tests from the comfort of one's home appeals to consumers (Gregory, 2019; Majumder et al., 2021; Phillips, 2016). The several popular inherited tests provide details on lineage, family relationships, and wellness. Recently, companies have launched in more unconventional fields like kid talent, pairings, and adultery. At the intersection of technological advancement, medical diagnosis, and test delivery models, DTC-GT holds a unique position (Hogarth & Saukko, 2017). Because of their unusual situation, there has been a lot of public attention and research into the ELSI (Ethical, Legal, and Social Implications) of both these companies methods and their products (Knoppers et al., 2021).

To define and develop standards, guidelines, and policies for exchanging biological information and data, several worldwide debates have taken place. These include the 1996 Bermuda Agreement, the 2003 Fort Lauderdale Agreement, the 2010 Nagoya Protocol on Access and Benefit Sharing, the 2013 (GA4GH) Guidelines of the Global Alliance for Genomics and Health Guidelines, and the 2016 General Data Protection Regulation of the European Union. The current system for sharing biotech resource information and data follows and preserves the ideals of international discourses and agreements, with particular emphasis on the idea that data should be published and shared as soon as possible after generation. This framework also complies with existing international agreements; nevertheless, if the event that national and international policies conflict, national rules and guidelines shall take precedence (Biotech - Pride Guidelines, 2021).

4. Current methodologies and tools in omics policy

As this chapter has abundantly accentuated, the biodata-driven omics research tends toward the birth of new protocols and tools (PARADIGM, iClusterPlus, LRAcluster, BCC, MDI, SNF, PFA, PINSPlus, NEMO, mixOmics, moCluster, MCIA, MFA, iNMF, MOFA, NetICS, FSMKL, PMA, Joint Bayeslan factor, CNAmet, Joint NMF) (Subramanian et al., 2020). Such that the data can be interpreted. There should be a development of methods or tools where there can be an association between clinical and omics data on a network-based platform even going as far as to web-surveying ongoing data-driven projects to find out their opinions on omics data, while also asking how the field could be improved as well as the potential barriers to progress within the field. Omics data integration can be advantageous in two ways: the creation database repositories can be shared and validate research being done, and secondly, produces new methods of analysis; quoting Gene Expression Omnibus (GEO) a data repository for next generation sequencing data and adhering according to MIAME rules (Minimum Information about a Microarray Experiment). Moreover, the web questionnaire showed that scientists were generally content with the availability of omics tools but there was an overall majority of scores that implied there was high demand for a more user friendly software, that tools were developed for more software prone people, and the highest average consensus of people voted that there needed to be a new integrative analysis of multiple data types (Gomez-Cabrero et al., 2014).

This calls for efficient computational platforms and knowledge management systems to gather, manage, analyze, and exchange clinical and experimental data as well as combine it with previously collected information from public databases (e.g., PubMed, BIND, Reactome, miRWalk, mirDB, Tarbase, miRBase, Gene Ontology, miRNAmap, and KEGG, SGD, MGD, YMDB, HMDB, Gene Bank, PMN, PRIDE, GENCODE, Gene Ontology, Chip-seq, methyl seq) (Mangul et al., 2019; Pinu et al., 2019; Subramanian et al., 2020). Software platforms seek to make data and knowledge accessible at any stage of the workflow, offer high interoperability, prevents errors in handling and analyzing data or models, and speed up the entire analysis. The creation of standards is crucial to enabling the integrated analysis of biological datasets because they are defined using very varied formats, nomenclatures, and data structure, for example The problems of minimal Information about a Microarray Experiment (MIAME), file format (i.e., how the information should be kept, typically XML-based), and ontologies are addressed by standards for data management (e.g., System Biology Ontology [SBO] and Gene Ontology [GO]). Users must be able to explore, query, and obtain data on genes, proteins, lipids, metabolites, or relevant pathways and networks using a functional interface. For this goal, a number of software systems have been created: (1) tab-delimited spreadsheet template files used with particular software interfaces for analysis; (2) safe data and research tools accessible online via wikis; (3) workflow management tools like Konstanz Information Miner (KNIME) and Galaxy for genetics; (4) laboratory information management systems (LIMS); and (5) genome browsers from Ensemb and UCSC. More recent initiatives include (1) integrating transcriptomics and protein—protein interactions, such as the Sage Bionetworks initiative; (2) integrating open source software, like the Garuda Alliance and the tranSMART platform; and (3) utilizing for-profit private tools from IDBS (Clinical Sense), Oracle (Translational Research Solution), and BioMax Informatics (BioXM) (Kuhn et al., 2013; Wheelock et al., 2013).

Multiomics data available to all people, it must first be "deconstructed" into separate data sets and then put into omics-specific databases. These challenges highlight the fact that high-quality multiomics studies require: (1) correct experimental design, (2) careful selection, organizing, and preservation of sufficient biological samples, (3) careful obtaining of quantitative multiomics data and correlated meta-data, (4) enhanced tools for combining and interpreting the data, (5) accepted minimum guidelines for multiomics techniques and meta-data, and (6) novel opportunities for the distribution of intact multiomics data. Surprisingly, a large number of the clinical, computational, and data integration needs required for

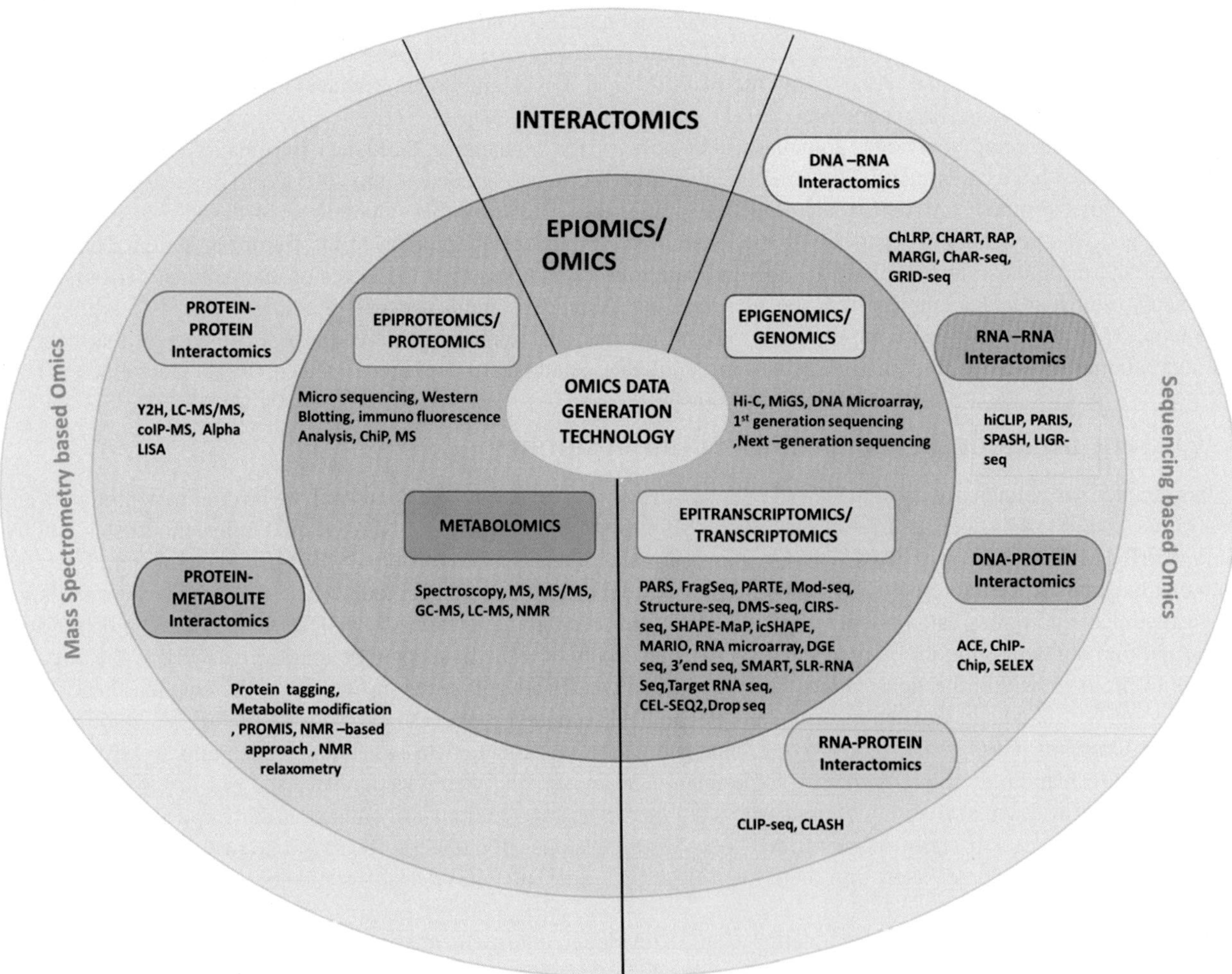

FIGURE 23.1 Classification of omics data generation technology.

metabolomics research are perfectly compatible with genomes, transcriptomics, and proteomics studies has been shown in Fig. 23.1 (Pinu et al., 2019).

A National Science, Technology, Innovation (STI) Observatory will be created as a result of Science, Technology, Innovation Policy (STIP), serving as a single repository for all types of data generated by and connected to the STI ecosystem. All financial plans, projects, grants, and incentives now in place in the ecosystem will be included in an open, centralized database platform. The Observatory will be dispersed, networked, and interoperable across important stakeholders, with central administration. All citizens of the nation and those interacting with the Indian STI ecosystem shall have equitable access to scientific data, knowledge, and tools through the development of a forward-thinking, all-encompassing Open Science Framework. The FAIR (findable, accessible, interoperable, and reusable) principles will apply to any data utilized in and produced by publicly funded research. Through the Indian Science and Technology Archive of Research (INDSTA), a dedicated portal to provide access to the results of such publicly funded research will be developed.

The following broad vision will direct the Science, Technology, and Innovation Policy.

(i) For India to become technologically self-sufficient and rank among the top three scientific superpowers in the upcoming 10 years.
(ii) Creating a science, technology, and innovation (STI) ecosystem that is "people centric" to create, develop, reinforce, and retain vital human capital.
(iii) To increase the Gross Domestic Expenditure on R&D (GERD), the private sector contribution to the GERD, and the number of Full-Time Equivalent (FTE) researchers by two every 5 years.
(iv) Building STI excellence at the individual and institutional levels to receive awards and recognition worldwide in the upcoming 10 years (Government of India, 2020).

5. Various types of data

Defining every biological data type produced by biotechnological technologies is impossible. Along with changing technologies, data types, in particular high-throughput data, also change.

5.1 Data on DNA sequences

These data may pertain to an entire genome, exomes, specific coding regions, DNA fragments, or single genes. That data may be a single sequence (for example, the sequence data produced by a Sanger sequencing) or several fragmented sequences from a genomic area with broad coverage (those produced by a highly parallel DNA sequencer).

5.2 Data on transcriptomic RNA sequences

Although cDNA synthesis is typically performed prior to sequencing, recent technological advancements allow single molecule direct RNA sequencing without cDNA synthesis, and as a result, the nature of the data is similar to those generated by a massively parallel DNA sequencer.

5.3 Data on a genotype

Individuals are genotyped using high-density microarrays that cover the complete genome at a large number of loci. For genome-wide association studies, particularly in plants, genotyping by sequencing (GBS) is being used progressively. However, small-scale genotyping using PCR-RFLP and other related technologies is still used for a variety of specific purposes.

5.4 Data on epigenomic

After appropriate preprocessing, these data are also primarily produced by high-throughput techniques similar to DNA microarrays or DNA sequencing.

5.5 Data on the microbiome

These data, which are also nucleic acid sequence data, currently come in three main subtypes: (a) Amplicon sequencing data, which can be used to identify specific groups of microorganisms present in any sample (such as human stool, soil, sediment, etc.); (b) Shotgun metagenomic sequence data, which enables thorough evaluation of all microbial organisms present in a sample; and (c) genome sequences of individual isolates. Additionally, there is information in the form of individual gene sequences used for taxonomic purposes, such as 16S rRNA, gyrase, and many other genes, or for multilocus sequence analysis and multilocus sequence typing.

5.6 Data on protein structure

Such data includes information on proteins and other significant biological macromolecules, such as their atomic coordinates. The 3D forms of proteins, nucleic acids, and complex assemblies provided by this data aid in the understanding of many aspects of protein production in different situations.

5.7 Data from mass spectrometry

Current proteomics relies heavily on mass spectrometry as an analytical technique, and mass spectrometers are frequently employed to produce data that enables protein identification, annotation of secondary modifications, and estimation of the relative or absolute abundance of specific proteins.

5.8 Data from flow cytometry

A method called flow cytometry is used to identify and quantify a population of cells or particles' physical and chemical properties. Counts and multiparameter profiles of various cells in a heterogeneous fluid mixture are the subject of flow cytometry data.

5.9 Data from imaging

Images of specific cells, organs, or bodily parts, such as X-rays of the chest or pictures of the eyes or oral cavity.

5.10 Data about metabolomics

With or without considering gene expression, metabolomics is being employed to understand metabolite levels. Additionally, it is used when combined with microbiome data to clarify host–microbiome interactions. The generation of small molecule metabolite patterns can be done by LC-MS, GC MS, or CE MS. Biological data also includes information on phenotypes (such as plant height, insulin level, morphology, and biochemical and chemotaxonomic characteristics for bacteria) on specific samples (Biotech - Pride Guidelines, 2021) (India).

Classification of Omics Data generation Technology are based on Technologies developed for comprehension the "Central dogma" and Omics data generation divided into three groups, i.e.,—Epiomics, Omics, and Interactomics. Epiomics/Omics (Epigenetics/Genomics, Epitranscriptomics/Transcriptomes, Epiproteomics/Proteomics, Metabolomics) and their Interactomics (DNA-RNA Interactomics, RNA-RNA Interactomics, DNA-Protein Interactomics, RNA-Protein Interactomics, Protein-Protein Interactomics, Protein-Metabolite Interactomics).

6. Genomics policy issues

National human genome research institute (NHGRI) increase knowledge and make sure that genomics improves the health of all people, promoting the appropriate use of genomics in society. To do this, the ethical, legal, and societal aspects of genomics research include these important challenges.

6.1 Genetic test coverage and reimbursement

The traditional methods of clinical medicine might be completely transformed by genomic medicine. Genetic testing to diagnose, predicted, and cure disease now has more options because of the mapping of the human genome. Payers like insurance companies and Medicare require standardized methods of determining genetic tests for reimbursement if patients are to fully benefit from the advantages of genetic testing. Insurance providers can't decide how much to pay for genetic tests without this information.

In addition, payers are finding it difficult to keep up with the number of novel genetic and next-generation sequencing tests that are being introduced to the market. Comprehensive statistics analyzing the economics of genetic testing are also lacking.

The Genetic Information Nondiscrimination Act (GINA), which was passed by the federal government to address these issues, prevents insurers from treating patients unfairly. Additionally, there are different state laws that provide additional protections against genetic discrimination by insurers. these laws state-by-state, visit NHGRI's Genome Statue and Legislation Database.

6.2 Discrimination based on genetics

Americans are protected from discrimination based on their genetic information in both health insurance and employment under the Genetic Information Nondiscrimination Act (GINA) of 2008. It prohibits health insurers from using genetic discrimination by updating the Social Security Act, the Employee Retirement Income Security Act (ERISA), the Public Health Service Act (PHSA), and the Internal Revenue Code (IRC) with the Health Insurance Portability and Accountability Act of 1996 (HIPAA). The Veterans Health Administration, Medicare, Medicaid, Federal Employees Health Benefits, and commercial health insurers are all covered by the health insurance provisions of GINA. GINA provides limited insurance coverage for the TRICARE program used by the US military. Although eligibility for TRICARE insurance is dependent on employment by the US military, GINA's employment prevents do not apply to the US military. TRICARE may not utilize genetic information for coverage, underwriting, or premium setting. It is legal for the US military to use medical and genetic data when employing fresh recruits. GINA's protections for health insurance do not apply to long-term care insurance, life insurance, or disability insurance.

6.3 Genome editing

With the use of genome editing tools, scientists may change DNA, modifying physical characteristics like eye color and risk of disease. Gene therapies, or treatments involving genome editing, are being developed by scientists to prevent and treat diseases in humans. Cystic fibrosis and diabetes are two conditions that may benefit from the use of genome editing techniques. Germline and somatic therapy are two distinct subcategories of gene therapies.

People believe that it is not now appropriate for scientists to edit the germline cell's DNA since doing so might have serious consequences for public health. Additionally, scientific communities throughout the world are carefully investigating germline treatment because mutations caused to a germline cell would be passed down through generations. To avoid germline editing, several nations and organizations have strong laws and restrictions. For instance, the NIH does not provide funding for studies involving human embryo editing.

6.4 Disparities in health

The National Human Genome Research Institute (NHGRI) is sure that all populations profit from the advancements in genomics research and encourages participation from research scientists, research participants, and decision-makers. Everybody should be aware of genomics research and understand the ethical, legal, and social impacts that result from it. NHGRI aims to reduce health disparities. To ensure that genomics benefits everyone, multiple efforts are being made at the intramural, extramural, and policy levels. For instance, the National Cancer Institute and NHGRI's Divisions of Genomic Medicine and Genomics and Society cooperated on the Clinical Sequencing Exploratory Research (CSER) program to support research into the adoption and use of genomic sequencing in clinical care. The program's main objectives include the development and advancement of best practices as well as the ethical, legal, and psychological effects of using genetic information in clinical decision-making. A U01 research project with a focus on Clinical Sites with Enhanced Diversity was recently released by CSER as part of the program's continuation (CSER2).

6.5 Research using human subjects

Human subjects are often used in biological, therapeutic, and social—behavioral studies because they can yield information and insights that cannot be gained through the use of model organisms or other techniques. Standards that prevent the rights and welfare of participants are applicable to all federally financed research involving human subjects, including those supported by the National Institutes of Health (NIH). These standards additionally require that researchers clarify the advantages and dangers of participation in the study.

The Common Rule specifies informed consent, under which an investigator must clearly outline the research study for each possible research participant, including any risks to participation, and get their written assent before engaging them in the study. The Institutional Review Board (IRB) must review all federally funded human subjects research projects to assure the safety of study participants, with Common Rules.

6.6 Informed agreement

Given that participants' contributions, such as samples and health-related information, are important for the success of the majority of human scientific research, it is important to protect their rights and interests. To understand the unique issucs related to providing genetic data, researchers and study participants should have a discussion. Genomic data may be personalized and unique to each individual.

- Can be kept and utilized indefinitely.
- Explain to people that they are susceptible to a wide range of conditions some of which may come as unexpectedly given their personal or family history.
- Involve dangers that are imprecize or unclear.
- Be reinterpreted and have their significance shift throughout time.
- Raise privacy issues (partially due to the possibility of reidentification).
- Be important to decisions about family and reproduction.

Additional components of the permission form may be needed by local or state laws, biobanks, data repositories, or Institutional Review Boards (IRBs) to ensure that research participants are given essential security.

6.7 Genetics-related intellectual property

Patents are granted to protect innovations and to promote innovation. Such provides investors the chance to get the most return on their money. Patents issued for genetic technology, such as novel DNA sequencing techniques, are not any different, and their licensing has proven very beneficial for companies creating goods based on genetic discoveries

However, when patents restrict how fundamental genetic information may be used, they pose a hazard of impeding or unfairly restricting scientific research and the conversion of research findings into therapeutic applications. Gene Patents and Licensing Practises and Their Impact on Patient Access to Genetic Tests is a paper written by the Secretary's Advisory Committee on Genetics, Health, and Society. The paper suggested that diagnostic (but not therapeutic) genetic testing be free from patent infringement, combined with a research use exemption, predicated on the underlying premise that patents on human genes were appropriate. At the time, there was debate around the decision to exclude diagnostic patents from infringement while acknowledging the possibility of diagnostic gene patents. This was especially true given that the case studies that followed the report provided conflicting evidence of the harm that gene patents had caused to patients.

6.8 Research in genomics

The Federal Bureau of Investigation's (FBI) program and software for supporting criminal justice DNA databases is called the Combined DNA Index System (CODIS). To identify criminal suspects, law enforcement utilizes CODIS to match DNA samples from crime scenes with DNA samples from suspects who have been arrested or convicted of crimes, as well as with DNA found at other crime sites. The method can only analyze 13–20 specific microsatellite DNA indices. If a match occurs in CODIS, the suspect can next have a confirmatory sample extracted from him or her. In court, the outcomes of a comparison between that sample and the sample from the crime scene might be presented as proof.

The term "investigative genetic genealogy" (IGG), also known as "forensic genetic genealogy," refers to a new investigative technique that combines the genetic analysis of crime scene samples with the mining of data from publicly accessible genetic genealogy databases and conventional genealogical records.

In November 2019, the Department of Justice's interim policy on investigative genetic ancestry went into legal effect. The policy restricts the beginning of IGG to situations involving an unsolved violent crime or unidentified remains of a suspected murder victim, and it applies to all law enforcement authorities (federal, state, and local). A conclusive policy on investigative genetic ancestry is anticipated from the Department of Justice.

6.9 Genomic research privacy

6.9.1 Research privacy

Two fundamental principles of scientific research should be balanced while conducting genomics research: the need to save investigation participants' privacy and the need to distribute data widely to maximize its use for future scientific research. Efforts to advance scientific research and save privacy for patients are balanced by federal laws like the Health Insurance Portability and Accountability Act (HIPAA) and the Common Rule. In the case of genomic data, this creates a problem since, except for the exception of identical twins, every person's DNA sequence varies, making it impossible to effectively anonymize a DNA sample.

6.9.2 Clinic patients' privacy

The use of genetic testing has changed from being uncommon to common in many therapeutic settings as a result of impressive developments in genomics research in recent years. Although this kind of testing has obvious clinical advantages for the patient, it also poses new privacy risks and puts the patient at risk of having their genetic information misused. The Genetic Information Nondiscrimination Act (GINA), which prohibits genetic discrimination and limits the access of health insurance providers and employers to an individual's genetic information.

6.10 Genetic testing regulations

6.10.1 Governmental regulation

The Food and Drug Administration (FDA), the Centers for Medicare and Medicaid Services (CMS), and the government Trade Commission (FTC) are among the government authorities that control genetic testing.

Adapted from the Genetics Home Reference at the National Library of Medicine, the following three criteria can be used to evaluate and regulate genetic and genomic testing, much like other diagnostic test types. Through its CLIA program, CMS controls clinical labs, including those that perform clinical genetic testing. The "Clinical Laboratory Improvement Amendments" (CLIA) set up a certification procedure that laboratories had to complete to properly conduct clinical testing.

6.10.2 FDA guidelines

Due to the widespread use of clinical genetic testing today, the expansion of direct-to-consumer (DTC) genomic testing, the rapid advancements in next-generation sequencing (NGS) technology, and FDA's growing concern that unregulated tests pose a threat to public health, FDA is changing its approach to LDTs T. FDA has created new guidance that specifies its strategies for regulating NGS genetic evaluations and verifying their analytic and clinical validity (National Institutes of Health, 2022) (India).

7. DNA technology (use and application) in India

The Minister of Science and Technology, Government of India introduced the DNA Technology (Use and Application) in India Regulation Bill, 2019. The Bill promotes regulating the use of DNA technologies for proving the identify of certain people.

7.1 Offenses

The bill provides punishments for various violations, including (i) disclosing DNA information or (ii) utilizing a DNA sample without permission. For example, releasing DNA information will result in a maximum 3-year imprisonment punishment and a maximum one lakh rupee fine (The DNA Technology, 2019), (India).

8. Limitations of omics technology policy

Despite the abundance of standards within the field, there is still a call for omics technology and data standards policies, as there have been reports from some funding authorities that research can be unclear in their methodologies or inconsistent for example: in their methods of preservation or managing biodata, thus all the more punctuating it's necessity in the research world (Field et al., 2009).

In order to harmonize the use of omics technology and data, terms of service privacy policy should be written in layman's terms for the average reader to understand and imperatively, there is to be a clear outlining of consumer informed consent procedures. The social and ethical implications that novel omics companies have such as clauses that they release about "sharing data with law enforcement when necessary." While their intentions may be, as they say, to stop illegal or unethical actions, it does raise a markedly serious question as to a company's level of discretion which can be very questionable (Knoppers et al., 2021).

While there is an apparent need for standardization and governance of omics technology in the form of policies, another downfall can be the social and technical long-term consequences. For instance, market and political conditions can skew the path that funding, and any potential research may take. Moreover, as discussed previously, there appears to be friction between globally appointed standards and the local implementation of them on a much more micro scale, as there are reports of some scientific literature where scientists are appearing to be stubborn and having "their own way" of doing things, compounded by the fact that other research shows that the looser that some standards are, the more likely people will be willing to conform to them (Holmes et al., 2010).

Another limitation faced by omics companies is that the diverse range of omic domains and formats are conforming to their own individualized, bespoke standard, which can result in "data heterogeneity." This can furthermore be a source of error, overlap and perhaps even biases that can affect downstream analytics of biodata and so suggests that catalogs of some kind must be implemented in order to ensure conformity to standards, or can have a system which makes overlaps redundant (Gomez-Cabrero et al., 2014).

Data heterogeneity essentially references the varying compatibility of certain omics technologies (for example, different sampling or sequencing techniques) and platforms (such as the varying resolutions and at which level single cellular or multi-organ), which, when combined, can often pose more of a problem than a solution. While it is noteworthy that there is a common difficulty in preprocessing this data so that they are all grouped and comprehendible under the same network, some tools such as mix Omics and Multi-Omics Factor Analysis (MOFA), etc., are the current management strategies use to line arise and extract the omics data for its scientific purpose (Krassowski et al., 2020). This heterogeneity is what warrants the cataloging of older and novel omics databases which is challenging but is another root of potential problem and can be strenuous when making omics technology policies. However, the heterogeneity of omics data sets and the integrative incompatibility is not just a source of varied preprocessing, as a recent study shows that the strategies for integrating metabolomics datasets were neither mentioned nor included despite the presence of genomic and proteomic results (Huang et al., 2017).

The experimental challenges as highlighted by this study is not exhaustive, as these include, but are not limited to (i) understanding the statistical behavior of readouts from each omics regime separately; (ii) recognizing nonobvious relationships between omics regimes within their original biological context; and (iii) utilizing time resolution in omics data, such as time course studies, to inform directionality (Buescher & Driggers, 2016).

That being said, not every single omics dataset possesses the four, i.e., the integration of "big data" is linked with volume, diversity, velocity, and truth. they all present similar difficulties, particularly in research with large sample sizes. The "curse of dimensionality," a condition where variances between samples become big and sparse and useless for cluster analysis, causes high-dimensional datasets with more than 1000 variables, further challenging the application of integrated omics datasets (Ronan et al., 2016). To be explicit, refer to the major omics data types, such as genomics, transcriptomics, proteomics, and metabolomics, as "integrated omics" and involve multiomics methodologies that combine three or more omics datasets (Misra et al., 2019).

9. Omics technology for human health

In addition to improving our knowledge of various physiological and pathophysiological processes, Next-Generation Sequencing (NGS), an omics technology, has a wide range of potential applications, including testing, diagnosing, and evolutionary history, response to therapy, and/or predictions of various disorders. It is sometimes possible to better direct our understanding of pathogenetic mechanisms at the molecular level by using these new technologies for the characterization and analysis of various biological levels. This will enable us to recognize patient subgroups or biomarkers to help with diagnosis or the development of customized treatments. Due to the abundance of high-throughput technologies accessible to the omics sciences and their fast evolution, the scientific community demands greater harmonization and standards in data production and analysis techniques. New data mining tools have a lot of possibilities to be implemented at the same moment. In consideration of this, one of the current challenges is to overcome the gap between the generation of omics data, the advancement of high-throughput technologies, and our ability to handle, integrate, analyze, and understand this vast quantity of data. As a result, over the next 10 years, genomics and other omics fields will play a vital role in the Big Data space (Alyass et al., 2015).

Given the intricate nature of omics studies and associated uses like personalized medical treatments, the healthcare systems must undoubtedly provide governance. An evolution in the way health services are set up will also be necessary to implement a tailored approach to healthcare. To guarantee that genomics and other technologies are exploited to their fullest potential, public health leaders have a larger responsibility for helping to catalyze change in the way that public policy and health services are set up. In order for genomics to be properly integrated into healthcare, knowledge advisors, influencers, and policy-makers in the health system need to be informed of the benefits and drawbacks of using data from genomics for disease risk identification, diagnosis, and therapy. They can formulate appropriate policies responses. Consequently, it is crucial to involve present health care system leaders, experts, and people of embracing the benefits and overcoming the drawbacks of a personalized health system. A key factor in determining whether or not the personalization of healthcare may ultimately result in larger improvements in population health will be how In response to this evolving health environment, the public health care system changes (Boccia et al., 2017; Ricciardi & Boccia, 2017).

The use of omics technology for human health also refers to the bioethical considerations toward the use and protection of data from Human Subjects. One of the earlier forms of this dates back to 1981: "The Federal Policy for the Protection of Human Subjects," or now more commonly known as the "Common Rule." The current policy makers and governing bodies have proposed revisions to the Common rule—based on the Belmont Report conference findings of 1979. The last decade has seen the Office for Human research Protections enforce a step-by-step improvement plan to the Common rule, including the respect for human privacy, confidentiality and dignity, and the obligatory advance in human health rights with regard to social justice within omics research to increase the protection of human subjects (National Human Genome Research Institute, 2019).

The process of discovering novel pharmaceutical medications, together with drug development (the process of validating, testing, and marketing a new drug), is one of the most significant activities in the field of pharmaceutical science and is also a way in which omics technology can be targeted for human health. According to a 2018 analysis, 210 novel molecular entities were discovered and developed between the years 2010 and 2016 with funding from around 20% of the US National Institutes of Health (NIH) budget. Since the advent of modern medical science, small molecule candidates have been the focus of the majority of systematic drug discovery efforts; for instance, approximately 86% of the medications (both authorized and investigational) in the Drug Bank database are made up of small molecules. This is because of a variety of factors, such as the relative simplicity of the synthesis, the generally high chemical stability, and the more obvious characterization of reactivity. In addition, bioinformatics tools may be used to identify the biological pathways and activities that are impacted by potential pharmaceuticals, predict the interactions between drugs and proteins, and identify genetic variations that can change a drug's response, among other things (Galkina Cleary et al., 2018; Romano & Tatonetti, 2019; Wishart et al., 2018).

10. Omics technology for livestock and crop improvement

Breeding and genetic selection are essential methods for improving animals. They have produced animals in a variety of livestock species that are genetically superior and disease-free, with increased productivity and efficiency. In the past, breeding animals were genetically selected mostly based on their phenotypic qualities, such as production attributes and estimated breeding value (BV). Eventually, additional economic qualities such as those related to reproduction and survival, animal health, illness resistance to stress, and animal welfare, among others, also played a crucial role in genetic enhancement projects. Rapid genetic development of production efficiency qualities to the following generation was assured by the selective breeding of genetically better animals. In order to fulfill consumer demand for production and animal welfare, a variety of breeding procedures been developed to validate the desired characteristic in genetically selected animals (Brito et al., 2020; Erasmus & van Marle-Köster, 2021; Plieschke et al., 2016; Rexroad et al., 2019). Recent developments in high-speed omics technologies have led to a growing acceptance of genome selection for the picking of reproductive animals. It has become a focus of study to use omics tools to enhance livestock because they may offer a more precise technique for breeding and animal selection. For use in cattle selection and improvement initiatives, this manuscript offers an overview of different omics methods and technologies (Chakraborty et al., 2022; Pedrosa et al., 2021; Ruan et al., 2021; Yang et al., 2021).

Omics is a term that refers to a group of recently created high-throughput technologies, such as metabolomics, transcriptomics, and genomes. These cutting-edge technologies may be used to analyze metabolism at the molecular level in a whole organism, tissue, or cell. The use of omics techniques to livestock offers a tremendous range of potential for understanding out what happened and how to increase the productivity of animals. Many studies on the heat stress gene expression, proteomics, and metabolomics in dairy cows have been conducted recently. The molecular pathways of HS in dairy cows are now better understood thanks to these results, which also present a fresh angle for future study (Min et al., 2017).

Many details on nucleotide and protein sequences are available from the huge molecular databases of the EMBL (Europe), DDBJ (Japan), and NCBI (United States). These datasets have been used in omics technology to comprehend the molecular and physiological underpinnings of economic features, as well as the genetic diversity. However, there is relatively little data on the specific regulatory networks through which these genes and proteins control the expression of economic features in phenotypes. Therefore, there is still cause for worry regarding a sizable unexplained source of variance among the phenotypes of several economic features in livestock. For a better understanding of gene regulatory networks (GRNs) and the discovery of functional genes via a systems biology approach, more modern machine learning (ML) methods have been created. These tools may be used to evaluate high throughput omics data that is accessible in databases (Chakraborty et al., 2022; Guttula et al., 2020; Ng et al., 2021; Wu et al., 2018).

Traditional food crops (TFPs) should be explored, exploited, and mainstreamed for nutritional food security, according to a number of recent research (Wolter et al., 2019). In epidemic situations or when the world's supply chains are broken down as a result of man-made or natural calamities, TFPs can serve as emergency meals as well as supplements to regular diets. An indigenous plant species that are either native to a particular region of the world or was introduced there from somewhere else centuries ago and has since integrated into the culture of that community or region is referred to as a traditional food product.

Such traditional crops that have been used for generations are still used and relied upon by several local indigenous populations across the world. Such traditional crops, which have been used for many generations but are now underutilized, restricted to specific geographical areas, and not in widespread use, are still used and relied upon by a number of local indigenous people across the world. However, consumers' preferences for these old traditional varieties have grown over the past several years, and there is a greater emphasis on reintroducing and mainstreaming these traditionally utilized ancient food crop kinds (Adhikari et al., 2019; Dwivedi et al., 2017; Longin & Würschum, 2016; Muthamilarasan et al., 2019, p. 1–38).

It is crucial to investigate the use of contemporary omics technologies for dissecting the molecular processes driving those features, especially in light of the nutritional, economic, and agronomic importance of TFPs and their possible application as future crops with climate resilience. Furthermore, to address the susceptibility of crop plants, significant genetic diversity exploitation is needed. The fragility of food plants due to their limited genetic variety must also be addressed, which calls for broad utilization of genetic diversity. For improved and more sustainable food production and supply, modern technologies may be utilized to describe agricultural germplasm collections. Utilizing genotyping and informatics technologies, researchers assessed more than 20,000 wild and cultivated barley genotypes to demonstrate the range of use of genetic resources in agricultural development (Langridge & Waugh, 2019; Milner et al., 2019).

It is useful to investigate the chromosomal organization, sequence polymorphism, and genome structure of plants by using structural genomics techniques and building genetic and physical maps of genomic regions regulating a particular characteristic of an organism. Transcriptomics enables the investigation of total mRNA translation in a cell, tissue, or organism under a particular circumstance. With the use of proteomics and metabolomics techniques, it is possible to quantify the quantity and quality of protein metabolite content. In a manner similar to this, ionomics tools can be used to fully comprehend the mineral and elemental composition of a plant species, and the incorporation of other omics tools like genomics, proteomics, and transcriptomics can help to establish the correlation between the elemental composition, transport, and storage, as well as the genes regulating these processes. Therefore, finding the genes accountable for a particular agricultural plant trait of interest requires omics tools (Hu, Scheben, & Edwards, 2018; Lowe et al., 2017; Van Emon, 2016; Yokoyama & Yura, 2016).

For agricultural enhancement efforts, such as the creation of high yielding, nutritionally superior, disease-resistant, and stress-tolerant crops, omics techniques can be applied. Now that genomics and gene editing techniques are integrated, it is feasible to change critical genes more quickly and accurately. Finding plants with superior features and desired qualities is a crucial first step. Priority should be given to the plants that exhibit one or more desired characteristics, such as higher nutritional content, large yields, and biotic and abiotic stress tolerance (Chen et al., 2019; Fernie & Yan, 2019; Fraser et al., 2020; Kumar et al., 2021).

11. Challenges in omics technology and policy framing

Although omics didn't develop a veggie like the Minnesota Cuke from Veggie Tales, it may offer consumers potential health benefits, support farms, help keep a steady supply of food and energy, and eventually reduce suffering. The development of more effective and less hazardous pesticides has been made possible by advances in our understanding of pesticide biodegradation, as well as the processes underpinning pesticide tolerance and metabolism, thanks to the use of genomics as well as the discovery of biomarkers that can be used to evaluate exposure levels in both humans and the environment. The expanded knowledge and insights generated by plant genomics are leading to unexpected findings and logical improvements in our understanding of plant biology. Programs like the Department of Energy's Office of Biological and Environmental Research provide a systems-level, anticipatory knowledge of vegetation. The development of omics technology has made it possible to collect large-scale collections of proteins (proteomes), interactions between proteins (interactomes), metabolites (metabolomes), and collections of observable characteristics (phenomes), which enable a systems biology approach for understanding plants from the single cell to the mature plant, not only during development but also, and this is crucial, under changing environmental conditions (Van Emon, 2016).

Omics has increased our ability to support the famished in the world, particularly those who live in less-than-ideal agricultural environments. Authorities must, however, find a compromise between customers and public interest groups who advocate prudence and oppose modern foods that have experienced genetic engineering, and researchers who are ready for their new products to be introduced to the market right away. The United States uses a coordinated, risk-based framework to evaluate biotechnology goods in order to guarantee public safety and the ongoing development of biotech technologies. The policy considers products that have been modified genetically as being on a continuum with existing products and emphasizes the outcome of genetic modification rather than the process itself (Van Emon, 2016).

Enthusiasm among consumers in agriculture techniques needs with to be handled facts rather than opinions, and it needs to be leveraged to further research on agriculture for both industrialized and dependent on agriculture civilizations lacking impeding efforts to feed a hungry world. Understanding and appreciating the results of omics research requires a systems approach that takes strong science, social repercussions, and economic factors into account. Social consequences vary between industrialized and agricultural societies, and these differences must be identified and handled. Between researchers, authorities, and the general, society, there must be tight communication and coordination. Several scientists have expressed worries about the fact that their work is not always accepted and, regrettably, blame regulators while doing little to nothing to inform the general public, their prospective customers. By disproving assumptions with evidence, omics researchers may do their share to advance the field. Every procedure that modifies biological creatures has the potential to pose safety problems, which raises questions that need to be clarified and solved with in a way that satisfies customers. Ironically, part of a thorough risk assessment that ensures bio safety may be achieved by omics research is the examination of transgenic crops, their genes, and predicted environmental interactions. Cooperation between researchers has been substantially aided by international organizations. Due to extensive partnerships, the sequencing of the rice genome and the Human Genome Project were both completed faster than anticipated. Several of websites, including www.gmoanswers.com and www.biotechbenefits.croplife.org, provide open access opportunities for discussion and information verification about biotech products. Involving the general public and academics in a public debate about the safety of biotech products is essential. For decisions to be made at all levels of participation with knowledge, a respected facility for researching the safety of transgenic crops may also provide academics with the opportunity to interact with the general public (Mehrotra & Goyal, 2013; Van Emon, 2016).

As a subject that is quickly developing, omics research faces special ethical challenges. In order to implement omics research, institutional review boards, doctors, researchers, and the general public will need to become more knowledgeable about omics and bioethics research. Researchers studying omics must reveal the elements of their studies that adhere to the principles of regard for people, justice, and kindness. Researchers may add to discussions of policies at the local level by volunteering their time to serve on the institution's IRB and engaging in multidisciplinary conversations about institutional bioethics policies (Williams & Anderson, 2018).

A variety of policies the frames of health policy, science, technology, and innovation policy, as well as more generally—are required to encourage innovation for healthy aging. Sweden, for instance, has developed a challenge-driven innovation policy strategy that encourages the use of technology as a solution to societal and economic problems. Sweden has a significant social and economic issue as a result of its aging population. With the number of persons 85 and older predicted to nearly increase by 2050, its population is aging quickly, driving up the expense of health and care responsibility (Report, 2014).

11.1 Privacy and confidentiality protection

Human-generated shared data must not contain any personally identifiable information and must have been obtained with knowledge and approval, including permission for sharing data after proper deidentification/anonymization. After anonymization or deidentification, reidentification cannot be performed unless there is a court order mandating it. Additionally, caution must be exercised to guarantee that the data resource is not employed to discriminate against any community, whether it is racial, religious, geographical in nature, or other. Before collecting data utilizing human samples, the information submitted must get the necessary ethical approval(s).

11.2 Transparency of policies

Guidelines for data transfer inside and beyond national boundaries, with private as well as public organizations, for information and commercial use, must be transparent and publicly available.

11.3 Public participation and criticism

Data-sharing rules and mechanisms should be developed with citizen input. The involvement should lead to an improvement in future policies. Additionally, there has to be a systematic system for registering and handling complaints about data misuse (Biotech - Pride Guidelines, 2021).

12. Ethical concern in omics technology and policy framing

Academics, research councils, groups, and society have been the primary factors of ethical concerns regarding human genetic changes for a very long time. See the most important materials concerning the ethical guidelines on human genetics provided by several international organizations in this area. The findings and comments issued by each of these organizations agree on one key issue: the need for human genome editing rules. They also focus on the requirement for the formation of a particular committee made up of a wide range of experts who can assess new technologies in a multidisciplinary and open-minded manner, taking into account its ethical and social implications (Rojas & Rosa, 2019).

Although the integration of multiomics, Machine learning (ML), Artificial intelligence (AI), and huge data into the clinical area care has considerable potential, now, limitations and ethical problems prevent its wider application. In particular, there is a basic lack of agreement over the clinically important genetic variants to concentrate on 18 as well as a lack of understanding by healthcare professionals, hospitals, and funding organizations of the potential benefits of genomic-guided treatment for patient outcomes. Due to these problems, concerns about the cost of omics in medicine, and the general resistance to change in healthcare systems, significant amounts of proof are required before these methods will be used more frequently (Kuchenbaecker et al., 2017).

The ethical concern surrounding the use of these technologies are just as significant as the logistical ones. Informed permission for testing and the management, distribution, and reporting of the significant quantity of data created for each individual continue to be of essential significance. Because of the vast number of genetic loci being examined and their possible impact on illness onset, obtaining consent is incredibly difficult. These ethical and practical difficulties must be considered against the huge possibility found in all areas of clinical medicine, including prognostics, treatment, and diagnosis. For the next 10 years, there will only be a clearer route toward more clinical omics utilization, thanks to official policy direction based on stronger evidence and physician education. An phenomenal evidence of such a development is the recent expansion of requirements for public works whole genome sequencing in Australian children, which was characterized by the introduction of genetic learning through a variety of methods (D'Adamo et al., 2021).

The Federal Policy for the Protection of Human Subjects, also known as the Common Rule, was first established in 1981 in reaction to the negative effects of improper oversight of human research subjects. According to conference conclusions from 1979, known as the Belmont Report, the basic ethical principles for the United States Health and Human Services (HHS) clinical study protection regulations were established (Belmont Report). Office for human research protections (OHRP) released a comprehensive plan to update the Common Rule in 2015; it is now awaiting adoption. The duty of the Institutional Review Board (IRB) is to ensure the safety of people participating in specific research projects, including those applying omics methods and technologies, is governed by the Common Rule. Human subject research is covered by the professional standards and guidelines found in the 2015 edition of the ANA Code of Ethics. Respect for human integrity, focusing on the patient's interests, maintaining the right to privacy and secrecy, the obligation to advance health and human rights and reduce inequalities, and incorporating social justice are all pertinent considerations in omics research. These components share values with the Common Rule's principles of beneficence, fairness, and respect for individuals. In order to ensure the safety of human subjects, it is further necessary to ensure that fundamental ethical standards are included into the study design and implementation due to the specific issues for omics research (Hodge & Gostin, 2017; Williams & Anderson, 2018).

SIENNA (Stakeholder-Informed Ethics for New technologies with high socio-ecoNomics and human rights impAct) it focuses on the ethical, legal, and social challenges faced by human genomics, human development, and human–machine interactions (such as AI and robots). Similar to other technological fields, human genomics is an innovative science that promises advantages to both people and society, but it also poses important ethical concerns, such as those related to autonomy, equality, liberty, privacy, and responsibility. Identification and evaluation of these challenges in cooperation with many stakeholders is the target of SIENNA (Genomics et al., 2019, p. 741716) and has been shown in Table 23.1.

Clinical genomics legal and bioethical researchers have emphasized complexity based on by several overlapping regulatory structures. Clinical applications of proteomics, like clinical genomics, are to support in both treatment and diagnostic of specific patients and the development of scientific knowledge. Similar to genomics, proteomics may be used in both the commercial sector or public health settings. This overlap is necessary for the application of various ethical rules and regulations in the areas of business, public health, research, and therapy. In cases of uncertainty or disagreement, it is generally recommended to use the rule that is most protective of individual rights (Wolf et al., 2020).

The bioethical paradigm, which encourages beneficence and justice while recognizing the liberties and independence of individuals, may be useful when predicting prospects and problems. For instance, avoiding numerous ethical concerns about the return of accidental discoveries is as easy as not looking for them or not analyzing the proteome profile's regions known to be associated with frequent incidental findings. Yet, chances to benefit specific individuals are lost when data that may be utilized to enhance their health or general well-being is refused to be analyzed or returned. Avoiding some ethical dilemmas may appear to be a smart strategy to prevent errors, but if doing so means passing up opportunities to significantly improve the lives of others for every the social price earned could be too high while maintaining their rights at a low cost (Mann et al., 2021).

TABLE 23.1 Important organizations that influence the ethics of human genomics and genetics.

S. no.	Organization name	URL	Founded	Objectives	Services
1.	BBMRI-ERIC is a European research ELSI	https://www.bbmri-eric.eu/services/elsi/	2009	A biobanking research facility in Europe is called BBMRI-ERIC. To promote biomedical research, the biobanking sector as an entire, including researchers, biobanks, industry, and patients.	A variety of online tools and software activities are available for biobankers and researchers, as well as quality control services, that assist with ethical, legal, and societal challenges.
2.	Strategic Initiative for Developing Capacity in Ethical Review (SIDCER)	https://www.sidcer-fercap.org/pages/whats-sidcer.html	2001	Enhancing human subject security through ethics for research in health, supporting quality enhancement, and maintaining international norms, To improve the Ethics Committees' survey and evaluation methodology for ongoing enhancement of quality.	Assisting with the development of National accreditation systems, EC/IRB member capacity building, mentoring EC/IRB members regarding International ethical review standards, and Networking with International Research Institutions, The SIDCER Recognition Program.
3.	The European Academies' Science Advisory Council (EASAC)	https://easac.eu/	2001	(i) Affect the formulation of EU policy; (ii) Specify the EU and Member States' involvement in international policy frameworks; (iii) Build a progressive and more durable network of science academies; (iv) Offer an independent forum for discussion; (v) Strengthen the capacity for science policy.	Legal counsel on public policy that is objective, competent, and supported by evidence should be given to decision-makers in European institutions.
4.	UNESCO, International Bioethics Committee (IBC)	https://www.unesco.org/en/ethics-science-technology/ibc	1993	When violent, antisemitic, and racist ideologies were promoted through systematic propaganda and taught in schools using contemporary methods of culture, communication, and information.	To advance intercultural communication and a better understanding of one another's life, UNESCO encourages the sharing of knowledge and the free exchange of ideas.
5.	The Nuffield council on bioethics	https://www.nuffieldbioethics.org/	1991	The advantages to society are realized in a manner that is aligned with the values of the public, policy and public debate should be informed through timely examination of the ethical issues generated by scientific and medical investigations.	Build on our knowledge, reputation, and collection of work while promoting social change, international cooperation, and biomedical research advancements. Connect and engage key stakeholders.
6.	American college of Medical Genetics and Genomoics (ACMG)	https://www.acmg.net/ACMG/About_ACMG/History.aspx	1991	To improve individual and social health, ACMG to demonstrate the lead in integrating genetics and genomics into all aspects of healthcare and medical treatment.	Clinical Genome Resource, The Newborn Screening Translational Research Network (NBSTRN), the Regional Genetics Network (RGN), and National Coordinating Centre (NCC).
7.	Japanese Organization Eubios Ethics Institute	https://www.eubios.info/	1990	The examination of ethical questions on a global scale, to encourage ethical behavior in science	Encourage practical actions based on research findings to put policy into place that is in keeping with the objectives and requirements of many communities throughout the world.

Continued

TABLE 23.1 Important organizations that influence the ethics of human genomics and genetics.—cont'd

S. no.	Organization name	URL	Founded	Objectives	Services
8.	National Center for Human Genome Research (NCHGR)	https://www.genome.gov/	1989	To advance genomics research, encourage cutting-edge research, create new technologies, and examine the social impacts of genomics in order to improve the health of everyone.	Create more effective methods for the eradication of genetic and hereditary diseases, as well as their early identification and treatment.
9.	Academia Europaea	https://www.ae-info.org/ae/	1988	European research and scholarship, advice to international organizations and governments, encourage cross-disciplinary and international research across all disciplines, especially in connection to European challenges.	Excellence in research in the humanities, legal, economic, social, medicine, political sciences, and mathematics all branches of the scientific and technological sciences is promoted.
10.	The National Academies of Sciences, Engineering, and Medicine	https://www.nationalacademies.org/	NAS (1863), NAE (1964), NAM (1970)	A country and a society that relies on scientific facts to make decisions that are in the best interests of humanity, as well as one that offers impartial, reliable counsel and facilitates the finding of answers to difficult problems by leveraging expertize, experience, and knowledge in the fields of science, engineering, and medicine.	Ingenuity, research, and evidence, as well as collaboration across various professions and fields of knowledge, are required to address the major challenges that society faces.
11.	Wellcome Trust (United Kingdom)	https://wellcome.org	1936	Studies on the three global health concerns of mental health, infectious illness, and climate change in addition to life, health, and wellbeing.	To learn more about life, health, and wellbeing, do research across a wide range of fields.
12.	Waters Corporation	https://www.waters.com/nextgen/us/en.html	1958	Finding new medicines, maintaining the security of the world's food and water resources, or preserving the quality of a chemical entity during manufacture.	Payment solutions, instrument services, software services, and compliance services.

13. Conclusion

To summate, omics technology has a wide range of social, ethical, and political considerations stemming from various levels of socioeconomic and research based origins. One of the earlier forms of this dates back to 1981: The "Common Rule," also known as "The Federal Policy for the Protection of Human Subjects," is now more widely used. The current policy makers and governing bodies have proposed revisions to the Common rule—based on the Belmont Report conference findings of 1979. The last decade has seen the Office for Protection of Human Subjects in Research enforce a step-by-step improvement plan to the standard law, preserving individual privacy is among them confidentiality and dignity, and the obligatory advance in human health rights with regard to social justice within omics research to increase the protection of human subjects.

Since the Common rule and the exponential emergence of various omics fields and data, the challenges of multidisciplinary research, outlined in this chapter, while having slight changes in the past few years is in dire need of updating. As we have outlined, the research barriers to omics technology include: the harmonization across sophisticated yet diverse systems, minimum information datasets, disintegrative properties of data resulting in an incomplete picture; whereas there can also be confounding variables such as the timeliness of accessing data across information systems, as well as the ethical and social impacts on the general public. That being said, this chapter highlights and acknowledges the current policy makers, whether they be governing bodies or independent companies, and their establishment of a more standardized community of practice including the harmonization of omics, cataloging of resources and interoperable systems.

References

Adhikari, T., & Hussain, A. (2019). Are traditional food crops really 'future smart foods?' A sustainability perspective. *Sustainability, 11*(19), 5236. https://doi.org/10.3390/su11195236

Alyass, A., Turcotte, M., & Meyre, D. (2015). From big data analysis to personalized medicine for all: Challenges and opportunities. *BMC Medical Genomics, 8*(1), 33. https://doi.org/10.1186/s12920-015-0108-y

Biotech - Pride Guidelines. (2021). *Major data types*. https://dbtindia.gov.in/sites/default/files/Biotech Pride Guidelines July 2021_0.pdf.

Boccia, S., Federici, A., Siliquini, R., Calabrò, G. E., & Ricciardi, W. (2017). Implementation of genomic policies in Italy: The new national plan for innovation of the health system based on omics sciences. *Epidemiology Biostatistics and Public Health, 14*(4), 1278221—1278223. https://doi.org/10.2427/12782

Bock, C., Farlik, M., & Sheffield, N. C. (2016). Multi-omics of single cells: Strategies and applications. *Trends in Biotechnology, 34*(8), 605—608. https://doi.org/10.1016/j.tibtech.2016.04.004

Brito, L. F., Oliveira, H. R., McConn, B. R., Schinckel, A. P., Arrazola, A., Marchant-Forde, J. N., & Johnson, J. S. (2020). Large-scale phenotyping of livestock welfare in commercial production systems: A new frontier in animal breeding. *Frontiers in Genetics, 11*. https://doi.org/10.3389/fgene.2020.00793

Buescher, J. M., & Driggers, E. M. (2016). Integration of omics: More than the sum of its parts. *Cancer and Metabolism, 4*(1), 4. https://doi.org/10.1186/s40170-016-0143-y

Chakraborty, D., Sharma, N., Kour, S., Sodhi, S. S., Gupta, M. K., Lee, S. J., & Son, Y. O. (2022). Applications of omics technology for livestock selection and improvement. *Frontiers in Genetics, 13*. https://doi.org/10.3389/fgene.2022.774113

Chen, K., Wang, Y., Zhang, R., Zhang, H., & Gao, C. (2019). CRISPR/Cas genome editing and precision plant breeding in agriculture. *Annual Review of Plant Biology, 70*(1), 667—697. https://doi.org/10.1146/annurev-arplant-050718-100049

Chervitz, S. A., Deutsch, E. W., Field, D., Parkinson, H., Quackenbush, J., Rocca-Serra, P., Sansone, S.-A., Stoeckert, C. J., Taylor, C. F., Taylor, R., & Ball, C. A. (2011). *Data standards for omics data: The basis of data sharing and reuse*. https://doi.org/10.1007/978-1-61779-027-0_2

D'Adamo, G. L., Widdop, J. T., & Giles, E. M. (2021). The future is now? Clinical and translational aspects of "omics" technologies. *Immunology and Cell Biology, 99*(2), 168—176. https://doi.org/10.1111/imcb.12404

Dwivedi, S. L., Lammerts van Bueren, E. T., Ceccarelli, S., Grando, S., Upadhyaya, H. D., & Ortiz, R. (2017). Diversifying food systems in the pursuit of sustainable food production and healthy diets. *Trends in Plant Science, 22*(10), 842—856. https://doi.org/10.1016/j.tplants.2017.06.011

Egea, R. R., Puchalt, N. G., Escrivá, M. M., & Varghese, A. C. (2014). Omics: Current and future perspectives in reproductive medicine and technology. *Journal of Human Reproductive Sciences, 7*(2), 73—92. https://doi.org/10.4103/0974-1208.138857

Erasmus, L. M., & van Marle-Köster, E. (2021). Moving towards sustainable breeding objectives and cow welfare in dairy production: A South African perspective. *Tropical Animal Health and Production, 53*(5). https://doi.org/10.1007/s11250-021-02914-w

Fernie, A. R., & Yan, J. (2019). De novo domestication: An alternative route toward new crops for the future. *Molecular Plant, 12*(5), 615—631. https://doi.org/10.1016/j.molp.2019.03.016

Field, D., Sansone, S.-A., Collis, A., Booth, T., Dukes, P., Gregurick, S. K., Kennedy, K., Kolar, P., Kolker, E., Maxon, M., Millard, S., Mugabushaka, A.-M., Perrin, N., Remacle, J. E., Remington, K., Rocca-Serra, P., Taylor, C. F., Thorley, M., Tiwari, B., & Wilbanks, J. (2009). 'Omics data sharing. *Science, 326*(5950), 234—236. https://doi.org/10.1126/science.1180598

Fraser, P. D., Aharoni, A., Hall, R. D., Huang, S., Giovannoni, J. J., Sonnewald, U., & Fernie, A. R. (2020). Metabolomics should be deployed in the identification and characterization of gene-edited crops. *The Plant Journal, 102*(5), 897—902. https://doi.org/10.1111/tpj.14679

Galkina Cleary, E., Beierlein, J. M., Khanuja, N. S., McNamee, L. M., & Ledley, F. D. (2018). Contribution of NIH funding to new drug approvals 2010–2016. *Proceedings of the National Academy of Sciences, 115*(10), 2329–2334. https://doi.org/10.1073/pnas.1715368115

Genomics, W. P., Soulier, A., Niemiec, E., & Howard, H. C. (2019). *Ethical analysis of human genetics and genomics.*

Gomez-Cabrero, D., Abugessaisa, I., Maier, D., Teschendorff, A., Merkenschlager, M., Gisel, A., Ballestar, E., Bongcam-Rudloff, E., Conesa, A., & Tegnér, J. (2014). Data integration in the era of omics: Current and future challenges. *BMC Systems Biology, 8*(Suppl. 2), I1. https://doi.org/10.1186/1752-0509-8-S2-I1

Government of India. (2020). *Science, technology, and innovation policy.* https://dst.gov.in/sites/default/files/STIP_Doc_1.4_Dec2020.pdf.

Gregory, K. (2019). Contestable kinship: User experience and engagement on DTC genetic testing sites. *New Genetics and Society, 38*(4), 387–409. https://doi.org/10.1080/14636778.2019.1677148

Guttula, P. K., Monteiro, P. T., & Gupta, M. K. (2020). A boolean logical model for reprogramming of testes-derived male germline stem cells into germline pluripotent stem cells. *Computer Methods and Programs in Biomedicine, 192*, 105473. https://doi.org/10.1016/j.cmpb.2020.105473

Hamdi, Y., Zass, L., Othman, H., Radouani, F., Allali, I., Hanachi, M., Okeke, C. J., Chaouch, M., Tendwa, M. B., Samtal, C., Mohamed Sallam, R., Alsayed, N., Turkson, M., Ahmed, S., Benkahla, A., Romdhane, L., Souiai, O., Tastan Bishop, O., Ghedira, K., … Kamal Kassim, S. (2021). Human OMICs and computational biology research in Africa: Current challenges and prospects. *OMICS: A Journal of Integrative Biology, 25*(4), 213–233. https://doi.org/10.1089/omi.2021.0004

Hodge, J. G., & Gostin, L. O. (2017). Revamping the US federal common rule. *JAMA, 317*(15), 1521. https://doi.org/10.1001/jama.2017.1633

Hogarth, S., & Saukko, P. (2017). A market in the making: The past, present and future of direct-to-consumer genomics. *New Genetics and Society, 36*(3), 197–208. https://doi.org/10.1080/14636778.2017.1354692

Holmes, C., McDonald, F., Jones, M., Ozdemir, V., & Graham, J. E. (2010). Standardization and omics science: Technical and social dimensions are inseparable and demand symmetrical study. *OMICS: A Journal of Integrative Biology, 14*(3), 327–332. https://doi.org/10.1089/omi.2010.0022

Hu, H., Scheben, A., & Edwards, D. (2018). Advances in integrating genomics and bioinformatics in the plant breeding pipeline. *Agriculture, 8*(6), 75. https://doi.org/10.3390/agriculture8060075

Hu, Y., An, Q., Sheu, K., Trejo, B., Fan, S., & Guo, Y. (2018). Single cell multi-omics technology: Methodology and application. *Frontiers in Cell and Developmental Biology, 6*(April). https://doi.org/10.3389/fcell.2018.00028

Huang, S., Chaudhary, K., & Garmire, L. X. (2017). More is better: Recent progress in multi-omics data integration methods. *Frontiers in Genetics, 8.* https://doi.org/10.3389/fgene.2017.00084

Jao, I., Kombe, F., Mwalukore, S., Bull, S., Parker, M., Kamuya, D., Molyneux, S., & Marsh, V. (2015). Involving research stakeholders in developing policy on sharing public health research data in Kenya. *Journal of Empirical Research on Human Research Ethics, 10*(3), 264–277. https://doi.org/10.1177/1556264615592385

Kalokairinou, L., Howard, H. C., Slokenberga, S., Fisher, E., Flatscher-Thöni, M., Hartlev, M., Van Hellemondt, R., Juškevičius, J., Kapelenska-Pregowska, J., Kováč, P., Lovrečić, L., Nys, H., De Paor, A., Phillips, A., Prudil, L., Rial-Sebbag, E., Casabona, C. M. R., Sándor, J., Schuster, A., … Borr, P. (2018). *Legislation of direct-to-consumer genetic testing in Europe: A fragmented regulatory landscape.* https://doi.org/10.1007/s12687-017-0344-2

Knoppers, T., Beauchamp, E., Dewar, K., Kimmins, S., Bourque, G., Joly, Y., & Dupras, C. (2021). The omics of our lives: Practices and policies of direct-to-consumer epigenetic and microbiomic testing companies. *New Genetics and Society, 40*(4), 541–569. https://doi.org/10.1080/14636778.2021.1997576

Krassowski, M., Das, V., Sahu, S. K., & Misra, B. B. (2020). State of the field in multi-omics research: From computational needs to data mining and sharing. *Frontiers in Genetics, 11.* https://doi.org/10.3389/fgene.2020.610798

Kuchenbaecker, K. B., Hopper, J. L., Barnes, D. R., Phillips, K.-A., Mooij, T. M., Roos-Blom, M.-J., Jervis, S., van Leeuwen, F. E., Milne, R. L., Andrieu, N., Goldgar, D. E., Terry, M. B., Rookus, M. A., Easton, D. F., Antoniou, A. C., McGuffog, L., Evans, D. G., Barrowdale, D., Frost, D., … Olsson, H. (2017). Risks of Breast, ovarian, and contralateral Breast cancer for BRCA1 and BRCA2 mutation carriers. *JAMA, 317*(23), 2402. https://doi.org/10.1001/jama.2017.7112

Kuhn, R. M., Haussler, D., & Kent, W. J. (2013). The UCSC genome browser and associated tools. *Briefings in Bioinformatics, 14*(2), 144–161. https://doi.org/10.1093/bib/bbs038

Kumar, A., Anju, T., Kumar, S., Chhapekar, S. S., Sreedharan, S., Singh, S., Choi, S. R., Ramchiary, N., & Lim, Y. P. (2021). Integrating omics and gene editing tools for rapid improvement of traditional food plants for diversified and sustainable food security. *International Journal of Molecular Sciences, 22*(15), 1–51. https://doi.org/10.3390/ijms22158093

Langridge, P., & Waugh, R. (2019). Harnessing the potential of germplasm collections. *Nature Genetics, 51*(2), 200–201. https://doi.org/10.1038/s41588-018-0340-4

Li, C.-X., Gao, J., Zhang, Z., Chen, L., Li, X., Zhou, M., & Wheelock, Å. M. (2022). Multiomics integration-based molecular characterizations of COVID-19. *Briefings in Bioinformatics, 23*(1). https://doi.org/10.1093/bib/bbab485

Lombardi, A., de Lange, P., Glinni, D., Senese, R., Cioffi, F., Lanni, A., Goglia, F., & Moreno, M. (2011). Studies of complex biological systems with applications to molecular medicine: The need to integrate transcriptomic and proteomic approaches. *Journal of Biomedicine and Biotechnology, 2011*, 1–19. https://doi.org/10.1155/2011/810242

Longin, C. F. H., & Würschum, T. (2016). Back to the future—Tapping into ancient grains for food diversity. *Trends in Plant Science, 21*(9), 731–737. https://doi.org/10.1016/j.tplants.2016.05.005

Lowe, R., Shirley, N., Bleackley, M., Dolan, S., & Shafee, T. (2017). Transcriptomics technologies. *PLoS Computational Biology, 13*(5), e1005457. https://doi.org/10.1371/journal.pcbi.1005457

Majumder, M. A., Guerrini, C. J., & McGuire, A. L. (2021). Direct-to-Consumer genetic testing: Value and risk. *Annual Review of Medicine, 72*(1), 151–166. https://doi.org/10.1146/annurev-med-070119-114727

Mangul, S., Martin, L. S., Hill, B. L., Lam, A. K. M., Distler, M. G., Zelikovsky, A., Eskin, E., & Flint, J. (2019). Systematic benchmarking of omics computational tools. In *Nature communications* (Vol. 10)Nature Publishing Group. https://doi.org/10.1038/s41467-019-09406-4, 1.

Mann, S. P., Treit, P. V., Geyer, P. E., Omenn, G. S., & Mann, M. (2021). Ethical principles, constraints, and opportunities in clinical proteomics. *Molecular and Cellular Proteomics, 20*, 100046. https://doi.org/10.1016/J.MCPRO.2021.100046

McQuilton, P., Gonzalez-Beltran, A., Rocca-Serra, P., Thurston, M., Lister, A., Maguire, E., & Sansone, S.-A. (2016). BioSharing: Curated and crowd-sourced metadata standards, databases and data policies in the life sciences. *Database*, baw075. https://doi.org/10.1093/database/baw075, 2016.

Mehrotra, S., & Goyal, V. (2013). Evaluation of designer crops for biosafety—A scientist's perspective. *Gene, 515*(2), 241–248. https://doi.org/10.1016/j.gene.2012.12.029

Milner, S. G., Jost, M., Taketa, S., Mazón, E. R., Himmelbach, A., Oppermann, M., Weise, S., Knüpffer, H., Basterrechea, M., König, P., Schüler, D., Sharma, R., Pasam, R. K., Rutten, T., Guo, G., Xu, D., Zhang, J., Herren, G., Müller, T., … Stein, N. (2019). Genebank genomics highlights the diversity of a global barley collection. *Nature Genetics, 51*(2), 319–326. https://doi.org/10.1038/s41588-018-0266-x

Min, L., Zhao, S., Tian, H., Zhou, X., Zhang, Y., Li, S., Yang, H., Zheng, N., & Wang, J. (2017). Metabolic responses and "omics" technologies for elucidating the effects of heat stress in dairy cows. *International Journal of Biometeorology, 61*(6), 1149–1158. https://doi.org/10.1007/s00484-016-1283-z

Misra, B. B., Langefeld, C., Olivier, M., & Cox, L. A. (2019). Integrated omics: Tools, advances and future approaches. *Journal of Molecular Endocrinology, 62*(1), R21–R45. https://doi.org/10.1530/JME-18-0055

Muthamilarasan, M., Singh, N. K., & Prasad, M. (2019). *Multi-omics approaches for strategic improvement of stress tolerance in underutilized crop species: A climate change perspective*. https://doi.org/10.1016/bs.adgen.2019.01.001

National Biotechnology Development Strategy, and Department of Biotechnology. (2021). *Policy enablers*. https://dbtindia.gov.in/sites/default/files/NATIONAL BIOTECHNOLOGY DEVELOPMENT STRATEGY_01.04.pdf.

National Human Genome Research Institute. (September, 2019). *Human subjects research*. https://www.genome.gov/about-genomics/policy-issues/Human-Subjects-Research-in-Genomics.

National Institutes of Health. (January 6, 2022). *Policy in genome—National Institutes of Health search results*. https://www.genome.gov/about-genomics/policy-issues.

Ng, B., Casazza, W., Kim, N. H., Wang, C., Farhadi, F., Tasaki, S., Bennett, D. A., De Jager, P. L., Gaiteri, C., & Mostafavi, S. (2021). Cascading epigenomic analysis for identifying disease genes from the regulatory landscape of GWAS variants. *PLoS Genetics, 17*(11), e1009918. https://doi.org/10.1371/journal.pgen.1009918

Pedrosa, V. B., Schenkel, F. S., Chen, S.-Y., Oliveira, H. R., Casey, T. M., Melka, M. G., & Brito, L. F. (2021). Genomewide association analyses of lactation persistency and milk production traits in holstein cattle based on imputed whole-genome sequence data. *Genes, 12*(11), 1830. https://doi.org/10.3390/genes12111830

Phillips, A. M. (2016). Only a click away—DTC genetics for ancestry, health, love…and more: A view of the business and regulatory landscape. *Applied and Translational Genomics, 8*, 16–22. https://doi.org/10.1016/j.atg.2016.01.001

Pinu, F. R., Beale, D. J., Paten, A. M., Kouremenos, K., Swarup, S., Schirra, H. J., & Wishart, D. (2019). Systems biology and multi-omics integration: Viewpoints from the metabolomics research community. *Metabolites, 9*(4). https://doi.org/10.3390/metabo9040076

Pirih, N., & Kunej, T. (2018). An updated taxonomy and a graphical summary tool for optimal classification and comprehension of omics research. *OMICS: A Journal of Integrative Biology, 22*(5), 337–353. https://doi.org/10.1089/omi.2017.0186

Plieschke, L., Edel, C., Pimentel, E. C. G., Emmerling, R., Bennewitz, J., & Götz, K.-U. (2016). Systematic genotyping of groups of cows to improve genomic estimated breeding values of selection candidates. *Genetics Selection Evolution, 48*(1), 73. https://doi.org/10.1186/s12711-016-0250-9

Raupach, M. J., Amann, R., Wheeler, Q. D., & Roos, C. (2016). The application of "-omics" technologies for the classification and identification of animals. *Organisms, Diversity and Evolution, 16*(1), 1–12. https://doi.org/10.1007/s13127-015-0234-6

Report, S. (2014). *Workshop on integrating omics and policy for healthy ageing* (Vol. 12, pp. 1–21).

Rexroad, C., Vallet, J., Matukumalli, L. K., Reecy, J., Bickhart, D., Blackburn, H., Boggess, M., Cheng, H., Clutter, A., Cockett, N., Ernst, C., Fulton, J. E., Liu, J., Lunney, J., Neibergs, H., Purcell, C., Smith, T. P. L., Sonstegard, T., Taylor, J., … Wells, K. (2019). Genome to phenome: Improving animal health, production, and well-being—A new USDA blueprint for animal genome research 2018–2027. *Frontiers in Genetics, 10*. https://doi.org/10.3389/fgene.2019.00327

Ricciardi, W., & Boccia, S. (2017). New challenges of public health: Bringing the future of personalised healthcare into focus. *The European Journal of Public Health, 27*(Suppl. 4), 36–39. https://doi.org/10.1093/eurpub/ckx164

Rojas, C. R., & Rosa, P. (2019). *Omics in society social, legal and ethical aspects of human genomics*. https://doi.org/10.2760/33084

Romano, J. D., & Tatonetti, N. P. (2019). Informatics and computational methods in natural product drug discovery: A review and perspectives. *Frontiers in Genetics, 10*(April), 1–16. https://doi.org/10.3389/fgene.2019.00368

Ronan, T., Qi, Z., & Naegle, K. M. (2016). Avoiding common pitfalls when clustering biological data. *Science Signaling, 9*(432). https://doi.org/10.1126/scisignal.aad1932

Ruan, D., Zhuang, Z., Ding, R., Qiu, Y., Zhou, S., Wu, J., Xu, C., Hong, L., Huang, S., Zheng, E., Cai, G., Wu, Z., & Yang, J. (2021). Weighted single-step GWAS identified candidate genes associated with growth traits in a duroc pig population. *Genes, 12*(1), 117. https://doi.org/10.3390/genes12010117

Sokolov-Mladenović, S., Cvetanović, S., & Mladenović, I. (2016). R& D expenditure and economic growth: EU28 evidence for the period 2002–2012. *Economic Research-Ekonomska Istraživanja, 29*(1), 1005–1020. https://doi.org/10.1080/1331677X.2016.1211948

Subramanian, I., Verma, S., Kumar, S., Jere, A., & Anamika, K. (2020). Multi-omics data integration, interpretation, and its application. In *Bioinformatics and biology insights* (Vol. 14)SAGE Publications Inc. https://doi.org/10.1177/1177932219899051

The DNA Technology (Use And Application) Regulation Bill, 2019 of Lok Sabha. (2019). *Testimony of the ministry of science and technology Mr. Harsh Vardhan).*

Van Emon, J. M. (2016). The omics revolution in agricultural research. *Journal of Agricultural and Food Chemistry, 64*(1), 36–44. https://doi.org/10.1021/acs.jafc.5b04515

Wheelock, C. E., Goss, V. M., Balgoma, D., Nicholas, B., Brandsma, J., Skipp, P. J., Snowden, S., Burg, D., D'Amico, A., Horvath, I., Chaiboonchoe, A., Ahmed, H., Ballereau, S., Rossios, C., Chung, K. F., Montuschi, P., Fowler, S. J., Adcock, I. M., Postle, A. D., ... Djukanović, R. (2013). Application of 'omics technologies to biomarker discovery in inflammatory lung diseases. *European Respiratory Journal, 42*(3), 802–825. https://doi.org/10.1183/09031936.00078812

Wilkinson, M. D., Dumontier, M., Aalbersberg, Ij J., Appleton, G., Axton, M., Baak, A., Blomberg, N., Boiten, J.-W., da Silva Santos, L. B., Bourne, P. E., Bouwman, J., Brookes, A. J., Clark, T., Crosas, M., Dillo, I., Dumon, O., Edmunds, S., Evelo, C. T., Finkers, R., ... Mons, B. (2016). The FAIR Guiding Principles for scientific data management and stewardship. *Scientific Data, 3*(1), 160018. https://doi.org/10.1038/sdata.2016.18

Williams, J. K., & Anderson, C. M. (2018). Omics research ethics considerations. *Nursing Outlook, 66*(4), 386–393. https://doi.org/10.1016/j.outlook.2018.05.003

Wishart, D. S., Feunang, Y. D., Guo, A. C., Lo, E. J., Marcu, A., Grant, J. R., Sajed, T., Johnson, D., Li, C., Sayeeda, Z., Assempour, N., Iynkkaran, I., Liu, Y., Maciejewski, A., Gale, N., Wilson, A., Chin, L., Cummings, R., Le, D., ... Wilson, M. (2018). DrugBank 5.0: A major update to the DrugBank database for 2018. *Nucleic Acids Research, 46*(D1), D1074–D1082. https://doi.org/10.1093/nar/gkx1037

Wolf, S. M., Ossorio, P. N., Berry, S. A., Greely, H. T., McGuire, A. L., Penny, M. A., & Terry, S. F. (2020). Integrating rules for genomic research, clinical care, public health screening and DTC testing: Creating translational law for translational genomics. *Journal of Law Medicine and Ethics, 48*(1), 69–86. https://doi.org/10.1177/1073110520916996

Wolter, F., Schindele, P., & Puchta, H. (2019). Plant breeding at the speed of light: The power of CRISPR/Cas to generate directed genetic diversity at multiple sites. *BMC Plant Biology, 19*(1), 176. https://doi.org/10.1186/s12870-019-1775-1

Wu, Y., Zeng, J., Zhang, F., Zhu, Z., Qi, T., Zheng, Z., Lloyd-Jones, L. R., Marioni, R. E., Martin, N. G., Montgomery, G. W., Deary, I. J., Wray, N. R., Visscher, P. M., McRae, A. F., & Yang, J. (2018). Integrative analysis of omics summary data reveals putative mechanisms underlying complex traits. *Nature Communications, 9*(1), 918. https://doi.org/10.1038/s41467-018-03371-0

Yang, C., Chowdhury, D., Zhang, Z., Cheung, W. K., Lu, A., Bian, Z., & Zhang, L. (2021). A review of computational tools for generating metagenome-assembled genomes from metagenomic sequencing data. *Computational and Structural Biotechnology Journal, 19*, 6301–6314. https://doi.org/10.1016/j.csbj.2021.11.028

Yokoyama, S., & Yura, K. (2016). Special issue: Big data analyses in structural and functional genomics. *Journal of Structural and Functional Genomics, 17*(4), 67. https://doi.org/10.1007/s10969-016-9213-1

Zhang, X., Xia, P., Wang, P., Yang, J., & Baird, D. J. (2018). Omics advances in ecotoxicology. *Environmental Science and Technology, 52*(7), 3842–3851. https://doi.org/10.1021/acs.est.7b06494

Index

Note: 'Page numbers followed by *f* indicate figures, *t* indicates tables.'

N

O

P